Jean-Pierre Fouque   Josselin Garnier
George Papanicolaou   Knut Sølna

Wave Propagation and Time Reversal in Randomly Layered Media

Springer

Authors

Jean-Pierre Fouque
Department of Statistics and
 Applied Probability
University of California
Santa Barbara, CA 93106-3110
USA
fouque@pstat.ucsb.edu

Josselin Garnier
UFR de Mathématiques
Université Paris VII
2 Place Jussieu
75251 Paris Cedex 05
France
garnier@math.jussieu.fr

George Papanicolaou
Mathematics Department
Stanford University
Stanford, CA 94305
USA
papanicolaou@stanford.edu

Knut Sølna
Department of Mathemathics
University of California at Irvine
Irvine, CA 92697
USA
ksolna@math.uci.edu

Managing Editors

B. Rozovskii
Division of Applied Mathematics
Brown University
182 George Street
Providence, RI 02912
rozovsky@dam.brown.edu

G. Grimmett
Centre for Mathematical Sciences
Wilberforce Road, Cambridge CB3 0WB, UK
G.R.Grimmett@statslab.cam.ac.uk

Mathematics Subject Classification (2000): 76Q05, 35L05, 35R30, 60G, 62M40, 73D35

ISSN: 0172-4568

ISBN-13: 978-1-4419-2162-8   eISBN-13: 978-0-387-49808-9

Printed on acid-free paper.

9 8 7 6 5 4 3 2 1

springer.com

To our families

Preface

Our motivation for writing this book is twofold: First, the theory of waves propagating in randomly layered media has been studied extensively during the last thirty years but the results are scattered in many different papers. This theory is now in a mature state, especially in the very interesting regime of separation of scales as introduced by G. Papanicolaou and his coauthors and described in [8], which is a building block for this book. Second, we were motivated by the time-reversal experiments of M. Fink and his group in Paris. They were done with ultrasonic waves and have attracted considerable attention because of the surprising effects of enhanced spatial focusing and time compression in random media. An exposition of this work and its applications is presented in [56]. Time reversal experiments were also carried out with sonar arrays in shallow water by W. Kuperman [113] and his group in San Diego. The enhanced spatial focusing and time compression of signals in time reversal in random media have many diverse applications in detection and in focused energy delivery on small targets as, for example, in the destruction of kidney stones. Enhanced spatial focusing is also useful in sonar and wireless communications for reducing interference. Time reversal ideas have played an important role in the development of new methods for array imaging in random media as presented in [19]. A quantitative mathematical analysis is crucial in the understanding of these phenomena and for the development of new applications. In a series of recent papers by the authors and their coauthors, starting with [40] in the one-dimensional case and [16] in the multidimensional case, a complete analysis of time reversal in random media has been proposed in the two extreme cases of strongly scattering layered media, and weak fluctuations in the parabolic approximation regime. These results are important in the understanding of the intermediate situations and will contribute to future applications of time reversal.

Wave propagation in three-dimensional random media has been studied mostly by perturbation techniques when the random inhomogeneities are small. The main results are that the amplitude of the mean waves decreases with distance traveled, because coherent wave energy is converted into incoherent fluctuations, while the mean energy propagates diffusively or by radiative transport. These phenomena are analyzed extensively from a physical and engineering point of view in the book of Ishimaru [90]. It was first noted by Anderson [5] that for electronic waves in strongly disordered materials there is wave localization. This means that wave energy does not propagate, because the random inhomogeneities trap it in finite regions. What is different and special in one-dimensional random media is that wave localization always occurs, even when the inhomogeneities are weak. This means that there is never a diffusive or transport regime in one-dimensional random media. This was first proved by Goldsheid, Molchanov, and Pastur in [79]. It is therefore natural that the analysis of waves in one-dimensional or strongly anisotropic layered media presented in this book should rely on methods and techniques that are different from those used in general, multidimensional random media.

The content of this book is multidisciplinary and presents many new physically interesting results about waves propagating in randomly layered media as well as applications in time reversal. It uses mathematical tools from probability and stochastic processes, partial differential equations, and asymptotic analysis, combined with the physics of wave propagation and modeling of time-reversal experiments. It addresses an interdisciplinary audience of students and researchers interested in the intriguing phenomena related to waves propagating in random media. We have tried to gradually bring together ideas and tools from all these areas so that no special background is required. The book can also be used as a textbook for advanced topics courses in which random media and related homogenization, averaging, and diffusion approximation methods are involved. The analytical results discussed here are proved in detail, but we have chosen to present them with a series of explanatory and motivating steps instead of a "theorem-proof" format. Most of the results in the book are illustrated with numerical simulations that are carefully calibrated to be in the regimes of the corresponding asymptotic analysis. At the end of each chapter we give references and additional comments related to the various results that are presented.

Acknowledgments

George Papanicolaou would like to thank his colleagues Joe Keller and Ragu Varadhan and his coauthors in the early work that is the basis of this book: Mark Asch, Bob Burridge, Werner Kohler, Pawel Lewicki, Marie Postel, Ping Sheng, Sophie Weinryb, and Ben White. The authors would like to thank their collaborators in developing the recent theory of time reversal presented in this book, in particular Jean-François Clouet, for early work on time reversal; André Nachbin, for numerous and fruitful recent collaborations on the subject; and Liliana Borcea and Chrysoula Tsogka for our extended collaboration on imaging. We also thank Mathias Fink and his group in Paris for many discussions of time-reversal experiments. We have benefited from numerous constructive discussions with our colleagues: Guillaume Bal, Peter Blomgren,

Grégoire Derveaux, Albert Fannjiang, Marteen de Hoop, Arnold Kim, Roger Maynard, Miguel Moscoso, Arogyaswami Paulraj, Lenya Ryzhik, Bill Symes, Bart Van Tiggelen, and Hongkai Zhao. We also would like to thank our students and postdoctoral fellows who have read earlier versions of the book: Petr Glotov, Renaud Marty, and Oleg Poliannikov.

Most of this book was written while the authors were visiting the Departments of Mathematics at North Carolina State University, University of California Irvine, Stanford University, Toulouse University, University Denis Diderot in Paris, IHES in Bures-sur-Yvette, and IMPA in Rio de Janeiro. The authors would like to acknowledge the hospitality of these places.

Santa Barbara, California
Paris, France
Stanford, California
Irvine, California

Jean-Pierre Fouque
Josselin Garnier
George Papanicolaou
Knut Sølna

December 19, 2006

Contents

1

Introduction and Overview of the Book

We begin by describing the organization of the book as shown in the diagram in Figure 1.1.

The basic theory of wave propagation in one-dimensional random media is contained in Chapters 2–9. Background for waves in deterministic, layered media is given in Chapters 2 and 3. In Chapters 4 and 5 we introduce the modeling of random media and describe in detail the scaling regimes that we consider in this book. In Chapter 6 we give a self-contained presentation of the asymptotic theory of random differential equations in a form that can be applied directly to the analysis of waves in random media in the following chapters. The asymptotic theory of reflection and transmission of waves in one-dimensional random media is presented in Chapters 7–9. Monochromatic reflection and transmission is analyzed in Chapter 7, which contains the well-known results of exponential decay of transmitted energy as the size of the random medium increases. In Chapter 8 we analyze the propagation of wave fronts and in Chapter 9 we characterize the statistical properties of wave fluctuations in the time domain.

The theory of time reversal in one-dimensional random media, both for reflected and for transmitted waves, along with applications to detection and communications, is presented in Chapters 10–13.

The extension of the theory of Chapters 8 and 9 to wave propagation in three-dimensional randomly layered media is given in Chapter 14. Time reversal in such media is analyzed in Chapter 15, where we derive analytical formulas that characterize the enhanced spatial focusing. An application to echo-mode energy refocusing on a passive scatterer is presented in Chapter 16.

Chapters 17–19 contain special topics and various generalizations to other asymptotic regimes and other types of waves. In Chapter 20 we analyze in detail wave propagation in randomly perturbed waveguides. This chapter is self-contained and could be read right after Chapter 6.

We now describe in more detail the contents of the chapters.

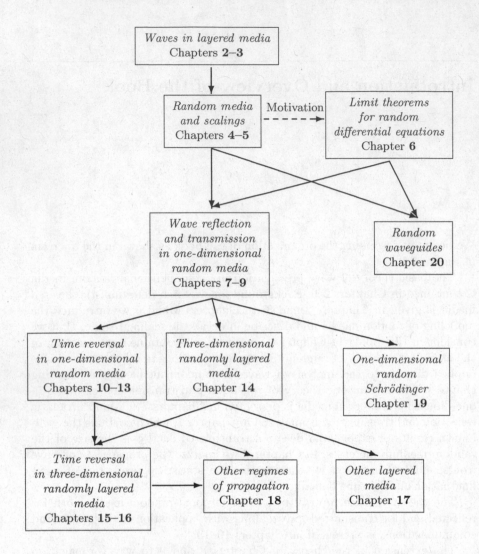

Fig. 1.1. Interdependence of the chapters.

Basic facts about wave propagation in **homogeneous media** are presented in **Chapter 2**.

In **Chapter 3** we consider one-dimensional **piecewise constant layered media**, and we introduce the usual formulation of reflection and transmission in terms of products of matrices.

Starting with **Chapter 4** we consider **randomly layered media**. We introduce the linear system of acoustic equations for waves propagating in one dimension, and then carefully describe the sequence of transformations that will be carried out throughout the rest of the book. We pay particular attention to boundary conditions and their interpretation, and to the reflected and transmitted waves in both the frequency and time domains. The concepts of random media and correlation lengths are introduced in this chapter. Our point of view is that randomness is closely associated with small-scale inhomogeneities leading naturally to the regime of homogenization and the notion of effective medium. This is done with an application of the law of large numbers, in the context of differential equations with random coefficients. This regime corresponds to waves propagating over distances of a few wavelengths, which are, however, much larger than the correlation length of the inhomogeneities.

We go a step further in **Chapter 5** by considering waves propagating over distances much larger than wavelengths. The fluctuations due to the multiple scattering by the random inhomogeneities accumulate and create *"noisy"* reflected and transmitted waves. We introduce important scaling regimes in which **diffusion approximations** are valid, leading to differential equations with random coefficients that are white noise. Even though the equations are linear, the probability distribution of the "noisy" wave field is a highly nonlinear function of the distribution of the random coefficients that model the random inhomogeneities. For a given frequency the random differential equations that enter are finite-dimensional, but in the time domain the problems become infinite-dimensional. Asymptotic approximations greatly simplify the analysis in the scaling regimes, and enable us to obtain useful information about the statistics of the reflected and transmitted waves.

In **Chapter 6** we present concepts and results about **stochastic processes** needed in the modeling of one-dimensional wave propagation and its asymptotic analysis. It is important to note that distance along the one-dimensional direction of propagation plays the role of the usual time parameter for these stochastic processes. The physical time is transformed by going into the frequency domain. In this chapter we present briefly the elements of the theory of Markov processes used for modeling randomly layered media and for describing the limit processes arising in the regime of diffusion approximations. A summary of the **stochastic calculus** is given at the end of the chapter, including Itô's formula, stochastic differential equations, the link with parabolic partial differential equations through the Feynman–Kac formula, and applications to the study of Lyapunov exponents of linear random differential equations.

A detailed analysis of the reflection and transmission of monochromatic waves in a one-dimensional random medium is given in **Chapter 7**. In one-dimensional random media all the wave energy is eventually converted into fluctuations, giving rise to the phenomenon of wave **localization**. This means that the energy is trapped by the random medium. It is entirely reflected back in the case of a random half-space. We show that the **exponential decay** of the transmitted energy through a random slab of random medium is closely related to the stability of the random harmonic oscillator, studied in this chapter. We also compute the moments of the transmitted energy, quantifying

the exponential decay, as well as the almost-sure exponential decay that is
related to the usual localization theory.

In **Chapter 8** we study the transmitted **wave front** in one-dimensional
random media, in the regimes of the diffusion approximation introduced in
the previous chapters. A pulse is sent from one end of a one-dimensional random medium and it is observed at the other end (see Figure 1.2). When the
pulse exits the slab it looks like a smeared and faded version of the original
one, followed by a noisy, incoherent coda. It is quite remarkable that in these
asymptotic regimes, the front of the transmitted pulse has a simple description: (i) its deterministic shape is given by the convolution of the original pulse
with a deterministic kernel that depends only on the second-order statistics
of the random medium, and (ii) the transmitted wave front is centered at a
random arrival time whose probability distribution is explicitly given in terms
of a single Brownian motion. In this chapter we also describe the wave front
reflected from a strong interface in a random medium.

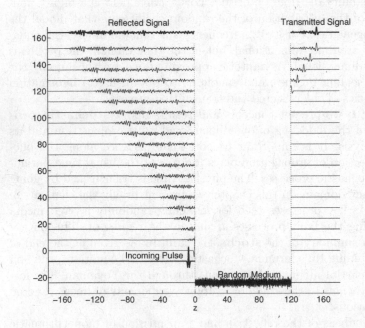

Fig. 1.2. Propagation of a pulse through a slab of random medium $(0, L)$. A right-
going wave is incoming from the left. Snapshots of the wave profile (here the pressure)
at different times are plotted from bottom to top. The reflected and transmitted
signals at the last time of the numerical simulation are plotted at the top.

In **Chapter 9** we characterize the **statistics of the reflected and transmitted waves**, including the coda, in both the frequency and time domains (see the wave signals plotted at the top of Figure 1.2). This is done by a careful asymptotic analysis of the moments of the reflection and transmission coefficients. They satisfy a system of differential equations with random coefficients and are scaled so that the diffusion approximation can be applied. The limiting moments are obtained as solutions of systems of transport equations, which play a central role in the analysis of time reversal with incoherent waves, discussed in the following chapters. The solutions of these deterministic transport equations admit a probabilistic representation in terms of jump Markov processes, which is particularly convenient for Monte Carlo simulations and, in some cases, for deriving explicit formulas.

In **Chapter 10** we analyze **time reversal in reflection** where the incoherent reflected waves are recorded and sent time-reversed back into the medium. We show that stable refocusing takes place at the original source point. This is observed in physical experiments and illustrated in numerical simulations in Figures 1.3 and 1.4. Time-reversal refocusing can be used to estimate power spectral densities of reflected waves. They contain information about the medium. In this chapter we also compare, with a detailed analysis of **signal-to-noise ratios**, the **spectral estimation** method using time reversal with a direct estimation of cross-correlations of the reflected signal.

In **Chapter 11** we present two applications of time reversal to detection. In the first application, we use time reversal to detect the presence of a **weak reflector** buried in the many random layers. In this case the refocusing kernel of the time-reversal process has a jump that is related to the depth and strength of the reflector, and we exploit this to identify the reflector. In the second application, we introduce **absorption** in the one-dimensional model and show that refocusing still takes place after time reversal. We apply this to the **detection** and characterization of a dissipative region embedded in the random medium. In the presence of a dissipative region the refocusing kernel is modified and has a jump in its derivative. The time of this jump is related to the depth of the dissipative region, and its amplitude to the strength of absorption.

In **Chapter 12** we study time reversal of waves in randomly layered media described in the previous chapters. In this chapter we analyze **time reversal in transmission**, which means that a pulse is emitted at one end of a random slab, recorded at a time-reversal mirror at the other end, and then sent back. The wave refocuses at the original source point and the quality of the refocusing depends on how much of the transmitted wave has been recorded. In particular, it is shown that recording some part of the incoherent coda wave improves refocusing.

Applications to **communications** are presented in **Chapter 13**, where we analyze **signal-to-interference ratios** with and without using time reversal for communications through a one-dimensional random channel.

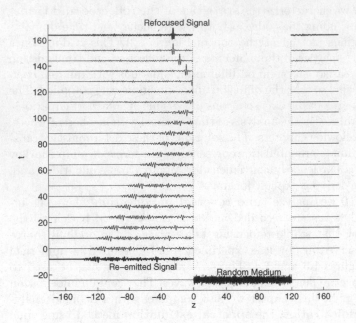

Fig. 1.3. We use the same random medium as in Figure 1.2 and send back, to the right, the time-reversed reflected signal (the one plotted at the top left corner of Figure 1.2). Snapshots of the wave profile (here the pressure) at different times are plotted from bottom to top. The refocused pulse is seen emerging from the random medium at the top.

Starting with **Chapter 14** we analyze waves propagating in a randomly layered **three-dimensional medium**. By taking Fourier transforms with respect to time and along the layers, the problem can be formulated as infinitely many one-dimensional problems. We model a physical source located at the surface of the random medium. Using a stationary phase analysis, we show that in the regime of diffusion approximations, and because of the separation of scales as in previous chapters, the stable wave front can again be described with an explicit formula that we derive.

Time reversal of waves propagating in three-dimensional randomly layered media is discussed in **Chapter 15**, where we consider a time-reversal mirror that records the signals generated by a source embedded in the random layers. We show that the time-reversed waves refocus around the original source point. We give a detailed analytical description of the refocused pulse in time and space. We compare this refocusing with **diffraction-limited** refocusing in homogeneous media and show that there is **superresolution** from

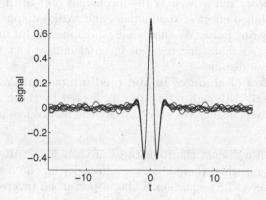

Fig. 1.4. We plot the refocused pulses generated by 10 independent simulations of time reversal (we follow the same procedure as in Figures 1.2–1.3, and we magnify the refocused pulse seen at the top line of Figure 1.3). The initial pulse is the second derivative of a Gaussian. We see here the remarkable statistical stability of the refocused pulse. Its shape and center do not depend on the realization of the medium, in contrast to the small-amplitude random wave fluctuations before and after the refocusing time.

multipathing. This means that the focusing is much tighter, as well as stable, in the random medium.

In **Chapter 16** we present an application of time reversal in three-dimensional randomly layered media to **echo-mode energy refocusing** on a passive scatterer. This means that when the reflected signals received at the time-reversal mirror from a scatterer in a randomly layered medium are time-reversed and suitably reemitted, they tend to focus on the scatterer.

In **Chapter 17** we present an extension of the theory of wave propagation and time reversal to more **general randomly layered media**. We analyze models in which the effective parameters of the random medium do not match those of the adjacent homogeneous medium. We also analyze the case in which the effective parameters of the random medium vary smoothly at the macroscopic scale. The case in which both the bulk modulus and the density of the medium are randomly fluctuating is analyzed in Section 17.3.

Chapter 18 is devoted to several extensions and generalizations including the following ones.

- We reconsider the analysis for a **different regime** of scale separation, in which the amplitude of the fluctuations of the medium parameters is small and the typical wavelength is comparable to the small correlation length of the random medium.
- We extend the analysis to **dispersive or weakly nonlinear** random media. In the dispersive case, time reversal succeeds in recompressing the

dispersive oscillatory tail as well as the incoherent part of the waves. We analyze the combined effect of randomness and weak nonlinearity on the front of a propagating pulse. We show that randomness helps in preventing shock formation, so that time reversal in transmission can be done for longer propagation distances.

- We study the effect of **changes in the medium** parameters before and after time reversal. Although refocusing is affected by these changes, we still have partial refocusing. We also quantify the partial loss of statistical stability.

In **Chapter 19** we discuss the robustness of wave localization in a randomly layered medium when there is also nonlinearity, in the context of the nonlinear Schrödinger (NLS) equation. Using a perturbed inverse scattering transform, we show in this chapter that a **soliton** can overcome the exponential decay experienced by linear waves propagating through a slab in random medium.

Wave propagation in **waveguides** is analyzed in **Chapter 20**. We consider the case in which the waveguide supports a finite number of propagating modes and the random fluctuations of the medium are three-dimensional. We analyze only transmitted waves through a randomly perturbed waveguide, in the forward-scattering approximation, and the space-time refocusing of these waves after time reversal. We show that stable refocusing does occur, especially when the number of modes is large. This chapter may be considered as a link with the theory of wave propagation in three-dimensional random media.

2

Waves in Homogeneous Media

In this chapter we present some basic aspects of acoustic wave propagation in a homogeneous medium with uniform properties. Solutions of the wave equation can be written explicitly in the form of integrals of elementary wave solutions. A time-dependent solution can be represented as a superposition of Fourier modes that are the solutions of the time-harmonic wave equation. In space, a general solution can be represented as a superposition of plane or spherical waves. We focus our attention on a representation in terms of time-dependent or time-harmonic plane waves since this formulation turns out to be particularly convenient in three-dimensional layered media that are considered in the following chapters.

2.1 Acoustic Wave Equations

The purpose of the first two sections, 2.1.1 and 2.1.2, is to briefly show how the acoustic wave equations can be obtained from the linearization of the conservation laws of fluid dynamics. The following sections consider these equations in a homogeneous medium.

2.1.1 Conservation Equations in Fluid Dynamics

The state of a fluid is characterized by macroscopic quantities such as the density ρ, the three-dimensional fluid velocity $\mathbf{U}$, the pressure P, and the temperature T. These quantities are functions of time t and space $\mathbf{r}$ and their evolution equations can be deduced from first principles, such as the conservation of mass, momentum, and energy. The conservation laws of mass and momentum have the form

$$\frac{\partial \rho}{\partial t} + \nabla \cdot (\rho \mathbf{U}) = 0 \,, \tag{2.1}$$

$$\frac{\partial \rho \mathbf{U}}{\partial t} + \nabla \cdot (\rho \mathbf{U} \otimes \mathbf{U}) + \nabla P = \mathbf{F} \,. \tag{2.2}$$

Here $\mathbf{U} \otimes \mathbf{U}$ is the matrix $(U_i U_j)_{i,j=1,2,3}$ and $\mathbf{F}(t, \mathbf{r})$ is an external force acting on the fluid. It can be an extended force, such as gravity, or a localized one. The energy-conservation equation is given in Appendix 2.3.2. It will be used in Section 2.1.8.

The system (2.1–2.2) is complemented by an equation of state that gives the pressure as a function of the density and the temperature. It can be determined by thermodynamic considerations, or estimated from experimental data. When the flow is isentropic, the pressure is a function of the density only, $P = P(\rho)$. The flow is isentropic if it is adiabatic, that is, no heat is transferred to or from the fluid, and reversible, that is, the flow conditions can return to their original values.

2.1.2 Linearization

The acoustic wave equations are obtained by linearizing the fluid dynamics equations for small disturbances around a fluid at rest. We denote by p_0 and ρ_0 the unperturbed pressure and density, with the unperturbed velocity equal to $\mathbf{0}$, and we consider small perturbations of the pressure, density, and velocity

$$P = p_0 + p, \qquad \rho = \rho_0 (1 + s), \qquad \mathbf{U} = \mathbf{u}, \tag{2.3}$$

where $p_0 = P(\rho_0)$ and s is sometimes refereed to as the condensation. The bulk modulus of the fluid is defined in terms of the equation of state by

$$K_0 = \rho_0 \left(\frac{\partial P}{\partial \rho} \right) (\rho_0). \tag{2.4}$$

The linearization of the equation of state $P = P(\rho)$ by (2.3) gives

$$p = K_0 s. \tag{2.5}$$

By linearizing the mass and momentum conservation equations (2.1–2.2) we obtain the acoustic wave equations

$$\frac{1}{K_0} \frac{\partial p}{\partial t} + \nabla \cdot \mathbf{u} = 0, \tag{2.6}$$

$$\rho_0 \frac{\partial \mathbf{u}}{\partial t} + \nabla p = \mathbf{F}. \tag{2.7}$$

These equations describe the evolution of the acoustic pressure p and velocity $\mathbf{u}$ in the fluid, with an external force $\mathbf{F}$. By taking the time derivative of (2.6) and using (2.7) we get the standard wave equation for the pressure

$$\frac{1}{c_0^2} \frac{\partial^2 p}{\partial t^2} - \Delta p = -\nabla \cdot \mathbf{F}, \tag{2.8}$$

where c_0 is the speed of sound defined by

$$c_0 = \sqrt{\frac{K_0}{\rho_0}} \,. \tag{2.9}$$

If the fluid is at rest at time 0, that is, $\mathbf{u}(t = 0, \mathbf{r}) = \mathbf{0}$, and if the external force comes from a potential ψ, that is, $\mathbf{F} = -\nabla\psi$, then integration of the momentum conservation equation (2.7) gives

$$\mathbf{u}(t, \mathbf{r}) = -\nabla \int_0^t \frac{(p + \psi)(s, \mathbf{r})}{\rho_0} ds \,.$$

This shows that the acoustic velocity field $\mathbf{u}$ is the gradient of a scalar field ϕ

$$\mathbf{u} = \nabla\phi \,.$$

Substituting into equation (2.7), we find that the velocity potential ϕ also satisfies the wave equation

$$\frac{1}{c_0^2}\frac{\partial^2 \phi}{\partial t^2} - \Delta\phi = \frac{-1}{K_0}\frac{\partial \psi}{\partial t} \,. \tag{2.10}$$

The velocity potential is unique only up to a function that depends only on time, and it has been chosen in a particular way here.

2.1.3 Hyperbolicity

By introducing the four-dimensional vector

$$\mathbf{w} = \begin{bmatrix} p \\ \mathbf{u} \end{bmatrix},$$

we can write the acoustic wave equations (2.6–2.7) as

$$\mathbf{M}\frac{\partial \mathbf{w}}{\partial t} + \sum_{j=1}^{3} \mathbf{D}_j \frac{\partial \mathbf{w}}{\partial x_j} = \begin{bmatrix} 0 \\ \mathbf{F} \end{bmatrix},$$

where $\mathbf{M}$ is the 4×4 diagonal matrix with entries $(K_0^{-1}, \rho_0 \mathbf{I})$, and the 4×4 symmetric matrices $\mathbf{D}_j$ are defined by

$$\mathbf{D}_1 = \begin{bmatrix} 0 & 1 & 0 & 0 \\ 1 & 0 & 0 & 0 \\ 0 & 0 & 0 & 0 \\ 0 & 0 & 0 & 0 \end{bmatrix}, \quad \mathbf{D}_2 = \begin{bmatrix} 0 & 0 & 1 & 0 \\ 0 & 0 & 0 & 0 \\ 1 & 0 & 0 & 0 \\ 0 & 0 & 0 & 0 \end{bmatrix}, \quad \mathbf{D}_3 = \begin{bmatrix} 0 & 0 & 0 & 1 \\ 0 & 0 & 0 & 0 \\ 0 & 0 & 0 & 0 \\ 1 & 0 & 0 & 0 \end{bmatrix}.$$

The matrix $\mathbf{M}^{-1}\mathbf{D}(\mathbf{k}) = \sum_{j=1}^{3} \mathbf{M}^{-1}\mathbf{D}_j k_j$ is diagonalizable for any $\mathbf{k} \in \mathbb{R}^3$, with eigenvalues $0, 0, -c_0|\mathbf{k}|$, and $c_0|\mathbf{k}|$, where c_0 is the speed of sound defined by (2.9). This means that the acoustic wave equations (2.6–2.7) form a symmetric hyperbolic system. However, we will not use the general theory here. We will rather give a self-contained analysis of the acoustic wave equations in homogeneous media.

2.1.4 The One-Dimensional Wave Equation

In this section we study briefly the one-dimensional wave equation. It plays an important role in layered media and in the analysis of the three-dimensional wave equation. We consider the partial differential equation

$$\frac{1}{c_0^2}\frac{\partial^2 \tilde{p}}{\partial t^2} - \frac{\partial^2 \tilde{p}}{\partial z^2} = 0\,, \qquad (t,z) \in (0,\infty) \times \mathbb{R}\,, \tag{2.11}$$

with smooth initial conditions $\tilde{p}(t=0,z) = \tilde{p}_0(z)$ and $\partial_t \tilde{p}(t=0,z) = \tilde{p}_1(z)$. By the change of variables $\alpha = z - c_0 t$, $\beta = z + c_0 t$ it can be written in the form

$$\frac{\partial^2 \tilde{p}}{\partial \alpha \partial \beta} = 0\,,$$

whose general solution is the sum $f(\alpha) + g(\beta)$ with arbitrary functions f and g. Therefore, the solution $\tilde{p}$ has the form

$$\tilde{p}(t,z) = f(z - c_0 t) + g(z + c_0 t)\,.$$

The identification of the functions f and g is obtained by inspection from the initial conditions, which gives the d'Alembert formula

$$\tilde{p}(t,z) = \frac{1}{2}\left[\tilde{p}_0(z + c_0 t) + \tilde{p}_0(z - c_0 t)\right] + \frac{1}{2c_0}\int_{z-c_0 t}^{z+c_0 t} \tilde{p}_1(z')dz'\,. \tag{2.12}$$

This representation shows that:

- The initial conditions split into two parts, one moving to the right with velocity c_0 and the other one moving to the left with velocity $-c_0$. This can clearly be seen in Figure 2.1.

- The regularity of the solution is determined by the regularity of the initial conditions. If $\tilde{p}_0 \in \mathcal{C}^k(\mathbb{R})$ and $\tilde{p}_1 \in \mathcal{C}^{k-1}(\mathbb{R})$, $k \geq 2$, then $u \in \mathcal{C}^k([0,\infty)\times\mathbb{R})$, but it is not smoother in general. This is typical for hyperbolic equations.

- The solution at time t and point z depends only on the initial data in the interval $[z - c_0 t, z + c_0 t]$. The finite speed of propagation, giving a finite range of influence, is also typical of hyperbolic equations.

Below we will need a representation formula for the solution of the partial differential equation

$$\frac{1}{c_0^2}\frac{\partial^2 \tilde{p}}{\partial t^2} - \frac{\partial^2 \tilde{p}}{\partial z^2} = 0\,, \qquad (t,z) \in (0,\infty)^2\,, \tag{2.13}$$

with the boundary condition $\tilde{p}(t,z=0) = 0$ and the initial conditions $\tilde{p}(t=0,z) = \tilde{p}_0(z)$ and $\partial_t \tilde{p}(t=0,z) = \tilde{p}_1(z)$. To solve it, we apply the method of

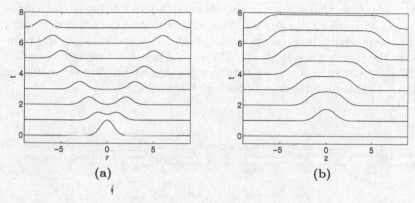

Fig. 2.1. Waves generated by the initial conditions $\tilde{p}(t = 0, z) = \tilde{p}_0(z)$, $\partial_t\tilde{p}(t = 0, z) = \tilde{p}_1(z)$ in a one-dimensional homogeneous medium with a constant speed of sound $c_0 = 1$. In picture (a), we choose $\tilde{p}_0(z) = \exp(-z^2)$ and $\tilde{p}_1(z) = 0$. In picture (b), we choose $\tilde{p}_0(z) = 0$ and $\tilde{p}_1(z) = \exp(-z^2)$. The spatial profiles of the field $\tilde{p}(t, z)$ are plotted at different times.

images, which here means extending the solution and the initial conditions to all of $\mathbb{R}$ by odd reflection. More precisely, for $(t, z) \in [0, \infty) \times \mathbb{R}$ we define

$$\check{p}(t, z) = \begin{cases} \tilde{p}(t, z) & \text{if } z \geq 0, \\ -\tilde{p}(t, -z) & \text{if } z \leq 0, \end{cases} \qquad \check{p}_j(z) = \begin{cases} \tilde{p}_j(z) & \text{if } z \geq 0, \\ -\tilde{p}_j(-z) & \text{if } z \leq 0, \end{cases} \quad j = 0, 1,$$
$$(2.14)$$

so that (2.13) becomes

$$\frac{1}{c_0^2}\frac{\partial^2\check{p}}{\partial t^2} - \frac{\partial^2\check{p}}{\partial z^2} = 0, \qquad (t, z) \in (0, \infty) \times \mathbb{R},$$

with the initial conditions $\check{p}(t = 0, z) = \check{p}_0(z)$ and $\partial_t\check{p}(t = 0, z) = \check{p}_1(z)$. By d'Alembert's formula (2.12) we get for $(t, z) \in [0, \infty) \times \mathbb{R}$,

$$\check{p}(t, z) = \frac{1}{2}\left[\check{p}_0(z + c_0 t) + \check{p}_0(z - c_0 t)\right] + \frac{1}{2c_0}\int_{z - c_0 t}^{z + c_0 t}\check{p}_1(z')dz'.$$

Using the definitions (2.14), we obtain the following expression for the solution $\tilde{p}(t, z)$ for $(t, z) \in [0, \infty)^2$:
If $z \geq c_0 t \geq 0$, then

$$\tilde{p}(t, z) = \frac{1}{2}\left[\tilde{p}_0(c_0 t + z) + \tilde{p}_0(z - c_0 t)\right] + \frac{1}{2c_0}\int_{z - c_0 t}^{z + c_0 t}\tilde{p}_1(z')dz'. \qquad (2.15)$$

If $0 \leq z \leq c_0 t$, then

$$\tilde{p}(t, z) = \frac{1}{2}\left[\tilde{p}_0(c_0 t + z) - \tilde{p}_0(c_0 t - z)\right] + \frac{1}{2c_0}\int_{c_0 t - z}^{c_0 t + z}\tilde{p}_1(z')dz'. \qquad (2.16)$$

This formula shows that the initial conditions split into two parts, one moving to the right with velocity c_0 and the other one moving to the left with velocity $-c_0$. The second part then reflects off the point $z = 0$, and subsequently propagates to the right with velocity c_0. This means that the Dirichlet boundary condition $\tilde{p}(t, z = 0) = 0$ corresponds to reflection with reflection coefficient equal to -1. This reflection can be seen in Figure 2.2.

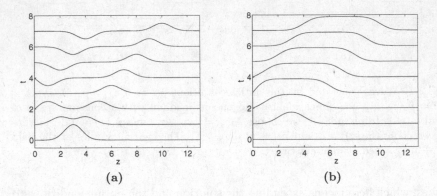

(a)   (b)

Fig. 2.2. Waves generated by the initial conditions $\tilde{p}(t = 0, z) = \tilde{p}_0(z)$, $\partial_t\tilde{p}(t = 0, z) = \tilde{p}_1(z)$ in a one-dimensional homogeneous medium with a constant speed of sound $c_0 = 1$ and a Dirichlet boundary condition $\tilde{p}(t, z = 0) = 0$. In picture (a) we choose $\tilde{p}_0(z) = \exp(-z^2)$ and $\tilde{p}_1(z) = 0$. In picture (b) we choose $\tilde{p}_0(z) = 0$ and $\tilde{p}_1(z) = \exp(-z^2)$. The spatial profiles of the field $\tilde{p}(t, z)$ are plotted at different times.

2.1.5 Solution of the Three-Dimensional Wave Equation by Spherical Means

In this section we obtain an integral representation of the solution of the wave equation

$$\frac{1}{c_0^2}\frac{\partial^2 p}{\partial t^2} - \Delta p = 0, \quad (t, \mathbf{r}) \in (0, \infty) \times \mathbb{R}^3, \quad (2.17)$$

with initial conditions $p(t = 0, \mathbf{r}) = p_0(\mathbf{r})$ and $\partial_t p(t = 0, \mathbf{r}) = p_1(\mathbf{r})$. We reduce the three-dimensional wave equation to a one-dimensional problem using spherical means. Let us assume that $p \in \mathcal{C}^2([0, \infty) \times \mathbb{R}^3)$ is a solution. We define the normalized average $\tilde{p}$ of p over the sphere $\partial B(\mathbf{r}, s)$ centered at $\mathbf{r}$ and with radius $s > 0$ by

$$\tilde{p}(t, \mathbf{r}, s) = \frac{1}{4\pi s}\int_{\partial B(\mathbf{r}, s)} p(t, \mathbf{r}')d\sigma(\mathbf{r}').$$

Note that the area of $\partial B(\mathbf{r}, s)$ is $4\pi s^2$ and consequently

$$\lim_{s \to 0} \tilde{p}(t, \mathbf{r}, s) = 0, \qquad \lim_{s \to 0} \frac{\tilde{p}(t, \mathbf{r}, s)}{s} = p(t, \mathbf{r}). \tag{2.18}$$

The change of variable $\mathbf{r}' \mapsto \mathbf{r} + s\mathbf{r}'$ gives

$$\tilde{p}(t, \mathbf{r}, s) = \frac{s}{4\pi} \int_{\partial B(\mathbf{0}, 1)} p(t, \mathbf{r} + s\mathbf{r}') d\sigma(\mathbf{r}').$$

By differentiating in s, we then obtain

$$\frac{\partial}{\partial s} \left[\frac{\tilde{p}(t, \mathbf{r}, s)}{s} \right] = \frac{1}{4\pi} \int_{\partial B(\mathbf{0}, 1)} \mathbf{r}' \cdot \nabla p(t, \mathbf{r} + s\mathbf{r}') d\sigma(\mathbf{r}')$$

$$= \frac{1}{4\pi s^2} \int_{\partial B(\mathbf{r}, s)} \frac{\mathbf{r}' - \mathbf{r}}{s} \cdot \nabla p(t, \mathbf{r}') d\sigma(\mathbf{r}').$$

Since $\frac{\mathbf{r}' - \mathbf{r}}{s}$ is the unit outward normal to the ball $B(\mathbf{r}, s)$ we can apply the Gauss–Green theorem (see Appendix 2.3.1), which gives

$$\frac{\partial}{\partial s} \left[\frac{\tilde{p}(t, \mathbf{r}, s)}{s} \right] = \frac{1}{4\pi s^2} \int_{B(\mathbf{r}, s)} \Delta p(t, \mathbf{r}') d\mathbf{r}' = \frac{1}{4\pi c_0^2 s^2} \int_{B(\mathbf{r}, s)} \frac{\partial^2 p}{\partial t^2}(t, \mathbf{r}') d\mathbf{r}'.$$

Next, we multiply by s^2 and differentiate with respect to s:

$$\frac{\partial}{\partial s} \left\{ s^2 \frac{\partial}{\partial s} \left[\frac{\tilde{p}(t, \mathbf{r}, s)}{s} \right] \right\} = \frac{1}{4\pi c_0^2} \frac{\partial}{\partial s} \left[\int_{B(\mathbf{r}, s)} \frac{\partial^2 p}{\partial t^2}(t, \mathbf{r}') d\mathbf{r}' \right]$$

$$= \frac{1}{4\pi c_0^2} \lim_{\delta s \to 0} \frac{1}{\delta s} \left[\int_{B(\mathbf{r}, s+\delta s) \setminus B(\mathbf{r}, s)} \frac{\partial^2 p}{\partial t^2}(t, \mathbf{r}') d\mathbf{r}' \right].$$

By the continuity of $\partial_t^2 p$ we find that

$$\frac{\partial}{\partial s} \left\{ s^2 \frac{\partial}{\partial s} \left[\frac{\tilde{p}(t, \mathbf{r}, s)}{s} \right] \right\} = \frac{1}{4\pi c_0^2} \int_{\partial B(\mathbf{r}, s)} \frac{\partial^2 p}{\partial t^2}(t, \mathbf{r}') d\sigma(\mathbf{r}') = \frac{s}{c_0^2} \frac{\partial^2 \tilde{p}(t, \mathbf{r}, s)}{\partial t^2}.$$

Since

$$\frac{\partial}{\partial s} \left\{ s^2 \frac{\partial}{\partial s} \left[\frac{\tilde{p}}{s} \right] \right\} = s \frac{\partial^2 \tilde{p}}{\partial s^2},$$

we see that $\tilde{p}(t, \mathbf{r}, s)$ is a solution of the one-dimensional wave equation as a function of (t, s),

$$\frac{1}{c_0^2} \frac{\partial^2 \tilde{p}(t, \mathbf{r}, s)}{\partial t^2} - \frac{\partial^2 \tilde{p}(t, \mathbf{r}, s)}{\partial s^2} = 0, \qquad (t, s) \in (0, \infty)^2.$$

It also satisfies the boundary condition $\tilde{p}(t, \mathbf{r}, s = 0) = 0$ and the initial conditions $\tilde{p}(t = 0, \mathbf{r}, s) = \tilde{p}_0(\mathbf{r}, s)$ and $\partial_t \tilde{p}(t = 0, \mathbf{r}, s) = \tilde{p}_1(\mathbf{r}, s)$. Here $\mathbf{r}$ is a frozen parameter and $\tilde{p}_j$, $j = 0, 1$, are defined as the normalized averages over the sphere $\partial B(\mathbf{r}, s)$ of p_j, $j = 0, 1$:

$$\tilde{p}_j(\mathbf{r}, s) = \frac{1}{4\pi s} \int_{\partial B(\mathbf{r}, s)} p_j(\mathbf{r}') d\sigma(\mathbf{r}') . \qquad (2.19)$$

By (2.16), if $0 \le s \le c_0 t$, then $\tilde{p}$ has the form

$$\tilde{p}(t, \mathbf{r}, s) = \frac{1}{2} \left[\tilde{p}_0 (\mathbf{r}, c_0 t + s) - \tilde{p}_0 (\mathbf{r}, c_0 t - s) \right] + \frac{1}{2c_0} \int_{c_0 t - s}^{c_0 t + s} \tilde{p}_1(\mathbf{r}, s') ds' .$$

Using (2.18) and (2.19) we finally get the representation

$$p(t, \mathbf{r}) = \frac{\partial \tilde{p}_0}{\partial s} (\mathbf{r}, c_0 t) + \frac{1}{c_0} \tilde{p}_1 (\mathbf{r}, c_0 t)$$

$$= \frac{\partial}{\partial t} \left[\frac{1}{4\pi c_0^2 t} \int_{\partial B(\mathbf{r}, c_0 t)} p_0(\mathbf{r}') d\sigma(\mathbf{r}') \right] + \frac{1}{4\pi c_0^2 t} \int_{\partial B(\mathbf{r}, c_0 t)} p_1(\mathbf{r}') d\sigma(\mathbf{r}') . \, (2.20)$$

This is known as Kirchhoff's representation formula. There is an equivalent form obtained by computing the time derivative

$$\frac{\partial}{\partial t} \left[\frac{1}{4\pi c_0^2 t} \int_{\partial B(\mathbf{r}, c_0 t)} p_0(\mathbf{r}') d\sigma(\mathbf{r}') \right]$$

$$= \frac{\partial}{\partial t} \left[\frac{t}{4\pi} \int_{\partial B(\mathbf{0}, 1)} p_0(\mathbf{r} + c_0 t \mathbf{r}') d\sigma(\mathbf{r}') \right]$$

$$= \frac{1}{4\pi} \int_{\partial B(\mathbf{0}, 1)} \left[p_0(\mathbf{r} + c_0 t \mathbf{r}') + c_0 t \mathbf{r}' \cdot \nabla p_0(\mathbf{r} + c_0 t \mathbf{r}') \right] d\sigma(\mathbf{r}')$$

$$= \frac{1}{4\pi c_0^2 t^2} \int_{\partial B(\mathbf{r}, c_0 t)} \left[p_0(\mathbf{r}') + (\mathbf{r}' - \mathbf{r}) \cdot \nabla p_0(\mathbf{r}') \right] d\sigma(\mathbf{r}') . \quad (2.21)$$

Kirchhoff's formula shows that the solution of the three-dimensional wave equation at a point $\mathbf{r}$ and at time t depends only on the initial data on the sphere $\partial B(\mathbf{r}, c_0 t)$. We will use this property of the wave equation in the context of acoustic waves in Section 2.1.7.

As an example we consider the initial data $p_0(\mathbf{r}) = \exp(-|\mathbf{r}|^2/r_0^2)$ and $p_1(\mathbf{r}) = 0$. The solution is a spherical wave since it depends only on $|\mathbf{r}|$ and has the form

$$p(t, \mathbf{r}) = \exp \left(-\frac{c_0^2 t^2 + |\mathbf{r}|^2}{r_0^2} \right) \left[\cosh \left(\frac{2 c_0 t |\mathbf{r}|}{r_0^2} \right) - \frac{c_0 t}{|\mathbf{r}|} \sinh \left(\frac{2 c_0 t |\mathbf{r}|}{r_0^2} \right) \right] .$$

As seen in Figure 2.3, the initial conditions localized at $\mathbf{0}$ give rise to a wave that propagates with speed c_0. Its amplitude decays as $1/|\mathbf{r}|$ and it takes an asymptotic form. For times associated with travel distances that are long compared to the support of the initial data, the field takes the asymptotic form

$$p(t, \mathbf{r}) \overset{c_0 t \gg r_0}{\sim} \frac{r_0}{2|\mathbf{r}|} \frac{|\mathbf{r}| - c_0 t}{r_0} \exp \left[-\frac{(|\mathbf{r}| - c_0 t)^2}{r_0^2} \right] .$$

We can also consider the case in which the initial data are $p_0(\mathbf{r}) = 0$ and $p_1(\mathbf{r}) = (1/t_0)\exp(-|\mathbf{r}|^2/r_0^2)$. The solution is then given by

$$p(t,\mathbf{r}) = \frac{r_0^2}{2c_0 t_0 |\mathbf{r}|} \exp\left(-\frac{c_0^2 t^2 + |\mathbf{r}|^2}{r_0^2}\right) \sinh\left(\frac{2c_0 t |\mathbf{r}|}{r_0^2}\right),$$

and it is plotted in Figure 2.3. Its asymptotic form is

$$p(t,\mathbf{r}) \overset{c_0 t \gg r_0}{\approx} \frac{r_0^2}{4c_0 t_0 |\mathbf{r}|} \exp\left[-\frac{(|\mathbf{r}| - c_0 t)^2}{r_0^2}\right].$$

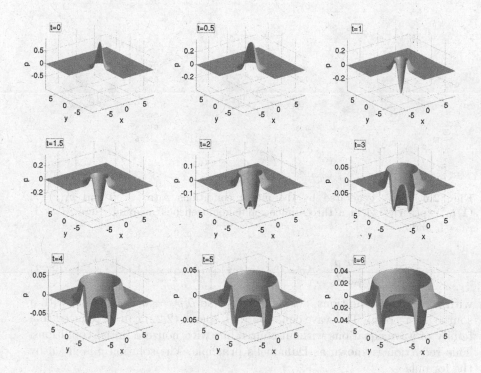

Fig. 2.3. Wave generated by the initial conditions $p_0(\mathbf{r}) = \exp(-|\mathbf{r}|^2/r_0^2)$ and $p_1(\mathbf{r}) = 0$ in a three-dimensional homogeneous medium. Here $c_0 = 1$ and $r_0 = 1$. The spatial profiles of the field $p(t, \mathbf{r} = (x, y, z))$ in the plane (x, y) (i.e., $z = 0$) are plotted at different times.

2.1.6 The Three-Dimensional Wave Equation With Source

In this section we obtain an integral representation for the solution of the wave equation

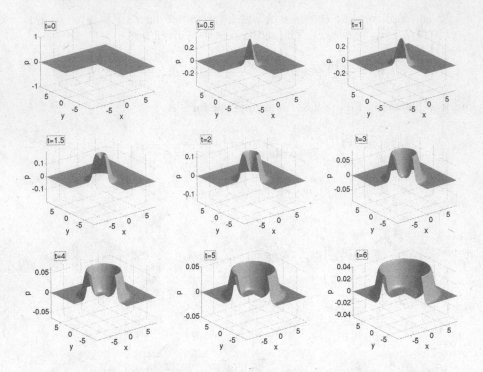

Fig. 2.4. Wave generated by the initial conditions $p_0(\mathbf{r}) = 0$ and $p_1(\mathbf{r}) = (1/t_0)\exp(-|\mathbf{r}|^2/r_0^2)$ in a three-dimensional homogeneous medium. Here $c_0 = 1$, $r_0 = 1$, and $t_0 = 1$.

$$\frac{1}{c_0^2}\frac{\partial^2 p}{\partial t^2} - \Delta p = f, \quad (t,\mathbf{r}) \in (0,\infty) \times \mathbb{R}^3, \tag{2.22}$$

with zero initial conditions, $p(t = 0,\mathbf{r}) = 0$ and $\partial_t p(t = 0,\mathbf{r}) = 0$, and with a source term $f(t,\mathbf{r})$. The wave equation with source (2.22) can be reduced to a family of wave equations without source but with nonzero initial conditions. This reduction is known as Duhamel's principle. The solution p is given by the formula

$$p(t,\mathbf{r}) = \int_0^t \tilde{p}(t,\mathbf{r},s)\,ds, \tag{2.23}$$

where $\tilde{p}(t,\mathbf{r},s)$, for any fixed $s \geq 0$, is a solution of

$$\frac{1}{c_0^2}\frac{\partial^2 \tilde{p}}{\partial t^2} - \Delta\tilde{p} = 0, \quad (t,\mathbf{r}) \in (s,\infty) \times \mathbb{R}^3,$$

with the initial conditions $\tilde{p}(t = s,\mathbf{r},s) = 0$ and $\partial_t\tilde{p}(t = s,\mathbf{r},s) = c_0^2 f(s,\mathbf{r})$. The function p defined by (2.23) satisfies

$$\frac{\partial p}{\partial t}(t,\mathbf{r}) = \tilde{p}(t,\mathbf{r},t) + \int_0^t \frac{\partial\tilde{p}}{\partial t}(t,\mathbf{r},s)ds = \int_0^t \frac{\partial\tilde{p}}{\partial t}(t,\mathbf{r},s)\,ds,$$

and

$$\frac{\partial^2 p}{\partial t^2}(t, \mathbf{r}) = \frac{\partial \tilde{p}}{\partial t}(t, \mathbf{r}, t) + \int_0^t \frac{\partial^2 \tilde{p}}{\partial t^2}(t, \mathbf{r}, s)\, ds = c_0^2 f(t, \mathbf{r}) + c_0^2 \int_0^t \Delta \tilde{p}(t, \mathbf{r}, s)\, ds$$
$$= c_0^2 f(t, \mathbf{r}) + c_0^2 \Delta p(t, \mathbf{r}).$$

This shows that p satisfies the wave equation with source (2.22). It is straightforward to check that it also satisfies the zero initial conditions. This verifies Duhamel's principle. Using the representation (2.20) for $\tilde{p}$ in (2.23), we get

$$p(t, \mathbf{r}) = \int_0^t \frac{1}{4\pi(t - s)} \int_{\partial B(\mathbf{r}, c_0(t-s))} f(s, \mathbf{r}') d\sigma(\mathbf{r}')\, ds$$
$$= \int_{B(\mathbf{r}, c_0 t)} \frac{1}{4\pi|\mathbf{r} - \mathbf{r}'|} f\left(t - \frac{|\mathbf{r} - \mathbf{r}'|}{c_0}, \mathbf{r}'\right) d\mathbf{r}', \qquad (2.24)$$

where $B(\mathbf{r}, c_0 t)$ is the ball centered at $\mathbf{r}$ and with radius $c_0 t$. This representation is known as the retarded potential representation.

2.1.7 Green's Function for the Acoustic Wave Equations

Using the results for the three-dimensional wave equation presented in the two previous sections, it is possible to obtain integral representations for the solution $(p, \mathbf{u})$ of the acoustic wave equations (2.6–2.7).

Let us assume that the fluid is initially at rest, that is, $\mathbf{u}(t = 0, \mathbf{r}) = \mathbf{0}$ and $p(t = 0, \mathbf{r}) = 0$, and that $\mathbf{F} = -\nabla\psi$ with $\psi \in \mathcal{C}^4([0, \infty) \times \mathbb{R}^3)$. By (2.24), there is a unique solution $(\mathbf{u}, p)$ in $\mathcal{C}^2([0, \infty) \times \mathbb{R}^3)$ that has the form $\mathbf{u} = \nabla\phi$ with

$$\phi(t, \mathbf{r}) = \mathcal{K}\left(\frac{-1}{K_0} \frac{\partial \psi}{\partial t}\right)(t, \mathbf{r}), \qquad p(t, \mathbf{r}) = \mathcal{K}\left(\Delta\psi\right)(t, \mathbf{r}). \qquad (2.25)$$

Here the operator $\mathcal{K}$ is defined by

$$\mathcal{K}(f)(t, \mathbf{r}) = \int_{B(\mathbf{r}, c_0 t)} \frac{1}{4\pi|\mathbf{r} - \mathbf{r}'|} f\left(t - \frac{|\mathbf{r} - \mathbf{r}'|}{c_0}, \mathbf{r}'\right) d\mathbf{r}'. \qquad (2.26)$$

This formula can be written in convolution form as

$$\mathcal{K}(f)(t, \mathbf{r}) = \iint f(t', \mathbf{r}') G(t - t', \mathbf{r} - \mathbf{r}')\, dt'\, d\mathbf{r}',$$

where G is the Green function

$$G(t, \mathbf{r}) = \frac{1}{4\pi|\mathbf{r}|} \delta\left(\frac{|\mathbf{r}|}{c_0} - t\right). \qquad (2.27)$$

It is the fundamental solution of the wave equation in $[0, \infty) \times \mathbb{R}^3$, that is, it solves the equation

$$\frac{1}{c_0^2}\frac{\partial^2 G}{\partial t^2} - \Delta G = \delta(t)\delta(\mathbf{r}), \qquad (t,\mathbf{r}) \in (0,\infty) \times \mathbb{R}^3,$$

with zero initial conditions. Here δ is the Dirac distribution. The Green function is the response of the system to a point source at position $\mathbf{0}$ that emits an impulse at time zero. From the form of the Green function we see that the response is a spherical wave, centered at $\mathbf{0}$, that propagates outward with time. In order for the wave to reach position $\mathbf{r}$ at time t it must have been emitted from the source at $\mathbf{0}$ at the time $t - |\mathbf{r}|/c_0$.

Let us consider the case in which there is no source inside the medium but the fluid is not initially at rest. Using (2.6), (2.7), (2.20), and (2.21) we find that the solution without sources and with nonzero initial conditions can be written as follows: If $\mathbf{u}(t=0,\mathbf{r}) = \mathbf{u}_0(\mathbf{r})$ and $p(t=0,\mathbf{r}) = p_0(\mathbf{r})$, and $\mathbf{F} = \mathbf{0}$, with $\mathbf{u}_0$ and $p_0 \in \mathcal{C}^2(\mathbb{R})$, then

$$p(t,\mathbf{r}) = \frac{\partial}{\partial t}\left[\frac{1}{4\pi c_0^2 t}\int_{\partial B(\mathbf{r},c_0 t)} p_0(\mathbf{r}')d\sigma(\mathbf{r}')\right]$$

$$-\frac{K_0}{4\pi c_0^2 t}\int_{\partial B(\mathbf{r},c_0 t)} \nabla \cdot \mathbf{u}_0(\mathbf{r}')\,d\sigma(\mathbf{r}')$$

$$= \frac{1}{4\pi c_0^2 t^2}\int_{\partial B(\mathbf{r},c_0 t)}\left[p_0(\mathbf{r}') + (\mathbf{r}'-\mathbf{r})\cdot\nabla p_0(\mathbf{r}') - K_0 t \nabla\cdot\mathbf{u}_0(\mathbf{r}')\right]d\sigma(\mathbf{r}'),$$

$$\mathbf{u}(t,\mathbf{r}) = \frac{\partial}{\partial t}\left[\frac{1}{4\pi c_0^2 t}\int_{\partial B(\mathbf{r},c_0 t)} \mathbf{u}_0(\mathbf{r}')\,d\sigma(\mathbf{r}')\right]$$

$$-\frac{1}{4\pi\rho_0 c_0^2 t}\int_{\partial B(\mathbf{r},c_0 t)} \nabla p_0(\mathbf{r}')\,d\sigma(\mathbf{r}')$$

$$= \frac{1}{4\pi c_0^2 t^2}\int_{\partial B(\mathbf{r},c_0 t)}\left[\mathbf{u}_0(\mathbf{r}') + [(\mathbf{r}'-\mathbf{r})\cdot\nabla]\mathbf{u}_0(\mathbf{r}') - \frac{t}{\rho_0}\nabla p_0(\mathbf{r}')\right]d\sigma(\mathbf{r}').$$

These formulas can be written in terms of the Green function (2.27) as

$$p(t,\mathbf{r}) = \frac{\partial}{\partial t}\left[c_0^2\int_{\mathbb{R}^3} p_0(\mathbf{r}')G(t,\mathbf{r}-\mathbf{r}')\,d\mathbf{r}'\right]$$

$$-c_0^2 K_0\int_{\mathbb{R}^3} \nabla\cdot\mathbf{u}_0(\mathbf{r}')G(t,\mathbf{r}-\mathbf{r}')\,d\mathbf{r}', \qquad (2.28)$$

$$\mathbf{u}(t,\mathbf{r}) = \frac{\partial}{\partial t}\left[c_0^2\int_{\mathbb{R}^3} \mathbf{u}_0(\mathbf{r}')G(t,\mathbf{r}-\mathbf{r}')\,d\mathbf{r}'\right]$$

$$-\frac{c_0^2}{\rho_0}\int_{\mathbb{R}^3} \nabla p_0(\mathbf{r}')G(t,\mathbf{r}-\mathbf{r}')\,d\mathbf{r}'. \qquad (2.29)$$

This shows that the solution at a given time t and point $\mathbf{r}$ depends only on the initial data on the sphere of radius $c_0 t$. It does not depend on data in the interior of this sphere. The interior of the sphere is a *lacuna* for the solution. This phenomenon is called Huygens's principle. It is true for waves propagating in odd dimensions, except for one dimension. The d'Alembert formula (2.12) shows that the solution of the one-dimensional wave equation depends on the initial data inside the "sphere." Huygens's principle is not satisfied in even dimensions.

By superposition, it is also possible to write the solution in the general case in which the fluid is not initially at rest and there is a source inside the medium.

2.1.8 Energy Density and Energy Flux

The wave energy density is defined by

$$e(t, \mathbf{r}) = \frac{\rho_0}{2} |\mathbf{u}(t, \mathbf{r})|^2 + \frac{1}{2K_0} p^2(t, \mathbf{r}) \,. \tag{2.30}$$

The time partial derivative of this quantity is obtained from (2.6–2.7):

$$\frac{\partial e}{\partial t} = -\mathbf{u} \cdot \nabla p - p \nabla \cdot \mathbf{u} + \mathbf{F} \cdot \mathbf{u} = -\nabla \cdot (p\mathbf{u}) + \mathbf{F} \cdot \mathbf{u} \,. \tag{2.31}$$

The wave energy in a domain V enclosed by a smooth surface ∂V satisfies the identity

$$\frac{d}{dt} \left(\int_V e(t, \mathbf{r}) d\mathbf{r} \right) = \int_V -\nabla \cdot (p\mathbf{u}) + \mathbf{F} \cdot \mathbf{u} \, d\mathbf{r}$$

$$= -\int_{\partial V} p\mathbf{u} \cdot \mathbf{n} \, d\sigma(\mathbf{r}) + \int_V \mathbf{F} \cdot \mathbf{u} \, d\mathbf{r} \,,$$

where $\mathbf{n}$ is the outward unit normal to V. Here we have used the Gauss–Green theorem. This identity shows that the energy flux is

$$\mathbf{m}(t, \mathbf{r}) = p\mathbf{u}(t, \mathbf{r}) \,. \tag{2.32}$$

It is also possible to get this energy equation from the linearization of the full energy conservation equation for fluid flows (see Appendix 2.3.2).

If the fluid is initially at rest with a source term $\mathbf{F}$ that is compactly supported in time and space, then the total wave energy

$$\mathcal{E}(t) = \int_{\mathbb{R}^3} e(t, \mathbf{r}) \, d\mathbf{r}$$

satisfies

$$\frac{d\mathcal{E}}{dt} = \int_{\mathbb{R}^3} \mathbf{F} \cdot \mathbf{u}(t, \mathbf{r}) \, d\mathbf{r} \,.$$

This identity shows that the energy is constant when there is no source in the medium.

2.2 Wave Decompositions in Three-Dimensional Media

2.2.1 Time Harmonic Waves

In this section we introduce wave field decompositions that can be used to reduce various wave propagation problems into simpler ones. We consider waves propagating in three spatial dimensions and in a homogeneous medium. The governing equations are the acoustic wave equations (2.6–2.7) with constant bulk modulus K_0 and density ρ_0. The source is modeled by the forcing term $\mathbf{F}(t, \mathbf{r})$, which we assume is smooth and compactly supported in space and time. The pressure p solves the scalar wave equation

$$\frac{1}{c_0^2}\frac{\partial^2 p}{\partial t^2} - \Delta p = -\nabla \cdot \mathbf{F}(t, \mathbf{r}) \, , \tag{2.33}$$

with $c_0 = \sqrt{K_0/\rho_0}$ the speed of sound in the homogeneous medium. Using the Green function in (2.27) we find that the pressure field can be expressed as

$$p(t, \mathbf{r}) = -\int \frac{\nabla \cdot \mathbf{F}(t - |\mathbf{r} - \mathbf{r}'|/c_0, \mathbf{r}')}{4\pi|\mathbf{r} - \mathbf{r}'|}d\mathbf{r}' \, , \tag{2.34}$$

and that $p(\cdot, \mathbf{r})$ is compactly supported in time.

We are interested in harmonic or monochromatic waves corresponding to excitation at a particular frequency. By the Fourier transform in time we can write

$$\hat{p}(\omega, \mathbf{r}) = \int e^{i\omega t}p(t, \mathbf{r})dt \, ,$$

$$p(t, \mathbf{r}) = \frac{1}{2\pi}\int e^{-i\omega t}\hat{p}(\omega, \mathbf{r})d\omega \, ,$$

so that the pressure field p is a superposition of wave components with time dependence $\exp(-i\omega t)$, which are called monochromatic waves. The harmonic component

$$e^{-i\omega t}\hat{p}(\omega, \mathbf{r})$$

results from harmonic forcing at the angular frequency ω. From (2.33) we find that $\hat{p}$ solves the Helmholtz equation with source term

$$k^2\hat{p} + \Delta\hat{p} = \nabla \cdot \hat{\mathbf{F}}(\omega, \mathbf{r}) \, ,$$

where the wave number k is defined by $k = \omega/c_0$. It follows from the representation (2.34) that we can associate the harmonic Green function

$$\hat{G}(\omega, \mathbf{r}) = \frac{e^{i\omega|\mathbf{r}|/c_0}}{4\pi|\mathbf{r}|} \tag{2.35}$$

with the three-dimensional Helmhotz equation. It is the Fourier transform of the time-dependent Green function (2.27) and it solves

$$k^2 \hat{G} + \Delta \hat{G} = -\delta(\mathbf{r}).$$

The sign in the phase of the exponential in (2.35) means that wave energy propagates away from the source as time increases. This is a form of causality. The function $\hat{G}$ is the outgoing free-space Green function of the Helmholtz equation.

We can now write

$$\hat{p}(\omega, \mathbf{r}) = -\int \nabla \cdot \hat{\mathbf{F}}(\omega, \mathbf{r}') \hat{G}(\omega, \mathbf{r} - \mathbf{r}') \, d\mathbf{r}'.$$

For the monochromatic spatial point source

$$-\nabla \cdot \hat{\mathbf{F}}(\omega, \mathbf{r}) = \delta(\mathbf{r} - \mathbf{r}_s)\hat{f}(\omega),$$

where $\mathbf{r}_s$ is the source location, we find that

$$\hat{p}(\omega, \mathbf{r}) = \hat{f}(\omega)\frac{e^{i\omega|\mathbf{r} - \mathbf{r}_s|/c_0}}{4\pi|\mathbf{r} - \mathbf{r}_s|}. \tag{2.36}$$

The corresponding progressing harmonic pressure field

$$p(t, \mathbf{r}) = \exp(-i\omega t)\hat{p}(\omega, \mathbf{r})$$

is a wave of decaying amplitude, temporal frequency ω, and wavelength $\lambda = 2\pi/k$, propagating away from the source with speed c_0. It follows from (2.36) that

$$\lim_{|\mathbf{r}| \to \infty} |\mathbf{r}| \left(\frac{\partial}{\partial|\mathbf{r}|} - i\frac{\omega}{c_0} \right) \hat{p}(\omega, \mathbf{r}) = 0,$$

uniformly in all directions $\hat{\mathbf{r}} = \mathbf{r}/|\mathbf{r}|$. This property is the Sommerfeld's radiation condition. It holds also for a source that is compactly supported in space. The pressure field behaves like an outgoing spherical wave far away from the source region. In homogeneous media, Sommerfeld's radiation condition is a consequence of the explicit form (2.35) of the Green function.

2.2.2 Plane Waves

We consider next the case in which the pressure has the form

$$p(t, \mathbf{r}) = f(t, \mathbf{r} \cdot \mathbf{q}),$$

for $\mathbf{q} \in \mathbb{R}^3$ a fixed vector and f differentiable. That is, the pressure field is a plane wave, because it is constant in planes perpendicular to $\mathbf{q}$, and it propagates in the direction $\mathbf{q}$. In the absence of a source we find from (2.33) that $f = f(t, r)$ solves the one-dimensional wave equation

$$\frac{1}{c_0^2}\frac{\partial^2 f}{\partial t^2} - q^2\frac{\partial^2 f}{\partial r^2} = 0\,, \tag{2.37}$$

for $q = |\mathbf{q}|$. By d'Alembert's formula the general solution is

$$f(t,r) = f_1\left(t - \frac{r}{qc_0}\right) + f_2\left(t + \frac{r}{qc_0}\right)\,, \tag{2.38}$$

with f_1 and f_2 arbitrary functions. We then see that

$$p(t,\mathbf{r}) = f_1\left(t - \frac{\mathbf{r}\cdot\mathbf{q}}{qc_0}\right) + f_2\left(t + \frac{\mathbf{r}\cdot\mathbf{q}}{qc_0}\right)$$

is the general form of a plane wave pressure field. The corresponding harmonic plane waves take the form

$$\hat{p}(\omega,\mathbf{r})e^{-i\omega t} = \hat{f}_1(\omega)e^{-i\omega(t-\mathbf{r}\cdot\mathbf{q}/(qc_0))} + \hat{f}_2(\omega)e^{-i\omega(t+\mathbf{r}\cdot\mathbf{q}/(qc_0))}\,.$$

They are characterized by the complex amplitudes $\hat{f}_1$ and $\hat{f}_2$, their direction of propagation $\pm\mathbf{q}$, their speed c_0, their angular frequency ω, and their wavelength $\lambda = 2\pi c_0/\omega = 2\pi/k$.

2.2.3 Spherical Waves

We consider now waves of the form

$$p(t,\mathbf{r}) = f(t,|\mathbf{r}|)\,,$$

and with f again differentiable. This pressure field is a spherical wave centered at the origin. Writing the Laplacian in spherical coordinates, we find that the equation satisfied by $f = f(t,r)$ is

$$\frac{1}{c_0^2}\frac{\partial^2 f}{\partial t^2} - \frac{1}{r}\frac{\partial^2(rf)}{\partial r^2} = 0\,,$$

which we rewrite as

$$\frac{1}{c_0^2}\frac{\partial^2(rf)}{\partial t^2} - \frac{\partial^2(rf)}{\partial r^2} = 0\,.$$

This is the one-dimensional wave equation in $r \in [0,\infty)$, with the Dirichlet boundary condition $rf(t,r)\,|_{r=0} = 0$. It then follows from the analysis of Section 2.1.4 that f has the form

$$rf(t,r) = f_1\left(t - \frac{r}{c_0}\right) + f_2\left(t + \frac{r}{c_0}\right)\,,$$

with f_1 and f_2 arbitrary odd functions, and

$$p(t, \mathbf{r}) = \frac{1}{|\mathbf{r}|} f_1 \left(t - \frac{|\mathbf{r}|}{c_0} \right) + \frac{1}{|\mathbf{r}|} f_2 \left(t + \frac{|\mathbf{r}|}{c_0} \right). \tag{2.39}$$

The pressure component f_1 corresponds to a spherical wave emanating from the origin, and the component f_2 to a spherical wave converging toward the origin, both of them propagating with speed c_0. In (2.36) we considered a harmonic point source that gives rise to a diverging spherical wave of the form given by the f_1 term. It follows from (2.39) that a spherical harmonic wave diverging from $\mathbf{r}_s$ has in general the form (2.36).

2.2.4 Weyl's Representation of Spherical Waves

In later chapters we consider a three-dimensional point source located above a layered medium. We give here a decomposition into plane waves of the spherical-wave field from a point source, when the parameterization is in terms of a horizontal slowness vector. Such decompositions are used frequently throughout the book.

Consider the outgoing free-space Green function $\hat{G}$, that is the solution of the Helmholtz equation corresponding to radiation from a point source. It solves the equation

$$k^2 \hat{G} + \Delta \hat{G} = -\delta(\mathbf{r}), \tag{2.40}$$

along with Sommerfeld's radiation condition. We then decompose the three-dimensional space coordinate $\mathbf{r}$ into the vertical component z and two-dimensional horizontal components $\mathbf{x}$. In layered media considered in the following chapters the vertical direction is in the direction of the layering so that the medium parameters are independent of the horizontal space coordinates. This motivates transforming (2.40) in the horizontal directions. We define

$$\check{G}(\omega, \boldsymbol{\kappa}, z) = \int \hat{G}(\omega, \mathbf{r}) e^{-i\omega \boldsymbol{\kappa} \cdot \mathbf{x}} d\mathbf{x}.$$

We shall see that the use of this particular Fourier transform with dual space variable $\omega\boldsymbol{\kappa}$ is convenient, since it leads to expressions that are plane waves. The harmonic Green function can be recovered by the inverse transform

$$\hat{G}(\omega, \mathbf{x}, z) = \frac{1}{4\pi^2} \int \check{G}(\omega, \boldsymbol{\kappa}, z) e^{i\omega \boldsymbol{\kappa} \cdot \mathbf{x}} \omega^2 d\boldsymbol{\kappa}.$$

From (2.40) we find that $\check{G}$ solves

$$\omega^2 (c_0^{-2} - |\boldsymbol{\kappa}|^2) \check{G} + \frac{\partial^2 \check{G}}{\partial z^2} = -\delta(z), \tag{2.41}$$

which is the Helmholtz equation for the free-space Green function in one dimension.

Let us first consider the case $|\boldsymbol{\kappa}|c_0 < 1$. The general solution of

$$q^2 \check{g} + \frac{\partial^2 \check{g}}{\partial z^2} = 0$$

is

$$\check{g}(z) = g_1 e^{-iqz} + g_2 e^{iqz}.$$

By (2.41) the harmonic Green function must be of this form in, respectively, the upper half-plane $z > 0$ and the lower half-plane $z < 0$:

$$\check{G}(\omega, \boldsymbol{\kappa}, z) = \begin{cases} g_1 e^{-i\omega\sqrt{c_0^{-2} - |\boldsymbol{\kappa}|^2}z} + g_2 e^{i\omega\sqrt{c_0^{-2} - |\boldsymbol{\kappa}|^2}z} & \text{for } z > 0, \\ g_3 e^{-i\omega\sqrt{c_0^{-2} - |\boldsymbol{\kappa}|^2}z} + g_4 e^{i\omega\sqrt{c_0^{-2} - |\boldsymbol{\kappa}|^2}z} & \text{for } z < 0. \end{cases}$$

The radiation condition enforces $g_1 = 0$ and $g_4 = 0$. Moreover, the Green function must satisfy the jump condition

$$\frac{\partial \check{G}}{\partial z}(z = 0^+) - \frac{\partial \check{G}}{\partial z}(z = 0^-) = -1.$$

This condition is obtained by integrating (2.41) between 0^+ and 0^-. Integrating one more time, we obtain the continuity condition

$$\check{G}(z = 0^+) - \check{G}(z = 0^-) = 0.$$

From these two conditions we conclude that $\check{G}$ is given by

$$\check{G}(\omega, \boldsymbol{\kappa}, z) = \frac{ie^{i\omega\sqrt{c_0^{-2} - |\boldsymbol{\kappa}|^2}|z|}}{2\omega\sqrt{c_0^{-2} - |\boldsymbol{\kappa}|^2}}, \tag{2.42}$$

for $z \neq 0$.

The same analysis in the case $|\boldsymbol{\kappa}|c_0 > 1$ gives

$$\check{G}(\omega, \boldsymbol{\kappa}, z) = \frac{e^{-|\omega|\sqrt{|\boldsymbol{\kappa}|^2 - c_0^{-2}}|z|}}{2|\omega|\sqrt{|\boldsymbol{\kappa}|^2 - c_0^{-2}}}. \tag{2.43}$$

Note the exponential decay away from the point source, so that Sommerfeld's radiation condition is satisfied. These exponentially damped modes are called *evanescent* modes. The case $c_0|\boldsymbol{\kappa}| = 1$ corresponds to plane waves traveling in the horizontal direction with speed c_0. For $c_0|\boldsymbol{\kappa}| < 1$ we have plane waves traveling obliquely relative to the horizontal direction, while for $c_0|\boldsymbol{\kappa}| > 1$ we have exponentially damped or evanescent modes due to the finite speed of propagation.

To obtain the Weyl representation, we inverse-transform in the lateral dimensions:

$$\hat{G}(\omega, \mathbf{x}, z) = \frac{i\omega}{8\pi^2} \int_{c_0|\boldsymbol{\kappa}|<1} \frac{e^{i\omega(\mathbf{x}\cdot\boldsymbol{\kappa}+\sqrt{c_0^{-2}-|\boldsymbol{\kappa}|^2}|z|)}}{\sqrt{c_0^{-2}-|\boldsymbol{\kappa}|^2}} d\boldsymbol{\kappa}$$

$$+\frac{|\omega|}{8\pi^2} \int_{c_0|\boldsymbol{\kappa}|>1} \frac{e^{i\omega\mathbf{x}\cdot\boldsymbol{\kappa}-|\omega|\sqrt{|\boldsymbol{\kappa}|^2-c_0^{-2}}|z|}}{\sqrt{|\boldsymbol{\kappa}|^2-c_0^{-2}}} d\boldsymbol{\kappa}$$

$$=\frac{i\omega}{8\pi^2} \int_{c_0|\boldsymbol{\kappa}|<1} \frac{e^{i\omega S(\boldsymbol{\kappa},\mathbf{x},z)}}{\sqrt{c_0^{-2}-|\boldsymbol{\kappa}|^2}} d\boldsymbol{\kappa}$$

$$+\frac{|\omega|}{8\pi^2} \int_{c_0|\boldsymbol{\kappa}|>1} \frac{e^{i\omega\mathbf{x}\cdot\boldsymbol{\kappa}-|\omega|\sqrt{|\boldsymbol{\kappa}|^2-c_0^{-2}}|z|}}{\sqrt{|\boldsymbol{\kappa}|^2-c_0^{-2}}} d\boldsymbol{\kappa}. \tag{2.44}$$

Here $\boldsymbol{\kappa}$ is the two-dimensional slowness vector and we have introduced the travel time

$$S(\boldsymbol{\kappa}, \mathbf{x}, z) = \mathbf{x} \cdot \boldsymbol{\kappa} + \sqrt{c_0^{-2} - |\boldsymbol{\kappa}|^2}|z|.$$

It follows from this representation that the modes parameterized by $(\omega, \boldsymbol{\kappa})$, with $c_0|\boldsymbol{\kappa}| < 1$, correspond to plane waves with frequency ω and three-dimensional wave vector

$$\mathbf{k} = \omega \left(\boldsymbol{\kappa}, \pm\sqrt{c_0^{-2} - |\boldsymbol{\kappa}|^2} \right)$$

in the upper and lower z half-planes, respectively. Thus, the plane-wave mode

$$e^{i\omega S(\boldsymbol{\kappa},\mathbf{x},z)}$$

propagates in the direction $\left(\boldsymbol{\kappa}, \sqrt{c_0^{-2} - |\boldsymbol{\kappa}|^2} \right)$ for $z > 0$ and in the direction $\left(\boldsymbol{\kappa}, -\sqrt{c_0^{-2} - |\boldsymbol{\kappa}|^2} \right)$ for $z < 0$.

2.2.5 The Acoustic Wave Generated by a Point Source

Let us return to the velocity field $\mathbf{u}$ and pressure field p in a homogeneous medium with constant K_0 and ρ_0 when there is a point source. The governing equations are

$$\rho_0 \frac{\partial \mathbf{u}}{\partial t} + \nabla p = f(t)\delta(\mathbf{r})\mathbf{e}, \tag{2.45}$$

$$\frac{1}{K_0} \frac{\partial p}{\partial t} + \nabla \cdot \mathbf{u} = 0, \tag{2.46}$$

with $\mathbf{e} \in \mathbb{R}^3$ being a source directivity vector. It is again convenient to carry out a specific joint Fourier transform in time and space. This decomposes the three-dimensional problem into plane-wave modes that satisfy one-dimensional problems. The transformed quantities are

$$\hat{\mathbf{u}}(\omega, \boldsymbol{\kappa}, z) = \iint e^{i\omega(t - \boldsymbol{\kappa} \cdot \mathbf{x})} \mathbf{u}(t, \mathbf{x}, z) \, dt \, d\mathbf{x},$$

$$\hat{p}(\omega, \boldsymbol{\kappa}, z) = \iint e^{i\omega(t - \boldsymbol{\kappa} \cdot \mathbf{x})} p(t, \mathbf{x}, z) \, dt \, d\mathbf{x},$$

where $\boldsymbol{\kappa}$ again is the two-dimensional slowness vector. The inverse transform is given by

$$p(t, \mathbf{x}, z) = \frac{1}{(2\pi)^3} \iint e^{-i\omega(t - \boldsymbol{\kappa} \cdot \mathbf{x})} \hat{p}(\omega, \boldsymbol{\kappa}, z) \omega^2 \, d\omega \, d\boldsymbol{\kappa},$$

$$\mathbf{u}(t, \mathbf{x}, z) = \frac{1}{(2\pi)^3} \iint e^{-i\omega(t - \boldsymbol{\kappa} \cdot \mathbf{x})} \hat{\mathbf{u}}(\omega, \boldsymbol{\kappa}, z) \omega^2 \, d\omega \, d\boldsymbol{\kappa}.$$

We denote by $\mathbf{v}$ the horizontal components of the velocity field and by u its vertical component, with $\hat{\mathbf{v}}, \hat{u}$ being the corresponding Fourier transformed quantities. The Fourier transformed acoustic equations (2.45) are then

$$-i\omega\rho_0 \hat{\mathbf{v}} + i\omega\boldsymbol{\kappa}\hat{p} = \delta(z)\hat{f}(\omega)\mathbf{e}_\perp,$$

$$-i\omega\rho_0 \hat{u} + \frac{\partial \hat{p}}{\partial z} = \delta(z)\hat{f}(\omega)\mathbf{e}_{\shortparallel},$$

$$-\frac{i\omega}{K_0}\hat{p} + i\omega\boldsymbol{\kappa} \cdot \hat{\mathbf{v}} + \frac{\partial \hat{u}}{\partial z} = 0,$$

using the notation $\mathbf{e} = (\mathbf{e}_\perp, \mathbf{e}_{\shortparallel})$. By eliminating $\hat{\mathbf{v}}$ we deduce that $(\hat{u}, \hat{p})$ satisfies the following system:

$$-i\omega\rho_0 \hat{u} + \frac{\partial \hat{p}}{\partial z} = \delta(z)\hat{f}(\omega)\mathbf{e}_{\shortparallel},$$

$$-i\omega \frac{1}{K_0}(1 - |\boldsymbol{\kappa}|^2 c_0^2)\hat{p} + \frac{\partial \hat{u}}{\partial z} = \delta(z)\hat{f}(\omega)\frac{\boldsymbol{\kappa} \cdot \mathbf{e}_\perp}{\rho_0}.$$

We have thus reduced the three-dimensional acoustic equations to a family of one-dimensional mode problems. The density and bulk modulus of the reduced modal problem are ρ_0 and

$$K(\boldsymbol{\kappa}) = \frac{K_0}{1 - |\boldsymbol{\kappa}|^2 c_0^2},$$

respectively. We further eliminate $\hat{u}$ to obtain

$$\omega^2(c_0^{-2} - |\boldsymbol{\kappa}|^2)\hat{p} + \frac{\partial^2 \hat{p}}{\partial z^2} = \hat{f}(\omega)\mathbf{e} \cdot \left(i\omega\boldsymbol{\kappa}, \frac{\partial}{\partial z}\right)\delta(z).$$

The field $\hat{p}$ can be expressed in terms of the fundamental solution $\check{G}$ of the one-dimensional wave equation (2.41):

$$\hat{p}(\omega, \boldsymbol{\kappa}, z) = -\hat{f}(\omega)\mathbf{e} \cdot \left(i\omega\boldsymbol{\kappa}, \frac{\partial}{\partial z}\right)\check{G}(\omega, \boldsymbol{\kappa}, z).$$

Using (2.42) and (2.43) we find for $z \neq 0$ the explicit formula

$$\hat{p}(\omega, \boldsymbol{\kappa}, z) = \frac{1}{2}\hat{f}(\omega)\mathbf{e} \cdot \left(\frac{\boldsymbol{\kappa}}{\sqrt{c_0^{-2} - |\boldsymbol{\kappa}|^2}}, \operatorname{sgn}(z)\right) e^{i\omega\sqrt{c_0^{-2} - |\boldsymbol{\kappa}|^2}|z|},$$

if $c_0|\boldsymbol{\kappa}| < 1$ and

$$\hat{p}(\omega, \boldsymbol{\kappa}, z) = \frac{1}{2}\hat{f}(\omega)\mathbf{e} \cdot \left(-i\operatorname{sgn}(\omega)\frac{\boldsymbol{\kappa}}{\sqrt{|\boldsymbol{\kappa}|^2 - c_0^{-2}}}, \operatorname{sgn}(z)\right) e^{-|\omega|\sqrt{|\boldsymbol{\kappa}|^2 - c_0^{-2}}|z|},$$

if $c_0|\boldsymbol{\kappa}| > 1$. Taking the inverse transform gives the representation

$$p(t, \mathbf{r}) = \frac{1}{16\pi^3} \int\!\!\int_{c_0|\boldsymbol{\kappa}|<1} e^{-i\omega(t - S(\boldsymbol{\kappa}, \mathbf{x}, z))} \hat{f}(\omega)$$

$$\times \mathbf{e} \cdot \left(\frac{\boldsymbol{\kappa}}{\sqrt{c_0^{-2} - |\boldsymbol{\kappa}|^2}}, \operatorname{sgn}(z)\right) \omega^2 \, d\omega \, d\boldsymbol{\kappa}$$

$$+ \frac{1}{16\pi^3} \int\!\!\int_{c_0|\boldsymbol{\kappa}|>1} e^{-i\omega(t - \boldsymbol{\kappa}\cdot\mathbf{x}) - |\omega|\sqrt{c_0^{-2} - |\boldsymbol{\kappa}|^2}|z|} \hat{f}(\omega)$$

$$\times \mathbf{e} \cdot \left(-i\operatorname{sgn}(\omega)\frac{\boldsymbol{\kappa}}{\sqrt{|\boldsymbol{\kappa}|^2 - c_0^{-2}}}, \operatorname{sgn}(z)\right) \omega^2 \, d\omega \, d\boldsymbol{\kappa} \,.$$

After integrating in ω and suppressing the evanescent modes, which is valid when $|z|$ is much larger than the typical wavelength of the source, we have

$$p(t, \mathbf{r}) \sim -\frac{1}{8\pi^2} \int_{c_0|\boldsymbol{\kappa}|<1} f''(t - S(\boldsymbol{\kappa}, \mathbf{x}, z)) \left(\frac{\mathbf{e}_\perp \cdot \boldsymbol{\kappa}}{\sqrt{c_0^{-2} - |\boldsymbol{\kappa}|^2}} + \operatorname{sgn}(z)\mathbf{e}_{\shortparallel}\right) d\boldsymbol{\kappa} \,.$$

The velocity field can be recovered from the relations

$$\hat{u} = \frac{1}{i\omega\rho_0}\frac{\partial \hat{p}}{\partial z}, \quad \hat{\mathbf{v}} = \frac{\boldsymbol{\kappa}}{\rho_0}\hat{p},$$

for $z \neq 0$.

The horizontal Fourier transform can also be carried out when the medium is layered to obtain a reduction to one-dimensional mode problems. In subsequent chapters we use such reductions of three-dimensional layered problems to simplify the analysis.

2.3 Appendix

2.3.1 Gauss–Green Theorem

The Gauss–Green theorem is a basic identity that is used in this chapter. Let V be a bounded open subset of $\mathbb{R}^n$ whose boundary ∂V is $\mathcal{C}^1$. Let $f \in \mathcal{C}^1(\overline{V})$, where $\overline{V} = V \cup \partial V$ is the closure of V. Then

$$\int_V \nabla f(\mathbf{r}) d\mathbf{r} = \int_{\partial V} f(\mathbf{r}) \mathbf{n}(\mathbf{r}) d\sigma(\mathbf{r}),$$

where $\mathbf{n}(\mathbf{r})$ is the outward unit normal to ∂V at $\mathbf{r} \in \partial V$.

2.3.2 Energy Conservation Equation

In this section we show that the energy density and flux described in Section 2.1.8 for the acoustic wave equations can also be obtained from the linearization of the energy-conservation equation for fluid dynamics.

The energy-conservation equation for a fluid in the absence of viscosity, heat conduction, and relaxation effects is

$$\frac{\partial}{\partial t} \left(\frac{1}{2} \rho |\mathbf{U}|^2 + \rho e \right) + \nabla \cdot \left[\left(\frac{1}{2} \rho |\mathbf{U}|^2 + \rho e + P \right) \mathbf{U} \right] = \mathbf{F} \cdot \mathbf{U}. \qquad (2.47)$$

Here e is the internal energy, which is a function of P and ρ. The conservation laws of mass, momentum, and energy in the form (2.1), (2.2), and (2.47) are the Euler equations. From thermodynamic arguments we have that

$$de + P(\rho) d \left(\frac{1}{\rho} \right) = 0$$

if the flow is isentropic. This implies that e is a function of ρ with

$$\frac{de}{d\rho} = \frac{P(\rho)}{\rho^2}.$$

By considering small perturbations around a fluid at rest, of the form (2.3), we can expand to second order,

$$\rho e(\rho) = \rho_0 e(\rho_0) + [\rho_0 e(\rho_0) + P(\rho_0)]s + \frac{\rho_0}{2} \frac{dP}{d\rho}(\rho_0) s^2 + \cdots$$

$$= \rho_0 e(\rho_0) + \frac{\rho_0 e(\rho_0) + P_0}{K_0} p + \frac{1}{2K_0} p^2 + \cdots,$$

where we have used the definition (2.4) of K_0 and the linearized equation of state (2.5). The linearization of the energy-conservation equation (2.47) gives to first order

$$\frac{\partial}{\partial t} \left(\frac{\rho_0 e(\rho_0) + P_0}{K_0} p \right) + \nabla \cdot [(\rho_0 e(\rho_0) + P_0) \mathbf{u}] = 0,$$

which is the linearized mass-conservation equation (2.6).

At second order we obtain

$$\frac{\partial}{\partial t} \left(\frac{\rho_0}{2} |\mathbf{u}|^2 + \frac{1}{2K_0} p^2 \right) + \nabla \cdot (p\mathbf{u}) = \mathbf{F} \cdot \mathbf{u},$$

which is the energy-conservation relation for the acoustic wave equations discussed in Section 2.1.8.

Notes

The results presented in this chapter are in the context of the acoustic wave equations but they can be extended to any physical system that can be reduced to the wave equation. This is the case in acoustics, in elasticity [167], and in electromagnetics, in homogeneous media [20]. The basic existence, uniqueness, and regularity results for solutions of the homogeneous three-dimensional wave equation can be found in the book by Evans [51]. A detailed analysis of the wave equation in $\mathbb{R}^n$ or in bounded domains can be found in [43], where different types of initial and boundary conditions are addressed and numerical schemes are discussed. For an introduction to nonlinear and dispersive waves we refer to the book by Whitham [167].

3

Waves in Layered Media

Throughout the book we use a number of essential transformations of the wave equation that are specific to layered media. In this chapter we consider the particular case in which the parameters of the medium vary in a piecewise-constant manner; in other words, we consider a stack of layers made of homogeneous media. We study the propagation of a normally incident plane wave, which enables us to reduce the problem to the one-dimensional acoustic wave equations. We will see that the problem can be recast as a product of matrices corresponding to the scattering of the wave by the successive interfaces between the layers. This is a classical setup for waves propagating in this particular type of layered media, and it is extremely useful for direct numerical simulations.

3.1 Reduction to a One-Dimensional System

The equations for the three-dimensional velocity $\mathbf{u}$ and pressure p are

$$\rho \frac{\partial \mathbf{u}}{\partial t} + \nabla p = \mathbf{0}, \tag{3.1}$$

$$\frac{1}{K} \frac{\partial p}{\partial t} + \nabla \cdot \mathbf{u} = 0, \tag{3.2}$$

where ρ is the density of the medium and K the bulk modulus of the medium. As seen in the previous chapter, these two equations correspond respectively to conservation of momentum and mass. The density and bulk modulus are assumed to be spatially varying along the z-coordinate. The restriction to a medium whose variations occur only in one direction is central in this book, and it is consistent with a range of applications, in geophysics for instance. If the initial conditions correspond to a plane wave that is normally incident to the layered medium, then the solution of the equations remains independent of the transverse variables, and the transverse velocity is zero. The system can then be reduced to the one-dimensional wave equations. Note that more general initial conditions, corresponding in particular to point sources, require a more general three-dimensional framework, and these problems will be fully addressed in Chapter 14. We will see there that it is always possible to decompose the wave solution into plane waves and to reduce the three-dimensional problem to an infinite set of one-dimensional problems. In this chapter we focus our attention to the one-dimensional case. However, the treatment applies in general to the plane-wave modes, which follows from other initial conditions.

In a one-dimensional medium the equations for the velocity u and pressure p are

$$\rho(z)\frac{\partial u(t,z)}{\partial t} + \frac{\partial p(t,z)}{\partial z} = 0, \tag{3.3}$$

$$\frac{1}{K(z)}\frac{\partial p(t,z)}{\partial t} + \frac{\partial u(t,z)}{\partial z} = 0,$$

with ρ being the density and K the bulk modulus of the medium, which are both functions of the spatial coordinate z. We write this system of equations in matrix form:

$$\frac{\partial}{\partial z}\begin{bmatrix} p(t,z) \\ u(t,z) \end{bmatrix} = -\begin{bmatrix} 0 & \rho(z) \\ K(z)^{-1} & 0 \end{bmatrix}\frac{\partial}{\partial t}\begin{bmatrix} p(t,z) \\ u(t,z) \end{bmatrix}.$$

A diagonalization of the 2×2 matrix gives

$$\begin{bmatrix} 0 & \rho(z) \\ K(z)^{-1} & 0 \end{bmatrix} = \mathbf{M}(z)^{-1}\begin{bmatrix} c(z)^{-1} & 0 \\ 0 & -c(z)^{-1} \end{bmatrix}\mathbf{M}(z),$$

where

$$\mathbf{M}(z) = \begin{bmatrix} \zeta(z)^{-1/2} & \zeta(z)^{1/2} \\ -\zeta(z)^{-1/2} & \zeta(z)^{1/2} \end{bmatrix}, \quad \mathbf{M}(z)^{-1} = \frac{1}{2}\begin{bmatrix} \zeta(z)^{1/2} & -\zeta(z)^{1/2} \\ \zeta(z)^{-1/2} & \zeta(z)^{-1/2} \end{bmatrix},$$

with $c(z) = \sqrt{K(z)/\rho(z)}$ and $\zeta(z) = \sqrt{K(z)\rho(z)}$ being respectively the local **speed of sound** and **impedance**. The system can then be written as

$$\frac{\partial}{\partial z}\begin{bmatrix} p(t,z) \\ u(t,z) \end{bmatrix} = -\frac{1}{c(z)}\mathbf{M}(z)^{-1}\begin{bmatrix} 1 & 0 \\ 0 & -1 \end{bmatrix}\mathbf{M}(z)\frac{\partial}{\partial t}\begin{bmatrix} p(t,z) \\ u(t,z) \end{bmatrix}.$$

In this representation the material parameters ρ and K may vary with respect to the space coordinate z. In the next section we consider the special case with constant coefficients when this decomposition can be used to decouple the wave into right- and left-going modes.

3.2 Right- and Left-Going Waves

We consider the special case with a **homogeneous** medium in which the coefficients ρ and K are constant. Consequently the speed of sound c and the impedance ζ are constant, and the system can be written:

$$\frac{\partial}{\partial z}\left(\mathbf{M}\begin{bmatrix} p(t,z) \\ u(t,z) \end{bmatrix}\right) = -\frac{1}{c}\begin{bmatrix} 1 & 0 \\ 0 & -1 \end{bmatrix}\frac{\partial}{\partial t}\left(\mathbf{M}\begin{bmatrix} p(t,z) \\ u(t,z) \end{bmatrix}\right).$$

Then, if we define

$$\begin{bmatrix} A(t,z) \\ B(t,z) \end{bmatrix} = \mathbf{M}\begin{bmatrix} p(t,z) \\ u(t,z) \end{bmatrix} = \begin{bmatrix} \zeta^{-1/2}p(t,z) + \zeta^{1/2}u(t,z) \\ -\zeta^{-1/2}p(t,z) + \zeta^{1/2}u(t,z) \end{bmatrix}, \quad (3.4)$$

it follows that

$$\frac{\partial}{\partial z}\begin{bmatrix} A(t,z) \\ B(t,z) \end{bmatrix} = -\frac{1}{c}\begin{bmatrix} 1 & 0 \\ 0 & -1 \end{bmatrix}\frac{\partial}{\partial t}\begin{bmatrix} A(t,z) \\ B(t,z) \end{bmatrix}. \quad (3.5)$$

The equations for A and B decouple,

$$\frac{\partial A(t,z)}{\partial z} + \frac{1}{c}\frac{\partial A(t,z)}{\partial t} = 0,$$
$$\frac{\partial B(t,z)}{\partial z} - \frac{1}{c}\frac{\partial B(t,z)}{\partial t} = 0,$$

and the waves can be written $A(t,z) = a(t - z/c)$ and $B(t,z) = b(t + z/c)$ for some wave-shape functions a and b. Thus, in the constant medium case we have decomposed the wave into the **right- and left-going waves** A and B, which do not interact.

To fully specify the problem we have to prescribe initial conditions, for instance the velocity and pressure profiles at time $t = 0$:

$$u(t=0,z) = u_0(z), \quad p(t=0,z) = p_0(z).$$

We then translate these initial conditions for u and p into initial conditions for the modes A and B:

$$A_0(-z) := A(t=0,z) = \zeta^{-1/2}p_0(z) + \zeta^{1/2}u_0(z),$$
$$B_0(z) := B(t=0,z) = -\zeta^{-1/2}p_0(z) + \zeta^{1/2}u_0(z),$$

which gives the expressions for the modes

$$A(t,z) = A_0(ct - z), \quad B(t,z) = B_0(ct + z),$$

and finally the expressions for the wave

$$p(t,z) = \zeta^{1/2}\frac{A(t,z) - B(t,z)}{2},$$
$$u(t,z) = \zeta^{-1/2}\frac{A(t,z) + B(t,z)}{2}.$$

The initial conditions determine the mode decomposition and can be chosen to generate a pure right-going wave (if $p_0 \equiv \zeta u_0$) or a pure left-going wave (if $p_0 \equiv -\zeta u_0$).

A more physical way to generate a wave is to assume that the wave vanishes as $t \to -\infty$ and to introduce a source term in the acoustic wave equations:

$$\rho \frac{\partial u(t,z)}{\partial t} + \frac{\partial p(t,z)}{\partial z} = F(t,z),$$

$$K^{-1} \frac{\partial p(t,z)}{\partial t} + \frac{\partial u(t,z)}{\partial z} = 0.$$

By assuming a point source $F(t,z) = \zeta^{1/2} f(t) \delta(z)$, the system for A and B becomes

$$\frac{\partial A(t,z)}{\partial z} + \frac{1}{c} \frac{\partial A(t,z)}{\partial t} = \delta(z) f(t),$$

$$\frac{\partial B(t,z)}{\partial z} - \frac{1}{c} \frac{\partial B(t,z)}{\partial t} = -\delta(z) f(t),$$

whose solutions are

$$A(t,z) = \begin{cases} f(t - z/c) & \text{if } z > 0, \\ 0 & \text{if } z < 0, \end{cases} \qquad B(t,z) = \begin{cases} 0 & \text{if } z > 0, \\ f(t + z/c) & \text{if } z < 0. \end{cases}$$

As a result, the velocity and pressure fields are

$$u(t,z) = \frac{\zeta^{-1/2}}{2} \begin{cases} f(t - z/c) & \text{if } z > 0, \\ f(t + z/c) & \text{if } z < 0. \end{cases} \quad p(t,z) = \frac{\zeta^{1/2}}{2} \begin{cases} f(t - z/c) & \text{if } z > 0, \\ -f(t + z/c) & \text{if } z < 0. \end{cases}$$

This means that the source term generates two waves with equal energy that propagate to the right and to the left (see Figure 3.1).

3.3 Scattering by a Single Interface

In this section we consider the case in which two homogeneous half-spaces are separated by an interface at $z = 0$:

$$\rho(z) = \begin{cases} \rho_0 & \text{if } z < 0, \\ \rho_1 & \text{if } z > 0, \end{cases} \qquad K(z) = \begin{cases} K_0 & \text{if } z < 0, \\ K_1 & \text{if } z > 0. \end{cases}$$

This section presents a new point of view on this simple scattering problem already discussed in Chapter 2. By diagonalizing the acoustic wave equations in the two half-spaces, we find that the waves can be decomposed into right- and left-going modes in each half-space. However, due to the mismatch of the medium parameters, the definition of the modes in terms of the pressure and velocity fields is not the same in the two half-spaces. As a result, a pure right-going wave incoming from the left half-space and impinging on the interface cannot be simply transmitted as a pure right-going wave propagating in the right half-space, because this would violate the continuity of the pressure and

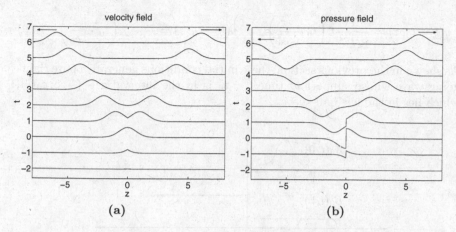

Fig. 3.1. Waves generated by a point source $F(t,z) = f(t)\delta(z)$ in a homogeneous medium. Here $f(t) = \exp(-t^2)$, $K = \rho = 1$. The spatial profiles of the velocity field (a) and of the pressure field (b) are plotted at times $t = -2$, $t = -1$, $t = 0$, $t = 1$, ..., $t = 6$.

velocity fields. The goal of this section is to analyze the scattering problem in terms of right- and left-going modes.

We introduce the local velocities $c_j = \sqrt{K_j/\rho_j}$ and impedances $\zeta_j = \sqrt{K_j\rho_j}$ and the right- and left-going modes defined by

$$z < 0: \begin{cases} A_0(t,z) = \zeta_0^{-1/2}p(t,z) + \zeta_0^{1/2}u(t,z)\,, \\ B_0(t,z) = -\zeta_0^{-1/2}p(t,z) + \zeta_0^{1/2}u(t,z)\,, \end{cases} \quad (3.6)$$

$$z > 0: \begin{cases} A_1(t,z) = \zeta_1^{-1/2}p(t,z) + \zeta_1^{1/2}u(t,z)\,, \\ B_1(t,z) = -\zeta_1^{-1/2}p(t,z) + \zeta_1^{1/2}u(t,z)\,. \end{cases} \quad (3.7)$$

For $j = 0, 1$, the pairs (A_j, B_j) satisfy the following system in their respective half-spaces:

$$\frac{\partial}{\partial z}\begin{bmatrix} A_j \\ B_j \end{bmatrix} = \frac{1}{c_j}\begin{bmatrix} -1 & 0 \\ 0 & 1 \end{bmatrix}\frac{\partial}{\partial t}\begin{bmatrix} A_j \\ B_j \end{bmatrix}\,, \quad (3.8)$$

which means that $A_j(t,z)$ is a function of $t - z/c_j$ only, and $B_j(t,z)$ is a function of $t + z/c_j$ only.

We assume that a right-going wave with the time profile f is incoming from the left and is partly reflected by the interface. We also assume a **radiation condition** in the right half-space so that no wave is coming from the right. Assume that f is compactly supported in $(0, \infty)$. We next introduce two ways to define proper **boundary conditions**:

(I) We can consider an initial value problem with initial conditions given at some time $t_0 < 0$ by

$$u(t = t_0, z) = \frac{1}{2\zeta_0^{1/2}} f\left(t_0 - \frac{z}{c_0}\right), \qquad p(t = t_0, z) = \frac{\zeta_0^{1/2}}{2} f\left(t_0 - \frac{z}{c_0}\right). \quad (3.9)$$

As shown in the previous section, these initial conditions generate a pure right-going wave whose support at time $t = t_0$ is in the interval $z \in (-\infty, c_0 t_0)$, which lies in the left half-space.

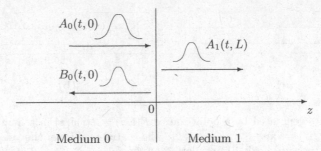

Fig. 3.2. Scattering of a pulse by an interface.

(II) We can consider a point source located at some point $z_0 < 0$ and generating a forcing term of the form

$$F(t, z) = \zeta_0^{1/2} f(t - z_0/c_0)\delta(z - z_0). \quad (3.10)$$

As seen in the previous section, this point source generates two waves. The left-going wave is propagating into the negative z-direction and will never interact with the interface, so we will ignore it. The right-going wave first propagates in the homogeneous left half-space and it eventually interacts with the interface $z = 0$.

In terms of the right- and left-going waves, these two formulations give the same descriptions. We have $A_0(t, z) = f(t - z/c_0)$ for $z < 0$, and $B_1(t, z) = 0$ for $z > 0$, and consequently, at the interface $z = 0$,

$$A_0(t, 0) = f(t), \qquad B_1(t, 0) = 0. \quad (3.11)$$

Note that the delays introduced in the initial conditions (3.9) and in the forcing term (3.10) have been chosen so that the boundary conditions (3.11) have a very simple form.

The pairs (A_0, B_0) and (A_1, B_1) are coupled by the jump conditions at $z = 0$ corresponding to the **continuity of the velocity and pressure fields**:

$$u(t, 0) = \zeta_0^{-1/2}\left(\frac{A_0(t, 0) + B_0(t, 0)}{2}\right) = \zeta_1^{-1/2}\left(\frac{A_1(t, 0) + B_1(t, 0)}{2}\right),$$

$$p(t, 0) = \zeta_0^{1/2}\left(\frac{A_0(t, 0) - B_0(t, 0)}{2}\right) = \zeta_1^{1/2}\left(\frac{A_1(t, 0) - B_1(t, 0)}{2}\right),$$

which gives

$$\begin{bmatrix} A_1(t,0) \\ B_1(t,0) \end{bmatrix} = \mathbf{J} \begin{bmatrix} A_0(t,0) \\ B_0(t,0) \end{bmatrix}, \qquad \mathbf{J} = \begin{bmatrix} r^{(+)} & r^{(-)} \\ r^{(-)} & r^{(+)} \end{bmatrix}, \qquad (3.12)$$

with $r^{(\pm)} = \frac{1}{2}\left(\sqrt{\zeta_1/\zeta_0} \pm \sqrt{\zeta_0/\zeta_1}\right)$. Note that $(r^{(+)})^2 - (r^{(-)})^2 = 1$. The matrix $\mathbf{J}$ can be interpreted as a propagator, since it "propagates" the right- and left-going modes from the left side of the interface to the right side. Such a propagator matrix will be called an **interface propagator** in the following.

Taking into account the boundary conditions (3.11) yields

$$\begin{bmatrix} A_1(t,0) \\ 0 \end{bmatrix} = \mathbf{J} \begin{bmatrix} f(t) \\ B_0(t,0) \end{bmatrix},$$

and solving this equation gives

$$B_0(t,0) = \mathcal{R}f(t), \qquad A_1(t,0) = \mathcal{T}f(t),$$

where $\mathcal{R}$ and $\mathcal{T}$ are the **reflection and transmission coefficients** of the interface:

$$\mathcal{R} = -\frac{r^{(-)}}{r^{(+)}} = \frac{\zeta_0 - \zeta_1}{\zeta_0 + \zeta_1}, \qquad \mathcal{T} = \frac{1}{r^{(+)}} = \frac{2\sqrt{\zeta_0\zeta_1}}{\zeta_0 + \zeta_1}.$$

These coefficients satisfy the energy-conservation relation

$$\mathcal{R}^2 + \mathcal{T}^2 = 1,$$

meaning that the sum of the energies of the reflected and transmitted waves is equal to the energy of the incoming wave. Finally, the complete solution for $z < 0$ in terms of the right- and left-going modes is

$$A_0(t,z) = f(t - z/c_0), \qquad B_0(t,z) = \mathcal{R}f(t + z/c_0),$$

and for $z > 0$,

$$A_1(t,z) = \mathcal{T}f(t - z/c_1), \qquad B_1(t,z) = 0.$$

Using (3.6–3.7) we can then obtain the pressure and velocity fields (see Figure 3.3). Note that the reflection coefficient is nonzero as soon as there is an impedance mismatch between the two half-spaces. Consequently, the interface between two different media generates no reflections if the impedances are equal, even though the speeds of sound may differ in the two media.

3.4 Single-Layer Case

3.4.1 Mathematical Setup

We consider in this section the case of a homogeneous slab with thickness L embedded between two homogeneous half-spaces:

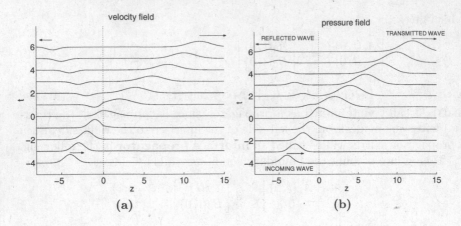

Fig. 3.3. Scattering of a pulse by an interface separating two homogeneous half-spaces $(c_0, \zeta_0, z < 0)$ and $(c_1, \zeta_1, z > 0)$. Here the incoming right-going wave has a Gaussian profile, $c_0 = \zeta_0 = 1$, and $c_1 = \zeta_1 = 2$. The spatial profiles of the velocity field (a) and of the pressure field (b) are plotted at times $t = -4$, $t = -3, \ldots, t = 6$.

$$\rho(z) = \begin{cases} \rho_0 & \text{if } z < 0, \\ \rho_1 & \text{if } z \in [0, L], \\ \rho_2 & \text{if } z > L, \end{cases} \qquad K(z) = \begin{cases} K_0 & \text{if } z < 0, \\ K_1 & \text{if } z \in [0, L]. \\ K_2 & \text{if } z > L. \end{cases}$$

We introduce the local velocities $c_j = \sqrt{K_j/\rho_j}$ and impedances $\zeta_j = \sqrt{K_j \rho_j}$ and the local right- and left-going modes defined by

$$A_j(t, z) = \zeta_j^{-1/2} p(t, z) + \zeta_j^{1/2} u(t, z), \qquad B_j(t, z) = -\zeta_j^{-1/2} p(t, z) + \zeta_j^{1/2} u(t, z),$$

with $j = 0$ for $z < 0$, $j = 1$ for $z \in [0, L]$, and $j = 2$ for $z > L$. The boundary conditions correspond to an impinging pulse at the interface $z = 0$ and a radiation condition at $z = L_2$:

$$A_0(t, 0) = f(t), \qquad B_2(t, L) = 0.$$

The propagation equations (3.8) in each homoegeneous region show that A_j is a function of $t - z/c_j$ only and B_j is a function of $t + z/c_j$ only. The waves inside the slab $[0, L]$ are therefore of the form

$$A_1(t, z) = a_1 \left(t - \frac{z}{c_1} \right), \qquad B_1(t, z) = b_1 \left(t + \frac{z}{c_1} \right),$$

while the reflected wave for $z < 0$ is of the form

$$B_0(t, z) = b_0 \left(t + \frac{z}{c_0} \right),$$

and the transmitted wave for $z > L$ is of the form

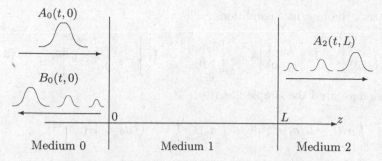

Fig. 3.4. Scattering of a pulse by a single layer.

$$A_2(t,z) = a_2\left(t - \frac{z-L}{c_2}\right).$$

We want to identify the functions b_0 and a_2, which give the shapes of the reflected and transmitted waves.

3.4.2 Reflection and Transmission Coefficient for a Single Layer

The unknown functions b_0 and a_2 can be obtained from the continuity conditions for the velocity and pressure at the two interfaces. At $z = 0$ we have

$$\begin{bmatrix} A_1(t,0) \\ B_1(t,0) \end{bmatrix} = \mathbf{J}_0 \begin{bmatrix} A_0(t,0) \\ B_0(t,0) \end{bmatrix}, \qquad \mathbf{J}_0 = \begin{bmatrix} r_0^{(+)} & r_0^{(-)} \\ r_0^{(-)} & r_0^{(+)} \end{bmatrix},$$

with $r_0^{(\pm)} = \frac{1}{2}\left(\sqrt{\zeta_1/\zeta_0} \pm \sqrt{\zeta_0/\zeta_1}\right)$. Similarly, at $z = L$,

$$\begin{bmatrix} A_2(t,L) \\ B_2(t,L) \end{bmatrix} = \mathbf{J}_1 \begin{bmatrix} A_1(t,L) \\ B_1(t,L) \end{bmatrix}, \qquad \mathbf{J}_1 = \begin{bmatrix} r_1^{(+)} & r_1^{(-)} \\ r_1^{(-)} & r_1^{(+)} \end{bmatrix},$$

with $r_1^{(\pm)} = \frac{1}{2}\left(\sqrt{\zeta_2/\zeta_1} \pm \sqrt{\zeta_1/\zeta_2}\right)$. We can write these relations in terms of the functions a_j, b_j as

$$\begin{bmatrix} a_1(t) \\ b_1(t) \end{bmatrix} = \mathbf{J}_0 \begin{bmatrix} f(t) \\ b_0(t) \end{bmatrix}, \qquad \begin{bmatrix} a_2(t) \\ 0 \end{bmatrix} = \mathbf{J}_1 \begin{bmatrix} a_1(t - L/c_1) \\ b_1(t + L/c_1) \end{bmatrix},$$

which can be solved to get the reflected and transmitted waves. The situation is more complicated than in the case of a single interface, because of the time delays $\pm L/c_1$. A convenient and general way to handle these delays is by going to the **frequency** domain, so that the time shifts are replaced by phase factors. The Fourier transforms of the modes are defined by

$$\hat{a}_j(\omega) \doteq \int a_j(t)e^{i\omega t}dt, \qquad \hat{b}_j(\omega) = \int a_j(t)e^{i\omega t}dt.$$

They satisfy the interface conditions

$$
\begin{bmatrix} \hat{a}_1(\omega) \\ \hat{b}_1(\omega) \end{bmatrix} = \mathbf{J}_0 \begin{bmatrix} \hat{f}(\omega) \\ \hat{b}_0(\omega) \end{bmatrix}, \qquad \begin{bmatrix} \hat{a}_2(\omega) \\ 0 \end{bmatrix} = \mathbf{J}_1 \begin{bmatrix} \hat{a}_1(\omega) e^{i\frac{\omega L}{c_1}} \\ \hat{b}_1(\omega) e^{-i\frac{\omega L}{c_1}} \end{bmatrix}, \qquad (3.13)
$$

where we have used the simple identity

$$
\int a_1(t - L/c_1) e^{i\omega t} dt = \int a_1(s) e^{i\omega\left(s + \frac{L}{c_1}\right)} ds = \hat{a}_1(\omega) e^{i\frac{\omega L}{c_1}}.
$$

Introducing the frequency-dependent matrix

$$
\hat{\mathbf{J}}_1(\omega) = \begin{bmatrix} r_1^{(+)} e^{i\frac{\omega L}{c_1}} & r_1^{(-)} e^{-i\frac{\omega L}{c_1}} \\ r_1^{(-)} e^{i\frac{\omega L}{c_1}} & r_1^{(+)} e^{-i\frac{\omega L}{c_1}} \end{bmatrix},
$$

the second equation of (3.13) can be rewritten as

$$
\begin{bmatrix} \hat{a}_2(\omega) \\ 0 \end{bmatrix} = \hat{\mathbf{J}}_1(\omega) \begin{bmatrix} \hat{a}_1(\omega) \\ \hat{b}_1(\omega) \end{bmatrix}. \qquad (3.14)
$$

The symplectic matrix $\hat{\mathbf{J}}_1(\omega)$ is a propagator in the frequency domain. It propagates the right- and left-going modes from the right side of the interface 0 to the right side of the interface 1, and it depends on the layer thickness L. Finally, combining the first equation of (3.13) and (3.14), we obtain the relation

$$
\begin{bmatrix} \hat{a}_2(\omega) \\ 0 \end{bmatrix} = \hat{\mathbf{K}}_0(\omega) \begin{bmatrix} \hat{f}(\omega) \\ \hat{b}_0(\omega) \end{bmatrix}, \qquad (3.15)
$$

where the frequency-dependent symplectic matrix

$$
\hat{\mathbf{K}}_0(\omega) = \hat{\mathbf{J}}_1(\omega) \mathbf{J}_0 = \begin{bmatrix} \hat{U}(\omega) & \overline{\hat{V}(\omega)} \\ \hat{V}(\omega) & \overline{\hat{U}(\omega)} \end{bmatrix}
$$

is the overall propagator of the slab. Equation (3.15) shows that $\hat{\mathbf{K}}_0(\omega)$ propagates the right- and left-going modes from the left side of the interface 0 to the right side of the interface 1. We find explicitly

$$
\hat{U}(\omega) = r_0^{(+)} r_1^{(+)} e^{i\frac{\omega L}{c_1}} + r_0^{(-)} r_1^{(-)} e^{-i\frac{\omega L}{c_1}},
$$
$$
\hat{V}(\omega) = r_0^{(+)} r_1^{(-)} e^{i\frac{\omega L}{c_1}} + r_0^{(-)} r_1^{(+)} e^{-i\frac{\omega L}{c_1}}.
$$

By solving equation (3.15), whose unknowns are $\hat{a}_2(\omega)$ and $\hat{b}_0(\omega)$, and using the expressions of $r_j^{(\pm)}$, we obtain

$$
\hat{b}_0(\omega) = \hat{\mathcal{R}}(\omega) \hat{f}(\omega), \qquad \hat{a}_2(\omega) = \hat{\mathcal{T}}(\omega) \hat{f}(\omega),
$$

where the frequency-dependent reflection and transmission coefficients are

$$\hat{\mathcal{R}}(\omega) = -\frac{\hat{V}(\omega)}{\hat{U}(\omega)} = \frac{R_1 e^{2i\frac{\omega L}{c_1}} + R_0}{1 + R_0 R_1 e^{2i\frac{\omega L}{c_1}}}, \tag{3.16}$$

$$\hat{T}(\omega) = \frac{1}{\hat{U}(\omega)} = \frac{T_0 T_1 e^{i\frac{\omega L}{c_1}}}{1 + R_0 R_1 e^{2i\frac{\omega L}{c_1}}}, \tag{3.17}$$

using that $|\hat{U}(\omega)|^2 - |\hat{V}(\omega)|^2 = 1$. Here $R_0 = \frac{\zeta_0 - \zeta_1}{\zeta_0 + \zeta_1}$, $R_1 = \frac{\zeta_1 - \zeta_2}{\zeta_1 + \zeta_2}$, $T_0 = \frac{2\sqrt{\zeta_0 \zeta_1}}{\zeta_0 + \zeta_1}$, and $T_1 = \frac{2\sqrt{\zeta_1 \zeta_2}}{\zeta_1 + \zeta_2}$ are the reflection and transmission coefficients of the two interfaces. The reflection and transmission coefficients of the layer satisfy the energy conservation relation $|\hat{\mathcal{R}}(\omega)|^2 + |\hat{T}(\omega)|^2 = 1$ for all ω, which means that the individual energies of the frequency components of the incoming pulse are preserved by the scattering process. The main qualitative difference between the scattering by a single interface and the scattering by a single layer is that the reflection and transmission coefficients in the layer case are frequency-dependent. This frequency dependence originates from interference effects between the waves that are scattered back and forth by the two interfaces of the layer. Constructive interferences, destructive interferences, and resonances can build a complicated picture, as we will see below.

3.4.3 Frequency-Dependent Reflectivity and Antireflection Layer

As an application let us consider a layer embedded between two homogeneous half-spaces that have the same material properties, i.e., the situation in which $\rho_2 = \rho_0$ and $K_2 = K_0$. We then have $R_1 = -R_0$ and $T_1 = T_0$, which implies that the global reflectivity of the layer can be written as

$$|\hat{\mathcal{R}}(\omega)|^2 = 1 - \frac{1 + R_0^4 - 2R_0^2}{1 + R_0^4 - 2R_0^2 \cos(\frac{2\omega L}{c_1})}.$$

The reflectivity is periodic with respect to the angular frequency ω with the period $\omega_c = \pi c_1 / L$. As a function of the angular frequency the reflectivity goes from the minimal value

$$|\hat{\mathcal{R}}|_{\min}^2 = 0 \text{ for } \omega = k\omega_c, \quad k \in \mathbb{Z},$$

to the maximal value

$$|\hat{\mathcal{R}}|_{\max}^2 = 1 - \left(\frac{1 - R_0^2}{1 + R_0^2}\right)^2 \text{ for } \omega = \left(k + \frac{1}{2}\right)\omega_c, \quad k \in \mathbb{Z}.$$

This shows that for any value of the reflection coefficient R_0 of a single interface, there exist frequencies that are fully transmitted by the layer. If we consider the case of strong scattering $T_0^2 \ll 1$, then the transmitted frequency bands have a width of the order of $\omega_c T_0^2$ around the fully transmitted frequencies $k\omega_c$. Outside of these bands the typical reflectivity is large, of order

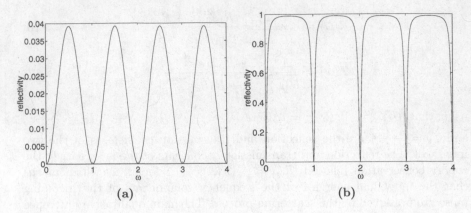

Fig. 3.5. Reflectivity $|\hat{\mathcal{R}}(\omega)|^2$ versus frequency for a single layer with $R_0 = -R_1 = 0.1$ (a) and $R_0 = -R_1 = 0.9$ (b). The period is $\omega_c = \pi c_1/L$.

$1 - T_0^4/4$. The plot of the reflectivity versus frequency clearly exhibits this phenomenon (Figure 3.5).

The total transmission phenomenon is also encountered in situations in which the two half-spaces are different. Indeed, if we now assume that the two homogeneous half spaces have different impedances $\zeta_0 \neq \zeta_2$, then it is possible to choose the thickness L and the impedance ζ_1 of the layer so that a given frequency ω will be fully transmitted from one half-space to the other one, which would not be the case in absence of such a layer. From the analysis of the reflectivity function

$$|\hat{\mathcal{R}}(\omega)|^2 = 1 - \frac{1 - R_0^2 - R_1^2 + R_0^2 R_1^2}{1 + 2R_0 R_1 \cos(\frac{2\omega L}{c_1}) + R_0^2 R_1^2},$$

one can show that a necessary and sufficient condition for $|\hat{\mathcal{R}}(\omega)|^2$ to be zero is that $R_0^2 + R_1^2 = -2R_0 R_1 \cos(2\omega L/c_1)$. In the case $\zeta_0 \neq \zeta_2$ this in turn enforces one to choose the impedance of the layer to be $\zeta_1 = \sqrt{\zeta_0 \zeta_2}$ (so that $R_0 = R_1$) and the thickness L to be chosen so that $\omega L/(\pi c_1)$ is half an integer (so that $\cos(2\omega L/c_1) = -1$). Usually the thickness is chosen to be equal to a quarter of the wavelength, meaning $\omega L/(\pi c_1) = 1/2$. The insertion of such an "antireflection" layer is often used in optics in order to transmit a laser beam from air to glass, for instance, or in echographic imaging in order to transmit an ultrasound beam from a transducer to a human body with the minimum loss of energy from the beam at the interface.

3.4.4 Scattering by a Single Layer in the Time Domain

In the time domain, we can write the reflected wave in the form

$$b_0(t) = \mathcal{R} * f(t),$$

where $\mathcal{R}$ is the inverse Fourier transform of $\hat{\mathcal{R}}$. If we exclude the degenerate situations in which the impedances can be 0 or $+\infty$, then $|R_j| < 1$ and we can expand the denominator of (3.16) as an infinite series

$$\hat{\mathcal{R}}(\omega) = R_0 + \sum_{n=1}^{\infty} \left[(-1)^n R_0^{n+1} R_1^n + (-1)^{n-1} R_0^{n-1} R_1^n\right] e^{2in\frac{\omega L}{c_1}}.$$

Using $T_0^2 + R_0^2 = 1$, this sum can also be written as

$$\hat{\mathcal{R}}(\omega) = R_0 + \sum_{n=1}^{\infty} (-1)^{n-1} R_0^{n-1} T_0^2 R_1^n e^{2in\frac{\omega L}{c_1}},$$

so that in the time domain,

$$\mathcal{R}(t) = R_0 \delta(t) + \sum_{n=1}^{\infty} (-1)^{n-1} R_0^{n-1} T_0^2 R_1^n \delta\left(t - 2n\frac{L}{c_1}\right), \qquad (3.18)$$

which is the reflected impulse response of the layer from the left side. It gives the train of impulses reflected by the layer when an impulse incoming from the left impinges on the layer at time 0. Each term of the expansion can be intuitively interpreted and associated with a particular scattering sequence (see Figures 3.6 and 3.7):

- The first term $R_0 \delta(t)$ corresponds to the reflection of the incident pulse by the interface $z = 0$.
- The term $n = 1$ corresponds to a transmission through the interface $z = 0$, a reflection by the interface $z = L$, and a transmission though the interface $z = 0$. The round trip takes the time $2L/c_1$.
- The general term n corresponds to a transmission through the interface $z = 0$, a sequence of $n - 1$ reflections by the interface $z = L$ followed by reflections by the interface $z = 0$, and finally a reflection by the interface $z = L$ and a transmission through the interface $z = 0$. The n round trips take time $2nL/c_1$.

In the case of a layer embedded between two half-spaces with the same properties, the expression (3.18) simplifies to

$$\mathcal{R}(t) = R_0 \delta(t) - \sum_{n=1}^{\infty} R_0^{2n-1} T_0^2 \delta\left(t - 2n\frac{L}{c_1}\right).$$

This decomposition of the reflection process can be used to explain the existence of reflectionless scattering at some particular frequencies as discussed in the previous paragraph. If the input wave is of the form $e^{ik\omega_c t}$, with $\omega_c = \pi c_1/L$ and $k \in \mathbb{Z}$, then the n-times multiplyreflected wave components exit the layer with a phase that is exactly shifted by an integer multiple of 2π with respect to the phase of the primary scattered wave $R_0 e^{ik\omega_c t}$, and these multiplyreflected waves fully cancel the primary component:

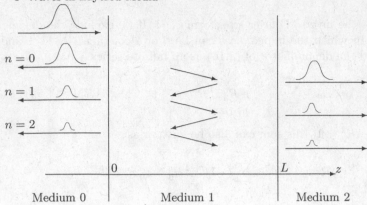

Fig. 3.6. Sketch of scattering sequences.

$$\mathcal{R} * e^{ik\omega_c t} = \left[R_0 - \sum_{n=1}^{\infty} R_0^{2n-1} T_0^2 \right] e^{ik\omega_c t} = 0 \,.$$

As a result, no wave is reflected at all, and by conservation of energy the wave is fully transmitted. Another way to understand this phenomenon involves analyzing the successive waves arriving at the interface $z = L$. These waves have the same phase and therefore they interfere constructively, which enhances the transmittivity of the layer to the point where it becomes equal to 1.

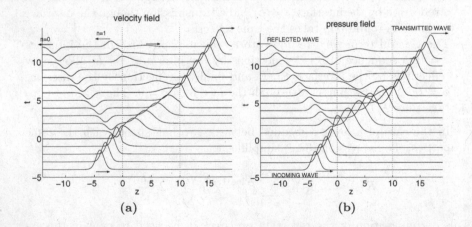

Fig. 3.7. Scattering of a pulse by a layer with parameters (c_1, ζ_1) separating two homogeneous half-spaces with the same parameters (c_0, ζ_0). Here the incoming right-going wave has a Gaussian profile, $c_0 = \zeta_0 = 1$, $c_1 = \zeta_1 = 2$, and the thickness of the layer is 10. The spatial profiles of the velocity field (a) and of the pressure field (b) are plotted at times $t = -4$, $t = -3$,..., $t = 12$. One can observe the first two terms ($n = 0$ and $n = 1$) of the reflected-impulse response of the layer.

3.4.5 Propagator and Scattering Matrices

We end this section by introducing a new object that characterizes the scattering process. We have seen that the overall propagator $\hat{\mathbf{K}}_0(\omega)$ transforms the pair of modes $(\hat{a}_0, \hat{b}_0)$ at the left side of the slab $[0, L]$ into the pair of modes $(\hat{a}_2, \hat{b}_2)$ at the right side of the slab $[0, L]$:

$$\begin{bmatrix} \hat{a}_2(\omega) \\ \hat{b}_2(\omega) \end{bmatrix} = \hat{\mathbf{K}}_0(\omega) \begin{bmatrix} \hat{a}_0(\omega) \\ \hat{b}_0(\omega) \end{bmatrix} .$$

This matrix is symplectic, of the form

$$\hat{\mathbf{K}}_0(\omega) = \begin{bmatrix} \hat{U}(\omega) & \overline{\hat{V}(\omega)} \\ \hat{V}(\omega) & \overline{\hat{U}(\omega)} \end{bmatrix} ,$$

with $|\hat{U}(\omega)|^2 - |\hat{V}(\omega)|^2 = 1$. Note that the moduli of $\hat{U}$ and $\hat{V}$ can take arbitrarily large values for general medium parameter variations, and this may in numerical simulations lead to some instabilities. There exists an alternative and equivalent way to characterize the scattering process through the scattering matrix $\hat{\mathbf{S}}_0(\omega)$ satisfying

$$\begin{bmatrix} \hat{a}_2(\omega) \\ \hat{b}_0(\omega) \end{bmatrix} = \hat{\mathbf{S}}_0(\omega) \begin{bmatrix} \hat{a}_0(\omega) \\ \hat{b}_2(\omega) \end{bmatrix} .$$

This matrix transports what is coming into the slab, namely the right-going mode $\hat{a}_0$ at 0 and the left-going mode $\hat{b}_2$ at L, into what is going out of the slab, the left-going mode $\hat{b}_0$ at 0 and the right-going mode $\hat{a}_2$ at L. We have computed the first column of the scattering matrix in this section, and the computation of the second column can be carried out in a similar way. We then obtain that the scattering matrix is of the form

$$\hat{\mathbf{S}}_0(\omega) = \begin{pmatrix} \hat{\mathcal{T}}(\omega) & \widetilde{\mathcal{R}}(\omega) \\ \hat{\mathcal{R}}(\omega) & \hat{\mathcal{T}}(\omega) \end{pmatrix} ,$$

where the entries of $\hat{\mathbf{S}}_0(\omega)$ and $\hat{\mathbf{K}}_0(\omega)$ are related to each other through the relations

$$\hat{\mathcal{R}}(\omega) = -\frac{\hat{V}(\omega)}{\hat{U}(\omega)} , \qquad \hat{\mathcal{T}}(\omega) = \frac{1}{\hat{U}(\omega)} , \qquad \widetilde{\mathcal{R}}(\omega) = \frac{\overline{\hat{V}(\omega)}}{\hat{U}(\omega)} .$$

Here $\hat{\mathcal{T}}$ is the transmission coefficient and $\hat{\mathcal{R}}$ the reflection coefficient for a wave incoming from the left, whereas the adjoint reflection coefficient corresponding to waves incoming from the right is denoted by $\widetilde{\mathcal{R}}$. The scattering picture is illustrated in Figure 3.8 and Figure 3.9. Using $|\hat{U}(\omega)|^2 - |\hat{V}(\omega)|^2 = 1$, it is easy to check that $\hat{\mathbf{S}}_0(\omega)^{-1} = \overline{\hat{\mathbf{S}}_0(\omega)}^T$, meaning that the scattering matrix is unitary. We also have the energy-conservation relation $|\hat{\mathcal{R}}(\omega)|^2 + |\hat{\mathcal{T}}(\omega)|^2 = 1$, which ensures the boundedness of the reflection and transmission coefficients and enhanced stability in the numerical simulations when one computes the coefficients of the scattering matrix (compared to those of the propagator matrix).

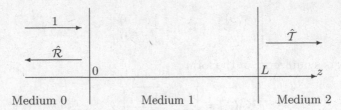

Fig. 3.8. Reflection and transmission coefficients of a layer.

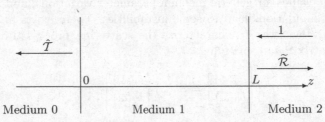

Fig. 3.9. Adjoint reflection and transmission coefficients of a layer.

3.5 Multilayer Piecewise-Constant Media

3.5.1 Propagation Equations

Let us consider a heterogeneous slab consisting of a stack of N layers. The medium is homogeneous inside each layer. The jth layer corresponds to the interval $[L_{j-1}, L_j)$ with $L_0 = 0$:

$$\rho(z) = \begin{cases} \rho_0 & \text{if } z < 0, \\ \rho_j & \text{if } z \in [L_{j-1}, L_j), \quad j = 1, \ldots, N, \\ \rho_{N+1} & \text{if } z > L_N, \end{cases} \tag{3.19}$$

$$K(z) = \begin{cases} K_0 & \text{if } z < 0, \\ K_j & \text{if } z \in [L_{j-1}, L_j), \quad j = 1, \ldots, N, \\ K_{N+1} & \text{if } z > L_N. \end{cases} \tag{3.20}$$

We introduce the local velocities $c_j = \sqrt{K_j/\rho_j}$ and impedances $\zeta_j = \sqrt{K_j \rho_j}$, and the local modes defined by

$$\begin{cases} A_0(t,z) = \zeta_0^{-1/2}p(t,z) + \zeta_0^{1/2}u(t,z)\,, \\ B_0(t,z) = -\zeta_0^{-1/2}p(t,z) + \zeta_0^{1/2}u(t,z)\,, \end{cases} \quad z < 0\,,$$

$$\begin{cases} A_j(t,z) = \zeta_j^{-1/2}p(t,z) + \zeta_j^{1/2}u(t,z)\,, \\ B_j(t,z) = -\zeta_j^{-1/2}p(t,z) + \zeta_j^{1/2}u(t,z)\,, \end{cases} \quad z \in (L_{j-1}, L_j)\,, \quad j = 1, \dots, N\,,$$

$$\begin{cases} A_{N+1}(t,z) = \zeta_{N+1}^{-1/2}p(t,z) + \zeta_{N+1}^{1/2}u(t,z)\,, \\ B_{N+1}(t,z) = -\zeta_{N+1}^{-1/2}p(t,z) + \zeta_{N+1}^{1/2}u(t,z)\,, \end{cases} \quad z > L_N\,.$$

In this section we analyze the wave propagation by studying the modes defined in terms of the local impedance and speed of sound. This is convenient because the wave equations for the modes (A_j, B_j) become very simple in each layer, as in the case of homogeneous medium addressed in Section 3.2:

$$\frac{\partial}{\partial z}\begin{bmatrix} A_j \\ B_j \end{bmatrix} = -\frac{1}{c_j}\begin{bmatrix} 1 & 0 \\ 0 & -1 \end{bmatrix}\frac{\partial}{\partial t}\begin{bmatrix} A_j \\ B_j \end{bmatrix}. \tag{3.21}$$

These equations can be readily integrated, and as a consequence the problem can be reduced to a product of matrices (one for each interface), as explained below.

The Boundary Conditions

The boundary conditions correspond to an impinging pulse at the interface $z = 0$ and a radiation condition at $z = L_N$:

$$A_0(t,0) = f(t)\,, \quad B_{N+1}(t, L_N) = 0\,.$$

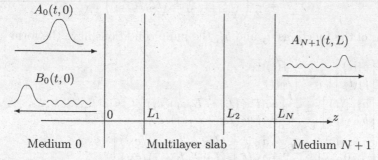

Fig. 3.10. Scattering of a pulse by a multilayer slab (with $N = 3$ layers).

The propagation equations in the two half-spaces dictates the form of the reflected wave for $z < 0$,

$$B_0(t,z) = b_0\left(t + \frac{z}{c_0}\right),$$

and the form of the transmitted wave for $z > L_N$,

$$A_{N+1}(t,z) = a_{N+1}\left(t - \frac{z - L_N}{c_N}\right).$$

We want to express the functions b_0 and a_{N+1} in terms of the medium's properties and the incoming waves. We have to study the propagation of the modes inside the medium $(0, L_N)$.

Propagator Formulation

Inside the layer (L_{j-1}, L_j), $j = 1, \ldots, N$, the pair (A_j, B_j) satisfies the system (3.21), which shows that A_j is a function of $t - z/c_j$ only and B_j is a function of $t + z/c_j$ only. Thus there exist functions $a_j(t)$ and $b_j(t)$ such that for $j = 1, \ldots, N$ and $L_{j-1} \leq z \leq L_j$,

$$A_j(t,z) = a_j(t - (z - L_{j-1})/c_j), \qquad B_j(t,z) = b_j(t + (z - L_{j-1})/c_j).$$

These are the right- and left-going modes moving with the local speed observed in the frame centered at the beginning of the local layer. The equations (3.8) are complemented by the jump conditions at the interfaces $z = L_j$, $j = 0, \ldots, N$, corresponding to the continuity of the velocity and pressure fields. Using the interface propagators gives

$$\begin{bmatrix} A_{j+1} \\ B_{j+1} \end{bmatrix}(t, L_j) = \mathbf{J}_j \begin{bmatrix} A_j \\ B_j \end{bmatrix}(t, L_j), \qquad \mathbf{J}_j = \begin{bmatrix} r_j^{(+)} & r_j^{(-)} \\ r_j^{(-)} & r_j^{(+)} \end{bmatrix},$$

with

$$r_j^{(\pm)} = \frac{1}{2}\left(\sqrt{\zeta_{j+1}/\zeta_j} \pm \sqrt{\zeta_j/\zeta_{j+1}}\right). \tag{3.22}$$

In terms of the functions a_j and b_j, the jump conditions have the form

$$\begin{bmatrix} a_1(t) \\ b_1(t) \end{bmatrix} = \mathbf{J}_0 \begin{bmatrix} f(t) \\ b_0(t) \end{bmatrix},$$

$$\begin{bmatrix} a_{j+1}(t) \\ b_{j+1}(t) \end{bmatrix} = \mathbf{J}_j \begin{bmatrix} a_j(t - (L_j - L_{j-1})/c_j) \\ b_j(t + (L_j - L_{j-1})/c_j) \end{bmatrix}, \qquad j = 1, \ldots, N-1,$$

$$\begin{bmatrix} a_{N+1}(t) \\ 0 \end{bmatrix} = \mathbf{J}_N \begin{bmatrix} a_N(t - (L_N - L_{N-1})/c_N) \\ b_N(t + (L_N - L_{N-1})/c_N) \end{bmatrix},$$

where we have taken into account the boundary conditions at $z = 0$ and $z = L_N$.

As in the single-layer case, it is convenient to go to the Fourier domain to handle the delays $(L_j - L_{j-1})/c_j$. Introducing the frequency dependent matrices

$$\hat{\mathbf{J}}_j(\omega) = \begin{bmatrix} r_j^{(+)} e^{i\frac{\omega(L_j - L_{j-1})}{c_j}} & r_j^{(-)} e^{-i\frac{\omega(L_j - L_{j-1})}{c_j}} \\ r_j^{(-)} e^{i\frac{\omega(L_j - L_{j-1})}{c_j}} & r_j^{(+)} e^{-i\frac{\omega(L_j - L_{j-1})}{c_j}} \end{bmatrix}, \qquad j = 0, \ldots, N, \quad (3.23)$$

where we have used the convention $L_{-1} = 0$, we can now concatenate the jump conditions and write

$$\begin{bmatrix} \hat{a}_{N+1}(\omega) \\ 0 \end{bmatrix} = \hat{\mathbf{J}}_N(\omega) \cdots \hat{\mathbf{J}}_1(\omega) \hat{\mathbf{J}}_0(\omega) \begin{bmatrix} \hat{f}(\omega) \\ \hat{b}_0(\omega) \end{bmatrix}. \qquad (3.24)$$

The solution of the scattering problem involves the computation of the product of matrices $\hat{\mathbf{J}}_N(\omega) \cdots \hat{\mathbf{J}}_0(\omega)$, and then the inversion of (3.24) to get the unknowns $\hat{a}_{N+1}(\omega)$ (transmitted wave) and $\hat{b}_0(\omega)$ (reflected wave).

3.5.2 Reflected and Transmitted Waves

The Frequency-Dependent Reflection and Transmission Coefficients

We discuss here the computation of the product in (3.24). First, we define the family of 2×2 frequency-dependent matrices

$$\hat{\mathbf{K}}_j(\omega) = \hat{\mathbf{J}}_N(\omega) \cdots \hat{\mathbf{J}}_j(\omega), \qquad j = 0, \ldots, N,$$

with the convention that $\hat{\mathbf{K}}_{N+1}(\omega) = \mathbf{I}$, where $\mathbf{I}$ is the 2×2 identity matrix. The matrix $\hat{\mathbf{K}}_j(\omega)$ is the overall propagator for the successive layers $j, \ldots, N$:

$$\begin{bmatrix} \hat{a}_{N+1}(\omega) \\ \hat{b}_{N+1}(\omega) = 0 \end{bmatrix} = \hat{\mathbf{K}}_j(\omega) \begin{bmatrix} \hat{a}_j(\omega) \\ \hat{b}_j(\omega) \end{bmatrix}.$$

Using the relation

$$\hat{\mathbf{K}}_j(\omega) = \hat{\mathbf{K}}_{j+1}(\omega) \hat{\mathbf{J}}_j(\omega)$$

and the expression (3.23) for $\hat{\mathbf{J}}_j(\omega)$, we can then show recursively that the matrix $\hat{\mathbf{K}}_j$ is of the form

$$\hat{\mathbf{K}}_j(\omega) = \begin{bmatrix} \hat{U}_j(\omega) & \overline{\hat{V}_j}(\omega) \\ \hat{V}_j(\omega) & \overline{\hat{U}_j}(\omega) \end{bmatrix}, \qquad j = 0, \ldots, N,$$

where the coefficients $(\hat{U}_j(\omega), \hat{V}_j(\omega))_{j=0,\ldots,N+1}$ satisfy the backward recursive linear system of equations

$$\hat{U}_j(\omega) = \left[r_j^{(+)} \hat{U}_{j+1}(\omega) + r_j^{(-)} \overline{\hat{V}_{j+1}}(\omega) \right] e^{i\frac{\omega(L_j - L_{j-1})}{c_j}}, \qquad (3.25)$$

$$\hat{V}_j(\omega) = \left[r_j^{(+)} \hat{V}_{j+1}(\omega) + r_j^{(-)} \overline{\hat{U}_{j+1}}(\omega) \right] e^{i\frac{\omega(L_j - L_{j-1})}{c_j}}, \qquad (3.26)$$

starting from $\hat{U}_{N+1}(\omega) = 1$ and $\hat{V}_{N+1}(\omega) = 0$, and where $r_j^{(\pm)}$ are given by (3.22). Note that these coefficients satisfy the conservation relation $|\hat{U}_j(\omega)|^2 - |\hat{V}_j(\omega)|^2 = 1$, which shows that $\hat{\mathbf{K}}_j(\omega)$ is symplectic. From the identity

$$
\begin{bmatrix} \hat{a}_{N+1}(\omega) \\ 0 \end{bmatrix} = \begin{bmatrix} \hat{U}_0(\omega) & \overline{\hat{V}_0}(\omega) \\ \hat{V}_0(\omega) & \overline{\hat{U}_0}(\omega) \end{bmatrix} \begin{bmatrix} \hat{f}(\omega) \\ \hat{b}_0(\omega) \end{bmatrix},
$$

we finally get

$$
\hat{b}_0(\omega) = \hat{\mathcal{R}}_0(\omega)\hat{f}(\omega), \qquad \hat{a}_{N+1}(\omega) = \hat{\mathcal{T}}_0(\omega)\hat{f}(\omega),
$$

where the reflection and transmission coefficients of the multilayer slab are

$$
\hat{\mathcal{R}}_0(\omega) = -\frac{\hat{V}_0(\omega)}{\overline{\hat{U}_0}(\omega)}, \qquad \hat{\mathcal{T}}_0(\omega) = \frac{1}{\overline{\hat{U}_0}(\omega)}.
$$

The system (3.25–3.26) allows us to get the reflection and transmission coefficients, but it turns out that the sequence $(\hat{U}_j, \hat{V}_j)$ required to compute $(\hat{\mathcal{R}}_0(\omega), \hat{\mathcal{T}}_0(\omega))$ can take large values within the conservation relation $|\hat{U}_j(\omega)|^2 - |\hat{V}_j(\omega)|^2 = 1$. This is an important numerical issue, and there is another way to compute the reflection and transmission coefficients, which we introduce now.

Linear Fractional Relations for the Local Reflection Coefficients

We introduce the local reflection and transmission coefficients

$$
\hat{\mathcal{R}}_j(\omega) = -\frac{\hat{V}_j(\omega)}{\hat{U}_j(\omega)}, \qquad \hat{\mathcal{T}}_j(\omega) = \frac{1}{\overline{\hat{U}_j}(\omega)}, \qquad j = 0, \ldots, N+1.
$$

Then we can write a nonlinear recursive relation satisfied by $\hat{\mathcal{R}}_j(\omega)$ from the linear recursive relation satisfied by $(\hat{U}_j, \hat{V}_j)$:

$$
\hat{\mathcal{R}}_j(\omega) = \frac{\hat{\mathcal{R}}_{j+1}(\omega) + R_j}{1 + \hat{\mathcal{R}}_{j+1}(\omega)R_j} e^{2i\frac{\omega(L_j - L_{j-1})}{c_j}}, \qquad j = 0, \ldots, N. \tag{3.27}
$$

Here $R_j = \frac{\zeta_j - \zeta_{j+1}}{\zeta_j + \zeta_{j+1}}$ is the reflection coefficient of the interface $z = L_j$. The "final" condition for this recursive system is given at $j = N+1$ by

$$
\hat{\mathcal{R}}_{N+1}(\omega) = 0. \tag{3.28}
$$

Equations (3.27–3.28) form a backward system that has to be solved recursively from $j = N$ to $j = 0$. This closed-form recursive system allows us to compute directly the reflection coefficient $\hat{\mathcal{R}}_0(\omega)$. Recall that $\hat{\mathcal{R}}_j(\omega)$ always belongs to the unit complex disk, so that no large values can be encountered along the recursive computation. This is in contrast to the linear system (3.25–3.26), where the pairs $(\hat{U}_j(\omega), \hat{V}_j(\omega))$ can take very large values. This remark turns out to be very important from the numerical point of view developed below.

Note also that each step in the recurrence (3.27) consists in applying a fractional linear transform to $\hat{\mathcal{R}}_{j+1}$ in order to get $\hat{\mathcal{R}}_j$. Since the set of fractional linear transforms forms a group under function composition, the relation between any pair of local reflection coefficients $\hat{\mathcal{R}}_j$ and $\hat{\mathcal{R}}_{j'}$ is a fractional linear transform. The four coefficients of these transforms can be computed recursively by (3.27). This remark allows the quick computation of the global reflection coefficient when two or several stacks of layers are merged.

Similarly, we can get the transmission coefficient $\hat{T}_0(\omega)$ from the backward recursive relation

$$\hat{T}_j(\omega) = \frac{T_j \hat{T}_{j+1}(\omega)}{1 + \hat{\mathcal{R}}_{j+1}(\omega)R_j} e^{i\frac{\omega(L_j - L_{j-1})}{c_j}}, \qquad j = 0, \ldots, N, \tag{3.29}$$

starting from $\hat{T}_{N+1}(\omega) = 1$. Here $T_j = \frac{2\sqrt{\zeta_j \zeta_{j+1}}}{\zeta_j + \zeta_{j+1}}$ is the transmission coefficient of the interface $z = L_j$.

Representations of the Reflected and Transmitted Waves in the Time Domain

Taking an inverse Fourier transform yields the integral representation of the reflected wave in the time domain

$$B(t, 0) = b_0(t) = \frac{1}{2\pi} \int \hat{\mathcal{R}}_0(\omega)\hat{f}(\omega)e^{-i\omega t} d\omega \,,$$

that is,

$$b_0(t) = \mathcal{R} * f(t) \,,$$

where $\mathcal{R}$ is the inverse Fourier transform of $\hat{\mathcal{R}}$. Using equation (3.27), we find that $\hat{\mathcal{R}}$ can be expanded as a series of the form

$$\hat{\mathcal{R}}(\omega) = \sum_{j=1}^{N} \sum_{k_j=0}^{\infty} \alpha_{k_1,\ldots,k_n} \exp\left(2i\omega \sum_{j=1}^{N} k_j \frac{L_j - L_{j-1}}{c_j}\right),$$

where $\alpha_{k_1,\ldots,k_n}$ is a coefficient that depends only on $R_0, \ldots, R_N$ (for instance, $\alpha_{0,0,\ldots,0} = R_0$). As a result, in the time domain,

$$\mathcal{R}(t) = \sum_{j=1}^{N} \sum_{k_j=0}^{\infty} \alpha_{k_1,\ldots,k_n} \delta\left(t - 2\sum_{j=1}^{N} k_j \frac{L_j - L_{j-1}}{c_j}\right).$$

Each term of this series can be associated with a scattering sequence involving reflections and transmissions by the different interfaces and that determines the value of the coefficients $\alpha_{k_1,\ldots,k_n}$. In this interpretation the time delay is simply the sum of travel times from an interface to another corresponding to the particular scattering sequence.

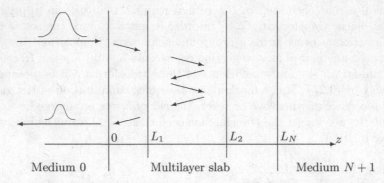

Medium 0   Multilayer slab   Medium $N+1$

Fig. 3.11. A particular scattering sequence inside the multilayer slab.

In the same way, the transmitted wave can be represented by

$$A_{N+1}(L_N,t) = a_{N+1}(t) = \frac{1}{2\pi}\int \hat{\mathcal{T}}_0(\omega)\hat{f}(\omega)e^{-i\omega t}d\omega\,,$$

that is,

$$a_{N+1}(t) = \mathcal{T} * f(t)\,,$$

where $\mathcal{T}$ is the inverse Fourier transform of $\hat{\mathcal{T}}$. Expanding $\hat{\mathcal{T}}$ yields that $\mathcal{T}$ is of the form

$$\mathcal{T}(t) = \sum_{j=1}^{N}\sum_{k_j=0}^{\infty}\beta_{k_1,\ldots,k_n}\delta\left(t - 2\sum_{j=1}^{N}k_j\frac{L_j - L_{j-1}}{c_j} - \sum_{j=1}^{N}\frac{L_j - L_{j-1}}{c_j}\right)\,,$$

where $\beta_{k_1,\ldots,k_n}$ is a coefficient that depends only on $R_0,\ldots,R_N$ (for instance, $\beta_{0,0,\ldots,0} = T_0 T_1 \cdots T_N = \sqrt{1 - R_0^2}\sqrt{1 - R_1^2}\cdots\sqrt{1 - R_N^2}$). Note that the first impulse of the transmitted impulse reponse $\mathcal{T}$ exits the layer at time $t = \sum_{j=1}^{N}\frac{L_j - L_{j-1}}{c_j}$, which corresponds to the total travel time through the slab. This first impulse describes the ballistic wave, that is to say the wave that has not been reflected at all, and its amplitude is $\prod_{j=0}^{N}T_j$.

3.5.3 Reflectivity Pattern and Bragg Mirror for Periodic Layers

Let us consider a very particular case, in which the multilayer slab consists of a stack of alternating layers of two different materials a and b with two different impedances $\zeta_a \neq \zeta_b$, with two thicknesses chosen so that the travel times through the layers are equal $L_a/c_a = L_b/c_b \equiv \tau$:

$$\zeta(z) = \begin{cases} \zeta_b & \text{if } z \in [L_{2j}, L_{2j+1}), \quad j = 0, \ldots, N-1, \\ \zeta_a & \text{otherwise}, \end{cases}$$

$$c(z) = \begin{cases} c_b & \text{if } z \in [L_{2j}, L_{2j+1}), \quad j = 0, \ldots, N-1, \\ c_a & \text{otherwise}, \end{cases}$$

where $L_{2j} = j(L_a + L_b)$ and $L_{2j+1} = L_{2j} + L_b$ (see Figure 3.12).

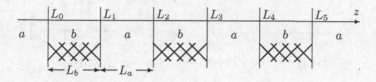

Fig. 3.12. $N = 3$-period Bragg mirror.

The periodic variation of the slab on a length scale comparable to the wavelength has a dramatic effect. In particular, there exist some frequency bands that are almost completely reflected. These bands are centered at the frequencies $\omega_k = (k + \frac{1}{2})\frac{\pi}{\tau}$. The recursive relation (3.27) takes the form

$$\hat{\mathcal{R}}_{2j-2}(\omega_k) = -\frac{\hat{\mathcal{R}}_{2j-1}(\omega_k) + R}{1 + \hat{\mathcal{R}}_{2j-1}(\omega_k)R}, \qquad \hat{\mathcal{R}}_{2j-1}(\omega_k) = -\frac{\hat{\mathcal{R}}_{2j}(\omega_k) - R}{1 - \hat{\mathcal{R}}_{2j}(\omega_k)R},$$

starting from $\hat{\mathcal{R}}_{2N}(\omega_k) = 0$. Here $R = \frac{\zeta_a - \zeta_b}{\zeta_a + \zeta_b}$. Grouping these relations by pairs yields

$$\hat{\mathcal{R}}_{2j-2}(\omega_k) = \frac{\alpha + \hat{\mathcal{R}}_{2j}(\omega_k)}{1 + \alpha\hat{\mathcal{R}}_{2j}(\omega_k)},$$

where $\alpha = \frac{-2R}{1 + R^2} = \frac{\zeta_b^2 - \zeta_a^2}{\zeta_a^2 + \zeta_b^2}$. Taking into account the initial condition $\hat{\mathcal{R}}_{2N}(\omega_k) = 0$, we obtain recursively that the local reflection coefficient is

$$\hat{\mathcal{R}}_{2j}(\omega_k) = \frac{1 - (\frac{\zeta_a}{\zeta_b})^{2(N-j)}}{1 + (\frac{\zeta_a}{\zeta_b})^{2(N-j)}}.$$

Finally, we get that the reflectivities of the frequencies ω_k are

$$\left| \hat{\mathcal{R}} \left((k + \frac{1}{2})\omega_0 \right) \right|^2 = \left(\frac{1 - (\frac{\zeta_a}{\zeta_b})^{2N}}{1 + (\frac{\zeta_a}{\zeta_b})^{2N}} \right)^2,$$

where N is the number of pairs of alternating layers. This shows that the reflectivity goes to one at an exponential rate as the number of layers increases. This phenomenon is referred to as Bragg resonance, which explains that a very high reflectivity can be achieved for some particular frequencies

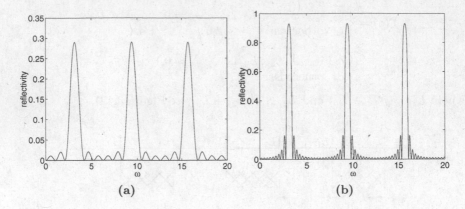

Fig. 3.13. Reflectivity $|\hat{\mathcal{R}}(\omega)|^2$ versus frequency for a N-period Bragg mirror with $N = 3$ (a), $N = 10$ (b). The two materials have impedances $\zeta_a = 0.9$ and $\zeta_b = 1.1$ and the travel time through each layer is 1. The reflected frequencies are $\omega_k = (2k+1)\pi$.

as a result of the periodicity of the structure, even for very small impedance contrast. The contrast between the two impedances controls the widths of the reflected frequency bands. The smaller the contrast, the narrower the reflected bands. Outside the reflected bands, the reflectivity is small, on the order of the reflectivity of one single interface. In conclusion, in the case of a periodic multilayer slab with many layers and small impedance contrasts, waves are either completely reflected back if their frequencies fulfill the Bragg resonance conditions, or waves are completely transmitted (see Figure 3.13).

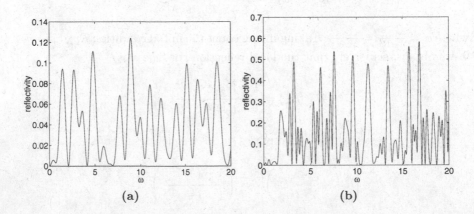

Fig. 3.14. Reflectivity versus frequency for a multilayer slab. We have perturbed the $N = 3$ (a) and $N = 10$ (b) Bragg mirror described in Figure 3.13 by changing randomly the thicknesses of the layers.

In the general framework of a multilayer slab, with different contrast impedances and layer thicknesses, the reflectivity $\omega \mapsto |\hat{\mathcal{R}}(\omega)|^2$ becomes a very complicated function as the number of layers N increases. As shown in Figure 3.14, the reflectivity as a function of the frequency has many local maxima and minima. Although the overall picture seems unpredictable, we will see in the next chapters that it is possible to describe the behavior of the reflectivity as the number of layers goes to infinity through an asympotic analysis based on limit theorems for stochastic processes.

3.5.4 Goupillaud Medium

A multilayer medium where layers have equal travel times is called a Goupillaud medium. This model is famous because it allows for an exact discretization of the pulse propagation in the time domain, as we show below. It is therefore often used in numerical simulations.

A Goupillaud medium consists of a stack of N layers of the type (3.19–3.20), where the local velocities and thicknesses of the layers are such that

$$\frac{L_j - L_{j-1}}{c_j} \equiv \tau \quad \forall j = 1, \ldots, N,$$

where τ is the travel time from one interface to another one. We aim at studying the impulse response of the medium, that is, how a right-going Dirac pulse incoming from the left half-space is transformed as it propagates into the multilayer slab. We are particularly interested in the reflected pulse at $z = 0$ (the reflected impulse response) and the transmitted pulse at $z = L_N$ (the transmitted impulse response). The impulse response is the analogy of the Green's function. The reflected (respectively transmitted) pulse when an arbitrary input pulse is considered is simply the convolution of the input pulse shape with the reflected (respectively transmitted) impulse response.

We assume that the slab is probed with a right-going propagating impulse:

$$A(t, z) = \delta(z - t/c_0), \quad B(t = 0, z) = 0, \quad \text{for } t < 0.$$

At time 0 the first scattering occurs as the impulse arrives at the interface $z = 0$. Just after $t = 0$ we have one reflected pulse and one transmitted pulse

$$A(t = 0^+, z) = T_0 \delta(z), \quad B(t = 0^+, z) = R_0 \delta(z),$$

where the support of the right-going impulse A is just to the right of the interface $z = 0$, and the support of the left-going impulse A is just to the left of the interface $z = 0$ (see top of Figure 3.15). Multiple-wave scattering leads to a set of right- and left-going impulses. The scattering events occur only at times that are multiples of τ due to the fact that the travel times from one interface to another are constant. If we observe the wave at times $k\tau^+$, $k \in \mathbb{N}$, then these impulses are located at the interfaces, more precisely just to the right of the interfaces for the right-going impulses and just to the left for the left-going impulses (see Figure 3.15). There are other impulse components that have exited the slab, but they no longer interact with the slab, since they propagate with constant speed, so they can be ignored. The right- and left-going waves inside the slab are of the form

$$A(t = k\tau^+, z) = \sum_{j=0}^{N} A_j^k \delta(z - L_j), \quad B(t = k\tau^+, z) = \sum_{j=0}^{N} B_j^k \delta(z - L_j),$$

with k being the time index and j the interface index of the impulse amplitudes A_j^k and B_j^k. The initial conditions at time 0^+ give

$$A_j^0 = T_0 \delta_{j0}, \qquad B_j^0 = R_0 \delta_{j0}, \tag{3.30}$$

where δ_{jk} is the Kronecker symbol: $\delta_{jk} = 1$ if $j = k$ and 0 if $j \neq k$.

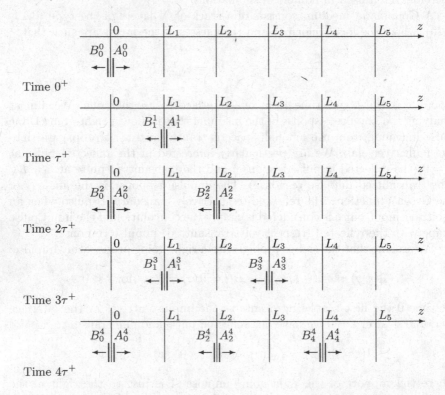

Fig. 3.15. Discretization of the scattering process in a Goupillaud medium. The nonzero impulse amplitudes are depicted for the times indices $k = 0, 1, 2, 3, 4$.

In this modeling, $(B_0^k)_{k \in \mathbb{N}}$ is the set of amplitudes of the reflected wave; that is, the reflected impulse response is

$$\mathcal{R}(t) = \sum_{k=0}^{\infty} B_0^k \delta(t - k\tau).$$

Similarly, the transmitted impulse response is

$$\mathcal{T}(t) = \sum_{k=0}^{\infty} A_N^k \delta(t - k\tau).$$

From the interface conditions at $z = L_0, \ldots, L_N$ we obtain the matrix relations

$$\begin{bmatrix} A_j^{k+1} \\ B_j^{k+1} \end{bmatrix} = \begin{bmatrix} T_j & -R_j \\ R_j & T_j \end{bmatrix} \begin{bmatrix} A_{j-1}^k \\ B_{j+1}^k \end{bmatrix}, \quad \text{for } j = 0, \ldots, N, \quad (3.31)$$

with the radiation conditions $A_{-1}^k = 0$ and $B_{N+1}^k = 0$. This system complemented with the initial conditions (3.30) allows us to compute recursively with respect to the time index k the right and left impulse amplitudes.

The initial conditions and the unit speed of propagation enforce that $A_j^k = B_j^k = 0$ for all $j > k$. This can be proved recursively with respect to k from (3.31) and is a manifestation of the hyperbolicity of the system. After a time $k\tau$, the wave has probed only the first k layers of the medium. This remark allows us to push the analysis forward by showing that the system (3.30–3.31) can be reduced to a triangular system. This is especially relevant for important applications, such as in geophysics, where the multilayer slab consists of a very large number N of layers or even an infinite number of layers; that is, the heterogeneous slab is a heterogeneous half-space. Note that the N-layer case and the infinite-layer case are strictly identical as long as the time interval for the computation is smaller than $N\tau$, according to the above remark.

We first note that $j = k$ is the position of the front impulse at time index k. Introducing the impulse amplitudes at distance l from the front,

$$\tilde{A}_l^k = A_{k-l}^k, \qquad \tilde{B}_l^k = B_{k-l}^k,$$

we can write a triangular system for $0 \le l \le k + 1$,

$$\begin{bmatrix} \tilde{A}_l^{k+1} \\ \tilde{B}_l^{k+1} \end{bmatrix} = \begin{bmatrix} T_{k+1-l} & -R_{k+1-l} \\ R_{k+1-l} & T_{k+1-l} \end{bmatrix} \begin{bmatrix} \tilde{A}_l^k \\ \tilde{B}_{l-2}^k \end{bmatrix},$$

with the initial conditions

$$\tilde{A}_l^0 = T_0 \delta_{l0}, \qquad \tilde{B}_l^0 = R_0 \delta_{l0},$$

at $k = 0$ and the convention $\tilde{B}_{-2}^k = \tilde{B}_{-1}^k = 0$, $\tilde{A}_{k+1}^k = 0$. Note that only the amplitudes $\tilde{A}_l^k$ and $\tilde{B}_l^k$ with even indices l are nonzero. Substituting the second equation into the first one, we get the equivalent system

$$\tilde{A}_{2l}^{k+1} = \frac{1}{T_{k+1-2l}} \tilde{A}_{2l}^k - \frac{R_{k+1-2l}}{T_{k+1-2l}} \tilde{B}_{2l}^{k+1},$$

$$\tilde{B}_{2l}^{k+1} = R_{k+1-2l} \tilde{A}_{2l}^k + T_{k+1-2l} \tilde{B}_{2(l-1)}^k.$$

Introducing the normalized amplitudes $0 \leq 2l \leq k$,

$$a_l^k = \left(\prod_{m=0}^{k-2l} T_m \right) \tilde{A}_{2l}^k, \qquad b_l^k = \left(\prod_{m=0}^{k-2l-1} T_m \right) \tilde{B}_{2l}^k,$$

with the convention $T_{-1} = 1$, we get the normalized equations

$$a_l^{k+1} = a_l^k - R_{k+1-2l} b_l^{k+1},$$
$$b_l^{k+1} = b_{l-1}^k + R_{k+1-2l} a_l^k.$$

By successive substitution of the second equation into the first one, it is possible to eliminate the variable b to get a closed system for a:

$$\mathbf{a}^{k+1} = \mathbf{a}^k + \mathbf{P}^k \mathbf{a}^k, \qquad\qquad (3.32)$$

where $\mathbf{a}_l^k = a_l^{k+l}$ and the matrix $\mathbf{P}^k$ is lower triangular with

$$\mathbf{P}_{l,m}^k = -R_{k-l+1} R_{k-m+1} \text{ for } l \geq m.$$

This triangular system can be easily integrated with the initial condition $\mathbf{a}_l^0 = T_0^2 \delta_{l0}$, and this gives the principle of a simple and efficient way to simulate wave propagation in Goupillaud layered media in the time domain. Equation (3.32) shows how a right-going wave changes as it propagates into the heterogeneous slab. Note that the relevant information about the medium for the analysis of the propagation is the product of two reflection coefficients. This remark shows that if the medium is described in terms of random coefficients, the statistics of the random propagating wave will be determined by the statistics of the products of reflection coefficients.

Notes

The various transforms carried out in this chapter on the wave equations are classical in the theory of hyperbolic systems. Our presentation is self-contained in the case of the one-dimensional acoustic equations. For a general introduction to wave propagation phenomena we refer to the book by Whitham (1974) [167], and for the particular case of deterministic layered media to the book by Brekhovskikh (1980) [24]. Efficient numerical schemes for wave propagating in multilayer media are proposed and discuseed in [8].

4

Effective Properties of Randomly Layered Media

In this chapter we study pulses traveling through finely layered media. We first consider in Section 4.1 the particular case of piecewise-constant media as introduced in the previous chapter. Using the formulation in terms of matrix products, we show with two examples, periodic and random, that the transmitted and reflected waves converge when the size of the layers goes to zero and their number tends to infinity. In fact, homogenization takes place and in the limit the pulse "sees" an effective homogeneous medium. The material properties of this effective medium are computed from the material properties of the heterogeneous medium through an averaging procedure. We then introduce in Section 4.2 more general models of finely layered media, for which the material parameters can vary continuously and/or with jumps. In Section 4.3 we present an alternative approach to the description of the propagating modes, based on differential equations, where a suitable reference medium is used. It requires the following sequence of transformations, some of which have already been introduced in the previous chapter, now discussed in Sections 4.3–4.4:

- Decomposition into **right-** and **left-going** waves.
- Specification of **boundary conditions** and identification of the **quantities of interest**.
- **Centering** along the **characteristics** of the wave equations in a reference medium.
- Performing Fourier transform in time, that is, passing to the **frequency domain**.
- Converting a boundary value problem into an initial value problem using a **propagator** matrix.

This approach will be the one exploited further in the rest of the book. We then show in Section 4.5 that in the case of random media, the homogenization problem is reduced to the averaging of random differential equations. The averaging procedure derives from three important concepts:

- **Stochastic modeling**: description of the fluctuations in the medium parameters, the density, and the bulk modulus, in terms of ergodic random processes.
- **Scale separation**: exploiting the assumption that the typical wavelength of the incoming wave is much longer than the typical scale of variation of the medium parameters.
- **Homogenization**: use of the law of large numbers or the ergodic theorem for ordinary differential equations with random coefficients.

In this chapter and in the rest of the book we are mainly concerned with waves propagating in randomly layered media, so we present here homogenization in this particular case. However, this theory applies to more general multi-dimensional random media.

4.1 Finely Layered Piecewise-Constant Media

In this section we consider the case of a pulse traveling across many small layers. We use the propagator formulation introduced in the previous chapter in Section 3.5.1. A pulse, whose Fourier transform is assumed to be compactly supported in $[-\omega_0, \omega_0]$, is incoming from the left homogeneous half-space and is impinging onto a multilayer slab $[0, L]$. We assume that the multilayer slab is formed by $2N$ layers made of two materials whose densities (respectively bulk moduli) are denoted by ρ_a and ρ_b (respectively K_a and K_b):

$$
(\rho, K)(z) = \begin{cases} (\rho_e, K_e) & \text{if } z < L_0 = 0, \\ (\rho_a, K_a) & \text{if } z \in [L_{2j}, L_{2j+1}), \quad j = 0, \ldots, N-1, \\ (\rho_b, K_b) & \text{if } z \in [L_{2j+1}, L_{2j+2}), \quad j = 0, \ldots, N-1, \\ (\rho_{e'}, K_{e'}) & \text{if } z > L = L_{2N+1}. \end{cases} \quad (4.1)
$$

The thickness of the jth layer will be denoted by $\Delta_j = L_j - L_{j-1}$, $j = 1, \ldots, 2N+1$. The homogeneous half-spaces $z < 0$ (respectively $z > L$) have material properties identified by the index e (respectively e'). The scattering problem (3.24) for the transmitted wave a_{2N+1}, the reflected wave b_0, and the incoming wave f can be written in the Fourier domain in the form

$$
\begin{bmatrix} \hat{a}_{2N+1}(\omega) \\ 0 \end{bmatrix} = \hat{\mathbf{K}}_{2N+1}(\omega) \begin{bmatrix} \hat{f}(\omega) \\ \hat{b}_0(\omega) \end{bmatrix}, \quad (4.2)
$$

$$
\hat{\mathbf{K}}_{2N+1}(\omega) = \hat{\mathbf{J}}_{2N+1}(\omega) \left(\hat{\mathbf{J}}_{2N}(\omega) \hat{\mathbf{J}}_{2N-1}(\omega) \right) \cdots
$$
$$
\cdots \left(\hat{\mathbf{J}}_2(\omega) \hat{\mathbf{J}}_1(\omega) \right) \hat{\mathbf{J}}_{ea}(\omega), \quad (4.3)
$$

where straightforward computations give the following:

- $\hat{\mathbf{J}}_{ea}$ corresponds to the interface between the left homogeneous medium and the first layer that is of type a:

$$\hat{\mathbf{J}}_{ea}(\omega) = \begin{bmatrix} r_{ea}^{(+)} & r_{ea}^{(-)} \\ r_{ea}^{(-)} & r_{ea}^{(+)} \end{bmatrix}, \tag{4.4}$$

with $r_{ea}^{(\pm)} = \frac{1}{2}\left(\sqrt{\zeta_a/\zeta_e} \pm \sqrt{\zeta_e/\zeta_a}\right)$.

- $\hat{\mathbf{J}}_{2N+1}$ corresponds to the interface with the right homogeneous medium:

$$\hat{\mathbf{J}}_{2N+1}(\omega) = \begin{bmatrix} r_{ae'}^{(+)} e^{i\frac{\omega\Delta_{2N+1}}{c_a}} & r_{ae'}^{(-)} e^{-i\frac{\omega\Delta_{2N+1}}{c_a}} \\ r_{ae'}^{(-)} e^{i\frac{\omega\Delta_{2N+1}}{c_a}} & r_{ae'}^{(+)} e^{i\frac{\omega\Delta_{2N+1}}{c_a}} \end{bmatrix}, \tag{4.5}$$

with $r_{ae'}^{(\pm)} = \frac{1}{2}\left(1\sqrt{\zeta_{e'}/\zeta_a} \pm \sqrt{\zeta_a/\zeta_{e'}}\right)$.

- $\hat{\mathbf{J}}_j^{(2)}(\omega) := \hat{\mathbf{J}}_{2j}(\omega)\hat{\mathbf{J}}_{2j-1}(\omega)$ corresponds to the propagation over the two successive layers $[L_{2j-2}, L_{2j-1}] \cup [L_{2j-1}, L_{2j}]$, and these composite propagators have elements

$$\hat{\mathbf{J}}_j^{(2)}(1,1) = r^{(+)2} e^{i\omega\left(\frac{\Delta_{2j-1}}{c_a} + \frac{\Delta_{2j}}{c_b}\right)} - r^{(-)2} e^{i\omega\left(\frac{\Delta_{2j-1}}{c_a} - \frac{\Delta_{2j}}{c_b}\right)}, \tag{4.6}$$

$$\hat{\mathbf{J}}_j^{(2)}(1,2) = r^{(+)}r^{(-)}\left(e^{i\omega\left(-\frac{\Delta_{2j-1}}{c_a} + \frac{\Delta_{2j}}{c_b}\right)} - e^{i\omega\left(-\frac{\Delta_{2j-1}}{c_a} - \frac{\Delta_{2j}}{c_b}\right)}\right), \tag{4.7}$$

$$\hat{\mathbf{J}}_j^{(2)}(2,1) = \overline{\hat{\mathbf{J}}_j^{(2)}(1,2)}, \tag{4.8}$$

$$\hat{\mathbf{J}}_j^{(2)}(2,2) = \overline{\hat{\mathbf{J}}_j^{(2)}(1,1)}, \tag{4.9}$$

with $r^{(\pm)} = \frac{1}{2}\left(\sqrt{\zeta_b/\zeta_a} \pm \sqrt{\zeta_a/\zeta_b}\right)$.

4.1.1 Periodic Case

We assume in this subsection that the layers have the same size: $\Delta_j \equiv \Delta$. Moreover, the thickness Δ is small, so that the total thickness $L = (2N+1)\Delta$ of the heterogeneous slab is of order one. In this case, the matrix $\hat{\mathbf{J}}_j^{(2)}$ is independent of the layer index j, and the matrix $\hat{\mathbf{J}}_{2N+1}(\omega)$ does not depend on N, but both depend on the layer thickness Δ:

$$\hat{\mathbf{J}}_j^{(2)}(\omega) \equiv \hat{\mathbf{J}}^{(2)}(\omega, \Delta), \quad j = 1, \dots, N,$$
$$\hat{\mathbf{J}}_{2N+1}(\omega) = \hat{\mathbf{J}}_{ae'}(\omega, \Delta).$$

The propagator $\hat{\mathbf{K}}_{2N+1}$ is given by

$$\hat{\mathbf{K}}_{2N+1}(\omega) = \hat{\mathbf{J}}_{ae'}(\omega, \Delta)\left(\hat{\mathbf{J}}^{(2)}(\omega, \Delta)\right)^N \hat{\mathbf{J}}_{ea}(\omega).$$

The matrix $\hat{\mathbf{J}}_{ea}(\omega)$ defined by (4.4) does not depend on Δ, and

$$\hat{\mathbf{J}}_{ae'}(\omega, \Delta) \xrightarrow{\Delta \to 0} \hat{\mathbf{J}}_{ae'} := \begin{bmatrix} r_{ae'}^{(+)} & r_{ae'}^{(-)} \\ r_{ae'}^{(-)} & r_{ae'}^{(+)} \end{bmatrix},$$

where the convergence is uniform for ω in the support $[-\omega_0, \omega_0]$ of the pulse $\hat{f}$.

We now study the convergence of $\left(\hat{\mathbf{J}}^{(2)}(\omega, \Delta)\right)^N$ as Δ goes to zero, with $N = (L - \Delta)/(2\Delta)$, and $L > 0$ fixed. Note that N goes to infinity as Δ goes to zero.

Remark 4.1. In the very particular case with no impedance contrast, $\zeta_a = \zeta_b$, we have $r^{(+)} = 1$ and $r^{(-)} = 0$, and the matrix $\hat{\mathbf{J}}^{(2)}(\omega, \Delta)$ is diagonal. As a result,

$$\left(\hat{\mathbf{J}}^{(2)}(\omega, \Delta)\right)^N = \begin{bmatrix} e^{i\frac{2\omega \Delta N}{\bar{c}}} & 0 \\ 0 & e^{-i\frac{2\omega \Delta N}{\bar{c}}} \end{bmatrix} \xrightarrow{\Delta \to 0} \begin{bmatrix} e^{i\frac{\omega L}{\bar{c}}} & 0 \\ 0 & e^{-i\frac{\omega L}{\bar{c}}} \end{bmatrix},$$

where the **effective velocity** $\bar{c}$ is the harmonic average of the individual velocities:

$$\frac{1}{\bar{c}} = \frac{1}{2}\left(\frac{1}{c_a} + \frac{1}{c_b}\right). \tag{4.10}$$

In the general case, $\zeta_a \neq \zeta_b$, we can expand the matrix $\hat{\mathbf{J}}^{(2)}(\omega, \Delta)$ to second order in Δ,

$$\hat{\mathbf{J}}^{(2)}(\omega, \Delta) = \mathbf{I} + i(\omega \Delta)\hat{\mathbf{J}}_1^{(2)} + \mathbf{O}(\Delta^2), \tag{4.11}$$

where

$$\hat{\mathbf{J}}_1^{(2)}(1,1) = r^{(+)2}\left(\frac{1}{c_a} + \frac{1}{c_b}\right) - r^{(-)2}\left(\frac{1}{c_a} - \frac{1}{c_b}\right), \tag{4.12}$$

$$\hat{\mathbf{J}}_1^{(2)}(1,2) = 2r^{(+)}r^{(-)}\frac{1}{c_b}, \tag{4.13}$$

$$\hat{\mathbf{J}}_1^{(2)}(2,1) = -\hat{\mathbf{J}}_1^{(2)}(1,2), \tag{4.14}$$

$$\hat{\mathbf{J}}_1^{(2)}(2,2) = -\hat{\mathbf{J}}_1^{(2)}(1,1). \tag{4.15}$$

Therefore we deduce the limit

$$\left(\hat{\mathbf{J}}^{(2)}(\omega, \Delta)\right)^N \xrightarrow{\Delta \to 0} \exp\left(i\omega\frac{L}{2}\hat{\mathbf{J}}_1^{(2)}\right).$$

The matrix $\hat{\mathbf{J}}_1^{(2)}$ can be written in the diagonal form

$$\hat{\mathbf{J}}_1^{(2)} = 2\mathbf{M}^{-1}\begin{bmatrix} \bar{c}^{-1} & 0 \\ 0 & -\bar{c}^{-1} \end{bmatrix}\mathbf{M},$$

where

$$\mathbf{M} = \begin{bmatrix} \bar{\zeta}^{-1/2} & \bar{\zeta}^{1/2} \\ -\bar{\zeta}^{-1/2} & \bar{\zeta}^{1/2} \end{bmatrix}.$$

The effective impedance and velocity are $\bar{c} = \sqrt{\bar{K}/\bar{\rho}}$ and $\bar{\zeta} = \sqrt{\bar{K}\bar{\rho}}$, with the **effective density and bulk modulus** defined by

$$\bar{\rho} = \frac{1}{2}\left(\rho_a + \rho_b\right), \qquad \frac{1}{\bar{K}} = \frac{1}{2}\left(\frac{1}{K_a} + \frac{1}{K_b}\right). \qquad (4.16)$$

The overall propagator as $\Delta \to 0$ takes the limit form

$$\hat{K}_{2N+1}(\omega) \xrightarrow{\Delta \to 0} \hat{\mathbf{J}}_{ae'}\mathbf{M}^{-1}\begin{bmatrix} e^{i\frac{\omega L}{\bar{c}}} & 0 \\ 0 & e^{-i\frac{\omega L}{\bar{c}}} \end{bmatrix}\mathbf{M}\hat{\mathbf{J}}_{ea}.$$

After some calculations, we see that

$$\hat{\mathbf{J}}_{ae'}\mathbf{M}^{-1} = \begin{bmatrix} r_{he'}^{(+)} & r_{he'}^{(-)} \\ r_{he'}^{(-)} & r_{he'}^{(+)} \end{bmatrix},$$

where $r_{he'}^{(\pm)}$ are the reflection coefficients of the interface between the **homogenized slab** characterized by the parameters $\bar{c}$ and $\bar{\zeta}$ and the right homogeneous half-space. They are given by $r_{he'}^{(\pm)} = \frac{1}{2}\left(\sqrt{\zeta_{e'}/\bar{\zeta}} \pm \sqrt{\bar{\zeta}/\zeta_{e'}}\right)$. Similarly,

$$\mathbf{M}\hat{\mathbf{J}}_{ea} = \begin{bmatrix} r_{eh}^{(+)} & r_{eh}^{(-)} \\ r_{eh}^{(-)} & r_{eh}^{(+)} \end{bmatrix},$$

where $r_{eh}^{(\pm)} = \frac{1}{2}\left(\sqrt{\bar{\zeta}/\zeta_e} \pm \sqrt{\zeta_e/\bar{\zeta}}\right)$ are the reflection coefficients of the interface between the left homogeneous half-space and the homogenized slab. Note that the reflection coefficients in the limit $\Delta \to 0$ do not depend on the choice of the type of the first and last thin layers, but only on the contrast impedances between the homogeneous half-spaces and the homogenized slab.

In conclusion, the heterogeneous slab $[0, L]$ in the limit $\Delta \to 0$ behaves like a homogeneous slab characterized by the parameters $\bar{\zeta}$ and $\bar{c}$ embedded between the two homogeneous half-spaces $z < 0$ and $z > L$. The values of the effective parameters $\bar{\rho}$ and $\bar{K}$ do not depend on the frequency ω. We note also that the parameters of the homogenized slab do not depend on those of the homogeneous half-spaces surrounding it, which means that homogenization is a local averaging process. The numerical experiment shown in Figure 4.1 illustrates homogenization in the case that the two half-spaces have material parameters that coincide with the limit effective parameters of the heterogeneous slab, so that in the limit, the pulse travels without perturbation as in an infinite homogeneous medium.

4.1.2 Random Case

We now look at a randomized version of the previous periodic case. We illustrate the role of randomness with a very particular example at this point. A more complete analysis will be presented later in this chapter.

We still consider the setup (4.1) describing a concatenation of small layers of two alternating materials. We assume here that the layer sizes Δ_j, $j = 1, \ldots, 2N+1$, are given by

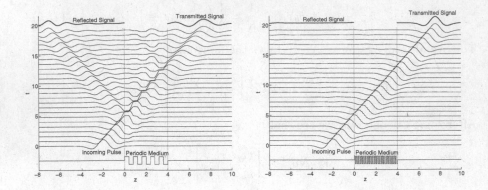

Fig. 4.1. Transmission of a pulse though a piecewise-constant periodic medium occupying the slab $[0, L]$ with $L = 4$. The medium is described by (4.1) with $\rho_e = \rho_a = \rho_b = \rho_{e'} = 1$, $K_e = K_{e'} = 1$, and $1/K_a = 0.2$, $1/K_b = 1.8$, so that $\bar{K} = 1$ (matched medium). The initial pulse is the second derivative of a Gaussian, with Fourier transform $\hat{f}(\omega) = \omega^2 \exp(-\omega^2/5)$. The (root-mean squared) time pulse width is $T_{\mathrm{rms}} = \sqrt{7/15} \sim 0.68$. We plot the pressure field. In the left picture (respectively right picture), the size of the layers is $\Delta = 0.4$ (respectively $\Delta = 0.08$). In the left picture, a significant backscattering can be observed, and the transmitted pulse is distorted. In the right picture, the backscattered wave is negligible, and the transmitted pulse is very close to the incoming pulse.

$$\Delta_j = \delta U_j, \tag{4.17}$$

where the U_j's are independent and identically distributed random variables with the common distribution being uniform over $[1/2, 3/2]$, and $\delta > 0$ is a small parameter. This particular choice is not essential in the analysis. Note that in this case the layer size is bounded and bounded away from zero, and its average is equal to δ.

We consider δ as a small parameter and take the number of layers $2N + 1$ of order δ^{-1}. This is achieved by setting $L'/\delta = 2N + 1$ with a fixed $L' > 0$ and restricting δ to values such that L'/δ is an odd integer. Note that the size L of the random slab is the random variable $L = \sum_{j=1}^{2N+1} \Delta_j$ with expected value

$$\mathbb{E}[L] = (2N + 1)\delta = L'$$

and variance

$$\mathbb{E}\left[(L - \mathbb{E}[L])^2\right] = (2N + 1)\frac{\delta^2}{12} = \frac{\delta L'}{12} \xrightarrow{\delta \to 0} 0.$$

In other words, L converges to L' in quadratic mean, and thus in probability.

The propagator $\hat{K}_{2N+1}(\omega)$ defined by (4.3) can now be written as the **product of random matrices**

$$\hat{K}_{2N+1}(\omega) = \hat{\mathbf{J}}_{2N+1}(\omega) \left(\hat{\mathbf{J}}_N^{(2)}(\omega) \cdots \hat{\mathbf{J}}_1^{(2)}(\omega) \right) \hat{\mathbf{J}}_{ea}(\omega),$$

where the matrices $\hat{\mathbf{J}}_{2N+1}(\omega)$, $\hat{\mathbf{J}}_{ea}(\omega)$, and $\hat{\mathbf{J}}_j^{(2)}(\omega)$, $j = 1, \ldots, N$ are given by (4.4–4.9) As in the periodic case, the first and last matrices in the product converge almost surely in the limit $\Delta \to 0$, so the problem is reduced to the study of the convergence of the product of the matrices $\hat{\mathbf{J}}_j^{(2)}(\omega)$.

In the very particular case in which there is no contrast of impedance, $\zeta_a = \zeta_b$, the matrices $\hat{\mathbf{J}}_j^{(2)}(\omega)$ are diagonal and the product is given by

$$\hat{\mathbf{J}}_N^{(2)}(\omega) \cdots \hat{\mathbf{J}}_1^{(2)}(\omega) = \begin{bmatrix} e^{i\omega S_N(\delta)} & 0 \\ 0 & e^{-i\omega S_N(\delta)} \end{bmatrix},$$

with

$$S_N(\delta) = \delta \sum_{j=1}^{N} \left(\frac{U_{2j-1}}{c_a} + \frac{U_{2j}}{c_b} \right).$$

Using $\delta = L'/(2N+1)$, an application of the strong law of large numbers shows the almost sure convergence

$$S_N(\delta) \xrightarrow{N \to \infty} \frac{L'}{\bar{c}}, \text{ with } \frac{1}{\bar{c}} = \frac{1}{2} \left(\frac{1}{c_a} + \frac{1}{c_b} \right).$$

The conclusion is as in the periodic case without impedance contrast.

In the case with contrast of impedance $\zeta_a \neq \zeta_b$, the expansion (4.11) is still valid and takes the form

$$\hat{\mathbf{J}}_j^{(2)}(\omega, \Delta) = \mathbf{I} + i(\omega\delta)\hat{\mathbf{J}}_{j,1}^{(2)} + \mathbf{O}(\delta^2), \tag{4.18}$$

where

$$\hat{\mathbf{J}}_{j,1}^{(2)}(1,1) = r^{(+)2} \left(\frac{U_{2j-1}}{c_a} + \frac{U_{2j}}{c_b} \right) - r^{(-)2} \left(\frac{U_{2j-1}}{c_a} - \frac{U_{2j}}{c_b} \right),$$

$$\hat{\mathbf{J}}_{j,1}^{(2)}(1,2) = 2r^{(+)}r^{(-)} \frac{U_{2j}}{c_b},$$

$$\hat{\mathbf{J}}_{j,1}^{(2)}(2,1) = -\hat{\mathbf{J}}_{j,1}^{(2)}(1,2),$$

$$\hat{\mathbf{J}}_{j,1}^{(2)}(2,2) = -\hat{\mathbf{J}}_{j,1}^{(2)}(1,1).$$

It is important to underline that an individual propagator matrix $\hat{\mathbf{J}}_j$ cannot be put in the form $\mathbf{I} + \delta\hat{\mathbf{J}}_{j,1} + O(\delta^2)$, because of the contrast of impedance appearing in $r^{(+)}$ and $r^{(-)}$, as in the periodic case. However, the model with two alternate materials allows us to obtain this form by pairing the propagators. Once we have this expansion, a combinatorics argument shows that with $\delta = L'/(2N+1)$,

$$\left\| \prod_{j=N}^{j=1} \left(\mathbf{I} + i(\omega\delta)\hat{\mathbf{J}}_{j,1}^{(2)} \right) - \exp \left(i\omega\delta \sum_{j=1}^{N} \hat{\mathbf{J}}_{j,1}^{(2)} \right) \right\| \xrightarrow{N \to \infty} 0.$$

We do not go into the technical details of this estimate since a different approach will be proposed and discussed in the rest of the book, starting with the next section. An application of the law of large numbers then gives

$$\delta \sum_{j=1}^{N} \hat{\mathbf{J}}_{j,1}^{(2)} \overset{N\to\infty}{\longrightarrow} \frac{L'}{2}\mathbb{E}[\hat{\mathbf{J}}_{j,1}^{(2)}] = \frac{L'}{2}\hat{\mathbf{J}}_{1}^{(2)} \,,$$

where $\hat{\mathbf{J}}_{1}^{(2)}$ is defined by (4.12–4.15). This establishes the convergence of the product of propagators

$$\hat{\mathbf{J}}_{N}^{(2)}(\omega) \cdots \hat{\mathbf{J}}_{1}^{(2)}(\omega) \overset{N\to\infty}{\longrightarrow} \exp\left(i\omega \frac{L'}{2}\hat{\mathbf{J}}_{1}^{(2)} \right) .$$

The rest of the argument is the same as in the periodic case.

In Figure 4.2 we present numerical evidence for the homogenization limit in a random medium. The situation is the same as in Figure 4.1, where all layers have the same size equal to 0.4 (left picture) or 0.08 (right picture), but in Figure 4.2 the layers have random sizes with means 0.4 and 0.08.

4.1.3 Conclusion

The method used so far to describe the transmission problem through a slab of a finely layered medium has been to solve locally the wave equation and to apply the interface conditions. This has naturally led to a formulation of the propagation problem in terms of a product of many matrices scaled so that the limit corresponds to a homogeneous effective medium. The two particular examples (periodic and random) we have addressed are relatively simple, and the analysis can be extended to more general media, for example to the case in which the impedances are independent and identically distributed random variables. Note, however, that in this case the expansion (4.18) is not valid, so the analysis is necessarily more involved. Moreover, random matrix formulation is well adapted to the case of piecewise-constant media. In the rest of the book we propose a different approach based on the analysis of the asymptotic behavior of differential equations. We will be mostly interested in the random case, in which these differential equations have random coefficients. In the following section we introduce this method in the regime of homogenization where the pulse travels through many inhomogeneities.

4.2 Random Media Varying on a Fine Scale

We extend the modeling of the layered medium from the piecewise-constant case addressed in the previous sections to the general situation in which the density and bulk modulus are spatially varying in a continuous way and/or

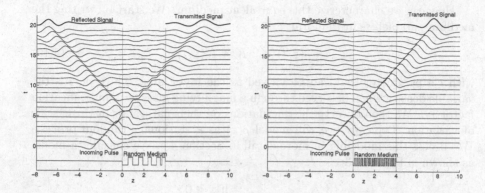

Fig. 4.2. Transmission of a pulse though a piecewise-constant random medium occupying the slab $[0, L]$ with $L = 4$. The medium is described by (4.1) with $\rho_e = \rho_a = \rho_b = \rho_{e'} = 1$, $K_e = K_{e'} = 1$, and $1/K_a = 0.2$, $1/K_b = 1.8$, so that $\bar{K} = 1$ (matched medium). The initial pulse is the second derivative of a Gaussian pulse function with width $T_{\mathrm{rms}} = \sqrt{7/15} \approx 0.68$. The sizes of the layers Δ_j are random and as described by (4.17). The Δ_j's are independent and identically distributed random variables with uniform distribution over $[0.2, 0.6]$ with mean $\delta = 0.4$ (left picture), and uniform distribution over $[0.04, 0.12]$ with mean $\delta = 0.08$ (right picture). Homogenization is seen clearly, although the convergence (as the size of the layers goes to zero) is not as rapid as in the periodic case (compare the reflected signals in the two right pictures of Figures 4.1–4.2). If we compare the simulations in the periodic and in the random cases, we see that fluctuations behind the main pulse are more important in the random case than in the periodic case, where they are practically nonexistent for $\delta = 0.08$. This is one of our motivations for introducing, as we do in the next sections, a different approach to the asymptotic analysis that is more suitable for long-distance propagation in random media.

with jumps. This can serve as a model for waves propagating through sedimentary layers of the earth's crust. In this case the layers are formed by a deposition process that results in a thin horizontally layered structure. We consider the idealized situation in which the parameters vary only with depth, and moreover, we make the important assumption that the variations are on a relatively **fine scale**. We assume that the scale of variation is small compared to the distance traveled by the pulse, as well as compared to the wavelength of the pulse. One may then expect that the waves are not strongly affected by the impedance in any particular layer. When a pulse propagates through such fine layers, the interaction with each layer is small, and propagation is not much affected. The pulse therefore travels as if the medium were homogeneous with the layers replaced by "averaged" ones. In general, we refer to this homogeneous medium as the **homogenized medium**. It is also referred to as an **effective, average, or equivalent medium**.

How can we characterize this equivalent medium? We start by writing the medium parameters in the form

$$\rho = \rho(z/l), \qquad K = K(z/l), \tag{4.19}$$

with l a parameter that can be viewed as the **layer size**. Thus, $\rho(z)$ is the variable density when observed through a magnifying glass with magnification factor $1/l$. We then observe the fluctuations on their natural or intrinsic scale of variation. Typically, we will model $\rho(z)$ as a stationary random process. We discuss this modeling in more detail in Section 4.5. We consider a model as shown in Figure 4.3:

$$\rho = \begin{cases} \rho_0 & \text{if } z < 0, \\ \rho_l(z) := \rho(z/l) & \text{if } z \in [0, L], \\ \rho_1 & \text{if } z > L, \end{cases} \tag{4.20}$$

$$K = \begin{cases} K_0 & \text{if } z < 0, \\ K_l(z) := K(z/l) & \text{if } z \in [0, L], \\ K_1 & \text{if } z > L. \end{cases} \tag{4.21}$$

In later chapters we will also consider wave propagation in three-dimensional layered media. An example is shown schematically in Figure 4.4. where a point source emits a spherical pulse that is incident on a heterogeneous, layered section. Before addressing the limit problem $l \to 0$, we introduce several important transformations of the wave equations.

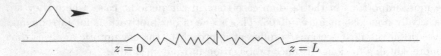

Fig. 4.3. Schematic of a heterogeneous slab embedded between two homogeneous half-spaces, with a pulse incident from the left.

4.3 Boundary Conditions and Equations for Right- and Left-Going Modes

As shown before, in Section 3.2, in each homogeneous half-space the wave can be decomposed into a right-going wave A_j and a left-going wave B_j ($j = 0$ for the left half-space and $j = 1$ for the right half-space):

$$u(t, z) = \zeta_0^{-1/2} \frac{A_0(t, z) + B_0(t, z)}{2}, \quad p(t, z) = \zeta_0^{1/2} \frac{A_0(t, z) - B_0(t, z)}{2}, \quad z < 0,$$

$$u(t, z) = \zeta_1^{-1/2} \frac{A_1(t, z) + B_1(t, z)}{2}, \quad p(t, z) = \zeta_1^{1/2} \frac{A_1(t, z) - B_1(t, z)}{2}, \quad z > L,$$

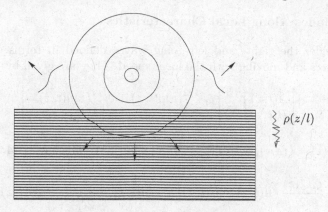

Fig. 4.4. A sperical wave from a point source is incident on a layered medium that varies on a fine scale.

where the impendance is $\zeta_j = \sqrt{K_j \rho_j}$. The right- and left-going modes travel with speed $c_0 = \sqrt{K_0/\rho_0}$ in the half-space $z < 0$ and with speed $c_1 = \sqrt{K_1/\rho_1}$ in the half-space $z > L$. This means that A_j is a function only of $t - z/c_j$, and B_j is a function only of $t + z/c_j$.

A right-going acoustic pulse incident from the left half-space enters the slab and no waves enter from the right. This is expressed by

$$A_0(t,z) = f\left(t - \frac{z}{c_0}\right) \quad \text{for} \quad z < 0\,, \tag{4.22}$$
$$B_1(t,z) = 0 \quad \text{for} \quad z > L\,.$$

The **pulse shape** function $f(t)$ is assumed to be smooth and with bounded support. The width of the pulse is an important time scale. For homogenization, this pulse width should be long compared to the time it takes to pass over an inhomogeneity, but it should not be short compared to the time it takes to traverse the slab. Propagation phenomena when the pulse width is short compared to the travel time of the slab cannot be described with homogenization or effective medium theory. They require more-elaborate statistical theories, discussed in subsequent chapters.

Next we decompose the wave inside the heterogeneous slab. There are two possibilities. We can either consider the right- and left-going waves defined in terms of the local impedances and moving with the local speed, or we can consider the right- and left-going waves defined in terms of reference values for the impedance and speed. This leads to two formulations that are essentially equivalent. However, depending on the situation, one approach can turn out to be more suitable than the other one for the asymptotic analysis. We present these two approaches in the next two subsections.

4.3.1 Modes Along Local Characteristics

We consider the right- and left-going waves defined in terms of the local impedances and moving with the local speed. They are given by

$$\begin{bmatrix} A(t,z) \\ B(t,z) \end{bmatrix} = \begin{bmatrix} \zeta_l^{-1/2}(z)p(t,z) + \zeta_l^{1/2}(z)u(t,z) \\ -\zeta_l^{-1/2}(z)p(t,z) + \zeta_l^{1/2}(z)u(t,z) \end{bmatrix}, \qquad (4.23)$$

where $\zeta_l(z) = \zeta(z/l)$, with $\zeta(z) = \sqrt{K(z)\rho(z)}$. We first invert (4.23):

$$p(t,z) = \frac{\zeta_l^{1/2}(z)}{2}\left(A(t,z) - B(t,z)\right), \qquad u(t,z) = \frac{1}{2\zeta_l^{1/2}(z)}\left(A(t,z) + B(t,z)\right).$$

Susbtituting these expressions into the wave equations gives a coupled system of two partial differential equations for the modes A and B. Let us assume first that the impedance function $\zeta(z)$ is differentiable. We compute the partial derivative of A:

$$\begin{aligned}
\frac{\partial A}{\partial z} &= \frac{1}{\zeta_l^{1/2}(z)}\frac{\partial p}{\partial z} + \zeta_l^{1/2}(z)\frac{\partial u}{\partial z} - \frac{\zeta_l'(z)}{2\zeta_l^{3/2}(z)}p + \frac{\zeta_l'(z)}{2\zeta_l^{1/2}(z)}u \\
&= -\frac{\rho_l(z)}{\zeta_l^{1/2}(z)}\frac{\partial u}{\partial t} - \frac{\zeta_l^{1/2}(z)}{K_l(z)}\frac{\partial p}{\partial t} + \frac{\zeta_l'(z)}{2\zeta_l(z)}\left(-\zeta_l^{-1/2}(z)p + \zeta_l^{1/2}(z)u\right) \\
&= \frac{1}{c_l(z)}\frac{\partial}{\partial t}\left(-\zeta_l^{1/2}(z)u - \zeta_l^{-1/2}(z)p\right) + \frac{\zeta_l'(z)}{2\zeta_l(z)}\left(-\zeta_l^{-1/2}(z)p + \zeta_l^{1/2}(z)u\right) \\
&= -\frac{1}{c_l(z)}\frac{\partial A}{\partial t} + \frac{\zeta_l'(z)}{2\zeta_l(z)}B.
\end{aligned}$$

Here ζ_l' stands for the z-derivative of ζ_l, and we have used the relation $c_l(z) = c(z/l)$ with $c(z) = \sqrt{K(z)/\rho(z)}$. A similar computation for $\partial B/\partial z$ leads to the system

$$\frac{\partial}{\partial z}\begin{bmatrix} A \\ B \end{bmatrix} = -\frac{1}{c(z/l)}\begin{bmatrix} 1 & 0 \\ 0 & -1 \end{bmatrix}\frac{\partial}{\partial t}\begin{bmatrix} A \\ B \end{bmatrix} + \frac{1}{l}\frac{\zeta'(z/l)}{2\zeta(z/l)}\begin{bmatrix} 0 & 1 \\ 1 & 0 \end{bmatrix}\begin{bmatrix} A \\ B \end{bmatrix}. \quad (4.24)$$

These are the linear equations that describe the propagation and coupling of the right- and left-going waves in the heterogeneous slab. The first term of the right-hand side describes the propagation of the modes with the local speed of sound $c(z/l)$. The second term of the right-hand side describes the coupling between the two modes due to the impedance variations $\zeta(z/l)$.

This method can be applied when the parameters ρ and K are piecewise-differentiable with discontinuities at isolated interfaces $0 = z_0 < z_1 < \cdots < z_{N-1} < z_N = L$, $j = 1, \ldots, N$. In this case, the modes satisfy (4.24) in each interval $z \in (z_j, z_{j+1})$, $j = 0, \ldots, N-1$. At each interface $z = z_j$, the continuity conditions for the pressure and velocity fields become interface relations for the modes A and B, which are similar to those of a single interface (3.12):

$$\begin{bmatrix} A \\ B \end{bmatrix}(t, z_j^+) = \mathbf{J}_j \begin{bmatrix} A \\ B \end{bmatrix}(t, z_j^-), \qquad \mathbf{J}_j = \begin{bmatrix} r_j^{(+)} & r_j^{(-)} \\ r_j^{(-)} & r_j^{(+)} \end{bmatrix}. \qquad (4.25)$$

Here $r_j^{(\pm)} = \frac{1}{2}\left(\sqrt{\zeta_l(z_j^+)/\zeta_l(z_j^-)} \pm \sqrt{\zeta_l(z_j^-)/\zeta_l(z_j^+)} \right)$. The partial differential equations (4.24) for $z \in (z_j, z_{j+1})$, the interface relations (4.25) at $z = z_j$, and the two interface relations at $z = 0$ and $z = L$ form a system that uniquely defines the modes A and B.

This formulation of scattering by a slab has been developed in the particular case of a piecewise-constant medium in Section 3.5, and the asymptotic limit $l \to 0$ has been studied for two particular examples in Section 4.1. More generally, this approach is appropriate when the fluctuations of the local speed are on a scale that is long compared to the pulse width times a characteristic propagation speed. In the frequency domain this is the setup for applications of the Wentzel–Kramer–Brillouin (WKB) method. However, in the regimes for random media analyzed in the following sections and chapters, the fluctuations of the local speed of sound are strong and rapid relative to the pulse width times a characteristic propagation speed. They cannot be captured by the WKB method. That is why another approach is necessary. We introduce it in the next subsection.

4.3.2 Modes Along Constant Characteristics

Another way to decompose the field inside the slab is to introduce the right- and left-going waves for constant values for the impedance and speed of sound. We shall see in the next section that there is a convenient choice for these two parameters, but in the present section we carry out the analysis with general, unspecified values $\bar{\zeta}$ and $\bar{c}$. Accordingly, we make the ansatz

$$\begin{bmatrix} A(t, z) \\ B(t, z) \end{bmatrix} = \begin{bmatrix} \bar{\zeta}^{-1/2} p(t, z) + \bar{\zeta}^{1/2} u(t, z) \\ -\bar{\zeta}^{-1/2} p(t, z) + \bar{\zeta}^{1/2} u(t, z) \end{bmatrix}. \qquad (4.26)$$

In order to derive equations for A and B we invert (4.26):

$$p = \frac{\bar{\zeta}^{1/2}}{2}(A - B), \qquad u = \frac{1}{2\bar{\zeta}^{1/2}}(A + B).$$

Computing $\partial A / \partial z$ gives

$$\frac{\partial A}{\partial z} = \frac{1}{\bar{\zeta}^{1/2}} \frac{\partial p}{\partial z} + \bar{\zeta}^{1/2} \frac{\partial u}{\partial z}$$

$$= \frac{1}{\bar{\zeta}^{1/2}} \left(-\rho_l(z) \frac{\partial u}{\partial t} \right) + \bar{\zeta}^{1/2} \left(-\frac{1}{K_l(z)} \frac{\partial p}{\partial t} \right)$$

$$= -\frac{\rho_l(z)}{2\bar{\zeta}}\frac{\partial(A+B)}{\partial t} - \frac{\bar{\zeta}}{2K_l(z)}\frac{\partial(A-B)}{\partial t}$$

$$= -\frac{1}{\bar{c}}\left(\Delta_l^{(+)}(z)\frac{\partial A}{\partial t} + \Delta_l^{(-)}(z)\frac{\partial B}{\partial t}\right),$$

where we have introduced the notation

$$\Delta_l^{(\pm)}(z) = \Delta^{(\pm)}(z/l), \qquad \Delta^{(\pm)}(z) = \frac{1}{2}\left(\frac{\rho(z)}{\bar{\rho}} \pm \frac{\bar{K}}{K(z)}\right). \tag{4.27}$$

The constant parameters $\bar{\rho}$ and $\bar{K}$ are defined through the relations $\bar{\zeta} = \sqrt{\bar{K}\bar{\rho}}$ and $\bar{c} = \sqrt{\bar{K}/\bar{\rho}}$. A similar computation for $\partial B/\partial z$ leads to the system

$$\frac{\partial}{\partial z}\begin{bmatrix} A \\ B \end{bmatrix} = -\frac{1}{\bar{c}}\begin{bmatrix} \Delta^{(+)}(z/l) & \Delta^{(-)}(z/l) \\ -\Delta^{(-)}(z/l) & -\Delta^{(+)}(z/l) \end{bmatrix}\frac{\partial}{\partial t}\begin{bmatrix} A \\ B \end{bmatrix}. \tag{4.28}$$

These are the linear equations that describe the propagation and **coupling** of the right- and left-going waves in the heterogeneous slab. This system is a generalization of (3.5) to the case of arbitrary density and bulk modulus. Indeed, if the slab has constant parameters $\rho(z) \equiv \bar{\rho}$ and $K(z) \equiv \bar{K}$, then we recover the decoupled system (3.5). It is important to underline here that the right-hand side of (4.28) does not have a term of order $1/l$, in contrast to the right-hand side of the system (4.24). This is one of our motivations for the introduction and the use of this second approach.

The relations (4.22) and the continuity of the velocity and pressure fields at the interfaces give boundary conditions prescribed at the endpoints of the heterogeneous slab for the modes A and B. More precisely, at $z = 0$ we have

$$\begin{bmatrix} A(t,0) \\ B(t,0) \end{bmatrix} = \mathbf{J}_0\begin{bmatrix} A_0(t,0) \\ B_0(t,0) \end{bmatrix}, \qquad \mathbf{J}_0 = \begin{bmatrix} r_0^{(+)} & r_0^{(-)} \\ r_0^{(-)} & r_0^{(+)} \end{bmatrix},$$

with $r_0^{(\pm)} = \frac{1}{2}\left(\sqrt{\zeta/\zeta_0} \pm \sqrt{\zeta_0/\zeta}\right)$. Similarly, at $z = L$,

$$\begin{bmatrix} A_1(t,L) \\ B_1(t,L) \end{bmatrix} = \mathbf{J}_1\begin{bmatrix} A(t,L) \\ B(t,L) \end{bmatrix}, \qquad \mathbf{J}_1 = \begin{bmatrix} r_1^{(+)} & r_1^{(-)} \\ r_1^{(-)} & r_1^{(+)} \end{bmatrix},$$

with $r_1^{(\pm)} = \frac{1}{2}\left(\sqrt{\zeta_1/\zeta} \pm \sqrt{\zeta/\zeta_1}\right)$. Using the boundary conditions (4.22) we get

$$\begin{bmatrix} A(t,0) \\ B(t,0) \end{bmatrix} = \mathbf{J}_0\begin{bmatrix} f(\omega_0 t) \\ B_0(t,0) \end{bmatrix}, \qquad \begin{bmatrix} A_1(t,L) \\ 0 \end{bmatrix} = \mathbf{J}_1\begin{bmatrix} A(t,L) \\ B(t,L) \end{bmatrix}.$$

By eliminating $B_0(t,0)$ and $A_1(t,L)$ we obtain the boundary conditions for the modes A and B:

$$r_0^{(+)}A(t,0) - r_0^{(-)}B(t,0) = f(t), \qquad r_1^{(-)}A(t,L) + r_1^{(+)}B(t,L) = 0.$$

In terms of the interface reflection and transmission coefficients

$$R_j = -\frac{r_j^{(-)}}{r_j^{(+)}}, \qquad T_j = \frac{1}{r_j^{(+)}}, \qquad j = 0, 1,$$

these boundary conditions read

$$A(t,0) + R_0 B(t,0) = T_0 f(t), \qquad T_1 A(t,L) - B(t,L) = 0. \qquad (4.29)$$

As we will see, the system (4.28) along with the boundary conditions (4.29) is sufficient to determine the modes A and B. Once the modes A and B are known, it is straightforward to extract the quantities of interest that are the unknown scattered waves:

- **Transmitted wave**, given by $A_1(t, L)$, the part of the wave escaping to the right:

$$A_1(t, L) = T_1 A(t, L).$$

- **Reflected wave**, given by $B_0(t, 0)$, the wave scattered back by the heterogeneities in the slab:

$$B_0(t, 0) = T_0 B(t, 0) + R_0 f(t).$$

Note that knowing $A_1(t, L)$ for all times gives us the right propagating wave for $z > L$, since the medium is constant. Therefore $A_1(t, z) = A_1(t - (z - L)/c_1, L)$. Similarly, we have $B_0(t, z) = B_0(t + z/c_0, 0)$ for all $z < 0$. One of the main objectives of this book is to analyze and describe these quantities in various asymptotic regimes and for various geometries.

4.4 Centering the Modes and Propagator Equations

4.4.1 Characteristic Lines

In the homogenized medium, as seen in the two examples given in Section 4.1, we expect that the right- and left-going waves are unchanged when we observe them in frames that move with the effective wave speeds $\pm \bar{c}$. We denote by (a, b) the waves in these moving frames, so that

$$a(s, z) = A(s + z/\bar{c}, z), \qquad b(s, z) = B(s - z/\bar{c}, z), \qquad (4.30)$$

where $t = s \pm z/\bar{c}$ are the two families of **characteristic lines** parameterized by s. We similarly define

$$a_1(s, L) = A_1(s + L/\bar{c}, L), \qquad b_0(s, 0) = B_0(s, 0). \qquad (4.31)$$

Note that a, b depend only on s if the medium is homogeneous with constant speed and impedance $\bar{c}$ and $\bar{\zeta}$, since in these coordinates the waves are **centered**. In the heterogeneous medium, scattering will couple the two waves. The governing equations for a and b given in (4.30) are nonlocal in time because of the shifts in the time variable. Rather than writing and analyzing these nonlocal equations we go into the Fourier domain, where these shifts are converted into phase factors.

4.4.2 Modes in the Fourier Domain

Since the medium is time-independent and our problem is linear, we can use the **Fourier transform** with respect to the time variable by defining

$$\hat{a}(\omega, z) = \int e^{i\omega s} a(s, z)\, ds\,, \qquad \hat{b}(\omega, z) = \int e^{i\omega s} b(s, z)\, ds\,.$$

We can introduce similarly the Fourier transforms $\hat{b}_0(\omega, 0)$ and $\hat{a}_1(\omega, L)$ of $b_0(t, 0)$ and $a_1(t, L)$. The frequency ω becomes a parameter. The only variable with respect to which we take derivatives is the space variable z. In other words, the partial differential equations in (t, z) have been transformed into an infinite family of ordinary differential equations parameterized by ω. Therefore we use the symbol d/dz instead of $\partial/\partial z$. We compute

$$\frac{d\hat{a}}{dz} = \int e^{i\omega s} \frac{\partial}{\partial z} A(s + z/\bar{c}, z)\, ds$$
$$= \int e^{i\omega s} \left(\frac{1}{\bar{c}} \frac{\partial A}{\partial s}(s + z/\bar{c}, z) + \frac{\partial A}{\partial z}(s + z/\bar{c}, z) \right) ds\,.$$

Using the first equation in (4.28) we express $\partial A/\partial z$ with time-differentiated terms, which gives a multiplicative factor $-i\omega$ in the frequency domain:

$$\frac{d\hat{a}}{dz} = -\frac{i\omega}{\bar{c}} \int e^{i\omega s} \left(a(s, z) - \Delta_l^{(+)} a(s, z) - \Delta_l^{(-)} b(s + 2z/\bar{c}, z) \right) ds$$
$$= -\frac{i\omega}{\bar{c}} \left((1 - \Delta_l^{(+)})\hat{a}(\omega, z) - \Delta_l^{(-)} e^{-2i\omega z/\bar{c}} \, \hat{b}(\omega, z) \right)\,.$$

Note that a **phase** appears in the last term because of the centering change of coordinates. Combined with a similar computation for the derivative of $\hat{b}$ we obtain the following system of ordinary differential equations for $(\hat{a}, \hat{b})$,

$$\frac{d}{dz} \begin{bmatrix} \hat{a} \\ \hat{b} \end{bmatrix} = \frac{i\omega}{\bar{c}} \begin{bmatrix} (\Delta_l^{(+)} - 1) & \Delta_l^{(-)} e^{-2i\omega z/\bar{c}} \\ -\Delta_l^{(-)} e^{+2i\omega z/\bar{c}} & (1 - \Delta_l^{(+)}) \end{bmatrix} \begin{bmatrix} \hat{a} \\ \hat{b} \end{bmatrix}\,, \qquad 0 < z < L\,, \quad (4.32)$$

with boundary conditions obtained by Fourier transforming (4.29):

$$\hat{a}(\omega, 0) + R_0 \hat{b}(\omega, 0) = T_0 \hat{f}(\omega)\,, \qquad R_1 e^{2i\frac{\omega L}{\bar{c}}} \hat{a}(\omega, L) - \hat{b}(\omega, L) = 0\,, \quad (4.33)$$

where

$$\hat{f}(\omega) = \int f(s)e^{i\omega s}\,ds\,.$$

Our problem now is to solve the system (4.32) with the boundary conditions (4.33). The problem (4.32–4.33) is a two-point boundary value problem and is not, at first, well suited for the asymptotic analysis that we develop. We now describe how the quantities of interest can be obtained from the solution of **initial value problems** in z.

4.4.3 Propagator

One way to transform the above two-point boundary value problem into an initial value problem is to introduce for each frequency ω the **propagator** $\mathbf{P}_\omega$. It is defined as the fundamental solution of the system (4.32), that is, the 2×2 complex matrix function satisfying

$$\frac{d}{dz}\mathbf{P}_\omega(0,z) = \mathbf{H}_\omega(z,z/l)\mathbf{P}_\omega(0,z)\,, \qquad \mathbf{P}_\omega(0,0) = \mathbf{I}\,, \qquad (4.34)$$

where we have denoted the identity matrix by $\mathbf{I}$ and we have introduced

$$\mathbf{H}_\omega(z,z') = \frac{i\omega}{\bar{c}}\left[\begin{array}{cc} (\Delta^{(+)}(z')-1) & \Delta^{(-)}(z')e^{-2i\omega z/\bar{c}} \\ -\Delta^{(-)}(z')e^{+2i\omega z/\bar{c}} & (1-\Delta^{(+)}(z')) \end{array}\right]\,. \qquad (4.35)$$

The matrix $\mathbf{P}_\omega(0,z)$ "propagates" the wave components from $z = 0$ to any other location $z > 0$, since the linearity of (4.32) implies that

$$\left[\begin{array}{c} \hat{a}(\omega,z) \\ \hat{b}(\omega,z) \end{array}\right] = \mathbf{P}_\omega(0,z)\left[\begin{array}{c} \hat{a}(\omega,0) \\ \hat{b}(\omega,0) \end{array}\right]\,, \qquad (4.36)$$

for any $z \in [0,L]$.

Note that the matrix $\mathbf{H}_\omega$ depends on both the "fast variable" $z' = z/l$ and the "slow" space variable z. We have carefully articulated this variation on different scales, since this will be essential in our asymptotic analysis.

Applying the Jacobi's formula for the derivative of a determinant,

$$\frac{d\det(\mathbf{P}_\omega)}{dz} = \mathrm{Tr}\left(\mathrm{Adj}(\mathbf{P}_\omega)\frac{d\mathbf{P}_\omega}{dz}\right),$$

where $\mathrm{Adj}(\mathbf{P}_\omega)$ is the adjugate of $\mathbf{P}_\omega$, which satisfies $\mathbf{P}_\omega\mathrm{Adj}(\mathbf{P}_\omega) = \det(\mathbf{P}_\omega)\mathbf{I}$, and using (4.34) we get

$$\frac{d\det(\mathbf{P}_\omega)}{dz} = \mathrm{Tr}\left(\mathrm{Adj}(\mathbf{P}_\omega)\mathbf{H}_\omega\mathbf{P}_\omega\right) = \mathrm{Tr}\left(\mathbf{H}_\omega\mathbf{P}_\omega\mathrm{Adj}(\mathbf{P}_\omega)\right),$$

where we use $\mathrm{Tr}(\mathbf{MN}) = \mathrm{Tr}(\mathbf{NM})$. Using the above relation between $\mathbf{P}_\omega$ and $\mathrm{Adj}(\mathbf{P}_\omega)$ we have

$$\frac{d \det(\mathbf{P}_\omega)}{dz} = \mathrm{Tr}\,(\mathbf{H}_\omega) \det(\mathbf{P}_\omega)\,.$$

Observe that the trace of the matrix $\mathbf{H}_\omega$ is zero. Thus the determinant of $\mathbf{P}_\omega$ is constant in z. The initial condition being the identity then gives

$$\det(\mathbf{P}_\omega) = 1\,.$$

If $(\alpha_\omega, \beta_\omega)^T$ satisfies (4.32) with initial condition $(1,0)$, then a simple computation shows that $(\overline{\beta_\omega}, \overline{\alpha_\omega})^T$ satisfies the same equation with initial condition $(0,1)^T$, which gives two linearly independent solutions. We deduce that the propagator $\mathbf{P}_\omega$ has the representation

$$\mathbf{P}_\omega = \begin{bmatrix} \alpha_\omega & \overline{\beta_\omega} \\ \beta_\omega & \overline{\alpha_\omega} \end{bmatrix}, \tag{4.37}$$

with

$$|\alpha_\omega|^2 - |\beta_\omega|^2 = 1\,.$$

The propagator and the boundary conditions (4.33) contain together all the information necessary to analyze the quantities of interest in the scattering problem. We first note that

$$\mathbf{P}_\omega(0,L) \begin{bmatrix} \hat{a}(\omega,0) \\ \hat{b}(\omega,0) \end{bmatrix} = \begin{bmatrix} \hat{a}(\omega,L) \\ \hat{b}(\omega,L) \end{bmatrix}.$$

Using the representation (4.37) we get

$$\hat{a}(\omega,L) = \alpha_\omega \hat{a}(\omega,0) + \overline{\beta_\omega}\hat{b}(\omega,0)\,,$$
$$\hat{b}(\omega,L) = \beta_\omega \hat{a}(\omega,0) + \overline{\alpha_\omega}\hat{b}(\omega,0)\,,$$

where we write $\alpha_\omega = \alpha_\omega(0,L)$ and $\beta_\omega = \beta_\omega(0,L)$ for simplicity. Substituting these relations into the second equation of (4.33) and solving the 2×2 linear system that results together with the first equation of (4.33), we find the expressions for the modes

$$\hat{b}(\omega,0) = \frac{T_0 \left[\beta_\omega(0,L) - R_1 e^{2i\frac{\omega L}{\bar{c}}} \alpha_\omega \right]}{(R_0 \beta_\omega - \overline{\alpha_\omega}) - R_1 e^{2i\frac{\omega L}{\bar{c}}} (R_0 \alpha_\omega - \overline{\beta_\omega})} \hat{f}(\omega)\,,$$

$$\hat{a}(\omega,L) = \frac{-T_0}{(R_0 \beta_\omega - \overline{\alpha_\omega}) - R_1 e^{2i\frac{\omega L}{\bar{c}}} (R_0 \alpha_\omega - \overline{\beta_\omega})} \hat{f}(\omega)\,.$$

Accordingly, the transmitted wave is given by

$$\hat{a}_1(\omega,L) = \frac{-T_0 T_1}{(R_0 \beta_\omega - \overline{\alpha_\omega}) - R_1 e^{2i\frac{\omega L}{\bar{c}}} (R_0 \alpha_\omega - \overline{\beta_\omega})} \hat{f}(\omega)\,, \tag{4.38}$$

and the reflected wave is given by

$$\hat{b}_0(\omega,0) = \frac{(\beta_\omega - R_0\overline{\alpha_\omega}) - R_1 e^{2i\frac{\omega L}{c}}(\alpha_\omega - R_0\overline{\beta_\omega})}{(R_0\beta_\omega - \overline{\alpha_\omega}) - R_1 e^{2i\frac{\omega L}{c}}(R_0\alpha_\omega - \overline{\beta_\omega})}\,\hat{f}(\omega)\,. \tag{4.39}$$

These representations of the exiting waves are useful, but there is an alternative way to get closed-form expressions for the same quantities that will be very convenient for the analysis in the next chapters, and which we discuss next.

4.4.4 The Riccati Equation for the Local Reflection Coefficient

Let us consider the reflection coefficient for a slab occupying the interval $(z, L]$, with $0 \leq z \leq L$, with waves incident from a homogeneous medium on the left. We call this the **local** reflection coefficient, and it is defined by

$$\hat{R}(\omega, z) = \frac{\hat{b}(\omega, z)}{\hat{a}(\omega, z)}\,.$$

Note that the coefficients $\hat{a}(\omega, z)$ and $\hat{b}(\omega, z)$ are global quantities associated with the full random medium between 0 and L. However, their ratio can be expressed as the local reflection coefficient $\hat{R}(\omega, z)$, which depends only on the section between z and L. This is a consequence of the geometry of the experiment, where the source is outside the random medium between 0 and L, on the negative z-axis. The fact that we terminate the random medium at $z = L$ enables us to identify a terminal condition for the local reflection condition at $z = L$. From the boundary condition (4.33) at $z = L$, we have

$$\hat{R}(\omega, L) = R_1 e^{2i\frac{\omega L}{c}}\,. \tag{4.40}$$

Differentiating the ratio $\hat{R} = \hat{b}/\hat{a}$ with respect to z and using (4.32) we see that the local reflection coefficient satisfies the **Riccati equation** for $z \in [0, L]$:

$$\frac{d\hat{R}}{dz} = \frac{i\omega}{c}\left[-\Delta_l^{(-)}(z)e^{2i\frac{\omega z}{c}} + 2(1 - \Delta_l^{(+)}(z))\hat{R} - \Delta_l^{(-)}(z)e^{-2i\frac{\omega z}{c}}\hat{R}^2\right]\,. \tag{4.41}$$

Taking into account the boundary condition (4.33) at $z = 0$, we get

$$\hat{b}(\omega, 0) = \frac{T_0\hat{R}(\omega, 0)}{1 + R_0\hat{R}(\omega, 0)}\,\hat{f}(\omega)\,,$$

and the reflected wave is

$$\hat{b}_0(\omega, 0) = \hat{\mathcal{R}}(\omega)\hat{f}(\omega)\,,$$

where $\hat{\mathcal{R}}$ is the reflection coefficient of the heterogeneous slab

$$\hat{\mathcal{R}}(\omega) = \frac{R_0 + \hat{R}(\omega, 0)}{1 + R_0\hat{R}(\omega, 0)}\,, \tag{4.42}$$

and R_0 is the interface reflection coefficient at $z = 0$. The Riccati equation (4.41) is a nonlinear terminal value problem for the local reflection coefficient. It is an alternative way to obtain the reflection coefficient for the slab, which replaces the linear two-point boundary value problem (4.32–4.33). It is a continuous analogue of the recursive linear fractional transformations (3.27) that were obtained for a piecewise-constant medium, without a special centering as is the case in (4.41). Note also that it is a backward Riccati equation that must be solved from $z = L$ to $z = 0$, starting from the terminal condition (4.40) at $z = L$. For a random heterogeneous slab this is problematic, because the application of standard stochastic analysis tools deals with forward random equations. One easy way to deal with this issue is to consider reflection from waves incident from the right of a slab occupying the interval $[-L, 0]$. The local reflection coefficient for the interval $[-L, z)$, with $-L \leq z \leq 0$, satisfies a forward Riccati equation with initial condition at $z = -L$. We will use this setup in Chapter 9.

The reflected wave is given in terms of the local reflection coefficient that solves (4.40–4.41). A nonlinear terminal value problem can also be obtained for the local transmission coefficient, from which the transmitted wave is determined. The local transmission coefficient is defined by

$$\hat{T}(\omega, z) = \frac{T_1 \hat{a}(\omega, L)}{\hat{a}(\omega, z)}, \quad 0 \leq z \leq L.$$

It satisfies the equation

$$\frac{d\hat{T}}{dz} = \frac{i\omega}{c} \hat{T} \left[(1 - \Delta_l^{(+)}(z)) - \Delta_l^{(-)}(z) e^{-2i \frac{\omega z}{c}} \hat{R} \right] \tag{4.43}$$

for $z \in [0, L]$, with terminal condition

$$\hat{T}(\omega, L) = T_1. \tag{4.44}$$

Taking into account the interface condition (4.33) at $z = 0$, we get

$$\hat{a}(\omega, L) = \frac{T_0 T_1^{-1} \hat{T}(\omega, 0)}{1 + R_0 \hat{R}(\omega, 0)} \hat{f}(\omega).$$

The transmitted wave is therefore

$$\hat{a}_1(\omega, L) = \hat{\mathcal{T}}(\omega) \hat{f}(\omega),$$

where $\hat{\mathcal{T}}$ is the transmission coefficient of the heterogeneous slab

$$\hat{\mathcal{T}}(\omega) = \frac{T_0 \hat{T}(\omega, 0)}{1 + R_0 \hat{R}(\omega, 0)}. \tag{4.45}$$

Here T_0 and R_0 are the interface transmission and reflection coefficients at $z = 0$.

By differentiating $|\hat{R}|^2 + |\hat{T}|^2$ with respect to z and using (4.41) and (4.43), we find that it is independent of z. From the terminal conditions at $z = L$ it follows that it is equal to one. This identity is preserved by the linear fractional transformations (4.42) and (4.45), and we obtain

$$|\hat{\mathcal{R}}(\omega)|^2 + |\hat{\mathcal{T}}(\omega)|^2 = 1 \,. \tag{4.46}$$

This is the **energy flux** conservation relation expressing the equality of the energy entering the slab with the energy exiting the slab. In particular, this implies that the complex-valued reflection and transmission coefficients, $\hat{\mathcal{R}}(\omega)$ and $\hat{\mathcal{T}}(\omega)$, are uniformly bounded in absolute value by one.

4.4.5 Reflection and Transmission in the Time Domain

We return to the time domain with an inverse Fourier transform. We have the following integral representations.

- The transmitted wave:

$$A_1(t, L) = a_1(t - L/\bar{c}, L) = \frac{1}{2\pi} \int e^{-i\omega(t - L/\bar{c})} \hat{\mathcal{T}}(\omega) \hat{f}(\omega) \, d\omega$$

$$= \frac{1}{2\pi} \int e^{-i\omega(t - L/\bar{c})} \left(\frac{T_0 \hat{T}(\omega, 0)}{1 + R_0 \hat{R}(\omega, 0)} \right) \hat{f}(\omega) \, d\omega \,. \tag{4.47}$$

- The reflected wave:

$$B_0(t, 0) = b_0(t, 0) = \frac{1}{2\pi} \int e^{-i\omega t} \hat{\mathcal{R}}(\omega) \hat{f}(\omega) \, d\omega$$

$$= \frac{1}{2\pi} \int e^{-i\omega t} \left(\frac{R_0 + \hat{R}(\omega, 0)}{1 + R_0 \hat{R}(\omega, 0)} \right) \hat{f}(\omega) \, d\omega \,. \tag{4.48}$$

In Section 4.5 we analyze the asymptotic behavior of the propagator matrices, from which we get the asymptotic behavior of the quantities of interest using (4.47) and (4.48). This is done in the regime in which the layer size l is small.

4.4.6 Matched Medium

When the two homogeneous half-spaces have the same material properties $\rho_0 = \rho_1$ and $K_0 = K_1$, it is natural to choose $\bar{K} = K_0$ and $\bar{\rho} = \rho_0$, so that $R_0 = R_1 = 0$ and $T_0 = T_1 = 1$. The reflection and transmission coefficients of the heterogeneous slab are then given by

$$\hat{\mathcal{R}}(\omega) = \hat{R}(\omega, 0) = -\frac{\beta_\omega}{\alpha_\omega} \,, \qquad \hat{\mathcal{T}}(\omega) = \hat{T}(\omega, 0) = \frac{1}{\alpha_\omega} \,.$$

They are also solutions of the Riccati equations (4.41)–(4.43) with the following terminal conditions at $z = L$: $\hat{R}(\omega, L) = 0$ and $\hat{T}(\omega, L) = 1$.

The local reflection coefficient $\hat{R}(\omega, z)$ characterizes reflection into the left half space from the random slab (z, L) at a fixed frequency ω. The transmission coefficient $\hat{T}(\omega, z)$ characterizes transmission by the slab (z, L) into the right half-space.

The transmitted and reflected waves (4.38) and (4.39) in the frequency domain are

$$\hat{a}_1(\omega, L) = \hat{a}(\omega, L) = \hat{T}(\omega)\hat{f}(\omega), \qquad \hat{b}_0(\omega, 0) = \hat{b}(\omega, 0) = \hat{R}(\omega)\hat{f}(\omega).$$

In the time domain they are

$$A_1(t, L) = a_1(t - L/\bar{c}, L) = \frac{1}{2\pi} \int e^{-i\omega(t - L/\bar{c})} \hat{T}(\omega, 0)\hat{f}(\omega) \, d\omega,$$

$$B_0(t, 0) = b_0(t, 0) = \frac{1}{2\pi} \int e^{-i\omega t} \hat{R}(\omega, 0)\hat{f}(\omega) \, d\omega.$$

4.5 Homogenization and the Law of Large Numbers

4.5.1 A Simple Discrete Random Medium

We illustrate in a simple setting the concept of homogenization from the point of view of differential equations and its connection to the law of large numbers. We first assume that the local propagation speed c varies with z but the impedance is constant. We choose the value $\bar{\zeta}$ to be equal to this constant. We consider a more general case in the next section. With these assumptions we have

$$\frac{\rho(z)}{\bar{\rho}} = \frac{\bar{K}}{K(z)} = \frac{\bar{c}}{c(z)},$$

and equation (4.34) for the propagator becomes

$$\frac{d}{dz}\mathbf{P}_\omega(0, z) = i\omega \left(\frac{1}{c(z/l)} - \frac{1}{\bar{c}} \right) \begin{bmatrix} 1 & 0 \\ 0 & -1 \end{bmatrix} \mathbf{P}_\omega(0, z).$$

This equation is diagonal and can be integrated by exponentiation:

$$\mathbf{P}_\omega(0, z) = \begin{bmatrix} \exp(i\omega S_l(z)) & 0 \\ 0 & \exp(-i\omega S_l(z)) \end{bmatrix}, \tag{4.49}$$

$$S_l(z) = \int_0^z \left(\frac{1}{c(y/l)} - \frac{1}{\bar{c}} \right) dy. \tag{4.50}$$

To determine the effective medium that emerges in the limit of fine layering $l \to 0$, we see from (4.50) that we need to study the behavior of $\int_0^z c^{-1}(y/l)dy$

as $l \to 0$. Homogenization can be illustrated using the simple model in which the medium is made up of independent and identically distributed layers of equal width $l \to 0$. The medium is defined by one sequence of **independent and identically distributed** positive random variables C_n that are bounded and bounded away from zero. Figure 4.5 is a schematic of the random layering in the slab. In this model, at a given location $z \in [0, L]$ in the slab, the local speed of propagation is given by

$$c(z/l) = C_{[z/l]},$$

where $[x]$ denotes the integer part of x. Since $[z/l] \to \infty$ as $l \to 0$, we can apply the law of large numbers to obtain

$$
\int_0^z c^{-1}(y/l)dy = l \int_0^{z/l} c^{-1}(\tilde{y}) \, d\tilde{y}
$$

$$
= \underbrace{l\,[z/l]}_{\downarrow \atop z} \times \underbrace{\frac{1}{[z/l]} \left(\sum_{j=0}^{[z/l]-1} \frac{1}{C_j} \right)}_{\text{a.s.} \downarrow \atop \mathbb{E}\left[\frac{1}{C_1}\right]} + \underbrace{l\,(z/l - [z/l]) \frac{1}{C_{[z/l]}}}_{\downarrow \atop 0}
$$

$$
\xrightarrow{l \to 0} z\mathbb{E}\left[\frac{1}{C_1}\right]. \tag{4.51}
$$

The convergence is in the almost sure sense, for almost all realizations of the medium, or with probability one with respect to the randomness. Note that the frequency ω does not appear in (4.51), so the set of probability one on which the convergence holds is independent of ω. A more general version of this result is presented in Section 4.5.2.

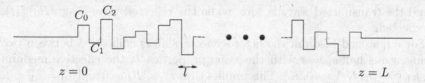

Fig. 4.5. A **piecewise-constant** heterogeneous slab is embedded in a homogeneous medium. The width of each small section is l. The propagation speed in the nth section is C_n where (C_n) is a sequence of independent and identically distributed random variables. The impedance is the same for all sections.

In this setting, homogenization in the frequency domain means that we should choose $\bar{c}$ such that in the limit that $l \to 0$ the propagator $\mathbf{P}_\omega(0, z)$ becomes the identity for all z. Using (4.49–4.50), we see that we must have

$$
\bar{c} = \left(\mathbb{E}\left[\frac{1}{C_1}\right] \right)^{-1}.
$$

Thus, the harmonic mean of local propagation speeds is the homogenized or **effective** propagation speed. This effective propagation speed is frequency independent in this example, and therefore it is also the effective propagation speed in the time domain. We can also consider homogenization in dispersive systems, as is done in Chapter 18.

In the limit $l \to 0$, the propagator $\mathbf{P}_\omega(0, L)$ becomes the identity, which means that $\alpha_\omega \to 1$ and $\beta_\omega \to 0$. Substituting into (4.38–4.39) we get the transmitted and reflected waves in the Fourier domain:

$$\hat{a}_1(\omega, L) = \frac{T_0 T_1}{1 + R_0 R_1 e^{2i \frac{\omega L}{\bar{c}}}} \hat{f}(\omega),$$

$$\hat{b}_0(\omega, 0) = \frac{R_0 + R_1 e^{2i \frac{\omega L}{\bar{c}}}}{1 + R_0 R_1 e^{2i \frac{\omega L}{\bar{c}}}} \hat{f}(\omega).$$

These coincide with the expressions (3.16) for the transmitted and reflected waves from a homogeneous layer embedded between two homogeneous half-spaces.

By Fourier transforming (4.31), we get the time domain limit $l \to 0$ of the transmitted and reflected waves:

$$\lim_{l \to 0} A_1(t, L) = \frac{1}{2\pi} \int e^{-i\omega(t - L/\bar{c})} \frac{T_0 T_1}{1 + R_0 R_1 e^{2i \frac{\omega L}{\bar{c}}}} \hat{f}(\omega) \, d\omega, \qquad (4.52)$$

$$\lim_{l \to 0} B(t, 0) = \frac{1}{2\pi} \int e^{-i\omega t} \frac{R_0 + R_1 e^{2i \frac{\omega L}{\bar{c}}}}{1 + R_0 R_1 e^{2i \frac{\omega L}{\bar{c}}}} \hat{f}(\omega) \, d\omega. \qquad (4.53)$$

Here we can take the limit $l \to 0$ inside the integral, since all quantities are bounded. The pulse function $f(t)$ is assumed to be a smooth function of compact support and so $\hat{f}$ is integrable. We also have $|R_0 R_1| \leq 1$. If $|R_0 R_1| < 1$, then the integrands in (4.52–4.53) are bounded. If $|R_0 R_1| = 1$ then $T_0 T_1 = 0$, and the transmitted wave is zero, while the reflected wave is $\operatorname{sgn}(R_0) f(t)$, as expected.

For a matched medium the heterogeneous slab is embedded between two homogeneous half-spaces with the same properties as the effective medium $(\rho_0, K_0) = (\rho_1, K_1) = (\bar{\rho}, \bar{K})$. This implies that $R_0 = R_1 = 0$ and $T_0 = T_1 = 1$, and the transmitted and reflected waves are given by

$$\lim_{l \to 0} A_1(t, L) = f\left(t - \frac{L}{\bar{c}}\right), \qquad \lim_{l \to 0} B(t, 0) = 0.$$

This is what happens in a homogeneous medium. The pulse is propagating to the right with constant speed $\bar{c}$ and there is no wave scattered back to the left.

The homogenized medium emerges as a consequence of the law of large numbers. Many independent small scattering events associated with thin layers are averaged. In homogenization the random matrix $\mathbf{H}_\omega$ in (4.35) is replaced by its average. This averaging makes the effective speed of propagation equal to the harmonic mean of the local propagation speeds. The averaged matrix $\mathbf{H}_\omega$ vanishes since the correct centering gives pure shift and no interaction of the right- and left-traveling waves.

4.5.2 Random Differential Equations

What is important in the analysis of the previous section is the existence of the limit

$$\lim_{l \to 0} \frac{1}{z} \int_0^z \frac{1}{c(y/l)} dy = \frac{1}{\bar{c}}.$$

This limit is well defined for a wide class of heterogeneous media. It is well defined for deterministic periodic media with small period l, as it is for media modeled with ergodic stochastic processes. We introduce below a general class of differential equations with random coefficients for which averaging or homogenization can be carried out. We then apply the general averaging theorem to the propagator equation (4.34).

Equation (4.34) for the propagator $\mathbf{P}_\omega$ is a prototype of differential equations with random coefficients. It has the general form

$$\frac{dX}{dz} = F(z, Y(z/l), X(z)), \qquad (4.54)$$

where:

- The variable z is one-dimensional, and in many applications of random differential equations this is a continuous time variable. The d-dimensional vector function F defines a random dynamical system with rapidly fluctuating coefficients represented by the process $Y(z/l)$, with l a small parameter. The trajectory of a random system along with that of the **averaged system** is shown in Figure 4.6. The closeness of the two trajectories is a manifestation of the averaging theorem.
- The solution $X(z)$ is the d-dimensional state vector of the system under study.
- The process $Y(z)$ is the **driving** random process, which takes values in an auxiliary space S. This process fluctuates on the **fine** scale z/l. We assume that it is ergodic. In the wave homogenization example discussed above, the process Y is defined by the piecewise-constant medium parameters (ρ, K), which are modeled as independent and identically distributed random variables.
- The function $F(X, Y)$ is a d-dimensional smooth function that is at most linearly growing in X, so that existence and uniqueness properties hold for equation (4.54).
- We will be interested in initial value problems with $X(0) = x_0$ given.

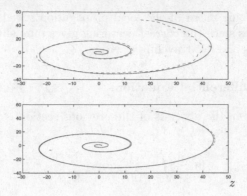

Fig. 4.6. The trajectory of a two-dimensional random differential equation and that of the averaged differential equation. The right-hand side of the random differential equation is $F_1(z, Y(z/l), X(z)) = (1 + Y_1(z/l)) \, X_1(z)/5$ and $F_2(z, Y(z/l), X(z)) = (1 + Y_2(z/l))$. The initial condition is $X(0) = (1, 0)$. The driving term Y is piecewise-constant and in each section, Y_1 and Y_2 are independent random variables, uniformly distributed in the interval $(-1.5, 1.5)$. In the top plot $l = 1/20$, in the bottom $l = 1/1000$. The first component X_1 is the radius and the second component X_2 the angle. The dashed lines show $X(z)$ for one realization of the random medium Y. The solid lines are the trajectories of the averaged equations.

In the equation (4.34) for the propagator, the dimension d is 8, since the propagator is a complex 2×2 matrix where each entry is decomposed into its real and imaginary parts. In this case the equation is linear in the state X and the function F is periodic in its first argument z due to the form of the phases in $\mathbf{H}_\omega$ defined in (4.35). As already noted, the driving process is

$$Y(z/l) = (\rho(z/l), K(z/l)) \, .$$

It takes its values in a compact subspace S of $(0, \infty) \times (0, \infty)$.

We consider the limit of l small and this corresponds to averaging the equation (4.54) with respect to the driving randomness Y. We now state a basic form of the **averaging theorem** for systems of random differential equations.

Theorem 4.2. *Under the ergodicity assumption for Y, the solution X of the random differential equation (4.54) converges almost surely to $\bar{X}$ given by*

$$\frac{d\bar{X}}{dz} = \bar{F}(z, \bar{X}) \quad with \quad \bar{X}(0) = x_0 \, , \tag{4.55}$$

and where

$$\bar{F}(z, x) = \lim_{Z \to \infty} \frac{1}{Z} \int_0^Z F(z, Y(y), x) \, dy = \mathbb{E}[F(z, Y, x)] \, . \tag{4.56}$$

It is important here to note that the expectation is taken when the arguments z and x are frozen (or fixed with respect to the randomness) and we average with respect to the invariant or "steady-state" distribution of Y.

We outline next the intuitive ideas of the proof of this result. A general proof, using perturbed test functions or correctors, is introduced in Chapter 6. Here we give an elementary direct proof. We consider the difference between the exact solution $X(z)$ and its small l limit $\bar{X}(z)$. From the differential equations satisfied by X and $\bar{X}$ we have

$$X(z) - \bar{X}(z) = \int_0^z F(y, Y(y/l), X(y))dy - \int_0^z \bar{F}(z, \bar{X}(y))dy$$

$$= \int_0^z \left(F(y, Y(y/l), X(y)) - F(y, Y(y/l), \bar{X}(y)) \right) dy + g(z),$$

where

$$g(z) := \int_0^z F(y, Y(y/l), \bar{X}(y)) - \bar{F}(y, \bar{X}(y))dy.$$

Taking the modulus and assuming that F is globally Lipschitz with respect to the X-variable. we have

$$|X(z) - \bar{X}(z)| \leq \int_0^z \left| F(y, Y(y/l), X(y)) - F(y, Y(y/l), \bar{X}(y)) \right| dy + |g(z)|$$

$$\leq C \int_0^z |X(y) - \bar{X}(y)| dy + |g(z)|.$$

We note that if

$$\lim_{l \to 0} |g(z)| = \lim_{l \to 0} \left| \int_0^z \left\{ F(y, Y(y/l), \bar{X}(y)) - \bar{F}(y, \bar{X}(y)) \right\} dy \right| = 0, \quad (4.57)$$

then Gronwall's lemma allows us to conclude that the difference $X(z) - \bar{X}(z)$ becomes small in the limit of l small, or

$$\lim_{l \to 0} X(z) = \bar{X}(z).$$

The convergence of an integral of the type (4.57) was carried out in the special case of a discrete driving process Y in Section 4.5.1. It is proven in general by discretizing the integral and applying the law of large numbers, hypothesis (4.56), on subintervals.

The Gronwall inequality in its simplest form is as follows. If $Z(t)$ satisfies for A and B positive constants the integral inequality

$$Z(t) \leq A + B \int_0^t Z(s)ds, \quad t \geq 0,$$

then

$$Z(t) \leq Ae^{Bt}, \quad t \geq 0.$$

4.5.3 The Effective Medium

We return to the form (4.34) without assuming that the impedance is constant and we use the averaging Theorem 4.2, introduced in the previous section, to find the effective medium parameters in the general case. The complex matrix-valued random dynamical system (4.34) for the propagator $\mathbf{P}_\omega \in \mathbb{C}^{2\times 2}$, with $\Delta^{(\pm)}$ given by (4.27), has the form

$$\frac{d\mathbf{P}_\omega}{dz} = F(z, Y(z/l), \mathbf{P}_\omega(z)),\tag{4.58}$$

where

$$F(z, Y(z/l), \mathbf{P}_\omega(z))\tag{4.59}$$

$$= \frac{i\omega}{\bar{c}} \left[\begin{matrix} \left(\frac{Y_1(z/l)+Y_2(z/l)}{2}\right) - 1 & \left(\frac{Y_1(z/l)-Y_2(z/l)}{2}\right) e^{-2i\omega z/\bar{c}} \\ -\left(\frac{Y_1(z/l)-Y_2(z/l)}{2}\right) e^{+2i\omega z/\bar{c}} & 1 - \left(\frac{Y_1(z/l)-Y_2(z/l)}{2}\right) \end{matrix} \right] \mathbf{P}_\omega(z),$$

and

$$Y_1(z/l) = \frac{\rho(z/l)}{\bar{\rho}}, \qquad Y_2(z/l) = \frac{\bar{K}}{K(z/l)}.$$

The effective medium approximation is now obtained by choosing $\bar{\rho}$ and $\bar{K}$ such that the "effective propagator" $\bar{\mathbf{P}}_\omega$, that is, the limit for the propagator when $l \to 0$ according to the averaging theorem, becomes the identity for any z:

$$\frac{d\bar{\mathbf{P}}_\omega}{dz} = 0.$$

This gives

$$\frac{1}{2}\left(\frac{\mathbb{E}[\rho]}{\bar{\rho}} + \bar{K}\mathbb{E}\left[\frac{1}{K}\right]\right) = 1, \qquad \frac{1}{2}\left(\frac{\mathbb{E}[\rho]}{\bar{\rho}} - \bar{K}\mathbb{E}\left[\frac{1}{K}\right]\right) = 0.$$

Thus, the homogenized or **effective medium** is given by

$$\bar{\rho} = \mathbb{E}[\rho], \qquad \frac{1}{\bar{K}} = \mathbb{E}\left[\frac{1}{K}\right].\tag{4.60}$$

The effective parameters are here frequency-independent and therefore it follows that they are also the effective parameters in the time domain. In the homogenization limit, the transmitted and reflected waves are given by (4.52) and (4.53), respectively, with the effective wave speed

$$\bar{c} = \sqrt{\bar{K}/\bar{\rho}}.\tag{4.61}$$

The effective wave speed is obtained by averaging the density and the *reciprocal* of the bulk modulus.

Example 4.3. Bubbles in water. This is not a typical one-dimensional random medium, but since homogenization is valid for general three-dimensional random media, the results that we get from the above elementary theory are physically correct. Air and water have the following density and bulk modulus:

$$\rho_a = 1.2 \ 10^3 \ \text{g/m}^3, \ K_a = 1.4 \ 10^8 \ \text{g/s}^2/\text{m}, \ c_a = 340 \ \text{m/s}.$$
$$\rho_w = 1.0 \ 10^6 \ \text{g/m}^3, \ K_w = 2.0 \ 10^{18} \ \text{g/s}^2/\text{m}, \ c_w = 1425 \ \text{m/s}.$$

If we consider a pulse whose bandwidth is in the range 10 Hz–30 kHz, then the wavelengths lie in the range 1 cm–100 m. Air bubbles in water are typically much smaller, so the effective-medium theory can be applied. Let us denote by ϕ the volume fraction of air in the mixture. The averaged density and bulk modulus are then

$$\bar{\rho} = \mathbb{E}[\rho] = \phi \rho_a + (1-\phi)\rho_w = \begin{cases} 9.9 \ 10^5 \ \text{g/m}^3 & \text{if } \phi = 1\%, \\ 9 \ 10^5 \ \text{g/m}^3 & \text{if } \phi = 10\%, \end{cases}$$

$$\bar{K} = \left(\mathbb{E}[K^{-1}]\right)^{-1} = \left(\frac{\phi}{K_a} + \frac{1-\phi}{K_w}\right)^{-1} = \begin{cases} 1.4 \ 10^{10} \ \text{g/s}^2/\text{m} & \text{if } \phi = 1\%, \\ 1.4 \ 10^9 \ \text{g/s}^2/\text{m} & \text{if } \phi = 10\%. \end{cases}$$

Accordingly, $\bar{c} = 120$ m/s if $\phi = 1\%$ and $\bar{c} = 37$ m/s if $\phi = 10\%$.

This important and physically relevant example shows that the average velocity may be much smaller than the minimum of the component velocities of the medium. However, it cannot happen in such a configuration that the velocity is larger than the maximum (or the essential supremum) of the component velocities. Indeed, by the Cauchy–Schwarz inequality,

$$\mathbb{E}[c^{-1}] = \mathbb{E}\left[K^{-1/2}\rho^{1/2}\right] \leq \mathbb{E}[K^{-1}]^{1/2}\mathbb{E}[\rho]^{1/2} = \bar{c}^{-1}.$$

Thus $\bar{c} \leq \mathbb{E}[c^{-1}]^{-1} \leq \|c\|_\infty$.

Notes

The theory and the results presented in this book rely heavily on modeling and analysis with separation of scales, which has been developed in the past thirty-five years. The main probabilistic tool for the homogenization theory of the equations considered in this book is the law of large numbers or, more generally, the ergodic theorem. We introduce this basic result in Section 4.5. We refer to the book of Breiman [23] for a more complete introduction to probabilistic tools at the level used in this chapter. In Section 4.5.2 we reformulate homogenization as an averaging theorem for random differential equations. Such averaging theorems were first given by Khasminskii [97]. A review of different averaging techniques can be found in the book by Holmes [89]. Multi-dimensional homogenization theory for periodic media is extensively treated by Milton [122] and Bensoussan–Lions–Papanicolaou [13]. A review of results on homogenization for random media is presented in [130]. Acoustic waves in bubbly liquids were analyzed in [34]. Electromagnetic waves in composite materials are discussed in [159].

5

Scaling Limits

In the previous chapter we considered a pulse propagating in a one-dimensional randomly layered medium in the homogenization or effective-medium regime. In this regime the typical wavelength of the propagating pulse λ_0 is comparable to the propagation distance L, while the size l of the layers is small. The typical wavelength is taken to be the pulse width times a reference propagation speed. In this homogenization regime, propagation in a random medium is asymptotically equivalent to propagation in a homogeneous effective medium obtained by averaging the density and the reciprocal of the bulk modulus. In many applications the propagation distance is large compared with the size of the pulse, and wave fluctuations build up behind it as it travels deep into the random medium. In order to model this regime, we take the propagation distance L to be large compared to the typical wavelength λ_0, and the typical layer size l small compared to λ_0,

$$l \ll \lambda_0 \ll L, \tag{5.1}$$

as illustrated in Figure 5.1.

We refer to this scaling as the *high-frequency white-noise regime*. It is a particularly interesting one because it is a high-frequency regime with respect to the large-scale variations of the medium, $L/\lambda_0 \gg 1$, but it is a low-frequency regime with respect to the small-scale random fluctuations, $l/\lambda_0 \ll 1$. As a result, the effect of the random fluctuations takes a canonical form, the white-noise limit, which is independent of the small-scale details. The high-frequency white-noise regime is one of the scaling regimes that have remarkably complete asymptotic theory, as we will see in the following chapters, but other interesting regimes can also be analyzed. In this chapter we introduce the different scales that are relevant to wave propagation in a random medium and we identify several interesting scaling regimes. The identification of scaling regimes provides small dimensionless parameters that quantify the separation of scales that are exploited by the asymptotic theory.

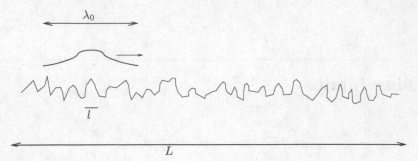

Fig. 5.1. This figure illustrates the high-frequency white-noise scaling regime. This is a regime in which the typical wavelength is much smaller than the propagation distance, that is, $\lambda_0 \ll L$, and in which the medium fluctuations are on a fine scale, $l \ll \lambda_0$. This scaling is typical of the applications that we have in mind in this book.

5.1 Identification of the Scaling Regimes

We consider again the acoustic wave equations

$$\rho(z)\frac{\partial u}{\partial t} + \frac{\partial p}{\partial z} = F(t, z)\,, \tag{5.2}$$

$$\frac{1}{K(z)}\frac{\partial p}{\partial t} + \frac{\partial u}{\partial z} = 0\,, \tag{5.3}$$

with a source term on the right side of the linearized momentum equation (5.2). We consider first the modeling of the random medium through the density ρ and the bulk modulus K.

5.1.1 Modeling of the Medium Fluctuations

For simplicity we assume that there are no random fluctuations in the density, that is, $\rho(z) = \bar{\rho}$, a constant for all z. This is not essential for the analysis of propagation in one-dimensional random media, but it greatly simplifies the analysis of wave propagation in three-dimensional randomly layered media, presented in Chapter 14. We consider media with randomly varying density in Chapter 17.

As we saw in Chapter 4, it is the reciprocal of the bulk modulus that is averaged over distances of propagation comparable with the width of the pulse. It is therefore natural to model the fluctuations in the form

$$\frac{1}{K(z)} = \begin{cases} \frac{1}{\bar{K}}(1 + \nu_K(z)) & \text{for } z \in [0, L]\,, \\ \frac{1}{\bar{K}} & \text{for } z \in (-\infty, 0) \cup (L, \infty)\,, \end{cases} \tag{5.4}$$

$$\rho(z) = \bar{\rho} \text{ for all } z\,,$$

where $\bar{\rho}$ and $\bar{K}$ are given positive constants. The relative fluctuations in the reciprocal of the bulk modulus are modeled by the zero-mean stationary random process ν_K defined on $(-\infty, \infty)$. In the language of the previous chapter, the effective bulk modulus is $\bar{K}$. Note that we consider here matched homogeneous media on either side of the random slab, which means that the properties of the two homogeneous half-spaces are the same as those of the homogenized slab. This matched-medium assumption simplifies the analysis. We consider in detail nonmatched media in Chapter 17.

We introduce the **standard deviation** σ_K and the **correlation length** l_K of the dimensionless fluctuations:

$$\sigma_K^2 = \mathbb{E}[\nu_K(z_0)^2], \qquad \sigma_K^2 l_K = \int_{-\infty}^{\infty} \mathbb{E}[\nu_K(z_0)\nu_K(z_0 + z)]dz. \qquad (5.5)$$

These quantities do not depend on z_0, since the fluctuation process ν_K is statistically stationary. Both are nonnegative, because the first one is a variance and the second one is the integral of the autocorrelation function proportional to the power spectral density of the stationary process ν_K at 0 frequency, which is nonnegative by the Wiener–Khintchine theorem. A general form of this positivity is presented in Chapter 6.

We now write the random process ν_K in scaled form

$$\nu_K(z) = \sigma\nu(z/l), \qquad (5.6)$$

where σ and l are two positive parameters and ν is a dimensionless stationary, zero-mean random function of a dimensionless argument. Using (5.5) we obtain the identities

$$\sigma_K^2 = \sigma^2 \mathbb{E}[\nu(z_0)^2],$$

$$\sigma_K^2 l_K = \sigma^2 l \int_{-\infty}^{\infty} \mathbb{E}[\nu(z_0)\nu(z_0 + z)]dz,$$

and we assume that $\mathbb{E}[\nu(z_0)^2]$ and $\int_{-\infty}^{\infty} \mathbb{E}[\nu(z_0)\nu(z_0 + z)]dz$ remain of order one in the various scaling regimes that we consider. We do not assume that $\mathbb{E}[\nu(z_0)^2]$ and $\int_{-\infty}^{\infty} \mathbb{E}[\nu(z_0)\nu(z_0 + z)]dz$ are equal to one, as seems natural at first, because we wish to consider applications in which the normalized fluctuation process ν is not globally stationary but only piecewise stationary or slowly varying. The first example of such a situation is in Chapter 8. From now on we use σ and l as the reference scales for the strength and the correlation length of the random fluctuations in the medium.

We also assume that $\sigma|\nu(z)| \leq C$ for some positive constant C such that $C < 1$. This ensures that the bulk modulus remains bounded and bounded away from zero. We shall further assume **ergodicity** and **mixing conditions** for the process ν, as discussed in detail in Chapter 6. For a specific example we may take $\nu(z)$ a piecewise-constant stationary process taking values that are independent and identically distributed random variables as in Section 4.5.1. Writing the fluctuations as $\sigma\nu(z/l)$ corresponds to having layers of typical size l, with typical impedance and velocity contrast of order σ.

5.1.2 Modeling of the Source Term

We consider a point source located in the homogeneous left half-space at some position $z_0 < 0$:

$$F(t, z) = \bar{\zeta}^{1/2} g(t) \delta(z - z_0) .$$

Such a source generates a wave that propagates to the left and never interacts with the random medium and a wave that propagates to the right. This wave interacts with the random slab. It has the form

$$A(t, z) = g\left(t - \frac{z - z_0}{\bar{c}}\right), \quad z < 0 .$$

Note that the factor $\bar{\zeta}^{1/2}$ has been included in the expression for the source F, so that the right-going mode is equal to g. We define the **pulse width** T_0 by

$$T_0^2 = \frac{\int_{-\infty}^{\infty} (t - \bar{T})^2 g^2(t) dt}{\int_{-\infty}^{\infty} g^2(t) dt}, \quad \text{where } \bar{T} = \frac{\int_{-\infty}^{\infty} t g^2(t) dt}{\int_{-\infty}^{\infty} g^2(t) dt},$$

which is the root mean square (rms) of the pulse g. We define the **typical frequency**, or more accurately the typical angular frequency, ω_0, of the incoming wave by

$$\omega_0 = \frac{2\pi}{T_0} .$$

The corresponding **typical wavelength** is $\lambda_0 = 2\pi \bar{c}/\omega_0$.

With these definitions we can write the source term and the corresponding incoming wave in the form

$$F(t, z) = \bar{\zeta}^{1/2} f(\omega_0 t) \delta(z - z_0) , \qquad (5.7)$$

$$A(t, z) = f(\omega_0(t - (z - z_0)/\bar{c})) , \qquad (5.8)$$

where f is the normalized **pulse shape** function, whose rms pulse width is one. Note that since the wave equation is linear, the order of magnitude of the pulse amplitude plays no role.

In our dimensional analysis we assume that the input pulse is characterized by a single, typical frequency ω_0, which is defined in terms of the rms pulse width. In many applications, as in communications discussed in Chapter 13, there are two frequencies that are naturally associated with a pulse. One is the **bandwidth**, which is the inverse of the pulse duration, and the other is the **carrier frequency**, around which the spectral energy is concentrated. If we consider a pulse of the form $g(t) = \cos(\omega_{\mathrm{HF}} t) \exp(-t^2/T_0^2)$, then the bandwidth is $1/T_0$, while the carrier frequency is ω_{HF}. In communications applications, we often have $\omega_{\mathrm{HF}} T_0 \gg 1$. In the main applications considered in this book, in geophysics, or ultrasound remote sensing, the carrier frequency and the bandwidth are of the same order. That is why we introduce only one quantity in the scaling theory, and call it the typical frequency.

5.1.3 The Dimensionless Wave Equations

We now put the wave equations in dimensionless form using the dimensionless space and time variables

$$\tilde{z} = \frac{z}{L_0}, \qquad \tilde{t} = \frac{c_0 t}{L_0}. \tag{5.9}$$

Here L_0 is a typical propagation distance and c_0 is a reference speed of propagation. A natural choice for the reference speed is the effective propagation speed $\bar{c}$, given by (4.61), but we will consider in the next chapters situations in which the half-spaces and the random slab have different speeds of sound, so it is preferable to do the dimensional analysis with a reference speed c_0. We introduce similarly a reference impedance ζ_0, so that the normalized pressure and velocity fields have the form

$$\tilde{p}(\tilde{t}, \tilde{z}) = \zeta_0^{-1/2} p\left(\tilde{t}\frac{L_0}{c_0}, \tilde{z}L_0\right), \qquad \tilde{u}(\tilde{t}, \tilde{z}) = \zeta_0^{1/2} u\left(\tilde{t}\frac{L_0}{c_0}, \tilde{z}L_0\right),$$

and the normalized source and fluctuations terms are given by

$$\tilde{F}(\tilde{t}, \tilde{z}) = \zeta_0^{-1/2} F\left(\tilde{t}\frac{L_0}{c_0}, \tilde{z}L_0\right), \qquad \tilde{\nu}(\tilde{z}) = \nu(\tilde{z}L_0).$$

In these dimensionless and normalized quantities, the wave equations (5.2–5.3) are given by

$$\tilde{\bar{\rho}}\frac{\partial \tilde{u}}{\partial \tilde{t}} + \frac{\partial \tilde{p}}{\partial \tilde{z}} = \tilde{F}(\tilde{t}, \tilde{z}), \tag{5.10}$$

$$\frac{1}{\tilde{\bar{K}}}\left(1 + \sigma\tilde{\nu}\left(\tilde{z}\frac{L_0}{l}\right)\right)\frac{\partial \tilde{p}}{\partial \tilde{t}} + \frac{\partial \tilde{u}}{\partial \tilde{z}} = 0, \tag{5.11}$$

where we have introduced $\tilde{\bar{\rho}} = c_0(\bar{\rho}/\zeta_0)$ and $\tilde{\bar{K}} = (\bar{K}/\zeta_0)/c_0$. Using (5.7) and the identity $\delta(az) = a^{-1}\delta(z)$ for $a > 0$ gives

$$\tilde{F}(\tilde{t}, \tilde{z}) = \tilde{\zeta}^{1/2} f\left(\tilde{t}\frac{\omega_0 L_0}{c_0}\right)\delta(\tilde{z} - \tilde{z}_0),$$

for the source in terms of the dimensionless parameters

$$\tilde{\bar{\zeta}} = \sqrt{\tilde{\bar{K}}\tilde{\bar{\rho}}} = \bar{\zeta}/\zeta_0, \quad \tilde{\bar{c}} = \sqrt{\tilde{\bar{K}}/\tilde{\bar{\rho}}} = \sqrt{\bar{K}/\bar{\rho}}/c_0, \quad \tilde{z}_0 = z_0/L_0.$$

We see that only three independent dimensionless groups of parameters appear: the **amplitude parameter** σ, and the **two scaling groups** L_0/l and $\omega_0 L_0/c_0$. We define two dimensionless parameters ε and θ by

$$\frac{L_0}{l} = \frac{1}{\varepsilon^2}, \qquad \frac{\omega_0 L_0}{c_0} = \frac{\theta}{\varepsilon}. \tag{5.12}$$

The important ratio θ/ε is the propagation distance measured in units of the wavelength. These relations can be inverted so that ε and θ are given by

$$\varepsilon = \sqrt{\frac{l}{L_0}}, \qquad \theta = \frac{\omega_0}{c_0}\sqrt{lL_0}.$$

From now on, we drop the tildes and write the scaled and dimensionless wave equations in the form

$$\bar{\rho}\frac{\partial u^\varepsilon}{\partial t} + \frac{\partial p^\varepsilon}{\partial z} = \bar{\zeta}^{1/2}f\left(\frac{\theta t}{\varepsilon}\right)\delta(z - z_0), \qquad (5.13)$$

$$\frac{1}{\bar{K}}\left(1 + \sigma\nu\left(\frac{z}{\varepsilon^2}\right)\right)\frac{\partial p^\varepsilon}{\partial t} + \frac{\partial u^\varepsilon}{\partial z} = 0. \qquad (5.14)$$

5.1.4 Scaling Limits

We introduce the wave number $k_0 = \omega_0/c_0$ and express ε and θ in terms of k_0L_0 and k_0l:

$$k_0L_0 = \frac{\omega_0 L_0}{c_0} = \frac{\theta}{\varepsilon},$$

$$k_0l = \frac{\omega_0 l}{c_0} = \frac{\omega_0 L_0}{c_0}\frac{l}{L_0} = \theta\varepsilon.$$

The orders of magnitude of these two parameters have the following physical interpretation.

(1) If k_0L_0 is large, then we are in a *high-frequency regime*, since the wavelength is much smaller than the propagation distance.

(2) If k_0l is small, then we are in a *white-noise regime*, because the scale of the inhomogeneities is smaller than the wavelength.

From the two dimensionless parameters k_0l and k_0L_0 and the amplitude parameter σ we can define a third scaling parameter that turns out to be very important in the asymptotic theory. It is the ratio of the propagation distance to the **localization length**, which is defined by

$$\frac{L_0}{L_{\text{loc}}} = \sigma^2(k_0l)(k_0L_0) = \sigma^2\theta^2.$$

The localization length $L_{\text{loc}} = 1/(\sigma^2 k_0^2 l)$ is discussed in detail in Chapter 7. The ratio L_0/L_{loc} plays an essential role in wave propagation in one-dimensional random media, because when it is comparable to one, multiple scattering leads to significant energy transfer between right- and left-going modes.

In the regime in which the three parameters k_0l, k_0L, and σ are all of order one, the wave field will interact with the details of the particular realization of the random medium. Such a regime is not of theoretical interest, because one of the main reasons for modeling complicated wave propagation problems with random media is the possible existence and identification of regimes in which the details of the medium fluctuations are captured in a canonical way. Moreover, it is in regimes in which numerical simulations become very cumbersome that the asymptotic analysis is interesting and useful. In certain scaling regimes we will find that a surprisingly simple description of the wave field and its statistics emerges. The detailed analysis of these regimes is the focus of this book. This simplified description arises often in regimes in which the wave field interacts strongly with the randomness.

We describe next some relevant scaling regimes. Note first that $\sigma \gg 1$ leads to a negative index of refraction, which is an unphysical regime, and so we will not discuss it further.

1. **Effective medium** or homogenization regime:

$$k_0 l \ll 1, \quad k_0 L_0 \sim 1, \quad \sigma \ll 1 \text{ or } \sigma \sim 1.$$

This is a low-frequency regime, since the propagation distance is on the order of the wavelength. It is a white-noise regime in the sense that the correlation length of the medium is much smaller than the wavelength. In this regime the wave propagation is described by a deterministic effective wave equation, random scattering is weak, and there is no backscattering since $L_0/L_{\text{loc}} \ll 1$ in this case.

· In terms of θ, ε, and σ this regime corresponds to

$$\varepsilon \ll 1, \quad \theta \sim \varepsilon, \quad \sigma \ll 1 \text{ or } \sigma \sim 1. \tag{5.15}$$

2. **Weakly heterogeneous** regime:

$$k_0 l \sim 1, \quad k_0 L_0 \gg 1, \quad \sigma \ll 1.$$

In this high-frequency regime, the coupling between the wave and the medium is weak because the strength of the fluctuations σ is small. As a result, the propagation distance must be large enough for the wave to experience significant scattering. The regime in which σ is small and $k_0 L_0$ is large, so that $\sigma^2 k_0 L_0 \sim 1$, is of particular interest because $L_0/L_{\text{loc}} \sim 1$. This means that mode coupling and backscattering are of order one. In terms of θ, ε, and σ this regime corresponds to

$$\varepsilon \ll 1, \quad \sigma \sim \varepsilon, \quad \theta \sim \varepsilon^{-1}. \tag{5.16}$$

3. **Strongly heterogeneous white-noise** regime:

$$k_0 l \ll 1, \quad k_0 L_0 \gg 1, \quad \sigma \sim 1.$$

In this high-frequency regime the coupling between the wave and the medium might be expected to be strong because the strength of the fluctuations σ is of order one. However, since the wavelength is much larger than the scale of variations of the medium, the wave cannot probe the small scales efficiently. The fluctuations of the medium tend to be averaged by the low sensitivity of the wave at these scales. As a result, a long propagation distance is necessary to build up sufficient backscattering. The interesting regime is that in which $k_0 l$ is small and $k_0 L_0$ is large, so that $k_0 l k_0 L_0 \sim 1$. We then have $L_0 / L_{\text{loc}} \sim 1$.

In terms of θ, ε, and σ this regime corresponds to

$$\varepsilon \ll 1, \quad \sigma \sim 1, \quad \theta \sim 1. \tag{5.17}$$

We already examined the effective medium or homogenization regime (5.15) in Chapter 4. We have seen that the key issue is the computation of the effective medium parameters, and we have shown how simple limit theorems such as the strong law of large numbers can help us to obtain closed-form formulas for the effective parameters. The relative simplicity of homogenization in one-dimensional random media is understandable because so much of it can also be carried out in several dimensions.

In the following chapters we address the last two regimes, (5.16) and (5.17), with special emphasis on the strongly heterogeneous white-noise regime, (5.17), which is the one that is encountered in geophysical applications. Typical scaling parameters in exploration seismology [168] can be taken to be as follows: the probing wavelength $\lambda_0 \approx 150$ m, which is small compared with the penetration depth $L_0 \approx 10$–15 km, but large compared with the correlation length that is estimated in the range $l \approx 2$–3 m. The weakly heterogeneous regime (5.16) is considered in Chapter 18.

The asymptotic analysis of the two regimes (5.16) and (5.17) is mathematically very similar. However, there are some important differences that should be kept in mind in using one or the other regime in applications. In the weakly heterogeneous regime (5.16), correlation lengths of the medium fluctuations are comparable to typical wavelengths, and so the asymptotic theory depends on the specific autocorrelation function of these fluctuations. In the strongly heterogeneous white-noise regime (5.17), typical wavelengths are much larger than correlation lengths, and so the asymptotic theory is not sensitive to the detailed structure of the autocorrelation of the fluctuations.

5.1.5 Right- and Left-Going Waves

The transformations and integral representations of the reflected and transmitted waves presented in this subsection hold for any values of the parameters ε, θ, and σ. However, the choice of these transformations is motivated by the intent to analyze the scaling regimes in which $\varepsilon \ll 1$, $\theta \gg 1$ or $\theta \sim 1$, and $\sigma \ll 1$ or $\sigma \sim 1$. We consider again the decomposition of the wave field $(p^\varepsilon, u^\varepsilon)$ into right-going and left-going waves. This is done with the transformation (3.4):

$$A^\varepsilon = \frac{p^\varepsilon}{\bar{\zeta}^{1/2}} + \bar{\zeta}^{1/2} u^\varepsilon, \qquad B^\varepsilon = -\frac{p^\varepsilon}{\bar{\zeta}^{1/2}} + \bar{\zeta}^{1/2} u^\varepsilon,$$

where $\bar{\zeta} = \sqrt{\bar{K}\bar{\rho}}$ is the constant impedance outside of the random medium and the effective impedance inside the medium. The constant background speed is $\bar{c} = \sqrt{\bar{K}/\bar{\rho}}$. Our boundary conditions are again those for an incident wave from the right,

$$A^{\varepsilon}(t,z) = f\left(\frac{\theta}{\varepsilon}\left(t - \frac{z}{\bar{c}}\right)\right), \quad z < 0,$$

and the radiation condition $B^{\varepsilon}(t,z) = 0$ in the right half-space $z > L$. The equations for A^{ε} and B^{ε} are given by (4.28) with the previous boundary conditions. The main difference with Chapter 4 is that the width of the incoming pulse is now small, of order ε/θ. It is natural to look at the quantities of interest, transmitted and reflected waves, on the same time scale as that of the incoming pulse. This is done by looking at the waves along the characteristics on a time scale of order ε/θ:

$$a^{\varepsilon}(s,z) = A^{\varepsilon}\left(\frac{\varepsilon}{\theta}s + \frac{z}{\bar{c}}, z\right), \tag{5.18}$$

$$b^{\varepsilon}(s,z) = B^{\varepsilon}\left(\frac{\varepsilon}{\theta}s - \frac{z}{\bar{c}}, z\right).$$

We now have $(\varepsilon/\theta)s$ instead of s in (4.30).

We next take the Fourier transform with respect to the time variable s:

$$\hat{a}^{\varepsilon}(\omega, z) = \int e^{i\omega s}a^{\varepsilon}(s,z)\, ds,$$

$$\hat{b}^{\varepsilon}(\omega, z) = \int e^{i\omega s}b^{\varepsilon}(s,z)\, ds.$$

The main difference with the Fourier transform in Section 4.4.2 is that now it is with respect to the short time scale s, rather than the original time scale. This scaling resolves short time scales and takes us into a **high-frequency** regime when ε/θ is small, as is seen by writing

$$\dot{\omega}s = \left(\frac{\omega}{\varepsilon/\theta}\right)((\varepsilon/\theta)s). \tag{5.19}$$

It is convenient also to write the velocity and pressure in terms of the right- and left-going waves in the Fourier domain, since shifts in the time variable become multiplication by phase factors,

$$\hat{p}^{\varepsilon}(\omega, z) = \frac{\sqrt{\bar{\zeta}}}{2}\left(\hat{a}^{\varepsilon}(\omega,z)e^{i\theta\omega z/(\varepsilon\bar{c})} - \hat{b}^{\varepsilon}(\omega,z)e^{-i\theta\omega z/(\varepsilon\bar{c})}\right), \tag{5.20}$$

$$\hat{u}^{\varepsilon}(\omega, z) = \frac{1}{2\sqrt{\bar{\zeta}}}\left(\hat{a}^{\varepsilon}(\omega,z)e^{i\theta\omega z/(\varepsilon\bar{c})} + \hat{b}^{\varepsilon}(\omega,z)e^{-i\theta\omega z/(\varepsilon\bar{c})}\right). \tag{5.21}$$

Equation (4.32) now becomes

$$\frac{d}{dz}\begin{bmatrix} \hat{a}^\varepsilon \\ \hat{b}^\varepsilon \end{bmatrix} = \frac{i\theta\omega}{\varepsilon\bar{c}}\begin{bmatrix} (\Delta^{(+)} - 1) & \Delta^{(-)}\,e^{-2i\theta\omega z/(\varepsilon\bar{c})} \\ -\Delta^{(-)}\,e^{+2i\theta\omega z/(\varepsilon\bar{c})} & (1 - \Delta^{(+)}) \end{bmatrix}\begin{bmatrix} \hat{a}^\varepsilon \\ \hat{b}^\varepsilon \end{bmatrix}. \qquad (5.22)$$

Using (5.4) we see that the quantities $\Delta^{(\pm)}$ in (4.27) become

$$\Delta^\pm = \frac{1}{2}\left\{1 \pm (1 + \sigma\nu(z/\varepsilon^2))\right\}. \qquad (5.23)$$

Substituting these quantities into (5.22) gives the new ordinary differential equations for the centered and transformed waves

$$\frac{d}{dz}\begin{bmatrix} \hat{a}^\varepsilon \\ \hat{b}^\varepsilon \end{bmatrix} = \frac{i\theta\omega\sigma}{2\bar{c}\varepsilon}\nu\left(\frac{z}{\varepsilon^2}\right)\begin{bmatrix} 1 & -e^{-2i\theta\omega z/(\varepsilon\bar{c})} \\ e^{+2i\theta\omega z/(\varepsilon\bar{c})} & -1 \end{bmatrix}\begin{bmatrix} \hat{a}^\varepsilon \\ \hat{b}^\varepsilon \end{bmatrix}. \qquad (5.24)$$

The boundary conditions are

$$\hat{a}^\varepsilon(\omega, 0) = \int e^{i\omega s} A^\varepsilon\left(\frac{\varepsilon}{\theta}s, 0\right)\,ds = \int e^{i\omega s} f(s)\,ds = \hat{f}(\omega), \qquad (5.25)$$

$$\hat{b}^\varepsilon(\omega, L) = 0. \qquad (5.26)$$

In the deterministic case ($\nu = 0$) the right-hand side in (5.24) vanishes. This corresponds to a constant medium and perfect transport of the right- and left-going wave components. In the random case the wave components couple because of the scattering from the fine-scale fluctuations in the random medium. The rate of change of the wave amplitudes is now random, of order ε^{-1}, and varies on the fine scale ε^2. It is the off-diagonal terms in (5.24) that couple right- and left-going waves, and they contain a phase factor that oscillates rapidly when θ/ε is large. This is important in the asymptotic analysis.

5.1.6 Propagator and Reflection and Transmission Coefficients

As in Section 4.4.3 we introduce the propagator matrix in order to convert the boundary value problem (5.24) for the wave amplitudes into an initial value problem. The propagator is defined by (4.36) and satisfies the matrix system

$$\frac{d}{dz}\mathbf{P}_\omega^\varepsilon(0, z) = \frac{\theta\sigma}{\varepsilon}\mathbf{H}_\omega\left(\frac{\theta z}{\varepsilon}, \nu\left(\frac{z}{\varepsilon^2}\right)\right)\mathbf{P}_\omega^\varepsilon(0, z), \qquad (5.27)$$

with initial condition $\mathbf{P}_\omega^\varepsilon(0, 0) = \mathbf{I}$. The right side depends on the fast variable $\theta z/\varepsilon$ through the phases and also on the fast variable z/ε^2 through the random fluctuation process ν,

$$\mathbf{H}_\omega(z, \nu) = \frac{i\omega}{2\bar{c}}\nu\begin{bmatrix} 1 & -e^{-2i\omega z/\bar{c}} \\ e^{+2i\omega z/\bar{c}} & -1 \end{bmatrix}. \qquad (5.28)$$

We can again express the propagator in the form

$$\mathbf{P}_\omega^\varepsilon(0,L) = \begin{bmatrix} \alpha_\omega^\varepsilon(0,L) & \overline{\beta_\omega^\varepsilon(0,L)} \\ \beta_\omega^\varepsilon(0,L) & \overline{\alpha_\omega^\varepsilon(0,L)} \end{bmatrix},$$

with $|\alpha_\omega^\varepsilon|^2 - |\beta_\omega^\varepsilon|^2 = 1$. Using this notation for the components of the propagator we find the following **integral representation** for the transmitted and reflected waves.

- The transmitted wave has the form

$$A^\varepsilon(t,L) = a^\varepsilon\left(\theta(t - L/\bar{c})/\varepsilon, L\right) = \frac{1}{2\pi}\int e^{-i\theta\omega(t-L/\bar{c})/\varepsilon}\hat{a}^\varepsilon(\omega, L)\, d\omega$$

$$= \frac{1}{2\pi}\int e^{-i\theta\omega(t-L/\bar{c})/\varepsilon}\left(\frac{1}{\alpha_\omega^\varepsilon(0,L)}\right)\hat{f}(\omega)\, d\omega. \tag{5.29}$$

The width of the incident pulse is of order ε/θ, and $t = z/\bar{c}$ is the travel time from the origin to location z in the effective homogeneous medium. Therefore we should observe the wave in a time window of the form

$$t = \frac{L}{\bar{c}} + \frac{\varepsilon}{\theta}s.$$

In this time window the transmitted wave has the form

$$A^\varepsilon\left(\frac{L}{\bar{c}} + \frac{\varepsilon}{\theta}s, L\right) = a^\varepsilon(s,L)$$

$$= \frac{1}{2\pi}\int e^{-i\omega s}\left(\frac{1}{\alpha_\omega^\varepsilon(0,L)}\right)\hat{f}(\omega)\, d\omega. \tag{5.30}$$

Note that the fast phase in the integral has been removed. This is important for the asymptotic analysis.

- The reflected wave has the form

$$B^\varepsilon(t,0) = b^\varepsilon\left(\frac{\theta}{\varepsilon}t, 0\right) = \frac{1}{2\pi}\int e^{-i\theta\omega t/\varepsilon}\hat{b}^\varepsilon(\omega, 0)\, d\omega$$

$$= \frac{1}{2\pi}\int e^{-i\theta\omega t/\varepsilon}\left(-\frac{\beta_\omega^\varepsilon(0,L)}{\alpha_\omega^\varepsilon(0,L)}\right)\hat{f}(\omega)\, d\omega. \tag{5.31}$$

There is no natural travel time associated with the reflected wave as there is for the transmitted wave. This is consistent with the fact that in the effective-medium approximation there is no reflected wave, since we assume that there is no mismatch at $z = 0$ and $z = L$.

The main objective now is to analyze the asymptotic behavior of the propagator matrix $\mathbf{P}_\omega^\varepsilon$ and to use it for the asymptotic analysis of the transmitted and reflected waves. From (5.30) and (5.31) we see that we must analyze the transmission and reflection coefficients in the frequency domain, which we now define.

- The transmission coefficient is

$$T_\omega^\varepsilon(0, L) = \frac{1}{\alpha_\omega^\varepsilon(0, L)}.$$
(5.32)

- The reflection coefficient is

$$R_\omega^\varepsilon(0, L) = -\frac{\beta_\omega^\varepsilon(0, L)}{\alpha_\omega^\varepsilon(0, L)}.$$
(5.33)

5.2 Diffusion Scaling

In order to focus on the probabilistic part of the asymptotic analysis of (5.27) we consider the following simple example of a scalar random equation,

$$\frac{dX^\varepsilon(z)}{dz} = \frac{\theta\sigma}{\varepsilon} \nu\left(\frac{z}{\varepsilon^2}\right) \quad \text{with} \quad X^\varepsilon(0) = 0,$$

where the right side does not depend on $X^\varepsilon(z)$. As in Section 4.5.1, we consider this equation with a discrete random process ν,

$$\nu(z/\varepsilon^2) = \nu_{[z/\varepsilon^2]}.$$

Here $\{\nu_n\}$ is a sequence of independent and identically distributed random variables, with zero mean ($\mathbb{E}[\nu_n] = 0$), variance one ($\mathbb{E}[\nu_n^2] = 1$). We also assume that they are bounded ($|\nu_n| < C$, for some positive constant C). The solution $X^\varepsilon(z)$ is the integral

$$X^\varepsilon(z) = \frac{\theta\sigma}{\varepsilon} \int_0^z \nu\left(\frac{y}{\varepsilon^2}\right) dy = \theta\sigma\varepsilon \int_0^{z/\varepsilon^2} \nu(y')\, dy' = \theta\sigma\varepsilon \sum_{n=0}^{[z/\varepsilon^2]-1} \nu_n,$$

where we assume for simplicity that z is an integer multiple of ε^2, for otherwise there is an additional $\mathcal{O}(\varepsilon)$ term. This is a sum of many independent and identically distributed centered random variables scaled by the (small) parameter $\theta\varepsilon$. The scaling that is appropriate for the **central limit theorem** is obtained by calculating the variance

$$\mathbb{E}[X^\varepsilon(L)^2] = \sigma^2\theta^2\varepsilon^2 \mathbb{E}\left[\left(\sum_{n=0}^{[L/\varepsilon^2]-1} \nu_n\right)^2\right]$$

$$= \sigma^2\theta^2\varepsilon^2 \left(\sum_{n=0}^{[L/\varepsilon^2]-1} \sum_{m=0}^{[L/\varepsilon^2]-1} \mathbb{E}[\nu_n\nu_m]\right)$$

$$= \sigma^2\theta^2\varepsilon^2 \left(\sum_{n=0}^{[L/\varepsilon^2]-1} \mathbb{E}[\nu_n^2]\right)$$

$$= \sigma^2\theta^2 L\mathbb{E}[\nu_n^2] = \sigma^2\theta^2 L.$$
(5.34)

Therefore, the variance of $X^\varepsilon(z)$ has nontrivial limit if $\sigma\theta$ is of order one, since L is a dimensionless order-one variable. This is indeed the case in the weakly heterogeneous regime, where $\sigma \sim \varepsilon$ and $\theta \sim \varepsilon^{-1}$, and also in the strongly heterogeneous white-noise regime where $\sigma \sim 1$ and $\theta \sim 1$. In the following sections we consider only the **strongly heterogeneous white-noise** regime. In this regime, ε is the only small parameter, and all other quantities are of order 1. In the next section we discuss the limit of $X^\varepsilon(z)$ as a family of random variables indexed by $z \geq 0$.

5.2.1 White-Noise Regime and Brownian Motion

We assume in this section that $\sigma = 1$ and $\theta = 1$ and we consider the process

$$X^\varepsilon(z) = \varepsilon \int_0^{z/\varepsilon^2} \nu(y)\,dy\,. \tag{5.35}$$

This is a family of random variables indexed by $z \geq 0$ and is an example of a stochastic process, with z playing here the role of what is usually a time variable. The assumption that the random medium is layered or **one-dimensional** is essential at this point because we can fully exploit the **stochastic calculus** associated with time-indexed stochastic processes. A key fact in stochastic calculus is that $X^\varepsilon(z)$ converges in distribution to the **Brownian motion**. In order to keep the discussion at a simple and intuitive level we will also assume here that the medium is discrete and defined by the independent and identically distributed sequence (ν_n).

We want to characterize this family of random variables in the small-ε limit. We look first at the increment of the process over the interval (z_1, z_2),

$$X^\varepsilon(z_2) - X^\varepsilon(z_1) = \varepsilon \sum_{n=[z_1/\varepsilon^2]}^{[z_2/\varepsilon^2]} \nu_n + \mathcal{O}(\varepsilon)\,.$$

The $\mathcal{O}(\varepsilon)$ term is due to the fact that the z_i's may not be integer multiples of ε^2, but this does not affect the limit. The main term is a sum of centered and normalized independent and identically distributed random variables with the scaling of the central limit theorem. This means that the increment converges in distribution to a centered Gaussian random variable with variance equal to $(z_2 - z_1)$, the length of the increment. A computation similar to the one in (5.34) shows that

$$\lim_{\varepsilon\to 0} \mathbb{E}\left[(X^\varepsilon(z_2) - X^\varepsilon(z_1))^2\right] = (z_2 - z_1)\mathbb{E}[\nu_n^2] = z_2 - z_1\,.$$

Let $0 < z_1 < z_2 < z_3 < \cdots < z_k$ and consider the increments over the disjoint intervals (z_i, z_{i+1}) for $1 \leq i \leq k-1$. Since these increments involve sums of disjoint sets of ν_n's, up to an $\mathcal{O}(\varepsilon)$ term they are independent, and this property carries over to any limit process. Thus, a limit process $X(z)$ has centered independent Gaussian increments with variance equal to the length of the z-increment. The family of random increments is the **white-noise** process. The limit process, $X(z)$, is continuous in z and has continuous trajectories. This is shown by proving a tightness property on the set of continuous trajectories, which in our case is a generalization of the computation (5.34) to the fourth moments

$$\lim_{\varepsilon \to 0} \mathbb{E}\left[(X^\varepsilon(z_2) - X^\varepsilon(z_1))^4 \right] = 3\mathbb{E}\left[\nu_n^2 \right]^2 (z_2 - z_1)^2 = 3(z_2 - z_1)^2 \, .$$

The Kolmogorov criterion for tightness can now be applied, giving convergence to Brownian motion.

We summarize the properties of the limit process, the Brownian motion $X(z)$:

- The trajectories $(z \mapsto X(z))$ are continuous and $X(0) = 0$.
- The process $X(z)$ has independent increments.
- The increment $X(z_2) - X(z_1)$ is a $\mathcal{N}(0, z_2 - z_1)$-distributed random variable (i.e., Gaussian with mean zero and variance $z_2 - z_1$).

The process $X(z)$ is called a **standard** Brownian motion. It is the simplest example of a **diffusion process**, a process that behaves in a "diffusive" way, since $\mathbb{E}[X(z)^2] = z$. This is in contrast to a "ballistic" behavior, where the mean square displacement is proportional to z^2.

5.2.2 Diffusion Approximation

We consider the strongly heterogeneous white-noise regime, that is, the regime in which $\varepsilon \ll 1$ while all other quantities are of order 1. Our main objective is to characterize the asymptotic behavior of the transmitted and reflected waves given by the integral representations (5.30) and (5.31). These expressions involve functionals of the propagator matrix $\mathbf{P}_\omega^\varepsilon(0, L)$ through the transmission and reflection coefficients given in (5.32) and (5.33). We will now characterize their asymptotic behavior in the diffusion limit introduced above. The equation for the propagator is

$$\frac{d}{dz}\mathbf{P}_\omega^\varepsilon(0, z) = \frac{i\theta\omega\sigma}{2\bar{c}\varepsilon}\nu\left(\frac{z}{\varepsilon^2}\right) \begin{bmatrix} 1 & -e^{-2i\theta\omega z/(\bar{c}\varepsilon)} \\ e^{+2i\theta\omega z/(\bar{c}\varepsilon)} & -1 \end{bmatrix} \mathbf{P}_\omega^\varepsilon(0, z), \quad (5.36)$$

with initial condition $\mathbf{P}_\omega^\varepsilon(0, 0) = \mathbf{I}$. By considering separately the real and imaginary parts of this system in vector form we find that it is a particular example of an equation of the general type

$$\frac{dX^\varepsilon}{dz} = \frac{1}{\varepsilon}F\left(\frac{z}{\varepsilon}, Y\left(\frac{z}{\varepsilon^2}\right), X^\varepsilon(z)\right) \, . \quad (5.37)$$

Here

- The solution X^ε is a d-dimensional real vector with initial condition $X^\varepsilon(0) = x_0$.
- The random process Y is stationary with good ergodic properties (discussed later).
- The function $F(z, y, x)$ is a smooth d-dimensional function at most linearly growing in x and has the important **centering** property

$$\mathbb{E}[F(z, Y(z_0), x)] = 0, \tag{5.38}$$

where z and x are fixed and the expectation is taken with respect to the law of the stationary process Y. The argument z_0 in the centering condition is arbitrary, since the process Y is assumed to be statistically stationary.
- The driving process $Y(z/\varepsilon^2)$ varies on the fine scale ε^2.

Equation (5.37) is similar to equation (4.54) (with $l = \varepsilon$) analyzed in the homogenization regime. The main difference is that F is now scaled by the large parameter $1/\varepsilon$ and we require F to be centered (5.38). The scaling and centering imply that the solution will exhibit diffusive stochastic behavior, in contrast to the deterministic ballistic behavior in the homogenization scaling. This is illustrated in Figures 4.6 (ballistic case) and Figure 5.2 (diffusive case). We note that if in (4.54), F is centered, then the homogenized solution is constant. However, for large z, that is, $z \to z/\varepsilon$, we have exactly the diffusive behavior exhibited by the solution of (5.37).

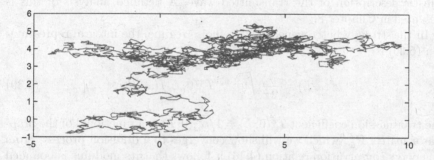

Fig. 5.2. Illustration of diffusive behavior. The coordinates of the trajectory shown are two independent Brownian motions, with time being the path parameter. The mean square displacement is proportional to the time.

The **diffusion approximation theorem** for (5.37) states that the process $X^\varepsilon(z)$ converges in distribution to a diffusion process $X(z)$. We discussed a particular simple example with convergence to Brownian motion in the previous section. The intuitive idea behind the diffusion approximation is that (5.37) has the scaling of the central limit theorem,

$$X^\varepsilon(z) - x_0 = \varepsilon \int_0^{z/\varepsilon^2} F(\varepsilon y, Y(y), X^\varepsilon(\varepsilon^2 y)) \, dy.$$

This is the integrated form of (5.37) with a change of integration variable so as to make it similar to (5.35). The characterization of the limit process for the random differential equation (5.37) is naturally more complicated than the simple example (5.35). It is presented in detail in Chapter 6 after we introduce the necessary background and tools from stochastic calculus. We end this chapter with a discussion of the meaning of **convergence in distribution** and the consequences it has for the transmitted and reflected waves.

Convergence in distribution is also referred to as weak convergence or convergence in law. If the process $X^\varepsilon(z)$ is weakly convergent to $X(z)$ then in particular, its finite-dimensional distributions are convergent. This means that for every $0 \le z_1 < z_2 < \cdots < z_k$ and any continuous bounded function ϕ we have

$$\lim_{\varepsilon \to 0} \mathbb{E}[\phi(X^\varepsilon(z_1), \ldots, X^\varepsilon(z_k))] = \mathbb{E}[\phi(X(z_1), \ldots, X(z_k))].$$

If the processes X^ε have moments, which is the case for the transmission and reflection coefficients, the function ϕ can be unbounded and we therefore have convergence of moments.

5.2.3 Finite-Dimensional Distributions of the Transmitted Wave

We show briefly how the diffusion approximation can be used to give a probabilistic description of the transmitted wave. A detailed analysis of this is presented in Chapter 8.

In the strongly heterogeneous white-noise regime, the integral representation (5.30) is

$$a^\varepsilon(s, L) = \frac{1}{2\pi} \int e^{-i\omega s} T_\omega^\varepsilon(0, L) \hat{f}(\omega) \, d\omega. \tag{5.39}$$

The transmission coefficient $T_\omega^\varepsilon(0, L) = 1/\overline{\alpha_\omega^\varepsilon(0, L)}$ is a functional of the propagator matrix $\mathbf{P}_\omega^\varepsilon$, which we will show converges to a diffusion process. From the energy conservation relation (4.46) it follows that its modulus is bounded by one, which implies that the transmitted wave is also bounded independently of ε by $(1/2\pi) \int |\hat{f}(\omega)| d\omega$. We see from (5.39) that the probability distribution of the wave in **time** depends on the **joint distribution** of the transmission coefficients at different frequencies. The joint moments of the transmission coefficient at different frequencies characterize the moments of the wave in time. From the joint moments we can determine finite-dimensional distributions, since the transmitted waves are bounded random variables. For $0 \le s_1 < \cdots < s_k$ we compute the joint moments of orders $m_1, \ldots, m_k$:

$$\mathbb{E}\left[(a^\varepsilon(s_1, L)^{m_1} \cdots a^\varepsilon(s_k, L)^{m_k})\right]$$
$$= \mathbb{E}\left[\frac{1}{(2\pi)^m} \prod_{h=1}^{k} \left(\int e^{-i\omega s_h} T_\omega^\varepsilon(0, L) \hat{f}(\omega) \, d\omega\right)^{m_h}\right],$$

where $m = \sum_{h=1}^{k} m_h$. This expression can be rewritten as a multiple integral with respect to the frequencies $\omega_{j,h}$ for $1 \leq h \leq k$ and $1 \leq j \leq m_h$:

$$\mathbb{E}\left[(a^{\varepsilon}(s_1, L)^{m_1} \cdots a^{\varepsilon}(s_k, L)^{m_k})\right]$$

$$= \frac{1}{(2\pi)^m} \int e^{-i\sum_{h,j} s_h \omega_{j,h}} \prod_{h,j} \hat{f}(\omega_{j,h}) \mathbb{E}\left[\prod_{h,j} T^{\varepsilon}_{\omega_{j,h}}(0, L)\right] \prod_{h,j} d\omega_{j,h}.$$

Here the sum and the products are taken over all frequencies. Therefore, in order to characterize the finite-dimensional distributions of the transmitted wave we need to know the joint moments of the transmission coefficients for a finite number of different frequencies, which we relabel as $\omega_1, \ldots, \omega_m$. This means that if we know the limits of moments of transmission coefficients

$$\lim_{\varepsilon \to 0} \mathbb{E}\left[\prod_{j=1}^{m} T^{\varepsilon}_{\omega_j}(0, L)\right], \tag{5.40}$$

then we can characterize all the finite-dimensional distributions of the transmitted wave in the time domain.

The diffusion approximation will be the main tool for the computation of these moments. In the following chapter we analyze in detail the diffusion approximation for random differential equations in a general framework, along with the associated stochastic calculus. The detailed discussion of applications to reflection and transmission of waves starts in Chapter 8. We show there that we can identify the moments (5.40) from the asymptotic distribution of the joint propagator matrix $\mathbf{P}^{\varepsilon}_{(\omega_1, \ldots, \omega_m)}(0, L)$, associated with the frequencies $(\omega_1, \ldots, \omega_m)$.

Notes

In this chapter we introduced characteristic parameters for the waves in randomly layered media, and we have put the wave equations in dimensionless form. We identified different scaling regimes in which multiple scattering is significant. We have also shown how the asymptotic analysis of wave propagation in these regimes reduces to the study of diffusion approximations for random ordinary differential equations. Limit theorems for random differential equations were first given by Khasminskii in 1966 [98]. The first application of such limit theorems to monochromatic waves in random media is due to Papanicolaou in 1971 [129]. The theory of waves in randomly layered media with the systematic use of diffusion approximations has been developed in the series of articles by Papanicolaou and his coauthors [8, 31, 32, 33, 104, 107, 132, 133, 152, 153, 169, 170]. In the physical literature, waves in randomly layered media were analyzed in [81, 82, 85, 154, 155, 156] and by Klyatskin [101], whose book contains many more references.

6

Asymptotics for Random Ordinary Differential Equations

In this chapter we give a self-contained presentation of the asymptotic analysis of random differential equations in the form that they have in models of wave propagation in randomly layered media. The chapter serves three purposes:

- It provides an introduction to Markovian models of random media.
- It gives a full treatment of the theory of diffusion approximations for random differential equations in a form that can be readily used for the asymptotic analysis of reflected and transmitted waves in randomly layered media.
- It gives an introduction to stochastic calculus as needed for computations in the asymptotic analysis of wave reflection and transmission.

We will carry out the asymptotic analysis of random differential equations of the form (5.37), where the process X^ε satisfies

$$\frac{dX^\varepsilon}{dz} = \frac{1}{\varepsilon} F\left(X^\varepsilon(z), Y\left(\frac{z}{\varepsilon^2}\right), \frac{z}{\varepsilon}\right), \quad z > 0, \quad X^\varepsilon(0) = x_0, \qquad (6.1)$$

with $\varepsilon > 0$ a small parameter. The process X^ε is d-dimensional. The function $F(x, y, \tau)$ is smooth and at most linearly growing in x and it is uniformly bounded in τ and y. We also assume that F is centered with respect to the driving random process $Y(z)$,

$$\mathbb{E}[F(x, Y(z_0), \tau)] = 0, \qquad (6.2)$$

for all x and τ, and the argument z_0 plays no role in this condition since $Y(z)$ is assumed statistically stationary. Although the asymptotic theory for X^ε can be carried out in great generality, we will assume in this book that the driving process $Y(z)$ is an ergodic Markovian process. This means that the randomly layered media that we consider here are Markovian.

We outline first, in the next section, some general notions on Markov processes and illustrate them with the standard Brownian motion $W(z)$, introduced in Section 5.2.1. We then give examples of Markovian models for the driving process $Y(z)$ in Section 6.2. The diffusion approximation theorems are stated and proved in Sections 6.3 through 6.5, and an introduction to stochastic calculus is given in Section 6.6.

References to general, more detailed treatments of limit theorems for random differential equations are give in the notes at the end of this chapter.

6.1 Markov Processes

The driving process $Y(z)$ is assumed to be Markovian, taking values in a state space S. The Markov property is defined by the condition that the σ-algebras (or information) generated by the two families of random variables $\{Y(s), s \geq z\}$ and $\{Y(s), s \leq z\}$ are independent given the value of $Y(z)$, for any z. When z is thought of as a time variable we say that "the future is independent of the past knowing the present." In the context of waves, z is a one-dimensional space variable and the Markov property means that "the right is independent from the left knowing the value at the boundary."

6.1.1 Semigroups

For bounded real-valued functions ϕ defined on S we consider the family of operators (P_s) defined by the expectations

$$(P_s\phi)(x) = \mathbb{E}[\phi(Y(z+s))|Y(z) = x], \tag{6.3}$$

where we have assumed that the Markovian process Y is **homogeneous** in the sense that the expectation in (6.3) depends on the interval $(z, z+s)$ only through its length s. The Markov property of Y implies that the family of operators $(P_s)_{s\geq0}$ is a **semigroup**:

$$P_{s+h} = P_s P_h,$$

with $P_0 = I$. As an example we consider the standard Brownian motion $W(z)$ defined in Section 5.2.1. It is a homogeneous Markov process, a property that comes from the independence of its increments:

$$\mathbb{E}[\phi(W(z+s))|W(z'), z' \leq z] = \mathbb{E}[\phi(W(z+s) - W(z) + W(z))|W(z'), z' \leq z]$$
$$= \mathbb{E}[\phi(W(z+s) - W(z) + W(z))|W(z)]$$
$$= (P_s\phi)(W(z)).$$

Here the notation

$$\mathbb{E}[X|W(z'), z' \leq z]$$

means conditional expectation of X given $W(z')$ for $z' \in [0, z]$, and we used the independence of the increment $W(z+s) - W(z)$ from $\{W(z'), z' \leq z\}$, also generated by increments on the left of z. The Brownian semigroup $(P_s\phi)(x)$ is computed using the Gaussian property of the increments

$$(P_s\phi)(x) = \int \phi(x+w)\frac{1}{\sqrt{2\pi s}}e^{-\frac{w^2}{2s}}\,dw = \int \phi(w)\frac{1}{\sqrt{2\pi s}}e^{-\frac{(w-x)^2}{2s}}\,dw. \quad (6.4)$$

6.1.2 Infinitesimal Generators

Semigroups have symbolically the form of exponentials of operators

$$P_s = e^{s\mathcal{L}}.$$

The operator $\mathcal{L}$ is called the **infinitesimal generator**. It is defined by

$$\mathcal{L} = \frac{dP_s}{ds}\bigg|_{s=0},$$

when the derivative on the right exists. It satisfies

$$\frac{dP_s}{ds} = \mathcal{L}P_s = P_s\mathcal{L}, \quad (6.5)$$

or more precisely,

$$\frac{dP_s\phi(x)}{ds} = \lim_{h\downarrow 0}\frac{P_{s+h}\phi(x) - P_s\phi(x)}{h} = \lim_{h\downarrow 0}\frac{P_h(P_s\phi)(x) - (P_s\phi)(x)}{h}$$
$$= P_s\lim_{h\downarrow 0}\frac{P_h\phi(x) - \phi(x)}{h} = P_s\mathcal{L}\phi(x).$$

These limits are considered for bounded functions ϕ for which they are well defined, that is, for functions ϕ in the **domain** of the infinitesimal generator $\mathcal{L}$. For the Brownian semigroup, we deduce from (6.4) that

$$\mathcal{L}\phi(x) = \lim_{h\downarrow 0}\frac{1}{h}\int (\phi(w) - \phi(x))\frac{1}{\sqrt{2\pi h}}e^{-\frac{(w-x)^2}{2h}}\,dw = \frac{1}{2}\phi''(x),$$

which is obtained with a Taylor expansion of ϕ around x when it is twice differentiable. More generally, in the multidimensional case, a Brownian motion is a vector with components that are independent one-dimensional Brownian motions, and its infinitesimal generator is the second-order partial differential operator $\frac{1}{2}\Delta$, the Laplace operator.

6.1.3 Martingales and Martingale Problems

Let Y be a homogeneous Markovian process with infinitesimal generator $\mathcal{L}$. For functions ϕ in the domain of $\mathcal{L}$ (which are bounded) we define the process

$$M(z) = \phi(Y(z)) - \phi(Y(0)) - \int_0^z \mathcal{L}\phi(Y(s))\,ds. \quad (6.6)$$

It satisfies the **martingale property**

$$\mathbb{E}[M(z+h)|Y(z'), z' \leq z] = M(z) , \qquad (6.7)$$

for all z and $h \geq 0$. This is seen from the following calculation:

$$\mathbb{E}[M(z+h)|Y(z'), z' \leq z]$$

$$= \mathbb{E}\left[M(z) + \phi(Y(z+h)) - \phi(Y(z)) - \int_z^{z+h} \mathcal{L}\phi(Y(s)) \, ds | Y(z'), z' \leq z \right]$$

$$= M(z) + P_h\phi(Y(z)) - \phi(Y(z)) - \int_0^h \frac{d}{ds} P_s\phi(Y(z)) \, ds$$

$$= M(z) ,$$

where we have used the identity

$$P_h\phi(x) - \phi(x) - \int_0^h \frac{d}{ds} P_s\phi(x) \, ds = 0 ,$$

which is simply the fundamental theorem of calculus. We have also used z-homogeneity, the Markov property, and the differentiation property (6.5). The martingale property implies in particular that $\mathbb{E}[M(z)]$ is constant, equal to zero here since $M(0) = 0$; but in fact, this is not enough to ensure the martingale property (6.7).

Conversely, if under a probability measure $\mathbb{P}$, $M(z)$ defined by (6.6) is a martingale for a collection of test functions ϕ that is large enough, then $\mathbb{P}$ characterizes uniquely the probability distribution of a Markov process Y with infinitesimal generator $\mathcal{L}$. This is known as the theory of **martingale problems**, for which we refer to [163]. Martingale formulations are particularly convenient for showing convergence in distribution in the diffusion approximation that we are considering here, as we will see in Section 6.3.

The square of a continuous martingale is not a martingale (we limit here the discussion to continuous martingales), but it can be made into one by subtracting an appropriate integral term, which is called its quadratic variation. We use the notation $\langle M, M \rangle (z)$ and we have that $M^2(z) - \langle M, M \rangle (z)$ is a martingale, so that

$$\mathbb{E}[\langle M, M \rangle (z+h) - \langle M, M \rangle (z)|\mathcal{F}_z] = \mathbb{E}[M^2(z+h) - M^2(z)|\mathcal{F}_z],$$

where $\mathcal{F}_z$ is the σ-algebra (or information) generated by $\{Y(z'), z' \leq z\}$. If ϕ, $\mathcal{L}\phi$, and $\mathcal{L}\phi^2$ are bounded functions, then the quadratic variation of $M(z)$ has the form

$$\langle M, M \rangle (z) = \int_0^z \mathcal{L}\phi^2(Y(s)) - 2\phi(Y(s))\mathcal{L}\phi(Y(s)) \, ds . \qquad (6.8)$$

The derivation of (6.8) is given in Appendix 6.9.1.

Finally, we note that (6.6) can be easily generalized by adding explicit dependence on z for the function ϕ. If the bounded function $\phi(z, x)$ is smooth in z and in the domain of $\mathcal{L}$ with respect to x, then the following process M is a martingale:

$$M(z) = \phi(z, Y(z)) - \phi(0, Y(0)) - \int_0^z \left(\frac{\partial}{\partial z} + \mathcal{L} \right) \phi(s, Y(s)) \, ds. \qquad (6.9)$$

6.1.4 Kolmogorov Backward and Forward Equations

Consider the **backward equation**

$$\frac{\partial u}{\partial z} + \mathcal{L}u = 0, \qquad (6.10)$$
$$u(Z, x) = U(x),$$

solved for $z < Z$ with the **terminal** condition at $z = Z$ given by the function $U(x)$, which we assume to be bounded. The solution $u(z, x)$ has the **probabilistic representation**

$$u(z, x) = \mathbb{E}[U(Y(Z))|Y(z) = x], \qquad (6.11)$$

where Y is the Markov process with infinitesimal generator $\mathcal{L}$. This can be easily understood by using (6.9) with $\phi = u$, assuming that u is sufficiently smooth, and noticing that

$$M(z) = u(z, Y(z)) - u(0, Y(0)),$$

since with this choice of ϕ, the integral in (6.9) is equal to zero. The martingale property of M between z and Z implies that

$$\mathbb{E}[M(Z)|Y(z'), z' \leq z] = M(z),$$

or equivalently

$$\mathbb{E}[u(Z, Y(Z))|Y(z'), z' \leq z] = u(z, Y(z)).$$

We deduce the representation (6.11) using the terminal condition

$$u(Z, Y(Z)) = U(Y(Z))$$

and the Markov property of Y.

Let us now consider the probability distribution $p(z, dx)$ of the process $Y(z)$ with an initial value $Y(0)$ distributed according to the given distribution $p(0, dx) = p_0(dx)$. That is, $p(z, dx)$, for $z \geq 0$, are probability distributions on the state space S on which the process Y takes its values, such that for every bounded function ϕ defined on S,

$$\mathbb{E}[\phi(Y(z))] = \int_S \phi(x)\, p(z, dx)\,.$$

When the state space S is a finite or countable set, then $p(z, dx)$ is a vector with finite or countably many components $p(z, i) = \mathbb{P}(Y(z) = S_i)$ with $p(0, i) = p_0(i) = \mathbb{P}(Y(0) = S_i)$. In a Euclidean space $S = \mathbb{R}^n$ we assume that $p(z, dx) = p(z, x)\, dx$, that is, we assume that the distribution of $Y(z)$ has a density. We derive the forward Kolmogorov equation in the Euclidean case and then give it also for a finite or countable state space.

The backward representation (6.11) integrated with respect to $p(z, x)$ gives

$$
\begin{aligned}
\int_S u(z, x)p(z, x)\, dx &= \int_S \mathbb{E}[u(Z, Y(Z))|Y(z) = x]p(z, x)\, dx \\
&= \mathbb{E}[u(Z, Y(Z)] \\
&= \int_S u(Z, x)p(Z, x)\, dx\,,
\end{aligned}
$$

which shows that the quantity $\int_S u(z, x)p(z, x)\, dx$ is independent of z. Differentiating it with respect to z, using equation (6.10) satisfied by u, and using the **adjoint** operator $\mathcal{L}^\star$ (defined by $\int \phi\mathcal{L}\psi dx = \int \psi\mathcal{L}^\star\phi dx$) gives

$$
\begin{aligned}
0 &= \int_S \frac{\partial u}{\partial z}(z, x)p(z, x)\, dx + \int_S u(z, x)\frac{\partial p}{\partial z}(z, x)\, dx \\
&= -\int_S (\mathcal{L}u)(z, x)p(z, x)\, dx + \int_S u(z, x)\frac{\partial p}{\partial z}(z, x)\, dx \\
&= -\int_S u(z, x)\mathcal{L}^\star p(z, x)\, dx + \int_S u(z, x)\frac{\partial p}{\partial z}(z, x)\, dx \\
&= \int_S u(z, x)\left(\frac{\partial p}{\partial z} - \mathcal{L}^\star p(z, x)\right) dx\,.
\end{aligned}
$$

At $z = Z$ we get

$$\int_S U(x)\left(\frac{\partial p}{\partial z} - \mathcal{L}^\star p(Z, x)\right) dx = 0\,,$$

which holds for all bounded functions U. Consequently, $\dfrac{\partial p}{\partial z} - \mathcal{L}^\star p(Z, x) = 0$ at any $Z > 0$, which means that the probability density $p(z, x)$ is solution of the **forward** equation

$$\frac{\partial p}{\partial z} = \mathcal{L}^\star p\,, \tag{6.12}$$

$$p(0, x) = p_0(x)\,.$$

In the case of a d-dimensional standard Brownian motion, starting from the origin, p solves the **heat** equation

$$\frac{\partial p}{\partial z} = \frac{1}{2}\Delta p \,,$$

with the initial condition $p_0 = \delta_0$, a Dirac delta function at the origin.

When the state space $S = \{S_i, i = 1, 2, \ldots\}$ is finite or countable, then the infinitesimal generator is a matrix $\mathcal{L}_{ij}$. It is defined by

$$\mathcal{L}_{ij} = \lim_{h \to 0} \frac{\mathbb{P}(Y(h) = S_i | Y(0) = S_j) - \delta_{ij}}{h} \,.$$

The forward Kolmogorov equation for $p(z, i) = \mathbb{P}(Y(z) = S_i)$ is a system of ordinary differential equations

$$\frac{dp(z, i)}{dz} = \sum_j \mathcal{L}_{ij}^\star p(z, j) \,, \quad i = 1, 2, 3, \ldots,$$

with the initial condition $p(0, i) = p_0(i)$. Here $\mathcal{L}_{ij}^\star$ is the adjoint of the matrix $\mathcal{L}_{ij}$.

In the Euclidean case, the existence of an **invariant probability density** means that $p(z, x)$ does not depend on z and consequently satisfies

$$\mathcal{L}^\star p = 0 \,. \tag{6.13}$$

In the finite or countable case the invariant probability $p(i)$ satisfies the matrix equation

$$\sum_j \mathcal{L}_{ij}^\star p(j) = 0 \,, \quad i = 1, 2, 3, \ldots.$$

In the next section we will see three examples with three types of infinitesimal generators: matrix, integral, and differential operators.

6.1.5 Ergodicity

The ergodic theorem states that under suitable hypotheses on a process $X(z)$, the z-average of a function of X tends to its ensemble average as $Z \to \infty$:

$$\frac{1}{Z} \int_0^Z \phi(X(z)) \, dz \overset{Z \to \infty}{\longrightarrow} \mathbb{E}[\phi(X(0))] \quad \mathbb{P} \text{ a.s.}$$

The mathematical formulation of hypotheses for the validity of the ergodic theorem is quite cumbersome in a general framework. In the case of a Markov process, under the assumption that the process can visit any neighborhood of the state space S with positive probability, in a finite time and from any starting point (irreducibility), then the existence of an invariant probability distribution ensures that the process is ergodic.

The d-dimensional Brownian motion has an invariant distribution that is the uniform distribution over $\mathbb{R}^d$. The total mass of this distribution is $+\infty$, so that it cannot be normalized to a probability distribution, and so the Brownian motion is not ergodic. However, the three models that we shall present in the next section have invariant probability distributions and they are ergodic.

Ergodicity is related to properties of the null spaces of $\mathcal{L}$ and its adjoint $\mathcal{L}^\star$. Recall that the functions in the domain of the infinitesimal generator $\mathcal{L}$ are bounded. If the process is irreducible, then ergodicity is equivalent to the infinitesimal generator having a null space reduced to the constant functions. In this case, the Markov process has a unique invariant probability distribution p, which belongs to the null space of the adjoint of the infinitesimal generator, and the distribution of $Y(z)$ converges to p as $z \to \infty$ for any initial distribution for $Y(0)$. We refer to [112] for the ergodic properties of Markov processes.

6.2 Markovian Models of Random Media

In this section we present examples of three types of Markovian models for the driving process Y.

- The first model is for a composite material with two components that alternate. The layer sizes are chosen randomly and independently, so that the resulting two-valued process is Markovian. Its infinitesimal generator is a 2×2 matrix. This model can be generalized easily to a Markovian model of a composite with any finite number of components.
- The second example is a **pure jump** Markov process, which can take a continuum of values. Its infinitesimal generator is an integral operator.
- The third example is an **ergodic diffusion** process, the Ornstein–Uhlenbeck process, whose infinitesimal generator is a differential operator.

In each of the three models we look carefully at their ergodic properties and their invariant distributions. We do not assume that these processes are centered, since the centering condition (6.2) applies to a function $F(x, Y(z_0), \tau)$ of this process.

6.2.1 Two-Component Composite Media

Our first example is a process $Y(z)$ taking alternately the values y_1 and y_2 representing two different materials. In this case the real-valued bounded functions ϕ defined on $S = \{y_1, y_2\}$ are simply two-dimensional real vectors with components $\phi_i = \phi(y_i)$ for $i = 1, 2$. The semigroup (P_s) is a family of 2×2 matrices acting on ϕ, which are given, according to (6.3), by

$$P_s = \begin{bmatrix} \mathbb{P}[Y(z+s) = y_1 | Y(z) = y_1] & \mathbb{P}[Y(z+s) = y_2 | Y(z) = y_1] \\ \mathbb{P}[Y(z+s) = y_1 | Y(z) = y_2] & \mathbb{P}[Y(z+s) = y_2 | Y(z) = y_2] \end{bmatrix}.$$

The construction of this Markov process is as follows. In a small interval of medium of length Δz, the process switches to the other material with probability proportional to Δz, say $\lambda \Delta z$, and remains the same with probability

$1 - \lambda \Delta z$. We assume that λ is a positive constant, which means that $Y(z)$ is homogeneous and symmetric in the two materials. With this description the semigroup has the form

$$P_{\Delta z} = \begin{bmatrix} 1 - \lambda \Delta z & \lambda \Delta z \\ \lambda \Delta z & 1 - \lambda \Delta z \end{bmatrix} + o(\Delta z),$$

and the infinitesimal generator is obtained by subtracting the identity matrix, dividing by Δz, and taking a limit as $\Delta z \downarrow 0$:

$$\mathcal{L} = \lambda \begin{bmatrix} -1 & 1 \\ 1 & -1 \end{bmatrix}.$$

The trajectories of this process can be constructed as follows. We introduce the increasing sequence $0 = Z_0 < Z_1 < Z_2 < \cdots < Z_n < \cdots$ of successive random transition points where the medium switches from one material to the other. In this model the layer sizes

$$(Z_1, Z_2 - Z_1, \ldots, Z_n - Z_{n-1}, \ldots)$$

form a sequence of independent random variables with the common **exponential distribution** with parameter λ, that is,

$$\mathbb{P}[Z_n - Z_{n-1} \leq z] = 1 - e^{-\lambda z},$$

for any $n \geq 1$. Equivalently, the sequence (Z_n) forms a **Poisson process** with intensity λ. The number of jumps in an interval $(z_1, z_2]$ is distributed according to the Poisson distribution with parameter $\lambda(z_2 - z_1)$,

$$\mathbb{P}\left[k \text{ jumps in } (z_1, z_2]\right] = e^{-\lambda(z_2 - z_1)} \frac{[\lambda(z_2 - z_1)]^k}{k!},$$

and the numbers of random jumps over disjoint intervals are independent random variables. Exponential distributions are naturally associated with the Markov property because of their **memoryless** property. We refer to [54] for a full treatment of this subject.

The parameter λ is here simply a scaling factor, with $1/\lambda$ the average size of a layer. In the asymptotic analysis the driving process is assumed to be normalized and dimensionless. This is done by taking $\lambda = 1$ here. We note, however, that for the rescaled process $Y(z/\varepsilon^2)$ the mean distance between transitions is ε^2, which is equivalent to setting $\lambda = \varepsilon^{-2}$. We consider now the normalized process $Y(z)$ with dimensionless argument and $\lambda = 1$. Its infinitesimal generator is

$$\mathcal{L} = \begin{bmatrix} -1 & 1 \\ 1 & -1 \end{bmatrix}. \tag{6.14}$$

The infinitesimal generator of the rescaled process $Y(z/\varepsilon^2)$ is simply $(1/\varepsilon^2)\mathcal{L}$, as can be seen from the general computation

$$\lim_{h \to 0} \frac{\mathbb{E}[\phi(Y(h/\varepsilon^2))|Y(0) = y] - \phi(y)}{h} = \frac{1}{\varepsilon^2} \lim_{h' \to 0} \frac{\mathbb{E}[\phi(Y(h'))|Y(0) = y] - \phi(y)}{h'}$$

$$= \frac{1}{\varepsilon^2} \mathcal{L}\phi(y),$$

where the rescaling corresponds to speeding up the transitions of the process Y. Realizations of this two-state random medium are shown in Figure 6.1 for two values for ε.

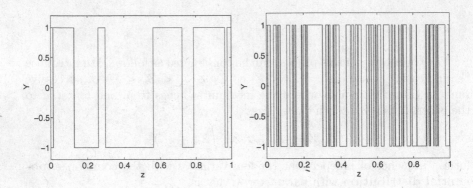

Fig. 6.1. Two realizations of the two-state random medium with $\varepsilon^2 = 0.1$ (left picture) and $\varepsilon^2 = 0.01$ (right picture). Here $y_1 = -1$ and $y_2 = 1$.

The invariant distribution of the two-state Markov process is obtained easily by solving equation (6.13), which is now

$$\mathcal{L}^\star p = \begin{bmatrix} -1 & 1 \\ 1 & -1 \end{bmatrix} \begin{bmatrix} p_1 \\ p_2 \end{bmatrix} = 0.$$

Here $p = (p_1, p_2)$ is a probability on $S = \{y_1, y_2\}$. The unique solution is the uniform distribution $(1/2, 1/2)$ on S, which is intuitively clear from the description of the process. As discussed in Section 6.1.5, the process is ergodic, and moreover, the distribution of $Y(z)$ converges exponentially fast to the uniform distribution as $z \to \infty$.

6.2.2 Multicomponent Composite Media

This example is a generalization of the previous one to a continuous state space S from a two-state space S. In each z interval the process $Y(z)$ takes values that are chosen randomly over an interval $[y_1, y_2]$ with a probability density $\mu(y)$. The layer widths are independent normalized exponential random variables with mean one, as in the previous section. Equivalently, the interfaces between layers are distributed according to a Poisson process with intensity normalized to one. This means that

$$\mathbb{E}[\phi(Y(z))|Y(0) = y] = \phi(y)\mathbb{P}[\text{no jump on } [0, z]]$$

$$+ \left(\int \phi(x)\mu(x)\,dx \right) \mathbb{P}[\text{at least one jump in } [0, z]]$$

$$= e^{-z}\phi(y) + (1 - e^{-z}) \left(\int \phi(x)\mu(x)\,dx \right).$$

Subtracting $\phi(y)$, dividing by z, and taking a limit as $z \downarrow 0$, we deduce that the infinitesimal generator is given by the following integral operator $\mathcal{L}$:

$$\mathcal{L}\phi(y) = \int \left(\phi(x) - \phi(y) \right) \mu(x)\,dx\,,$$

where we have used the fact that the integral of $\mu(x)$ is one, since it is a probability density. Scaling the process Y, that is, replacing it by $Y(z/\varepsilon^2)$, will again have the effect of multiplying $\mathcal{L}$ by the scaling factor $1/\varepsilon^2$.

To compute the adjoint of $\mathcal{L}$ we note that for any probability density $p(y)$,

$$\int p(y)\mathcal{L}\phi(y)\,dy = \int \int p(y) \left(\phi(x) - \phi(y) \right) \mu(x)\,dx\,dy$$

$$= \int \phi(x)\mu(x)\,dx - \int p(y)\phi(y)\,dy$$

$$= \int \phi(x) \left(\mu(x) - p(x) \right) dx$$

$$= \int \phi(x)\mathcal{L}^* p(x)\,dx\,,$$

which identifies the adjoint operator

$$\mathcal{L}^* p = \mu - p\,.$$

The unique solution of equation (6.13) for the invariant distribution is simply $p = \mu$. This is intuitively clear since in this model we distribute Y according to μ and independently layer by layer. Realizations of such a medium are shown in Figure 6.2. This model is similar to the one used in Chapter 4, where the random coefficients are distributed randomly and independently layer by layer but the size of the layers is constant. The model considered in Chapter 4 is, however, not Markovian. This is easily seen by noting that the information in $(Y(z'), z' \le z)$ contains the present value $Y(z)$ but also the position of z within the layer relative to the jump on the left. This in turn fixes the position of the jump on the right since the layers have constant width. By considering jointly the process $Y(z)$ and the position within a layer $\tau(z)$, we obtain the process $(Y(z), \tau(z))$, which is Markovian on $S \times [0, 1)$. Although this Markov process is not irreducible and memorizes its initial conditions, the general methods of Markov processes that we need can be applied. In particular, the invariant distribution is obtained as the product of the distribution of the parameter over S by the uniform distribution over $[0, 1)$. This example is important for the applications in this book because it shows that Markovian models are not as special as they might appear at first.

> *Many non-Markovian models of random media can be made Markovian by enlarging the state space of the driving process.*

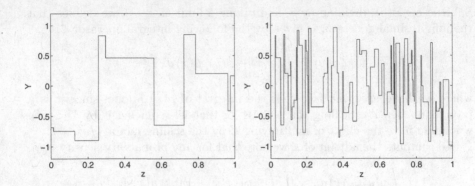

Fig. 6.2. Two realizations of a multicomponent process with $\varepsilon^2 = 0.1$ (left picture) and $\varepsilon^2 = 0.01$ (right picture). Here the invariant distribution is the uniform one over $[-1, 1]$.

6.2.3 A Continuous Random Medium

The third example is different from the other two because the process $Y(z)$ varies continuously with z. One-dimensional media with continuously varying parameters are also layered media even though there are no sharp interfaces separating in them. Wave propagation in such media is analyzed in the same way as in piecewise-constant ones. In particular, the asymptotic analysis can be used with or without sharp interfaces in the random medium because of the scaling regime that we are considering here.

The simplest example of a continuous Markovian and ergodic process $Y(z)$ for modeling the properties of the medium is the Ornstein–Uhlenbeck process. It is constructed by adding white noise to a damped scalar equation

$$dY(z) = -Y(z)\,dz + \sqrt{2}\,dW(z)\,, \ z > 0\,, \tag{6.15}$$

with $Y(0)$ given. Here the level of the white noise dW has been normalized to $\sqrt{2}$ and the damping rate to one, so that the invariant density of the process is a standard Gaussian, as is shown below. Equation (6.15) is an example of a **stochastic differential equation**, which will be discussed in greater detail in Section 6.6.3. However, we do not need the full theory of stochastic differential equations for this example. We can use the usual variation of constants method for linear differential equations to obtain the process

$$Y(z) = Y(0)e^{-z} + \sqrt{2}e^{-z} \int_0^z e^s \, dW(s) \,. \tag{6.16}$$

The integral of the deterministic function e^z with respect to the Brownian motion W, known as a Wiener integral, can be defined by integration by parts,

$$\int_0^z e^s \, dW(s) = e^z W(z) - \int_0^z W(s)e^s \, ds \,,$$

where the integral on the right-hand side is a usual integral of continuous functions. This integral can also be looked at as a random variable obtained as a limit in the mean-square sense (in L^2) of the centered Gaussian random variables

$$\sum_{j=0}^{n-1} e^{z_j} \left(W(z_{j+1}) - W(z_j) \right) \,,$$

where $z_j = (j/n)z$. This implies that the limiting integral is Gaussian, centered, and has a variance given by

$$\mathbb{E}\left[\left(\int_0^z e^s \, dW(s) \right)^2 \right] = \lim_{n \to \infty} \sum_{j=0}^{n-1} e^{2z_j} \left(z_{j+1} - z_j \right) = \int_0^z e^{2s} \, ds \,,$$

where we have used the independence of increments of the Brownian motion W and the variance $\mathbb{E}[(W(z_{j+1}) - W(z_j))^2] = z_{j+1} - z_j$.

For a given initial state $Y(0)$, the continuous process (6.16) is Gaussian with mean $Y(0)e^{-z}$ and variance $1 - e^{-2z}$. Therefore its long-time distribution, as $z \to \infty$, is the standard Gaussian distribution $\mathcal{N}(0,1)$. This is also its invariant distribution, since if $Y(0)$ is $\mathcal{N}(0,1)$-distributed and independent of W, then we get from the sum of the Gaussian random variables in (6.16) that $Y(z)$ is also $\mathcal{N}(0,1)$-distributed for all $z > 0$.

We obtain from (6.16) that

$$Y(z+h) = Y(z)e^{-h} + \sqrt{2}e^{-(z+h)} \int_z^{z+h} e^s \, dW(s) \,,$$

from which we infer the Markov property of the process Y, since $Y(z+h)$ is a function of the present value $Y(z)$ and the increments of W in the "future" $[z, z+h]$. The semigroup of the Ornstein–Uhlenbeck process is given by

$$P_z \phi(x) = \mathbb{E}[\phi(Y(z))|Y(0) = x] = \int \phi(y) \frac{1}{\sqrt{2\pi(1 - e^{-2z})}} e^{-\frac{(y - xe^{-z})^2}{2(1 - e^{-2z})}} \, dy \,,$$

and its infinitesimal generator is obtained by taking the derivative of the semigroup at $z = 0$ leading to

$$\mathcal{L} = -y\frac{\partial}{\partial y} + \frac{\partial^2}{\partial y^2},$$

which is a partial differential operator. The adjoint operator $\mathcal{L}^\star$ is defined by

$$\int \psi(y)\mathcal{L}\phi(y)\,dy = \int \phi(y)\mathcal{L}^\star\psi(y)\,dy.$$

Integration by parts gives

$$\mathcal{L}^\star\psi = \frac{\partial}{\partial y}(y\psi) + \frac{\partial^2\psi}{\partial y^2}.$$

It is easily seen that $\mathcal{L}^\star\mathcal{N} = 0$, where $\mathcal{N}(y) = e^{-y^2/2}/\sqrt{2\pi}$ is the standard Gaussian density. As in the other models, in the asymptotic analysis we speed up the process Y by replacing it with $Y(z/\varepsilon^2)$. Its infinitesimal generator is $(1/\varepsilon^2)\mathcal{L}$, and its invariant density is the standard Gaussian.

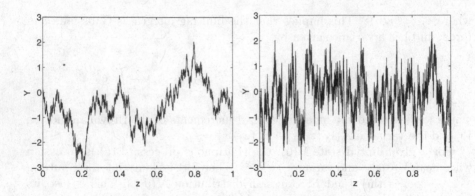

Fig. 6.3. Two realizations of the rescaled Ornstein–Uhlenbeck process $Y(z/\varepsilon^2)$ with $\varepsilon^2 = 0.1$ (left picture) and $\varepsilon^2 = 0.01$ (right picture).

6.3 Diffusion Approximation Without Fast Oscillation

We have just discussed in detail the Markovian models for the random processes driving the ordinary differential equations (6.1) that we want to analyze. Now we return to this problem, and we consider first the simple case in which there is no fast oscillation in the equation for X^ε, that is,

$$\frac{dX^\varepsilon}{dz} = \frac{1}{\varepsilon}F\left(X^\varepsilon(z), Y^\varepsilon(z)\right) + G\left(X^\varepsilon(z), Y^\varepsilon(z)\right), \tag{6.17}$$

where we have used the notation $Y^\varepsilon(z) = Y(z/\varepsilon^2)$. There is no explicit periodic dependence of F and G on z as in (6.1), and that is what we mean by no
fast oscillation. The state space for X^ε is $\mathbb{R}^d$, and Y is a Markov process with
state space S. The functions $F(x,y)$ and $G(x,y)$ are defined on $\mathbb{R}^d \times S$ and
take values in $\mathbb{R}^d$. We assume that F and G are smooth and at most linearly
growing in x. We also assume that F is centered, that is,

$$\mathbb{E}[F(x, Y(z))] = 0 \,, \tag{6.18}$$

for every x, where the expectation is taken with respect to the invariant
distribution of Y. In many applications the function F is of the form $F(x,y) = \tilde{F}(x,y) - \mathbb{E}[\tilde{F}(x, Y(z))]$, which ensures the important **centering condition**.

We want to analyze the asymptotic behavior of the process $X^\varepsilon(z)$ as $\varepsilon \to 0$
in the sense of convergence of distributions, and in particular, to characterize
its limit process $X(z)$. We will use for this purpose the perturbed-test-function
method, which identifies the generator of the limit process. We have to use a
family of perturbed or modified test functions that provide correctors for the
rapid fluctuations.

6.3.1 Markov Property

We note first that the process X^ε is not Markovian by itself. This is because
the driving process $Y^\varepsilon(z)$ is also needed to determine the increments dX^ε.
Because of the assumed Markov property of Y^ε, it is therefore clear that the
pair $(X^\varepsilon, Y^\varepsilon)$ is Markovian.

Consider a test function $\phi(x,y)$ on $\mathbb{R}^d \times S$ that is smooth in x and in the
domain of the infinitesimal generator $\mathcal{L}_Y$ of the Markov process Y in the y
variable. For $\varepsilon > 0$ fixed we then have

$$\frac{\mathbb{E}\left[\phi(X^\varepsilon(h), Y^\varepsilon(h)) \mid X^\varepsilon(0) = x, Y^\varepsilon(0) = y\right] - \phi(x,y)}{h}$$

$$= \mathbb{E}\left[\frac{\phi(X^\varepsilon(h), Y^\varepsilon(h)) - \phi(x, Y^\varepsilon(h))}{h} \,\Big|\, X^\varepsilon(0) = x, Y^\varepsilon(0) = y\right]$$

$$+ \frac{\mathbb{E}\left[\phi(x, Y^\varepsilon(h)) \mid Y^\varepsilon(0) = y\right] - \phi(x,y)}{h}$$

$$= \mathbb{E}\left[\left(\frac{X^\varepsilon(h) - x}{h}\right) \nabla_x \phi(x, Y^\varepsilon(h)) \mid X^\varepsilon(0) = x, Y^\varepsilon(0) = y\right]$$

$$+ \frac{\mathbb{E}\left[\phi(x, Y^\varepsilon(h)) \mid Y^\varepsilon(0) = y\right] - \phi(x,y)}{h} + \mathcal{O}(h) \,,$$

where we have expanded the smooth function ϕ in the variable x. Taking a
limit as $h \downarrow 0$ we obtain the infinitesimal generator of the Markovian pair
$(X^\varepsilon, Y^\varepsilon)$, which we denote by $\mathcal{L}^\varepsilon$:

$$\mathcal{L}^\varepsilon \phi(x,y) = \frac{1}{\varepsilon^2} \mathcal{L}_Y \phi(x,y) + \frac{1}{\varepsilon} F(x,y) \cdot \nabla_x \phi(x,y) + G(x,y) \cdot \nabla_x \phi(x). \tag{6.19}$$

Here we have used the differential equation (6.17) for X^ε, the assumed right continuity of Y^ε, the definition of the infinitesimal generator $\mathcal{L}_Y$ of the Markov process Y, and the definition of the scaled process $Y^\varepsilon(z) = Y(z/\varepsilon^2)$.

6.3.2 Perturbed Test Functions

Since we want to get the asymptotic behavior of the process X^ε, it is natural to consider test functions that depend only on the variable x. Applying the infinitesimal generator $\mathcal{L}^\varepsilon$, given by (6.19), to such test functions gives

$$\mathcal{L}^\varepsilon \phi(x) = \frac{1}{\varepsilon} F(x,y) \cdot \nabla_x \phi(x) + G(x,y) \cdot \nabla_x \phi(x) \,, \qquad (6.20)$$

since $\phi(x)$ is a constant in the y variable and $\mathcal{L}1 = 0$ for any infinitesimal generator, so that $\mathcal{L}_Y \phi(x) = 0$. The fact that $\mathcal{L}1 = 0$ is a direct consequence of $P_s 1 = 1$ for general semigroups defined by (6.3). With this choice of test functions the quantity $\mathcal{L}^\varepsilon \phi(x)$ in (6.20) contains diverging terms in ε. However, they can be canceled approximately by **correcting** the test function $\phi(x)$ by adding a small term of the form $\varepsilon \phi_1(x,y)$. To see this, we write

$$\mathcal{L}^\varepsilon \left(\phi + \varepsilon \phi_1 \right)(x,y) = \frac{1}{\varepsilon} \left\{ F(x,y) \cdot \nabla_x \phi(x) + \mathcal{L}_Y \phi_1(x,y) \right\}$$
$$+ G(x,y) \cdot \nabla_x \phi(x) + F(x,y) \cdot \nabla_x \phi_1(x,y) + O(\varepsilon) \,, (6.21)$$

from which we see that we should choose the correction ϕ_1 so that the diverging term in $1/\varepsilon$ vanishes. In order to accomplish this we must solve the equation for ϕ_1:

$$\mathcal{L}_Y \phi_1 + F \cdot \nabla_x \phi = 0 \,. \qquad (6.22)$$

For each fixed $x \in \mathbb{R}^d$ this is a **Poisson equation** with respect to $\mathcal{L}_Y$ and the variable y. This important equation is discussed in the next section. The centering condition (6.18),

$$\mathbb{E}[F(x,Y(z)) \cdot \nabla_x \phi(x)] = \mathbb{E}[F(x,Y(z))] \cdot \nabla_x \phi(x) = 0 \,,$$

is an essential part of the solvability theory for (6.22).

6.3.3 The Poisson Equation and the Fredholm Alternative

We assume from now on that Y is an ergodic Markov process and that its infinitesimal generator satisfies the Fredholm alternative, which we will introduce and discuss in this section. The models presented in Section 6.2 belong to this class.

   We want to find a solvability condition on the function $f(y)$ for the Poisson equation

$$\mathcal{L}_Y u(y) = f(y) \,, \qquad (6.23)$$

and characterize its solutions $u(y)$. We note first that for any function $v(y)$ in the domain of the generator,

$$\mathbb{E}[\mathcal{L}_Y v(Y(z))] = 0 \,,$$

where the expectation is with respect to the invariant distribution of Y. This is just the weak form of the equation $\mathcal{L}_Y^* p = 0$ for the invariant distribution. When S is a subset of $\mathbb{R}^n$ and p has a density, then we have

$$\mathbb{E}[\mathcal{L}_Y v(Y(z))] = \int_S p(y)\mathcal{L}_Y v(y) \, dy = \int_S v(y)\mathcal{L}_Y^* p(y) \, dy = 0 \,;$$

for all test functions $v(y)$ in the domain of $\mathcal{L}_Y$. We conclude that if the Poisson equation has a solution $u(y)$, then the inhomogeneous term $f(y)$ must have mean zero with respect to the invariant distribution

$$\mathbb{E}[f(Y(z))] = 0 \,.$$

In the Euclidean case this has the form

$$\int_S f(y)p(y) \, dy = 0 \,.$$

This means that the centering condition on f is a necessary condition for the solvability of the Poisson equation. However, the existence of a unique invariant probability distribution for Y and the centering condition for f are not sufficient for the Poisson equation to have a solution. If it has a solution $u(y)$, then it is clearly not unique, since $u(y) + C$ is also a solution for any constant C.

The infinitesimal generator $\mathcal{L}_Y$ is not invertible, since it has the nontrivial function $u(y) = 1$ in its null space, that is, $\mathcal{L}_Y 1 = 0$. The Markov process Y is ergodic, so that the null space of $\mathcal{L}_Y$ is reduced to the constant functions, as explained in Section 6.1.5. Moreover, the one-dimensional space Null_1 spanned by the invariant probability distribution p is contained in the null space of $\mathcal{L}_Y^*$. The Fredholm alternative is the statement that the Poisson equation (6.23) admits a solution if f satisfies the orthogonality condition $f \perp \mathrm{Null}_1$, which is here the centering condition on f, that is, $\mathbb{E}[f(Y(0))] = 0$. In this case, a particular solution of the Poisson equation $\mathcal{L}_Y u = f$ is given by

$$u_0(y) = -\int_0^\infty P_s f(y) \, ds \,, \tag{6.24}$$

assuming that the integral exists, since we have formally

$$\begin{aligned} \mathcal{L}_Y u_0(y) &= -\int_0^\infty \mathcal{L}_Y e^{s\mathcal{L}_Y} f(y) \, ds \\ &= -\left[e^{s\mathcal{L}_Y} f(y)\right]_0^\infty = f(y) - \mathbb{E}[f(Y(0))] = f(y) \,. \end{aligned}$$

Another bounded solution u_1 of the Poisson equation is such that the difference $\tilde{u}_1 = u_1 - u_0$ belongs to the null space of the infinitesimal generator $\mathcal{L}_Y$. However, $\text{Null}(\mathcal{L}_Y)$ is generated by 1, so $\tilde{u}_1$ is a constant. The particular solution (6.24) satisfies $\mathbb{E}[u_0(Y(0))] = 0$, since $\mathbb{E}[f(Y(s))] = 0$, so that u_0 is the unique solution with zero mean of the Poisson equation.

The question that needs to be addressed now is under what conditions on the state space S and on the Markov process Y, beyond ergodicity, does the Fredholm alternative hold for the generator $\mathcal{L}_Y$? This is a difficult question that does not have a general or a simple answer. It depends, for example, on what class of solutions to the Poisson equation we wish to consider for a given class of centered inhomogeneous terms f. We will discuss here two particular cases in which a complete analysis can be given in an elementary way.

The first case is that in which the space S is a finite set and the transition probabilities of the Markov process Y are positive. In this case the infinitesimal generator is a matrix, the Poisson equation is a linear system of equations, and the Fredholm alternative is the well-known one from linear algebra. This finite dimensional case extends easily to generators that are integral operators on a compact set, like the one we considered in Section 6.2.2.

The second case is that of an ergodic diffusion process in a compact submanifold of a Euclidean space without a boundary, such as the sphere or the torus, in which case the infinitesimal generator is an elliptic second-order partial differential operator. If it is a uniformly elliptic operator, then the Fredholm alternative holds for any function f that is bounded and centered and with a solution to the Poisson equation that is also bounded. This follows from the basic theory of second-order elliptic partial differential equations. Brownian motion with reflection in a bounded region of $\mathbb{R}^n$ is not in this class but does have the Fredholm alternative property. Its infinitesimal generator is the Laplacian with Neumann boundary conditions. The Ornstein–Uhlenbeck process of Section 6.2.3 does not fall in this class either, since its state space is not compact. However, the simple form of its generator allows for a full treatment of the validity of the Fredholm alternative.

6.3.4 Limiting Infinitesimal Generator

After this discussion on the Fredholm alternative we now return to the analysis of the limiting problem. The process $(X^\varepsilon, Y^\varepsilon)$ is Markov with generator $\mathcal{L}^\varepsilon$ given by (6.19). This implies that for any smooth and bounded test function ϕ, as above, the process

$$\phi(X^\varepsilon(z), Y^\varepsilon(z)) - \int_0^z \mathcal{L}^\varepsilon \phi(X^\varepsilon(u), Y^\varepsilon(u))\, du$$

is a martingale. Convergence in distribution of X^ε is obtained by showing that the martingale problem associated with $\mathcal{L}^\varepsilon$ converges to the martingale problem associated with the limiting generator $\mathcal{L}$. It is based on the perturbed-test-function method.

Step 1. Perturbed-test-function method.

For any smooth test function ϕ and any $\varepsilon > 0$ there exists a test function ϕ^ε and a generator $\mathcal{L}$ such that

$$\sup_{x\in K, y\in S} |\phi^\varepsilon(x,y)-\phi(x)| \xrightarrow{\varepsilon\to 0} 0, \qquad \sup_{x\in K, y\in S} |\mathcal{L}^\varepsilon\phi^\varepsilon(x,y)-\mathcal{L}\phi(x)| \xrightarrow{\varepsilon\to 0} 0, \quad (6.25)$$

for any compact subset K of $\mathbb{R}^d$.

We look for a perturbed test function ϕ^ε of the form

$$\phi^\varepsilon(x,y) = \phi(x) + \varepsilon\phi_1(x,y) + \varepsilon^2\phi_2(x,y). \qquad (6.26)$$

Applying $\mathcal{L}^\varepsilon$ to this ϕ^ε we get

$$\begin{aligned}\mathcal{L}^\varepsilon\phi^\varepsilon &= \frac{1}{\varepsilon}\left(\mathcal{L}_Y\phi_1(x,y) + F(x,y)\cdot\nabla_x\phi(x)\right) \\ &\quad + \left(\mathcal{L}_Y\phi_2(x,y) + F(x,y)\cdot\nabla_x\phi_1(x,y) + G(x,y)\cdot\nabla_x\phi(x)\right) \\ &\quad + O(\varepsilon).\end{aligned} \qquad (6.27)$$

We define the first corrector ϕ_1 to cancel the ε^{-1} term (6.27). This gives a Poisson equation for ϕ_1 as a function of $y\in S$ with $x\in\mathbb{R}^d$ a parameter:

$$\mathcal{L}_Y\phi_1(x,y) + F(x,y)\cdot\nabla_x\phi(x) = 0.$$

Since $F(x,y)$ is centered by hypothesis (6.18) and we assume that the Fredholm alternative holds for $\mathcal{L}_Y$, the Poisson equation has a solution bounded in y and smooth in x. It has the representation (6.24), which is here

$$\phi_1(x,y) = \int_0^\infty \mathbb{E}[F(x,Y(z))\cdot\nabla_x\phi(x) \mid Y(0)=y]\,dz.$$

In all examples considered here, the convergence of the integral is actually exponential.

We cannot define the second corrector ϕ_2 by canceling the order-one terms in (6.27) because that would require solving a Poisson equation with an inhomogeneous term that is not centered. To center this term we subtract its mean relative to the invariant distribution of Y. This gives the Poisson equation

$$\begin{aligned}\mathcal{L}_Y\phi_2(x,y) &+ F(x,y)\cdot\nabla_x\phi_1(x,y) + G(x,y)\cdot\nabla_x\phi(x) \\ &- \mathbb{E}[F(x,Y(0))\cdot\nabla_x\phi_1(x,Y(0)) + G(x,Y(0))\cdot\nabla_x\phi(x)] = 0.\end{aligned}$$

As with the Poisson equation for ϕ_1, this equation has a solution that is bounded in y and smooth in x. It therefore follows that

$$\mathcal{L}^\varepsilon\phi^\varepsilon = \mathbb{E}[F(x,Y(0))\cdot\nabla_x\phi_1(x,Y(0))] + \mathbb{E}[G(x,Y(0))\cdot\nabla_x\phi(x)] + O(\varepsilon).$$

Both parts of (6.25) will now be satisfied if we define the limit generator $\mathcal{L}$ by

$$\mathcal{L}\phi(x) = \int_0^\infty \mathbb{E}\left[F(x, Y(0)) \cdot \nabla_x \left(F(x, Y(z)) \cdot \nabla_x \phi(x)\right)\right] dz$$
$$+ \mathbb{E}[G(x, Y(0)) \cdot \nabla_x \phi(x)], \tag{6.28}$$

where the expectation is relative to the invariant distribution of Y.

The limit operator (6.28) is a second-order elliptic partial differential operator, a diffusion operator of the form

$$\mathcal{L} = \frac{1}{2} \sum_{i,j=1}^d a_{ij}(x) \frac{\partial^2}{\partial x_i \partial x_j} + \sum_{i=1}^d b_i(x) \frac{\partial}{\partial x_i},$$

with

$$a_{ij}(x) = 2 \int_0^\infty \mathbb{E}\left[F_i(x, Y(0)) F_j(x, Y(z))\right] dz \tag{6.29}$$

and

$$b_i(x) = \sum_{j=1}^d \int_0^\infty \mathbb{E}\left[F_j(x, Y(0)) \frac{\partial F_i}{\partial x_j}(x, Y(z))\right] dz + \mathbb{E}[G_i(x, Y(0))]. \tag{6.30}$$

The ellipticity is seen by noting that only the symmetric part of the diffusion coefficients $\{a_{ij}(x)\}$ enters into the operator. The symmetric part of these coefficients can be written in the form, also denoted by $a_{ij}(x)$,

$$a_{ij}(x) = \int_{-\infty}^\infty \mathbb{E}\left[F_i(x, Y(0)) F_j(x, Y(z))\right] dz.$$

The ellipticity condition is

$$\sum_{i,j=1}^d a_{ij}(x)\xi_i\xi_j \geq 0,$$

for any vector $\xi = (\xi_1, \ldots, \xi_d)$. Using the symmetric form of the a_{ij} in this ellipticity condition, we have

$$\sum_{i,j=1}^d a_{ij}(x)\xi_i\xi_j = \int_{-\infty}^\infty \mathbb{E}\left[Y_{\xi,x}(0)Y_{\xi,x}(z)\right] dz \geq 0,$$

where we have introduced the zero-mean stationary process $Y_{\xi,x}(z) = \xi \cdot F(x, Y(z))$ with $x \in \mathbb{R}^d$ and $\xi \in \mathbb{R}^d$ fixed. The positivity condition follows by noting that the right side is the integral of the autocorrelation function of $Y_{\xi,x}$, which equals the power spectral density of $Y_{\xi,x}$,

$$\int_{-\infty}^\infty e^{ikz} \mathbb{E}\left[Y_{\xi,x}(0)Y_{\xi,x}(z)\right] dz,$$

at zero frequency, $k = 0$. It is therefore nonnegative by Bochner's theorem. We discuss this further in Section 6.3.6.

Step 2. Convergence of martingale problems. The martingale property (6.6) of the process $(X^\varepsilon, Y^\varepsilon)$ applied to the perturbed test function ϕ^ε, (6.26), that we have just constructed gives

$$
\mathbb{E}\left[\; \phi^\varepsilon(X^\varepsilon(z'), Y^\varepsilon(z')) - \phi^\varepsilon(X^\varepsilon(z), Y^\varepsilon(z)) \right.
$$
$$
\left. - \int_z^{z'} \mathcal{L}^\varepsilon \phi^\varepsilon(X^\varepsilon(s), Y^\varepsilon(s))\, ds \; \middle| \; (X^\varepsilon(z''), Y^\varepsilon(z'')), z'' \le z \right] = 0 \,,
$$

for all $0 \le z \le z'$. Since we are interested only in the limit of the process X^ε, we use this identity only with conditioning with respect to X^ε. The properties of conditional expectations allow us to write the reduced martingale property in the form

$$
\mathbb{E}\left[\left(\; \phi^\varepsilon(X^\varepsilon(z'), Y^\varepsilon(z')) - \phi^\varepsilon(X^\varepsilon(z), Y^\varepsilon(z)) \right.\right.
$$
$$
\left.\left. - \int_z^{z'} \mathcal{L}^\varepsilon \phi^\varepsilon(X^\varepsilon(s), Y^\varepsilon(s))\, ds \; \right) h_1(X^\varepsilon(z_1)) \cdots h_m(X^\varepsilon(z_m)) \right] = 0 \,,
$$

for any bounded and continuous functions $h_1, h_2, \ldots, h_m$ and $0 \le z_1 < z_2 < \cdots < z_m \le z \le z'$. Using the approximation properties of ϕ^ε and $\mathcal{L}^\varepsilon \phi^\varepsilon$ given by (6.25), we have

$$
\mathbb{E}\left[\left(\; \phi(X^\varepsilon(z')) - \phi(X^\varepsilon(z)) \right.\right.
$$
$$
\left.\left. - \int_z^{z'} \mathcal{L}\phi(X^\varepsilon(s))\, ds \; \right) h_1(X^\varepsilon(z_1)) \cdots h_m(X^\varepsilon(z_m)) \right] = O(\varepsilon), \quad (6.31)
$$

with many terms incorporated in the $O(\varepsilon)$ term on the right. If the right side $F(x, y)$ in the differential equation (6.17) is bounded along with its x derivatives, then the correctors ϕ_1 and ϕ_2 will also be bounded along with their x derivatives. In this case the $O(\varepsilon)$ estimate on the right side is elementary and (6.25) holds with $K = \mathbb{R}^d$. When $F(x, y)$ has at most linear growth in x and is smooth, then similar properties hold for the correctors, and ε-independent moment estimates are needed for X^ε to get the $O(\varepsilon)$ estimate. We refer to the specialized literature for this in the notes at the end of the chapter.

The next step is to pass to the limit $\varepsilon \to 0$ in (6.31). Only the probability distribution of X^ε depends on ε on the left in (6.31). The functional of the trajectory of the process whose expectation is taken on the left is fixed and independent of ε. If we know that the distribution of X^ε is convergent, or that it has a convergent subsequence, then we can pass to the limit in (6.31) and obtain

$$
\mathbb{E}\left[\left(\; \phi(X(z')) - \phi(X(z)) - \int_z^{z'} \mathcal{L}\phi(X(s))\, ds \right) h_1(X(z_1)) \cdots h_m(X(z_m)) \right] = 0 \,,
$$

for any bounded and continuous test functions $h_1, h_2, \ldots, h_m$ and any $0 \leq z_1 < z_2 < \cdots < z_m \leq z \leq z'$. The expectation here is with respect to any limit distribution for X^ε. By the properties of conditional expectation, this implies that the functional of the trajectory

$$\phi(X(z)) - \phi(X(0)) - \int_0^z \mathcal{L}\phi(X(s))\, ds, \quad z \geq 0, \tag{6.32}$$

is a martingale for any limit distribution of X^ε and for any smooth and bounded test function $\phi(x)$. If for the limit diffusion generator (6.28) the distribution for which this functional is a martingale is unique, then we know that the distribution of X^ε must converge weakly to this distribution. This is the general way in which the process X^ε defined by (6.17) is shown to converge to the diffusion process whose generator $\mathcal{L}$ is given by (6.28).

There are two items that need to be addressed in order to complete the above step-by-step asymptotic analysis of $X^\varepsilon(z)$. The first is showing that the probability distributions of $X^\varepsilon(z)$ have convergent subsequences, and the second is showing that the martingale problem for the limit generator $\mathcal{L}$ defines the distribution of a limit process $X(z)$ uniquely. We address the existence of convergent subsequences for X^ε in the next section.

The uniqueness of the probability distribution of the process for which (6.32) is a martingale for any test function $\phi(x)$, with the generator $\mathcal{L}$ given by (6.28), depends on the properties of this elliptic operator. If the diffusion and drift coefficients given by (6.29) and (6.30) are smooth and bounded, then this martingale problem has a unique solution. Uniform ellipticity is not necessary. If the diffusion coefficients (6.29) grow at most quadratically in x and are smooth, and the drift coefficients (6.30) grow at most linearly in x and are smooth, then again the martingale problem defines the distribution uniquely.

We summarize the convergence analysis of this section in the following theorem.

Theorem 6.1. *Let $X^\varepsilon(z)$ for $z \geq 0$ be the process in $\mathbb{R}^d$ defined by the random differential equation*

$$\frac{dX^\varepsilon}{dz}(z) = \frac{1}{\varepsilon} F\left(X^\varepsilon(z), Y\left(\frac{z}{\varepsilon^2}\right)\right) + G\left(X^\varepsilon(z), Y\left(\frac{z}{\varepsilon^2}\right)\right),$$

starting from $X^\varepsilon(0) = x_0 \in \mathbb{R}^d$. Assume that $Y(z)$ is a z-homogeneous Markov ergodic process on a state space S with generator $\mathcal{L}_Y$ satisfying the Fredholm alternative, and the $\mathbb{R}^d$-valued function F satisfies the centering condition $\mathbb{E}[F(x, Y(0))] = 0$, where $\mathbb{E}[\cdot]$ denotes expectation with respect to the invariant probability distribution of $Y(z)$. Assume also that $F(x, y)$ and $G(x, y)$ are at most linearly growing and smooth in x. Then the random processes $X^\varepsilon(z)$ converge in distribution to the Markov diffusion process $X(z)$ with generator

$$\mathcal{L}\phi(x) = \int_0^\infty \mathbb{E}\left[F(x, Y(0)) \cdot \nabla_x \left(F(x, Y(z)) \cdot \nabla_x \phi(x)\right)\right] dz$$
$$+ \mathbb{E}\left[G(x, Y(0)) \cdot \nabla_x \phi(x)\right]. \tag{6.33}$$

The infinitesimal generator $\mathcal{L}$ has the form

$$\mathcal{L} = \frac{1}{2} \sum_{i,j=1}^d a_{ij}(x) \frac{\partial^2}{\partial x_i \partial x_j} + \sum_{i=1}^d b_i(x) \frac{\partial}{\partial x_i},$$

with

$$a_{ij}(x) = 2 \int_0^\infty \mathbb{E}\left[F_i(x, Y(0)) F_j(x, Y(z))\right] dz,$$

$$b_i(x) = \sum_{j=1}^d \int_0^\infty \mathbb{E}\left[F_j(x, Y(0)) \frac{\partial F_i}{\partial x_j}(x, Y(z))\right] dz + \mathbb{E}[G_i(x, Y(0))].$$

The symmetric part of $a_{ij}(x)$ is nonnegative definite, as is shown in the discussion following (6.29) and (6.30).

Theorem 6.1 can be readily extended to include a small perturbation on the right side of the random differential equation. We assume that the process X^ε in $\mathbb{R}^d$ is defined as the solution of the random differential equation

$$\frac{dX^\varepsilon}{dz}(z) = \frac{1}{\varepsilon} F\left(X^\varepsilon(z), Y\left(\frac{z}{\varepsilon^2}\right)\right) + G\left(X^\varepsilon(z), Y\left(\frac{z}{\varepsilon^2}\right)\right) + R^\varepsilon\left(X^\varepsilon(z), Y\left(\frac{z}{\varepsilon^2}\right)\right),$$

where the differentiable in x vector function $R^\varepsilon(x, y)$ satisfies

$$\sup_{x \in K, y \in S} |R^\varepsilon(x, y)| \xrightarrow{\varepsilon \to 0} 0,$$

for any compact set $K \subset \mathbb{R}^d$. Then the random processes X^ε converge in distribution to the diffusion process with the same generator (6.33). This can be established by applying the perturbed-test-function method with the same family of perturbed test functions as in the case $R^\varepsilon = 0$. This is because they satisfy the key properties (6.25) even in the presence of R^ε.

6.3.5 Relative Compactness of the Laws of the Processes

In this section we introduce and discuss a method that can be used to obtain the relative compactness, or tightness, of the laws of the processes $\{X^\varepsilon(z), \ 0 \le z \le Z\}$, $\varepsilon > 0$, and therefore the existence of convergent subsequences. A sufficient condition for the existence of convergent subsequences of the distributions of $(X^\varepsilon)_{\varepsilon > 0}$, in the space $\mathcal{C}$ of continuous paths, is the Kolmogorov moment estimate

$$\mathbb{E}[|X^\varepsilon(z) - X^\varepsilon(z')|^4] \le C(z - z')^2, \tag{6.34}$$

for all $0 \leq z' \leq z \leq Z$, and with C a constant independent of ε. We do not, in fact, have an estimate of this form for the family of processes $(X^\varepsilon)_{\varepsilon>0}$, but we do have it for a family of processes $(\tilde{X}^\varepsilon)_{\varepsilon>0}$, defined below, which is uniformly close to $(X^\varepsilon)_{\varepsilon>0}$, in probability

$$\limsup_{\varepsilon \to 0} \mathbb{P}\left(\sup_{0 \leq z \leq Z} |X^\varepsilon(z) - \tilde{X}^\varepsilon(z)| > \delta\right) = 0, \tag{6.35}$$

for all $\delta > 0$. This is enough for the existence of convergent subsequences for the laws of X^ε.

We obtain moment estimates uniform in ε using the perturbed-test-function method. Using vector notation, let $\phi(x) = x$ and let the associated perturbed test function $\phi^\varepsilon(x, y) = x + \varepsilon\phi_1(x, y)$, where $\phi_1(x, y)$ is a solution of the Poisson equation (6.22): $\mathcal{L}_Y\phi_1(x, y) = -F(x, y)$. In this equation, the coordinate vector x plays the role of a frozen parameter, and ϕ_1 inherits the boundedness and smoothness properties of F with respect to x. If F is bounded, then the function ϕ_1 is bounded. If F has linear growth in x, then ϕ_1 also has linear growth. We assume in the following that F has bounded x-derivatives and linear growth in x, uniformly in y.

In order to check the Kolmogorov criterion (6.34), we first need estimates uniform in ε for moments of $X^\varepsilon(z)$. They can be obtained using perturbed test functions ϕ^ε of the above form $\phi^\varepsilon(x, y) = x + \varepsilon\phi_1(x, y)$ and with the vector-valued martingale M^ε defined by

$$M^\varepsilon(z) = \phi^\varepsilon(X^\varepsilon(z), Y^\varepsilon(z)) - \phi^\varepsilon(x_0, Y^\varepsilon(0)) - \int_0^z \mathcal{L}^\varepsilon\phi^\varepsilon(X^\varepsilon(s), Y^\varepsilon(s)) \, ds.$$

We can now represent $X^\varepsilon(z)$ in the form

$$X^\varepsilon(z) = x_0 - \varepsilon(\phi_1(X^\varepsilon(z), Y^\varepsilon(z)) - \phi_1(x_0, Y^\varepsilon(0)))$$
$$+ \int_0^z \mathcal{L}^\varepsilon\phi^\varepsilon(X^\varepsilon(s), Y^\varepsilon(s)) \, ds + M^\varepsilon(z).$$

Here x_0 is the starting point $X^\varepsilon(0) = x_0$. The functions $\phi_1(x, y)$ and $\mathcal{L}^\varepsilon\phi^\varepsilon(x, y) = F \cdot \nabla_x\phi_1(x, y)$ have linear growths in x uniformly in y. We have therefore the inequality, in $0 \leq z \leq Z$,

$$|X^\varepsilon(z)| \leq |x_0| + \varepsilon c_1(1 + |X^\varepsilon(z)| + |x_0|) + c_2 \int_0^z (1 + |X^\varepsilon(s)|) \, ds + \sup_{0 \leq s \leq Z} |M^\varepsilon(s)|.$$

Here c_1 and c_2 are constants independent of ε. We can rewrite this in the form

$$(1 - c_1\varepsilon)|X^\varepsilon(z)| \leq c_Z \left(1 + |x_0| + \sup_{0 \leq s \leq Z} |M^\varepsilon(s)|\right) + c_2 \int_0^z |X^\varepsilon(s)| \, ds,$$

where c_Z is a constant depending on Z but not on ε. For $\varepsilon < 1/(2c_1)$ we can apply Gronwall's lemma to obtain

$$\sup_{z \in [0,Z]} |X^\varepsilon(z)| \le C_Z \left[1 + |x_0| + \sup_{z \in [0,Z]} |M^\varepsilon(z)| \right], \qquad (6.36)$$

where C_Z is another constant that depends on Z but not on ε. As we saw in Section 6.1.3, the quadratic variation of the martingale M^ε is given by

$$\langle M^\varepsilon, M^\varepsilon \rangle (z) = \int_0^z (\mathcal{L}^\varepsilon \phi^{\varepsilon 2} - 2\phi^\varepsilon \mathcal{L}^\varepsilon \phi^\varepsilon)(X^\varepsilon(s), Y^\varepsilon(s)) \, ds.$$

Using the expressions $\phi^\varepsilon(x,y) = x + \varepsilon \phi_1(x,y)$ and $\mathcal{L}^\varepsilon = \frac{1}{\varepsilon^2} \mathcal{L}_Y + \frac{1}{\varepsilon} F \cdot \nabla_x$, we can verify that

$$\mathcal{L}^\varepsilon \phi^{\varepsilon 2} - 2\phi^\varepsilon \mathcal{L}^\varepsilon \phi^\varepsilon = \mathcal{L}_Y \phi_1^2 - 2\phi_1 \mathcal{L}_Y \phi_1,$$

which shows that this term is independent of ε and has quadratic growth in x. Using Doob's inequality, which holds for any zero-mean martingale, we have for any $p \ge 2$,

$$\mathbb{E} \left[\sup_{z \in [0,Z]} |M^\varepsilon(z)|^p \right] \le c_p \mathbb{E} \left[\langle M^\varepsilon, M^\varepsilon \rangle (Z)^{p/2} \right] \le C_p \int_0^Z 1 + \mathbb{E} \left[|X^\varepsilon(z)|^p \right] dz.$$

Substituting into the pth power of (6.36) and using once again Gronwall's lemma, we obtain the ε-independent moment estimate

$$\mathbb{E} \left[\sup_{z \in [0,Z]} |X^\varepsilon(z)|^p \right] \le C_{p,Z}(1 + |x_0|^p). \qquad (6.37)$$

Let the process $\tilde{X}^\varepsilon(z)$ be defined by

$$\tilde{X}^\varepsilon(z) = x_0 + \int_0^z \mathcal{L}^\varepsilon \phi^\varepsilon(X^\varepsilon(s), Y^\varepsilon(s)) \, ds + M^\varepsilon(z). \qquad (6.38)$$

Then

$$X^\varepsilon(z) - \tilde{X}^\varepsilon(z) = -\varepsilon(\phi_1(X^\varepsilon(z), Y^\varepsilon(z)) - \phi_1(x_0, Y^\varepsilon(0))).$$

We deduce from the ε-independent moment estimate (6.37) and the linear growth of $\phi_1(x,y)$ in x, uniformly in y, that

$$\mathbb{E} \left[\sup_{0 \le z \le Z} |X^\varepsilon(z) - \tilde{X}^\varepsilon(z)| \right] \le C\varepsilon,$$

where C is a constant that depends on Z but not on ε. This implies the uniform-in-probability closeness (6.35) of the two processes X^ε and $\tilde{X}^\varepsilon$. So it is enough to show the validity of the Kolmogorov condition (6.34) for the process $\tilde{X}^\varepsilon$.

We now consider the increments of $\tilde{X}^\varepsilon$, and from (6.38) we have

$$\tilde{X}^{\varepsilon}(z) - \tilde{X}^{\varepsilon}(z') = \int_{z'}^{z} \mathcal{L}^{\varepsilon} \phi^{\varepsilon}(X^{\varepsilon}(s), Y^{\varepsilon}(s)) \, ds + M^{\varepsilon}(z) - M^{\varepsilon}(z'). \quad (6.39)$$

We estimate the fourth moment of this increment in the following way. The fourth power of the first term on the right of (6.39) can be bounded by $C_1(1+\sup_{0 \leq z'' \leq Z} |X^{\varepsilon}(z'')|^4)(z - z')^4$, where C_1 is a constant. In view of (6.37), the expectation of $1 + \sup_{0 \leq z'' \leq Z} |X^{\varepsilon}(z'')|^4$ is bounded uniformly in ε. The second term on the right in (6.39) is a martingale, and its quadratic variation can be bounded by $C_2(1 + \sup_{0 \leq z'' \leq Z} |X^{\varepsilon}(z'')|^2)$, where C_2 is a constant independent of ε. Using Doob's inequality we have

$$\mathbb{E}\left[|M^{\varepsilon}(z) - M^{\varepsilon}(z')|^4\right] \leq c_4 \mathbb{E}[(\langle M^{\varepsilon}, M^{\varepsilon}\rangle(z) - \langle M^{\varepsilon}, M^{\varepsilon}\rangle(z'))^2] \leq C_4(z - z')^2.$$

This shows that the Kolmogorov criterion (6.34) is fulfilled for $(\tilde{X}^{\varepsilon})_{\varepsilon > 0}$, and in view of its uniform proximity (6.34) to $(X^{\varepsilon})_{\varepsilon > 0}$ in probability, the laws of the processes $(X^{\varepsilon})_{\varepsilon > 0}$ have weakly convergent subsequences in $\mathcal{C}$.

6.3.6 The Multiplicative-Noise Case

We give now a simple example to illustrate the asymptotic analysis of the previous sections. We consider a special $F(x, y)$ in which the noise is multiplicative,

$$F(x, y) = F_1(x) F_2(y).$$

Here F_1 is a bounded smooth function from $\mathbb{R}^d$ to itself and $F_2(y)$ is a scalar function on S that is centered with respect to the invariant distribution of $Y(z)$,

$$\mathbb{E}[F_2(Y(0))] = 0.$$

The limiting infinitesimal generator given in (6.33) becomes

$$\mathcal{L}\phi(x) = \int_0^{\infty} \mathbb{E}\left[F(x, Y(0)) \cdot \nabla_x \left(F(x, Y(z)) \cdot \nabla_x \phi(x)\right)\right] dz$$
$$= \frac{\gamma}{2} F_1(x) \cdot \nabla_x \left(F_1(x) \cdot \nabla_x \phi(x)\right),$$

where the parameter γ is the integrated autocorrelation of the stationary process $F_2(Y(z))$. It is given by

$$\gamma = 2 \int_0^{\infty} \mathbb{E}\left[F_2(Y(0)) F_2(Y(z))\right] dz = 2 \int_0^{\infty} \mathbb{E}\left[\nu(0)\nu(z)\right] dz, \quad (6.40)$$

where we have used the notation $F(Y(z)) = \nu(z)$. The parameter γ is nonnegative. This follows from the general considerations of Section 6.3.4. It can also be shown by noting that it is the limit of the variance of

$$\frac{1}{\sqrt{L}} \int_0^L \nu(z)\, dz \, .$$

More generally, we have, for any real k,

$$0 \leq \frac{1}{L}\mathbb{E}\left[\left|\int_0^L \nu(z)e^{ikz}\, dz\right|^2\right] = \frac{1}{L}\int_0^L \int_0^L \mathbb{E}\left[\nu(z_1)\nu(z_2)\right] e^{ik(z_1-z_2)}\, dz_1\, dz_2$$

$$= \frac{2}{L}\int_0^L \int_0^{z_1} \mathbb{E}\left[\nu(z_1-z_2)\nu(0)\right]\cos(k(z_1-z_2))\, dz_2\, dz_1$$

$$= \frac{2}{L}\int_0^L \int_0^{z_1} \mathbb{E}\left[\nu(z)\nu(0)\right]\cos(kz)\, dz\, dz_1$$

$$= \frac{2}{L}\int_0^L \mathbb{E}\left[\nu(z)\nu(0)\right]\cos(kz)\left(\int_z^L dz_1\right)\, dz$$

$$= 2\int_0^L \mathbb{E}\left[\nu(z)\nu(0)\right]\cos(kz)\, dz - \frac{2}{L}\int_0^L z\mathbb{E}\left[\nu(z)\nu(0)\right]\cos(kz)\, dz$$

$$\rightarrow 2\int_0^\infty \mathbb{E}\left[\nu(z)\nu(0)\right]\cos(kz)\, dz \quad \text{as} \quad L \to \infty,$$

assuming exponential decay of the autocorrelation function (for instance). This calculation shows that the Fourier transform of the autocorrelation function of the stationary process ν is proportional to the power spectral density of the process and is nonnegative. In particular, the integrated autocorrelation γ is the Fourier transform of the autocorrelation function at 0 frequency and is consequently nonnegative.

6.4 The Averaging and Fluctuation Theorems

The perturbed-test-function method can be used for many types of limit theorems. In this section we will use it for the averaging Theorem 4.2 that we discussed in Section 4.5.2. We will then use it again to analyze the fluctuations about the deterministic limit process.

6.4.1 Averaging

We consider the random differential equation

$$\frac{dX^\varepsilon}{dz} = F\left(X^\varepsilon(z), Y\left(\frac{z}{\varepsilon^2}\right)\right), \quad X^\varepsilon(0) = x_0,$$

where now we do not assume that $F(x, y)$ is centered and, for simplicity, we assume that F does not depend on z explicitly as it does in Section 4.5.2. We denote the mean of F with respect to the invariant distribution of $Y(z)$ by

$$\bar{F}(x) = \mathbb{E}[F(x, Y(0))].$$

The joint process $(X^\varepsilon(\cdot), Y(\cdot/\varepsilon^2))$ is Markovian with generator

$$\mathcal{L}^\varepsilon = \frac{1}{\varepsilon^2}\mathcal{L}_Y + F(x, y) \cdot \nabla_x.$$

Let $\phi(x)$ be a test function and define the perturbed test function $\phi^\varepsilon(x, y) = \phi(x) + \varepsilon^2\phi_1(x, y)$, where the corrector ϕ_1 solves the Poisson equation

$$\mathcal{L}_Y\phi_1(x, y) + F(x, y) \cdot \nabla_x\phi(x) - \bar{F}(x) \cdot \nabla_x\phi(x) = 0.$$

We see as before that $\mathcal{L}^\varepsilon\phi^\varepsilon(x, y) = \bar{F}(x) \cdot \nabla_x\phi(x) + O(\varepsilon^2)$. Therefore the processes $X^\varepsilon(z)$ converge to the solution of the martingale problem associated with the generator $\mathcal{L}\phi(x) = \bar{F}(x) \cdot \nabla_x\phi(x)$. The solution is the deterministic process $\bar{X}(z)$ defined as the solution of the ordinary differential equation

$$\frac{d\bar{X}}{dz} = \bar{F}(\bar{X}(z)), \quad \bar{X}(0) = x_0. \tag{6.41}$$

Convergence in distribution to a deterministic limit process implies in general convergence in probability. Therefore we have the following theorem.

Theorem 6.2. *Let $X^\varepsilon(z)$ be the process defined by the random ordinary differential equation*

$$\frac{dX^\varepsilon}{dz} = F\left(X^\varepsilon(z), Y\left(\frac{z}{\varepsilon^2}\right)\right), \quad X^\varepsilon(0) = x_0,$$

where $\nabla_x F(x, y)$ is bounded in x and y. Let $\bar{F}(x) = \mathbb{E}[F(x, Y(0))]$, where the expectation is with respect to the invariant law of the Markov process $Y(z)$, which satisfies the hypotheses of Theorem 6.1. Let $\bar{X}(z)$ be the solution of the ordinary equation (6.41). Then

$$\mathbb{P}\left(\sup_{z \in [0, Z]} |X^\varepsilon(z) - \bar{X}(z)| > \delta\right) \xrightarrow{\varepsilon \to 0} 0,$$

for all $\delta > 0$.

In view of this it is natural to consider the behavior of the fluctuations about $\bar{X}(z)$, as we do in the next section.

6.4.2 Fluctuation Theory

We will consider the behavior of the fluctuation process $X^\varepsilon(z) - \bar{X}(z)$ as $\varepsilon \to 0$. It is of order ε, so anticipating this we introduce the rescaled fluctuation process defined by

$$U^\varepsilon(z) = \frac{X^\varepsilon(z) - \bar{X}(z)}{\varepsilon}. \tag{6.42}$$

It is the solution of the random differential equation

$$\frac{dU^\varepsilon}{dz}(z) = \frac{1}{\varepsilon}\left[F\left(\bar{X}(z) + \varepsilon U^\varepsilon(z), Y\left(\frac{z}{\varepsilon^2}\right)\right) - \bar{F}(\bar{X}(z))\right], \tag{6.43}$$

starting from $U^\varepsilon(0) = 0$. We want to find the limit in distribution of the process $U^\varepsilon(z)$ as $\varepsilon \to 0$.

We first compute the form of the limit process by a simple formal expansion in ε and then apply Theorem 6.1 to justify it. Expanding the right side of (6.43), we obtain

$$\frac{dU^\varepsilon}{dz}(z) = \frac{1}{\varepsilon}\left[F\left(\bar{X}(z), Y\left(\frac{z}{\varepsilon^2}\right)\right) - \bar{F}(\bar{X}(z))\right]$$
$$+ \nabla_x F\left(\bar{X}(z), Y\left(\frac{z}{\varepsilon^2}\right)\right) U^\varepsilon(z) + O(\varepsilon).$$

We neglect the $O(\varepsilon)$ term and consider the integral form of this differential equation linear in U^ε:

$$U^\varepsilon(z) = V^\varepsilon(z) + \int_0^z \nabla_x F\left(\bar{X}(s), Y\left(\frac{s}{\varepsilon^2}\right)\right) U^\varepsilon(s)\,ds. \tag{6.44}$$

Here we have defined

$$V^\varepsilon(z) = \frac{1}{\varepsilon}\int_0^z F^{(c)}\left(\bar{X}(s), Y\left(\frac{s}{\varepsilon^2}\right)\right) ds, \quad \text{with } F^{(c)}(x, y) = F(x, y) - \bar{F}(x).$$

The vector function $F^{(c)}(x, y)$ satisfies the centering condition

$$\mathbb{E}[F^{(c)}(x, Y(0))] = 0$$

for all x. Therefore, the limiting distribution of the process V^ε is given by a slight variant of Theorem 6.1, which is extended to include a slow variation through $\bar{X}(z)$ in $F^{(c)}$. We see that the processes V^ε converge in distribution to a diffusion process whose infinitesimal generator is inhomogeneous in z,

$$\mathcal{L}_z = \frac{1}{2}\sum_{i,j=1}^d a_{ij}(\bar{X}(z))\frac{\partial^2}{\partial v_i \partial v_j},$$

where $\mathbf{a}(x) = (a_{ij}(x))_{i,j=1,\dots,d}$ is the nonnegative definite diffusion matrix

$$a_{ij}(x) = \int_{-\infty}^{\infty} \mathbb{E}[F_i^{(c)}(x, Y(0))F_j^{(c)}(x, Y(s))]\,ds. \tag{6.45}$$

If we denote by $\boldsymbol{\sigma}(x) = (\sigma_{ij}(x))_{i,j=1,\dots,d}$ a symmetric square root of the matrix $\mathbf{a}(x)$, then we can identify the limit V of V^ε as

$$V(z) = \int_0^z \boldsymbol{\sigma}(\bar{X}(s))\,dW(s),$$

and in coordinate form

$$V_i(z) = \sum_{j=1}^{d} \int_0^z \sigma_{ij}(\bar{X}(s)) \, dW_j(s),$$

where $W = (W_j)_{j=1,\dots,d}$ and the W_j's are independent standard Brownian motions.

If we can apply the averaging theorem of the previous section to the second term on the right in (6.44), then we get the following effective equation for the limit U of U^ε:

$$U(z) = V(z) + \int_0^z \nabla_x \bar{F}(\bar{X}(s)) U(s) \, ds.$$

The solution can be written in the form

$$U(z) = \int_0^z \mathbf{P}(s, z) \, dV(s) = \int_0^z \mathbf{P}(s, z) \boldsymbol{\sigma}(\bar{X}(s)) \, dW(s), \qquad (6.46)$$

where we have introduced the propagator or fundamental solution $\mathbf{P}$ of the linear variational equation of the deterministic system (6.41):

$$\frac{d\mathbf{P}}{dz}(z_0, z) = \nabla_x \bar{F}(\bar{X}(z)) \mathbf{P}(z_0, z), \qquad z \geq z_0, \qquad (6.47)$$

starting from $\mathbf{P}(z_0, z = z_0) = \mathbf{I}$. The limit process U is therefore a zero-mean Gaussian process with autocorrelation function

$$\mathbb{E}[U(z') U^T(z)] = \int_0^{\min(z, z')} \mathbf{P}(s, z') \mathbf{a}(\bar{X}(s)) \mathbf{P}^T(s, z) \, ds, \qquad (6.48)$$

for all $z, z' \geq 0$.

We now give a proof of this fluctuation theorem using Theorem 6.1. We first introduce the $\mathbb{R}^{2d}$-valued process

$$\tilde{U}^\varepsilon(z) = \begin{bmatrix} \bar{X}(z) \\ U^\varepsilon(z) \end{bmatrix}.$$

We have included the deterministic trajectory $\bar{X}(z)$ in this joint process so as to have z-homogeneous equations. Using (6.41) and (6.43), the process $\tilde{U}^\varepsilon(z)$ is a solution of the random differential equation

$$\frac{d\tilde{U}^\varepsilon}{dz}(z) = \frac{1}{\varepsilon} \tilde{F}\left(\tilde{U}^\varepsilon(z), Y\left(\frac{z}{\varepsilon^2}\right)\right) + \tilde{G}\left(\tilde{U}^\varepsilon(z), Y\left(\frac{z}{\varepsilon^2}\right)\right) + \tilde{R}^\varepsilon\left(\tilde{U}^\varepsilon(z), Y\left(\frac{z}{\varepsilon^2}\right)\right).$$

Here

$$\tilde{F}\left(\tilde{u}, y\right) = \begin{bmatrix} 0 \\ F\left(\bar{x}, y\right) - \bar{F}(\bar{x}) \end{bmatrix},$$

$$\tilde{G}\left(\tilde{u}, y\right) = \begin{bmatrix} \bar{F}(\bar{x}) \\ \nabla_x F\left(\bar{x}, y\right) u \end{bmatrix},$$

$$\tilde{R}^\varepsilon\left(\tilde{u}, y\right) = \begin{bmatrix} 0 \\ \frac{1}{\varepsilon} F(\bar{x} + \varepsilon u, y) - \frac{1}{\varepsilon} F(\bar{x}, y) - \nabla_x F\left(\bar{x}, y\right) u \end{bmatrix},$$

where

$$\tilde{u} = \begin{bmatrix} \bar{x} \\ u \end{bmatrix}.$$

We note that $\tilde{R}^\varepsilon(\tilde{u}, y)$ is of order ε uniformly in y and for $\tilde{u}$ in compact subsets. Furthermore, the function $\tilde{F}(\tilde{u}, y)$ satisfies the centering condition $\mathbb{E}[\tilde{F}(\tilde{u}, Y(0))] = 0$ for all $\tilde{u}$. Therefore, by applying Theorem 6.1, we conclude that the processes $\tilde{U}^\varepsilon$ converge in distribution to a diffusion process whose infinitesimal generator is

$$\mathcal{L} = \int_0^\infty \mathbb{E}\left[\tilde{F}(\tilde{u}, Y(0)) \cdot \nabla_{\tilde{u}}\left(\tilde{F}(\tilde{u}, Y(z)) \cdot \nabla_{\tilde{u}}\right)\right] dz + \mathbb{E}\left[\tilde{G}(\tilde{u}, Y(0)) \cdot \nabla_{\tilde{u}}\right]$$

$$= \frac{1}{2} \sum_{i,j=1}^d a_{ij}(\bar{x}) \frac{\partial^2}{\partial u_i \partial u_j} + \sum_{i=1}^d \bar{F}_i(\bar{x}) \frac{\partial}{\partial \bar{x}_i} + \sum_{i,j=1}^d \frac{\partial \bar{F}_i}{\partial \bar{x}_j}(\bar{x}) u_j \frac{\partial}{\partial u_i}, \qquad (6.49)$$

with the diffusion matrix given by (6.45). The symmetric square root of the nonnegative definite matrix $(a_{ij}(\bar{x}))$ is denoted by $\sigma_{ij}(\bar{x})$.

The process $\tilde{U}(z) = (\bar{X}(z), U(z))^T$, with U being the process defined by (6.46), can now be identified as a diffusion with the generator (6.49). This shows that the limit process of $\tilde{U}^\varepsilon(z) = (\bar{X}(z), U^\varepsilon(z))^T$ is $\tilde{U}(z) = (\bar{X}(z), U(z))^T$, where U is the Gaussian process defined by (6.46). Its integral representation in terms of the propagator $\mathbf{P}$ is

$$U(z) = \int_0^z \mathbf{P}(s, z)\boldsymbol{\sigma}(\bar{X}(s)) \, dW(s). \qquad (6.50)$$

We summarize the results in the following theorem.

Theorem 6.3. *Let $X^\varepsilon(z)$ and $\bar{X}(z)$ be defined as in Theorem 6.2. Let the fluctuation process $U^\varepsilon(z)$ be defined by (6.42). In addition to the hypotheses of Theorem 6.2 we assume that the second derivatives in x of $F(x, y)$ are uniformly bounded in x and y. Then the processes $U^\varepsilon(z)$ converge in distribution as $\varepsilon \to 0$ to the mean-zero Gaussian process $U(z)$ given by (6.50) and whose covariance is given by (6.48).*

6.5 Diffusion Approximation with Fast Oscillations

6.5.1 Semifast Oscillations

In applications to waves in randomly layered media in the strongly heterogeneous white-noise regime (5.17), the random differential equations that come up have the form (5.37) for X^ε, which is written again below, (6.51). The right side of this differential equation has a rapidly varying argument, z/ε, along with the random driving term $Y(z/\varepsilon^2)$. A diffusion approximation can be obtained in this more general framework by extending the analysis of Section 6.3. The result is summarized in the following theorem.

Theorem 6.4. *Let the process $X^\varepsilon(z)$ be defined by the system of random ordinary differential equations*

$$\frac{dX^\varepsilon}{dz}(z) = \frac{1}{\varepsilon}F\left(X^\varepsilon(z), Y\left(\frac{z}{\varepsilon^2}\right), \frac{z}{\varepsilon}\right) + G\left(X^\varepsilon(z), Y\left(\frac{z}{\varepsilon^2}\right), \frac{z}{\varepsilon}\right), \qquad (6.51)$$

starting from $X^\varepsilon(0) = x_0 \in \mathbb{R}^d$. We assume the same hypotheses as in Theorem 6.1. We also assume that $F(x, y, \tau)$ and $G(x, y, \tau)$ are periodic in the variable τ with period Z_0 and $F(x, y, \tau)$ satisfies the centering condition

$$\mathbb{E}[F(x, Y(0), \tau)] = 0\,,$$

for all x and τ. Then the random processes $X^\varepsilon(z)$ converge in distribution to the diffusion Markov process $X(z)$ with generator

$$\mathcal{L}\phi(x) = \frac{1}{Z_0}\int_0^{Z_0}\int_0^\infty \mathbb{E}\left[F(x, Y(0), \tau)\cdot\nabla_x\left(F(x, Y(z), \tau)\cdot\nabla_x\phi(x)\right)\right] dz\, d\tau$$

$$+\frac{1}{Z_0}\int_0^{Z_0} \mathbb{E}\left[G(x, Y(0), \tau)\cdot\nabla_x\phi(x)\right] d\tau\,. \qquad (6.52)$$

This limit generator is simply the average over the periodic argument τ of the generators of Theorem 6.1 with fixed τ. In fact, this theorem holds also when the dependence of $F(x, y, \tau)$ on τ is more general than periodic such as quasiperiodic or almost periodic. In this case the generator of the limit process has the form

$$\mathcal{L}\phi(x) =$$

$$\lim_{Z_0\to\infty}\frac{1}{Z_0}\int_0^{Z_0}\int_0^\infty \mathbb{E}\left[F(x, Y(0), \tau)\cdot\nabla_x\left(F(x, Y(z), \tau)\cdot\nabla_x\phi(x)\right)\right] dz\, d\tau$$

$$+ \lim_{Z_0\to\infty}\frac{1}{Z_0}\int_0^{Z_0} \mathbb{E}\left[G(x, Y(0), \tau)\cdot\nabla_x\phi(x)\right] d\tau\,.$$

This generalization of the periodic case is discussed at the end of this section. The proof of Theorem 6.4 is similar to that of Theorem 6.1. We need only explain how to construct perturbed test functions in this case.

It is convenient to introduce the linear motion on the torus $[0, Z_0]$ defined by $\tau(z) := z \mod Z_0$. The joint process $(X^\varepsilon(z), Y(z/\varepsilon^2), \tau(z/\varepsilon))$ on the state space $\mathbb{R}^d \times S \times [0, Z_0]$ is now a z-homogeneous Markov process with infinitesimal generator

$$\mathcal{L}^\varepsilon = \frac{1}{\varepsilon^2}\mathcal{L}_Y + \frac{1}{\varepsilon}F(x, y, \tau)\cdot\nabla_x + \frac{1}{\varepsilon}\frac{\partial}{\partial\tau} + G(x, y, \tau)\cdot\nabla_x\,.$$

The reason we introduced the motion on the torus $\tau(z)$ is to make the new joint process z-homogeneous. As a result the formalism of generators used in Theorem 6.1 can be carried over with minor adjustments. The main step that requires adjustment here is the construction of the perturbed test functions.

Let $\phi(x)$ be a smooth test function and as with (6.26) we look for a perturbed test function of the form

$$\phi^\varepsilon(x, y, \tau) = \phi(x) + \varepsilon\phi_1(x, y, \tau) + \varepsilon^2\phi_2(x, y, \tau).$$

As in Section 6.3 we apply $\mathcal{L}^\varepsilon$ to ϕ^ε and get

$$\mathcal{L}^\varepsilon\phi^\varepsilon = \frac{1}{\varepsilon}\Big[\mathcal{L}_Y\phi_1 + F(x, y, \tau)\cdot\nabla_x\phi(x)\Big] + \Big[\mathcal{L}_Y\phi_2 + F(x, y, \tau)\cdot\nabla_x\phi_1(x, y, \tau)$$

$$+ G(x, y, \tau)\cdot\nabla_x\phi(x) + \frac{\partial\phi_1}{\partial\tau}(x, y, \tau)\Big] + O(\varepsilon). \tag{6.53}$$

We choose the first corrector ϕ_1 in order to cancel the $O(\varepsilon^{-1})$ terms. This means that ϕ_1 must satisfy the Poisson equation

$$\mathcal{L}_Y\phi_1 + F(x, y, \tau)\cdot\nabla_x\phi(x) = 0.$$

Since F satisfies the centering condition and $\mathcal{L}_Y$ the Fredholm alternative, this equation has a solution. We write it as $\phi_1(x, y, \tau) = \phi_{11}(x, y, \tau) + \phi_{12}(x, \tau)$, where ϕ_{11} is the same as before,

$$\phi_{11}(x, y, \tau) = \int_0^\infty \mathbb{E}[F(x, Y(z), \tau)\cdot\nabla_x\phi(x)|Y(0) = y]\,dz,$$

while ϕ_{12} does not depend on y, so that $\mathcal{L}_Y\phi_{12} = 0$. It will be determined so that the limit generator is independent of τ.

We determine the second corrector ϕ_2 so that after centering, the $O(1)$ terms (6.53) cancel,

$$\mathcal{L}_Y\phi_2(x, y, \tau) + \phi_3(x, y, \tau) - \mathbb{E}[\phi_3(x, Y(0), \tau)] = 0,$$

where we have defined

$$\phi_3(x, y, \tau) = F(x, y, \tau)\cdot\nabla_x\phi_1(x, y, \tau) + \frac{\partial\phi_1}{\partial\tau}(x, y, \tau) + G(x, y, \tau)\cdot\nabla_x\phi(x).$$

With this determination of the correctors, (6.53) has the form

$$\mathcal{L}^\varepsilon\phi^\varepsilon = \mathbb{E}[\phi_3(x, Y(0), \tau)] + O(\varepsilon)$$

$$= \mathbb{E}[F(x, Y(0), \tau)\cdot\nabla_x\phi_{11}(x, Y(0), \tau)] + \frac{\partial\phi_{12}}{\partial\tau}(x, \tau)$$

$$+ \mathbb{E}[G(x, Y(0), \tau)\cdot\nabla_x\phi(x)] + O(\varepsilon), \tag{6.54}$$

where we use the decomposition $\phi_1(x, y, \tau) = \phi_{11}(x, y, \tau) + \phi_{12}(x, \tau)$.

We now determine the y-independent part of ϕ_1, which is ϕ_{12}, so that the right side of (6.54) is independent of τ. In order that it also be a bounded function of τ we have to write it as the integral over τ of a function that has mean zero in τ. This gives

$$\phi_{12}(x,\tau) = -\int_0^\tau \mathbb{E}[F(x,Y(0),s) \cdot \nabla_x \phi_{11}(x,Y(0),s)]$$
$$+\mathbb{E}[G(x,Y(0),s) \cdot \nabla_x \phi(x)] - \mathcal{L}\phi(x) \, ds \,,$$

where

$$\mathcal{L}\phi(x) = \frac{1}{Z_0}\int_0^{Z_0} \mathbb{E}\left[F(x,Y(0),\tau) \cdot \nabla_x \phi_{11}(x,Y(0),\tau)\right] d\tau$$
$$+\frac{1}{Z_0}\int_0^{Z_0} \mathbb{E}\left[G(x,Y(0),\tau) \cdot \nabla_x \phi(x)\right] d\tau$$
$$= \frac{1}{Z_0}\int_0^{Z_0}\int_0^\infty \mathbb{E}\left[F(x,Y(0),\tau) \cdot \nabla_x \left(F(x,Y(z),\tau) \cdot \nabla_x \phi(x)\right)\right] dz \, d\tau$$
$$+\frac{1}{Z_0}\int_0^{Z_0} \mathbb{E}\left[G(x,Y(0),\tau) \cdot \nabla_x \phi(x)\right] d\tau \,. \tag{6.55}$$

We therefore have

$$\mathcal{L}^\varepsilon \phi^\varepsilon(x,y,\tau) = \mathcal{L}\phi(x) + O(\varepsilon) \,.$$

With the perturbed test functions determined as we have just done, and with the limit generator defined by (6.55), the rest of the proof of Theorem 6.4 is similar to that of Theorem 6.1.

The extension to an $F(x,y,\tau)$ with almost-periodic dependence in τ does not require any changes in the construction of the perturbed test functions. However, the corrector $\phi_{12}(x,\tau)$ is not bounded in τ, but its growth is sublinear, so that $\varepsilon\phi_{12}(x,z/\varepsilon) \to 0$ as $\varepsilon \to 0$, and this is enough for the convergence proof.

6.5.2 Fast Oscillations

The random ordinary differential equations that come up in the study of waves in randomly layered media in the weakly heterogeneous regime (5.16) have the form (6.56) below. The right side of the differential equation (6.56) has a rapidly varying argument, z/ε^2. This rapid variation in z is on the same scale as that of the driving process Y. We analyze wave propagation in the weakly heterogeneous regime, in one-dimensional random media in the next two chapters, in three-dimensional randomly layered media in Chapter 18, and in random waveguides in Chapter 20. Theorems 6.1 and 6.4 extend to random differential equations with such fast oscillations. The results of the asymptotic analysis are summarized in the following theorem.

Theorem 6.5. *Let the process $X^\varepsilon(z)$ be defined by the system of random ordinary differential equations*

$$\frac{dX^\varepsilon}{dz}(z) = \frac{1}{\varepsilon} F\left(X^\varepsilon(z), Y\left(\frac{z}{\varepsilon^2}\right), \frac{z}{\varepsilon^2}\right) + G\left(X^\varepsilon(z), Y\left(\frac{z}{\varepsilon^2}\right), \frac{z}{\varepsilon^2}\right), \quad (6.56)$$

starting from $X^\varepsilon(0) = x_0 \in \mathbb{R}^d$. We assume the same hypotheses as in Theorem 6.1. We assume also that $F(x, y, \tau)$ and $G(x, y, \tau)$ are periodic with respect to τ with period Z_0 and that F satisfies the centering condition

$$\int_0^{Z_0} \mathbb{E}\left[F(x, Y(0), \tau)\right] d\tau = 0,$$

for all x. Then the random processes $X^\varepsilon(z)$ converge in distribution to the diffusion Markov process $X(z)$ with generator

$$\mathcal{L}\phi(x) = \frac{1}{Z_0} \int_0^{Z_0} \int_0^\infty \mathbb{E}\left[F(x, Y(0), \tau) \cdot \nabla_x \left(F(x, Y(z), \tau + z) \cdot \nabla_x \phi(x)\right)\right] dz\, d\tau$$

$$+ \frac{1}{Z_0} \int_0^{Z_0} \mathbb{E}\left[G(x, Y(0), \tau) \cdot \nabla_x \phi(x)\right] d\tau. \quad (6.57)$$

This theorem can be extended to functions $F(x, y, \tau)$ that are quasiperiodic or almost periodic in τ. The generator of the limit process has then the form

$$\mathcal{L}\phi(x) =$$

$$\lim_{Z_0 \to \infty} \frac{1}{Z_0} \int_0^{Z_0} \int_0^\infty \mathbb{E}\left[F(x, Y(0), \tau) \cdot \nabla_x \left(F(x, Y(z), \tau + z) \cdot \nabla_x \phi(x)\right)\right] dz\, d\tau$$

$$+ \lim_{Z_0 \to \infty} \frac{1}{Z_0} \int_0^{Z_0} \mathbb{E}\left[G(x, Y(0), \tau) \cdot \nabla_x \phi(x)\right] d\tau.$$

However, as we will see at the end of this section, the centering condition must be uniform in both x and τ,

$$\mathbb{E}[F(x, Y(0), \tau)] = 0. \quad (6.58)$$

For the proof of Theorem 6.5 it is enough to give the detailed construction of the perturbed test functions. The rest is similar to the proof of Theorem 6.1. To be able to deal with z-homogeneous Markov processes and generators we consider, as in Theorem 6.4, the enlarged joint Markov process $(X^\varepsilon(z), Y(z/\varepsilon^2), \tau(z/\varepsilon^2))$ whose infinitesimal generator is

$$\mathcal{L}^\varepsilon = \frac{1}{\varepsilon^2}\left(\mathcal{L}_Y + \frac{\partial}{\partial \tau}\right) + \frac{1}{\varepsilon} F(x, y, \tau) \cdot \nabla_x + G(x, y, \tau) \cdot \nabla_x.$$

The process $(Y(z), \tau(z))$ is Markov with infinitesimal generator $Q = \mathcal{L}_Y + \frac{\partial}{\partial \tau}$ in the state space $S \times [0, Z_0]$. It has an invariant probability distribution that is the product of the invariant distribution of Y times the uniform distribution in $[0, Z_0]$. The joint driving process (Y, τ) is ergodic and Q satisfies the Fredholm alternative if $\mathcal{L}_Y$ does.

Let $\phi(x)$ be a smooth test function and look for the perturbed test functions in the form

$$\phi^\varepsilon(x, y, \tau) = \phi(x) + \varepsilon\phi_1(x, y, \tau) + \varepsilon^2\phi_2(x, y, \tau).$$

Applying $\mathcal{L}^\varepsilon$ to ϕ^ε we get

$$\mathcal{L}^\varepsilon \phi^\varepsilon = \frac{1}{\varepsilon}\left[Q\phi_1 + F(x, y, \tau) \cdot \nabla_x\phi(x)\right]$$
$$+ \left[Q\phi_2 + F(x, y, \tau) \cdot \nabla_x\phi_1(x, y, \tau) + G(x, y, \tau) \cdot \nabla_x\phi(x)\right] + O(\varepsilon).$$

As before, we determine the first corrector ϕ_1 so that the $O(\varepsilon^{-1})$ terms vanish. This gives the Poisson equation

$$Q\phi_1(x, y, \tau) + F(x, y, \tau) \cdot \nabla_x\phi(x) = 0,$$

whose solution is well defined because the inhomogeneous term satisfies the centering condition with respect to the invariant distribution for Q, and it has the Fredholm alternative. The zero-mean solution of this Poisson equation admits the integral representation

$$\phi_1(x, y, \tau) = \int_0^\infty \mathbb{E}\left[F(x, Y(z), \tau + z) \cdot \nabla_x\phi(x) \mid Y(0) = y\right] dz. \tag{6.59}$$

Here we have used the fact that the two parts of $Q = \mathcal{L}_Y + \frac{\partial}{\partial\tau}$ commute, so that the semigroups for Y and τ act independently.

The second corrector is determined by the condition that the $O(1)$ terms in $\mathcal{L}^\varepsilon\phi^\varepsilon$ vanish after centering. This gives the Poisson equation

$$Q\phi_2(x, y, \tau) + F(x, y, \tau) \cdot \nabla_x\phi_1(x, y, \tau) + G(x, y, \tau) \cdot \nabla_x\phi(x) - \mathcal{L}\phi(x) = 0,$$

where

$$\mathcal{L}\phi(x) = \frac{1}{Z_0}\int_0^{Z_0} \mathbb{E}\left[F(x, Y(0), \tau) \cdot \nabla_x\phi_1(x, Y(0), \tau)\right] d\tau$$
$$+ \frac{1}{Z_0}\int_0^{Z_0} \mathbb{E}\left[G(x, Y(0), \tau) \cdot \nabla_x\phi(x)\right] d\tau$$
$$= \frac{1}{Z_0}\int_0^{Z_0}\int_0^\infty \mathbb{E}\left[F(x, Y(0), \tau) \cdot \nabla_x\left(F(x, Y(z), \tau + z) \cdot \nabla_x\phi(x)\right)\right] dz\, d\tau$$
$$+ \frac{1}{Z_0}\int_0^{Z_0} \mathbb{E}\left[G(x, Y(0), \tau) \cdot \nabla_x\phi(x)\right] d\tau.$$

The solution ϕ_2 of this Poisson equation is well defined since the inhomogeneous term is centered and Q has the Fredholm alternative property. We therefore have that

$$\mathcal{L}^\varepsilon \phi^\varepsilon = \mathcal{L}\phi + O(\varepsilon) \, ,$$

and the proof is completed as in Theorem 6.1.

The main change that occurs when $F(x, y, \tau)$ is almost periodic in τ is that the generator Q of the joint driving process $(Y(z), \tau(z))$ does not satisfy the Fredholm alternative. However, the generator $\mathcal{L}_Y$ of $Y(z)$ satisfies the Fredholm alternative, and the stronger centering condition (6.58) allows the first corrector ϕ_1 to be well defined by the integral representation (6.59). As in the almost-periodic extension of Theorem 6.4, the second corrector need only have sublinear growth in τ so that $\varepsilon^2 \phi_2(x, y, z/\varepsilon^2) \to 0$ as $\varepsilon \to 0$ uniformly in $y \in S$ on bounded sets in $x \in \mathbb{R}^d$.

6.6 Stochastic Calculus

A diffusion Markov process in $\mathbb{R}^d$ is defined by its infinitesimal generator, which is a second-order partial differential operator

$$\mathcal{L} = \frac{1}{2} \sum_{i,j=1}^d a_{ij}(x) \frac{\partial^2}{\partial x_i \partial x_j} + \sum_{i=1}^d b_i(x) \frac{\partial}{\partial x_i} \, . \tag{6.60}$$

If the coefficients of the generator are smooth and if $f(x)$ is a bounded smooth function, then

$$u(z, x) = \mathbb{E}_{z,x}[f(X(Z))] = \mathbb{E}[f(X(Z))|X(z) = x] \tag{6.61}$$

satisfies the backward Kolmogorov equation

$$\frac{\partial u}{\partial z} + \mathcal{L}u = 0 \, , \quad z < Z \, , \tag{6.62}$$

with terminal condition $u(Z, x) = f(x)$. When the diffusion process $X(z)$ is well defined and $f(x)$ is any bounded function, then the probabilistic representation (6.61) makes sense even though it may not be a classical solution of the backward Kolmogorov equation (6.62).

Starting from the generator $\mathcal{L}$, that is, from the diffusion and drift coefficients, $a_{ij}(x)$ and $b_i(x)$, there are many ways in which to construct the diffusion process $X(z)$. If the diffusion and drift coefficients are continuous and bounded and there is strong ellipticity

$$\sum_{i,j=1}^d a_{ij}(x)\xi_i\xi_j \geq \delta \sum_{i=1}^d \xi_i^2 \, , \quad x, \xi \in \mathbb{R}^d \, , \tag{6.63}$$

for some $\delta > 0$, then the martingale problem (6.6) determines the probability distribution of the diffusion process. In the martingale problem the distribution of the process in the space of continuous paths is uniquely determined when for any smooth and bounded function $\phi(x)$,

$$M_\phi(z) = \phi(X(z)) - \phi(x) - \int_0^z \mathcal{L}\phi(X(s))\,ds \qquad (6.64)$$

is a martingale. As we saw in Section 6.3, this is a particularly convenient characterization of diffusion processes for limit theorems. It is mostly the formalism of the martingale problem that is convenient in the asymptotic analysis, for we do not use the deeper unique characterization of the limit with minimal regularity, which is the main part of this theory. If the diffusion and drift coefficients have additional regularity properties then the transition probabilities of the process, and its semigroup (6.3), can be constructed from solutions of the Kolmogorov equation (6.62) using results from the theory of partial differential equations. From the semigroup the full distribution of the diffusion process in the space of continuous paths can be constructed using the Markov property.

When the drift coefficients $b_i(x)$ are smooth and the diffusion coefficients $a_{ij}(x)$ are smooth and elliptic,

$$\sum_{i,j=1}^d a_{ij}(x)\xi_i\xi_j \geq 0, \quad x,\xi \in \mathbb{R}^d, \qquad (6.65)$$

then the diffusion process can be constructed as a functional of the paths of Brownian motion by solving Itô stochastic differential equations. This is particularly useful for Monte Carlo simulations. It helps in some calculations in waves in layered media, and it often gives a quick and efficient way to get the asymptotic behavior in complicated situations, but it is not an essential tool for the analysis. The Itô theory of stochastic differential equations for diffusions can be thought of as a generalization of Langevin equations in which the diffusion coefficients $a_{ij}(x)$ do not depend on x. Itô stochastic differential equations appear very rarely in the physics literature.

The motivation for using stochastic differential equations to characterize diffusions with smooth coefficients is as follows. In the smooth case (6.65) the diffusion coefficients can be factored,

$$a_{ij}(x) = \sum_{k=1}^d \sigma_{ik}(x)\sigma_{jk}(x), \qquad (6.66)$$

and the matrix $\sigma_{ij}(x)$ is itself smooth but may be degenerate. We can then try to construct $X(z)$ by solving the differential equation

$$dX_i(z) = b_i(X(z))\,dz + \sum_{j=1}^d \sigma_{ij}(X(z))\,dW_j(z), \quad i = 1,\ldots,d, \qquad (6.67)$$

where $W_1(z), W_2(z), \ldots, W_d(z)$ are independent standard Brownian motions and $X_i(0) = x_i$, $i = 1, 2, \ldots, d$. The intuitive meaning of this equation is that the increments of the path of the diffusion process $X_i(z + \Delta z) - X_i(z)$ given

$X(z)$ are approximately Gaussian random variables with mean $b_i(X(z))\Delta z$ and covariance $a_{ij}(X(z))\Delta z$. We write this equation with differentials, rather than derivatives, because Brownian motion is not differentiable. The difficulty in constructing the diffusion as a functional of Brownian motion using (6.67) is that the paths of Brownian motion are continuous but not differentiable, and therefore the paths of any solution $X(z)$ of (6.67) are expected to be continuous but not differentiable. This means that in the integral form of the stochastic differential equation

$$X_i(z) = X_i(0) + \int_0^z b_i(X(s))\,ds + \int_0^z \sum_{j=1}^d \sigma_{ij}(X(s))\,dW_j(s)\,, \qquad (6.68)$$

the integral with respect to Brownian motion is not well defined by the usual integration theories. This is because neither the integrand nor the integrator in the Brownian integral is differentiable. A suitable definition of this integral, the stochastic integral, is not just a mathematical technicality but rather an essential part of dealing with the Markov property of diffusions, which instantaneously lose memory of the past. For Langevin equations the σ_{ij} are constants and the Brownian integral is elementary, for it is itself another nonstandard Brownian motion.

In this section we review briefly the basic facts from stochastic calculus, namely, Brownian stochastic integrals, Itô's formula, stochastic differential equations for diffusions, and connections with partial differential equations.

We then use the stochastic calculus to identify the limit diffusion processes of random differential equations as solutions of stochastic differential equations. This is particularly useful when the random differential equations are linear systems, which is the case for the propagator matrices in randomly layered media.

6.6.1 Stochastic Integrals

Our goal in this subsection is to define the stochastic integral with respect to the standard Brownian motion process $W(z)$, $z \geq 0$. The Itô stochastic integral can be defined for integrands $f(z, W)$ that are nonanticipating functionals of Brownian motion. Let us denote by $(\mathcal{F}_z)$ the filtration generated by Brownian motion W, that is, $\mathcal{F}_z$ is the σ-algebra (or information) generated by $\{W(s), 0 \leq s \leq z\}$. If for any z, $f(z, W)$ is a functional that depends only on $\mathcal{F}_z$, then it is called an adapted function or functional. We first define the stochastic integral for elementary nonanticipating functionals of Brownian motion and then in the general case with a limiting process.

An elementary nonanticipating functional of Brownian motion on $[0, 1]$ is defined by

$$f(z, W) = \sum_{k=0}^{n-1} \alpha_k(W)\mathbf{1}_{[z_k, z_{k+1})}(z)\,,$$

where the $\alpha_k(W)$ depend only on $\mathcal{F}_{z_k}$ for $0 \le k \le n-1$ and have finite variance, and $0 = z_0 < z_1 < \cdots < z_n = 1$. The integral $M_f(z) = \int_0^z f(s, W) \, dW(s)$ is defined by

$$M_f(z) = \sum_{k=0}^{n-1} \alpha_k(W)(W(z \wedge z_{k+1}) - W(z \wedge z_k)), \qquad (6.69)$$

where $a \wedge b = \min\{a, b\}$. The random process $M_f(z)$ is a zero-mean continuous, square-integrable martingale

$$\mathbb{E}\left[M_f(z')|\mathcal{F}_z\right] = M_f(z), \quad z' \ge z, \qquad (6.70)$$

whose increments have conditional variance

$$\mathbb{E}\left[(M_f(z') - M_f(z))^2|\mathcal{F}_z\right] = \mathbb{E}\left[\int_z^{z'} f^2(s, W) \, ds|\mathcal{F}_z\right]. \qquad (6.71)$$

We extend this definition to all square-integrable nonanticipating functionals f such that $\mathbb{E}[\int_0^1 f^2(s, W) \, ds] < \infty$ by approximating them with elementary ones in L^2. The martingale property and the Itô isometry (6.71) are preserved in this extension.

What is different and peculiar to the theory of the stochastic integral is that in the definition (6.69), the increments of the Brownian motion point forward. This is why the stochastic integral is a martingale and the Itô isometry holds. The definition of the integral is sensitive to the position of the Brownian increment relative to the position at which the integrand is evaluated. If, for example, the integrand is evaluated at the midpoint of the Brownian increment, then the resulting integral is different from the Itô integral, and it is called the Stratonovich integral. The Itô integral is the right one for developing a theory of stochastic differential equations of the form (6.67) because we want the diffusion Markov process $X(z)$ to have $\mathcal{L}$ in (6.60) as generator. For this to be the case, the stochastic integral must have the martingale property (6.70) and the Itô isometry property (6.71).

To see that stochastic integrals do not behave like ordinary integrals we consider the case $f(z, W) = W(z)$. By direct calculation from the definition we show that

$$\int_0^z W(s) \, dW(s) = \frac{1}{2}W^2(z) - \frac{z}{2}. \qquad (6.72)$$

Indeed, this is the integral of the functional $f(s, W) = W(s)$. We consider only the case $z = 1$. Setting $z_k = k/n$, the sequence of functions $f_n(s, W) = \sum_{k=0}^{n-1} W(z_k)\mathbf{1}_{[z_k, z_{k+1})}(s)$ approximates f in L^2:

$$\mathbb{E}\left[\int_0^1 |f(s, W) - f_n(s, W)|^2 \, ds\right] = \sum_{k=0}^{n} \int_{z_k}^{z_{k+1}} \mathbb{E}[(W(s) - W(z_k))^2] \, ds$$

$$= \sum_{k=0}^{n-1} \int_{z_k}^{z_{k+1}} (s - z_k) \, ds = \sum_{k=0}^{n-1} \frac{1}{2n^2} = \frac{1}{2n},$$

which goes to 0 as $n \to \infty$. The integral of f_n up to 1 is given by (6.69):

$$M_{f_n}(1) = \sum_{k=0}^{n-1} W(z_k)(W(z_{k+1}) - W(z_k)).$$

Using the fact that

$$W^2(1) = \sum_{k=0}^{n-1} \left(W(z_{k+1})^2 - W(z_k)^2\right)$$

$$= \sum_{k=0}^{n-1} (W(z_{k+1}) - W(z_k))(W(z_{k+1}) + W(z_k)),$$

we obtain the difference

$$M_{f_n}(1) - \frac{1}{2}W^2(1)$$

$$= \sum_{k=0}^{n-1} (W(z_{k+1}) - W(z_k)) \left(W(z_k) - \frac{1}{2}(W(z_{k+1}) + W(z_k))\right)$$

$$= -\frac{1}{2} \sum_{k=0}^{n-1} (W(z_{k+1}) - W(z_k))^2.$$

Therefore we get the first moment

$$\mathbb{E}\left[M_{f_n}(1) - \frac{1}{2}W^2(1)\right] = -\frac{1}{2} \sum_{k=0}^{n-1} \frac{1}{n} = -\frac{1}{2},$$

and the second moment

$$\mathbb{E}\left[\left(M_{f_n}(1) - \frac{1}{2}W^2(1)\right)^2\right] = \frac{1}{4} \sum_{k \neq l=0}^{n-1} \frac{1}{n^2} + \frac{1}{4} \sum_{k=0}^{n-1} \frac{3}{n^2} = \frac{1}{4} + \frac{2}{n},$$

where we have used the fact that $\mathbb{E}[(W(z_{k+1}) - W(z_k))^2] = 1/n$ and $\mathbb{E}[(W(z_{k+1}) - W(z_k))^4] = 3/n^2$. Finally, we deduce

$$\mathbb{E}\left[\left(M_{f_n}(1) - \left(\frac{1}{2}W^2(1) - \frac{1}{2}\right)\right)^2\right] = \frac{2}{n},$$

which goes to 0 as $n \to \infty$, which shows that $M_{f_n}(1)$ converges to $\frac{1}{2}W^2(1) - \frac{1}{2}$ in L^2 as $n \to \infty$. This is the desired result (6.72) at $z = 1$. The derivation for general z follows the same lines.

A similar computation with the Stratonovich integral shows that

$$\int_0^z W(s) \circ dW(s) = \frac{1}{2}W^2(z),$$

which is what ordinary integration gives. However, the Stratonovich integral is not a martingale and does not satisfy the Itô isometry. Stratonovich integrals will be discussed further in Section 6.7.2.

6.6.2 Itô's Formula

If the Brownian motion were differentiable and $\phi(z,x)$ is a smooth function then $\psi(z) = \phi(z, W(z))$ would be differentiable and $\psi'(z) = \phi_z(z, W(z)) + \phi_x(z, W(z))W'(z)$, which is the chain rule. But the Brownian motion is not differentiable, and this formula is not correct. Itô's formula provides a general way to extend the chain rule to functions of Brownian motion and of stochastic integrals. In the example just considered, Itô's formula gives $d\psi(z) = \phi_z(z, W(z))\,dz + \phi_x(z, W(z))\,dW(z) + \frac{1}{2}\phi_{xx}(z, W(z))\,dz$, which is written in terms of differentials, since Brownian motion is not differentiable. The unexpected, at first, new term in this formula is the one with the second x derivative. We will explain briefly how this formula arises and discuss its generalizations.

The reason why Itô's formula has an extra term is that while Brownian motion does not have finite variation, for otherwise it would be differentiable with probability one, it does have finite quadratic variation,

$$\inf_P \sum_k (W(z_{k+1}) - W(z_k))^2 = z, \tag{6.73}$$

where the infimum is taken over all finite partitions P of the interval $[0, z]$. We write this in compact notation as $(dW(z))^2 = dz$. Variations of order higher than two vanish for Brownian motion. If for a fixed partition we write

$$\psi(z) - \psi(0) = \sum_k [\psi(z_{k+1}) - \psi(z_k)],$$

with $\psi(z) = \phi(z, W(z))$, use a two-term Taylor expansion on each subinterval $[z_k, z_{k+1}]$, and use the quadratic variation property of Brownian motion, we get Itô's formula in integral form:

$$\phi(z, W(z)) = \phi(0,0) + \int_0^z \frac{\partial \phi}{\partial z}(s, W(s))\,ds + \int_0^z \frac{\partial \phi}{\partial x}(s, W(s))\,dW(s)$$

$$+ \frac{1}{2}\int_0^z \frac{\partial^2 \phi}{\partial x^2}(s, W(s))\,ds. \tag{6.74}$$

As an example consider $\phi(z, x) = x^2$. Itô's formula gives

$$W(z)^2 = 2\int_0^z W(s)\,dW(s) + z,$$

which agrees with the direct calculation of the quadratic stochastic integral (6.72).

Itô's formula can be generalized to functions of Brownian stochastic integrals, not only of Brownian motion. We consider the $\mathbb{R}$-valued stochastic integral

$$X(z) = X(0) + \int_0^z f(s, W)\,dW(s) + \int_0^z g(s, W)\,ds,$$

where f and g are nonanticipating square-integrable and integrable functionals, respectively. We call this a stochastic integral, even though it also contains an ordinary integral, because Itô's formula applied to such processes is also a stochastic integral of this form. We have just given in (6.74) Itô's formula for $\phi(z, X(z))$ for the case $g = 0$ and $f = 1$. Its generalization is

$$\phi(z, X(z)) = \phi(0, X(0)) + \int_0^z \frac{\partial \phi}{\partial z}(s, X(s))\, ds + \int_0^z \frac{\partial \phi}{\partial x}(s, X(s))\, dX(s)$$
$$+ \frac{1}{2} \int_0^z \frac{\partial^2 \phi}{\partial x^2}(s, X(s))\, d\langle X, X \rangle(s), \qquad (6.75)$$

where

$$dX(z) = f(z, W)\, dW(z) + g(z, W)\, dz \quad \text{and} \quad d\langle X, X \rangle(z) = f^2(z, W)\, dz.$$

The increasing process

$$\langle X, X \rangle(z) = \int_0^z f^2(s, W)\, ds$$

is the quadratic variation of the stochastic integral $X(z)$, defined in the same way as for Brownian motion (6.73):

$$\langle X, X \rangle(z) = \inf_P \sum_k (X(z_{k+1}) - X(z_k))^2.$$

It is characterized by the fact that

$$X^2(z) - \langle X, X \rangle(z)$$

is a martingale, which also follows from the Itô isometry (6.71). For Brownian motion, $f = 1$, $g = 0$, we have $\langle W, W \rangle(z) = z$.

Next we compute the multidimensional version of Itô's formula for the stochastic integrals of the form

$$X_i(z) = X_i(0) + \int_0^z \sum_{j=1}^n F_{ij}(s, W)\, dW_j(s) + \int_0^z G_i(s, W)\, ds, \quad i = 1, \ldots, d,$$

where $W_1(z), \ldots, W_n(z)$ are n independent standard Brownian motions. Here $(F_{ij}(s, W))_{i=1,\ldots,d, j=1,\ldots,n}$ is an adapted (nonanticipating) random $d \times n$ matrix that is square-integrable, and $(G_i(s, W))_{i=1,\ldots,d}$ is an adapted random $\mathbb{R}^d$ vector that is integrable. Let $\phi : \mathbb{R} \times \mathbb{R}^d \to \mathbb{R}$ be a smooth function. Then Itô's formula is given by

$$\phi(z, X(z)) = \phi(0, X(0)) + \int_0^z \frac{\partial \phi}{\partial z}(s, X(s))\, ds + \sum_{i=1}^d \int_0^z \frac{\partial \phi}{\partial x_i}(s, X(s))\, dX_i(s)$$
$$+ \frac{1}{2} \sum_{i,j=1}^d \int_0^z \frac{\partial^2 \phi}{\partial x_i \partial x_j}(s, X(s))\, d\langle X_i, X_j \rangle(s), \quad (6.76)$$

where

$$dX_i(z) = \sum_{j=1}^{n} F_{ij}(z, W) \, dW_j(z) + G_i(z, W) \, dz$$

and

$$d\langle X_i, X_j \rangle (z) = \sum_{k=1}^{n} F_{ik}(z, W) F_{jk}(z, W) \, dz \, .$$

This last expression is the cross quadratic variation of the stochastic integrals $X_i(z)$ and $X_j(z)$:

$$\langle X_i, X_j \rangle (z) = \inf_{P} \sum_{k} (X_i(z_{k+1}) - X_i(z_k))(X_j(z_{k+1}) - X_j(z_k)) \, .$$

6.6.3 Stochastic Differential Equations

With stochastic integrals well defined and with Itô's formula as an analytical tool, we can consider the solution of stochastic differential equations of the form (6.67).

We will use vector notation. Let $W(z)$ be the standard n-dimensional Brownian motion. Let $b : \mathbb{R}^d \times \mathbb{R} \to \mathbb{R}^d$ and $\sigma : \mathbb{R}^d \times \mathbb{R} \to \mathbb{R}^{d \times n}$ be smooth vector and matrix functions, respectively. We say that the random process $X(z)$ is a solution of the stochastic differential equation (SDE)

$$dX(z) = \sigma(z, X(z)) \, dW(z) + b(z, X(z)) \, dz$$

if $X(z)$ is $\mathcal{F}_z$-adapted and if

$$X(z) = X(0) + \int_0^z \sigma(s, X(s)) \, dW(s) + \int_0^z b(s, X(s)) \, ds \, . \tag{6.77}$$

The main result is that there exists a unique solution with continuous paths if b and σ are uniformly Lipschitz in x and grow at most linearly in x. The proof is an extension of the Picard iteration method for ordinary differential equations and uses the Kolmogorov inequality for martingales. The process $X(z)$ has finite moments of all orders on any finite z interval and starting from any point in $\mathbb{R}^d$ (see for instance [128]).

The Markov property of the solution process $X(z)$ follows from the initial value structure of the defining equation (6.77), from the independent increments property of Brownian motion, and from the definition of the stochastic integral. The solution $X(z')$ at any $z' > z$ is a functional of Brownian motion increments in $[z, z']$ and of $X(z)$. It does not depend on the past given $X(z)$. A Markov process with continuous paths is a diffusion process. Its generator is given in the next section.

We give two basic examples.

Example 1. Let $\lambda \in \mathbb{R}$ and consider the SDE

$$dX(z) = \lambda X(z)\, dW(z), \quad X(0) = 1.$$

Then the unique solution is the exponential martingale

$$X(z) = \exp\left(\lambda W(z) - \frac{\lambda^2}{2} z\right).$$

This can be seen by applying Itô's formula to $\phi(z, W(z))$ with $\phi(z,x) = \exp(\lambda x - \lambda^2 z/2)$.

Example 2. The solution of the SDE

$$dX(z) = -X(z)\, dz + \sqrt{2}\, dW(z)$$

is the Ornstein–Uhlenbeck process introduced in Section 6.2.3.

6.6.4 Diffusions and Partial Differential Equations

In the z-homogeneous case the smooth matrix field $\boldsymbol{\sigma} : \mathbb{R}^d \to \mathbb{R}^{d \times n}$ and the vector field $b : \mathbb{R}^d \to \mathbb{R}^d$ do not depend on z. The SDE for the $\mathbb{R}^d$-valued diffusion process $X(z)$ is

$$dX(z) = \boldsymbol{\sigma}(X(z))\, dW(z) + b(X(z))\, dz. \tag{6.78}$$

Let $\phi : \mathbb{R}^d \to \mathbb{R}$ be a smooth and bounded function. Using Itô's formula we get that

$$\phi(X(z)) = \phi(X(0)) + \int_0^z \mathcal{L}\phi(X(s))\, ds + \int_0^z \nabla_x \phi(X(s)) \cdot \boldsymbol{\sigma}(X(s))\, dW(s),$$

where $\mathcal{L}$ is the diffusion operator

$$\mathcal{L} = \frac{1}{2} \sum_{i,j=1}^d a_{ij}(x) \frac{\partial^2}{\partial x_i \partial x_j} + \sum_{i=1}^d b_i(x) \frac{\partial}{\partial x_i}, \tag{6.79}$$

with

$$a_{ij}(x) = \sum_{k=1}^n \sigma_{ik}(x) \sigma_{jk}(x). \tag{6.80}$$

The diffusion operator $\mathcal{L}$ is the generator of the diffusion process $X(z)$. This can be seen by letting $u(z, Z, x)$ be the solution of the partial differential equation

$$\frac{\partial u}{\partial z} + \mathcal{L}u = 0, \quad z < Z, \tag{6.81}$$

with terminal condition $u(z = Z, Z, x) = \phi(x)$. If the solution of this equation is smooth, then we can apply Itô's formula to $u(z, Z, X(z))$ and conclude that

$$u(Z, Z, X(Z)) - u(0, Z, X(0)) = \int_0^Z \nabla_x u(s, Z, X(s)) \cdot \sigma(X(s)) \, dW(s).$$

Under the present conditions the integrand of the stochastic integral is square-integrable, and so the mean of the stochastic integral is zero. Taking into account the terminal condition for u we have the probabilistic representation

$$u(0, Z, x) = \mathbb{E}_{0,x}[\phi(X(Z))] = \mathbb{E}[\phi(X(Z))|X(0) = x].$$

We conclude from this representation that $\mathcal{L}$ is the generator of the z-homogeneous diffusion process $X(z)$ and that (6.81) is its backward Kolmogorov equation.

Itô's formula shows also that the distribution of $X(z)$ is the solution of the martingale problem associated with the generator $\mathcal{L}$, that is,

$$\phi(X(z)) - \phi(X(0)) - \int_0^z \mathcal{L}\phi(X(s)) \, ds$$

is a martingale for any test function $\phi(x)$.

Clearly the study of diffusion processes is intimately connected with parabolic differential equations, even if in principle everything can be obtained from the SDE (6.78). The distribution of the process $X(z)$ can be obtained with partial differential equations methods. Let us assume that the diffusion coefficients a_{ij} are twice differentiable with bounded derivatives, the b_i's are continuously differentiable with bounded derivatives, and a satisfies the strong ellipticity condition (6.63). Then there exists a unique Green's function $p(z, x, y)$ from $\mathbb{R}^+ \times \mathbb{R}^d \times \mathbb{R}^d$ to $\mathbb{R}$ such that $p(0, x, y) = \delta(x - y)$ and:

1. $p(z, x, y) > 0 \ \forall z > 0, \ x, y \in \mathbb{R}^d$,
2. p is continuous on $\mathbb{R}^{+*} \times \mathbb{R}^d \times \mathbb{R}^d$, p is $\mathcal{C}^2$ in x and y, and $\mathcal{C}^1$ in z,
3. as a function of z and x, p satisfies the partial differential equation (PDE)

$$\frac{\partial p}{\partial z} = \mathcal{L}p,$$

4. as a function of z and y, p satisfies the PDE

$$\frac{\partial p}{\partial z} = \mathcal{L}^* p,$$

where $\mathcal{L}^*$ is the adjoint operator

$$\mathcal{L}^* p = -\sum_{i=1}^d \frac{\partial}{\partial y_i}(b_i(y)p) + \sum_{i,j=1}^d \frac{\partial^2}{\partial y_i \partial y_j}(a_{ij}(y)p).$$

The Green's function $p(z, x, y)$ is consequently the density of the kernel of the semigroup P_z of the Markov process with the generator $\mathcal{L}$:

$$(P_z\phi)(x) = \int \phi(y)p(z, x, y) \, dy.$$

6.6.5 Feynman–Kac Representation Formula

Itô's formula can be used to obtain a probabilistic representation for the solution $u(z, Z, x)$ of partial differential equations of the form

$$\frac{\partial u}{\partial z} + \mathcal{L}u - Vu = 0, \quad z < Z, \tag{6.82}$$

$$u(z = Z, Z, x) = \phi(x), \tag{6.83}$$

where in addition to the operator $\mathcal{L}$ defined in (6.79), $V = V(z, x)$ is a bounded function, a potential term in the equation; $\phi(x)$ is a smooth function in the terminal condition at $z = Z$; and the equation is to be solved for $z < Z$. This situation is a generalization of the backward Kolmogorov equation (6.81).

We consider the diffusion Markov process $X(z)$ that solves the SDE (6.78) and has infinitesimal generator $\mathcal{L}$. We now apply Itô's formula to $M(z)$ defined by

$$M(z) = u(z, Z, X(z))e^{-\int_0^z V(s, X(s))\, ds}.$$

The dz part of $dM(z)$ in Itô's formula is given by

$$\left(\frac{\partial u}{\partial z} + \mathcal{L}u - Vu\right)(z, Z, X(z))e^{-\int_0^z V(s, X(s))\, ds}\, dz,$$

and it is zero because u satisfies the partial differential equation (6.82). The martingale property of $M(z)$ between z and Z gives $\mathbb{E}\left(M(Z) \mid \mathcal{F}_z\right) = M(z)$, which can be written as

$$u(z, Z, X(z)) = \mathbb{E}\left[e^{-\int_z^Z V(s, X(s))\, ds}\phi(X(Z)) \mid \mathcal{F}_z\right],$$

where we have used $u(Z, Z, X(Z)) = \phi(X(Z))$ from the terminal condition (6.83) and the fact that $\exp\left(-\int_0^z V(s, X(s))\, ds\right)$ is $\mathcal{F}_z$-adapted. From the Markov property of $X(z)$ we obtain the *Feynman–Kac representation formula*

$$u(z, Z, x) = \mathbb{E}\left[e^{-\int_z^Z V(s, X(s))\, ds}\phi(X(Z)) \mid X(z) = x\right]. \tag{6.84}$$

Thus, the solution u of the PDE (6.82) has a representation as an expectation of a functional of the path of the process $X(z)$. This representation formula remains valid in the z-inhomogeneous case in which the coefficients of the diffusion $X(z)$ depend also on z, and the operator $\mathcal{L}$ depends also on z through $b(z, x)$ and $\sigma(z, x)$. In the homogeneous case with a potential independent of z, that is, $V(z, x) = V(x)$, the change of variable $Z - z = z'$ and with v defined by $v(z', x) = u(Z - z', x)$, we have that it satisfies

$$\frac{\partial v}{\partial z'} = \mathcal{L}v - V(x)v, \tag{6.85}$$

with the initial condition $v(0, x) = \phi(x)$. The form of the probabilistic representation for v is now

$$v(z', x) = \mathbb{E}\left[e^{-\int_{Z-z'}^{Z} V(X(s))\,ds}\phi(X(Z)) \mid X(Z-z') = x\right]$$
$$= \mathbb{E}\left[e^{-\int_{0}^{z'} V(X(s))\,ds}\phi(X(z')) \mid X(0) = x\right],$$

where we have used the homogeneity in z of the diffusion Markov process $X(z)$.

6.7 Limits of Random Equations and Stochastic Equations

We now consider the random differential equations of Section 6.3 and their diffusion limits from the viewpoint of stochastic calculus. The random differential equations have the form

$$\frac{dX^\varepsilon(z)}{dz} = \frac{1}{\varepsilon}F\left(X^\varepsilon(z), Y^\varepsilon(z)\right) + G\left(X^\varepsilon(z), Y^\varepsilon(z)\right), \qquad (6.86)$$

or those with fast oscillations (6.51) and (6.56). The limit generator for (6.86) is given by

$$\mathcal{L}\phi(x) = \int_{0}^{\infty} \mathbb{E}\left[F(x, Y(0)) \cdot \nabla_x \left(F(x, Y(z)) \cdot \nabla_x \phi(x)\right)\right] dz$$
$$+ \mathbb{E}[G(x, Y(0)) \cdot \nabla_x \phi(x)] \qquad (6.87)$$

as in Theorem 6.1. For the equations with fast oscillations the limit generators are given by (6.52) and (6.57) or their analogues in the almost-periodic case stated just below these expressions.

Next we introduce a special class of random differential equations, with or without fast oscillations, for which the limit process has a simple characterization by a stochastic differential equation.

6.7.1 Itô Form of the Limit Process

In all of the applications that we present in this book the random differential equations that we encounter have a right-hand side of the form

$$F(x, y, \tau) = \sum_{p=1}^{n} F^{(p)}(x) g^{(p)}(y, \tau), \qquad (6.88)$$

where $F^{(p)}(x)$ are smooth vector fields in $\mathbb{R}^d$ and $g^{(p)}(y, \tau)$ are real-valued scalar functions of $y \in S$ and $\tau \in \mathbb{R}$, and periodic or almost periodic in τ. We assume that the centering condition

$$\mathbb{E}[g^{(p)}(Y(0), \tau)] = 0 \qquad (6.89)$$

holds for all $\tau \in \mathbb{R}$ and $p = 1, 2, \ldots, n$. In applying the limit Theorem 6.1, where $g^{(p)}(y, \tau)$ is independent of τ, we have the correlation integrals

$$C_{pq} = 2 \int_0^\infty \mathbb{E}\left[g^{(p)}(Y(0))g^{(q)}(Y(z))\right] dz, \tag{6.90}$$

$p, q = 1, \ldots, n$. In applying the limit Theorem 6.4 we have the correlation integrals

$$C_{pq} = 2 \lim_{Z_0 \to \infty} \frac{1}{Z_0} \int_0^{Z_0} \int_0^\infty \mathbb{E}\left[g^{(p)}(Y(0), \tau)g^{(q)}(Y(z), \tau)\right] dz \, d\tau, \tag{6.91}$$

$p, q = 1, \ldots, n$, and for the case of Theorem 6.5 the correlation integrals

$$C_{pq} = 2 \lim_{Z_0 \to \infty} \frac{1}{Z_0} \int_0^{Z_0} \int_0^\infty \mathbb{E}\left[g^{(p)}(Y(0), \tau)g^{(q)}(Y(z), \tau + z)\right] dz \, d\tau, \tag{6.92}$$

$p, q = 1, \ldots, n$. The constant $n \times n$ matrix $\mathbf{C} = (C_{pq})_{p,q=1,\ldots,n}$ is not symmetric in general. However, its symmetric part $\mathbf{C}^{(S)} = \frac{1}{2}(\mathbf{C} + \mathbf{C}^T)$ is nonnegative, as was shown in Section 6.3.4.

With these definitions the limit generator has, in all cases, the form

$$\mathcal{L}\phi(x) = \frac{1}{2} \sum_{p,q=1}^n C_{pq} F^{(p)}(x) \cdot \nabla_x \left[F^{(q)}(x) \cdot \nabla_x \phi(x)\right] + \bar{G}(x) \cdot \nabla_x \phi(x), \tag{6.93}$$

where $\bar{G}(x) = \lim_{Z_0 \to \infty} \frac{1}{Z_0} \int_0^{Z_0} \mathbb{E}[G(x, Y(0), \tau)] \, d\tau$. We define by $\tilde{\sigma}$ the symmetric square root of $\mathbf{C}^{(S)} = \tilde{\sigma}^2$ and define further

$$\sigma_{il}(x) = \sum_{p=1}^n \tilde{\sigma}_{lp} F_i^{(p)}(x) \tag{6.94}$$

and

$$b_i(x) = \frac{1}{2} \sum_{p,q=1}^n \sum_{j=1}^d C_{pq} F_j^{(p)}(x) \frac{\partial F_i^{(q)}(x)}{\partial x_j} + \bar{G}_i(x). \tag{6.95}$$

Then the limit process $X(z)$ is identified with the solution of the Itô stochastic differential equation

$$dX_i(z) = \sum_{l=1}^n \sigma_{il}(X(z)) \, dW_l(z) + b_i(X(z)) \, dz, \quad i = 1, 2, \ldots, d, \tag{6.96}$$

with $X_i(0) = x_i$, and $W_1(z), \ldots, W_n(z)$ independent standard Brownian motions. We note that with the definitions (6.94) and (6.95) the drift coefficients are also given by

$$b_i(x) = \frac{1}{2} \sum_{l=1}^n \sum_{j=1}^d \sigma_{jl}(x) \frac{\partial \sigma_{il}(x)}{\partial x_j} + b_i^{(A)}(x) + \bar{G}_i(x), \tag{6.97}$$

where

$$b_i^{(A)}(x) = \frac{1}{2} \sum_{p,q=1}^{n} \sum_{j=1}^{d} C_{pq}^{(A)} F_j^{(p)}(x) \frac{\partial F_i^{(q)}(x)}{\partial x_j}, \tag{6.98}$$

and $\mathbf{C}^{(A)} = \frac{1}{2}(\mathbf{C} - \mathbf{C}^T)$ is the antisymmetric part of the matrix $\mathbf{C}$.

Stochastic differential equations in which the drift coefficients $b_i(x)$ are related to the $\sigma_{il}(x)$ by the relation (6.97) are special because they have a simpler form when written as Stratonovich stochastic differential equations

$$dX_i(z) = \sum_{l=1}^{n} \sigma_{ip}(X(z)) \circ dW_l(z) + (b_i^{(A)}(X(z)) + \bar{G}_i(X(z))) \, dz, \quad i = 1, 2, \ldots, d.$$

$$\tag{6.99}$$

Here the Brownian stochastic integral

$$\int_0^z \sigma_{il}(X(s)) \circ dW_l(s)$$

is not the Itô integral of Section 6.6.1 but the Stratonovich integral, which is denoted by the circle. We note that in many of the applications that follow the matrix $\mathbf{C}$ is symmetric, and therefore $b_i^{(A)}(x) = 0$.

We discuss the relation between Itô and Stratonovich stochastic integrals and stochastic differential equations in the next section.

6.7.2 Stratonovich Stochastic Integrals

Let $X(z)$ be the solution of the Itô stochastic differential equation (6.96) and let $\phi(x)$ be a smooth and bounded function on $\mathbb{R}^d$. We saw in Section 6.6.1 that Itô stochastic integrals have the Brownian increments pointing forward,

$$\int_0^z \phi(X(s)) \, dW_p(s) \approx \sum_k \phi(X(z_k))(W_p(z_{k+1}) - W_p(z_k)),$$

where $0 = z_0 < z_1 < \cdots < z_N = z$. The Stratonovich stochastic integral is defined with the midpoint rule

$$\int_0^z \phi(X(s)) \circ dW_p(s) \approx \sum_k \frac{\phi(X(z_k)) + \phi(X(z_{k+1}))}{2} (W_p(z_{k+1}) - W_p(z_k)).$$

By adding and subtracting $\phi(X(z_k))$ in this expression the Stratonovich integral can be written as an Itô integral and a correction

$$\int_0^z \phi(X(s)) \circ dW_p(s) \approx \int_0^z \phi(X(s)) \, dW_p(s)$$

$$+ \frac{1}{2} \sum_k (\phi(X(z_{k+1})) - \phi(X(z_k)))(W_p(z_{k+1}) - W_p(z_k)).$$

We can use Itô's formula to write the ϕ increments as

$$\phi(X(z_{k+1})) - \phi(X(z_k))$$

$$= \int_{z_k}^{z_{k+1}} \left(\sum_{i=1}^{d} b_i(X(s)) \frac{\partial \phi}{\partial x_i}(X(s)) + \frac{1}{2} \sum_{i,j=1}^{d} a_{ij}(X(s)) \frac{\partial^2 \phi}{\partial x_i \partial x_j}(X(s)) \right) ds$$

$$+ \sum_{q=1}^{n} \int_{z_k}^{z_{k+1}} \sum_{j=1}^{d} \sigma_{jq}(X(s)) \frac{\partial \phi}{\partial x_j}(X(s)) \, dW_q(s).$$

We use this expression to calculate the form of the correction term on the right side of the relation between the Stratonovich and the Itô integrals. As in the calculation of quadratic variations of stochastic integrals, the ds part of the ϕ increments does not contribute, and we get

$$\int_0^z \phi(X(s)) \circ dW_p(s) = \int_0^z \phi(X(s)) \, dW_p(s)$$

$$+ \frac{1}{2} \sum_{j=1}^{d} \int_0^z \sigma_{jp}(X(s)) \frac{\partial \phi}{\partial x_j}(X(s)) \, ds. \quad (6.100)$$

Note that the expectation of the Stratonovich integral is not zero as it is for the Itô integral. It is equal to the expectation of the ordinary integral on the right.

When we use this equality for $\int_0^z \sigma_{ip}(X(s)) \circ dW_p(s)$ we get

$$\int_0^z \sigma_{ip}(X(s)) \circ dW_p(s) = \int_0^z \sigma_{ip}(X(s)) \, dW_p(s)$$

$$+ \int_0^z \frac{1}{2} \sum_{j=1}^{d} \sigma_{jp}(X(s)) \frac{\partial \sigma_{ip}}{\partial x_j}(X(s)) \, ds.$$

This is the relation that we used in the previous section to write the Itô stochastic differential equation (6.96) is the Stratonovich form (6.99).

We can use the relation (6.100) between the Stratonovich and the Itô integrals to rewrite Itô's formula (6.76). Taking into account the explicit dependence of ϕ on z we have

$$\phi(z, X(z)) = \phi(0, X(0))$$

$$+ \int_0^z \left(\frac{\partial \phi}{\partial z}(s, X(s)) + \sum_{i=1}^{d} \frac{\partial \phi}{\partial x_i}(s, X(s)) \bar{G}_i(X(s)) \right) ds$$

$$+ \sum_{i=1}^{d} \sum_{p=1}^{n} \int_0^z \frac{\partial \phi}{\partial x_i}(s, X(s)) \sigma_{ip}(X(s)) \circ dW_p(s), \quad (6.101)$$

which looks superficially like the ordinary chain rule of calculus.

What is interesting about Stratonovich integrals here is that they appear naturally in the limit theorems for random differential equations. It is the form (6.93) of the limit generator as a sum of squares of vector fields that gives it.

6.7.3 Limits of Random Matrix Equations

Motivated by the random differential equations (5.27), (5.28) for the 2×2 propagator matrices $\mathbf{P}_\omega^\varepsilon(0, z)$ to which the limit theorems can be applied, we will now write the limit generators in a form that exhibits some important properties of the limit diffusion process. We are dealing with linear random matrix equations, and the property of the limit process that we want to identify is a form of space-time homogeneity of its increments.

The general form of the matrix equations is

$$\frac{d\mathbf{P}^\varepsilon(z)}{dz} = \frac{1}{\varepsilon}\mathbf{F}\left(\mathbf{P}^\varepsilon(z), Y^\varepsilon(z)\right), \tag{6.102}$$

or those with fast oscillations (6.51) and (6.56). Here $\mathbf{P}^\varepsilon$ is a $d \times d$ matrix, and the matrix-valued function $\mathbf{F}$ has the form

$$\mathbf{F}(\mathbf{P}, y, \tau) = \sum_{p=1}^n g^{(p)}(y, \tau)\mathbf{h}_p\mathbf{P}, \tag{6.103}$$

where the $\mathbf{h}_p$ are given constant $d \times d$ matrices. We assume for simplicity that the order-one term G in (6.86) is zero. As in Section 6.7.1 we assume that the scalar functions $g^{(p)}(y, \tau)$ satisfy the centering condition (6.89) and that the integrated correlations C_{pq}, $p, q = 1, \ldots, n$ are defined by (6.90), (6.91), and (6.92) depending on which limit theorem is appropriate. The generator of the limit matrix-valued diffusion is given by

$$\mathcal{L}\phi(\mathbf{P}) = \frac{1}{2}\sum_{p,q=1}^n C_{pq}\mathcal{D}_p\mathcal{D}_q\phi(\mathbf{P}). \tag{6.104}$$

Here the first-order partial differential operators $\mathcal{D}_p$ are defined by

$$\mathcal{D}_p\phi(\mathbf{P}) = \mathbf{h}_p\mathbf{P} \cdot \nabla_{\mathbf{P}}\phi(\mathbf{P}) = \lim_{\delta \to 0}\frac{1}{\delta}\left(\phi(e^{\delta\mathbf{h}_p}\mathbf{P}) - \phi(\mathbf{P})\right), \tag{6.105}$$

where $\mathbf{P} \cdot \mathbf{Q} = \sum_{i,j} P_{ij}Q_{ij}$. They do not commute with each other but satisfy the same commutation relations that the matrices $\mathbf{h}_p$ do. As we noted in Section 6.7.1, the matrix $\mathbf{C}$ is not symmetric in general. However, its symmetric part $\mathbf{C}^{(S)}$ is nonnegative definite.

As in Section 6.7.1, we introduce the symmetric square root $\tilde{\sigma}$ of the symmetric part $\mathbf{C}^{(S)}$ of the matrix $\mathbf{C}$. We then define the $d \times d$ matrices $\tilde{\mathbf{h}}_l = \sum_{p=1}^n \tilde{\sigma}_{lp}\mathbf{h}_p$. The Stratonovich form of the matrix-valued stochastic differential equation for the limit diffusion process $\mathbf{P}(z)$ is given by

$$d\mathbf{P}(z) = \sum_{l=1}^{n} \tilde{\mathbf{h}}_l \mathbf{P}(z) \circ dW_l(z) + \frac{1}{2} \sum_{p,q=1}^{n} C_{pq}^{(A)} \mathbf{h}_q \mathbf{h}_p \mathbf{P}(z)\, dz, \qquad (6.106)$$

where $W_l(z)$, $l = 1, \ldots, n$ are independent standard Brownian motions and $\mathbf{P}(0) = \mathbf{P}_0$. We denote by $\mathbf{C}^{(A)}$ the antisymmetric part of the matrix $\mathbf{C}$. If the initial matrix $\mathbf{P}_0$ is nonsingular, then the diffusion process $\mathbf{P}(z)$ is nonsingular for any $z \geq 0$ because by the Liouville identity and the Itô–Stratonovich formula we have that

$$d \det \mathbf{P}(z) = \mathrm{Tr} \left[\sum_{l=1}^{n} \tilde{\mathbf{h}}_l \circ dW_l(z) + \frac{1}{2} \sum_{p,q=1}^{n} C_{pq}^{(A)} \mathbf{h}_q \mathbf{h}_p\, dz \right] \det \mathbf{P}(z),$$

with $\det \mathbf{P}(0) = \det \mathbf{P}_0$.

The differential operators $\mathcal{D}_p$ have the property that they commute with right-multiplicative translations $\mathcal{T}_\mathbf{Q}\phi(\mathbf{P}) = \phi(\mathbf{PQ})$. That is, $\mathcal{T}_\mathbf{Q}\mathcal{D}_p = \mathcal{D}_p\mathcal{T}_\mathbf{Q}$. This property is inherited by the generator $\mathcal{L}$, and it implies that the matrix diffusion process $\mathbf{P}(z)$ has homogeneous in z independent multiplicative increments. This means that for any $0 < z_1 < z_2$ the distribution of $\mathbf{P}(z_2)\mathbf{P}^{-1}(z_1)$ starting from $\mathbf{P}_0$ at $z = 0$ is the same as the distribution of $\mathbf{P}(z_2 - z_1)$ starting from $\mathbf{I}$ at $z = 0$ and it is independent of $\{W_p(z),\ z \leq z_1,\ p = 1, \ldots, n\}$. In particular, the distribution of $\mathbf{P}(z_2)\mathbf{P}^{-1}(z_1)$ does not depend on the initial matrix $\mathbf{P}_0$, which is the spatial homogeneity property that we mentioned in the beginning of this subsection.

To prove this property we note that $\mathbb{E}_{0,\mathbf{PQ}}[\phi(\mathbf{P}(z))] = \mathbb{E}_{0,\mathbf{P}}[\phi(\mathbf{P}(z)\mathbf{Q})]$, because both sides satisfy the backward Kolmogorov equation and the generator commutes with right translations. Here $\mathbb{E}_{0,\mathbf{P}}[\cdot]$ is conditional expectation for the diffusion $\mathbf{P}(z)$ starting at $z = 0$ from $\mathbf{P}$. We now have that for any bounded random variable g_{z_1} that depends on $\{W_p(z),\ z \leq z_1,\ p = 1, \ldots, n\}$

$$\begin{aligned}
\mathbb{E}_{0,\mathbf{P}_0}[\phi(\mathbf{P}(z_2)\mathbf{P}^{-1}(z_1))g_{z_1}] &= \mathbb{E}_{0,\mathbf{P}_0}[\mathbb{E}_{z_1,\mathbf{P}(z_1)}[\phi(\mathbf{P}(z_2)\mathbf{P}^{-1}(z_1))]g_{z_1}] \\
&= \mathbb{E}_{0,\mathbf{P}_0}[\mathbb{E}_{z_1,\mathbf{I}}[\phi(\mathbf{P}(z_2))]g_{z_1}] \\
&= \mathbb{E}_{0,\mathbf{I}}[\phi(\mathbf{P}(z_2 - z_1))]\mathbb{E}_{0,\mathbf{P}_0}[g_{z_1}].
\end{aligned}$$

This proves the space-time homogeneity and the independence of the multiplicative increments of the matrix diffusion process $\mathbf{P}(z)$.

6.8 Lyapunov Exponent for Linear Random Differential Equations

All the limit theorems that we have given are for processes $X^\varepsilon(z)$ that are defined by random differential equations over a finite interval $0 \leq z \leq Z$, as ε tends to zero. This includes the averaging Theorem 6.2, the fluctuation Theorem 6.3, and the diffusion approximation Theorems 6.1, 6.4, and 6.5. Under what circumstances is the large-z behavior of the limit process characteristic of the behavior of the original process $X^\varepsilon(z)$ for fixed but small ε as z tends to infinity? The answer is that only under rather special circumstances is the behavior of $X^\varepsilon(z)$ for ε fixed and z large reflected by the large z behavior of the limit process. In this section we consider an important special configuration in which this is the case. More precisely, we study the exponential behavior of solutions of some linear systems of random differential equations for which the large-z and the small-ε limits can be interchanged.

6.8.1 Lyapunov Exponent of the Random Differential Equation

We consider the linear systems of random differential equations of the form

$$\frac{dX^\varepsilon}{dz}(z) = \frac{1}{\varepsilon}\mathbf{\Omega} X^\varepsilon(z) + \frac{1}{\varepsilon}g\left(Y\left(\frac{z}{\varepsilon^2}\right)\right)\mathbf{h}X^\varepsilon(z), \quad X^\varepsilon(0) = x_0. \tag{6.107}$$

In this and the next section we will assume that the following hypotheses hold:

- The driving process $Y(z)$ is Markovian as in Theorem 6.1.
- The real-valued function $g(y)$ is bounded and the centering condition $\mathbb{E}[g(Y(0))] = 0$ holds.
- $\mathbf{\Omega}$ and $\mathbf{h}$ are $d \times d$ constant matrices. The matrix $\mathbf{\Omega}$ is skew symmetric, $\mathbf{\Omega} = -\mathbf{\Omega}^T$, and it has no invariant vectors. The matrix $\mathbf{\Omega}$ generates a flow $e^{\mathbf{\Omega} z}$ on the sphere $\mathbb{S}^{d-1}$ such that for $j = 1, 2, 3$,

$$\lim_{Z \to \infty} \frac{1}{Z}\int_0^Z \phi_j(e^{\mathbf{\Omega} z}\hat{x})\,dz \text{ is independent of } \hat{x} \in \mathbb{S}^{d-1}, \tag{6.108}$$

where $\phi_1(\hat{x}) = (\mathbf{h}\hat{x}, \mathbf{h}\hat{x})$, $\phi_2(\hat{x}) = (\mathbf{h}^2\hat{x}, \hat{x})$, and $\phi_3(\hat{x}) = (\mathbf{h}\hat{x}, \hat{x})^2$.

We will show in Lemma 6.6 that the limits (6.108) always exist, but are not independent of $\hat{x}$ in general. Given $\mathbf{\Omega}$ and $\mathbf{h}$ we assume that the limit is independent of $\hat{x}$ for the three polynomial functions $\phi_j(\hat{x})$, $j = 1, 2, 3$, shown below (6.108). This hypothesis is satisfied for the random harmonic oscillator that we consider in Chapter 7.

We are interested in the large-z behavior of the process $X^\varepsilon(z)$ as characterized by the Lyapunov exponent

$$\Gamma^\varepsilon = \limsup_{z \to \infty} \frac{1}{z}\ln|X^\varepsilon(z)|. \tag{6.109}$$

We introduce polar coordinates for the solution $X^\varepsilon(z) \in \mathbb{R}^d$,

$$X^\varepsilon(z) = e^{R^\varepsilon(z)}\hat{X}^\varepsilon(z), \tag{6.110}$$

where

$$R^\varepsilon(z) = \log|X^\varepsilon(z)| \in \mathbb{R} \text{ and } \hat{X}^\varepsilon(z) = e^{-R^\varepsilon(z)}X^\varepsilon(z) \in \mathbb{S}^{d-1}. \tag{6.111}$$

We have

$$\frac{d\hat{X}^\varepsilon}{dz}(z) = \frac{1}{\varepsilon}\Omega\hat{X}^\varepsilon(z) + \frac{1}{\varepsilon}g\left(Y\left(\frac{z}{\varepsilon^2}\right)\right)h^\perp(\hat{X}^\varepsilon(z)), \quad \hat{X}^\varepsilon(0) = \hat{x}_0, \quad (6.112)$$

and

$$\frac{dR^\varepsilon}{dz}(z) = \frac{1}{\varepsilon}q\left(Y\left(\frac{z}{\varepsilon^2}\right), \hat{X}^\varepsilon(z)\right), \quad R^\varepsilon(0) = \log|x_0|. \quad (6.113)$$

Here we have used the fact that $(\Omega\hat{x}, \hat{x}) = 0$ for all $\hat{x} \in \mathbb{S}^{d-1}$. The real-valued function $q(y, \hat{x})$ and the vector-valued function $h^\perp(\hat{x})$ are defined by

$$q(y, \hat{x}) = g(y)(\hat{x}, \mathbf{h}\hat{x}), \quad h^\perp(\hat{x}) = \mathbf{h}\hat{x} - (\hat{x}, \mathbf{h}\hat{x})\hat{x}. \quad (6.114)$$

The joint process $\left(Y\left(\frac{z}{\varepsilon^2}\right), \hat{X}^\varepsilon(z)\right)$ is Markovian with state space $S \times \mathbb{S}^{d-1}$ and its infinitesimal generator is given by

$$\mathcal{L}^\varepsilon = \frac{1}{\varepsilon^2}\mathcal{L}_Y + \frac{1}{\varepsilon}\left[\Omega\hat{x} + g(y)h^\perp(\hat{x})\right] \cdot \nabla_{\hat{x}}. \quad (6.115)$$

We will assume that there is an $\varepsilon_0 > 0$ such that for all $0 < \varepsilon \leq \varepsilon_0$ the joint process $\left(Y\left(\frac{z}{\varepsilon^2}\right), \hat{X}^\varepsilon(z)\right)$ is ergodic, which means that there is a unique invariant probability law p^ε on the state space $S \times \mathbb{S}^{d-1}$ such that for any test function $\phi(y, \hat{x})$,

$$\int_{S \times \mathbb{S}^{d-1}} [\mathcal{L}^\varepsilon\phi(y, \hat{x})] \, p^\varepsilon(dy, d\hat{x}) = 0, \quad (6.116)$$

which has the form

$$\int_{S \times \mathbb{S}^{d-1}} \left(\mathcal{L}_Y + \varepsilon\left[\Omega\hat{x} + g(y)h^\perp(\hat{x})\right] \cdot \nabla_{\hat{x}}\right)\phi(y, \hat{x}) \, p^\varepsilon(dy, d\hat{x}) = 0. \quad (6.117)$$

The Lyapunov exponent Γ^ε, defined by (6.109), is the limit as z tends to infinity of $R^\varepsilon(z)/z$,

$$\Gamma^\varepsilon = \lim_{z\to\infty} \frac{1}{z}R^\varepsilon(z) = \lim_{z\to\infty} \frac{1}{z}\left[\log|x_0| + \int_0^z \frac{1}{\varepsilon}q\left(Y\left(\frac{s}{\varepsilon^2}\right), \hat{X}^\varepsilon(s)\right)ds\right],$$
$$(6.118)$$

so that under the assumed ergodicity of the joint process $\left(Y\left(\frac{z}{\varepsilon^2}\right), \hat{X}^\varepsilon(z)\right)$, Γ^ε is given by

$$\Gamma^\varepsilon = \frac{1}{\varepsilon}\int_{S \times \mathbb{S}^{d-1}} q(y, \hat{x}) \, p^\varepsilon(dy, d\hat{x}). \quad (6.119)$$

We will now use a variant of the perturbed-test-function method to show that indeed the limit of Γ^ε exists as ε tends to zero, and we will calculate the limit Lyapunov exponent explicitly.

Suppose that we can construct a perturbed test function $q^{\varepsilon,\lambda}(y, \hat{x}) = q_0^\lambda(y, \hat{x}) + \varepsilon q_1^\lambda(y, \hat{x})$, with $q_0^\lambda(y, \hat{x})$ and $q_1^\lambda(y, \hat{x})$ determined as solutions of Poisson equations. The parameter $\lambda > 0$ is introduced in order to regularize the solution of a Poisson equation below, and we will let $\lambda \to 0$ at the end. The perturbed test function will be constructed such that

$$\varepsilon^2 \mathcal{L}^\varepsilon q^{\varepsilon,\lambda} = q - \varepsilon \Gamma + \varepsilon \lambda q_{00}^\lambda + O(\varepsilon^2). \tag{6.120}$$

Here $q(y, \hat{x})$ is given by (6.114), and Γ is a constant to be determined, as is the function $q_{00}^\lambda(\hat{x})$. The error term is $O(\varepsilon^2)$ for each $\lambda > 0$. Then using (6.116) we have

$$
\begin{aligned}
0 &= \int_{S \times \mathbb{S}^{d-1}} \varepsilon \mathcal{L}^\varepsilon (q_0^\lambda(y, \hat{x}) + \varepsilon q_1^\lambda(y, \hat{x})) \, p^\varepsilon(dy, d\hat{x}) \\
&= \frac{1}{\varepsilon} \int_{S \times \mathbb{S}^{d-1}} q(y, \hat{x}) p^\varepsilon(dy, d\hat{x}) - \Gamma \int_{S \times \mathbb{S}^{d-1}} p^\varepsilon(dy, d\hat{x}) \\
&\qquad + \int_{S \times \mathbb{S}^{d-1}} \lambda q_{00}^\lambda(\hat{x}) \, p^\varepsilon(dy, d\hat{x}) + O(\varepsilon) \\
&= \Gamma^\varepsilon - \Gamma + \int_{S \times \mathbb{S}^{d-1}} \lambda q_{00}^\lambda(\hat{x}) \, p^\varepsilon(dy, d\hat{x}) + O(\varepsilon). \tag{6.121}
\end{aligned}
$$

This proves that

$$\limsup_{\varepsilon \to 0} |\Gamma^\varepsilon - \Gamma| \le \sup_{\hat{x} \in \mathbb{S}^{d-1}} |\lambda q_{00}^\lambda(\hat{x})|. \tag{6.122}$$

It remains therefore to show that the perturbed test function $q^{\varepsilon,\lambda}$ can be constructed with the desired properties and that

$$\lim_{\lambda \to 0} \sup_{\hat{x} \in \mathbb{S}^{d-1}} |\lambda q_{00}^\lambda(\hat{x})| = 0. \tag{6.123}$$

Expanding the left side of (6.120) in ε and equating to zero coefficients of ε leads first to the following Poisson equation for q_0^λ:

$$\mathcal{L}_Y q_0^\lambda = q. \tag{6.124}$$

Since $\mathbb{E}[g(Y(0))] = 0$ we have that $\mathbb{E}[q(Y(0), \hat{x})] = 0$ for all $\hat{x}$, so the Poisson equation for q_0^λ can be solved, and its solution has the form

$$q_0^\lambda(y, \hat{x}) = \chi(y)(\hat{x}, \mathbf{h}\hat{x}) + q_{00}^\lambda(\hat{x}). \tag{6.125}$$

Here $\chi(y)$ is the zero-mean solution for $\mathcal{L}_Y \chi = g$, and $q_{00}^\lambda(\hat{x})$ is a function that depends only on $\hat{x}$ and that will be determined from the second Poisson equation, which has the form

$$\mathcal{L}_Y q_1^\lambda(y, \hat{x}) = -\Gamma - \mathcal{L}_1 q_0^\lambda(y, \hat{x}) + \lambda q_{00}^\lambda(\hat{x}). \tag{6.126}$$

Here $\mathcal{L}_1$ is given by

$$\mathcal{L}_1 = [\mathbf{\Omega}\hat{x} + g(y)h^\perp(\hat{x})] \cdot \nabla_{\hat{x}}. \tag{6.127}$$

Note than in (6.126) we have added on the right side the term $\lambda q_{00}^\lambda(\hat{x})$, which makes the perturbed test functions q_0^λ and q_1^λ depend on λ. The solvability condition for this equation is that the right side should have zero mean with respect to the invariant probability distribution of the random process $Y(z)$, which means that we should have

$$-\Gamma - \mathbb{E}\left[\mathcal{L}_1 q_0^\lambda(Y(0), \hat{x})\right] + \lambda q_{00}^\lambda(\hat{x}) = 0\,,$$

which reduces to

$$(-\lambda + \mathbf{\Omega}\hat{x} \cdot \nabla_{\hat{x}}) q_{00}^\lambda(\hat{x}) = -\Gamma + \frac{\gamma}{2} h^\perp(\hat{x}) \cdot \nabla_{\hat{x}}(\hat{x}, \mathbf{h}\hat{x})\,, \tag{6.128}$$

for all $\hat{x} \in \mathbb{S}^{d-1}$. Here

$$\gamma = -2\mathbb{E}[\chi(Y(0))g(Y(0))] = 2 \int_0^\infty \mathbb{E}[g(Y(z))g(Y(0))]\,dz\,. \tag{6.129}$$

We define Γ so that the mean with respect to the uniform measure over the sphere of the right side of (6.128) is zero:

$$\Gamma = \frac{\gamma}{2} \int_{\mathbb{S}^{d-1}} h^\perp(\hat{x}) \cdot \nabla_{\hat{x}}(\hat{x}, \mathbf{h}\hat{x})\,d\hat{x},\,. \tag{6.130}$$

Equation (6.128) has the form of a regularized Poisson equation

$$(-\lambda + \mathbf{\Omega}\hat{x} \cdot \nabla_{\hat{x}}) q_{00}^\lambda(\hat{x}) = \phi(\hat{x})$$

for the generator $\mathbf{\Omega}\hat{x} \cdot \nabla_{\hat{x}}$ and for the function

$$\phi(\hat{x}) = -\Gamma + \frac{\gamma}{2} h^\perp(\hat{x}) \cdot \nabla_{\hat{x}}(\hat{x}, \mathbf{h}\hat{x})\,,$$

with the mean of ϕ equal to zero, $\int_{\mathbb{S}^{d-1}} \phi(\hat{x})\,d\hat{x} = 0$. Note that $\phi(\hat{x})$ is a polynomial in $\hat{x}$, and by substituting the expression for $h^\perp$ we can write it in the form

$$\phi(\hat{x}) = -\Gamma + \frac{\gamma}{2} \left[(\mathbf{h}\hat{x}, \mathbf{h}\hat{x}) + (\mathbf{h}^2\hat{x}, \hat{x}) - 2(\mathbf{h}\hat{x}, \hat{x})^2\right]\,, \tag{6.131}$$

which shows that ϕ is a linear combination of the functions ϕ_j, $j = 1, 2, 3$, in the hypothesis (6.108). The regularized Poisson equation (6.128) has a solution if λ is positive, and its properties are summarized in Lemma 6.7. We first establish the existence of Cesaro limits for the flow $e^{\mathbf{\Omega}z}$.

Lemma 6.6. *Let ϕ be a polynomial defined on $\mathbb{S}^{d-1}$. Then there exists a polynomial ϕ_0 such that*

$$\lim_{Z \to \infty} \sup_{\hat{x} \in \mathbb{S}^{d-1}} \left| \frac{1}{Z} \int_0^Z \phi(e^{\mathbf{\Omega}z}\hat{x})\,dz - \phi_0(\hat{x}) \right| = 0\,.$$

The function ϕ_0 is the $\mathbf{\Omega}$-average of ϕ and it has the same average over the sphere as ϕ,

$$\int_{\mathbb{S}^{d-1}} \phi_0(\hat{x})\,d\hat{x} = \int_{\mathbb{S}^{d-1}} \phi(\hat{x})\,d\hat{x}\,.$$

Proof. The flow $e^{\Omega z}$ has no invariant vectors. Therefore the dimension d is even, the eigenvalues of Ω form a set of the form $\{\pm i\omega_j, j = 1, \ldots, d/2\}$, where $\omega_j > 0$, and there exists an orthonormal basis such that the matrices Ω and $e^{\Omega z}$ have the form

$$
\Omega = \begin{bmatrix} \begin{pmatrix} 0 & \omega_1 \\ -\omega_1 & 0 \end{pmatrix} & & 0 \\ & \ddots & \\ 0 & & \begin{pmatrix} 0 & \omega_{d/2} \\ -\omega_{d/2} & 0 \end{pmatrix} \end{bmatrix},
$$

$$
e^{\Omega z} = \begin{bmatrix} \begin{pmatrix} \cos(\omega_1 z) & \sin(\omega_1 z) \\ -\sin(\omega_1 z) & \cos(\omega_1 z) \end{pmatrix} & & 0 \\ & \ddots & \\ 0 & & \begin{pmatrix} \cos(\omega_{d/2} z) & \sin(\omega_{d/2} z) \\ -\sin(\omega_{d/2} z) & \cos(\omega_{d/2} z) \end{pmatrix} \end{bmatrix}.
$$

The function $\phi(\hat{x})$ is assumed to be a polynomial of degree n, so that $\phi(e^{\Omega z}\hat{x})$ can be expanded as

$$
\phi(e^{\Omega z}\hat{x}) = \phi_0(\hat{x}) + \sum_{\omega \in \mathcal{O}} \phi_\omega^{(c)}(\hat{x}) \cos(\omega z) + \phi_\omega^{(s)}(\hat{x}) \sin(\omega z),
$$

where $\mathcal{O}$ is the finite set of nonzero sums and differences of up to n frequencies ω_j, $j = 1, \ldots, d/2$. The functions ϕ_0, $\phi_\omega^{(c)}$, and $\phi_\omega^{(s)}$ are polynomials, which implies that they are smooth and bounded on $\mathbb{S}^{d-1}$. For any $\hat{x} \in \mathbb{S}^{d-1}$, we have

$$
\frac{1}{Z} \int_0^Z \phi(e^{\Omega z}\hat{x}) \, dz = \phi_0(\hat{x}) + \sum_{\omega \in \mathcal{O}} \phi_\omega^{(c)}(\hat{x}) \frac{\sin(\omega Z)}{\omega Z} + \phi_\omega^{(s)}(\hat{x}) \frac{1 - \cos(\omega Z)}{\omega Z},
$$

so that

$$
\lim_{Z \to \infty} \sup_{\hat{x} \in \mathbb{S}^{d-1}} \left| \frac{1}{Z} \int_0^Z \phi(e^{\Omega z}\hat{x}) \, dz - \phi_0(\hat{x}) \right| = 0.
$$

Finally, the matrix $e^{\Omega z}$ is orthogonal, which means that the flow $e^{\Omega z}$ preserves the uniform measure over the sphere $\mathbb{S}^{d-1}$. Therefore

$$
\int_{\mathbb{S}^{d-1}} \phi(\hat{x}) \, d\hat{x} = \int_{\mathbb{S}^{d-1}} \frac{1}{Z} \int_0^Z \phi(e^{\Omega z}\hat{x}) \, dz \, d\hat{x} = \int_{\mathbb{S}^{d-1}} \phi_0(\hat{x}) \, d\hat{x},
$$

which completes the proof of the lemma. $\square$

In the next lemma we consider the behavior of solutions of the Poisson equation (6.128).

Lemma 6.7. *Let ϕ be a bounded function defined on $\mathbb{S}^{d-1}$. For any $\lambda > 0$ let the function $\xi_\lambda(\hat{x})$ be defined by*

$$\xi_\lambda(\hat{x}) = -\int_0^\infty \phi(e^{\Omega s}\hat{x})e^{-\lambda s}\,ds\,. \qquad (6.132)$$

Then $\xi_\lambda(\hat{x})$ is the unique bounded solution of the equation

$$(-\lambda + \Omega\hat{x}\cdot\nabla_{\hat{x}})\xi_\lambda = \phi\,. \qquad (6.133)$$

Moreover, if ϕ is a polynomial whose Ω-average is zero, that is, the asserted Cesaro limits satisfy

$$\phi_0(\hat{x}) \equiv 0\,, \quad \forall \hat{x} \in \mathbb{S}^{d-1}\,, \qquad (6.134)$$

then $\lambda\xi_\lambda$ tends to zero uniformly as $\lambda \to 0$:

$$\limsup_{\lambda\to 0}\ \sup_{\hat{x}\in\mathbb{S}^{d-1}} |\lambda\xi_\lambda(\hat{x})| = 0\,. \qquad (6.135)$$

Proof. For $\lambda > 0$ the function ξ_λ is well defined, since $|\phi(e^{\Omega s}\hat{x})e^{-\lambda s}| \leq e^{-\lambda s}\|\phi\|_\infty$, where $\|\cdot\|$ stands for the supremum norm over $\mathbb{S}^{d-1}$. Furthermore, we have the identity

$$\frac{\partial}{\partial s}\left(\phi(e^{\Omega s}\hat{x})e^{-\lambda s}\right) = -\lambda\phi(e^{\Omega s}\hat{x})e^{-\lambda s} + (\nabla_{\hat{x}}\phi)(e^{\Omega s}\hat{x})\cdot(e^{\Omega s}\Omega\hat{x})e^{-\lambda s}$$

$$= -\lambda\left[\phi(e^{\Omega s}\hat{x})e^{-\lambda s}\right] + \Omega\hat{x}\cdot\nabla_{\hat{x}}\left[\phi(e^{\Omega s}\hat{x})e^{-\lambda s}\right]\,.$$

By integrating this equality from $s = 0$ to $s = \infty$ we see that ξ_λ satisfies (6.133),

$$-\phi(\hat{x}) = \lambda\xi_\lambda(\mathbf{x}) - \Omega\hat{x}\cdot\nabla_{\hat{x}}\xi_\lambda(\hat{x})\,,$$

which proves the first part of the lemma. For the second part of the lemma, we assume that the condition (6.134) holds for ϕ. By Lemma 6.6, for any $\delta > 0$ there exists Z_δ such that

$$\sup_{\hat{x}\in\mathbb{S}^{d-1}}\left|\frac{1}{Z_\delta}\int_0^{Z_\delta}\phi(e^{\Omega s}\hat{x})\,ds\right| \leq \delta\,. \qquad (6.136)$$

Let us write

$$\xi_\lambda(\hat{x}) = -\sum_{k=0}^\infty \int_{kZ_\delta}^{(k+1)Z_\delta} \phi(e^{\Omega s}\hat{x})e^{-\lambda s}\,ds\,,$$

and define

$$\xi_\lambda^{(k)}(\hat{x}) = -\int_{kZ_\delta}^{(k+1)Z_\delta}\phi(e^{\Omega s}\hat{x})\,ds\,e^{-\lambda kZ_\delta}\,.$$

We then have the estimate

$$\left\| \xi_\lambda - \sum_{k=0}^\infty \xi_\lambda^{(k)} \right\|_\infty \leq \sum_{k=0}^\infty \|\phi\|_\infty e^{-\lambda k Z_\delta} \int_0^{Z_\delta} 1 - e^{-\lambda s} ds$$

$$\leq \frac{\|\phi\|_\infty}{\lambda} \frac{\lambda Z_\delta - 1 + e^{-\lambda Z_\delta}}{1 - e^{-\lambda Z_\delta}},$$

and by (6.136),

$$\left\| \sum_{k=0}^\infty \xi_\lambda^{(k)} \right\|_\infty \leq \sum_{k=0}^\infty e^{-\lambda k Z_\delta} \sup_{\hat{x} \in \mathbb{S}^{d-1}} \left| \int_0^{Z_\delta} \phi(e^{\Omega s} \hat{x}) \, ds \right| \leq \frac{\delta Z_\delta}{1 - e^{-\lambda Z_\delta}}.$$

As a consequence,

$$\limsup_{\lambda \to 0} \sup_{\hat{x} \in \mathbb{S}^{d-1}} |\lambda \xi_\lambda(\hat{x})| \leq \delta.$$

Taking the limit $\delta \to 0$ gives the desired result.   □

With the definition (6.130) of Γ and with the hypothesis (6.108), the function ϕ defined by (6.131) has a zero Ω-average. Therefore, Lemma 6.7 can be applied to the solution q_{400}^λ of the regularized Poisson equation (6.128), and we can take the limit $\lambda \to 0$ in (6.122) and use (6.123) to obtain the desired result $\lim_{\varepsilon \to 0} \Gamma^\varepsilon = \Gamma$. We summarize this result in the following theorem.

Theorem 6.8. *Assume that for fixed $\varepsilon > 0$ the joint Markov process*

$$\left(Y\left(\frac{z}{\varepsilon^2} \right), \hat{X}^\varepsilon(z) \right)$$

with generator (6.115) is ergodic on $S \times \mathbb{S}^{d-1}$. We also assume the set of hypotheses stated below (6.107). Then the Lyapunov exponent limit (6.109) exists and is deterministic and independent of x_0, for any fixed $\varepsilon > 0$. Moreover, the limit of the Lyapunov exponent for small ε exists,

$$\lim_{\varepsilon \to 0} \Gamma^\varepsilon = \Gamma, \tag{6.137}$$

and Γ is given by (6.130), which has also the form

$$\Gamma = \frac{\gamma}{2} \int_{\mathbb{S}^{d-1}} [(\mathbf{h}\hat{x}, \mathbf{h}\hat{x}) + (\mathbf{h}^2\hat{x}, \hat{x}) - 2(\mathbf{h}\hat{x}, \hat{x})^2] \, d\hat{x}. \tag{6.138}$$

The constant γ is nonnegative and is given by

$$\gamma = \int_{-\infty}^\infty \mathbb{E}[g(Y(0))g(Y(z))] \, dz. \tag{6.139}$$

In the next section we will show that if we first take the diffusion limit $\varepsilon \to 0$ and then calculate the Lyapunov exponent of the limit diffusion process, we get the same result as in (6.138). This implies that as far as the rate of growth or decay of solutions of the random linear system (6.107) is concerned, the limits $\varepsilon \to 0$ and $z \to \infty$ can be interchanged.

6.8.2 Lyapunov Exponent of the Limit Diffusion

We cannot get a diffusion limit for the process X^ε because it has a large deterministic drift of order $1/\varepsilon$. We therefore remove this drift by introducing the process $\widetilde{X}^\varepsilon$ in $\mathbb{R}^d$ defined by

$$\widetilde{X}^\varepsilon(z) = e^{-\Omega\frac{z}{\varepsilon}} X^\varepsilon(z).$$

We have $|X^\varepsilon(z)| = |\widetilde{X}^\varepsilon(z)|$, and the process $\widetilde{X}^\varepsilon(z)$ satisfies the random differential equation

$$\frac{d\widetilde{X}^\varepsilon}{dz}(z) = \frac{1}{\varepsilon}\widetilde{F}\left(\widetilde{X}^\varepsilon(z), Y\left(\frac{z}{\varepsilon^2}\right), \frac{z}{\varepsilon}\right), \tag{6.140}$$

where the function $\widetilde{F}(\widetilde{x}, y, \tau)$ has the form

$$\widetilde{F}(\widetilde{x}, y, \tau) = g(y) e^{-\Omega\tau} \mathbf{h} e^{\Omega\tau} \widetilde{x}.$$

We can write this as

$$\widetilde{F}(\widetilde{x}, y, \tau) = \sum_{i,j=1}^{d} g^{(ij)}(y, \tau) \mathbf{h}^{(ij)} \widetilde{x},$$

where the real-valued functions $g^{(ij)}$ and the constant matrices $\mathbf{h}^{(ij)}$ are given by

$$\mathbf{h}_{kl}^{(ij)} = \delta_{ik}\delta_{jl}, \qquad g^{(ij)}(y, \tau) = g(y)(e^{-\Omega\tau}\mathbf{h} e^{\Omega\tau})_{ij}.$$

The random differential equation is of the form (6.88), and the diffusion approximation theorem 6.4 can be applied if the following limits exist and are independent of z_0:

$$C_{ij,i'j'} = 2 \lim_{Z\to\infty} \frac{1}{Z} \int_{z_0}^{z_0+Z} \int_0^\infty \mathbb{E}\left[g^{(ij)}(Y(0), \tau) g^{(i'j')}(Y(z), \tau)\right] dz\, d\tau$$

$$= \gamma \lim_{Z\to\infty} \frac{1}{Z} \int_{z_0}^{z_0+Z} (e^{-\Omega\tau}\mathbf{h} e^{\Omega\tau})_{ij}(e^{-\Omega\tau}\mathbf{h} e^{\Omega\tau})_{i'j'} \, d\tau.$$

Here γ is defined by (6.139). By Lemma 6.6, these limits exist and the limit process is a diffusion with infinitesimal generator

$$\mathcal{L} = \frac{1}{2} \sum_{i,j,i',j'=1}^{d} C_{ij,i'j'} \mathbf{h}^{(ij)}\widetilde{x} \cdot \nabla_{\widetilde{x}}[\mathbf{h}^{(i'j')}\widetilde{x} \cdot \nabla_{\widetilde{x}}]$$

$$= \frac{1}{2} \sum_{i,j,i',j'=1}^{d} C_{ij,i'j'} \widetilde{x}_j \frac{\partial}{\partial\widetilde{x}_i}\left[\widetilde{x}_{j'}\frac{\partial}{\partial\widetilde{x}_{i'}}\right]$$

$$= \frac{1}{2}\sum_{i,i'=1}^{d} C_{ii'}(\widetilde{x}) \frac{\partial^2}{\partial\widetilde{x}_i\partial\widetilde{x}_{i'}} + \frac{1}{2}\sum_{i'=1}^{d} D_{i'}(\widetilde{x}) \frac{\partial}{\partial\widetilde{x}_{i'}},$$

where

$$C_{ii'}(\widetilde{x}) = \sum_{j,j'=1}^{d} C_{ij,i'j'} \widetilde{x}_j \widetilde{x}_{j'}$$

$$= \gamma \lim_{Z \to \infty} \frac{1}{Z} \int_0^Z (e^{-\Omega\tau} \mathbf{h} e^{\Omega\tau} \widetilde{x})_i (e^{-\Omega\tau} \mathbf{h} e^{\Omega\tau} \widetilde{x})_{i'} \, d\tau \,,$$

$$D_{i'}(\widetilde{x}) = \sum_{i,j=1}^{d} C_{ij,i'i} \widetilde{x}_j$$

$$= \gamma \lim_{Z \to \infty} \frac{1}{Z} \int_0^Z (e^{-\Omega\tau} \mathbf{h}^2 e^{\Omega\tau} \widetilde{x})_{i'} \, d\tau \,.$$

We introduce polar coordinates for the limit diffusion $\widetilde{X}(z) \in \mathbb{R}^d$,

$$\widetilde{X}(z) = e^{R(z)} \check{X}(z) \,, \tag{6.141}$$

where

$$R(z) = \log|\widetilde{X}(z)| \in \mathbb{R} \quad \text{and} \quad \check{X}(z) = e^{-R(z)} \widetilde{X}(z) \in \mathbb{S}^{d-1} \,. \tag{6.142}$$

The joint process $(R(z), \check{X}(z))$ in $\mathbb{R} \times \mathbb{S}^{d-1}$ is also a diffusion. Using Itô's formula we can find its infinitesimal generator, which has the form

$$\mathcal{L} = \frac{1}{2} a^{(R)}(R, \check{x}) \frac{\partial^2}{\partial R^2} + \frac{1}{2} b^{(R)}(R, \check{x}) \frac{\partial}{\partial R}$$

$$+ \frac{1}{2} \sum_{i,j=1}^{d} a_{ij}(R, \check{x}) \frac{\partial^2}{\partial \check{x}_i \partial \check{x}_j} + \frac{1}{2} \sum_{i=1}^{d} b_i(R, \check{x}) \frac{\partial}{\partial \check{x}_i} \,.$$

Here

$$a^{(R)}(R, \check{x}) = \sum_{i,i'=1}^{d} C_{ii'}(\check{x}) \check{x}_i \check{x}_j \,,$$

$$b^{(R)}(R, \check{x}) = \sum_{i=1}^{d} C_{ii}(\check{x}) + \sum_{i=1}^{d} D_i(\check{x}) \check{x}_i - 2 \sum_{ii'=1}^{d} C_{ii'}(\check{x}) \check{x}_i \check{x}_j \,,$$

and a_{ij} and b_i are given by similar expressions. From these expressions we see that the diffusion and drift coefficients of the limit process $R(z)$ do not depend on R. The following computations show that they do not depend on $\check{x}$ either:

$$\sum_{i,i'=1}^{d} C_{ii'}(\check{x}) \check{x}_i \check{x}_{i'} = \gamma \lim_{Z \to \infty} \frac{1}{Z} \int_0^Z (\check{x}, e^{-\Omega\tau} \mathbf{h} e^{\Omega\tau} \check{x})^2 \, d\tau$$

$$= \gamma \lim_{Z \to \infty} \frac{1}{Z} \int_0^Z (e^{\Omega\tau} \check{x}, \mathbf{h} e^{\Omega\tau} \check{x})^2 \, d\tau$$

$$= \gamma \int_{\mathbb{S}^{d-1}} (\hat{x}, \mathbf{h}\hat{x})^2 \, d\check{x} \,.$$

Here we have used the fact that $e^{\Omega \tau}$ is orthogonal, so that $(e^{\Omega \tau} \check{x}, e^{\Omega \tau} \check{y}) = (\check{x}, \check{y})$, and we have applied Lemma 6.6 with the hypothesis (6.108) in order to get the expression of the limit in terms of the mean over the sphere. We have similarly

$$\sum_{i=1}^{d} D_i(\check{x}) \check{x}_i = \gamma \lim_{Z \to \infty} \frac{1}{Z} \int_0^Z (\check{x}, e^{-\Omega \tau} \mathbf{h}^2 e^{\Omega \tau} \check{x}) \, d\tau$$

$$= \gamma \lim_{Z \to \infty} \frac{1}{Z} \int_0^Z (e^{\Omega \tau} \check{x}, \mathbf{h}^2 e^{\Omega \tau} \check{x}) \, d\tau$$

$$= \gamma \int_{\mathbb{S}^{d-1}} (\hat{x}, \mathbf{h}^2 \hat{x}) \, d\hat{x} \, ,$$

$$\sum_{i=1}^{d} C_{ii}(\check{x}) = \gamma \lim_{Z \to \infty} \frac{1}{Z} \int_0^Z (e^{-\Omega \tau} \mathbf{h} e^{\Omega \tau} \check{x}, e^{-\Omega \tau} \mathbf{h} e^{\Omega \tau} \check{x}) \, d\tau$$

$$= \gamma \lim_{Z \to \infty} \frac{1}{Z} \int_0^Z (\mathbf{h} e^{\Omega \tau} \check{x}, \mathbf{h} e^{\Omega \tau} \check{x}) \, d\tau$$

$$= \gamma \int_{\mathbb{S}^{d-1}} (\mathbf{h} \hat{x}, \mathbf{h} \hat{x}) \, d\hat{x} \, .$$

This means that the limit process $R(z)$ is a Gaussian process with constant diffusion and drift:

$$R(z) = \sqrt{\gamma a^{(r)}} W(z) + \frac{\gamma}{2} b^{(r)} z \, , \tag{6.143}$$

where

$$a^{(r)} = \int_{\mathbb{S}^{d-1}} (\mathbf{h} \hat{x}, \hat{x})^2 \, d\hat{x} \, ,$$

$$b^{(r)} = \int_{\mathbb{S}^{d-1}} \left[(\mathbf{h} \hat{x}, \mathbf{h} \hat{x}) + (\mathbf{h}^2 \hat{x}, \hat{x}) - 2(\mathbf{h} \hat{x}, \hat{x})^2 \right] d\hat{x} \, .$$

The Lyapunov exponent of the limit diffusion is

$$\lim_{z \to \infty} \frac{1}{z} R(z) = \frac{\gamma}{2} b^{(r)} = \Gamma \, ,$$

where Γ is again given by (6.138).

We summarize the results of this section in the following theorem.

Theorem 6.9. *Under the hypotheses of Theorem 6.8 the processes* $\log |X^\varepsilon(z)|$ *converge in distribution as* $\varepsilon \to 0$ *to the Gaussian process* $R(z)$ *given by* (6.143) *and*

$$\lim_{z \to \infty} \frac{1}{z} R(z) = \Gamma \, ,$$

where Γ *is the limit of the Lyapunov exponent* (6.138).

We have thus shown that if we first take the diffusion limit $\varepsilon \to 0$ and then calculate the Lyapunov exponent of the limit diffusion process we get the same result as in (6.138). Therefore, under the hypotheses of Theorems 6.8 and 6.9, the rate of growth or decay of solutions of the random linear system (6.107) is the same regardless of the order in which the limits $\varepsilon \to 0$ and $z \to \infty$ are taken.

We note here that the ergodicity hypothesis of the process $(Y\left(\frac{z}{\varepsilon^2}\right), \hat{X}^{\varepsilon}(z))$ is not used in Theorem 6.9.

The expression (6.138) can be used to determine the sign of the Lyapunov exponent in some special cases. If, for example, $\mathbf{h}$ is skew symmetric, then

$$\Gamma = \frac{\gamma}{2} \int_{\mathbb{S}^{d-1}} [(\mathbf{h}\hat{x}, \mathbf{h}\hat{x}) - (\mathbf{h}\hat{x}, \mathbf{h}\hat{x})] \, d\hat{x} = 0 \,.$$

If $\mathbf{h}$ is symmetric then

$$\Gamma = \gamma \int_{\mathbb{S}^{d-1}} [(\mathbf{h}\hat{x}, \mathbf{h}\hat{x}) - (\mathbf{h}\hat{x}, \hat{x})^2] \, d\hat{x} \,,$$

which is nonnegative by Cauchy–Schwarz inequality

$$(\mathbf{h}\hat{x}, \hat{x})^2 \le (\mathbf{h}\hat{x}, \mathbf{h}\hat{x})(\hat{x}, \hat{x}) = (\mathbf{h}\hat{x}, \mathbf{h}\hat{x}) \,.$$

We note that in the symmetric case the only way that we can have $\Gamma = 0$ (assuming that $\gamma > 0$) is if $\mathbf{h}\hat{x}$ is proportional to $\hat{x}$ for all $\hat{x}$. This means that if $\mathbf{h}$ is symmetric and not proportional to the identity matrix, and if $\gamma > 0$, then Γ is positive.

6.9 Appendix

6.9.1 Quadratic Variation of a Continuous Martingale

The quadratic variation of the continuous martingale $M(z)$ given by (6.6) has the form (6.8):

$$\langle M, M \rangle (z) = \int_0^z \mathcal{L}\phi^2(Y(s)) - 2\phi(Y(s))\mathcal{L}\phi(Y(s)) \, ds \,.$$

This is seen by the following calculations. We first note that by the martingale property of M, we have $\mathbb{E}[M(z+h)M(z)|\mathcal{F}_z] = M(z)^2$ and therefore

$$\mathbb{E}[M(z+h)^2 - M(z)^2|\mathcal{F}_z] = \mathbb{E}[(M(z+h) - M(z))^2|\mathcal{F}_z] \,.$$

The square increment of $M(z)$ can be written as

$$(M(z+h) - M(z))^2 = \left(\phi(Y(z+h)) - \phi(Y(z)) - \int_z^{z+h} \mathcal{L}\phi(Y(s)) \, ds \right)^2 \,.$$

Neglecting the integral term in this equation is possible under the assumption that ϕ and $\mathcal{L}\phi$ are bounded. Let $\Delta\phi = \phi(Y(z+h)) - \phi(Y(z))$ and $J = \int_z^{z+h} \mathcal{L}\phi(Y(s))\,ds$. We want to estimate $(\Delta\phi - J)^2 - \Delta\phi^2$ when h is small. We know that $J = O(h)$. We have then that $|(\Delta\phi - J)^2 - \Delta\phi^2| \le 2|\Delta\phi||J| + J^2$. But $2|\Delta\phi||J| + J^2 \le \delta\Delta\phi^2 + (1+\delta^{-1})J^2$. Choosing $\delta = \sqrt{h}$ gives the estimate

$$(M(z+h) - M(z))^2 = (\phi(Y(z+h)) - \phi(Y(z)))^2(1 + O(h^{1/2})) + O(h^{3/2}),$$

where the $O(\cdot)$ are deterministic. Continuing with the above calculation we have

$$
\begin{aligned}
(\phi(Y(z+h)) &- \phi(Y(z)))^2 \\
&= \phi^2(Y(z+h)) - \phi^2(Y(z)) - 2\phi(Y(z))(\phi(Y(z+h)) - \phi(Y(z))) \\
&= M_2(z+h) - M_2(z) + \int_z^{z+h} \mathcal{L}\phi^2(Y(s))\,ds \\
&\quad -2\phi(Y(z))\left(M(z+h) - M(z) + \int_z^{z+h} \mathcal{L}\phi(Y(s))\,ds\right),
\end{aligned}
$$

where $M_2(z)$ is the martingale of the form (6.6) with ϕ replaced by ϕ^2. Using the martingale property for M and M_2 we obtain that

$$
\begin{aligned}
\mathbb{E}[(\phi(Y(z+h)) &- \phi(Y(z)))^2|\mathcal{F}_z] \\
&= \mathbb{E}[\int_z^{z+h} \mathcal{L}\phi^2(Y(s))\,ds - 2\phi(Y(z))\int_z^{z+h} \mathcal{L}\phi(Y(s))\,ds|\mathcal{F}_z] \\
&= \int_0^h P_s\mathcal{L}\phi^2(Y(z)) - 2\phi(Y(z))P_s\mathcal{L}\phi(Y(z))\,ds \\
&= [\mathcal{L}\phi^2(Y(z)) - 2\phi(Y(z))\mathcal{L}\phi(Y(z))]h + o(h)\,.
\end{aligned}
$$

Here we have used the Markov property for $Y(z)$ and the continuity of the associated semigroup. This shows that

$$\mathbb{E}[\langle M, M\rangle\,(z+h) - \langle M, M\rangle\,(z)|\mathcal{F}_z] = [\mathcal{L}\phi^2(Y(z)) - 2\phi(Y(z))\mathcal{L}\phi(Y(z))]h + o(h)\,,$$

from which the integral form of $\langle M, M\rangle\,(z)$ is obtained by summing over small increments in z.

Notes

In this chapter we present a self-contained summary of the basic tools of the theory of stochastic processes needed for modeling randomly layered media and for carrying out asymptotic analysis in various scaling limits. For an introduction to Markov processes we refer to the book by Breiman [23]. An advanced treatment of the theory of Markov processes, associated semigroups, and limit theorems is in the book by Ethier and Kurtz [50]. The martingale approach to diffusions and limit theorems is in the book of Stroock and Varadhan [163]. An introduction to stochastic calculus with Brownian motions can be found in the book by Oksendal [128] and a more advanced treatment in the book by Karatzas and Shreve [92]. The first diffusion-approximation results for random differential equations were given by Khasminskii in 1966 [97, 98]. The martingale approach to limit theorems for random differential equations was presented by Papanicolaou–Stroock–Varadhan in 1976 [135] and in Blankenship–Papanicolaou [14], including the *perturbed-test-function method* that is used extensively in this chapter. Similar methods are used in homogenization [13] and in stochastic stability and control [114]. We also refer to a recent series of papers by Pardoux and Veretennikov [138] for an extended treatment of Poisson equations and diffusion approximation. We have only considered Markovian models of random equations here for simplicity. The results, however, can be extended to a large class of mixing processes as is done in [96] and in the books by Ethier–Kurtz [50] and by Kushner [114]. Finally, the theory of random dynamical systems, including Lyapunov exponents and the multiplicative ergodic theory, is presented in the book by Arnold [6]. The two theorems in this area in Section 6.8 are motivated by applications to wave localization and the random harmonic oscillator that are discussed in the next chapter. These theorems are presented here for the first time.

7

Transmission of Energy Through a Slab of Random Medium

In this chapter we consider the simplest problem of wave propagation in one-dimensional random media, which is the reflection and tranmission of wave energy by a slab. We focus on the exponential decay of the transmission coefficient as the size of the random slab goes to infinity. This exponential decay is characteristic of wave propagation in one-dimensional random media and holds regardless of the strength of the fluctuations in the medium properties. It is not true in three-dimensional isotropic random media unless the fluctuations are strong enough. It is a manifestation of the phenomenon of *wave localization* where wave energy does not propagate and is localized in space because of strong interference induced by the random medium.

We will use the limit theorems of the previous chapter to analyze quantitatively the reflection and transmission of energy by one-dimensional random media in the two scaling regimes of Chapter 5. They are the weakly heterogeneous regime (5.16), which is the one most often considered in the literature, and the strongly heterogeneous white-noise regime (5.17). We calculate in this chapter the exponential decay rate of transmitted energy by a slab of random medium.

We consider monochromatic waves in the weakly heterogeneous regime in Section 7.1 and study the exponential decay of the power transmission coefficient as a function of the size of the random medium. We show that the exponential decay rate defines the localization length (7.42) of the random medium in the asymptotic regime. By studying carefully the statistics of the power transmission coefficient we also show that its expected value is different from its most probable value (Propositions 7.3–7.4). We study the transmission of pulses and their energy decay in Section 7.2. We show that the transmitted energy of the pulse is a self-averaging quantity and analyze its decay rate (Proposition 7.5). Self-averaging means that the transmitted energy tends to its mean value in the asymptotic regime. We study in Section 7.3 the exponential decay of the power transmission coefficient before going into the asymptotic regime. This defines a localization length that converges to that of the asymptotic regime (Proposition 7.6). All the analysis is carried out in the weakly heterogeneous regime. In Section 7.4 we show that it can be extended to the strongly heterogeneous white-noise regime (Proposition 7.7).

7.1 Transmission of Monochromatic Waves

We consider the acoustic wave equations in one dimension (3.3) with a slab of random medium in $(0, L)$ and surrounded by a homogeneous medium. We assume matched medium boundary conditions at both ends of the slab, that is, the parameters of the homogeneous half-spaces are equal to the effective parameters of the random slab. We consider a right-going monochromatic wave incident from the homogeneous left half-space. We will analyze first reflection and transmission by the random slab in the *weakly heterogeneous* regime (5.16) introduced in Chapter 5. In this regime the correlation length of the fluctuations in the medium properties is of order ε^2, as is the wavelength. They are both much smaller than the size of the slab, which is of order 1. The typical amplitude of the fluctuations of the medium is small, of order ε in this regime. We assume that the medium parameters have the form

$$\frac{1}{K(z)} = \begin{cases} \dfrac{1}{\overline{K}} \left(1 + \varepsilon \nu \left(\dfrac{z}{\varepsilon^2}\right)\right) & \text{for } z \in [0, L], \\ \dfrac{1}{\overline{K}} & \text{for } z \in (-\infty, 0) \cup (L, \infty), \end{cases}$$

$$\rho(z) = \bar{\rho} \text{ for all } z,$$

where ν is a zero-mean, stationary random process satisfying strong decorrelation conditions. As in Chapter 6, we assume that the fluctuations have the form $\nu(z) = g(Y(z))$, where Y is a homogeneous in z Markov process with values in a compact space. We assume that it is strongly ergodic, satisfying the Fredholm alternative for solutions of the Poisson equation as in Section 6.3.3. The function g is a bounded real-valued function satisfying the centering condition $\mathbb{E}[g(Y(0))] = 0$. In the weakly heterogeneous scaling regime (5.16) the frequency of the monochromatic waves is ω/ε^2.

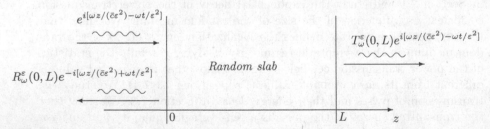

$e^{i[\omega z/(\bar{c}\varepsilon^2) - \omega t/\varepsilon^2]}$

$T_\omega^\varepsilon(0, L)e^{i[\omega z/(\bar{c}\varepsilon^2) - \omega t/\varepsilon^2]}$

$R_\omega^\varepsilon(0, L)e^{-i[\omega z/(\bar{c}\varepsilon^2) + \omega t/\varepsilon^2]}$

Random slab

0 L z

Fig. 7.1. Reflection and transmission of monochromatic waves.

We denote by $\hat{u}^\varepsilon$ and $\hat{p}^\varepsilon$ the time-harmonic complex velocity and pressure fields

$$\begin{bmatrix} u^{\varepsilon}(t,z) \\ p^{\varepsilon}(t,z) \end{bmatrix} = \exp\left(-\frac{i\omega t}{\varepsilon^2}\right) \begin{bmatrix} \hat{u}^{\varepsilon}(z) \\ \hat{p}^{\varepsilon}(z) \end{bmatrix},$$

which satisfy the system of random ordinary differential equations

$$-\frac{i\omega\bar{\rho}}{\varepsilon^2}\hat{u}^{\varepsilon} + \frac{d\hat{p}^{\varepsilon}}{dz} = 0, \tag{7.1}$$

$$-\frac{i\omega}{\bar{K}\varepsilon^2}\left(1 + \varepsilon\nu\left(\frac{z}{\varepsilon^2}\right)\right)\hat{p}^{\varepsilon} + \frac{d\hat{u}^{\varepsilon}}{dz} = 0. \tag{7.2}$$

As in Section 5.1.5, we introduce the right-going and left-going modes

$$\hat{a}^{\varepsilon}(\omega,z) = \left(\bar{\zeta}^{1/2}\hat{u}^{\varepsilon}(z) + \bar{\zeta}^{-1/2}\hat{p}^{\varepsilon}(z)\right)e^{\frac{-i\omega z}{\bar{c}\varepsilon^2}},$$

$$\hat{b}^{\varepsilon}(\omega,z) = \left(\bar{\zeta}^{1/2}\hat{u}^{\varepsilon}(z) - \bar{\zeta}^{-1/2}\hat{p}^{\varepsilon}(z)\right)e^{\frac{i\omega z}{\bar{c}\varepsilon^2}},$$

where the effective impedance and velocity are $\bar{\zeta} = \sqrt{\bar{K}\bar{\rho}}$ and $\bar{c} = \sqrt{\bar{K}/\bar{\rho}}$. The modes satisfy the differential equations

$$\frac{d}{dz}\begin{bmatrix} \hat{a}^{\varepsilon} \\ \hat{b}^{\varepsilon} \end{bmatrix} = \frac{1}{\varepsilon}\mathbf{H}_{\omega}\left(\frac{z}{\varepsilon^2}, \nu\left(\frac{z}{\varepsilon^2}\right)\right)\begin{bmatrix} \hat{a}^{\varepsilon} \\ \hat{b}^{\varepsilon} \end{bmatrix}, \tag{7.3}$$

$$\mathbf{H}_{\omega}(z,\nu) = \frac{i\omega}{2\bar{c}}\nu\begin{bmatrix} 1 & -e^{-2i\omega z/\bar{c}} \\ e^{2i\omega z/\bar{c}} & -1 \end{bmatrix},$$

which are the same as (5.27) with $\sigma = \varepsilon$ and $\theta = \varepsilon^{-1}$ for the propagator $\mathbf{P}_{\omega}^{\varepsilon}$ that we will introduce in the next section. The mode amplitudes satisfy the boundary conditions corresponding to a unit right-going monochromatic wave incident from the left at $z = 0$ and no wave incident from the right at $z = L$

$$\hat{a}^{\varepsilon}(\omega,0) = 1, \qquad \hat{b}^{\varepsilon}(\omega,L) = 0. \tag{7.4}$$

The reflection and transmission coefficients are given by

$$R_{\omega}^{\varepsilon}(0,L) = \hat{b}^{\varepsilon}(\omega,0), \qquad T_{\omega}^{\varepsilon}(0,L) = \hat{a}^{\varepsilon}(\omega,L), \tag{7.5}$$

see Figure 7.1. We will analyze the transmission coefficient in the limit $\varepsilon \to 0$ as a stochastic process in L for fixed frequency ω. We will study the joint statistics for two distinct ω in Section 7.2.3.

7.1.1 The Diffusion Limit for the Propagator

We first transform the boundary value problem (7.3–7.4) into an initial value problem. This step is similar to the analysis carried out in Section 5.1.6 (with $\sigma = \varepsilon$ and $\theta = \varepsilon^{-1}$). We introduce the propagator $\mathbf{P}_{\omega}^{\varepsilon}(0,z)$, i.e., the fundamental matrix solution of the linear system of random differential equations (7.3) with the initial condition $\mathbf{P}_{\omega}^{\varepsilon}(0, z = 0) = \mathbf{I}$. From symmetries in (7.3), as discussed in Section 4.4.3, $\mathbf{P}_{\omega}^{\varepsilon}$ has the form

$$\mathbf{P}_\omega^\varepsilon(0, z) = \begin{bmatrix} \alpha_\omega^\varepsilon(0, z) & \overline{\beta_\omega^\varepsilon(0, z)} \\ \beta_\omega^\varepsilon(0, z) & \overline{\alpha_\omega^\varepsilon(0, z)} \end{bmatrix}, \qquad (7.6)$$

where $(\alpha_\omega^\varepsilon, \beta_\omega^\varepsilon)^T$ is the solution of (7.3) with initial conditions

$$\alpha_\omega^\varepsilon(0, z = 0) = 1, \qquad \beta_\omega^\varepsilon(0, z = 0) = 0. \qquad (7.7)$$

The modes $\hat{a}^\varepsilon$ and $\hat{b}^\varepsilon$ can be expressed in terms of the propagator as

$$\begin{bmatrix} \hat{a}^\varepsilon(\omega, z) \\ \hat{b}^\varepsilon(\omega, z) \end{bmatrix} = \mathbf{P}_\omega^\varepsilon(0, z) \begin{bmatrix} \hat{a}^\varepsilon(\omega, 0) \\ \hat{b}^\varepsilon(\omega, 0) \end{bmatrix}. \qquad (7.8)$$

From (7.8) when applied at $z = L$ and from the boundary conditions (7.4) we deduce that

$$R_\omega^\varepsilon(0, L) = -\frac{\beta_\omega^\varepsilon(0, L)}{\alpha_\omega^\varepsilon(0, L)}, \quad T_\omega^\varepsilon(0, L) = \frac{1}{\alpha_\omega^\varepsilon(0, L)}. \qquad (7.9)$$

Since the trace of $\mathbf{H}_\omega$ in (7.3) is zero, we see as in Section 4.4.3 that

$$\det(\mathbf{P}_\omega^\varepsilon(0, L)) = |\alpha_\omega^\varepsilon(0, L)|^2 - |\beta_\omega^\varepsilon(0, L)|^2 = 1. \qquad (7.10)$$

From this and (7.9) we obtain the following energy conservation law for reflection and transmission:

$$|R_\omega^\varepsilon(0, L)|^2 + |T_\omega^\varepsilon(0, L)|^2 = 1. \qquad (7.11)$$

The meaning of this conservation law is that the input wave of unit energy splits into a reflected wave and a transmitted wave without any losses. The power transmission coefficient, $|T_\omega^\varepsilon(0, L)|^2$, gives the proportion of energy that is transmitted through the slab. It is equal to $1/|\alpha_\omega^\varepsilon(0, L)|^2$.

We now apply the diffusion-approximation Theorem 6.56 in its linear version, described in Section 6.7.3, to obtain the asymptotic distribution of the propagator $\mathbf{P}_\omega^\varepsilon$. This asymptotic distribution will then be used to find the asymptotic distribution of the transmission coefficient T_ω^ε using (7.9). We first rewrite the equation for the propagator in the expanded form

$$\frac{d}{dz}\mathbf{P}_\omega^\varepsilon(0, z) = \frac{i\omega}{2\varepsilon\bar{c}}\nu\left(\frac{z}{\varepsilon^2}\right)\begin{bmatrix} 1 & 0 \\ 0 & -1 \end{bmatrix}\mathbf{P}_\omega^\varepsilon(0, z)$$

$$-\frac{\omega}{2\bar{c}\varepsilon}\nu\left(\frac{z}{\varepsilon^2}\right)\sin\left(\frac{2\omega z}{\bar{c}\varepsilon^2}\right)\begin{bmatrix} 0 & 1 \\ 1 & 0 \end{bmatrix}\mathbf{P}_\omega^\varepsilon(0, z)$$

$$-\frac{i\omega}{2\bar{c}\varepsilon}\nu\left(\frac{z}{\varepsilon^2}\right)\cos\left(\frac{2\omega z}{\bar{c}\varepsilon^2}\right)\begin{bmatrix} 0 & 1 \\ -1 & 0 \end{bmatrix}\mathbf{P}_\omega^\varepsilon(0, z). \qquad (7.12)$$

This matrix-valued ordinary differential equation is of the form (6.102) with fast phase

$$\frac{d}{dz}\mathbf{P}_\omega^\varepsilon(0, z) = \frac{1}{\varepsilon}\mathbf{F}\left(\mathbf{P}_\omega^\varepsilon(0, z), \nu\left(\frac{z}{\varepsilon^2}\right), \frac{z}{\varepsilon^2}\right).$$

The decomposition (6.103) of the matrix field $\mathbf{F}$ is here given by

$$\mathbf{F}(\mathbf{P}, \nu, \tau) = \frac{\omega}{2\bar{c}} \sum_{p=0}^{2} g^{(p)}(\nu, \tau) \mathbf{h}_p \mathbf{P},$$

where

$$\mathbf{h}_0 = i\boldsymbol{\sigma}_3, \quad \mathbf{h}_1 = -\boldsymbol{\sigma}_1, \quad \mathbf{h}_2 = \boldsymbol{\sigma}_2,$$

with $\boldsymbol{\sigma}_1$, $\boldsymbol{\sigma}_2$, and $\boldsymbol{\sigma}_3$ the Pauli spin matrices

$$\boldsymbol{\sigma}_1 = \begin{bmatrix} 0 & 1 \\ 1 & 0 \end{bmatrix}, \quad \boldsymbol{\sigma}_2 = \begin{bmatrix} 0 & -i \\ i & 0 \end{bmatrix}, \quad \boldsymbol{\sigma}_3 = \begin{bmatrix} 1 & 0 \\ 0 & -1 \end{bmatrix}. \quad (7.13)$$

The real-valued functions $g^{(p)}$ are given by

$$g^{(0)}(\nu, \tau) = \nu, \quad g^{(1)}(\nu, \tau) = \nu \sin\left(\frac{2\omega z}{\bar{c}}\right), \quad g^{(2)}(\nu, \tau) = \nu \cos\left(\frac{2\omega z}{\bar{c}}\right).$$

The correlation matrix $\mathbf{C} = (C_{pq})_{p,q=0,1,2}$, defined by (6.92), can be computed explicitly in terms of the covariance of ν. Let us calculate its first two entries:

$$C_{00} = 2 \int_0^\infty \mathbb{E}[\nu(0)\nu(z)]\, dz,$$

$$C_{11} = 2\frac{1}{2\pi} \int_0^{2\pi} \int_0^\infty \mathbb{E}[\nu(0)\nu(z)]\, dz\, \sin(x) \sin\left(x + \frac{2\omega z}{\bar{c}}\right) dx$$

$$= \frac{1}{\pi} \int_0^{2\pi} \sin^2(x)\, dx \int_0^\infty \mathbb{E}[\nu(0)\nu(z)] \cos\left(\frac{2\omega z}{\bar{c}}\right) dz$$

$$+ \frac{1}{\pi} \int_0^{2\pi} \sin(x)\cos(x)\, dx \int_0^\infty \mathbb{E}[\nu(0)\nu(z)] \sin\left(\frac{2\omega z}{\bar{c}}\right) dz$$

$$= \int_0^\infty \mathbb{E}[\nu(0)\nu(z)] \cos\left(\frac{2\omega z}{\bar{c}}\right) dz.$$

The other entries of the matrix $\mathbf{C}$ in (6.92) can be calculated in a similar way, and we have

$$\mathbf{C} = \begin{bmatrix} \gamma(0) & 0 & 0 \\ 0 & \frac{1}{2}\gamma(\omega) & -\frac{1}{2}\gamma^{(s)}(\omega) \\ 0 & \frac{1}{2}\gamma^{(s)}(\omega) & \frac{1}{2}\gamma(\omega) \end{bmatrix},$$

with

$$\gamma(\omega) = 2 \int_0^\infty \mathbb{E}\left[\nu(0)\nu(z)\right] \cos\left(\frac{2\omega z}{\bar{c}}\right) dz, \quad (7.14)$$

$$\gamma^{(s)}(\omega) = 2 \int_0^\infty \mathbb{E}\left[\nu(0)\nu(z)\right] \sin\left(\frac{2\omega z}{\bar{c}}\right) dz. \quad (7.15)$$

The parameter $\gamma(\omega)$ is a nonnegative real number because it is proportional to the power spectral density of the stationary random process ν, as shown in Section 6.3.6. The symmetric and antisymmetric parts of $\mathbf{C}$ are given by

$$\mathbf{C}^{(S)} = \begin{bmatrix} \gamma(0) & 0 & 0 \\ 0 & \frac{1}{2}\gamma(\omega) & 0 \\ 0 & 0 & \frac{1}{2}\gamma(\omega) \end{bmatrix}, \qquad \mathbf{C}^{(A)} = \begin{bmatrix} 0 & 0 & 0 \\ 0 & 0 & -\frac{1}{2}\gamma^{(s)}(\omega) \\ 0 & \frac{1}{2}\gamma^{(s)}(\omega) & 0 \end{bmatrix}.$$

The symmetric part is a diagonal matrix, whose square root is also diagonal.

From the general results obtained in Section 6.7.3, we know that $\mathbf{P}_\omega^\varepsilon(0, z)$ converges in distribution to $\mathbf{P}_\omega(0, z)$, which is the solution of the Stratonovich stochastic differential equation (6.106),

$$dP_\omega(0, z) = \frac{\sqrt{\gamma(0)}\omega}{2\bar{c}} \mathbf{h}_0 \mathbf{P}_\omega(0, z) \circ dW_0(z) + \frac{\sqrt{\gamma(\omega)}\omega}{2\sqrt{2}\bar{c}} \mathbf{h}_1 \mathbf{P}_\omega(0, z) \circ dW_1(z)$$

$$+ \frac{\sqrt{\gamma(\omega)}\omega}{2\sqrt{2}\bar{c}} \mathbf{h}_2 \mathbf{P}_\omega(0, z) \circ d\tilde{W}_1(z) - \frac{\gamma^{(s)}(\omega)\omega^2}{8\bar{c}^2} \mathbf{h}_0 \mathbf{P}_\omega(0, z)\, dz,$$

where W_0, W_1, and $\tilde{W}_1$ are independent standard Brownian motions, and we have used the fact that $\mathbf{h}_2\mathbf{h}_1 = \mathbf{h}_0$. At $z = 0$, we have $\mathbf{P}_\omega(0, z = 0) = \mathbf{I}$. Using the explicit form of the matrices $\mathbf{h}_j$, $j = 0, 1, 2$, the limit propagator matrix is a linear diffusion process solution of the Stratonovich stochastic differential equation

$$dP_\omega(0, z) = \frac{i\omega\sqrt{\gamma(0)}}{2\bar{c}} \begin{bmatrix} 1 & 0 \\ 0 & -1 \end{bmatrix} \mathbf{P}_\omega(0, z) \circ dW_0(z)$$

$$- \frac{\omega\sqrt{\gamma(\omega)}}{2\sqrt{2}\bar{c}} \begin{bmatrix} 0 & 1 \\ 1 & 0 \end{bmatrix} \mathbf{P}_\omega(0, z) \circ dW_1(z)$$

$$- \frac{i\omega\sqrt{\gamma(\omega)}}{2\sqrt{2}\bar{c}} \begin{bmatrix} 0 & 1 \\ -1 & 0 \end{bmatrix} \mathbf{P}_\omega(0, z) \circ d\tilde{W}_1(z)$$

$$- \frac{i\omega^2\gamma^{(s)}(\omega)}{8\bar{c}^2} \begin{bmatrix} 1 & 0 \\ 0 & -1 \end{bmatrix} \mathbf{P}_\omega(0, z)\, dz. \qquad (7.16)$$

We can transform these stochastic differential equations into their Itô form by replacing in (7.16) the Stratonovich integrals by Itô integrals, and using (6.100) to compute the modified drift given by:

$$- \frac{i\omega^2\gamma^{(s)}(\omega)}{8\bar{c}^2} \begin{bmatrix} 1 & 0 \\ 0 & -1 \end{bmatrix} \mathbf{P}_\omega(0, z)\, dz - \frac{\omega^2[\gamma(0) - \gamma(\omega)]}{8\bar{c}^2} \mathbf{P}_\omega(0, z)\, dz. \qquad (7.17)$$

7.1.2 Polar Coordinates for the Propagator

By the symmetry of (7.16), the propagator matrix has the form

$$\mathbf{P}_\omega(0, z) = \begin{bmatrix} \alpha_\omega(0, z) & \overline{\beta_\omega(0, z)} \\ \beta_\omega(0, z) & \overline{\alpha_\omega(0, z)} \end{bmatrix},$$

which preserves in the limit $\varepsilon \to 0$ the form (7.6). The pair $(\alpha_\omega, \beta_\omega)$ satisfies the system of stochastic differential equations

$$d\alpha_\omega = \frac{\omega}{2\bar{c}} \left(i\sqrt{\gamma(0)}\alpha_\omega \circ dW_0(z) - \frac{\sqrt{\gamma(\omega)}}{\sqrt{2}}\beta_\omega \circ (dW_1(z) + id\tilde{W}_1(z)) \right)$$
$$-\frac{i\omega^2\gamma^{(s)}(\omega)}{8\bar{c}^2}\alpha_\omega \, dz,$$

$$d\beta_\omega = \frac{\omega}{2\bar{c}} \left(-i\sqrt{\gamma(0)}\beta_\omega \circ dW_0(z) - \frac{\sqrt{\gamma(\omega)}}{\sqrt{2}}\alpha_\omega \circ (dW_1(z) - id\tilde{W}_1(z)) \right)$$
$$+\frac{i\omega^2\gamma^{(s)}(\omega)}{8\bar{c}^2}\beta_\omega \, dz,$$

starting from $\alpha_\omega(0, z = 0) = 1$ and $\beta_\omega(0, z = 0) = 0$. All matrices that appear in the right side of (7.16) have trace zero, so that the determinant of $\mathbf{P}_\omega(0, z)$ is a conserved quantity. Therefore, the pair $(\alpha_\omega, \beta_\omega)$ satisfies the conservation of energy relation $|\alpha_\omega|^2 - |\beta_\omega|^2 = 1$ and can be parameterized as follows:

$$\alpha_\omega(0, z) = \cosh\left(\frac{\theta_\omega(z)}{2}\right)e^{i\phi_\omega(z)}, \tag{7.18}$$

$$\beta_\omega(0, z) = \sinh\left(\frac{\theta_\omega(z)}{2}\right)e^{i(\psi_\omega(z)+\phi_\omega(z))}, \tag{7.19}$$

with $\theta_\omega(z) \in [0, \infty)$, $\psi_\omega(z), \phi_\omega(z) \in \mathbb{R}$. Since we are in the Stratonovich framework, we find by the standard chain rule of differentiation that the process $(\theta_\omega, \psi_\omega, \phi_\omega)$ satisfies the system of Stratonovich stochastic differential equations given by

$$d\phi_\omega = -\frac{\omega\sqrt{\gamma(\omega)}}{2\sqrt{2}\bar{c}}\tanh\left(\frac{\theta_\omega}{2}\right)\left(\sin(\psi_\omega) \circ dW_1(z) + \cos(\psi_\omega) \circ d\tilde{W}_1(z)\right)$$
$$+\frac{\omega\sqrt{\gamma(0)}}{2\bar{c}}dW_0(z) - \frac{\omega^2\gamma^{(s)}(\omega)}{8\bar{c}^2}\,dz, \tag{7.20}$$

$$d\psi_\omega = \frac{\omega\sqrt{\gamma(\omega)}}{\sqrt{2}\bar{c}\tanh(\theta_\omega)}\left(\sin(\psi_\omega) \circ dW_1(z) + \cos(\psi_\omega) \circ d\tilde{W}_1(z)\right)$$
$$-\frac{\omega\sqrt{\gamma(0)}}{\bar{c}}dW_0(z) + \frac{\omega^2\gamma^{(s)}(\omega)}{4\bar{c}^2}\,dz, \tag{7.21}$$

$$d\theta_\omega = \frac{\omega\sqrt{\gamma(\omega)}}{\sqrt{2}\bar{c}}\left(-\cos(\psi_\omega) \circ dW_1(z) + \sin(\psi_\omega) \circ d\tilde{W}_1(z)\right). \tag{7.22}$$

We next transform these stochastic differential equations into their Itô form. We use (6.100) and we compute the Itô–Stratonovich corrections. As a result, the Itô form of (7.20–7.22) is

$$d\phi_\omega = -\frac{\omega\sqrt{\gamma(\omega)}}{2\sqrt{2}\bar{c}} \tanh\left(\frac{\theta_\omega}{2}\right)\left(\sin(\psi_\omega)\,dW_1(z) + \cos(\psi_\omega)\,d\tilde{W}_1(z)\right)$$

$$+\frac{\omega\sqrt{\gamma(0)}}{2\bar{c}}\,dW_0(z) - \frac{\omega^2\gamma^{(s)}(\omega)}{8\bar{c}^2}\,dz\,, \tag{7.23}$$

$$d\psi_\omega = \frac{\omega\sqrt{\gamma(\omega)}}{\sqrt{2}\bar{c}\tanh(\theta_\omega)}\left(\sin(\psi_\omega)\,dW_1(z) + \cos(\psi_\omega)\,d\tilde{W}_1(z)\right)$$

$$-\frac{\omega\sqrt{\gamma(0)}}{\bar{c}}\,dW_0(z) + \frac{\omega^2\gamma^{(s)}(\omega)}{4\bar{c}^2}\,dz\,, \tag{7.24}$$

$$d\theta_\omega = \frac{\omega\sqrt{\gamma(\omega)}}{\sqrt{2}\bar{c}}\left(-\cos(\psi_\omega)\,dW_1(z) + \sin(\psi_\omega)\,d\tilde{W}_1(z)\right)$$

$$+\frac{\omega^2\gamma(\omega)}{4\bar{c}^2\tanh(\theta_\omega)}\,dz\,. \tag{7.25}$$

If we introduce a new pair of processes $(W_1^*, \tilde{W}_1^*)$ by the orthogonal transformation

$$\begin{bmatrix} W_1^*(z) \\ \tilde{W}_1^*(z) \end{bmatrix} = \int_0^z \begin{bmatrix} \sin(\psi_\omega) & \cos(\psi_\omega) \\ -\cos(\psi_\omega) & \sin(\psi_\omega) \end{bmatrix} d \begin{bmatrix} W_1(z) \\ \tilde{W}_1(z) \end{bmatrix},$$

then these transformed processes are again independent standard Brownian motions. Therefore the stochastic processes $(\theta_\omega, \psi_\omega, \phi_\omega)$ can be written as the solution of the Itô stochastic differential equations

$$d\phi_\omega = -\frac{\omega\sqrt{\gamma(\omega)}}{2\sqrt{2}\bar{c}} \tanh\left(\frac{\theta_\omega}{2}\right) dW_1^*(z) + \frac{\omega\sqrt{\gamma(0)}}{2\bar{c}}\,dW_0(z)$$

$$-\frac{\omega^2\gamma^{(s)}(\omega)}{8\bar{c}^2}\,dz\,, \tag{7.26}$$

$$d\psi_\omega = \frac{\omega\sqrt{\gamma(\omega)}}{\sqrt{2}\bar{c}\tanh(\theta_\omega)}\,dW_1^*(z) - \frac{\omega\sqrt{\gamma(0)}}{\bar{c}}\,dW_0(z) + \frac{\omega^2\gamma^{(s)}(\omega)}{4\bar{c}^2}\,dz\,, \tag{7.27}$$

$$d\theta_\omega = \frac{\omega\sqrt{\gamma(\omega)}}{\sqrt{2}\bar{c}}\,d\tilde{W}_1^*(z) + \frac{\omega^2\gamma(\omega)}{4\bar{c}^2\tanh(\theta_\omega)}\,dz\,, \tag{7.28}$$

with the initial conditions $\theta_\omega(0) = 0$, $\psi_\omega(0) = 0$, $\phi_\omega(0) = 0$. Note that the problem (7.26–7.28) is well posed because the transformations (7.18–7.19) and the choice of the specific branch $\theta_\omega(z) \in [0, \infty)$ define uniquely the process $(\theta_\omega, \psi_\omega, \phi_\omega)$.

The generator of the Markov process $(\theta_\omega, \psi_\omega, \phi_\omega)$ is

$$\mathcal{L} = \frac{\gamma(\omega)\omega^2}{16\bar{c}^2}\left[\tanh^2\left(\frac{\theta_\omega}{2}\right)\frac{\partial^2}{\partial\phi_\omega^2} + 2\left(1 + \tanh^2\left(\frac{\theta_\omega}{2}\right)\right)\frac{\partial^2}{\partial\phi_\omega\partial\psi_\omega}\right]$$

$$+\frac{\gamma(\omega)\omega^2}{4\bar{c}^2}\left[\frac{\partial^2}{\partial\theta_\omega^2} + \frac{1}{\tanh(\theta_\omega)}\frac{\partial}{\partial\theta_\omega} + \frac{1}{\tanh^2(\theta_\omega)}\frac{\partial^2}{\partial\psi_\omega^2}\right]$$

$$+\frac{\gamma(0)\omega^2}{2\bar{c}^2}\left[\frac{\partial^2}{\partial\psi_\omega^2} + \frac{1}{4}\frac{\partial^2}{\partial\phi_\omega^2}\right] + \frac{\gamma^{(s)}(\omega)\omega^2}{4\bar{c}^2}\left[\frac{\partial}{\partial\psi_\omega} - \frac{1}{2}\frac{\partial}{\partial\phi_\omega}\right]. \tag{7.29}$$

The process $(\theta_\omega, \psi_\omega)$ is Markov by itself with generator

$$\mathcal{L} = \frac{\gamma(\omega)\omega^2}{4\bar{c}^2} \left[\frac{\partial^2}{\partial\theta_\omega^2} + \frac{1}{\tanh(\theta_\omega)} \frac{\partial}{\partial\theta_\omega} + \frac{1}{\tanh^2(\theta_\omega)} \frac{\partial^2}{\partial\psi_\omega^2} \right]$$
$$+ \frac{\gamma(0)\omega^2}{2\bar{c}^2} \frac{\partial^2}{\partial\psi_\omega^2} + \frac{\gamma^{(s)}(\omega)\omega^2}{4\bar{c}^2} \frac{\partial}{\partial\psi_\omega}. \tag{7.30}$$

The part of this generator in the square brackets is the Laplace–Beltrami operator on the hyperbolic disk, which is the space $(\theta, \psi) \in [0, \infty) \times [0, 2\pi)$ with the Riemannian metric

$$ds^2 = d\theta^2 + \sinh^2\theta\, d\psi^2.$$

We note in particular that the process $(\theta_\omega, \psi_\omega)$ is not quite a Brownian motion on the hyperbolic disk. However, the radial process θ_ω is by itself a diffusion Markov process with infinitesimal generator

$$\mathcal{L}_{rad} = \frac{\gamma(\omega)\omega^2}{4\bar{c}^2} \left[\frac{\partial^2}{\partial\theta_\omega^2} + \frac{1}{\tanh(\theta_\omega)} \frac{\partial}{\partial\theta_\omega} \right], \tag{7.31}$$

which is the radial part of the Laplace–Beltrami operator on the hyperbolic disk.

7.1.3 Martingale Representation of the Transmission Coefficient

From (7.9) and (7.18) we see that the transmission coefficient $T_\omega^\varepsilon(0, L)$ converges in distribution as $\varepsilon \to 0$ to a limit $T_\omega(0, L)$ that has the form

$$T_\omega(0, L) = \frac{e^{i\phi_\omega(L)}}{\cosh\left(\frac{\theta_\omega(L)}{2}\right)}. \tag{7.32}$$

Using the stochastic differential equations (7.26–7.28) and Itô's formula, we will show that the transmission coefficient $T_\omega(0, L)$ has the following martingale representation.

Proposition 7.1. *The transmission coefficient $T_\omega(0, L)$ has the representation*

$$T_\omega(0, L) = M_\omega(0, L)\tilde{T}_\omega(0, L), \tag{7.33}$$

where

$$\tilde{T}_\omega(0, L) = \exp\left[i\frac{\sqrt{\gamma(0)}\omega}{2\bar{c}} W_0(L) - i\frac{\gamma^{(s)}(\omega)\omega^2}{8\bar{c}^2} L - \frac{\gamma(\omega)\omega^2}{8\bar{c}^2} L \right], \tag{7.34}$$

and $M_\omega(0, L)$ is the complex martingale

$$M_\omega(0, L) = \exp\left[-\frac{\sqrt{\gamma(\omega)}\omega}{2\sqrt{2}\bar{c}} \int_0^L \tanh\left(\frac{\theta_\omega(z)}{2}\right) \left(d\tilde{W}_1^*(z) + idW_1^*(z) \right) \right], \tag{7.35}$$

whose mean is one.

We will use this representation in the next chapter, where we study the propagation of pulse fronts.

From (7.32) the power transmission coefficient $|T_\omega^\varepsilon(0, L)|^2$ converges in distribution as $\varepsilon \to 0$ to

$$|T_\omega(0, L)|^2 = \frac{1}{\cosh^2\left(\frac{\theta_\omega(L)}{2}\right)}. \tag{7.36}$$

From the representation (7.33) of $T_\omega(0, L)$, we obtain the following stochastic integral representation for the power transmission coefficient:

$$|T_\omega(0, L)|^2 = \exp\left[-\frac{\sqrt{\gamma(\omega)}\omega}{\sqrt{2}\bar{c}} \int_0^L \tanh\left(\frac{\theta_\omega(z)}{2}\right) d\tilde{W}_1^* - \frac{\gamma(\omega)\omega^2}{4\bar{c}^2}L\right]. \tag{7.37}$$

We will use (7.37) in the next section and in Section 7.1.6.

To show that the transmission coefficient has the martingale representation (7.33), we note the following. First, from (7.26) we deduce that

$$\exp(i\phi_\omega(L)) = \exp\left[-i\frac{\sqrt{\gamma(\omega)}\omega}{2\sqrt{2}\bar{c}} \int_0^L \tanh\left(\frac{\theta_\omega(z)}{2}\right) dW_1^*(z)\right.$$
$$\left. +i\frac{\omega\sqrt{\gamma(0)}}{2\bar{c}}W_0(L) - i\frac{\gamma^{(s)}(\omega)\omega^2}{8\bar{c}^2}L\right]. \tag{7.38}$$

Second, from (7.28) we write the Stratonovich differential equation

$$d\frac{1}{\cosh\left(\frac{\theta_\omega(z)}{2}\right)} = -\frac{1}{2}\frac{\tanh\left(\frac{\theta_\omega(z)}{2}\right)}{\cosh\left(\frac{\theta_\omega(z)}{2}\right)} \circ \left[\frac{\sqrt{\gamma(\omega)}\omega}{\sqrt{2}\bar{c}} d\tilde{W}_1^*(z) + \frac{\gamma(\omega)\omega^2}{2\bar{c}^2\tanh(\theta_\omega(z))} dz\right],$$

which we can integrate to obtain

$$\frac{1}{\cosh\left(\frac{\theta_\omega(L)}{2}\right)} = \exp\left[-\frac{\sqrt{\gamma(\omega)}\omega}{2\sqrt{2}\bar{c}} \int_0^L \tanh\left(\frac{\theta_\omega(z)}{2}\right) \circ d\tilde{W}_1^*(z)\right.$$
$$\left. -\frac{\gamma(\omega)\omega^2}{4\bar{c}^2} \int_0^L \frac{\tanh\left(\frac{\theta_\omega(z)}{2}\right)}{\tanh(\theta_\omega(z))} dz\right].$$

Writing the stochastic integral on the right side in Itô form and using the formula $\tanh(s/2)/\tanh(s) = (1/2)(1 + \tanh^2(s/2))$ for the drift term, we get

$$\frac{1}{\cosh\left(\frac{\theta_\omega(L)}{2}\right)} = \exp\left[-\frac{\sqrt{\gamma(\omega)}\omega}{2\sqrt{2}\bar{c}} \int_0^L \tanh\left(\frac{\theta_\omega(z)}{2}\right) d\tilde{W}_1^*(z) - \frac{\gamma(\omega)\omega^2}{8\bar{c}^2}L\right].$$

Multiplying this identity by (7.38) gives (7.33).

7.1.4 The Localization Length $L_{\text{loc}}(\omega)$

We have just established that the power transmission coefficient $|T_\omega^\varepsilon(0, L)|^2$ converges in distribution to $\tau_\omega(L) = \cosh^{-2}(\theta_\omega(L)/2)$, which is a diffusion Markov process. By (7.31) the infinitesimal generator of $\tau_\omega(L)$ is

$$\mathcal{L}_\omega = \frac{\gamma(\omega)\omega^2}{4\bar{c}^2} \left[\tau_\omega^2 (1 - \tau_\omega) \frac{\partial^2}{\partial \tau_\omega^2} - \tau_\omega^2 \frac{\partial}{\partial \tau_\omega} \right]. \tag{7.39}$$

From now on we assume that

$$\gamma(\omega) > 0, \tag{7.40}$$

which means that the power spectral density (7.14) of the fluctuation process ν is not zero at the frequencies of interest.

As we shall see in the next section, we can compute all moments of the limit power transmission coefficient τ_ω by solving the diffusion equation associated with the infinitesimal generator $\mathcal{L}_\omega$. In this section we prove the following proposition.

Proposition 7.2. *The logarithm of the power transmission coefficient converges with probability one as $L \to \infty$:*

$$\lim_{L \to \infty} \frac{1}{L} \ln \left[\tau_\omega(L) \right] = -\frac{1}{L_{\text{loc}}(\omega)}, \tag{7.41}$$

where

$$L_{\text{loc}}(\omega) = \frac{4\bar{c}^2}{\gamma(\omega)\omega^2}. \tag{7.42}$$

This result follows from

$$\frac{\theta_\omega(L)}{L} \xrightarrow{L \to \infty} \frac{1}{L_{\text{loc}}(\omega)}, \tag{7.43}$$

with probability one, which we show below. From (7.43) we have

$$\lim_{L \to \infty} \frac{1}{L} \ln \left[\tau_\omega(L) \right] = \lim_{L \to \infty} \frac{1}{L} \ln \left[\cosh^{-2} \left(\frac{\theta_\omega(L)}{2} \right) \right]$$

$$= -\lim_{L \to \infty} \frac{1}{L} \theta_\omega(L) = -\frac{1}{L_{\text{loc}}(\omega)},$$

with probability one, which is (7.41).

From the representation (7.37) and (7.41) we have, with probability one, the asymptotic equivalence in the sense of logarithms

$$\tau_\omega(L) \sim \exp \left(-\frac{L}{L_{\text{loc}}(\omega)} - \frac{\sqrt{2}}{\sqrt{L_{\text{loc}}(\omega)}} \tilde{W}_1^*(L) \right), \tag{7.44}$$

as $L \to \infty$. Since $\tilde{W}_1^*(L)/L$ tends to zero with probability one as $L \to \infty$, the frequency-dependent length $L_{\mathrm{loc}}(\omega)$ characterizes the exponential decay of the power transmission coefficient, at frequency ω, in the weak fluctuations regime as $\varepsilon \to 0$.

From (7.44) we note that the fluctuations of the log power transmission coefficient can be characterized in the limit $L \to \infty$. We have that

$$\sqrt{L}\left[\frac{1}{L}\ln \tau_\omega(L) - \frac{1}{L_{\mathrm{loc}}(\omega)}\right]$$

converges in distribution as $L \to \infty$ to a Gaussian random variable with mean zero and variance $2/L_{\mathrm{loc}}(\omega)$.

We call $L_{\mathrm{loc}}(\omega)$, defined by (7.42), the localization length of the random medium, and we will use this notation frequently throughout the book. In this section, we have shown that L_{loc} characterizes the exponential decay rate (as $L \to \infty$) of the limit in distribution (as $\varepsilon \to 0$) of the power transmission coefficient. In Section 7.3, we will show that the power transmission coefficient for fixed $\varepsilon > 0$ has exponential decay characterized by some ε-dependent localization length $L_{\mathrm{loc}}^\varepsilon$, and that $L_{\mathrm{loc}}^\varepsilon$ has the limit L_{loc} given by (7.42) as $\varepsilon \to 0$.

By **wave localization** we mean in this book exponential decay of the power transmission coefficient as a function of the size of the random medium. It is a special property of randomly layered media that the decay rate is exponential regardless of the strength of the fluctuations of the random medium. It is also known, as we comment in the notes, that the spectrum of the reduced wave equation in a random medium is pure point with exponentially decaying eigenfunctions.

To prove (7.43) we first note that the process θ_ω takes values in $(0, \infty)$, and so $\tanh(\theta_\omega(z)) \leq 1$, which implies that

$$\frac{1}{\tanh(\theta_\omega(z))} \geq 1,$$

for all $z > 0$. From the integral representation of θ_ω,

$$\theta_\omega(L) = \frac{1}{L_{\mathrm{loc}}(\omega)}\int_0^L \frac{1}{\tanh(\theta_\omega(s))}\,ds + \frac{\sqrt{2}}{\sqrt{L_{\mathrm{loc}}(\omega)}}\tilde{W}_1^*(L), \qquad (7.45)$$

we obtain therefore the inequality

$$\theta_\omega(L) \geq \frac{L}{L_{\mathrm{loc}}(\omega)} + \frac{\sqrt{2}}{\sqrt{L_{\mathrm{loc}}(\omega)}}\tilde{W}_1^*(L).$$

This shows that

$$\liminf_{L\to\infty}\frac{\theta_\omega(L)}{L} \geq \frac{1}{L_{\mathrm{loc}}(\omega)}, \qquad (7.46)$$

with probability one, since $\tilde{W}_1^*(L)/L \to 0$ as $L \to \infty$ with probability one. Since $\gamma(\omega) > 0$, we have $\liminf_{L\to\infty} \theta_\omega(L) = \infty$ with probability one, and

$$\limsup_{L\to\infty} \frac{\theta_\omega(L)}{L} \leq \frac{1}{L_{\mathrm{loc}}(\omega)} \tag{7.47}$$

with probability one. From (7.46) and (7.47) we deduce (7.43).

7.1.5 Mean and Fluctuations of the Power Transmission Coefficient

In the previous section we have shown that the power transmission coefficient converges to a diffusion Markov process with the infinitesimal generator (7.39). This result is summarized in the following proposition.

Proposition 7.3. *The power transmission coefficient $|T_\omega^\varepsilon(0,L)|^2$ converges in distribution as a continuous process in L to the Markov process $\tau_\omega(L)$ whose infinitesimal generator is*

$$\mathcal{L}_\omega = \frac{1}{L_{\mathrm{loc}}(\omega)} \left[\tau_\omega^2 (1-\tau_\omega) \frac{\partial^2}{\partial \tau_\omega^2} - \tau_\omega^2 \frac{\partial}{\partial \tau_\omega} \right]. \tag{7.48}$$

Here the localization length $L_{\mathrm{loc}}(\omega)$ is defined by (7.42). The generator (7.48) characterizes the statistical distribution of the process $\tau_\omega(L)$, and in particular its moments $\mathbb{E}[\tau_\omega(L)^n]$, $n \in \mathbb{N}$. It is clear from the form of the generator (7.48) that the moments of $\tau_\omega(L)$ are functions of $L/L_{\mathrm{loc}}(\omega)$ only. We compute these moments in Appendix 7.6, where we show that they have the integral representations

$$\mathbb{E}[\tau_\omega(L)^n] = \xi_n \left(\frac{L}{L_{\mathrm{loc}}(\omega)} \right). \tag{7.49}$$

Here the functions $\xi_n(l)$ are defined by

$$\xi_n(l) = \exp\left(-\frac{l}{4} \right) \int_0^\infty e^{-\mu^2 l} \frac{2\pi\mu \sinh(\pi\mu)}{\cosh^2(\pi\mu)} K^{(n)}(\mu)\, d\mu, \tag{7.50}$$

where $K^{(1)}(\mu) = 1$ and, for $n \geq 2$,

$$K^{(n)}(\mu) = \prod_{j=1}^{n-1} \frac{1}{j^2} \left[\mu^2 + (j - \frac{1}{2})^2 \right].$$

We state in particular the limiting form of the mean power transmission coefficient in the following proposition.

Proposition 7.4. *The mean power transmission coefficient $\mathbb{E}[|T_\omega^\varepsilon(0,L)|^2]$ converges as $\varepsilon \to 0$,*

$$\lim_{\varepsilon \to 0} \mathbb{E}[|T_\omega^\varepsilon(0,L)|^2] = \mathbb{E}[\tau_\omega(L)] = \xi_1 \left(\frac{L}{L_{\mathrm{loc}}(\omega)}\right), \tag{7.51}$$

where $\xi_1(l)$ is given by

$$\xi_1(l) = \exp\left(-\frac{l}{4}\right) \int_0^\infty e^{-\mu^2 l} \frac{2\pi\mu \sinh(\mu\pi)}{\cosh^2(\mu\pi)} \, d\mu . \tag{7.52}$$

The exponential decay rate of the mean power transmission coefficient is

$$\lim_{L \to \infty} \frac{1}{L} \ln\left(\mathbb{E}[\tau_\omega(L)]\right) = -\frac{1}{4L_{\mathrm{loc}}(\omega)} . \tag{7.53}$$

The meaning of (7.53) is that for $L \gg L_{\mathrm{loc}}(\omega)$, the expectation of $\tau_\omega(L)$ is given by

$$\mathbb{E}[\tau_\omega(L)] \sim \exp\left(-\frac{L}{4L_{\mathrm{loc}}(\omega)}\right) .$$

The asymptotic representation (7.53) is obtained with the Laplace asymptotic approximation method, which gives for $l \gg 1$,

$$\xi_1(l) \sim 2\pi^2 \exp\left(-\frac{l}{4}\right) \int_0^\infty e^{-\mu^2 l} \mu^2 \, d\mu = \frac{\pi^{5/2}}{2l^{3/2}} \exp\left(-\frac{l}{4}\right) . \tag{7.54}$$

More generally, we have for $l \gg 1$,

$$\xi_n(l) \sim 2\pi^2 K^{(n)}(0) \exp\left(-\frac{l}{4}\right) \int_0^\infty e^{-\mu^2 l} \mu^2 \, d\mu = \frac{\pi^{5/2} K^{(n)}(0)}{2l^{3/2}} \exp\left(-\frac{l}{4}\right),$$

where $K^{(n)}(0) = \prod_{j=1}^{n-1} (1 - \frac{1}{2j})^2$. This shows that for any $n \geq 1$,

$$\lim_{L \to \infty} \frac{1}{L} \ln\left(\mathbb{E}[\tau_\omega(L)^n]\right) = -\frac{1}{4L_{\mathrm{loc}}(\omega)}, \tag{7.55}$$

which means that the exponential decay rate of the limiting moments of the power transmission coefficient is the same for all $n \geq 1$ and is given by $1/(4L_{\mathrm{loc}}(\omega))$. We plot in Figure 7.2 the mean power transmission coefficient and its standard deviation as a function of $L/L_{\mathrm{loc}}(\omega)$. Note that the relative standard deviation plotted in picture (b) of Figure 7.2 tends to infinity as $L/L_{\mathrm{loc}}(\omega) \to \infty$, which shows that the power transmission is a very fluctuating process.

7.1.6 The Strongly Fluctuating Character of the Power Transmission Coefficient

Comparing (7.53) with (7.41), we see that the exponential behavior of the mean power transmission coefficient $\sim \exp[-L/(4L_{\mathrm{loc}}(\omega))]$ is different from its typical behavior $\sim \exp[-L/L_{\mathrm{loc}}(\omega)]$. This is also seen in the exponential

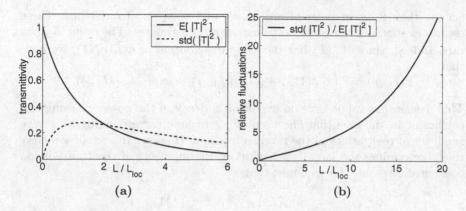

Fig. 7.2. Plots of the behavior of the power transmission coefficient as a function of $L/L_{\text{loc}}(\omega)$. The mean and the standard deviation of the limit power transmission coefficient $\tau_\omega(0)$ are plotted in picture (a). They decay exponentially with $L/L_{\text{loc}}(\omega)$ with rates $1/(4L_{\text{loc}}(\omega))$ and $1/(8L_{\text{loc}}(\omega))$, respectively. Picture (b) is a plot of the relative fluctuations of the power transmission coefficient, which grow exponentially with rate $1/(8L_{\text{loc}}(\omega))$.

growth of the normalized standard deviation of the power transmission coefficient plotted in Figure 7.2b.

This is a common phenomenon for strongly fluctuating random processes. We now give some heuristic arguments to complete the discussion. To analyze it, we use the representation of the limit power transmission coefficient

$$\tau_\omega(L) = \frac{1}{\cosh^2\left(\frac{\theta_\omega(L)}{2}\right)} \tag{7.56}$$

in terms of the process $\theta_\omega(L)$, which is the solution of the stochastic differential equation (7.28). The integral representation (7.45) of this process gives for $L \gg L_{\text{loc}}(\omega)$,

$$\theta_\omega(L) \sim \frac{\sqrt{2}}{\sqrt{L_{\text{loc}}}}\tilde{W}_1^*(L) + \frac{L}{L_{\text{loc}}}.$$

This asymptotic representation is an extension of the analysis leading to (7.43). Since $L \gg L_{\text{loc}}(\omega)$, with high probability we know that $\tilde{W}_1^*(L)$ is of order $\sqrt{L}$, which is negligible compared to L, so that $\cosh^2(\theta_\omega(L)/2) \sim \exp(L/L_{\text{loc}})$. By (7.56) this implies that $\tau_\omega(L) \sim \exp(-L/L_{\text{loc}})$ with high probability. This is the observable exponential decay of the power transmission coefficient, with decay rate equal to the reciprocal of the localization length L_{loc} given by (7.80).

If, however, the realization of the Brownian motion $\tilde{W}_1^*$ is such that the event

$$E_{\omega,L} := \left\{ \sqrt{2/L_{\text{loc}}}\tilde{W}_1^*(L) < -L/L_{\text{loc}} \right\}$$

occurs, then θ_ω is at most of order one, and so τ_ω is of order one. These estimates should be understood as exponential estimates. The event $E_{\omega,L}$ is rare. Indeed, since $\tilde{W}_1^*(L)$ has the same distribution as $\sqrt{L}\tilde{W}_1^*(1)$, we have that

$$\mathbb{P}(E_{\omega,L}) = \mathbb{P}\left(\tilde{W}_1^*(1) < -\sqrt{L/(2L_{\text{loc}})}\right) \sim \exp(-L/(4L_{\text{loc}})).$$

This means that we observe no exponential decay of the power transmission coefficient in the case that the event $E_{\omega,L}$ occurs, corresponding to a very small set of realizations of the random medium. However this small set effectively determines the large-L behaviors of the moments of the transmission coefficient. We note, for example, that

$$\begin{aligned}
\mathbb{E}[\tau_\omega(L)] &= \mathbb{E}[\tau_\omega(L)\mathbf{1}_{E_{\omega,L}}] + \mathbb{E}[\tau_\omega(L)\mathbf{1}_{E_{\omega,L}^c}] \\
&\overset{L\gg1}{\sim} 1 \times \mathbb{P}(E_{\omega,L}) + \exp(-L/L_{\text{loc}})\mathbb{P}(E_{\omega,L}^c) \\
&\sim \exp(-L/(4L_{\text{loc}})) + \exp(-L/L_{\text{loc}}) \\
&\sim \exp(-L/(4L_{\text{loc}})),
\end{aligned}$$

from which we recover the result (7.53). For the nth moment we have

$$\begin{aligned}
\mathbb{E}[\tau_\omega(L)^n] &= \mathbb{E}[\tau_\omega(L)^n\mathbf{1}_{E_{\omega,L}}] + \mathbb{E}[\tau_\omega(L)^n\mathbf{1}_{E_{\omega,L}^c}] \\
&\overset{L\gg1}{\sim} 1 \times \mathbb{P}(E_{\omega,L}) + \exp(-nL/L_{\text{loc}})\mathbb{P}(E_{\omega,L}^c) \\
&\sim \exp(-L/(4L_{\text{loc}})) + \exp(-nL/L_{\text{loc}}) \\
&\sim \exp(-L/(4L_{\text{loc}})),
\end{aligned}$$

from which we recover (7.55). Thus the large-L behavior of the moments of the power transmission coefficient is determined by exceptional realizations of the random medium. What is called the localization length is $L_{\text{loc}}(\omega)$, given by (7.42), which is what is observed for a typical realization of the random medium. We will see in the next section that this is true only for monochromatic waves. The reason for the difference between monochromatic and pulsed waves is that the rare event $E_{\omega,L}$ depends on the frequency ω, and with a continuum of frequencies in the pulse case it is likely that these events occur for some frequencies with high probability.

7.2 Exponential Decay of the Transmitted Energy for a Pulse

7.2.1 Transmission of a Pulse Through a Slab of Random Medium

We consider a right-going pulse incoming from the left homogeneous half-space

$$A^\varepsilon(t, z) = \frac{1}{2\pi} \int \hat{f}^\varepsilon(\omega) \exp\left(i\frac{\omega z}{\bar{c}} - i\omega t\right) d\omega, \quad z \leq 0, \qquad (7.57)$$

where $\hat{f}^\varepsilon$ is a function whose effective bandwidth is of order ε^{-2}, so that

$$\hat{f}^\varepsilon(\omega) = \varepsilon\hat{f}(\varepsilon^2\omega) \iff f^\varepsilon(t) = \frac{1}{\varepsilon}f\left(\frac{t}{\varepsilon^2}\right),$$

where $\hat{f}(\omega)$ is a rapidly decaying function. The pulse width is of order ε^{-2} and the pulse amplitude has been normalized so that the pulse energy is of

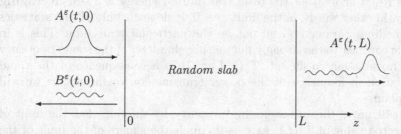

Fig. 7.3. Transmission of a pulse.

order one:

$$\mathcal{E}_{\text{inc}} := \int |A^\varepsilon(t,0)|^2\, dt = \frac{1}{2\pi}\int |\hat{f}^\varepsilon(\omega)|^2\, d\omega = \frac{1}{2\pi}\int |\hat{f}(\omega)|^2\, d\omega\,.$$

The total field in the region $z \leq 0$ is the superposition of the incoming wave A^ε and the reflected wave

$$B^\varepsilon(t,z) = \frac{1}{2\pi\varepsilon}\int \hat{f}(\omega)R_\omega^\varepsilon(0,L)\exp\left(-i\frac{\omega z}{\bar{c}\varepsilon^2} - i\frac{\omega t}{\varepsilon^2}\right)d\omega\,, \quad z \leq 0, \quad (7.58)$$

where $R_\omega^\varepsilon(0,L)$ is the reflection coefficient for the frequency ω/ε^2. The field in the region $z \geq L$ is only the transmitted right-going wave

$$A^\varepsilon(t,z) = \frac{1}{2\pi\varepsilon}\int \hat{f}(\omega)T_\omega^\varepsilon(0,L)\exp\left(i\frac{\omega z}{\bar{c}\varepsilon^2} - i\frac{\omega t}{\varepsilon^2}\right)d\omega\,, \quad z \geq L, \quad (7.59)$$

where $T_\omega^\varepsilon(0,L)$ is the transmission coefficient for the frequency ω/ε^2. The total transmitted energy is therefore

$$\mathcal{T}^\varepsilon(L) := \int |A^\varepsilon(t,L)|^2\, dt = \frac{1}{2\pi}\int |\hat{f}(\omega)|^2 |T_\omega^\varepsilon(0,L)|^2\, d\omega\,, \quad (7.60)$$

by the Parseval identity.

7.2.2 Self-Averaging Property of the Transmitted Energy

In this section we show that the random variable $\mathcal{T}^\varepsilon(L)$ converges in probability to its limit expectation.

Proposition 7.5. *The total transmitted pulse energy $\mathcal{T}^{\varepsilon}(L)$, given by (7.60), converges in probability to the deterministic quantity*

$$\mathcal{T}(L) = \frac{1}{2\pi} \int |\hat{f}(\omega)|^2 \xi_1 \left(\frac{L}{L_{\mathrm{loc}}(\omega)} \right) d\omega, \tag{7.61}$$

where $\xi_1(l)$ is given by (7.52).

This result shows that the total transmitted energy is a **self-averaging** quantity, in other words, in the limit $\varepsilon \to 0$, it depends only on the statistics of the medium parameters and not on the particular realization. This is in dramatic contrast to the strongly fluctuating character of the single-frequency power transmission coefficient $|T_{\omega}^{\varepsilon}(0, L)|^2$. It is a consequence of the rapid decorrelation in frequency of the power transmission coefficient, as we will now explain.

We will prove this self-averaging property by showing that the limit of the expected value of $\mathcal{T}^{\varepsilon}(L)^2$ as $\varepsilon \to 0$ equals the square of the limit of the expected value of $\mathcal{T}^{\varepsilon}(L)$. This implies that the fluctuations of $\mathcal{T}^{\varepsilon}(L)$ converge to zero as $\varepsilon \to 0$.

By linearity, the first moment of $\mathcal{T}^{\varepsilon}(L)$ has the integral representation

$$\mathbb{E}\left[\mathcal{T}^{\varepsilon}(L)\right] = \frac{1}{2\pi} \int |\hat{f}(\omega)|^2 \mathbb{E}\left[|T_{\omega}^{\varepsilon}(0, L)|^2\right] d\omega. \tag{7.62}$$

Therefore, by Proposition 7.4 we have that

$$\mathbb{E}\left[\mathcal{T}^{\varepsilon}(L)\right] \xrightarrow{\varepsilon \to 0} \mathcal{T}(L),$$

where $\mathcal{T}(L)$ is given by (7.61).

The second moment of $\mathcal{T}^{\varepsilon}(L)$ has the integral representation

$$\mathbb{E}\left[\mathcal{T}^{\varepsilon}(L)^2\right] = \frac{1}{4\pi^2} \int \int |\hat{f}(\omega)|^2 |\hat{f}(\omega')|^2 \mathbb{E}\left[|T_{\omega}^{\varepsilon}(0, L)|^2 |T_{\omega'}^{\varepsilon}(0, L)|^2\right] d\omega \, d\omega'. \tag{7.63}$$

It is therefore necessary to study the limit as $\varepsilon \to 0$ of the two-frequency process $(|T_{\omega}^{\varepsilon}(0, L)|^2, |T_{\omega'}^{\varepsilon}(0, L)|^2)$ for $\omega \neq \omega'$. We show in the next section that $|T_{\omega}^{\varepsilon}(0, L)|^2$ and $|T_{\omega'}^{\varepsilon}(0, L)|^2$ are asymptotically uncorrelated for any $\omega \neq \omega'$, so that

$$\mathbb{E}\left[\mathcal{T}^{\varepsilon}(L)^2\right] \xrightarrow{\varepsilon \to 0} \frac{1}{4\pi^2} \int \int |\hat{f}(\omega)|^2 |\hat{f}(\omega')|^2 \xi_1 \left(\frac{L}{L_{\mathrm{loc}}(\omega)} \right) \xi_1 \left(\frac{L}{L_{\mathrm{loc}}(\omega')} \right) d\omega \, d\omega'$$

$$= \left(\frac{1}{2\pi} \int |\hat{f}(\omega)|^2 \xi_1 \left(\frac{L}{L_{\mathrm{loc}}(\omega)} \right) d\omega \right)^2$$

$$= \mathcal{T}(L)^2.$$

Therefore $\mathcal{T}^{\varepsilon}(L)$ converges to $\mathcal{T}(L)$ in mean square:

$$\mathbb{E}\left[(\mathcal{T}^\varepsilon(L) - \mathcal{T}(L))^2\right] = \mathbb{E}\left[\mathcal{T}^\varepsilon(L)^2\right] - 2\mathbb{E}\left[\mathcal{T}^\varepsilon(L)\right]\mathcal{T}(L) + \mathcal{T}(L)^2 \xrightarrow{\varepsilon \to 0} 0.$$

By the Chebyshev inequality this implies convergence in probability. For any $\delta > 0$,

$$\mathbb{P}\left(|\mathcal{T}^\varepsilon(L) - \mathcal{T}(L)| > \delta\right) \leq \frac{\mathbb{E}\left[(\mathcal{T}^\varepsilon(L) - \mathcal{T}(L))^2\right]}{\delta^2} \xrightarrow{\varepsilon \to 0} 0,$$

which completes the proof of Proposition 7.5.

7.2.3 The Diffusion Limit for the Two-Frequency Propagator

In this section we show that the power transmission coefficients at two distinct frequencies are asymptotically independent in the limit $\varepsilon \to 0$. To do this we fix two frequencies $\omega_1 \neq \omega_2$ and apply the diffusion approximation theorem 6.56 to obtain the limit distribution of the two-frequency propagator. This limit distribution will then be used to show the asymptotic independence of the pair of power transmission coefficients $(|T^\varepsilon_{\omega_1}(0, L)|^2, |T^\varepsilon_{\omega_2}(0, L)|^2)$, using (7.9).

We introduce the 4×4 two-frequency propagator matrix

$$\mathbf{P}^\varepsilon_2(0, z) = \begin{bmatrix} \mathbf{P}^\varepsilon_{\omega_1}(0, z) & \mathbf{0} \\ \mathbf{0} & \mathbf{P}^\varepsilon_{\omega_2}(0, z) \end{bmatrix},$$

where $\mathbf{0}$ is the 2×2 zero matrix. From the equations (7.12) satisfied by the two propagators $\mathbf{P}^\varepsilon_{\omega_j}$, the random differential equation satisfied by $\mathbf{P}^\varepsilon_2$ is of the form (6.102) with fast phase

$$\frac{d}{dz}\mathbf{P}^\varepsilon_2(0, z) = \frac{1}{\varepsilon}\mathbf{F}\left(\mathbf{P}^\varepsilon_2(0, z), \nu\left(\frac{z}{\varepsilon^2}\right), \frac{z}{\varepsilon^2}\right).$$

Here the decomposition (6.103) of the matrix field $\mathbf{F}$ is

$$\mathbf{F}(\mathbf{P}_2, \nu, \tau) = \frac{1}{2\bar{c}} \sum_{p=0}^{4} g^{(p)}(\nu, \tau)\mathbf{\Omega}_2 \mathbf{h}^{(p)}\mathbf{P}_2,$$

where the constant 4×4 matrices $\mathbf{\Omega}_2$ and $\mathbf{h}^{(p)}$, $p = 0, ..., 4$ are given by

$$\mathbf{\Omega}_2 = \begin{bmatrix} \omega_1\mathbf{I} & 0 \\ 0 & \omega_2\mathbf{I} \end{bmatrix}, \qquad \mathbf{h}^{(0)} = i\begin{bmatrix} \sigma_3 & 0 \\ 0 & \sigma_3 \end{bmatrix}, \qquad \mathbf{h}^{(1)} = -\begin{bmatrix} \sigma_1 & 0 \\ 0 & 0 \end{bmatrix},$$

$$\mathbf{h}^{(2)} = \begin{bmatrix} \sigma_2 & 0 \\ 0 & 0 \end{bmatrix}, \qquad \mathbf{h}^{(3)} = -\begin{bmatrix} 0 & 0 \\ 0 & \sigma_1 \end{bmatrix}, \qquad \mathbf{h}^{(4)} = \begin{bmatrix} 0 & 0 \\ 0 & \sigma_2 \end{bmatrix},$$

with $\mathbf{I}$ the 2×2 identity matrix and σ_1, σ_2, and σ_3 the Pauli spin matrices (7.13). The real-valued functions $g^{(p)}$, $p = 0, \ldots, 4$ are given by

$$g^{(0)}(\nu, \tau) = \nu,$$

$$g^{(1)}(\nu, \tau) = \nu \sin\left(\frac{2\omega_1 z}{\bar{c}}\right), \qquad g^{(2)}(\nu, \tau) = \nu \cos\left(\frac{2\omega_1 z}{\bar{c}}\right),$$

$$g^{(3)}(\nu, \tau) = \nu \sin\left(\frac{2\omega_2 z}{\bar{c}}\right), \qquad g^{(4)}(\nu, \tau) = \nu \cos\left(\frac{2\omega_2 z}{\bar{c}}\right).$$

The correlation matrix $\mathbf{C} = (C_{pq})_{p,q=0,\ldots,4}$ defined by (6.92) has here the form

$$\mathbf{C} = \begin{bmatrix} \gamma(0) & 0 & 0 & 0 & 0 \\ 0 & \frac{1}{2}\gamma(\omega_1) & -\frac{1}{2}\gamma^{(s)}(\omega_1) & 0 & 0 \\ 0 & \frac{1}{2}\gamma^{(s)}(\omega_1) & \frac{1}{2}\gamma(\omega_1) & 0 & 0 \\ 0 & 0 & 0 & \frac{1}{2}\gamma(\omega_2) & -\frac{1}{2}\gamma^{(s)}(\omega_2) \\ 0 & 0 & 0 & \frac{1}{2}\gamma^{(s)}(\omega_2) & \frac{1}{2}\gamma(\omega_2) \end{bmatrix},$$

with $\gamma(\omega)$ and $\gamma^{(s)}(\omega)$ given by (7.14–7.15). From the general results obtained in Section 6.7.3, we get that $\mathbf{P}_2^\varepsilon(0, z)$ converges in distribution to $\mathbf{P}_2(0, z)$, which is the solution of the linear Stratonovich stochastic differential equation

$$d\mathbf{P}_2(0, z) = \frac{\sqrt{\gamma(0)}}{2\bar{c}} \boldsymbol{\Omega}_2 \mathbf{h}^{(0)} \mathbf{P}_2(0, z) \circ dW_0(z)$$

$$+ \frac{\sqrt{\gamma(\omega_1)}\omega_1}{2\sqrt{2}\bar{c}} \mathbf{h}^{(1)} \mathbf{P}_2(0, z) \circ dW_1(z) + \frac{\sqrt{\gamma(\omega_1)}\omega_1}{2\sqrt{2}\bar{c}} \mathbf{h}^{(2)} \mathbf{P}_2(0, z) \circ d\tilde{W}_1(z)$$

$$+ \frac{\sqrt{\gamma(\omega_2)}\omega_2}{2\sqrt{2}\bar{c}} \mathbf{h}^{(3)} \mathbf{P}_2(0, z) \circ dW_2(z) + \frac{\sqrt{\gamma(\omega_2)}\omega_2}{2\sqrt{2}\bar{c}} \mathbf{h}^{(4)} \mathbf{P}_2(0, z) \circ d\tilde{W}_2(z)$$

$$- \frac{\gamma^{(s)}(\omega_1)\omega_1^2}{8\bar{c}^2} \mathbf{h}^{(2)} \mathbf{h}^{(1)} \mathbf{P}_2(0, z)\, dz - \frac{\gamma^{(s)}(\omega_2)\omega_2^2}{8\bar{c}^2} \mathbf{h}^{(4)} \mathbf{h}^{(3)} \mathbf{P}_2(0, z)\, dz, \quad (7.64)$$

where W_0, W_1, $\widetilde{W}_1$, W_2, $\widetilde{W}_2$ are five independent standard Brownian motions.

The matrix $\mathbf{P}_2$ is made up of two 2×2 diagonal subblocks and zeros elsewhere, and we denote the diagonal 2×2 blocks by $\mathbf{P}_{\omega_1}$ and $\mathbf{P}_{\omega_2}$. The matrix $\mathbf{P}_{\omega_j}$ is the limit (in distribution) of the propagator $\mathbf{P}_{\omega_j}^\varepsilon$, for $j = 1, 2$. We can rewrite the stochastic differential equation (7.64) for the individual propagators:

$$d\mathbf{P}_{\omega_j}(0, z) = \frac{i\omega_j\sqrt{\gamma(0)}}{2\bar{c}} \begin{bmatrix} 1 & 0 \\ 0 & -1 \end{bmatrix} \mathbf{P}_{\omega_j}(0, z) \circ dW_0(z)$$

$$- \frac{\omega_j\sqrt{\gamma(\omega_j)}}{2\sqrt{2}\bar{c}} \begin{bmatrix} 0 & 1 \\ 1 & 0 \end{bmatrix} \mathbf{P}_{\omega_j}(0, z) \circ dW_j(z)$$

$$- \frac{i\omega_j\sqrt{\gamma(\omega_j)}}{2\sqrt{2}\bar{c}} \begin{bmatrix} 0 & 1 \\ -1 & 0 \end{bmatrix} \mathbf{P}_{\omega_j}(0, z) \circ d\tilde{W}_j(z)$$

$$- \frac{i\omega_j^2\gamma^{(s)}(\omega_j)}{8\bar{c}^2} \begin{bmatrix} 1 & 0 \\ 0 & -1 \end{bmatrix} \mathbf{P}_{\omega_j}(0, z)\, dz, \quad (7.65)$$

for $j = 1, 2$. We note here that the two limit propagators $\mathbf{P}_{\omega_1}$ and $\mathbf{P}_{\omega_2}$ satisfy dynamically distinct stochastic differential equations, which, however, are not statistically independent, because they both involve the same Brownian motion W_0.

By the symmetry property of (7.65), the propagator matrices have the form

$$\mathbf{P}_{\omega_j}(0, z) = \begin{bmatrix} \alpha_{\omega_j}(0, z) & \overline{\beta_{\omega_j}(0, z)} \\ \beta_{\omega_j}(0, z) & \overline{\alpha_{\omega_j}(0, z)} \end{bmatrix},$$

where the pairs $(\alpha_{\omega_j}, \beta_{\omega_j})$ satisfy the system of stochastic differential equations

$$d\alpha_{\omega_j} = \frac{\omega_j}{2\bar{c}} \left(i\sqrt{\gamma(0)} \alpha_{\omega_j} \circ dW_0(z) - \frac{\sqrt{\gamma(\omega_j)}}{\sqrt{2}} \beta_{\omega_j} \circ (dW_j(z) + id\tilde{W}_j(z)) \right)$$

$$- \frac{i\omega_j{}^2 \gamma^{(s)}(\omega_j)}{8\bar{c}^2} \alpha_{\omega_j} \, dz,$$

$$d\beta_{\omega_j} = \frac{\omega_j}{2\bar{c}} \left(-i\sqrt{\gamma(0)} \beta_{\omega_j} \circ dW_0(z) - \frac{\sqrt{\gamma(\omega_j)}}{\sqrt{2}} \alpha_{\omega_j} \circ (dW_j(z) - id\tilde{W}_j(z)) \right)$$

$$+ \frac{i\omega_j{}^2 \gamma^{(s)}(\omega_j)}{8\bar{c}^2} \beta_{\omega_j} \, dz,$$

starting from $\alpha_{\omega_j}(0, z = 0) = 1$ and $\beta_{\omega_j}(0, z = 0) = 0$, for $j = 1, 2$. We proceed as in Section 7.1.1 to parameterize the pairs $(\alpha_{\omega_j}, \beta_{\omega_j})$, $j = 1, 2$, by

$$\alpha_{\omega_j}(0, z) = \cosh\left(\frac{\theta_{\omega_j}(z)}{2}\right) e^{i\phi_{\omega_j}(z)},$$

$$\beta_{\omega_j}(0, z) = \sinh\left(\frac{\theta_{\omega_j}(z)}{2}\right) e^{i(\psi_{\omega_j}(z) + \phi_{\omega_j}(z))}.$$

The two processes $\tilde{W}_1^*$ and $\tilde{W}_2^*$ defined by

$$\tilde{W}_j^*(z) = \int_0^z -\cos(\psi_{\omega_j}) \, dW_j(s) + \sin(\psi_{\omega_j}) \, d\tilde{W}_j(s)$$

are independent standard Brownian motions. This is because they are orthogonal transformations of Brownian stochastic integrals with nonanticipating arguments. It can also be seen by direct computation of the cross-quadratic variation. As a consequence, the processes θ_{ω_j}, $j = 1, 2$, satisfy the decoupled and statistically independent one-dimensional stochastic differential equations

$$d\theta_{\omega_j} = \frac{\omega_j \sqrt{\gamma(\omega_j)}}{\sqrt{2}\bar{c}} d\tilde{W}_j^*(z) + \frac{\omega_j{}^2 \gamma(\omega_j)}{4\bar{c}^2 \tanh(\theta_{\omega_j})} \, dz, \qquad (7.66)$$

with the initial condition $\theta_{\omega_j}(0) = 0$. Since the power transmission coefficients are given by $|T_{\omega_j}^\varepsilon(0, L)|^2 = 1/|\alpha_{\omega_j}^\varepsilon(0, L)|^2$, we conclude that the pair $(|T_{\omega_1}^\varepsilon(0, L)|^2, |T_{\omega_2}^\varepsilon(0, L)|^2)$ converges in distribution to $(\tau_{\omega_1}(L), \tau_{\omega_2}(L))$, where

$$\tau_{\omega_j}(L) = \cosh^{-2}\left(\frac{\theta_{\omega_j}(L)}{2}\right).$$

The two processes $\tau_{\omega_1}(L)$ and $\tau_{\omega_2}(L)$ are therefore two independent Markov processes whose infinitesimal generators are respectively $\mathcal{L}_{\omega_1}$ and $\mathcal{L}_{\omega_2}$ defined by (7.48).

This result is sufficient for showing the statistical stability of the transmitted pulse energy in Proposition 7.5. In the following chapters we will see that the two-frequency autocorrelation function of the transmission coefficient is a key quantity that deserves a thorough study. We will see that it is only when $\omega_1 - \omega_2$ is of order ε^2 that $|T_{\omega_1}^\varepsilon(0,L)|^2$ and $|T_{\omega_2}^\varepsilon(0,L)|^2$ are correlated. In the strongly heterogeneous white-noise regime (5.17) this is true when $\omega_1 - \omega_2$ is of order ε. We study in detail these correlated processes in Chapter 9, where we present a deeper analysis of pulse propagation in random media.

7.3 Wave Localization in the Weakly Heterogeneous Regime

In this section we study the decay of the power transmission coefficient $|T_\omega^\varepsilon(0,L)|^2$ for fixed ε as $L \to \infty$. Subsequently we consider the limit $\varepsilon \to 0$. The analysis can be divided into two parts. First, we show that $|T_\omega^\varepsilon(0,L)|^2$ decays exponentially with the size L of the slab of the random medium and that the decay rate is the reciprocal of the localization length $L_{\mathrm{loc}}^\varepsilon(\omega)$. The reciprocal of the localization length is equal to the Lyapunov exponent of an associated random harmonic oscillator problem. Second the Lyapunov exponent of this random harmonic oscillator is computed in the asymptotic limit $\varepsilon \to 0$. We show that the limit of $L_{\mathrm{loc}}^\varepsilon(\omega)$ as $\varepsilon \to 0$ equals the localization length $L_{\mathrm{loc}}(\omega)$ given by (7.42) in Section 7.1.4. From the results of this section we conclude that wave localization in the sense of exponential decay of the power transmission coefficient does not occur only in the diffusion limit $\varepsilon \to 0$. There is no explicit formula for the localization length $L_{\mathrm{loc}}^\varepsilon(\omega)$ for $\varepsilon > 0$. It can be shown only that its limit as $\varepsilon \to 0$ is $L_{\mathrm{loc}}(\omega)$ given by (7.42). We also conclude that the limits $L \to \infty$ and $\varepsilon \to 0$ can be interchanged as far as the exponential decay rate of the power transmission coefficient is concerned.

7.3.1 Determination of the Power Transmission Coefficient from a Random Harmonic Oscillator

We first describe how the wave propagation problem is related to a **random harmonic oscillator**. Using the components of the propagator matrix $\mathbf{P}_\omega^\varepsilon(0,z)$ in (7.6), we define the process

$$v_\omega^\varepsilon(z) := \alpha_\omega^\varepsilon(0,z)e^{i\omega z/(\bar{c}\varepsilon^2)} - \beta_\omega^\varepsilon(0,z)e^{-i\omega z/(\bar{c}\varepsilon^2)}. \tag{7.67}$$

Differentiating with respect to z and using the fact that $(\alpha_\omega^\varepsilon, \beta_\omega^\varepsilon)$ satisfy (7.3), we get

$$\frac{dv_\omega^\varepsilon}{dz} = \frac{i\omega}{\bar{c}\varepsilon^2}\left(\alpha_\omega^\varepsilon(0,z)e^{i\omega z/(\bar{c}\varepsilon^2)} + \beta_\omega^\varepsilon(0,z)e^{-i\omega z/(\bar{c}\varepsilon^2)}\right). \tag{7.68}$$

Differentiating once again we see that $v_\omega^\varepsilon(z)$ satisfies the random harmonic oscillator equation

$$\frac{d^2v_\omega^\varepsilon}{dz^2} + \frac{\omega^2}{\bar{c}^2\varepsilon^4}\left(1 + \varepsilon\nu\left(\frac{z}{\varepsilon^2}\right)\right)v_\omega^\varepsilon = 0. \tag{7.69}$$

From (7.7), the initial conditions for v_ω^ε are

$$v_\omega^\varepsilon(0) = 1, \quad \frac{dv_\omega^\varepsilon}{dz}(0) = \frac{i\omega}{\bar{c}\varepsilon^2}. \tag{7.70}$$

Let us introduce the energy of the oscillator

$$r_\omega^\varepsilon(z) := \frac{1}{2}\left(|v_\omega^\varepsilon(z)|^2 + \frac{\bar{c}^2\varepsilon^4}{\omega^2}\left|\frac{dv_\omega^\varepsilon}{dz}(z)\right|^2\right).$$

From (7.67) and (7.68) we see that

$$r_\omega^\varepsilon(z) = |\alpha_\omega^\varepsilon(0,z)|^2 + |\beta_\omega^\varepsilon(0,z)|^2.$$

Using the relation (7.10), $|\alpha_\omega^\varepsilon|^2 - |\beta_\omega^\varepsilon|^2 = 1$, this can also be written as

$$r_\omega^\varepsilon(z) = 2|\alpha_\omega^\varepsilon(0,z)|^2 - 1,$$

and therefore, by (7.9), we get the following relation between r_ω^ε and T_ω^ε,

$$r_\omega^\varepsilon(z) = 2|T_\omega^\varepsilon(0,z)|^{-2} - 1, \tag{7.71}$$

or equivalently, at $z = L$,

$$|T_\omega^\varepsilon(0,L)|^2 = \frac{2}{1 + r_\omega^\varepsilon(L)}. \tag{7.72}$$

We have expressed the power transmission coefficient in terms of the energy of the random harmonic oscillator (7.69) with initial conditions (7.70). The asymptotic analysis of the oscillator problem as $\varepsilon \to 0$ is very singular compared to that for $(\alpha_\omega^\varepsilon, \beta_\omega^\varepsilon)$. This is because the fast phases in (7.67) and (7.68) are present in the oscillator, but they are not present in the equations (7.3) for $(\alpha_\omega^\varepsilon, \beta_\omega^\varepsilon)$. Therefore the limit $\varepsilon \to 0$ is more conveniently analyzed in the $(\alpha_\omega^\varepsilon, \beta_\omega^\varepsilon)$ formulation as we have done in the previous sections. Nevertheless, from (7.72) we see that the asymptotic distribution of $r_\omega^\varepsilon(L)$ as $\varepsilon \to 0$ is known and equals that of $\cosh(\theta_\omega(L))$, which is characterized by the stochastic differential equation (7.28).

If we want to analyze the limit behavior as $L \to \infty$ of the power transmission coefficient $|T_\omega^\varepsilon(0,L)|^2$ for $\varepsilon > 0$ fixed, then it is more convenient to use the oscillator formulation. We can deduce this limit behavior of $|T_\omega^\varepsilon(0, L)|^2$ from that of the energy $r_\omega^\varepsilon(L)$ of the random harmonic oscillator, since they are related by (7.72). This is because we can use the theory of Lyapunov exponents of linear random differential equations presented in Section 6.8.

We analyze the random harmonic oscillator in Section 7.5. The behavior of its Lyapunov exponent for small ε is given in Proposition 7.9 of that section. We show that the limit

$$\lim_{z \to \infty} \frac{1}{z} \ln[r_\omega^\varepsilon(z)]$$

exists with probability one and that it is equal to a deterministic quantity $\gamma_\omega^\varepsilon$ that is twice the Lyapunov exponent of the oscillator. We show also that this coefficient has the limit

$$\lim_{\varepsilon \to 0} \gamma_\omega^\varepsilon = \frac{\gamma(\omega)\omega^2}{4\bar{c}^2}, \qquad \gamma(\omega) = \int_{-\infty}^{\infty} \mathbb{E}[\nu(0)\nu(z)] \cos\left(\frac{2\omega z}{\bar{c}}\right) dz. \qquad (7.73)$$

The parameter $\gamma(\omega)$ is the same as (7.14), which is assumed, in (7.40), to be positive. Therefore, for ε small enough, $\gamma_\omega^\varepsilon > 0$. From (7.72) we deduce the existence with probability one of the limit

$$\lim_{L \to \infty} \frac{1}{L} \ln[|T_\omega^\varepsilon(0, L)|^2] = -\lim_{L \to \infty} \frac{1}{L} \ln[r_\omega^\varepsilon(L)] = -\gamma_\omega^\varepsilon.$$

Let

$$L_{\text{loc}}^\varepsilon(\omega) = \frac{1}{\gamma_\omega^\varepsilon}. \qquad (7.74)$$

We summarize these results in the following proposition.

Proposition 7.6. *For ε small enough and fixed, there exists a finite localization length $L_{\text{loc}}^\varepsilon(\omega)$, given by (7.74), such that with probability one,*

$$\lim_{L \to \infty} \frac{1}{L} \ln[|T_\omega^\varepsilon(0, L)|^2] = -\frac{1}{L_{\text{loc}}^\varepsilon(\omega)}. \qquad (7.75)$$

The localization length $L_{\text{loc}}^\varepsilon(\omega)$ has the limit

$$\lim_{\varepsilon \to 0} L_{\text{loc}}^\varepsilon(\omega) = L_{\text{loc}}(\omega), \qquad (7.76)$$

where $L_{\text{loc}}(\omega)$ is given by (7.42) and recalled in (7.73).

7.3.2 Comparisons of Decay Rates

The main point of the analysis presented in the previous section is that we get the same result for the exponential decay of the power transmission coefficient by taking first the limit $L \to \infty$, and then $\varepsilon \to 0$ (Proposition 7.6), and by taking first the limit $\varepsilon \to 0$, and then $L \to \infty$ (equation (7.41). This is a special case of the general result about Lyapunov exponents of linear random differential equations presented in Section 6.8.

We can also compare the exponential decay rate of the transmitted pulse energy with the almost-sure decay rate of the power transmission coefficient of a monochromatic wave, in the limit $\varepsilon \to 0$. They behave differently as functions of the size L of the random medium. Both decay exponentially, but the transmitted pulse energy has a much slower decay rate. Consider first the case in which the incoming pulse is narrowband, that is to say, the spectrum $\hat{f}$ is concentrated around the carrier wave number ω_0/ε^2 with a bandwidth that is small compared to $1/\varepsilon^2$, but large compared to 1 (in dimensionless variables). It follows from (7.61) and (7.54) that $\mathcal{T}(L)$ decays exponentially as $L \to \infty$:

$$\frac{1}{L}\ln[\mathcal{T}(L)] \stackrel{L \gg 1}{\approx} -\frac{1}{4L_{\mathrm{loc}}(\omega_0)}, \qquad \text{with } \frac{1}{L_{\mathrm{loc}}(\omega_0)} = \frac{\gamma(\omega_0)\omega_0^2}{4\bar{c}^2}. \qquad (7.77)$$

Thus, the decay rate of the transmitted pulse energy with carrier frequency ω_0/ε^2 equals that of the mean power transmission coefficient at frequency ω_0/ε^2 (see (7.53)). This is because the transmitted pulse energy is self-averaging. The decay rate of the monochromatic power transmission coefficient is four times that of the one for the pulse. For broadband pulses, that is, with bandwidth an interval of the form $[\omega_0/\varepsilon^2, \omega_1/\varepsilon^2]$, the decay rate of the transmitted pulse energy is

$$\frac{1}{L}\ln[\mathcal{T}(L)] \stackrel{L \gg 1}{\approx} -\frac{1}{4L_{\mathrm{loc}}(\omega^*)},$$

where ω^* is the frequency in the interval $[\omega_0, \omega_1]$ for which the localization length is maximal. Usually, the lowest frequency ω_0 has the maximal localization length. This can be seen from expression (7.42), which shows that $L_{\mathrm{loc}}(\omega) \sim 4\bar{c}^2/[\gamma\omega^2]$ if the power spectral density of the random medium is flat over the bandwidth.

7.4 Wave Localization in the Strongly Heterogeneous White-Noise Regime

In this section we revisit the results obtained in this chapter in the *strongly heterogeneous white-noise* regime (5.17) introduced in Chapter 5, that is to say we consider the regime in which the fluctuations of the medium are strong ~ 1, the correlation length of the medium $\sim \varepsilon^2$ is much smaller than the wavelength $\sim \varepsilon$, which is much smaller than the size of the slab ~ 1. Accordingly, we assume that the fluctuations of the medium are given by

$$\frac{1}{K(z)} = \begin{cases} \dfrac{1}{\overline{K}}\left(1 + \nu\left(\dfrac{z}{\varepsilon^2}\right)\right) & \text{for } z \in [0,L], \\ \dfrac{1}{\overline{K}} & \text{for } z \in (-\infty,0) \cup (L,\infty), \end{cases}$$

$$\rho(z) = \bar{\rho} \quad \text{for all } z,$$

and we parameterize the frequency of the incoming right-going monochromatic wave by ω/ε. Under these conditions, the exponential decay of the transmission coefficient is related to the Lyapunov exponent of the random harmonic oscillator:

$$\frac{d^2 v_\omega^\varepsilon}{dz^2} + \frac{\omega^2}{\bar{c}^2 \varepsilon^2} \left(1 + \nu\left(\frac{z}{\varepsilon^2}\right)\right) v_\omega^\varepsilon = 0. \tag{7.78}$$

The analysis of this equation goes along the same lines as in the weak fluctuations regime and is presented in Section 7.5. The results are summarized in Proposition 7.8. Using this proposition, we have the following result.

Proposition 7.7. *For ε small enough and fixed, there exists a finite localization length $L_{\mathrm{loc}}^\varepsilon(\omega)$ such that with probability one,*

$$\lim_{L\to\infty} \frac{1}{L}\ln[|T_\omega^\varepsilon(0,L)|^2] = -\frac{1}{L_{\mathrm{loc}}^\varepsilon(\omega)}. \tag{7.79}$$

The localization length $L_{\mathrm{loc}}^\varepsilon(\omega)$ has the limit

$$\lim_{\varepsilon\to 0} L_{\mathrm{loc}}^\varepsilon(\omega) = L_{\mathrm{loc}}(\omega), \tag{7.80}$$

where

$$L_{\mathrm{loc}}(\omega) = \frac{4\bar{c}^2}{\gamma\omega^2}, \qquad \gamma = \gamma(0) = \int_{-\infty}^{\infty} \mathbb{E}[\nu(0)\nu(z)]\,dz. \tag{7.81}$$

This result is simply the low-frequency limit of Proposition 7.6, where $\gamma(\omega)$ is replaced by its limit $\gamma = \gamma(0)$. All the results stated in the previous sections can be extended to the strongly heterogeneous white-noise regime by substituting for $L_{\mathrm{loc}}(\omega)$ the low-frequency limit (7.81). In the analysis we apply the diffusion-approximation Theorem 6.4 instead of Theorem 6.5, which we used in the weakly heterogeneous regime. As a consequence, the statements of Propositions 7.3 and 7.4 hold in the strongly heterogeneous white-noise regime with the localization length defined by (7.81).

Figure 7.4a shows the almost-sure convergence of the logarithm of the power transmission coefficient as $L \to \infty$. In the numerical simulations, the random medium is a concatenation of thin layers with constant density and alternating bulk modulus values $\kappa_K = 1/(1 \pm \sigma_K)$, with $\sigma_K = 0.8$. The effective density, bulk modulus, and speed of sound are all equal to one. The thicknesses of the layers are independent and identically distributed random variables with exponential distribution of mean $l_c = 0.02$. The integrated autocorrelation is $\gamma = \sigma_K^2 l_c = 0.0128$. The frequency ω is equal to 5, for which the localization length $L_{\mathrm{loc}}(\omega)$ equals 12.5. In dimensional variables, the localization length is

$$[L_{\mathrm{loc}}(\omega_0)]_{\mathrm{dim}} = \frac{4\bar{c}^2}{\sigma_K^2 l_K \omega_0^2}, \tag{7.82}$$

where σ_K and l_K are defined by (5.5). It increases as the frequency decreases, which means that high-frequency waves do not penetrate as deeply into the medium as the low-frequency ones. The frequency above which waves cannot be transmitted through a slab of size L_0 depends on the statistical properties of the medium and is obtained from (7.82):

$$[\omega_K]_{\text{dim}} = \frac{2\bar{c}}{\sqrt{\sigma_K^2 l_K L_0}}. \qquad (7.83)$$

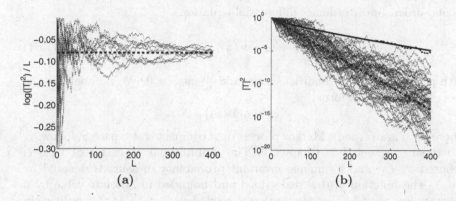

(a)   (b)

Fig. 7.4. Plot (a): The logarithm of the power transmission coefficient divided by the propagation distance L plotted as a function of L. The thin dotted lines correspond to 10 different realizations of the random medium. The thick dashed line is the theoretical limit $-1/L_{\text{loc}}(\omega)$. We can see clearly the almost-sure convergence to the asymptotic limit. Plot (b): the power transmission coefficient as a function of propagation distance. The thin dotted lines correspond to 50 different realizations of the random medium. The dashed line very near the solid line at the top corresponds to the power transmission coefficient averaged over 5×10^4 realizations. We see clearly that the behavior of the average power transmission coefficient is very different from its typical behavior. The thick solid line at the top comes from formula (7.52) for the mean power transmission coefficient. The thick dashed line below it comes from the asymptotic formula $\exp[-L/L_{\text{loc}}(\omega)]$.

7.5 The Random Harmonic Oscillator

In this section we analyze the Lyapunov exponent of the random harmonic oscillator. It is needed in Sections 7.3 and 7.4 of this chapter, where we have related the localization length of monochromatic waves to the Lyapunov exponent of the random harmonic oscillator. The presentation of this section is self-contained. However, it is a special case of the problem considered in Section 6.8. We addressed there a general multidimensional random linear system, which required some general conditions ensuring the validity of the results. In the case of the random harmonic oscillator that we consider here, each step in the analysis can be carried out without using the general theory as detailed in this section.

7.5.1 The Lyapunov Exponent of the Random Harmonic Oscillator

Since $\varepsilon > 0$ is fixed throughout the analysis in this section, for simplicity we set it equal to one.

The random harmonic oscillator process is the solution of the random second-order, linear ordinary differential equation

$$\frac{d^2 v}{dz^2} + (1 + \nu(z)) v = 0, \tag{7.84}$$

with prescribed initial conditions for v and $\frac{dv}{dz}$ at $z = 0$. We assume that the process $\nu(z)$ has the form

$$\nu(z) = g(Y(z)),$$

where $Y(z)$ is an ergodic Markov process on a compact state space S, following the theory presented in Chapter 6. The infinitesimal generator of $Y(z)$ is denoted by $\mathcal{L}_Y$ and its unique invariant probability measure is denoted by $\bar{p}(dy)$. The function $g(y)$ is real-valued and bounded in absolute value by a constant less than 1. This assumption is needed so that $1 + \nu(z)$ is positive in the oscillator equation (7.84). We also assume that $\nu(z)$ has mean zero,

$$\mathbb{E}[\nu(z)] = \mathbb{E}[g(Y(z))] = \int_S g(y) \, \bar{p}(dy) = 0.$$

Here $\mathbb{E}$ denotes the expectation with respect to the invariant distribution of $Y(z)$.

The Lyapunov exponent is defined by

$$\Gamma = \lim_{z \to \infty} \frac{1}{z} \ln[R(z)], \qquad R(z) = \sqrt{v(z)^2 + \frac{dv}{dz}(z)^2}, \tag{7.85}$$

when the limit exists. It is the exponential growth rate of the square root of the energy of the oscillator. We show below that the limit exists with probability one and is given by (7.90). The positivity of Γ is addressed in the next two sections.

We introduce polar coordinates $(R(z), \psi(z))$ by

$$v(z) = R(z) \cos(\psi(z)) \quad \text{and} \quad \frac{dv}{dz}(z) = R(z) \sin(\psi(z)),$$

in terms of which the harmonic oscillator equation (7.84) has the form

$$R(z) = R(0) \exp\left(\int_0^z q(\psi(s), Y(s)) \, ds \right), \tag{7.86}$$

$$\frac{d\psi}{dz}(z) = F(\psi(z), Y(z)). \tag{7.87}$$

The functions q and F are given by

$$q(\psi, y) = -g(y)\sin(\psi)\cos(\psi),$$
$$F(\psi, y) = -1 - g(y)\cos^2(\psi).$$

From (7.87) and the analysis of Section 6.3 (with $\varepsilon = 1$), we see that (ψ, Y) is a Markov process on the compact state space $\mathbb{S}^1 \times S$, where $\mathbb{S}^1$ denotes the unit circle. Its infinitesimal generator is given by

$$\mathcal{L} = \bar{\mathcal{L}}_Y + F(\psi, y)\frac{\partial}{\partial \psi}. \tag{7.88}$$

Because of the assumption that g is bounded by a constant less than 1, the function F is strictly negative. This implies that the Markov process (ψ, Y) is ergodic and its invariant probability distribution has the form $p(\psi, y)d\psi\bar{p}(dy)$, where p is the solution of

$$\mathcal{L}^\star p = 0. \tag{7.89}$$

The adjoint $\mathcal{L}^\star$ of the infinitesimal generator $\mathcal{L}$ was introduced in Section 6.1.4 and is given here by

$$\mathcal{L}^\star p(\psi, y) = \mathcal{L}_Y^\star p(\psi, y) - \frac{\partial}{\partial \psi}\left(F(\psi, y)p(\psi, y)\right).$$

From the ergodicity of (ψ, Y) and (7.86), we conclude that with probability one,

$$\lim_{z \to \infty} \frac{1}{z}\ln[R(z)] = \lim_{z \to \infty} \frac{1}{z}\left[\ln(R_0) + \int_0^z q(\psi(s), Y(s))\, ds\right] = \Gamma,$$

where the Lyapunov exponent Γ is the deterministic quantity given by

$$\Gamma = \int_{\mathbb{S}^1 \times S} q(\psi, y)p(\psi, y)\, d\psi\, \bar{p}(dy). \tag{7.90}$$

We will next analyze the Lyapunov exponent in two different asymptotic regimes.

7.5.2 Expansion of the Lyapunov Exponent in the Strongly Heterogeneous Regime.

We will first show how the theory of Section 6.8 can be used, and then derive an approximate formula for the Lyapunov exponent by direct calculation.

The scaled random harmonic oscillator in the strongly heterogeneous white-noise regime has the form

$$\frac{d^2 v_\omega^\varepsilon}{dz^2} + \frac{\omega^2}{\bar{c}^2 \varepsilon^2}\left(1 + \nu\left(\frac{z}{\varepsilon^2}\right)\right)v_\omega^\varepsilon = 0. \tag{7.91}$$

As shown in the previous subsection, the limit

$$\lim_{z \to \infty} \frac{1}{z} \log \sqrt{|v_\omega^\varepsilon(z)|^2 + \left|\frac{dv_\omega^\varepsilon}{dz}(z)\right|^2} \tag{7.92}$$

exists with probability one and is deterministic. It is the Lyapunov exponent Γ^ε, which is also equal to

$$\Gamma^\varepsilon = \frac{1}{2} \lim_{z \to \infty} \frac{1}{z} \log r_\omega^\varepsilon(z),$$

where r_ω^ε is the energy of the oscillator

$$r_\omega^\varepsilon(z) = \frac{1}{2}\left(|v_\omega^\varepsilon(z)|^2 + \frac{\bar{c}^2\varepsilon^2}{\omega^2}\left|\frac{dv_\omega^\varepsilon}{dz}(z)\right|^2\right).$$

This problem can be put in the form of the general linear system of random differential equations discussed in Section 6.8.

The random harmonic oscillator can be written in system form as

$$\frac{dX^\varepsilon}{dz}(z) = \frac{1}{\varepsilon}\Omega X^\varepsilon(z) + \frac{1}{\varepsilon}g\left(Y\left(\frac{z}{\varepsilon^2}\right)\right)\mathbf{h}X^\varepsilon(z),$$

where the vector $X^\varepsilon(z)$ is defined by

$$X^\varepsilon(z) = \begin{bmatrix} v_\omega^\varepsilon(z) \\ \frac{\varepsilon\bar{c}}{\omega}\frac{dv_\omega^\varepsilon}{dz}(z) \end{bmatrix}.$$

The matrices Ω and $\mathbf{h}$ are given by

$$\Omega = \frac{\omega}{\bar{c}}\begin{bmatrix} 0 & 1 \\ -1 & 0 \end{bmatrix}, \quad \mathbf{h} = \frac{\omega}{\bar{c}}\begin{bmatrix} 0 & 0 \\ -1 & 0 \end{bmatrix}.$$

In this formulation, the Lyapunov exponent Γ^ε, defined by (7.92), is equal to the Lyapunov exponent of X^ε defined by

$$\Gamma^\varepsilon = \lim_{z \to \infty} \frac{1}{z} \log |X^\varepsilon(z)|,$$

where $|\cdot|$ is the Euclidean norm.

We now verify that the hypotheses stated below (6.107) are satisfied. Here the flow generated by Ω is simply the rotation

$$e^{\Omega z} = \begin{bmatrix} \cos(\frac{\omega z}{\bar{c}}) & \sin(\frac{\omega z}{\bar{c}}) \\ -\sin(\frac{\omega z}{\bar{c}}) & \cos(\frac{\omega z}{\bar{c}}) \end{bmatrix}.$$

The functions $\phi_1(\hat{x}) = (\mathbf{h}\hat{x}, \mathbf{h}\hat{x}) = (\omega^2/\bar{c}^2)\hat{x}_1^2$, $\phi_2(\hat{x}) = (\mathbf{h}^2\hat{x}, \hat{x}) = 0$, and $\phi_3(\hat{x}) = (\mathbf{h}\hat{x}, \hat{x})^2 = (\omega^2/\bar{c}^2)\hat{x}_1^2\hat{x}_2^2$ have Ω-averages given by

$$\lim_{Z \to \infty} \frac{1}{Z} \int_0^Z \phi_1(e^{\Omega z}\hat{x}) \, dz = \frac{\omega^2}{\bar{c}^2} \lim_{Z \to \infty} \frac{1}{Z} \int_0^Z \left(\hat{x}_1 \cos(\frac{\omega z}{\bar{c}}) + \hat{x}_2 \sin(\frac{\omega z}{\bar{c}}) \right)^2 dz$$

$$= \frac{\omega^2}{2\bar{c}^2}\hat{x}_1^2 + \frac{\omega^2}{2\bar{c}^2}\hat{x}_2^2 = \frac{\omega^2}{2\bar{c}^2},$$

$$\lim_{Z \to \infty} \frac{1}{Z} \int_0^Z \phi_2(e^{\Omega z}\hat{x}) \, dz = 0,$$

$$\lim_{Z \to \infty} \frac{1}{Z} \int_0^Z \phi_3(e^{\Omega z}\hat{x}) \, dz = \frac{\omega^2}{\bar{c}^2} \lim_{Z \to \infty} \frac{1}{Z} \int_0^Z \left(\hat{x}_1 \cos(\frac{\omega z}{\bar{c}}) + \hat{x}_2 \sin(\frac{\omega z}{\bar{c}}) \right)^2$$

$$\times \left(\hat{x}_1 \sin(\frac{\omega z}{\bar{c}}) + \hat{x}_2 \cos(\frac{\omega z}{\bar{c}}) \right)^2 dz$$

$$= \frac{\omega^2}{8\bar{c}^2}\hat{x}_1^4 + \frac{\omega^2}{4\bar{c}^2}\hat{x}_1^2\hat{x}_2^2 + \frac{\omega^2}{8\bar{c}^2}\hat{x}_2^2$$

$$= \frac{\omega^2}{8\bar{c}^2}(\hat{x}_1^2 + \hat{x}_2^2)^2 = \frac{\omega^2}{8\bar{c}^2}.$$

These Ω-averages are independent of $\hat{x} \in \mathbb{S}^1$, as required by the theory of Section 6.8. Theorem 6.8 can now be used to show that the limit as $\varepsilon \to 0$ of the Lyapunov exponent exists,

$$\lim_{\varepsilon \to 0} \Gamma^\varepsilon = \Gamma, \tag{7.93}$$

and is given by (6.138), that is, $\Gamma = \omega^2\gamma/(8\bar{c}^2)$, where γ is defined by (6.139).

We now derive this result by a simple formal expansion. We introduce the polar coordinates $(R^\varepsilon, \psi^\varepsilon)$ for the scaled harmonic oscillator (7.91), which, as in (7.86–7.87), satisfy

$$R^\varepsilon(z) = R^\varepsilon(0) \exp\left[\frac{1}{\varepsilon} \int_0^z q\left(\psi^\varepsilon(s), Y\left(\frac{s}{\varepsilon^2}\right) \right) ds \right], \tag{7.94}$$

$$\frac{d\psi^\varepsilon}{dz}(z) = \frac{1}{\varepsilon} F\left(\psi^\varepsilon(z), Y\left(\frac{z}{\varepsilon^2}\right) \right). \tag{7.95}$$

The functions q and F are given here by

$$q(\psi, y) = -\frac{\omega}{\bar{c}} g(y) \sin(\psi) \cos(\psi),$$

$$F(\psi, y) = -\frac{\omega}{\bar{c}} \left(1 + g(y) \cos^2(\psi) \right),$$

and the Lyapunov exponent Γ^ε is given by

$$\Gamma^\varepsilon = \frac{1}{\varepsilon} \int_{\mathbb{S}^1 \times S} q(\psi, y) p^\varepsilon(\psi, y) \, d\psi \, \bar{p}(dy),$$

where p^ε is the solution of

$$\mathcal{L}^{\varepsilon \star} p^\varepsilon = 0. \tag{7.96}$$

The generator $\mathcal{L}^\varepsilon$ of the Markov process $(\psi^\varepsilon(z), Y(\frac{z}{\varepsilon^2}))$ is given by

$$\mathcal{L}^\varepsilon = \frac{1}{\varepsilon^2}\mathcal{L}_Y + \frac{1}{\varepsilon}\mathcal{L}_1 \text{ with } \mathcal{L}_1 = F(\psi, y)\frac{\partial}{\partial\psi}.$$

We will assume for simplicity that $\mathcal{L}_Y$ is self-adjoint and derive an expansion of the Lyapunov exponent Γ^ε in powers of ε. In this section we assume that the invariant probability density p^ε can be expanded in powers of ε,

$$p^\varepsilon = p_0 + \varepsilon p_1 + \varepsilon^2 p_2 + \cdots.$$

Substituting this expansion into (7.96) and collecting the terms with the same powers in ε, we get for p_0, p_1, and p_2 the following hierarchy of Poisson equations:

$$\mathcal{L}_Y p_0 = 0, \tag{7.97}$$

$$\mathcal{L}_Y p_1 = -\mathcal{L}_1^\star p_0, \tag{7.98}$$

$$\mathcal{L}_Y p_2 = -\mathcal{L}_1^\star p_1, \tag{7.99}$$

where we have used the fact that $\mathcal{L}_Y^\star = \mathcal{L}_Y$. The adjoint of $\mathcal{L}_1$ is given by $\mathcal{L}_1^\star p = -\partial_\psi(Fp)$. Once the expansion of p^ε has been determined, it can be used in (7.90) to obtain an expansion for Γ^ε:

$$\Gamma^\varepsilon = \frac{1}{\varepsilon}\int_{\mathbb{S}^1\times S} (qp_0 + \varepsilon qp_1)(\psi, y)\, d\psi\, \bar{p}(dy) + O(\varepsilon). \tag{7.100}$$

From (7.97), p_0 satisfies $\mathcal{L}_Y p_0 = 0$. Therefore we can choose p_0 to be independent of y, so that $p_0 = p_0(\psi)$. For p_1, (7.98) gives $\hat{\mathcal{L}}_Y p_1 = -\mathcal{L}_1^\star p_0 = \partial_\psi(Fp_0)$:

$$\mathcal{L}_Y p_1 = -\frac{\omega}{c}\left\{\partial_\psi[p_0(\psi)] + g(y)\partial_\psi[\cos^2(\psi)p_0(\psi)]\right\}. \tag{7.101}$$

This is an equation in which ψ plays the role of a frozen parameter. By the assumed Fredholm alternative, the solvability condition of the Poisson equation requires that the right-hand side of (7.101) have mean zero with respect to the invariant probability distribution $\bar{p}(dy)$ of $\mathcal{L}_Y$. Since $g(Y(z))$ has zero mean, the mean of the right-hand side is $-(\omega/\bar{c})\partial_\psi[p_0(\psi)]$, and consequently p_0 is constant, independent of both ψ and y. Since p^ε is a probability density, its integral with respect to $d\psi\, \bar{p}(dy)$ over $\mathbb{S}^1 \times S$ is equal to one. Therefore

$$p_0 \equiv \frac{1}{2\pi},$$

and p_j, $j \geq 1$, have integrals equal to zero. As a result the first term in the expansion (7.100) of the Lyapunov exponent vanishes:

$$\Gamma^\varepsilon = \frac{1}{\varepsilon}\int_{\mathbb{S}^1\times S} qp_0(\psi, y_0)\, d\psi\, \bar{p}(dy_0) + O(1)$$

$$= -\frac{\omega}{4\pi\bar{c}}\left(\int_{\mathbb{S}^1}\sin(2\psi)\, d\psi\right)\left(\int_S g(y)\, \bar{p}(dy)\right) + O(1)$$

$$= O(1).$$

The Poisson equation (7.101) for p_1 reduces to

$$\mathcal{L}_Y p_1 = \frac{\omega}{2\pi\bar{c}} g(y) \sin(2\psi).$$

Since we must choose the solution whose integral with respect to $d\psi \, \bar{p}(dy)$ is equal to zero, we get

$$p_1(\psi, y_0) = -\frac{\omega}{2\pi\bar{c}} \sin(2\psi) \int_0^\infty \mathbb{E}[g(Y(z))|Y(0) = y_0] \, dz.$$

Therefore, the expansion of the Lyapunov exponent becomes

$$
\begin{aligned}
\Gamma^\varepsilon &= \int_{\mathbb{S}^1 \times S} q p_1(\psi, y_0) \, d\psi \, \bar{p}(dy_0) + O(\varepsilon) \\
&= \frac{\omega^2}{4\pi\bar{c}^2} \left(\int_{\mathbb{S}^1} \sin^2(2\psi) \, d\psi \right) \left(\int_0^\infty \int_S g(y_0) \mathbb{E}[g(Y(z)) \mid Y(0) = y_0] \, \bar{p}(dy_0) \, dz \right) \\
&\quad + O(\varepsilon) \\
&= \frac{\omega^2}{4\bar{c}^2} \int_0^\infty \mathbb{E}[g(Y(0))g(Y(z))] \, dz + O(\varepsilon).
\end{aligned}
$$

This shows that the Lyapunov exponent is given by

$$\Gamma^\varepsilon = \frac{\gamma\omega^2}{8\bar{c}^2} + O(\varepsilon),$$

where

$$\gamma = \int_{-\infty}^\infty \mathbb{E}[\nu(0)\nu(z)] \, dz. \tag{7.102}$$

This agrees, of course, with the result (7.93) that we obtained with the general theory of Section 6.8.

We summarize the results of this section in the following proposition.

Proposition 7.8. *Let v_ω^ε be the solution of the random harmonic oscillator equation (7.91). Its Lyapunov exponent defined by (7.92) is positive for ε small enough, and has the limit*

$$\lim_{\varepsilon \to 0} \Gamma^\varepsilon = \frac{\gamma\omega^2}{8\bar{c}^2},$$

where γ is given by (7.102).

This result is used in Section 7.4 to obtain the limit (7.80) of the localization length of monochromatic waves in the strongly heterogeneous white-noise regime.

7.5.3 Expansion of the Lyapunov Exponent in the Weakly Heterogeneous Regime

In this section we extend the analysis of the previous section to the weakly heterogeneous regime. In this regime, the scaled random harmonic oscillator equation has the form

$$\frac{d^2 v_\omega^\varepsilon}{dz^2} + \frac{\omega^2}{\bar{c}^2 \varepsilon^4}\left(1 + \varepsilon\nu\left(\frac{z}{\varepsilon^2}\right)\right)v_\omega^\varepsilon = 0. \tag{7.103}$$

We will carry out only the formal expansion of the previous section. The general theory of Section 6.8 applies here too, when suitably extended. We will not consider it in detail in this section.

We obtain the expansion of the Lyapunov exponent Γ^ε from its definition (7.92), which in this scaling has the form

$$\Gamma^\varepsilon = \frac{1}{\varepsilon^2}\int_{\mathbb{S}^1 \times S} q(\psi, y)p^\varepsilon(\psi, y)\,d\psi\,\bar{p}(dy). \tag{7.104}$$

The invariant probability density p^ε is the solution of (7.96), and the generator $\mathcal{L}^\varepsilon$ has the form

$$\mathcal{L}^\varepsilon = \frac{1}{\varepsilon^2}\mathcal{L}_0 + \frac{1}{\varepsilon}\mathcal{L}_1,$$

where

$$\mathcal{L}_0 = \mathcal{L}_Y - \frac{\omega}{\bar{c}}\frac{\partial}{\partial\psi}, \qquad \mathcal{L}_1 = -\frac{\omega}{\bar{c}}g(y)\cos^2(\psi)\frac{\partial}{\partial\psi}.$$

As in the previous section, we expand the invariant probability density $p^\varepsilon = p_0 + \varepsilon p_1 + \varepsilon^2 p_2 + \cdots$, and find that p_0, p_1, and p_2 satisfy the hierarchy of Poisson equations

$$\mathcal{L}_0^\star p_0 = 0, \tag{7.105}$$

$$\mathcal{L}_0^\star p_1 = -\mathcal{L}_1^\star p_0, \tag{7.106}$$

$$\mathcal{L}_0^\star p_2 = -\mathcal{L}_1^\star p_1. \tag{7.107}$$

The adjoint of $\mathcal{L}_0$ is explicitly given by $\mathcal{L}_0^\star = \mathcal{L}_Y + (\omega/\bar{c})\partial_\psi$, where we assume, as in the formal expansion in the previous section, that $\mathcal{L}_Y$ is self-adjoint. As in the strong-fluctuations regime, we choose p_0 to be equal to $(2\pi)^{-1}$. For p_1, (7.106) gives

$$\left(\mathcal{L}_Y + \frac{\omega}{\bar{c}}\partial_\psi\right)p_1 = -\mathcal{L}_1^\star p_0 = -\frac{\omega}{\bar{c}}g(y)\partial_\psi(\cos^2(\psi)p_0) = \frac{\omega}{2\pi\bar{c}}g(y)\sin(2\psi).$$

The right-hand side has zero mean with respect to the invariant probability $p_0\,\bar{p}(dy)\,d\psi$ of $\mathcal{L}_0^\star$, and so the Poisson equation admits a solution p_1 that has zero mean

$$p_1(\psi, y_0) = -\frac{\omega}{2\pi\bar{c}}\int_0^\infty \mathbb{E}[g(Y(z))|Y(0) = y_0]\sin\left(2\psi + \frac{2\omega z}{\bar{c}}\right)dz.$$

Substituting the expansion of p^ε into (7.104), we get for the Lyapunov exponent

$$
\begin{aligned}
\Gamma^\varepsilon &= \int_{\mathbb{S}^1 \times S} q p_1(\psi, y_0)\, d\psi\, \bar{p}(dy_0) + O(\varepsilon^3) \\
&= \frac{\omega^2}{4\pi \bar{c}^2} \int_{\mathbb{S}^1} \sin(2\psi) \int_0^\infty \mathbb{E}[g(Y(0))g(Y(z))] \sin\left(2\psi + \frac{2\omega z}{\bar{c}}\right) dz\, d\psi + O(\varepsilon) \\
&= \frac{\omega^2}{4\bar{c}^2} \int_0^\infty \cos\left(\frac{2\omega z}{\bar{c}}\right) \mathbb{E}[g(Y(0))g(Y(z))] dz + O(\varepsilon) \\
&= \frac{\gamma(\omega)\omega^2}{8\bar{c}^2} + O(\varepsilon).
\end{aligned}
\tag{7.108}
$$

Here

$$
\gamma(\omega) = \int_{-\infty}^\infty \cos\left(\frac{2\omega z}{\bar{c}}\right) \mathbb{E}[\nu(0)\nu(z)]dz.
\tag{7.109}
$$

Without having provided all the mathematical details for the results of this section, we nevertheless state them in the following proposition.

Proposition 7.9. *Let v_ω^ε be the solution of the random harmonic oscillator equation (7.103). Its Lyapunov exponent defined by (7.92) is positive for ε small enough, and has the limit*

$$
\lim_{\varepsilon \to 0} \Gamma^\varepsilon = \frac{\gamma(\omega)\omega^2}{8\bar{c}^2},
$$

where $\gamma(\omega)$ is given by (7.109).

This result is used in Section 7.3 to derive the expansion (7.76) of the localization length of monochromatic waves in the weakly heterogeneous regime.

7.6 Appendix. Statistics of the Power Transmission Coefficient

In this section we derive explicit formulas for the moments of the Markov process $\tau_\omega(L)$ using the infinitesimal generator given by (7.48). The initial condition for the power transmission coefficient is $\tau_\omega(L = 0) = 1$.

7.6.1 The Probability Density of the Power Transmission Coefficient

The computations of the probability density of $\tau_\omega(L)$ are carried out using the Mehler–Fock transform. It is convenient to introduce the auxiliary process

$$
\eta_\omega = \frac{2 - \tau_\omega}{\tau_\omega},
$$

which is a Markov process taking values in $[1, \infty)$. Its infinitesimal generator is given by

$$\mathcal{L} = \frac{1}{L_{\text{loc}}}(\eta^2 - 1)\frac{\partial^2}{\partial \eta^2} + \frac{2}{L_{\text{loc}}}\eta\frac{\partial}{\partial \eta} = \frac{1}{L_{\text{loc}}}\frac{\partial}{\partial \eta}(\eta^2 - 1)\frac{\partial}{\partial \eta}. \qquad (7.110)$$

We omit the ω-dependence of L_{loc} for simplicity in this appendix. The infinitesimal generator of η_ω is self-adjoint, so the Fokker–Planck equation for the probability density is

$$\frac{\partial p}{\partial L}(L, \eta) = \frac{1}{L_{\text{loc}}}\frac{\partial}{\partial \eta}\left[(\eta^2 - 1)\frac{\partial p}{\partial \eta}(L, \eta)\right], \quad \eta > 1,$$

starting from $p(L = 0, \eta) = \delta(\eta - 1)$.

We denote by $P_{-1/2+i\mu}(\eta)$, $\eta \geq 1$, $\mu \geq 0$, the Legendre function of the first kind, which is the solution of

$$\frac{d}{d\eta}(\eta^2 - 1)\frac{d}{d\eta}P_{-1/2+i\mu}(\eta) = -\left(\mu^2 + \frac{1}{4}\right)P_{-1/2+i\mu}(\eta), \qquad (7.111)$$

starting from $P_{-1/2+i\mu}(1) = 1$. It has the integral representation

$$P_{-1/2+i\mu}(\eta) = \frac{\sqrt{2}}{\pi}\cosh(\pi\mu)\int_0^\infty \frac{\cos(\mu\tau)}{\sqrt{\cosh\tau + \eta}}\,d\tau. \qquad (7.112)$$

The Mehler–Fock transform of an integrable function f defined on $(1, \infty)$ is the function $\check{f}$ defined on $(0, \infty)$ given by

$$\check{f}(\mu) = \int_1^\infty f(\eta)P_{-1/2+i\mu}(\eta)\,d\eta.$$

Its inverse transform is

$$f(\eta) = \int_0^\infty \check{f}(\mu)\mu\tanh(\mu\pi)P_{-1/2+i\mu}(\eta)\,d\mu.$$

We apply the Mehler–Fock transform to the probability density $p(L, \eta)$:

$$\check{p}(L, \mu) = \int_1^\infty p(L, \eta)P_{-1/2+i\mu}(\eta)\,d\eta.$$

Differentiating with respect to L gives

$$\frac{\partial \check{p}}{\partial L}(L, \mu) = \frac{1}{L_{\text{loc}}}\int_1^\infty \frac{\partial}{\partial \eta}\left[(\eta^2 - 1)\frac{\partial p}{\partial \eta}(L, \eta)\right]P_{-1/2+i\mu}(\eta)\,d\eta.$$

Integrating twice by parts, we obtain

$$\frac{\partial \check{p}}{\partial L}(L, \mu) = \frac{1}{L_{\text{loc}}}\int_1^\infty p(L, \eta)\frac{\partial}{\partial \eta}\left[(\eta^2 - 1)\frac{\partial P_{-1/2+i\mu}}{\partial \eta}(\eta)\right]d\eta.$$

Using the differential equation (7.111) satisfied by the Legendre function, we see that the Mehler–Fock transform of p satisfies the ordinary differential equation

$$\frac{\partial \check{p}}{\partial L}(L, \mu) = -\frac{1}{L_{\text{loc}}} \left(\mu^2 + \frac{1}{4} \right) \check{p}(L, \mu),$$

starting from $\check{p}(L = 0, \mu) = 1$. Its solution is

$$\check{p}(L, \mu) = \exp\left[-\left(\mu^2 + \frac{1}{4} \right) \frac{L}{L_{\text{loc}}} \right].$$

By applying the inverse Mehler–Fock transform we get an integral representation of the probability density of $\eta_\omega(L)$:

$$p(L, \eta) = \int_0^\infty \mu \tanh(\mu\pi) P_{-1/2+i\mu}(\eta) \exp\left[-\left(\mu^2 + \frac{1}{4} \right) \frac{L}{L_{\text{loc}}} \right] d\mu. \quad (7.113)$$

The probability density of the power transmission coefficient $\tau_\omega(L)$ is therefore

$$p(L, \tau) = \frac{\tau^2}{2} \int_0^\infty \mu \tanh(\mu\pi) P_{-1/2+i\mu} \left(\frac{2}{\tau} - 1 \right) \exp\left[-\left(\mu^2 + \frac{1}{4} \right) \frac{L}{L_{\text{loc}}} \right] d\mu.$$

7.6.2 Moments of the Power Transmission Coefficient

The moments of the power transmission coefficient $\tau_\omega(L)$ are given by

$$\mathbb{E}\left[\tau_\omega(L)^n \right] = \int_1^\infty \left(\frac{2}{1+\eta} \right)^n p(L, \eta) \, d\eta.$$

Using the integral representation (7.113) and introducing

$$J^{(n)}(\mu) = 2^n \int_1^\infty \frac{P_{-1/2+i\mu}(\eta)}{(1+\eta)^n} \, d\eta,$$

we have

$$\mathbb{E}\left[\tau_\omega(L)^n \right] = \int_0^\infty \mu \tanh(\pi\mu) J^{(n)}(\mu) \exp\left[-\left(\mu^2 + \frac{1}{4} \right) \frac{L}{L_{\text{loc}}} \right] d\mu. \quad (7.114)$$

We now derive explicit formulas for the functions $J^{(n)}(\mu)$.

We first consider $J^{(1)}(\mu)$. Using the explicit expression (7.112) of the Legendre function, we have

$$J^{(1)}(\mu) = \frac{2\sqrt{2}}{\pi} \cosh(\mu\pi) \int_0^\infty \cos(\mu\tau) \left[\int_1^\infty \frac{d\eta}{(1+\eta)\sqrt{\cosh(\tau) + \eta}} \right] d\tau.$$

The integral within the square brackets can be computed using the change of variables $u = \sqrt{\cosh(\tau) + \eta}/\sqrt{\cosh(\tau) - 1}$, which gives

$$J^{(1)}(\mu) = \frac{2\cosh(\mu\pi)}{\pi} \int_0^\infty \frac{\tau\cos(\mu\tau)}{\sinh(\tau/2)}\,d\tau = \frac{\cosh(\mu\pi)}{\pi} \int_{-\infty}^\infty \frac{\tau\cos(\mu\tau)}{\sinh(\tau/2)}\,d\tau .$$

The integral on the right is computed by contour integration as shown in Figure 7.5. Note the presence of the pole at $z = i\pi$. The result of the integration is

$$J^{(1)}(\mu) = \frac{2\pi}{\cosh(\pi\mu)} . \tag{7.115}$$

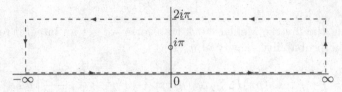

Fig. 7.5. Contour used for the integration of $J^{(1)}(\mu)$ in the complex plane.

We next consider $J^{(n)}(\mu)$, $n \geq 2$. Using the differential equation (7.111) satisfied by the Legendre function, we can write

$$J^{(n)}(\mu) = -\frac{1}{\mu^2 + 1/4} \int_1^\infty \frac{2^n}{(1+\eta)^n} \frac{d}{d\eta}\left[(\eta^2 - 1)\frac{dP_{-1/2+i\mu}}{d\eta}(\eta) \right] d\eta .$$

We then integrate twice by parts, and we get

$$\begin{aligned}
J^{(n)}(\mu) &= \frac{1}{\mu^2 + 1/4} \int_1^\infty \left[\frac{2^n n(1-n)}{(1+\eta)^n} + \frac{2^{n+1}n^2}{(1+\eta)^{n+1}} \right] P_{-1/2+i\mu}(\eta)\,d\eta \\
&= \frac{n(1-n)}{\mu^2 + 1/4} J^{(n)}(\mu) + \frac{n^2}{\mu^2 + 1/4} J^{(n+1)}(\mu) .
\end{aligned}$$

This establishes a recurrence relation for the functions $J^{(n)}(\mu)$:

$$J^{(n+1)}(\mu) = \frac{1}{n^2}\left[\mu^2 + (n - \tfrac{1}{2})^2 \right] J^{(n)}(\mu) .$$

Using (7.115) we see that

$$J^{(n)}(\mu) = \frac{2\pi}{\cosh(\pi\mu)} K^{(n)}(\mu) ,$$

where

$$K^{(n)}(\mu) = \prod_{j=1}^{n-1} \frac{1}{j^2}\left[\mu^2 + \left(j - \frac{1}{2} \right)^2 \right] , \qquad K^{(1)}(\mu) = 1 .$$

Substituting this into (7.114), we get

$$\mathbb{E}\left[\tau_\omega(L)^n\right] = \exp\left(-\frac{L}{4L_{\text{loc}}} \right) \int_0^\infty e^{-\mu^2 L/L_{\text{loc}}} \frac{2\pi\mu\sinh(\pi\mu)}{\cosh^2(\pi\mu)} K^{(n)}(\mu)\,d\mu .$$

$$\tag{7.116}$$

Notes

The presentation of power transmission through a slab of random medium in Section 7.1, using the limit theorems of Chapter 6, follows the treatment in [104]. A more physical approach to this problem is given in Klyatskin's book [101].

The self-averaging property presented in Section 7.2 does not seem to be well known. The analytical reason for the phenomenon is the decorrelation of the power transmission coefficients at distinct frequencies, which is well known [32].

The treatment of wave localization in Section 7.3 is limited to the analysis of the exponential decay of the power transmission coefficient. The phenomenon of wave localization was discovered in 1958 by Anderson [5] in connection with electron waves in semiconductors. The mathematical theory was developed only twenty years later, starting with the paper of Goldsheid–Molchanov–Pastur [79]. Since that paper there has been a great deal of research published on the subject, in particular in the one-dimensional case, for which the theory of products of random matrices is available. We cite here some books that also contain additional references: Carmona–Lacroix [36], Pastur–Figotin [140], and the review papers by Van Tiggelen, Lacroix, and Klein in the proceedings [58].

The strongly heterogeneous white-noise regime addressed in Section 7.4 was considered for the first time in [32] and in more detail in [8].

The analysis of the Lyapunov exponent of the random harmonic oscillator is presented in detail in Arnold's book [6]. The positivity of the Lyapunov exponent for a general class of random media is shown in [111]. The ε-expansion of the Lyapunov exponent follows that in [7] and is a special case of the one presented in Section 6.8.

A general reference for the Mehler–Fock transform used to obtain the integral formulas in the appendix is [49].

8

Wave-Front Propagation

In Chapter 4 we discussed the homogenization regime in which we can replace the medium parameters by their (deterministic) homogenized or averaged values. This enormously simplifies the analysis of wave propagation. When the homogenized parameters are constant, as in the uniform-background case treated in Chapter 4, the transmitted wave is simply the incident wave shifted in time. There are no reflections in the homogenized medium, and the transmitted wave arrives at the end of the slab at time $L/\bar{c}$. In this chapter, as in Chapter 7, we consider the *weakly heterogeneous* regime (5.16) and the *strongly heterogeneous white-noise* regime (5.17) when the wave travels a distance that is large relative to its wavelength. These are two regimes in which random effects build up and affect the wave. We have seen in Chapter 7 that the total transmitted energy of a pulse decays exponentially with the width of the slab, and is a self-averaging quantity. In this chapter, we analyze the **transmitted wave front**. The main results are stated in Propositions 8.1 and 8.3 for the two asymptotic regimes, the weakly and the strongly heterogeneous regimes. The effect of the random medium on the wave front can be described as follows: When we observe the wave front at its random arrival time it can be expressed as a convolution of the transmitted front in the effective medium with a deterministic smoothing kernel that depends on the statistics of the medium fluctuations and the observation and source points.

In this chapter we start by analyzing the pulse spreading of the front in the weakly heterogeneous regime in Section 8.1. The main analytical tool is a functional averaging theorem whose proof is given in the appendix. In Section 8.2 we derive the pulse-spreading formula in the strongly heterogeneous white-noise regime using moments and the diffusion approximation results presented in Chapter 6, applied in the frequency domain. We then study the energy of the transmitted front and compare it with the total transmitted energy analyzed in Chapter 7. We also explain how the correct travel time for the front is related to the random medium, and present some numerical simulations. In Section 8.3 we discuss the problem of reflection of the front by an interface.

The analysis of the one-dimensional wave front presented in this chapter is generalized to three-dimensional randomly layered media in Chapter 14.

8.1 The Transmitted Wave Front in the Weakly Heterogeneous Regime

We consider the acoustic wave equations in the *weakly heterogeneous* regime (5.16), introduced in Section 5.1. We assume that a point source located at $z_0 < 0$ emits a short pulse at time $z_0/\bar{c}$. We showed in Chapter 5 that the dimensionless acoustic equations have the form (5.13–5.14)

$$\rho(z)\frac{\partial u}{\partial t} + \frac{\partial p}{\partial z} = \bar{\zeta}^{1/2} f\left(\frac{t - z_0/\bar{c}}{\varepsilon^2}\right)\delta(z - z_0)\,, \tag{8.1}$$

$$\frac{1}{K(z)}\frac{\partial p}{\partial t} + \frac{\partial u}{\partial z} = 0\,, \tag{8.2}$$

with the medium parameters in the weakly heterogeneous regime given by

$$\frac{1}{K(z)} = \begin{cases} \frac{1}{\bar{K}}\left(1 + \varepsilon\nu(z/\varepsilon^2)\right) & \text{for } z \in [0, L], \\ \frac{1}{\bar{K}} & \text{for } z \in (-\infty, 0) \cup (L, \infty), \end{cases}$$

$$\rho(z) = \bar{\rho} \quad \text{for all } z\,.$$

Here, the effective impedance and speed of sound are $\bar{\zeta} = \sqrt{\bar{K}\bar{\rho}}$ and $\bar{c} = \sqrt{\bar{K}/\bar{\rho}}$, respectively. The pulse emitted at $z_0 < 0$ at time $z_0/\bar{c}$ impinges on the random medium in $[0, L]$ at time 0. We assume, as in the previous chapter, that the fluctuation process $\nu(z) = g(Y(z))$ is a bounded function of an ergodic Markov process $Y(z)$ on a compact state space S. It has zero mean $\mathbb{E}[\nu(z)] = 0$. These are the general hypotheses needed for applying the limit theorems of Chapter 6. As in the previous chapter we match the background medium outside of the slab with the homogenized parameters $(\bar{\rho}, \bar{K})$ inside, and assume that ρ is constant. Generalizations to varying unmatched backgrounds and fluctuating density are treated in Chapter 17.

In this section, we give an analytical description of the transmitted pulse front in the weakly heterogeneous regime. We use a formulation similar to those found in the physical literature as well as in several mathematical papers. It is a natural time-domain approach that does not use the Fourier representation of the waves. The essential probabilistic step involves a relatively simple functional averaging theorem (Proposition 8.2). In order to use this averaging theorem we need to assume, however, that the first and second derivatives of the fluctuation process $\nu(z)$ are uniformly bounded. We shall see in Section 8.2 that the approach based on diffusion approximation in the frequency domain does not require smoothness properties and can also be applied in the strongly heterogeneous white-noise regime. There does not appear to be a simple averaging argument that can be used to analyze front propagation in the strongly heterogeneous white-noise regime.

We consider the right- and left-going waves defined in terms of the local impedances and moving with the local sound speed, as defined in (4.23):

$$
\begin{bmatrix} A^\varepsilon(t,z) \\ B^\varepsilon(t,z) \end{bmatrix} = \begin{bmatrix} \zeta^{\varepsilon-1/2}(z)p(t,z) + \zeta^{\varepsilon 1/2}(z)u(t,z) \\ -\zeta^{\varepsilon-1/2}(z)p(t,z) + \zeta^{\varepsilon 1/2}(z)u(t,z) \end{bmatrix} . \tag{8.3}
$$

The local impedance is

$$
\zeta^\varepsilon(z) = \sqrt{K(z)\rho(z)} = \frac{\bar{\zeta}}{\sqrt{1 + \varepsilon\nu(z/\varepsilon^2)}} , \quad \text{with } \bar{\zeta} = \sqrt{\bar{K}\bar{\rho}}.
$$

The mode amplitudes satisfy (4.24), which now takes the form

$$
\frac{\partial}{\partial z}\begin{bmatrix} A^\varepsilon \\ B^\varepsilon \end{bmatrix} = -\frac{1}{c^\varepsilon(z)}\begin{bmatrix} 1 & 0 \\ 0 & -1 \end{bmatrix}\frac{\partial}{\partial t}\begin{bmatrix} A^\varepsilon \\ B^\varepsilon \end{bmatrix} + \frac{\zeta^{\varepsilon\prime}(z)}{2\zeta^\varepsilon(z)}\begin{bmatrix} 0 & 1 \\ 1 & 0 \end{bmatrix}\begin{bmatrix} A^\varepsilon \\ B^\varepsilon \end{bmatrix}. \tag{8.4}
$$

Here $\zeta^{\varepsilon\prime}$ is the z-derivative of ζ^ε and

$$
\frac{\zeta^{\varepsilon\prime}(z)}{\zeta^\varepsilon(z)} = -\frac{1}{2\varepsilon}\frac{\nu'(z/\varepsilon^2)}{1 + \varepsilon\nu(z/\varepsilon^2)} .
$$

The local sound speed is

$$
c^\varepsilon(z) = \sqrt{K(z)/\rho(z)} = \frac{\bar{c}}{\sqrt{1 + \varepsilon\nu(z/\varepsilon^2)}} , \quad \text{with } \bar{c} = \sqrt{\bar{K}/\bar{\rho}}. \tag{8.5}
$$

This system is completed with an initial condition corresponding to a right-going wave that is incoming from the homogeneous half-space $z < 0$ and is impinging on the random medium in $[0, L]$,

$$
A^\varepsilon(t,z) = f\left(\frac{t-z}{\varepsilon^2}\right), \quad B^\varepsilon(t,z) = 0, \quad t < 0 .
$$

The pulse function f is compactly supported in the interval $(-T_0, T_0)$. Equation (8.4) clearly exhibits the two relevant propagation mechanisms. The first term on the right describes transport along the random characteristics with the local sound speed $c^\varepsilon(z)$. The second term on the right describes coupling between the right- and left-going modes, which is proportional to the derivative $\zeta^{\varepsilon\prime}$ of the impedance.

8.1.1 Stabilization of the Transmitted Wave Front

We now state the main result that characterizes the wave front transmitted through the random medium.

Proposition 8.1. *The wave front observed in the frame moving with the sound speed $\bar{c}$ of the effective medium,*

$$A^\varepsilon \left(\frac{z}{\bar{c}} + \varepsilon^2 s, z \right), \quad z > 0 ,$$

converges in distribution as $\varepsilon \to 0$ to

$$a(s, z) = a_0 \left(s - \frac{\sqrt{\gamma(0)}}{2\bar{c}} W_0(z), z \right). \tag{8.6}$$

Here $W_0(z)$ is a standard Brownian motion and a_0 is the deterministic pulse profile given by

$$a_0(s, z) = \frac{1}{2\pi} \int \exp \left(-i\omega s - \frac{\gamma(\omega)\omega^2}{8\bar{c}^2} z - i \frac{\gamma^{(s)}(\omega)\omega^2}{8\bar{c}^2} z \right) \hat{f}(\omega) d\omega , \tag{8.7}$$

with $\hat{f}(\omega)$ the Fourier transform of the initial pulse and

$$\gamma(\omega) = 2 \int_0^\infty \mathbb{E}[\nu(0)\nu(z)] \cos \left(\frac{2\omega z}{\bar{c}} \right) dz ,$$

$$\gamma^{(s)}(\omega) = 2 \int_0^\infty \mathbb{E}[\nu(0)\nu(z)] \sin \left(\frac{2\omega z}{\bar{c}} \right) dz .$$

We see from this proposition that the frequency-dependent decay rate of the wave front in (8.7) is equal to one-half the reciprocal of the localization length of the power transmission coefficient analyzed in Section 7.1.4 (equation (7.42)):

$$\frac{\gamma(\omega)\omega^2}{8\bar{c}^2} = \frac{1}{2L_{\text{loc}}(\omega)} . \tag{8.8}$$

This is always nonnegative because $\gamma(\omega)$ is the power spectral density of the stationary fluctuations $\nu(z)$ of the random medium. The factor one-half in (8.8) can be explained by the fact that the localization length is the reciprocal of the decay rate of the power transmission coefficient, which is the squared modulus in the Fourier domain. It is, however, not immediate that the decay rate for the transmitted wave front is exactly equal to one-half of the localization length as in (8.8) because this is the *asymptotic* decay rate of the mean-power transmission for large z, as we saw in Section 7.1.4. The decay rate of the wave front in (8.7) is valid for *any* z. The explanation for this interesting phenomenon is clear from the martingale representation (7.33) of the limit transmission coefficient $T_\omega(0, L)$, as we now show.

By (5.30) (with $\sigma = \varepsilon$ and $\theta = 1/\varepsilon$ in the weakly heterogeneous regime) the wave front observed at the end of the random medium in an ε^2-neighborhood of the arrival time $L/\bar{c}$ of the effective medium has the Fourier representation

$$A^\varepsilon \left(\frac{L}{\bar{c}} + \varepsilon^2 s, L \right) = \frac{1}{2\pi} \int \hat{f}(\omega) T_\omega^\varepsilon(0, L) e^{-i\omega s} d\omega .$$

By (8.6) and (8.7) the limit wave front can be written as

$$a(s, L) = \frac{1}{2\pi} \int \hat{f}(\omega) \tilde{T}_\omega(0, L) e^{-i\omega s} d\omega ,$$

where $\tilde{T}_\omega(0, L)$ is the multiplicative factor (7.34) in the martingale representation (7.33), $T_\omega(0, L) = \tilde{T}_\omega(0, L) M_\omega(0, L)$, of the limit transmission coefficient

$$\tilde{T}_\omega(0, L) = \exp\left[i \frac{\sqrt{\gamma(0)}}{2c} W_0(L) - \frac{\gamma(\omega)\omega^2}{8\bar{c}^2} L - i \frac{\gamma^{(s)}(\omega)\omega^2}{8\bar{c}^2} L \right] .$$

The fact that the martingale part does not appear in the limit wave front (8.6) is the essential mathematical content of Proposition 8.1. The reason for this is that (a) for each ω the factor $\tilde{T}_\omega(0, L)$ and the martingale $M_\omega(0, L)$ are independent and (b) for distinct ω_1 and ω_2 the martingales $M_{\omega_1}(0, L)$ and $M_{\omega_2}(0, L)$ are independent, so they average out in the ω-integration. These facts will be used in Section 8.2 to prove the analogue of Proposition 8.1 in the strongly heterogeneous white-noise regime. As we have already noted, the proof of Proposition 8.1 that we give here is based on an averaging theorem for integral equations in the time domain.

The second term $\exp[-i\gamma^{(s)}(\omega)\omega^2 z/(8\bar{c}^2)]$ in (8.7) is a frequency-dependent phase modulation and $\gamma^{(s)}(\omega)$ is conjugate to $\gamma(\omega)$, which is a factor in the decay rate. This shows that the transmitted wave front propagates in a dispersive effective medium with frequency-dependent wave number, given by

$$k(\omega) = \frac{\omega}{\bar{c}} - \varepsilon^2 \frac{\gamma^{(s)}(\omega)\omega^2}{8\bar{c}^2} ,$$

up to higher-order terms.

Proposition 8.1 shows that the transmitted wave front in the random medium is modified in two ways compared to propagation in a homogeneous one. First, its arrival time at the end of the slab $z = L$ has a small random component of order ε^2. Second, if we observe the wave front near its random arrival time, then we see a pulse profile that, to leading order, is deterministic and is the original pulse shape convolved with a deterministic kernel that depends on the second-order statistics of the medium through the autocorrelation function of ν. From the integral equation formulation of the wave front problem that we derive next and is given by (8.18), we can see that only second-order scattering events enter into the asymptotic analysis. This explains why only second-order statistics of the fluctuations are involved.

We can interpret the presence of the random shift in (8.6) by analyzing the asymptotic behavior of the random travel time defined by

$$\tau_0^\varepsilon(z) = \frac{z}{\bar{c}} + \frac{\varepsilon}{2\bar{c}} \int_0^z \nu\left(\frac{y}{\varepsilon^2}\right) dy . \tag{8.9}$$

We note that $\tau_0^\varepsilon(z)$ given by (8.9) is not the travel time along the random characteristics of (8.4), which by (8.5) is given by

$$\tau^\varepsilon(z) = \int_0^z \frac{1}{c^\varepsilon(y)} dy = \frac{z}{\bar{c}} + \frac{\varepsilon}{2\bar{c}} \int_0^z \nu\left(\frac{y}{\varepsilon^2}\right) dy - \frac{\varepsilon^2}{8\bar{c}} \int_0^z \nu^2\left(\frac{y}{\varepsilon^2}\right) dy + O(\varepsilon^3),$$

$$(8.10)$$

corresponding to the first arrival time to depth z for a point source at the surface. The ε^2 term in (8.10) is not present in (8.9). It is one of the results of Proposition 8.1 that the travel time of the wave front has the form (8.9). It implies that the stable wave front arrives after $\tau^\varepsilon(z)$, with the delay $\varepsilon^2 \mathbb{E}[\nu(0)^2]z/(8\bar{c})$.

By the central limit theorem applied to the fluctuation process $\nu(z)$ we have that as $\varepsilon \to 0$,

$$\frac{1}{\varepsilon^2}\left(\tau_0^\varepsilon(z) - \frac{z}{\bar{c}}\right) = \frac{1}{2\varepsilon\bar{c}} \int_0^z \nu\left(\frac{y}{\varepsilon^2}\right) dy$$

converges in distribution to

$$\frac{\sqrt{\gamma(0)}}{2\bar{c}} W_0(z), \tag{8.11}$$

where $W_0(z)$ is a standard Brownian motion. This characterizes the weak limit of the fluctuations in the arrival time of the stable wave front around the deterministic arrival time associated with the homogenized medium.

If we compute the expectation of the pulse front $a(s, z)$ in (8.6) then we obtain the **coherent** or **mean wave front**

$$\mathbb{E}[a(s, z)] = \frac{1}{2\pi} \int \exp\left(-i\omega s - \frac{[\gamma(0) + \gamma(\omega)]\omega^2}{8\bar{c}^2} z - i \frac{\gamma^{(s)}(\omega)\omega^2}{8\bar{c}^2} z\right) \hat{f}(\omega) d\omega.$$

Thus, the decay rate of the coherent wave is

$$\frac{[\gamma(0) + \gamma(\omega)]\omega^2}{8\bar{c}^2}, \tag{8.12}$$

which is larger than the decay rate (8.8) for the randomly centered wave front. This is discussed further in the strongly heterogeneous white-noise regime in Proposition 8.3, where it can be related to the rate of spreading of the pulse by (8.46).

8.1.2 The Integral Equation for the Transmitted Field

In this section we will transform the initial value problem (8.4) into an integral equation for the transmitted field using travel time coordinates.

We carry out a series of transformations to rewrite the evolution equations for the mode amplitudes in centered coordinates along the characteristic of the right-going mode. This gives a lower-triangular system that can be analyzed more easily. In a second step we apply the averaging theorem to this system in order to get an asymptotic description of the front of the advancing pulse.

We introduce the characteristic random travel time (8.10) and consider the new reference frame

$$(z, t) \mapsto (\tau, s) , \quad \text{with } \tau = \tau^\varepsilon(z) \text{ and } s = \frac{t - \tau^\varepsilon(z)}{\varepsilon^2} , \qquad (8.13)$$

which moves with the right-going mode A^ε and is adjusted to be on the time scale of the incident pulse. In this new reference frame the equations for $(A^\varepsilon, B^\varepsilon)$ have the form

$$\frac{\partial}{\partial \tau} \begin{bmatrix} A^\varepsilon \\ B^\varepsilon \end{bmatrix} = \frac{1}{\varepsilon^2} \begin{bmatrix} 0 & 0 \\ 0 & 2 \end{bmatrix} \frac{\partial}{\partial s} \begin{bmatrix} A^\varepsilon \\ B^\varepsilon \end{bmatrix} - \frac{1}{4\varepsilon} M^\varepsilon \left(\frac{z^\varepsilon(\tau)}{\varepsilon^2} \right) \begin{bmatrix} 0 & 1 \\ 1 & 0 \end{bmatrix} \begin{bmatrix} A^\varepsilon \\ B^\varepsilon \end{bmatrix} , \qquad (8.14)$$

where

$$M^\varepsilon(z) = \bar{c} \frac{\nu'(z)}{(1 + \varepsilon \nu(z))^{3/2}} ,$$

and $z^\varepsilon(\tau)$ is the inverse function of the travel time $\tau^\varepsilon(z)$. This is a lower-triangular system that we can integrate. More precisely, the equation for A^ε can be integrated for $\tau > 0$:

$$A^\varepsilon(s, \tau) = -\frac{1}{4\varepsilon} \int_0^\tau M^\varepsilon \left(\frac{z^\varepsilon(y)}{\varepsilon^2} \right) B^\varepsilon(s, y) dy + f(s). \qquad (8.15)$$

For $\tau \leq 0$, we simply have $A^\varepsilon(s, \tau) = f(s)$. The integrated form of the equation for B^ε is

$$B^\varepsilon(s, \tau) = -\frac{\varepsilon^2}{2} \int_{-\infty}^s S_B^\varepsilon \left(u, \tau + \frac{\varepsilon^2}{2}(s - u) \right) du , \qquad (8.16)$$

where

$$S_B^\varepsilon(s, \tau) = -\frac{1}{4\varepsilon} M^\varepsilon \left(\frac{z^\varepsilon(\tau)}{\varepsilon^2} \right) A^\varepsilon(s, \tau). \qquad (8.17)$$

The integral in (8.16) is over the infinite range $(-\infty, s)$. However, the initial conditions restrict A^ε and B^ε to be zero for $s < -T_0$ and $\tau = 0$. From equations (8.14) we then see that A^ε and B^ε are zero for $s < -T_0$ for any $\tau \geq 0$. Thus the integral with respect to u in (8.16) is effectively limited to the range $(-T_0, s)$. If we now substitute the integral representation (8.16) for B^ε into the one (8.15) for A^ε we obtain

$$A^\varepsilon(s, \tau) = f(s) - \frac{1}{32} \int_0^\tau M^\varepsilon \left(\frac{z^\varepsilon(y)}{\varepsilon^2} \right)$$

$$\times \int_{-T_0}^s M^\varepsilon \left(\frac{z^\varepsilon(y + \varepsilon^2(s - u)/2)}{\varepsilon^2} \right) A^\varepsilon \left(u, y + \varepsilon^2 \frac{s - u}{2} \right) du \, dy. \ (8.18)$$

This is the closed integral equation for the advancing front of the transmitted wave. We will apply the averaging theorem to a somewhat simplified version of this equation.

8.1.3 Asymptotic Analysis of the Transmitted Wave Front

We first transform the integral equation (8.18) into a form that is asymptotically equivalent to it as $\varepsilon \to 0$ and that allows direct application of the averaging theorem.

From (8.18) we get the inequality

$$\sup_{\tau \in [0, \tau^\varepsilon(L)]} |A^\varepsilon(s, \tau)| \leq |f(s)| + \frac{M^2 \tau^\varepsilon(L)}{32} \int_{-T_0}^{s} \sup_{\tau \in [0, \tau^\varepsilon(L)]} |A^\varepsilon(u, \tau)| \, du \,,$$

where $M = \bar{c} \|\nu'\|_\infty / (1 - \|\nu\|_\infty)^{3/2}$ is an upper bound for M^ε valid for any $\varepsilon < 1$. We also have $\tau^\varepsilon(L) \leq L/[\bar{c}(1 - \|\nu\|_\infty)]$. Using Gronwall's lemma we then obtain for any $\varepsilon < 1$ and $T > 0$ the estimate

$$\sup_{\tau \in [0, \tau^\varepsilon(L)], s \leq T} |A^\varepsilon(s, \tau)| \leq e^{M_2 L(T + T_0)} \|f\|_\infty \,.$$

Here $M_2 = M^2 / [32 \bar{c}(1 - \|\nu\|_\infty)]$. Substituting this estimate into (8.16) and (8.15), we get the further estimates

$$\sup_{\tau \in [0, \tau^\varepsilon(L)], s \leq T} |B^\varepsilon(s, \tau)| \leq \varepsilon K_{T,L} \,, \qquad \sup_{\tau \in [0, \tau^\varepsilon(L)], s \leq T} \left| \frac{\partial A^\varepsilon}{\partial \tau}(s, \tau) \right| \leq K_{T,L} \,,$$

where $K_{T,L}$ is a constant that depends only on T and L. From the estimate for $\partial_\tau A^\varepsilon$, we see that we can replace the last term of the integral in (8.18) by $A^\varepsilon(u, y)$, with an error of order ε^2. After the change of variable $x = z^\varepsilon(y)$ we obtain the integral equation

$$A^\varepsilon(s, \tau) = f(s) - \frac{1}{32} \int_0^{z^\varepsilon(\tau)} M^\varepsilon\left(\frac{x}{\varepsilon^2}\right) \frac{1}{c^\varepsilon(x)}$$

$$\times \int_{-T_0}^{s} M^\varepsilon\left(\frac{z^\varepsilon(\tau^\varepsilon(x)) + \varepsilon^2(s - u)/2)}{\varepsilon^2}\right) A^\varepsilon(u, \tau^\varepsilon(x)) \, du \, dx \,. \tag{8.19}$$

We assume here that the second derivative of ν is bounded. Then we have

$$M^\varepsilon\left(\frac{z^\varepsilon(\tau^\varepsilon(x)) + \varepsilon^2(s - u)/2)}{\varepsilon^2}\right) = M^\varepsilon\left(\frac{x}{\varepsilon^2} + \bar{c}\frac{s - u}{2}\right) + O(\varepsilon^2)$$

$$= \bar{c}\nu'\left(\frac{x}{\varepsilon^2} + \bar{c}\frac{s - u}{2}\right) + O(\varepsilon) \,.$$

We also have that

$$c^\varepsilon(x) = \bar{c} + O(\varepsilon) \,, \qquad z^\varepsilon(\tau) = \bar{c}\tau + O(\varepsilon) \,, \qquad \tau^\varepsilon(x) = x/\bar{c} + O(\varepsilon) \,,$$

uniformly in $x \in [0, L]$ and $\tau \in [0, L/[\bar{c}(1 - \|\nu\|_\infty)]]$. Using once again the uniform bound on $\partial_\tau A^\varepsilon$, we see that

$$A^\varepsilon(u, \tau^\varepsilon(x)) = A^\varepsilon(u, x/\bar{c}) + O(\varepsilon) \,,$$

which allows us to simplify the integral equation (8.19) for A^ε,

$$A^\varepsilon(s,\tau) = f(s) - \frac{\bar{c}}{32} \int_0^{\bar{c}\tau} \nu'\left(\frac{x}{\varepsilon^2}\right) \int_{-T_0}^s \nu'\left(\frac{x}{\varepsilon^2} + \bar{c}\frac{s-u}{2}\right) A^\varepsilon\left(u, \frac{x}{\bar{c}}\right) du\, dx\,,$$

where we have neglected terms of order ε. With the change of variable $x = \bar{c}y$, this integral equation can also be written as

$$A^\varepsilon(s,\tau) = f(s) - \frac{\bar{c}^2}{32} \int_0^\tau \nu'\left(\bar{c}\frac{y}{\varepsilon^2}\right) \int_{-T_0}^s \nu'\left(\bar{c}\frac{y}{\varepsilon^2} + \bar{c}\frac{s-u}{2}\right) A^\varepsilon(u,y)\, du\, dy\,,$$

or, in a functional form,

$$A^\varepsilon(\cdot,\tau) = f(\cdot) + \int_0^\tau F\left(\frac{y}{\varepsilon^2}\right) A^\varepsilon(\cdot,y)dy\,, \tag{8.20}$$

where $F(y)$ is the random linear operator acting on functions $A(\cdot)$ with support in $(-T_0, \infty)$, defined by

$$[F(y)A](s) = -\frac{\bar{c}^2}{32}\nu'(\bar{c}y) \int_{-T_0}^s \nu'\left(\bar{c}y + \bar{c}\frac{s-u}{2}\right) A(u)du\,. \tag{8.21}$$

Using the ergodic properties of ν, the following averaging theorem holds.

Proposition 8.2. *The solution $A^\varepsilon(\cdot,\tau)$ of the integral equation (8.20) converges, as a process in the space of continuous functions, in probability as $\varepsilon \to 0$ to the solution of the averaged integral equation*

$$\tilde{A}(\cdot,\tau) = f(\cdot) + \int_0^\tau \tilde{F}\tilde{A}(\cdot,y)dy\,, \tag{8.22}$$

where $\tilde{F} = \mathbb{E}[F(y)]$, that is,

$$[\tilde{F}A](s) = -\frac{\bar{c}^2}{32} \int_{-T_0}^s \mathbb{E}\left[\nu'(\bar{c}y)\nu'\left(\bar{c}y + \bar{c}\frac{s-u}{2}\right)\right] A(u)du\,. \tag{8.23}$$

The proof of this averaging theorem is given in Appendix 8.4. If we denote by ϕ_1 the autocorrelation function of the stationary random process ν',

$$\phi_1(x) = \mathbb{E}[\nu'(z)\nu'(z+x)]\,,$$

then the operator $\tilde{F}$ acting on functions $A(\cdot)$ with support in $(-T_0, \infty)$ has the form

$$\tilde{F}A(s) = -\frac{\bar{c}^2}{32} \int_{-T_0}^s \phi_1\left(\frac{\bar{c}}{2}(s-u)\right) A(u)du = -\frac{\bar{c}^2}{32} \int_0^{T_0+s} \phi_1\left(\frac{\bar{c}}{2}u\right) A(s-u)du\,.$$

This operator can also be written as a convolution independently of the point $-T_0$ defining the left end of support of A:

$$\tilde{F}A(s) = -\frac{\bar{c}^2}{32} \int_0^\infty \phi_1\left(\frac{\bar{c}}{2}u\right) A(s-u)du \,.$$

In the Fourier domain the convolution operator $\tilde{F}$ is the multiplication operator

$$\int_{-\infty}^\infty \tilde{F}A(s)e^{i\omega s}ds = -\frac{\bar{c}}{16}b_1\left(\frac{2\omega}{\bar{c}}\right) \int_{-\infty}^\infty A(s)e^{i\omega s}ds \,, \qquad (8.24)$$

where

$$b_1(k) = \int_0^\infty \phi_1(x)e^{ikx}dx \,. \qquad (8.25)$$

We will now rewrite b_1 in terms of the autocorrelation function of the stationary random process ν. Let ϕ_0 be the autocorrelation function of ν,

$$\phi_0(x) = \mathbb{E}[\nu(z)\nu(z+x)] \,, \qquad (8.26)$$

and let

$$b_0(k) = \int_0^\infty \phi_0(x)e^{ikx}dx \,. \qquad (8.27)$$

First we note that $\partial_x^2\phi_0(x) = \mathbb{E}[\nu(z)\nu''(z+x)]$. We also note that ϕ_0 is independent of z, by stationarity, so that $0 = \partial_z\partial_x\phi_0(x) = \mathbb{E}[\nu(z)\nu''(z+x)] + \mathbb{E}[\nu'(z)\nu'(z+x)]$. As a result we have the identity

$$\phi_1(x) = -\phi_0''(x) \,. \qquad (8.28)$$

By integration by parts we get

$$b_1(k) = -\int_0^\infty \phi_0''(x)e^{ikx}dx = -\left[\phi_0'(x)e^{ikx}\right]_0^\infty + ik\int_0^\infty \phi_0'(x)e^{ikx}dz \,.$$

Since ϕ_0 is even and differentiable we have $\phi_0'(0) = 0$, and the first term on the right side vanishes. Integrating by parts once again we obtain

$$b_1(k) = ik\left[\phi_0(x)e^{ikx}\right]_0^\infty + k^2\int_0^\infty \phi_0(x)e^{ikx}dx \,,$$

which can be written as

$$b_1(k) = -ik\phi_0(0) + k^2b_0(k) \,. \qquad (8.29)$$

Using (8.29) in (8.24) the linear operator $\tilde{F}$ is therefore given by

$$\int_{-\infty}^\infty \tilde{F}A(s)e^{i\omega s}ds = \left[\frac{i\omega}{8}\phi_0(0) - \frac{\omega^2}{4\bar{c}}b_0\left(\frac{2\omega}{\bar{c}}\right)\right] \int_{-\infty}^\infty A(s)e^{i\omega s}ds \,. \qquad (8.30)$$

We have shown that the pulse front converges to a deterministic profile when it is observed in the frame moving to the right with the random local sound speed $c^\varepsilon(z)$.

If we observe the wave front in the deterministic frame moving with the average speed $\bar{c}$, then we have to account for the small random difference between $z/\bar{c}$ and the random characteristic travel time $\tau^\varepsilon(z)$ given by (8.10). The rescaled travel time correction is

$$\frac{1}{\varepsilon^2}\left(\tau^\varepsilon(z) - \frac{z}{\bar{c}}\right) = \frac{1}{2\varepsilon\bar{c}}\int_0^z \nu\left(\frac{x}{\varepsilon^2}\right) dx - \frac{1}{8\bar{c}}\int_0^z \nu\left(\frac{x}{\varepsilon^2}\right)^2 dx + O(\varepsilon). \quad (8.31)$$

By the central limit theorem applied to the fluctuation process $\nu(z)$ and the ergodic theorem to $\nu^2(z)$, we have that as $\varepsilon \to 0$,

$$\frac{1}{\varepsilon^2}\left(\tau^\varepsilon(z) - \frac{z}{\bar{c}}\right)$$

converges in distribution to

$$\frac{1}{\sqrt{2\bar{c}}}\sqrt{b_0(0)}W_0(z) - \frac{1}{8\bar{c}}\phi_0(0)z\,, \quad (8.32)$$

where $W_0(z)$ is a standard Brownian motion. The deterministic correction $-\phi_0(0)z/(8\bar{c})$ cancels with the first term on the right in (8.30), when written in the time domain and used in (8.22). That is why the travel time fluctuation of the wave front is simply the Brownian motion part of (8.32) in the limit $\varepsilon \to 0$. This completes the proof of Proposition 8.1.

8.2 The Transmitted Wave Front in the Strongly Heterogeneous Regime

We consider now the acoustic wave equations in the *strongly heterogeneous white-noise* regime (5.17) introduced in Section 5.1. As shown in Chapter 5, the dimensionless acoustic equations have the form (5.13–5.14),

$$\rho(z)\frac{\partial u}{\partial t} + \frac{\partial p}{\partial z} = \zeta^{1/2}f\left(\frac{t - z_0/\bar{c}}{\varepsilon}\right)\delta(z - z_0)\,, \quad (8.33)$$

$$\frac{1}{K(z)}\frac{\partial p}{\partial t} + \frac{\partial u}{\partial z} = 0\,, \quad (8.34)$$

with the medium parameters in the strongly heterogeneous regime given by

$$\frac{1}{K(z)} = \begin{cases} \frac{1}{\overline{K}}\left(1 + \nu(z/\varepsilon^2)\right) & \text{for } z \in [0, L], \\ \frac{1}{\overline{K}} & \text{for } z \in (-\infty, 0) \cup (L, \infty), \end{cases}$$

$$\rho(z) = \bar{\rho} \text{ for all } z\,.$$

Note that now the pulse width is of order ε and the medium fluctuations are of order one. The fluctuation process $\nu(z)$ satisfies the general hypotheses of Chapter 6, as stated in the beginning of Section 8.1.

8.2.1 Asymptotic Representation of the Transmitted Wave Front

We will analyze front propagation in the strongly heterogeneous white-noise regime using the Fourier representation of the wave front. From Section 5.1.6, it has the form (5.30) (with $\sigma = 1$ and $\theta = 1$),

$$A^\varepsilon(L/\bar{c} + \varepsilon s, L) = a^\varepsilon(s, L) = \frac{1}{2\pi} \int e^{-i\omega s} T_\omega^\varepsilon(0, L) \hat{f}(\omega) \, d\omega, \qquad (8.35)$$

where s is scaled time relative to the arrival time $L/\bar{c}$ in the effective medium. The transmission coefficient

$$T_\omega^\varepsilon(0, L) = \frac{1}{\alpha_\omega^\varepsilon(0, L)} \qquad (8.36)$$

is expressed in terms of the propagator $\mathbf{P}_\omega^\varepsilon(0, z)$, which has the form

$$\mathbf{P}_\omega^\varepsilon(0, L) = \begin{bmatrix} \alpha_\omega^\varepsilon(0, L) & \overline{\beta_\omega^\varepsilon(0, L)} \\ \beta_\omega^\varepsilon(0, L) & \overline{\alpha_\omega^\varepsilon(0, L)} \end{bmatrix}.$$

It satisfies the random ordinary differential equations (5.27), introduced in Section 5.1.6,

$$\frac{d}{dz} \mathbf{P}_\omega^\varepsilon(0, z) = \frac{1}{\varepsilon} \mathbf{H}_\omega\left(\frac{z}{\varepsilon}, \nu\left(\frac{z}{\varepsilon^2}\right)\right) \mathbf{P}_\omega^\varepsilon(0, z), \qquad (8.37)$$

with the initial condition $\mathbf{P}_\omega^\varepsilon(0, 0) = \mathbf{I}$. The 2×2 complex matrix $\mathbf{H}_\omega$ is defined by

$$\mathbf{H}_\omega(z, \nu) = \frac{i\omega}{2\bar{c}} \nu \begin{bmatrix} 1 & -e^{-2i\omega z/\bar{c}} \\ e^{2i\omega z/\bar{c}} & -1 \end{bmatrix}.$$

As ordinary differential equations, the systems (8.37) are dynamically decoupled for different frequencies ω but they are stochastically coupled through the common fluctuation process ν. Since $\mathbb{E}[\nu(z)] = 0$, the matrix $\mathbf{H}_\omega$ is centered:

$$\mathbb{E}[\mathbf{H}_\omega(\tau, \nu(z))] = \mathbf{0}.$$

This is the setup for the limit theorems in Section 6.7.3, which we use in this section to obtain the following result.

Proposition 8.3. *The wave front observed at the right end of the random medium in an ε-neighborhood of the arrival time $L/\bar{c}$ in the effective medium,*

$$A^\varepsilon\left(\frac{L}{\bar{c}} + \varepsilon s, L\right),$$

converges in distribution as $\varepsilon \to 0$ to

$$a(s, L) = a_0 \left(s - \frac{\sqrt{\gamma}}{2\bar{c}} W_0(L), L \right). \tag{8.38}$$

Here $W_0(z)$ is a standard Brownian motion and a_0 is the deterministic pulse profile given by

$$a_0(s, L) = \frac{1}{2\pi} \int \exp \left(-i\omega s - \frac{\gamma \omega^2 L}{8\bar{c}^2} \right) \hat{f}(\omega) d\omega, \tag{8.39}$$

with $\hat{f}(\omega)$ the Fourier transform of the initial pulse and

$$\gamma = \int_{-\infty}^{\infty} \mathbb{E}[\nu(0)\nu(z)] dz.$$

Comparing this proposition with Proposition 8.1 we see that the results are the same if we set $\omega = 0$ in $\gamma(\omega)$ and $\gamma^{(s)}(\omega)$ in (8.7), in which case $\gamma(0) = \gamma$ and $\gamma^{(s)}(0) = 0$. This is consistent with the two scaling limits that are involved. In the weakly heterogeneous regime (5.16) wavelengths are comparable to the correlation length, and this is the reason why $\gamma(\omega)$ and $\gamma^{(s)}(\omega)$ enter into the pulse front characterization (8.7). In the strongly heterogeneous white-noise regime (5.17), wavelengths are long compared to the correlation length, and so $\gamma(\omega)$ and $\gamma^{(s)}(\omega)$ are evaluated at $\omega = 0$.

Proposition 8.3 shows that the frequency-dependent decay rate is

$$\frac{\gamma \omega^2}{8\bar{c}^2} = \frac{1}{2L_{\mathrm{loc}}(\omega)}, \tag{8.40}$$

where $L_{\mathrm{loc}}(\omega)$ is the localization length (7.81). Using (8.39) we can write (8.38) as a convolution in the time domain,

$$\begin{aligned}
a(s, L) &= \frac{1}{2\pi} \int e^{-i\omega s} e^{\left(i\omega \frac{\sqrt{\gamma}}{2\bar{c}} W_0(L) - \omega^2 \frac{\gamma}{8\bar{c}^2} L \right)} \int e^{i\omega u} f(u) du \, d\omega \\
&= \int f(u) \frac{1}{2\pi} \int e^{-i\omega \left(s - \frac{\sqrt{\gamma}}{2\bar{c}} W_0(L) - u \right)} e^{-\omega^2 \frac{\gamma}{8\bar{c}^2} L} d\omega \, du \\
&= \int f(u) \left\{ \frac{1}{2\pi} \int e^{-i\omega(s - \Theta_L - u)} e^{-\frac{\omega^2 D_L^2}{2}} d\omega \right\} du \\
&= \int f(u) \left\{ \frac{e^{-\frac{(s - \Theta_L - u)^2}{2D_L^2}}}{\sqrt{2\pi D_L^2}} \right\} du, \tag{8.41}
\end{aligned}$$

where we have defined

$$D_L^2 = \frac{\gamma}{4\bar{c}^2} L, \tag{8.42}$$

$$\Theta_L = \frac{\sqrt{\gamma}}{2\bar{c}} W_0(L). \tag{8.43}$$

Here we have also used the inverse Fourier transform for the characteristic function of the centered Gaussian density

$$\frac{1}{2\pi} \int e^{-i\omega s} e^{-\frac{\omega^2 D_L^2}{2}} d\omega = \frac{1}{\sqrt{2\pi D_L^2}} e^{-\frac{s^2}{2D_L^2}} = \mathcal{N}_{D_L}(s). \qquad (8.44)$$

The front $a(s, L)$ is simply the convolution of the initial pulse shape with this density evaluated at the **randomly shifted time** $s - \Theta_L$,

$$a(s, L) = [f \star \mathcal{N}_{D_L}] (s - \Theta_L), \qquad (8.45)$$

where $\star$ denotes convolution.

As we have seen in Proposition 8.1, in connection with (8.10), the random shift Θ_L is the limit of the **travel time** fluctuation. This is discussed in more detail in Section 8.2.7. If we compute the expectation of the pulse front in (8.45) then we obtain the **coherent** or **mean wave front**

$$\mathbb{E}[a(s, L)] = [f \star \mathcal{N}_{2D_L}] (s). \qquad (8.46)$$

It is a pulse that is centered at the arrival time $L/\bar{c}$ in the effective medium. Its profile is the convolution of the original pulse with a Gaussian density whose variance $2D_L^2$ is twice that of the stable front profile in (8.45). This factor two is due to the combination of the spreading of the stable front in (8.45) and averaging over the random fluctuations in the arrival time. Note that the doubling of the spreading factor follows from (8.12) when we replace $\gamma(\omega)$ by $\gamma(0) = \gamma$.

The discussion presented in Section 5.2.3 shows that the finite dimensional time distributions of the process $a^\varepsilon(s, L)$ are characterized by the moments in (5.40),

$$\lim_{\varepsilon \to 0} \mathbb{E}\left[\prod_{j=1}^m T_{\omega_j}^\varepsilon(0, L) \right], \qquad (8.47)$$

with $\omega_j \neq \omega_k$ for $j \neq k$. We note here the special form of the products in these moments. They are linear in each $T_{\omega_j}^\varepsilon(0, L)$ and not all possible moments are needed.

Our objective is to describe the limiting distribution of the transmitted front $a^\varepsilon(s, L)$ with the following strategy. First, we characterize the limit of the specific moments in (8.47). Second, we identify a simple process $a(s, L)$ that has the **same specific moments** in the limit $\varepsilon \to 0$. This allows us to conclude that $a^\varepsilon(s, L)$ converges to $a(s, L)$ **in distribution**. For the identification of $a(s, L)$ we use the martingale representation, (7.33). This representation is given in the weak fluctuation regime. However, setting $\gamma^{(s)}(\omega) = \gamma^{(s)}(0) = 0$ and $\gamma(\omega) = \gamma(0) = \gamma$ in (7.34) and (7.35) gives also the representation in the strongly heterogeneous white-noise regime.

In order to describe **jointly** the transmission coefficients occurring in the moments (8.47), we consider the system of equations (8.37) that describes the propagators for the finite set of frequencies $(\omega_1, \ldots, \omega_m)$. This will be carried out in Section 8.2.4 after a discussion of the transmitted energy in Section 8.2.2 and the presentation of some numerical simulations in Section 8.2.3.

8.2.2 The Energy of the Transmitted Wave

We will discuss the transmission of energy only in the strongly heterogeneous white-noise regime. The same analysis can be applied in the weakly heterogeneous regime using the results of Proposition 8.1.

The energy of the transmitted wave front is nonrandom in the limit $\varepsilon \to 0$ and is given by

$$\mathcal{E}_{\mathrm{stab}}(L) = \int |f \star \mathcal{N}_{D_L}(s)|^2 ds.$$

By the Parseval identity and (8.39) it also has the form

$$\mathcal{E}_{\mathrm{stab}}(L) = \frac{1}{2\pi} \int |\hat{f}\omega)|^2 \exp\left(-\frac{L}{L_{\mathrm{loc}}(\omega)}\right) d\omega. \tag{8.48}$$

The wave front energy exiting the medium at $z = L$ is strictly less than $\int |f(s)|^2 ds$. We may ask, does a part of the missing energy exit the medium in a stable way somewhere else or at a different time? In other words, what is the limit in distribution of $A^\varepsilon(L/\bar{c} + t_0 + \varepsilon s, L)$ for $t_0 \neq 0$ (energy exiting at $z = L$) or $B^\varepsilon(t_0 + \varepsilon s, 0)$ (energy reflected at $z = 0$)? An analysis similar to that for Proposition 8.3 shows that these two processes (in s) vanish as ε goes to 0 due to the fast phase $e^{-i\omega t_0/\varepsilon}$ in their integral representations

$$A^\varepsilon(L/\bar{c} + t_0 + \varepsilon s, L) = \frac{1}{2\pi} \int e^{-i\omega s} e^{-i\omega t_0/\varepsilon} T_\omega^\varepsilon(0, L)\hat{f}(\omega) \, d\omega,$$

$$B^\varepsilon(t_0 + \varepsilon s, 0) = \frac{1}{2\pi} \int e^{-i\omega s} e^{-i\omega t_0/\varepsilon} R_\omega^\varepsilon(0, L)\hat{f}(\omega) \, d\omega.$$

This implies that there is no other stable energy, in the ε-limit, exiting the slab $[0, L]$.

There is no absorption in the medium, so that the energy of the impinging pulse is conserved. This means that part of the incoming energy is transformed into small incoherent fluctuations. As we shall see in Chapter 9, the incoherent waves have small amplitude, of order $\sqrt{\varepsilon}$, but large support, of order one, so that they are not captured by the analysis of this chapter.

Using the results of Section 7.2.2, we can now describe the energy content of the transmitted wave when the input pulse is narrowband with carrier frequency ω_0. It consists of a stable part, described by the wave front theory of this chapter, whose energy decays quickly as $\exp[-L/L_{\mathrm{loc}}(\omega_0)]$. There is also an incoherent part in the transmitted wave whose energy decays as $\exp[-L/(4L_{\mathrm{loc}}(\omega_0))]$, as seen in Section 7.2. Thus the incoherent wave contains most of the energy of the total transmitted wave in the regime $L \geq L_{\mathrm{loc}}(\omega_0)$.

8.2.3 Numerical Illustration of Pulse Spreading

In this section we present numerical results that illustrate the transmitted wave front theory. The numerical setup is that described at the end of Section 7.4. We give it again for completeness. The random medium is a concatenation of thin layers with constant density and alternating bulk modulus with values $\kappa_K = 1/(1 \pm \sigma_K)$, with $\sigma_K = 0.8$. The effective density, bulk modulus, and speed of sound are all equal to one. The thicknesses of the layers are independent and identically distributed random variables with exponential distribution with mean $l_c = 0.02$. The integrated covariance is $\gamma = \sigma_K^2 l_c = 0.0128$. The total thickness of the random slab is $L = 120$, so that the variance D_L^2 is equal to 0.384. The incoming pulse is the second derivative of a Gaussian, with Fourier transform $\hat{f}(\omega) = \omega^2 \exp(-\omega^2/5)$. The pulse width is 4.7 and the carrier (angular) frequency is 2.9.

In Figure 8.1a we compare the transmitted signals recorded at $z = L$ in the absence of randomness and the signals obtained with 10 different realizations of the random medium. The transmitted signals in the random cases contain a short stable front and a small-amplitude long **coda**, that is, incoherent wave fluctuations. We can first check the pulse front stabilization qualitatively since the shapes of the recorded signals are obviously deterministic, while the coda is changing when the medium changes. We can also observe a (positive or negative) time delay around the expected arrival time $t = 120$ that depends on the realization of the random medium. We can also verify the pulse front stabilization quantitatively. In Figure 8.1b we take the signal corresponding to the homogeneous case, convolve it with the deterministic kernel (8.45) with $\Theta_L = 0$, and compare the convolved signal with the transmitted signals time-shifted to allow for better comparison. In Figure 8.1c we plot the mean transmitted pulse obtained by averaging of the numerical signals, and compare it with the theoretical expectation (8.46). As pointed out above, we can see the increased spreading due to the averaging over the random time delay. The agreement between the theoretical predictions and the numerical simulations is excellent.

8.2.4 The Diffusion Limit for the Multifrequency Propagators

We now come back to the derivation of Proposition 8.3. We will use the diffusion approximations in the form of Section 6.7.3 to analyze the propagators jointly at several different frequencies. We first write the propagator equations for each frequency,

$$\frac{d}{dz}\mathbf{P}_{\omega_j}^\varepsilon(0,z) = \frac{i\omega_j}{2\bar{c}\varepsilon}\nu\left(\frac{z}{\varepsilon^2}\right)\begin{bmatrix} 1 & 0 \\ 0 & -1 \end{bmatrix}\mathbf{P}_{\omega_j}^\varepsilon(0,z)$$

$$-\frac{\omega_j}{2\bar{c}\varepsilon}\nu\left(\frac{z}{\varepsilon^2}\right)\sin\left(\frac{2\omega_j z}{\bar{c}\varepsilon}\right)\begin{bmatrix} 0 & 1 \\ 1 & 0 \end{bmatrix}\mathbf{P}_{\omega_j}^\varepsilon(0,z)$$

$$-\frac{i\omega_j}{2\bar{c}\varepsilon}\nu\left(\frac{z}{\varepsilon^2}\right)\cos\left(\frac{2\omega_j z}{\bar{c}\varepsilon}\right)\begin{bmatrix} 0 & 1 \\ -1 & 0 \end{bmatrix}\mathbf{P}_{\omega_j}^\varepsilon(0,z), \quad (8.49)$$

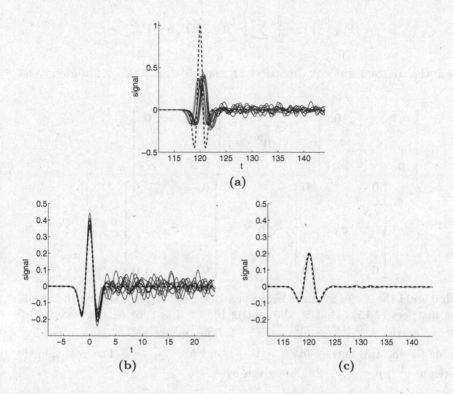

Fig. 8.1. Transmitted signals. Plot (a): comparison between the signal obtained with a homogeneous medium (thick dashed line) and the signals obtained from simulations with 10 different realizations of the random medium. Plot (b): Comparison between the theoretical stable pulse front obtained by the pulse front stabilization theory (thick dashed line) and the signals obtained from simulations (and time-shifted to remove the random time shifts). Plot (c): Comparison between the theoretical expected pulse (thick dashed line) and the mean transmitted signals obtained by averaging over 100 simulations.

for $j = 1, \ldots, m$. Introducing the $2m \times 2m$ multifrequency propagator matrix

$$
\mathbf{P}_m^\varepsilon(0, z) = \begin{bmatrix} \mathbf{P}_{\omega_1}^\varepsilon(0, z) & \cdots & 0 \\ & \ddots & \\ 0 & \cdots & \mathbf{P}_{\omega_m}^\varepsilon(0, z) \end{bmatrix},
$$

the system (8.49) is of the form (6.102),

$$
\frac{d}{dz} \mathbf{P}_m^\varepsilon(0, z) = \frac{1}{\varepsilon} \mathbf{F}\left(\mathbf{P}_m^\varepsilon(0, z), \nu\left(\frac{z}{\varepsilon^2}\right), \frac{z}{\varepsilon} \right).
$$

The decomposition (6.103) of the matrix field $\mathbf{F}$ is

$$\mathbf{F}(\mathbf{P}_m, \nu, \tau) = \frac{1}{2\bar{c}} \sum_{p=0}^{2m} g^{(p)}(\nu, \tau) \mathbf{\Omega}_m \mathbf{h}^{(p)} \mathbf{P}_m,$$

where the constant $2m \times 2m$ matrices $\mathbf{\Omega}_m$ and $\mathbf{h}^{(p)}$, $p = 0, \ldots, 2m$ are given by

$$\mathbf{\Omega}_m = \begin{bmatrix} \omega_1 \mathbf{I} & \cdots & \mathbf{0} \\ & \ddots & \\ \mathbf{0} & \cdots & \omega_m \mathbf{I} \end{bmatrix}, \qquad \mathbf{h}^{(0)} = i \begin{bmatrix} \boldsymbol{\sigma}_3 & \cdots & \mathbf{0} \\ & \ddots & \\ \mathbf{0} & \cdots & \boldsymbol{\sigma}_3 \end{bmatrix},$$

$$\mathbf{h}^{(2j-1)} = - \begin{bmatrix} \mathbf{0} \cdots & & \cdots \mathbf{0} \\ \vdots & & \vdots \\ & \boldsymbol{\sigma}_1 & \\ \vdots & & \vdots \\ \mathbf{0} \cdots & & \cdots \mathbf{0} \end{bmatrix}, \qquad \mathbf{h}^{(2j)} = \begin{bmatrix} \mathbf{0} \cdots & & \cdots \mathbf{0} \\ \vdots & & \vdots \\ & \boldsymbol{\sigma}_2 & \\ \vdots & & \vdots \\ \mathbf{0} \cdots & & \cdots \mathbf{0} \end{bmatrix}, \qquad j = 1, \ldots, m,$$

with $\mathbf{0}$ and $\mathbf{I}$ the 2×2 zero and identity matrices, and $\boldsymbol{\sigma}_1$, $\boldsymbol{\sigma}_2$, and $\boldsymbol{\sigma}_3$ the Pauli spin matrices (7.13). That is, the matrix $\mathbf{h}^{(2j-1)}$ has zero entries except for the two entries $h^{(2j-1)}_{2j-1,2j} = h^{(2j-1)}_{2j,2j-1} = -1$. The matrix $\mathbf{h}^{(2j)}$ has zero entries except for the two entries $h^{(2j)}_{2j-1,2j} = -i$ and $h^{(2j)}_{2j,2j-1} = i$. The real-valued functions $g^{(p)}$, $p = 0, \ldots, 2m$, are given by

$$g^{(0)}(\nu, \tau) = \nu,$$

$$g^{(2j-1)}(\nu, \tau) = \nu \sin\left(\frac{2\omega_j z}{\bar{c}}\right), \qquad j = 1, \ldots, m,$$

$$g^{(2j)}(\nu, \tau) = \nu \cos\left(\frac{2\omega_j z}{\bar{c}}\right), \qquad j = 1, \ldots, m.$$

The correlation matrix $\mathbf{C} = (C_{pq})_{p,q,=0,\ldots,2m}$ defined by (6.91) can be computed in terms of the integrated covariance of ν. It is diagonal here because of the orthogonality of $\cos(2\omega_j z/\bar{c})$ and $\sin(2\omega_j z/\bar{c})$,

$$C_{pq} = \lim_{Z_0 \to \infty} \frac{1}{Z_0} \int_0^{Z_0} \int_0^{\infty} \mathbb{E}[g^{(p)}(\nu(0), \tau) g^{(q)}(\nu(z), \tau)] dz \, d\tau = \frac{1}{2} \gamma_p \delta_{pq},$$

with

$$\gamma_0 = \gamma = \int_{-\infty}^{\infty} \mathbb{E}[\nu(0)\nu(z)] \, dz, \qquad \gamma_p = \frac{\gamma}{2}, \qquad p = 1, \ldots, 2m.$$

Therefore, the multifrequency propagator $\mathbf{P}^{\varepsilon}_m(0, z)$ converges in distribution to $\mathbf{P}_m(0, z)$, which is the solution of the Stratonovich stochastic differential equations

$$dP_m = \frac{\sqrt{\gamma}}{2\bar{c}}\Omega_m \mathbf{h}^{(0)}\mathbf{P}_m \circ dW_0(z) + \frac{\sqrt{\gamma}}{2\sqrt{2}\bar{c}}\sum_{j=1}^{m}\Omega_m \mathbf{h}^{(2j-1)}\mathbf{P}_m \circ dW_j(z)$$

$$+ \frac{\sqrt{\gamma}}{2\sqrt{2}\bar{c}}\sum_{j=1}^{m}\Omega_m \mathbf{h}^{(2j)}\mathbf{P}_m \circ d\tilde{W}_j(z)\,. \quad (8.50)$$

Here W_0, W_j, and $\tilde{W}_j$, $j = 1, \ldots, m$, are $2m+1$ independent standard Brownian motions. The matrix $\mathbf{P}_m$ is made up of 2×2 diagonal subblocks and zeros elsewhere, and we denote the jth diagonal 2×2 block by $\mathbf{P}_{\omega_j}$. The matrix $\mathbf{P}_{\omega_j}$ is the limit (in distribution) of the propagator $\mathbf{P}_{\omega_j}^{\varepsilon}$. We can rewrite the stochastic differential equation (8.50) for the individual propagators and in Itô's form

$$d\mathbf{P}_{\omega_j}(0,z) = \frac{i\omega_j\sqrt{\gamma}}{2\bar{c}}\begin{bmatrix} 1 & 0 \\ 0 & -1 \end{bmatrix}\mathbf{P}_{\omega_j}(0,z)dW_0(z)$$

$$-\frac{\omega_j\sqrt{\gamma}}{2\sqrt{2}\bar{c}}\begin{bmatrix} 0 & 1 \\ 1 & 0 \end{bmatrix}\mathbf{P}_{\omega_j}(0,z)dW_j(z)$$

$$-\frac{i\omega_j\sqrt{\gamma}}{2\sqrt{2}\bar{c}}\begin{bmatrix} 0 & 1 \\ -1 & 0 \end{bmatrix}\mathbf{P}_{\omega_j}(0,z)d\tilde{W}_j(z)\,, \quad (8.51)$$

for all $j = 1, \ldots, m$. Note here the remarkable cancellation of Itô's corrections. This is consistent with the analogous calculation in the weakly heterogeneous regime in Section 7.1.1, where by setting $\gamma(\omega) \to \gamma$ and $\gamma^{(s)}(\omega) \to 0$ in (7.17) we obtain (8.51).

8.2.5 Martingale Representation of the Multifrequency Transmission Coefficient

This section follows closely the lines of the analysis of Sections 7.1.2 and 7.1.3, which addressed the weakly heterogeneous regime. We will show that the transmission coefficients have martingale representations similar to (7.33), and that the martingale parts are independent for different frequencies.

By the symmetry of (8.51), the propagator matrices have the form

$$\mathbf{P}_{\omega_j}(0,z) = \begin{bmatrix} \alpha_{\omega_j}(0,z) & \overline{\beta_{\omega_j}(0,z)} \\ \beta_{\omega_j}(0,z) & \overline{\alpha_{\omega_j}(0,z)} \end{bmatrix}.$$

The pairs $(\alpha_{\omega_j}, \beta_{\omega_j})$, $j = 1, \ldots, m$, satisfy the conservation of energy relation $|\alpha_{\omega_j}|^2 - |\beta_{\omega_j}|^2 = 1$ and can be parameterized in polar coordinates as follows:

$$\alpha_{\omega_j}(0,z) = \cosh\left(\frac{\theta_{\omega_j}(z)}{2}\right)e^{i\phi_{\omega_j}(z)}, \quad (8.52)$$

$$\beta_{\omega_j}(0,z) = \sinh\left(\frac{\theta_{\omega_j}(z)}{2}\right)e^{i(\psi_{\omega_j}(z)+\phi_{\omega_j}(z))}. \quad (8.53)$$

If we introduce the new pairs of processes $(W_j^*, \tilde{W}_j^*)$ by the orthogonal transformation

$$\begin{bmatrix} W_j^*(z) \\ \tilde{W}_j^*(z) \end{bmatrix} = \int_0^z \begin{bmatrix} \sin(\psi_{\omega_j}) & \cos(\psi_{\omega_j}) \\ -\cos(\psi_{\omega_j}) & \sin(\psi_{\omega_j}) \end{bmatrix} d \begin{bmatrix} W_j(z) \\ \tilde{W}_j(z) \end{bmatrix},$$

then W_0, W_j^*, $\tilde{W}_j^*$, $j = 1, \ldots, m$, are $2m+1$ independent standard Brownian motions. The stochastic processes $(\theta_{\omega_j}, \psi_{\omega_j}, \phi_{\omega_j})$ can then be written as in (7.26–7.28), as solutions of the Itô stochastic differential equations

$$d\phi_{\omega_j} = -\frac{\omega_j \sqrt{\gamma}}{2\sqrt{2\bar{c}}} \tanh(\frac{\theta_{\omega_j}}{2}) dW_j^*(z) + \frac{\omega_j \sqrt{\gamma}}{2\bar{c}} dW_0(z), \qquad (8.54)$$

$$d\psi_{\omega_j} = \frac{\omega_j \sqrt{\gamma}}{\sqrt{2\bar{c}} \tanh(\theta_{\omega_j})} dW_j^*(z) - \frac{\omega_j \sqrt{\gamma}}{\bar{c}} dW_0(z), \qquad (8.55)$$

$$d\theta_{\omega_j} = \frac{\omega_j \sqrt{\gamma}}{\sqrt{2\bar{c}}} d\tilde{W}_j^*(z) + \frac{\omega_j^2 \gamma}{4\bar{c}^2 \tanh(\theta_{\omega_j})} dz, \qquad (8.56)$$

with the initial conditions $\theta_{\omega_j}(0) = 0$, $\psi_{\omega_j}(0) = 0$, $\phi_{\omega_j}(0) = 0$.

Using the same arguments as in Section 7.1.3, we see that the random vector $(T_{\omega_1}^\varepsilon(0, L), \ldots, T_{\omega_m}^\varepsilon(0, L))$ converges in distribution as $\varepsilon \to 0$ to the limit $(T_{\omega_1}(0, L), \ldots, T_{\omega_m}(0, L))$, where the limit transmission coefficients have the martingale representations

$$T_{\omega_j}(0, L) = M_{\omega_j}(0, L)\tilde{T}_{\omega_j}(0, L), \qquad j = 1, \ldots, m. \qquad (8.57)$$

Here

$$\tilde{T}_{\omega_j}(0, L) = \exp\left[i\frac{\sqrt{\gamma}\omega_j}{2\bar{c}} W_0(L) - \frac{\gamma \omega_j^2}{8\bar{c}^2} L \right], \qquad (8.58)$$

and $M_{\omega_j}(0, L)$ is the complex martingale

$$M_{\omega_j}(0, L) = \exp\left[-\frac{\sqrt{\gamma}\omega_j}{2\sqrt{2\bar{c}}} \int_0^L \tanh\left(\frac{\theta_{\omega_j}(z)}{2} \right) \left(d\tilde{W}_j^*(z) + i dW_j^*(z) \right) \right]. \qquad (8.59)$$

By (8.56) the process θ_{ω_j} is the solution of a stochastic differential equation driven by the Brownian motion $\tilde{W}_j^*$. Therefore, the martingale $M_{\omega_j}(0, L)$ depends only on the pair of Brownian motions $(W_j^*, \tilde{W}_j^*)$. This shows that these martingales are independent of each other, and independent of $\tilde{T}_{\omega_j}(0, L)$, which is a function of the Brownian motion W_0 only.

8.2.6 Identification of the Limit Wave Front

The transmitted front was shown in Section 5.2.3 to be characterized by the specific moments

$$\lim_{\varepsilon \to 0} \mathbb{E} \left[\prod_{j=1}^{m} T^{\varepsilon}_{\omega_j}(0, L) \right] = \mathbb{E} \left[\prod_{j=1}^{m} T_{\omega_j}(0, L) \right],$$

where T_{ω_j} has the martingale representation (8.57). The key facts about this representation are that

- the multiplicative factor $\tilde{T}_{\omega_j}(0, L)$ and the martingale $M_{\omega_j}(0, L)$ are independent
- the martingales $M_{\omega_j}(0, L)$ and $M_{\omega_k}(0, L)$ are independent for $j \neq k$.

It follows now that

$$\mathbb{E} \left[\prod_{j=1}^{m} T_{\omega_j}(0, L) \right] = \mathbb{E} \left[\prod_{j=1}^{m} \tilde{T}_{\omega_j}(0, L) M_{\omega_j}(0, L) \right]$$

$$= \mathbb{E} \left[\prod_{j=1}^{m} \tilde{T}_{\omega_j}(0, L) \right] \mathbb{E} \left[\prod_{j=1}^{m} M_{\omega_j}(0, L) \right]$$

$$= \mathbb{E} \left[\prod_{j=1}^{m} \tilde{T}_{\omega_j}(0, L) \right].$$

The expression for the wave front in (8.35) is

$$A^{\varepsilon}(L/\bar{c} + \varepsilon s, L) = a^{\varepsilon}(s, L) = \frac{1}{2\pi} \int e^{-i\omega s} T^{\varepsilon}_{\omega}(0, L) \hat{f}(\omega) \, d\omega.$$

We have shown that as a process with respect to the time variable s, $a^{\varepsilon}(s, L)$ converges in distribution to the process

$$a(s, L) = \frac{1}{2\pi} \int e^{-i\omega s} \tilde{T}_{\omega}(0, L) \hat{f}(\omega) \, d\omega,$$

where $\tilde{T}_{\omega}$ is the multiplicative factor (8.58) in the martingale representation (8.57),

$$\tilde{T}_{\omega}(0, L) = \exp \left(i\omega \frac{\sqrt{\gamma}}{2\bar{c}} W_0(L) - \omega^2 \frac{\gamma}{8\bar{c}^2} L \right). \tag{8.60}$$

Therefore the limit wave front is given by

$$a(s, L) = \frac{1}{2\pi} \int e^{-i\omega s} \tilde{T}_{\omega}(0, L) \hat{f}(\omega) d\omega$$

$$= \frac{1}{2\pi} \int e^{-i\omega s} e^{\left(i\omega \frac{\sqrt{\gamma}}{2\bar{c}} W_0(L) - \omega^2 \frac{\gamma L}{8\bar{c}^2} \right)} \hat{f}(\omega) d\omega$$

$$= \frac{1}{2\pi} \int e^{-i\omega \left(s - \frac{\sqrt{\gamma}}{2\bar{c}} W_0(L) \right)} e^{-\omega^2 \frac{\gamma L}{8\bar{c}^2}} \hat{f}(\omega) d\omega.$$

This is the result stated in Proposition 8.3.

We have characterized the limiting pulse front $a(s, L)$ through its finite-dimensional time distributions as explained in Section 5.2.3. In fact, this limit is also in the sense of convergence in distribution for continuous processes. For this we need an estimate on the modulus of continuity

$$M^\varepsilon(\delta) = \sup_{|s_2 - s_1| \leq \delta} |a^\varepsilon(s_1, L) - a^\varepsilon(s_2, L)|\,.$$

From the integral representation (8.35) of the transmitted wave front a^ε and the uniform bound

$$|T_\omega^\varepsilon(0, L)| \leq 1\,,$$

which follows from the conservation of energy relation (7.11), the modulus of continuity $M^\varepsilon(\delta)$ is uniformly bounded in ε by the deterministic quantity

$$M^\varepsilon(\delta) \leq \frac{1}{2\pi} \int \sup_{|s_2 - s_1| \leq \delta} \left|1 - e^{i\omega(s_2 - s_1)}\right| \left|\hat{f}(\omega)\right| d\omega.$$

When $\hat{f}(\omega)$ is rapidly decaying at infinity, then

$$M^\varepsilon(\delta) \leq \delta \frac{1}{2\pi} \int |\omega| \left|\hat{f}(\omega)\right| d\omega\,.$$

Relative compactness, or tightness, of the probability laws follows by taking the limit $\delta \to 0$. This property is important if, for instance, we want to study the convergence of the maximum value of the wave front over a given time window. The proof of Proposition 8.3 is complete.

8.2.7 Asymptotic Analysis of Travel Times

The random shift Θ_L in (8.45) is a **random travel time correction**. If we observe the transmitted signal at its random arrival time, then we will actually see a deterministic pulse shape in the asymptotic limit $\varepsilon \to 0$. This shape differs from the shape in a homogeneous medium because of the convolution with the Gaussian kernel $\mathcal{N}_{D_L}$ in (8.45). However, the shape does not depend on the particular realization of the random medium. The result (8.45) characterizes the transmitted pulse in distribution but does not describe how the travel time correction is related to the realization of the random medium. In order to understand this relation in more detail we observe the transmitted pulse at the random time

$$\tau_0^\varepsilon(L) = L/\bar{c} + \varepsilon\Theta_L^\varepsilon\,, \tag{8.61}$$

where the fluctuation Θ_L^ε is defined by

$$\Theta_L^\varepsilon = \frac{1}{2\bar{c}\varepsilon} \int_0^L \nu\left(\frac{z}{\varepsilon^2}\right) dz\,. \tag{8.62}$$

This travel time $\tau_0^\varepsilon(L)$ is different from that along the local characteristics given by

$$\tau^\varepsilon(L) = \int_0^L \frac{1}{c^\varepsilon(z)} dz, \quad \text{with } c^\varepsilon(z) = \frac{\bar{c}}{\sqrt{1 + \nu(z/\varepsilon^2)}}.$$

The calculations of this section show that the appropriate arrival time in the strongly heterogeneous white-noise regime is (8.61).

The transmitted wave observed around the *corrected* arrival time is given by

$$a_c^\varepsilon(s, L) := A(L/\bar{c} + \varepsilon(\Theta_L^\varepsilon + s), L) = \frac{1}{2\pi} \int e^{-i\omega s} T_\omega^\varepsilon(0, L) e^{-i\omega \Theta_L^\varepsilon} \hat{f}(\omega) d\omega.$$

(8.63)

Generalizing the method of identification of the limiting distribution used in the previous section, we see that we need to compute the limit of the corrected specific moments

$$\mathbb{E}\left[T_{\omega_1}^\varepsilon(0, L) \cdots T_{\omega_m}^\varepsilon(0, L) e^{-i(\sum_{j=1}^m \omega_j) \Theta_L^\varepsilon} \right], \qquad (8.64)$$

for any finite set of distinct frequencies $\omega_1, \ldots, \omega_m$. From this expression it is clear that we need to keep track of the joint distribution of the multi-frequency propagator $(\mathbf{P}_{\omega_1}^\varepsilon(0, z), \ldots, \mathbf{P}_{\omega_m}^\varepsilon(0, z))$ and the process Θ_z^ε. This is achieved by a straightforward generalization of the derivation given in Section 8.2.4. We see that $(\mathbf{P}_{\omega_1}^\varepsilon(0, z), \ldots, \mathbf{P}_{\omega_m}^\varepsilon(0, z), \Theta_z^\varepsilon)$ converges in distribution to $(\mathbf{P}_{\omega_1}(0, z), \ldots, \mathbf{P}_{\omega_m}(0, z), \Theta_z)$, where the $\mathbf{P}_{\omega_j}$'s satisfy the stochastic differential equations (8.51) and Θ_z is given by

$$\Theta_z = \frac{\sqrt{\gamma}}{2\bar{c}} W_0(z),$$

where indeed W_0 is the common Brownian motion driving the equations (8.51) for the propagators.

For the corrected specific moments (8.64) we have the convergence

$$\lim_{\varepsilon \to 0} \mathbb{E}\left[\prod_{j=1}^m T_{\omega_j}^\varepsilon(0, L) e^{-i(\sum_{j=1}^m \omega_j) \Theta_L^\varepsilon} \right] = \mathbb{E}\left[\prod_{j=1}^m T_{\omega_j}(0, L) e^{-i(\sum_{j=1}^m \omega_j) \frac{\sqrt{\gamma}}{2\bar{c}} W_0(L)} \right]$$

$$= \mathbb{E}\left[\prod_{j=1}^m \tilde{T}_{\omega_j}(0, L) e^{-i(\sum_{j=1}^m \omega_j) \frac{\sqrt{\gamma}}{2\bar{c}} W_0(L)} \right].$$

The last equality follows from the martingale representation (8.57) and its properties discussed in Section 8.2.5. We see from (8.58) that

$$\tilde{T}_{\omega_j}(0, L) e^{-i\omega_j \frac{\sqrt{\gamma}}{2\bar{c}} W_0(L)} = e^{-\frac{\gamma \omega_j^2}{8\bar{c}^2} L},$$

and therefore

$$a_c(s, L) = \frac{1}{2\pi} \int e^{-i\omega s} e^{-\frac{\gamma \omega^2}{8 \bar{c}^2} L} \hat{f}(\omega) d\omega = [f \star \mathcal{N}_{D_L}](s),$$

where D_L is given by (8.42). Thus, the transmitted wave front observed around its arrival time $\tau^\varepsilon(L)$ is asymptotically deterministic.

8.3 The Reflected Front in Presence of an Interface

In this section we analyze the reflected front when there is a discontinuity in the parameters of the background medium in the strongly heterogeneous case. In a homogeneous medium the resulting interface produces a reflected wave that is simply given as the incident pulse multiplied by the interface reflection coefficient and centered according to the two-way travel time to the interface. In the random case we show that a generalized pulse front theory can be used where the reflected wave is the convolution of the wave in the deterministic case with a kernel that depends on the medium statistics. This problem will be considered again in Section 11.1 when there is no impedance contrast and therefore no coherent reflected front. The case of smoothly varying background parameters is studied in Chapter 17.

8.3.1 Integral Representation of the Reflected Pulse

We have seen in Section 8.2.2 that no coherent energy part can be observed in the reflected wave. This is so because the background medium does not induce any coherent reflection. The incoherent reflected waves will be studied in detail in Chapter 9. We address in this section a case in which an interface is embedded into the random medium as sketched in Figure 8.2.

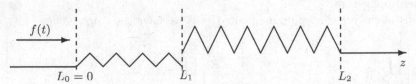

Fig. 8.2. We consider reflection and transmission of a pulse incoming from the left homogeneous half-space and impinging on two heteregeneous slabs with different background and statistical parameters.

We suppose that the distribution of the bulk modulus is not stationary but has a discontinuity at some location L_1,

$$\frac{1}{K(z)} = \begin{cases} \frac{1}{\bar{K}_1} & \text{for } z \in (-\infty, L_0), \\ \frac{1}{\bar{K}_1}\left(1 + \nu_1(z/\varepsilon^2)\right) & \text{for } z \in [L_0, L_1), \\ \frac{1}{\bar{K}_2}\left(1 + \nu_2(z/\varepsilon^2)\right) & \text{for } z \in [L_1, L_2), \\ \frac{1}{\bar{K}_2} & \text{for } z \in [L_2, \infty), \end{cases}$$

$$\rho(z) = \bar{\rho} \quad \text{for all } z \,,$$

where $L_0 = 0$. Our setup includes the general case $\bar{K}_1 \neq \bar{K}_2$ of a jump in the effective coefficients, as well as the particular case $\bar{K}_1 = \bar{K}_2$, where there is only a change in the statistics of the fluctuations. Here ν_1 and ν_2 are two independent zero-mean random processes. To simplify the presentation, we have assumed matched medium boundary conditions at both ends of the random medium. We consider an impinging pulse from the left.

In the absence of randomness, $\nu_1 = \nu_2 = 0$, a part of the wave is reflected by the interface. The analysis carried out in Chapter 3 shows that the reflection coefficient of the interface is

$$R_I = \frac{\bar{\zeta}_1 - \bar{\zeta}_2}{\bar{\zeta}_1 + \bar{\zeta}_2} \,, \tag{8.65}$$

where $\bar{\zeta}_j = \sqrt{\bar{K}_j \bar{\rho}}$, $j = 1, 2$, are the impedances of the two homogeneous media separated by the interface at $z = L_1$.

To analyze the effect of random fluctuations of the medium on the reflection by the interface, we introduce the local effective speeds $\bar{c}_j = \sqrt{\bar{K}_j / \bar{\rho}}$ and the right- and left-going modes defined by

$$\begin{cases} A_j(t, z) = \bar{\zeta}_j^{-1/2} p(t, z) + \bar{\zeta}_j^{1/2} u(t, z), \\ B_j(t, z) = -\bar{\zeta}_j^{-1/2} p(t, z) + \bar{\zeta}_j^{1/2} u(t, z), \end{cases} \quad \text{for } L_{j-1} \leq z \leq L_j \,.$$

The boundary conditions correspond to an impinging pulse at the interface $z = 0$ and a radiation condition at $z = L_2$:

$$A_1(t, 0) = f\left(\frac{t}{\varepsilon}\right), \quad B_2(t, L_2) = 0.$$

For $j = 1, 2$, inside the medium (L_{j-1}, L_j) the pair (A_j, B_j) satisfies the system

$$\frac{\partial}{\partial z}\begin{bmatrix} A_j \\ B_j \end{bmatrix} = \frac{1}{2\bar{c}_j}\begin{bmatrix} -2 - \nu_j(z/\varepsilon^2) & \nu_j(z/\varepsilon^2) \\ -\nu_j(z/\varepsilon^2) & 2 + \nu_j(z/\varepsilon^2) \end{bmatrix}\frac{\partial}{\partial t}\begin{bmatrix} A_j \\ B_j \end{bmatrix}.$$

For $j = 1$ and $j = 2$, the two systems are coupled by the jump conditions at $z = L_1$ corresponding to the continuity of the velocity and pressure fields

$$u(t, L_1) = \bar{\zeta}_1^{-1/2}\left(\frac{A_1(t, L_1) + B_1(t, L_1)}{2}\right) = \bar{\zeta}_1^{-1/2}\left(\frac{A_2(t, L_1) + B_2(t, L_1)}{2}\right),$$

$$p(t, L_1) = \bar{\zeta}_1^{1/2}\left(\frac{A_1(t, L_1) - B_1(t, L_1)}{2}\right) = \bar{\zeta}_2^{1/2}\left(\frac{A_2(t, L_1) - B_2(t, L_1)}{2}\right).$$

This gives the interface conditions

$$\begin{bmatrix} A_2 \\ B_2 \end{bmatrix}(t, L_1) = \mathbf{J} \begin{bmatrix} A_1 \\ B_1 \end{bmatrix}(t, L_1), \qquad \mathbf{J} = \begin{bmatrix} r^{(+)} & r^{(-)} \\ r^{(-)} & r^{(+)} \end{bmatrix},$$

where

$$r^{(\pm)} = \frac{1}{2}\left(\sqrt{\bar{\zeta}_2/\bar{\zeta}_1} \pm \sqrt{\bar{\zeta}_1/\bar{\zeta}_2} \right). \tag{8.66}$$

For $j = 1, 2$ and $L_{j-1} \leq z \leq L_j$ we define the centered waves by

$$a_j^\varepsilon(s, z) = A_j(\varepsilon s + (z - L_{j-1})/\bar{c}_j, z),$$
$$b_j^\varepsilon(s, z) = B_j(\varepsilon s - (z - L_{j-1})/\bar{c}_j, z).$$

These are the right- and left-going modes, respectively, in the frames moving with the local effective speed, centered at the beginning of the slabs ($L_0 = 0$ for $j = 1$ and L_1 for $j = 2$), and observed on the scale ε of the incoming pulse.

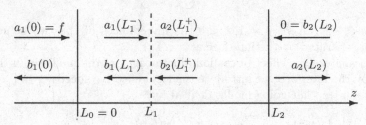

Fig. 8.3. Boundary conditions for the modes in presence of an interface.

In the frequency domain, the mode amplitudes obey the systems

$$\frac{d}{dz}\begin{bmatrix} \hat{a}_j^\varepsilon \\ \hat{b}_j^\varepsilon \end{bmatrix} = \frac{1}{\varepsilon}\mathbf{H}_{\omega,j}\left(\frac{z - L_{j-1}}{\varepsilon}, \nu_j\left(\frac{z}{\varepsilon^2}\right) \right)\begin{bmatrix} \hat{a}_j^\varepsilon \\ \hat{b}_j^\varepsilon \end{bmatrix}, \qquad L_{j-1} \leq z \leq L_j,$$

where

$$\mathbf{H}_{\omega,j}(z, \nu) = \frac{i\omega}{2\bar{c}_j}\nu \begin{bmatrix} 1 & -e^{-2i\omega z/\bar{c}_j} \\ e^{2i\omega z/\bar{c}_j} & -1 \end{bmatrix}.$$

The associated propagators satisfy the random differential equations

$$\frac{d}{dz}\mathbf{P}_{\omega,j}^\varepsilon(L_{j-1}, z) = \frac{1}{\varepsilon}\mathbf{H}_{\omega,j}\left(\frac{z - L_{j-1}}{\varepsilon}, \nu_j\left(\frac{z}{\varepsilon^2}\right) \right)\mathbf{P}_{\omega,j}^\varepsilon(L_{j-1}, z),$$

on $L_{j-1} \leq z \leq L_j$, with initial conditions $\mathbf{P}_{\omega,j}^\varepsilon(L_{j-1}, L_{j-1}) = \mathbf{I}$. Using the propagators we have

$$\begin{bmatrix} \hat{a}_j^\varepsilon \\ \hat{b}_j^\varepsilon \end{bmatrix}(\omega, L_j) = \mathbf{P}_{\omega,j}^\varepsilon(L_{j-1}, L_j)\begin{bmatrix} \hat{a}_j^\varepsilon \\ \hat{b}_j^\varepsilon \end{bmatrix}(\omega, L_{j-1}), \qquad j = 1, 2.$$

We also have the boundary and jump conditions

$$\hat{a}_1^\varepsilon(\omega,0) = \hat{f}(\omega), \qquad \hat{b}_2^\varepsilon(\omega,L_2) = 0, \qquad \begin{bmatrix} \hat{a}_2^\varepsilon \\ \hat{b}_2^\varepsilon \end{bmatrix}(\omega,L_1) = \mathbf{J}_\omega^\varepsilon \begin{bmatrix} \hat{a}_1^\varepsilon \\ \hat{b}_1^\varepsilon \end{bmatrix}(\omega,L_1),$$

where the jump matrix $\mathbf{J}_\omega^\varepsilon$ takes into account the fact that the time origins of a_1^ε and b_1^ε are shifted with respect to those of a_2^ε and b_2^ε:

$$\mathbf{J}_\omega^\varepsilon = \begin{bmatrix} r^{(+)}e^{i\omega L_1/(\bar{c}_1\varepsilon)} & r^{(-)}e^{-i\omega L_1/(\bar{c}_1\varepsilon)} \\ r^{(-)}e^{i\omega L_1/(\bar{c}_1\varepsilon)} & r^{(+)}e^{-i\omega L_1/(\bar{c}_1\varepsilon)} \end{bmatrix}.$$

This can be checked with the following computation:

$$\hat{a}_2^\varepsilon(\omega,L_1) = \int e^{i\omega s} a_2^\varepsilon(s,L_1)ds$$

$$= \int e^{i\omega s} A_2(\varepsilon s,L_1)ds$$

$$= \int e^{i\omega s}\left(r^{(+)}A_1(\varepsilon s,L_1) + r^{(-)}B_1(\varepsilon s,L_1)\right)ds$$

$$= r^{(+)}\int e^{i\omega s} a_1^\varepsilon\left(s - \frac{L_1}{\varepsilon\bar{c}_1},L_1\right)ds + r^{(-)}\int e^{i\omega s} b_1^\varepsilon\left(s + \frac{L_1}{\varepsilon\bar{c}_1},L_1\right)ds$$

$$= r^{(+)}e^{i\omega L_1/(\bar{c}_1\varepsilon)}\hat{a}_1^\varepsilon(\omega,L_1) + r^{(-)}e^{-i\omega L_1/(\bar{c}_1\varepsilon)}\hat{b}_1^\varepsilon(\omega,L_1),$$

and a similar computation for $\hat{b}_2^\varepsilon(\omega,L_1)$.

From all these relations, we get

$$\begin{bmatrix} \hat{a}_2^\varepsilon(\omega,L_2) \\ 0 \end{bmatrix} = \mathbf{P}_{\omega,2}^\varepsilon(L_1,L_2)\mathbf{J}_\omega^\varepsilon\mathbf{P}_{\omega,1}^\varepsilon(0,L_1)\begin{bmatrix} \hat{f}(\omega) \\ \hat{b}_1^\varepsilon(\omega,0) \end{bmatrix}.$$

Inverting this relation yields an expression for the left-going mode,

$$\hat{b}_1^\varepsilon(\omega,0) = -\frac{r^{(+)}\eta_\omega^\varepsilon + r^{(-)}\kappa_\omega^\varepsilon}{r^{(+)}\tilde{\eta}_\omega^\varepsilon + r^{(-)}\tilde{\kappa}_\omega^\varepsilon}\hat{f}(\omega), \tag{8.67}$$

where

$$\eta_\omega^\varepsilon = \alpha_{\omega,1}^\varepsilon(0,L_1)\beta_{\omega,2}^\varepsilon(L_1,L_2)e^{i\omega L_1/(\bar{c}_1\varepsilon)} + \beta_{\omega,1}^\varepsilon(0,L_1)\overline{\alpha_{\omega,2}^\varepsilon(L_1,L_2)}e^{-i\omega L_1/(\bar{c}_1\varepsilon)},$$

$$\kappa_\omega^\varepsilon = \beta_{\omega,1}^\varepsilon(0,L_1)\beta_{\omega,2}^\varepsilon(L_1,L_2)e^{-i\omega L_1/(\bar{c}_1\varepsilon)} + \alpha_{\omega,1}^\varepsilon(0,L_1)\overline{\alpha_{\omega,2}^\varepsilon(L_1,L_2)}e^{i\omega L_1/(\bar{c}_1\varepsilon)},$$

$$\tilde{\eta}_\omega^\varepsilon = \alpha_{\omega,1}^\varepsilon(0,L_1)\beta_{\omega,2}^\varepsilon(L_1,L_2)e^{-i\omega L_1/(\bar{c}_1\varepsilon)} + \overline{\beta_{\omega,1}^\varepsilon(0,L_1)}\,\overline{\alpha_{\omega,2}^\varepsilon(L_1,L_2)}e^{i\omega L_1/(\bar{c}_1\varepsilon)},$$

$$\tilde{\kappa}_\omega^\varepsilon = \overline{\beta_{\omega,1}^\varepsilon(0,L_1)}\beta_{\omega,2}^\varepsilon(L_1,L_2)e^{i\omega L_1/(\bar{c}_1\varepsilon)} + \overline{\alpha_{\omega,1}^\varepsilon(0,L_1)}\,\overline{\alpha_{\omega,2}^\varepsilon(L_1,L_2)}e^{-i\omega L_1/(\bar{c}_1\varepsilon)}.$$

The integral representation of the reflected wave around time t_0 on the ε-scale is

$$B_1(t_0 + \varepsilon s,0) = \frac{1}{2\pi}\int e^{-i\omega(s+\frac{t_0}{\varepsilon})}\hat{b}_1^\varepsilon(\omega,0)d\omega, \tag{8.68}$$

with $\hat{b}_1^\varepsilon(\omega,0)$ given by (8.67).

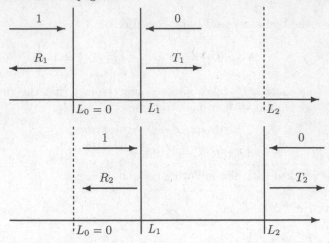

Fig. 8.4. Reflection and transmission coefficients.

8.3.2 The Limit for the Reflected Front

There are several ways in which to simplify the expression (8.67) for $\hat{b}_1^\varepsilon$. One of them is to use the reflection and transmission coefficients of the two random slabs. Let us define, for $j = 1, 2$, the transmission and reflection coefficients $T_{\omega,j}^\varepsilon$ and $R_{\omega,j}^\varepsilon$ for the slab $[L_{j-1}, L_j]$ as illustrated in Figure 8.4. In terms of $\alpha_{\omega,j}^\varepsilon$ and $\beta_{\omega,j}^\varepsilon$ they are given by

$$R_{\omega,j}^\varepsilon = -\frac{\beta_{\omega,j}^\varepsilon(L_{j-1}, L_j)}{\alpha_{\omega,j}^\varepsilon(L_{j-1}, L_j)}, \qquad T_{\omega,j}^\varepsilon = \frac{1}{\alpha_{\omega,j}^\varepsilon(L_{j-1}, L_j)}.$$

We also introduce $\tilde{R}_{\omega,1}^\varepsilon$ and $\tilde{T}_{\omega,1}^\varepsilon$, the reflection and transmission coefficients for a left-going incident wave coming from the right onto the medium 1 as illustrated in Figure 8.5. These **adjoint** coefficients are given in terms of $\alpha_{\omega,1}^\varepsilon$ and $\beta_{\omega,1}^\varepsilon$ by

$$\tilde{R}_{\omega,1}^\varepsilon = e^{\frac{2i\omega L_1}{\bar{c}_1\varepsilon}}\frac{\overline{\beta_{\omega,1}^\varepsilon(0, L_1)}}{\alpha_{\omega,1}^\varepsilon(0, L_1)}, \qquad \tilde{T}_{\omega,1}^\varepsilon = e^{\frac{2i\omega L_1}{\bar{c}_1\varepsilon}}\frac{1}{\alpha_{\omega,1}^\varepsilon(0, L_1)}.$$

The left-going mode amplitude $\hat{b}_1^\varepsilon$ given by (8.67) can be rewritten in terms of these reflection and transmission coefficients and their adjoints. Since the reflection coefficients are less than one, we can expand these expressions to obtain the series

$$\hat{b}_1^\varepsilon(\omega, 0) = \left(R_{\omega,1}^\varepsilon + \frac{T_{\omega,1}^\varepsilon}{\tilde{T}_{\omega,1}^\varepsilon}R_{\omega,2}^\varepsilon\right)\sum_{m=0}^{\infty}(R_{\omega,2}^\varepsilon\tilde{R}_{\omega,1}^\varepsilon)^m\hat{f}(\omega) + (T_{\omega,1}^\varepsilon)^2[1 - (R_{\omega,2}^\varepsilon)^2]$$

$$\times e^{\frac{2i\omega L_1}{\bar{c}_1\varepsilon}}\sum_{n=0}^{\infty}\left(-\frac{r^{(-)}}{r^{(+)}}\right)^{n+1}(\tilde{R}_{\omega,1}^\varepsilon - R_{\omega,2}^\varepsilon)^n\left[\sum_{m=0}^{\infty}(R_{\omega,2}^\varepsilon\tilde{R}_{\omega,1}^\varepsilon)^m\right]^n\hat{f}(\omega).$$

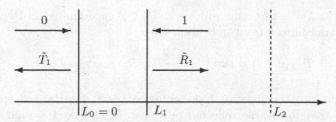

Fig. 8.5. Adjoint reflection and transmission coefficients.

The situation is now similar to the one encountered in Section 8.2.4. The reflected front is characterized by the moments of $\hat{b}_1^\varepsilon$,

$$\mathbb{E}\left[\prod_{j=1}^m \hat{b}_1^\varepsilon(\omega_j, 0)\right],$$

for m distinct frequencies $(\omega_j)_{1 \le j \le m}$. These moments involve sums of expectations of products of reflection and transmission coefficients. These expectations can be factored because the coefficients associated with the medium 1 are asymptotically independent from the coefficients associated to the medium 2. It follows from our analysis in Chapter 9 that an expectation involving a product of reflection and transmission coefficients vanishes as $\varepsilon \to 0$ as soon as the product contains reflection coefficients. Only one term in the series expansion of $\hat{b}_1^\varepsilon$ does not involve a reflection coefficient, and it is proportional to $(T_{\omega,1}^\varepsilon)^2(0, L_1)$. As a result, the problem is reduced to the identification of the limits of the moments

$$\mathbb{E}\left[\prod_{j=1}^m (T_{\omega_1,1}^\varepsilon)^2(0, L_1)\right],$$

for m distinct frequencies $(\omega_j)_{1 \le j \le m}$. This can be done in the same way as in Section 8.2.6. We need only to observe that the limit transmission coefficients $T_{\omega_j,1}^2(0, L_1)$ have the martingale representation

$$T_{\omega_j,1}^2(0, L_1) = M_{\omega_j,1}^2(0, L_1)\tilde{T}_{\omega_j,1}^2(0, L_1),$$

which is the square of (8.57). This is because $M_{\omega_j,1}^2(0, L_1)$, the square of (8.59), is also a complex martingale. The moments of the phase-compensated coefficient

$$\tilde{b}_1^\varepsilon(\omega) = \hat{b}_1^\varepsilon(\omega, 0)e^{\frac{-2i\omega L_1}{\bar{c}_1 \varepsilon}},$$

at different frequencies converge, and the limits are given by

$$\lim_{\varepsilon \to 0} \mathbb{E}\left[\prod_{j=1}^m \tilde{b}_1^\varepsilon(\omega_j)\right] = \left(-\frac{r^{(-)}}{r^{(+)}}\right)^m \mathbb{E}\left[\prod_{j=1}^m \tilde{T}_{\omega_j,1}^2(0, L_1)\right]\prod_{j=1}^m \hat{f}(\omega_j),$$

where $\tilde{T}_{\omega,1}(0, L_1)$ is the same process as (8.58), encountered in the study of the stable transmitted front in Section 8.2.5:

$$\tilde{T}_{\omega,1}(0, L_1) = \exp\left(i\omega \frac{\sqrt{\gamma_1}}{2\bar{c}_1} W_0(L_1) - \omega^2 \frac{\gamma_1}{8\bar{c}_1^2} L_1 \right).$$

Substituting into the integral representation (8.68) for the reflected wave, this shows that the stable reflected pulse can be observed around the time $t_0 = 2L_1/\bar{c}_1$ and has the form

$$B_1(2L_1/\bar{c}_1 + \varepsilon s, 0) \xrightarrow{\varepsilon \to 0} b(s) := \left(-\frac{r^{(-)}}{r^{(+)}} \right) \frac{1}{2\pi} \int e^{-i\omega s} \tilde{T}_{\omega,1}^2(0, L_1) \hat{f}(\omega) d\omega,$$

where we assume that there is impedance contrast, that is, $r^{(-)} \neq 0$ in (8.66). At any other observation times $t_0 \neq 2L_1/\bar{c}_1$ the reflected wave vanishes in the limit $\varepsilon \to 0$ because of the remaining rapid phase in the integral representation. This implies in particular that, even with random inhomogeneities in the medium, the arrival time of the stable reflection can be used to identify the depth of the jump in the background parameters with a precision of order ε, which is due to the random time shift.

The limiting **stable reflected front** has the form

$$b(s) = R_I f \star \mathcal{N}_{D_{2L_1}}(s - 2\Theta_{L_1}), \tag{8.69}$$

where D_z and Θ_z are defined as in (8.43) and (8.42),

$$D_z^2 = \frac{\gamma_1 z}{4\bar{c}_1^2}, \qquad \Theta_z = \frac{\sqrt{\gamma_1}}{2\bar{c}_1} W_0(z), \qquad \gamma_1 = \int_{-\infty}^{\infty} \mathbb{E}[\nu_1(0)\nu_1(z)] dz,$$

and $\mathcal{N}_D$ is the centered Gaussian density with variance D^2. The reflection coefficient

$$R_I = \left(-\frac{r^{(-)}}{r^{(+)}} \right) = \frac{\bar{\zeta}_1 - \bar{\zeta}_2}{\bar{\zeta}_1 + \bar{\zeta}_2}$$

is the one introduced in (8.65). It is the one corresponding to the case in which the interface separates two homogeneous media with impedances $\bar{\zeta}_1$ and $\bar{\zeta}_2$. The reflected pulse front has a deterministic shape imposed by the convolution with the Gaussian kernel $\mathcal{N}_{D_{2L_1}}$, and it is random only through the random time shift $2\Theta_{L_1}$.

The result that we obtain in the random case is not surprising once the behavior of a transmitted pulse front is understood. Indeed the reflected front does a round trip in the random medium in order to go from the surface $z = 0$ to the interface $z = L_1$ and back. The deterministic spreading thus corresponds to a travel distance of $2L_1$, and the random time shift is simply twice the one-way shift because the wave travels in the same medium.

Finally, we note that a stable transmitted pulse can be observed at the end of the slab ($z = L_2$) around the time $t_1 = L_1/\bar{c}_1 + (L_2 - L_1)/\bar{c}_2$. The limiting form of this **stable transmitted front** is

$$A(t_1 + \varepsilon s, L_2) \xrightarrow{\varepsilon \to 0} T_I f \star \mathcal{N}_D(s - \Theta), \qquad (8.70)$$

where the width of the convolution kernel is

$$D^2 = \frac{\gamma_1 L_1}{4\bar{c}_1^2} + \frac{\gamma_2(L_2 - L_1)}{4\bar{c}_2^2},$$

and the random time delay is given by

$$\Theta = \frac{\sqrt{\gamma_1}}{2\bar{c}_1} W_0(L_1) + \frac{\sqrt{\gamma_2}}{2\bar{c}_2}(W_0(L_2) - W_0(L_1)).$$

The transmission coefficient

$$T_I = \frac{2\sqrt{\bar{\zeta}_1 \bar{\zeta}_2}}{\bar{\zeta}_1 + \bar{\zeta}_2}$$

is the one for the interface between two homogeneous media.

8.4 Appendix. Proof of the Averaging Theorem

In this Appendix we give a proof of Proposition 8.2. We fix $T > 0$ and prove the convergence in the space of continuous functions over $[-T_0, T]$ with the supremum norm $\| \cdot \|_\infty$. We first list some properties of the operators F and $\tilde{F}$ defined by (8.21) and (8.23), respectively, in the following two lemmas.

Lemma 8.4. *Let $A(s)$ be a deterministic continuous function. Then*

$$\mathbb{E}\left[\left\| \frac{1}{Z} \int_0^Z [F(y)A]dy - \tilde{F}A] \right\|_\infty\right] \xrightarrow{Z \to \infty} 0.$$

Proof. Let us define

$$\Delta_Z(s) := \frac{1}{Z} \int_0^Z [F(y)A](s)dy - \tilde{F}A(s)$$

$$= -\frac{\bar{c}^2}{32} \int_{-T_0}^s \left[\frac{1}{Z} \int_0^Z \nu'(\bar{c}y)\nu'\left(\bar{c}y + \bar{c}\frac{s-u}{2}\right) \right.$$

$$\left. -\mathbb{E}\left[\nu'(\bar{c}y)\nu'\left(\bar{c}y + \bar{c}\frac{s-u}{2}\right) \right] dy \right] A(u)du.$$

By the ergodic theorem, for any $s, u \in [-T_0, T]$,

$$\mathbb{E}\left[\left| \frac{1}{Z} \int_0^Z \nu'(\bar{c}y)\nu'\left(\bar{c}y + \bar{c}\frac{s-u}{2}\right) - \mathbb{E}\left[\nu'(\bar{c}y)\nu'\left(\bar{c}y + \bar{c}\frac{s-u}{2}\right) \right] dy \right|\right] \xrightarrow{Z \to \infty} 0.$$

Therefore, by the dominated convergence theorem, for any $s \in [-T_0, T]$,

$$\mathbb{E}\left[|\Delta_Z(s)|\right] \xrightarrow{Z\to\infty} 0.$$

It remains to control the modulus of continuity to get a uniform in s estimate. From the uniform boundedness of the process ν', we have

$$\sup_{|s_1-s_2|\leq\delta} |\Delta_Z(s_1) - \Delta_Z(s_2)| \leq \frac{\bar{c}^2\|\nu'\|_\infty^2}{16} \sup_{|s_1-s_2|\leq\delta} |A(s_1) - A(s_2)|.$$

Therefore, setting $s_k = -T_0 + k(T+T_0)/N$, $k = 0, \ldots, N$, we have

$$\mathbb{E}\left[\|\Delta_Z\|_\infty\right] \leq \sum_{k=0}^{N} \mathbb{E}\left[|\Delta_Z(s_k)|\right] + \frac{\bar{c}^2\|\nu'\|_\infty^2}{16} \sup_{|s_1-s_2|\leq(T+T_0)/N} |A(s_1) - A(s_2)|.$$

Taking first the limit $Z \to \infty$ and then $N \to \infty$ gives the result from the uniform continuity of A over the compact interval $[-T_0, T]$.

Lemma 8.5. *(1) For any y, the operators $F(y)$ and $\tilde{F}$ are uniformly Lipschitz with a nonrandom Lipschitz constant c:*

$$\|F(y)A - F(y)B\|_\infty \leq c\|A - B\|_\infty, \quad \|\tilde{F}A - \tilde{F}B\|_\infty \leq c\|A - B\|_\infty.$$

(2) There exists $C > 0$ such that

$$\sup_{y\in\mathbb{R}} \|F(y)A\|_\infty + \|\tilde{F}A\|_\infty \leq C\|A\|_\infty.$$

Proof. The first part of the lemma follows from the uniform in s estimate

$$|[F(y)A](s) - [F(y)B](s)| \leq \frac{\bar{c}^2\|\nu'\|_\infty^2}{16}\|A - B\|_\infty,$$

which also holds true for $\tilde{F}$. The second part follows directly from the boundedness of the process ν'.

We can now give the proof of Proposition 8.2. It is enough to prove convergence in the mean of the supremum norm of the difference between A^ε and $\tilde{A}$, because this implies convergence in probability. From the integral equation formulations

$$A^\varepsilon(s,\tau) = f(s) + \int_0^\tau F\left(\frac{y}{\varepsilon^2}\right) A^\varepsilon(s,y)dy, \qquad \tilde{A}(s,\tau) = f(s) + \int_0^\tau \tilde{F}\tilde{A}(s,y)dy,$$

the difference between A^ε and $\tilde{A}$ satisfies

$$A^\varepsilon(s,\tau) - \tilde{A}(s,\tau) = \int_0^\tau \left(F\left(\frac{y}{\varepsilon^2}\right) A^\varepsilon(s,y) - F\left(\frac{y}{\varepsilon^2}\right) \tilde{A}(s,y)\right) dy + g^\varepsilon(s,\tau),$$

where $g^\varepsilon(s,\tau) := \int_0^\tau F(\frac{y}{\varepsilon^2})\tilde{A}(s,y) - \tilde{F}\tilde{A}(s,y)dy$. Taking the supremum norm, we obtain

$$\|A^\varepsilon(\cdot,\tau) - \tilde{A}(\cdot,\tau)\|_\infty \le \int_0^\tau \left\|F\left(\frac{y}{\varepsilon^2}\right) A^\varepsilon(\cdot,y) - F\left(\frac{y}{\varepsilon^2}\right)\tilde{A}(\cdot,y)\right\|_\infty dy$$
$$+\|g^\varepsilon(\cdot,\tau)\|_\infty$$
$$\le c\int_0^\tau \|A^\varepsilon(\cdot,y) - \tilde{A}(\cdot,y)\|_\infty dy + \|g^\varepsilon(\cdot,\tau)\|_\infty.$$

Taking the expectation and applying Gronwall's lemma, we obtain for any arbitrary $\tau_0 > 0$,

$$\sup_{\tau\in[0,\tau_0]}\mathbb{E}\left[\|A^\varepsilon(\cdot,\tau) - \tilde{A}(\cdot,\tau)\|_\infty\right] \le e^{c\tau_0}\sup_{\tau\in[0,\tau_0]}\mathbb{E}[\|g^\varepsilon(\cdot,\tau)\|_\infty].$$

It remains to show that the last term goes to 0 as $\varepsilon \to 0$. Let $\delta > 0$:

$$g^\varepsilon(s,\tau) = \sum_{k=0}^{[\tau/\delta]-1}\int_{k\delta}^{(k+1)\delta}\left(F\left(\frac{y}{\varepsilon^2}\right)\tilde{A}(s,y) - \tilde{F}\tilde{A}(s,y)\right) dy$$
$$+\int_{\delta[\tau/\delta]}^\tau\left(F\left(\frac{y}{\varepsilon^2}\right)\tilde{A}(s,y) - \tilde{F}\tilde{A}(s,y)\right) dy.$$

Set $M_{\tau_0} = \sup_{\tau\in[0,\tau_0]}\|\tilde{A}(\cdot,\tau)\|_\infty$. From Lemma 8.5, the last term of the right-hand side is bounded by $CM_{\tau_0}\delta$. Furthermore, F is Lipschitz, so that

$$\left\|F\left(\frac{y}{\varepsilon^2}\right)\tilde{A}(\cdot,y) - F\left(\frac{y}{\varepsilon^2}\right)\tilde{A}(\cdot,k\delta)\right\|_\infty \le c\left\|\tilde{A}(\cdot,y) - \tilde{A}(\cdot,k\delta)\right\|_\infty$$
$$\le cCM_{\tau_0}|y - k\delta|.$$

Similarly we have

$$\left\|\tilde{F}\tilde{A}(\cdot,y) - \tilde{F}\tilde{A}(\cdot,k\delta)\right\|_\infty \le cCM_{\tau_0}|y - k\delta|.$$

Therefore

$$\|g^\varepsilon(\cdot,\tau)\|_\infty \le \left\|\sum_{k=0}^{[\tau/\delta]-1}\int_{k\delta}^{(k+1)\delta}\left(F\left(\frac{y}{\varepsilon^2}\right)\tilde{A}(\cdot,k\delta) - \tilde{F}\tilde{A}(\cdot,k\delta)\right) dy\right\|_\infty$$
$$+2cCM_{\tau_0}\sum_{k=0}^{[\tau/\delta]-1}\int_{k\delta}^{(k+1)\delta}(y - k\delta)dy + 2cCM_{\tau_0}\delta$$
$$\le \varepsilon^2\sum_{k=0}^{[\tau/\delta]-1}\left\|\int_{k\delta/\varepsilon^2}^{(k+1)\delta/\varepsilon^2}\left(F(y)\tilde{A}(\cdot,k\delta) - \tilde{F}\tilde{A}(\cdot,k\delta)\right) dy\right\|_\infty$$
$$+cCM_{\tau_0}(\tau + 2)\delta.$$

Taking the expectation and the supremum over $\tau \in [0,\tau_0]$, we get

$$\sup_{\tau \in [0, \tau_0]} \mathbb{E}[\|g^\varepsilon(\cdot, \tau)\|_\infty]$$

$$\leq \delta \sum_{k=0}^{[\tau_0/\delta]-1} \mathbb{E}\left[\left\|\frac{\varepsilon^2}{\delta} \int_{k\delta/\varepsilon^2}^{(k+1)\delta/\varepsilon^2} \left(F(y)\tilde{A}(\cdot, k\delta) - \tilde{F}\tilde{A}(\cdot, k\delta)\right) dy\right\|_\infty\right]$$

$$+ cCM_{\tau_0}(\tau_0 + 2)\delta.$$

Taking the limit $\varepsilon \to 0$, we obtain from Lemma 8.4

$$\limsup_{\varepsilon \to 0} \sup_{\tau \in [0, \tau_0]} \mathbb{E}[\|g^\varepsilon(\cdot, \tau)\|_\infty] \leq cCM_{\tau_0}(\tau_0 + 2)\delta.$$

Letting $\delta \to 0$ completes the proof.

Notes

The stabilization of the wave front in randomly layered media was first noted by O'Doherty and Anstey in a geophysical context [126]. A time-domain integral equation approach to pulse stabilization is given in [28, 33]. The frequency-domain approach presented here in Proposition 8.3 follows [39]. The use of the martingale representation for the transmission coefficient is new. An approach using the Riccati equation of Chapter 9 is in [117]. The analysis of wave front reflection from an interface in Section 8.3 is new. Generalizations to three-dimensional randomly layered media are presented in Chapter 14.

9

Statistics of Incoherent Waves

This chapter is a self-contained statistical analysis of the time- and frequency-domain properties of the incoherent waves scattered by a randomly layered medium. We have shown in Chapter 8 that the energy of the transmitted wave front decays exponentially with the size of the random medium, which implies that the incoherent waves carry most of the energy. The analysis of the incoherent waves is therefore important in many applications, especially in time reversal, as we will see in the next chapters. We have also seen in Chapter 7 that the total transmitted energy decays exponentially with the size of the random medium, but that the decay rate is slower than that of the wave front. We will therefore focus attention on the incoherent reflected waves in a homogeneous effective medium. In the last section of this chapter we extend the analysis to the incoherent transmitted waves.

In Proposition 9.1 the limit of the frequency autocorrelation function of the reflection coefficient is described in terms of a system of transport equations. General moments of the reflection coefficient are computed in Proposition 9.2. The limit of the frequency autocorrelation function of the transmission coefficient is given in Proposition 9.4. Finally, in Propositions 9.3–9.5, we show that asymptotically both the reflected and the transmitted incoherent waves have Gaussian statistics, which means that their probability distributions are completely characterized by the time autocorrelation function.

9.1 The Reflected Wave

9.1.1 Reformulation of the Reflection and Transmission Problem

From the results of Chapter 7 (equations (7.9) and (7.18–7.19)), the limit reflection coefficient at frequency ω is given by

$$R_\omega(0, L) = -\tanh\left(\frac{\theta_\omega(L)}{2}\right) e^{i(\psi_\omega(L) + 2\phi_\omega(L))},$$

where $(\theta_\omega, \psi_\omega, \phi_\omega)$ satisfies the system of stochastic differential equations (7.26–7.28). As we have seen in Chapter 7, this representation is appropriate to study the single-frequency moments of the reflection coefficient. For the multifrequency analysis that we carry out in this chapter it turns out that it is more convenient to use directly the Riccati equation satisfied by the reflection coefficient. This requires a reformulation of the scattering problem as shown in Figure 9.1. The incident wave comes from the right, and the reflected wave exits into the homogeneous half-space on the right also. In the next section we derive a closed random differential equation for the reflection coefficient $R_\omega^\varepsilon(-L, z)$ for z going from $-L$ to 0.

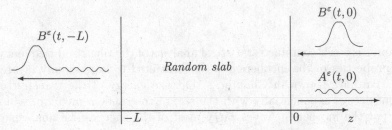

Fig. 9.1. The new scattering problem.

The asymptotic analysis of incoherent reflected waves will be carried out in the strongly heterogeneous white-noise regime (5.17) described in Chapter 5. The scaled acoustic equations (5.13–5.14) with $\theta = 1$ and $\sigma = 1$ and an incident wave from the right are

$$\rho(z)\frac{\partial u^\varepsilon}{\partial t} + \frac{\partial p^\varepsilon}{\partial z} = 0, \tag{9.1}$$

$$\frac{1}{K(z)}\frac{\partial p^\varepsilon}{\partial t} + \frac{\partial u^\varepsilon}{\partial z} = 0. \tag{9.2}$$

We assume that the medium parameters are

$$\frac{1}{K(z)} = \begin{cases} \frac{1}{\overline{K}}\left(1 + \nu(z/\varepsilon^2)\right) & \text{for } z \in [-L, 0], \\ \frac{1}{\overline{K}} & \text{for } z \in (-\infty, -L) \cup (0, \infty), \end{cases}$$

$$\rho(z) = \bar{\rho} \text{ for all } z,$$

and an incoming left-going wave impinges on the interface $z = 0$. Since we are in the strongly heterogeneous white-noise regime, the pulse width is of order ε, and its amplitude is scaled so that it has energy of order one. It is given by

$$\frac{1}{\sqrt{\varepsilon}}f\left(\frac{t}{\varepsilon}\right),$$

where f is square-integrable, so that

$$\int_{-\infty}^{\infty} \left[\frac{1}{\sqrt{\varepsilon}} f\left(\frac{t}{\varepsilon}\right) \right]^2 dt = \int_{-\infty}^{\infty} f(u)^2 du < \infty.$$

As in previous chapters, we introduce the right- and left-going modes

$$A^\varepsilon(t,z) = \bar{\zeta}^{1/2} u^\varepsilon(t,z) + \bar{\zeta}^{-1/2} p^\varepsilon(t,z),$$
$$B^\varepsilon(t,z) = \bar{\zeta}^{1/2} u^\varepsilon(t,z) - \bar{\zeta}^{-1/2} p^\varepsilon(t,z),$$

where the effective impedance is $\bar{\zeta} = \sqrt{\bar{K}\bar{\rho}}$. We consider these modes in coordinates moving with the effective speed $\bar{c}$ and on the time scale of the incoming pulse,

$$a^\varepsilon(s,z) = A^\varepsilon(\varepsilon s + z/\bar{c}, z),$$
$$b^\varepsilon(s,z) = B^\varepsilon(\varepsilon s - z/\bar{c}, z).$$

Here the effective speed is $\bar{c} = \sqrt{\bar{K}/\bar{\rho}}$. In the Fourier domain the modes satisfy the differential equations

$$\frac{d}{dz}\begin{bmatrix} \hat{a}^\varepsilon \\ \hat{b}^\varepsilon \end{bmatrix} = \frac{1}{\varepsilon} \mathbf{H}_\omega\left(\frac{z}{\varepsilon}, \nu\left(\frac{z}{\varepsilon^2}\right)\right) \begin{bmatrix} \hat{a}^\varepsilon \\ \hat{b}^\varepsilon \end{bmatrix}, \tag{9.3}$$

$$\mathbf{H}_\omega(z,\nu) = \frac{i\omega}{2\bar{c}}\nu \begin{bmatrix} 1 & -e^{-2i\omega z/\bar{c}} \\ e^{2i\omega z/\bar{c}} & -1 \end{bmatrix}.$$

The modes also satisfy boundary conditions corresponding to a left-going wave impinging at $z = 0$ and the radiation condition at $z = -L$,

$$\hat{b}^\varepsilon(\omega, 0) = \frac{1}{\sqrt{\varepsilon}} \hat{f}(\omega), \qquad \hat{a}^\varepsilon(\omega, -L) = 0. \tag{9.4}$$

We first transform the boundary value problem (9.3–9.4) into an initial value problem. This step is similar to the analysis carried out in Section 5.1. We introduce the propagator $\mathbf{P}_\omega^\varepsilon(-L, z)$, that is, the fundamental solution matrix of the linear system of differential equations (9.3) with initial condition $\mathbf{P}_\omega^\varepsilon(-L, z = -L) = \mathbf{I}$. From symmetries in (9.3), $\mathbf{P}_\omega^\varepsilon$ is of the form

$$\mathbf{P}_\omega^\varepsilon(-L, z) = \begin{bmatrix} \alpha_\omega^\varepsilon(-L, z) & \overline{\beta_\omega^\varepsilon(-L, z)} \\ \beta_\omega^\varepsilon(-L, z) & \overline{\alpha_\omega^\varepsilon(-L, z)} \end{bmatrix}, \tag{9.5}$$

where $(\alpha_\omega^\varepsilon, \beta_\omega^\varepsilon)^T$ is a solution of (9.3) with the initial conditions

$$\alpha_\omega^\varepsilon(-L, z = -L) = 1, \qquad \beta_\omega^\varepsilon(-L, z = -L) = 0. \tag{9.6}$$

The modes $\hat{a}^\varepsilon$ and $\hat{b}^\varepsilon$ can be expressed in terms of the propagator as

$$\begin{bmatrix} \hat{a}^\varepsilon(\omega, z) \\ \hat{b}^\varepsilon(\omega, z) \end{bmatrix} = \mathbf{P}_\omega^\varepsilon(-L, z) \begin{bmatrix} \hat{a}^\varepsilon(\omega, -L) \\ \hat{b}^\varepsilon(\omega, -L) \end{bmatrix}. \tag{9.7}$$

We can now define the transmission and reflection coefficients $T_\omega^\varepsilon(-L, z)$ and $R_\omega^\varepsilon(-L, z)$, respectively, for a slab $[-L, z]$ by (see Figure 9.2)

$$\mathbf{P}_\omega^\varepsilon(-L, z) \begin{bmatrix} 0 \\ T_\omega^\varepsilon(-L, z) \end{bmatrix} = \begin{bmatrix} R_\omega^\varepsilon(-L, z) \\ 1 \end{bmatrix}. \tag{9.8}$$

In terms of the propagator entries they are given by

$$R_\omega^\varepsilon(-L, z) = \frac{\overline{\beta_\omega^\varepsilon(-L, z)}}{\alpha_\omega^\varepsilon(-L, z)}, \qquad T_\omega^\varepsilon(-L, z) = \frac{1}{\alpha_\omega^\varepsilon(-L, z)}. \tag{9.9}$$

By (9.4) and (9.7) applied at $z = 0$, the reflected and transmitted mode amplitudes can be expressed in terms of the reflection and transmission coefficients as

$$\hat{a}^\varepsilon(\omega, 0) = \frac{1}{\sqrt{\varepsilon}} \hat{f}(\omega) R_\omega^\varepsilon(-L, 0), \qquad \hat{b}^\varepsilon(\omega, -L) = \frac{1}{\sqrt{\varepsilon}} \hat{f}(\omega) T_\omega^\varepsilon(-L, 0).$$

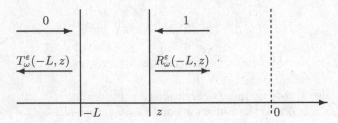

Fig. 9.2. Reflection and transmission coefficients.

9.1.2 The Riccati Equation for the Reflection Coefficient

We want to derive a closed equation for the reflection coefficient. By differentiating $R_\omega^\varepsilon(-L, z)$ and $T_\omega^\varepsilon(-L, z)$ with respect to z, we have

$$\frac{dR_\omega^\varepsilon}{dz} = \frac{1}{\alpha_\omega^\varepsilon} \frac{d\beta_\omega^\varepsilon}{dz} - \frac{\beta_\omega^\varepsilon}{(\alpha_\omega^\varepsilon)^2} \frac{d\alpha_\omega^\varepsilon}{dz},$$

$$\frac{dT_\omega^\varepsilon}{dz} = -\frac{1}{(\alpha_\omega^\varepsilon)^2} \frac{d\alpha_\omega^\varepsilon}{dz}.$$

From the equations (9.3) satisfied by $(\alpha_\omega^\varepsilon, \beta_\omega^\varepsilon)$, we get

$$\frac{dR_\omega^\varepsilon}{dz} = -\frac{i\omega}{2\bar{c}\varepsilon} \nu \left(\frac{z}{\varepsilon^2} \right) \left(e^{-2i\omega z/(\bar{c}\varepsilon)} - 2R_\omega^\varepsilon + (R_\omega^\varepsilon)^2 e^{2i\omega z/(\bar{c}\varepsilon)} \right), \tag{9.10}$$

$$\frac{dT_\omega^\varepsilon}{dz} = \frac{i\omega}{2\bar{c}\varepsilon} \nu \left(\frac{z}{\varepsilon^2} \right) \left(1 - R_\omega^\varepsilon e^{2i\omega z/(\bar{c}\varepsilon)} \right) T_\omega^\varepsilon. \tag{9.11}$$

The initial conditions for these nonlinear differential equations are

$$R_\omega^\varepsilon(-L, z = -L) = 0, \qquad T_\omega^\varepsilon(-L, z = -L) = 1,$$

at $z = -L$. This is because the medium is homogeneous for $z < -L$, and left-going waves simply travels at constant speed to the left. Equation (9.10) is the Riccati equation for the reflection coefficient, and (9.11) is the associated linear equation for the transmission coefficient, which depends on the reflection coeffcient.

9.1.3 Representation of the Reflected Field

The reflected wave at $z = 0$ admits the following representation in terms of the reflection coefficient:

$$A^\varepsilon(t, 0) = a^\varepsilon\left(\frac{t}{\varepsilon}, 0\right)$$

$$= \frac{1}{2\pi} \int \hat{a}^\varepsilon(\omega, 0) e^{-i\frac{\omega t}{\varepsilon}} \, d\omega$$

$$= \frac{1}{2\pi\sqrt{\varepsilon}} \int R_\omega^\varepsilon(-L, 0) \hat{f}(\omega) e^{-i\frac{\omega t}{\varepsilon}} \, d\omega. \qquad (9.12)$$

The statistical description of the reflected wave is thus closely related to the joint statistical distribution of the reflection coefficient at different frequencies. In this chapter we will focus attention on:

1. The mean amplitude $\mathbb{E}[A^\varepsilon(t, 0)]$, which describes the coherent reflected wave.
2. The mean intensity $\mathbb{E}[A^\varepsilon(t, 0)^2]$, which describes the energy distribution of the reflected wave in the time domain.
3. The correlation function $c_t^\varepsilon(s) = \mathbb{E}[A^\varepsilon(t + \varepsilon s, 0) A^\varepsilon(t, 0)]$, which describes the time fluctuations in a time window of size of the order of ε.

We will also give the complete statistical distribution of the reflected wave. These results will be derived from the integral representation (9.12).

The mean amplitude is

$$\mathbb{E}[A^\varepsilon(t, 0)] = \frac{1}{2\pi\sqrt{\varepsilon}} \int \mathbb{E}[R_\omega^\varepsilon(-L, 0)] \hat{f}(\omega) e^{-i\frac{\omega t}{\varepsilon}} \, d\omega. \qquad (9.13)$$

Higher-order moments of the reflected wave involve an expansion in multiple integrals and moments of products of reflection coefficients. Let us consider the second moment, that is, the mean intensity. Since $A^\varepsilon(t, 0)$ is real-valued,

$$A^\varepsilon(t, 0)^2 = A^\varepsilon(t, 0) \overline{A^\varepsilon(t, 0)}$$

$$= \frac{1}{4\pi^2\varepsilon} \left(\int R_{\omega_1}^\varepsilon(-L, 0) \hat{f}(\omega_1) e^{-\frac{i\omega_1 t}{\varepsilon}} \, d\omega_1 \right) \left(\int \overline{R_{\omega_2}^\varepsilon(-L, 0) \hat{f}(\omega_2)} e^{\frac{i\omega_2 t}{\varepsilon}} \, d\omega_2 \right)$$

$$= \frac{1}{4\pi^2\varepsilon} \int \int R_{\omega_1}^\varepsilon(-L, 0) \overline{R_{\omega_2}^\varepsilon(-L, 0)} \hat{f}(\omega_1) \overline{\hat{f}(\omega_2)} e^{i\frac{(\omega_2 - \omega_1)t}{\varepsilon}} \, d\omega_1 \, d\omega_2.$$

By taking the expectation we obtain an expression of the mean intensity in terms of the frequency autocorrelation function of the reflection coefficient:

$$\mathbb{E}[A^\varepsilon(t,0)^2] = \frac{1}{4\pi^2\varepsilon} \int\int \mathbb{E}\left[R^\varepsilon_{\omega_1}(-L,0)\overline{R^\varepsilon_{\omega_2}(-L,0)}\right]$$
$$\times \hat{f}(\omega_1)\overline{\hat{f}(\omega_2)}e^{i\frac{(\omega_2-\omega_1)t}{\varepsilon}}\,d\omega_1\,d\omega_2.$$

The presence of the fast phase $(\omega_2 - \omega_1)t/\varepsilon$ suggests the change of variables

$$\omega_1 = \omega + \varepsilon h/2\,, \qquad \omega_2 = \omega - \varepsilon h/2\,,$$

which leads to the representation

$$\mathbb{E}[A^\varepsilon(t,0)^2] = \frac{1}{4\pi^2} \int\int \mathbb{E}\left[R^\varepsilon_{\omega+\varepsilon h/2}(-L,0)\overline{R^\varepsilon_{\omega-\varepsilon h/2}(-L,0)}\right]$$
$$\times \hat{f}(\omega+\varepsilon h/2)\overline{\hat{f}(\omega-\varepsilon h/2)}e^{-iht}\,d\omega\,dh. \qquad (9.14)$$

This shows that the correlation between the reflection coefficients at two nearby frequencies plays an important role. Note in particular that for ε small,

$$\mathbb{E}[A^\varepsilon(t,0)^2] \sim \frac{1}{4\pi^2} \int\int \mathbb{E}\left[R^\varepsilon_{\omega+\varepsilon h/2}(-L,0)\overline{R^\varepsilon_{\omega-\varepsilon h/2}(-L,0)}\right]e^{-iht}dh$$
$$\times |\hat{f}(\omega)|^2 d\omega.$$

We shall thus carry out in the next section a careful analysis of the distribution of the reflection coefficient at two nearby frequencies in the asymptotic limit $\varepsilon \to 0$.

9.2 Statistics of the Reflected Wave in the Frequency Domain

9.2.1 Moments of the Reflection Coefficient

We aim at computing the moments of the reflection coefficient. In view of (9.13) and (9.14), we are particularly interested in the first and second moments. However, the Riccati equation (9.10) satisfied by the reflection coefficient is nonlinear. As a result, we need to introduce a complete family of moments in order to get a closed system of equations. We introduce for $p, q \in \mathbb{N}$,

$$U^\varepsilon_{p,q}(\omega,h,z) = \left(R^\varepsilon_{\omega+\varepsilon h/2}(-L,z)\right)^p \left(\overline{R^\varepsilon_{\omega-\varepsilon h/2}(-L,z)}\right)^q. \qquad (9.15)$$

The moments of interest to us are the first moment

$$\mathbb{E}[R^\varepsilon_\omega(-L,0)] = \mathbb{E}\left[U^\varepsilon_{1,0}(\omega,0,0)\right] \qquad (9.16)$$

and the two-frequency autocorrelation function

$$\mathbb{E}\left[R^{\varepsilon}_{\omega+\varepsilon h/2}(-L,z)\overline{R^{\varepsilon}_{\omega-\varepsilon h/2}(-L,z)}\right] = \mathbb{E}\left[U^{\varepsilon}_{1,1}(\omega,h,z)\right]. \tag{9.17}$$

Using the Riccati equation (9.10) satisfied by R^{ε}_{ω}, we see that the family $(U^{\varepsilon}_{p,q})_{p,q\in\mathbb{N}}$ satisfies

$$\frac{\partial U^{\varepsilon}_{p,q}}{\partial z} = \frac{i\omega}{\bar{c}}\nu^{\varepsilon}(p-q)U^{\varepsilon}_{p,q} + \frac{i\omega}{2\bar{c}}\nu^{\varepsilon}e^{\frac{2i\omega z}{\bar{c}\varepsilon}}\left(qe^{-\frac{ihz}{\bar{c}}}U^{\varepsilon}_{p,q-1} - pe^{\frac{ihz}{\bar{c}}}U^{\varepsilon}_{p+1,q}\right)$$

$$+ \frac{i\omega}{2\bar{c}}\nu^{\varepsilon}e^{-\frac{2i\omega z}{\bar{c}\varepsilon}}\left(qe^{\frac{ihz}{\bar{c}}}U^{\varepsilon}_{p,q+1} - pe^{\frac{-ihz}{\bar{c}}}U^{\varepsilon}_{p-1,q}\right), \qquad -L \le z \le 0,$$

starting from

$$U^{\varepsilon}_{p,q}(\omega,h,z=-L) = \mathbf{1}_0(p)\mathbf{1}_0(q).$$

Here $\mathbf{1}_0(p) = 1$ if $p = 0$ and is 0 otherwise, and we have set

$$\nu^{\varepsilon}(z) = \frac{1}{\varepsilon}\nu\left(\frac{z}{\varepsilon^2}\right).$$

The system of random ordinary differential equations for $U^{\varepsilon}_{p,q}$ has a form that is almost suitable for the application of the limit theorems of Chapter 6. One major problem is that we need an infinite-dimensional version of these theorems. This requires a weak formulation and the introduction of an appropriate space of test functions. We refer to [137] for the details, and here we simply apply the result as if it were in a finite-dimensional context. Another problem is the presence of slow components of the form $\exp(\pm ihz/\bar{c})$. We first remove these terms by taking a shifted and scaled Fourier transform with respect to h:

$$V^{\varepsilon}_{p,q}(\omega,\tau,z) = \frac{1}{2\pi}\int e^{-ih(\tau-(p+q)z/\bar{c})}U^{\varepsilon}_{p,q}(\omega,h,z)dh. \tag{9.18}$$

The system of equations satisfied by $(V^{\varepsilon}_{p,q})_{p,q\in\mathbb{N}}$ is

$$\frac{\partial V^{\varepsilon}_{p,q}}{\partial z} = -\frac{p+q}{\bar{c}}\frac{\partial V^{\varepsilon}_{p,q}}{\partial \tau} + \frac{i\omega}{\bar{c}}\nu^{\varepsilon}(p-q)V^{\varepsilon}_{p,q} + \frac{i\omega}{2\bar{c}}\nu^{\varepsilon}e^{\frac{2i\omega z}{\bar{c}\varepsilon}}\left(qV^{\varepsilon}_{p,q-1} - pV^{\varepsilon}_{p+1,q}\right)$$

$$+ \frac{i\omega}{2\bar{c}}\nu^{\varepsilon}e^{-\frac{2i\omega z}{\bar{c}\varepsilon}}\left(qV^{\varepsilon}_{p,q+1} - pV^{\varepsilon}_{p-1,q}\right), \tag{9.19}$$

starting from

$$V^{\varepsilon}_{p,q}(\omega,\tau,z=-L) = \delta(\tau)\mathbf{1}_0(p)\mathbf{1}_0(q).$$

We now apply the limit theorem of Section 6.7.3. This establishes that the process $(V^{\varepsilon}_{p,q})_{p,q\in\mathbb{N}}$ converges in distribution as $\varepsilon \to 0$ to a diffusion process $(V_{p,q})_{p,q\in\mathbb{N}}$. The limit diffusion process is identified as the solution of the Itô stochastic differential equation

$$dV_{p,q} = -\frac{q+p}{\bar{c}}\frac{\partial V_{p,q}}{\partial \tau}dz + \frac{i\sqrt{\gamma}\omega}{\bar{c}}(p-q)V_{p,q}dW_0(z)$$

$$+\frac{i\sqrt{\gamma}\omega}{2\sqrt{2}\bar{c}}\left(qV_{p,q-1} - pV_{p+1,q} + qV_{p,q+1} - pV_{p-1,q}\right)dW_1(z)$$

$$+\frac{\sqrt{\gamma}\omega}{2\sqrt{2}\bar{c}}\left(qV_{p,q-1} - pV_{p+1,q} - qV_{p,q+1} + pV_{p-1,q}\right)dW_2(z)$$

$$+\frac{\gamma\omega^2}{4\bar{c}^2}\left[pq(V_{p+1,q+1} + V_{p-1,q-1} - 2V_{p,q}) - 3(p-q)^2 V_{p,q}\right]dz, \quad (9.20)$$

where W_j, $j = 0,1,2$, are three independent Brownian motions and γ is the integrated covariance of the process ν. The form of these stochastic differential equations (9.20) can be derived from (9.19) by replacing the integrals of $\nu^\varepsilon(z)$, $\nu^\varepsilon(z)\cos(2\omega z/(\bar{c}\varepsilon))$ and $\nu^\varepsilon(z)\sin(2\omega z/(\bar{c}\varepsilon))$ by the three independent Brownian motions $\sqrt{\gamma}W_0$, $\sqrt{\gamma/2}W_1$, and $\sqrt{\gamma/2}W_2$. The last line in (9.20) is the Itô-Stratonovich correction (6.100).

Taking the expectation of the stochastic differential equation (9.20) yields a closed system satisfied by the moments

$$\frac{\partial \mathbb{E}[V_{p,q}]}{\partial z} = -\frac{q+p}{\bar{c}}\frac{\partial \mathbb{E}[V_{p,q}]}{\partial \tau} - \frac{3\gamma\omega^2}{4\bar{c}^2}(p-q)^2\mathbb{E}[V_{p,q}]$$

$$+\frac{\gamma\omega^2}{4\bar{c}^2}pq\left(\mathbb{E}[V_{p+1,q+1}] + \mathbb{E}[V_{p-1,q-1}] - 2\mathbb{E}[V_{p,q}]\right).$$

We now proceed with the computation of the moments.

Consider first the family of moments $f_p(\omega,\tau,z) = \mathbb{E}[V_{p+1,p}(\omega,\tau,z)]$, $p \in \mathbb{N}$. It satisfies the closed system

$$\frac{\partial f_p}{\partial z} = -\frac{2p+1}{\bar{c}}\frac{\partial f_p}{\partial \tau} + \frac{\gamma\omega^2}{4\bar{c}^2}\left[p(p+1)(f_{p+1} + f_{p-1} - 2f_p) - 3f_p\right],$$

starting from $f_p(\omega,\tau,z=-L) = 0$. This is a linear system of transport equations starting from a zero initial condition. As a result, the solution is $f_p \equiv 0$ for all p. From $f_0 = 0$ we see therefore that $\mathbb{E}[V_{1,0}^\varepsilon(\omega,\tau,0)]$ converges to zero as $\varepsilon \to 0$, so that $\mathbb{E}[U_{1,0}^\varepsilon(\omega,h,0)]$ also converges to zero as $\varepsilon \to 0$. Note that the last implication is rigorous in the weak formulation described in [137]. As a consequence, the first moment (9.16) converges to zero:

$$\mathbb{E}[R_\omega^\varepsilon(-L,0)] \xrightarrow{\varepsilon \to 0} 0. \quad (9.21)$$

This result can be generalized as follows. For a fixed positive integer n_0, consider the family of moments $f_p(\omega,\tau,z) = \mathbb{E}[V_{p+n_0,p}(\omega,\tau,z)]$, $p \in \mathbb{N}$. Proceeding as above, the family of functions $(f_p(\omega,\tau,z))_{p\in\mathbb{N}}$ is a solution of a system of transport equations with zero initial conditions. Thus $f_p \equiv 0$, and consequently

$$\mathbb{E}[U_{p,q}^\varepsilon(\omega,h,0)] \xrightarrow{\varepsilon \to 0} 0, \quad (9.22)$$

for $p \neq q$.

Consider now the **diagonal** family of moments $g_p(\omega, \tau, z) = \mathbb{E}[V_{p,p}(\omega, \tau, z)]$, $p \in \mathbb{N}$. It satisfies the closed system

$$\frac{\partial g_p}{\partial z} = -\frac{2p}{\bar{c}}\frac{\partial g_p}{\partial \tau} + \frac{\gamma \omega^2}{4\bar{c}^2}p^2(g_{p+1} + g_{p-1} - 2g_p),$$

starting from $g_p(\omega, \tau, z = -L) = \delta(\tau)\mathbf{1}_0(p)$. This is a linear system of transport equations that admits a nontrivial solution. We have thus identified the limits of the expectations $\mathbb{E}[V_{p,p}^\varepsilon(\omega, \tau, z)]$, $p \in \mathbb{N}$. They converge to $\mathcal{W}_p(\omega, \tau, -L, z)$, which obey the closed system of **transport equations**

$$\frac{\partial \mathcal{W}_p}{\partial z} + \frac{2p}{\bar{c}}\frac{\partial \mathcal{W}_p}{\partial \tau} = (\mathcal{L}_\omega \mathcal{W})_p, \quad z \geq -L, \ \tau \in \mathbb{R}, \ p \in \mathbb{N}, \qquad (9.23)$$

$$(\mathcal{L}_\omega \phi)_p = \frac{1}{L_{\mathrm{loc}}(\omega)}p^2 (\phi_{p+1} + \phi_{p-1} - 2\phi_p), \qquad (9.24)$$

starting from

$$\mathcal{W}_p(\omega, \tau, -L, z = -L) = \delta(\tau)\mathbf{1}_0(p).$$

Here $L_{\mathrm{loc}}(\omega)$ is the localization length defined by (7.81),

$$L_{\mathrm{loc}}(\omega) = \frac{4\bar{c}^2}{\gamma \omega^2}.$$

Using (9.17) and (9.18), we get the limit of the autocorrelation function of the reflection coefficient

$$\mathbb{E}\left[R_{\omega+\varepsilon h/2}^\varepsilon(-L, 0)\overline{R_{\omega-\varepsilon h/2}^\varepsilon(-L, 0)}\right] = \mathbb{E}\left[U_{11}^\varepsilon(\omega, h, 0)\right]$$

$$= \int \mathbb{E}\left[V_{11}^\varepsilon(\omega, \tau, 0)\right]e^{ih\tau}d\tau$$

$$\xrightarrow{\varepsilon \to 0} \int \mathcal{W}_1(\omega, \tau, -L, 0)e^{ih\tau}d\tau. \quad (9.25)$$

More generally, we get

$$\mathbb{E}[U_{p,p}^\varepsilon(\omega, h, 0)] \xrightarrow{\varepsilon \to 0} \int \mathcal{W}_p(\omega, \tau, -L, 0)e^{ih\tau}d\tau. \qquad (9.26)$$

We can summarize the results of this section in the following proposition.

Proposition 9.1. *The expectation of the product of two reflection coefficients at two nearby frequencies,*

$$\mathbb{E}\left[\left(R_{\omega+\varepsilon h/2}^\varepsilon(-L, 0)\right)^p \left(\overline{R_{\omega-\varepsilon h/2}^\varepsilon(-L, 0)}\right)^q\right],$$

has the following limit as $\varepsilon \to 0$:
(1) If $p \neq q$, then it converges to 0.
(2) If $p = q$, then it converges to

$$\int \mathcal{W}_p(\omega, \tau, -L, 0) e^{ih\tau} d\tau,$$

where $\mathcal{W}_p(\omega, \tau, -L, z)$ is the solution of the system of transport equations (9.23).

The solution $\mathcal{W}_p(\omega, \tau, -L, 0)$ of the transport equation can be written as a scaled function of two variables as follows. We introduce the canonical system of transport equations for $\tilde{\mathcal{W}}_p(\tilde{\tau}, \tilde{z})$,

$$\frac{\partial \tilde{\mathcal{W}}_p}{\partial \tilde{z}} + 2p\frac{\partial \tilde{\mathcal{W}}_p}{\partial \tilde{\tau}} = (\tilde{\mathcal{L}}\tilde{\mathcal{W}})_p, \quad \tilde{z} \geq 0, \ \tilde{\tau} \in \mathbb{R}, \ p \in \mathbb{N}, \tag{9.27}$$

$$\tilde{\mathcal{W}}_p(\tilde{\tau}, \tilde{z} = 0) = \delta(\tilde{\tau})\mathbf{1}_0(p), \tag{9.28}$$

where $\tilde{\mathcal{L}}$ is given by

$$(\tilde{\mathcal{L}}\phi)_p = p^2 (\phi_{p+1} + \phi_{p-1} - 2\phi_p). \tag{9.29}$$

We have that

$$\mathcal{W}_p(\omega, \tau, -L, 0) = \frac{\bar{c}}{L_{\text{loc}}(\omega)} \tilde{\mathcal{W}}_p\left(\frac{\bar{c}\tau}{L_{\text{loc}}(\omega)}, \frac{L}{L_{\text{loc}}(\omega)}\right). \tag{9.30}$$

Note that we have shifted the z-coordinate to start at $\tilde{z} = 0$ and that the ω-dependence of $\mathcal{W}_p(\omega, \tau, -L, 0)$ comes entirely from $L_{\text{loc}}(\omega)$. This reduction in the number of variables is important for numerical simulations.

9.2.2 Probabilistic Representation of the Transport Equations

In this section we give a probabilistic representation of the solution to the transport equations (9.23) in terms of a jump Markov process. This representation is helpful because it leads to explicit solutions in some particular cases, and in the general case it provides an efficient Monte Carlo method for numerical simulations.

We introduce the jump Markov process $(N_z)_{z \geq -L}$ with state space $\mathbb{N}$ and infinitesimal generator $\mathcal{L}_\omega$ given by (9.24). The construction of the jump process is as follows. When it reaches the state $n > 0$, a random clock with exponential distribution and parameter $2n^2/L_{\text{loc}}(\omega)$ starts running. When the clock strikes, the process jumps to $n + 1$ or $n - 1$ with probability $1/2$. Zero is an absorbing state. Define the process

$$\frac{\partial \mathcal{T}_z}{\partial z} = -\frac{2}{\bar{c}} N_z,$$

with $\mathcal{T}_{-L} = \tau$. The pair $(N_z, \mathcal{T}_z)_{z \geq -L}$ is Markovian with generator

$$\mathcal{L}_\omega - \frac{2n}{\bar{c}} \frac{\partial}{\partial \tau}.$$

The probabilistic representation of the solution of the Kolmogorov equation

$$\frac{\partial u}{\partial z} = \left(\mathcal{L}_\omega - \frac{2n}{\bar{c}} \frac{\partial}{\partial \tau} \right) u, \qquad z > -L, \qquad u(n, \tau, z = -L) = u_0(n, \tau), \quad (9.31)$$

is

$$u(n, \tau, z) = \mathbb{E}\left[u_0 \left(N_z, \mathcal{T}_z \right) \mid N_{-L} = n, \mathcal{T}_{-L} = \tau \right]$$

$$= \mathbb{E}\left[u_0 \left(N_z, \tau - \frac{2}{\bar{c}} \int_{-L}^{z} N_{z'} dz' \right) \mid N_{-L} = n \right]. \qquad (9.32)$$

The solution of the transport equations (9.23) is exactly of the form (9.31), so we can use the probabilistic representation in terms of the jump Markov process $(N_z)_{z \geq -L}$. Taking $u_0(n, \tau) = \mathbb{1}_0(n)\delta(\tau)$, we obtain $u(p, \tau, 0) = \mathcal{W}_p(\omega, \tau, -L, 0)$, which gives

$$\int_{\tau_0}^{\tau_1} \mathcal{W}_p(\omega, \tau, -L, 0) d\tau = \mathbb{P}\left(N_0 = 0 , \ \frac{2}{\bar{c}} \int_{-L}^{0} N_{z'} dz' \in [\tau_0, \tau_1] \mid N_{-L} = p \right),$$

$$(9.33)$$

after integrating in τ between τ_0 and τ_1.

From this probabilistic representation of the solution $\mathcal{W}_p$ of the system of transport equations (9.23), we deduce the following hyperbolicity property. If $\tau_1 < 2L/\bar{c}$, then the only paths that can contribute to the probability (9.33) should satisfy

$$\frac{2}{\bar{c}} \int_{-L}^{0} N_z dz \leq \tau_1 < \frac{2}{\bar{c}} L,$$

and thus N_z, which takes only integer values, has to vanish before reaching 0. We recall that zero is an absorbing state, so that the process stays at zero afterwards. As a result, $\mathcal{W}_p(\omega, \tau, -L, 0)$ does not depend on the value of L for $L \geq \bar{c}\tau/2$. This result, derived from the probabilistic representation of the transport equations, is consistent with the hyperbolic nature of the acoustic wave equations in the homogenized medium with finite speed of propagation $\bar{c}$.

We plot in Figure 9.3 the function $\tau \mapsto \mathcal{W}_1(\omega, \tau, -L, 0)$ for different values of L. We use a Monte Carlo method to compute $\mathcal{W}_1$ from its probabilistic representation (9.33). In the next section we will see that in the limit $L \to \infty$, $\mathcal{W}_1$ has a simple explicit form.

We can give another application of the probabilistic representation (9.33). If we take $h = 0$ in (9.25), then we obtain

$$\mathbb{E}\left[|R_\omega^\varepsilon|^2(-L, 0) \right] \overset{\varepsilon \to 0}{\longrightarrow} \int \mathcal{W}_1(\omega, \tau, -L, 0) d\tau.$$

We thus have a simple probabilistic representation of the limit of the mean-square reflection coefficient

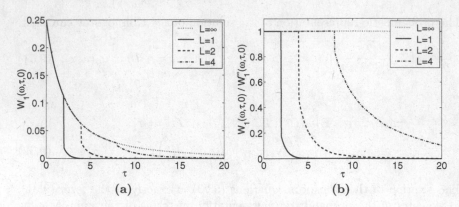

Fig. 9.3. Plots of profiles of the function $\tau \mapsto \mathcal{W}_1(\omega, \tau, -L, 0)$ for different values of the width of the slab L. Here $\bar{c} = 1$, $L_{\mathrm{loc}}(\omega) = 2$. We see that the profiles for two different values $L_0 < L_1$ are identical for τ smaller than $2L_0$. Plot (b) shows the function $\tau \mapsto \mathcal{W}_1(\omega, \tau, -L, 0)/\mathcal{W}_1^\infty(\omega, \tau, 0)$, where $\mathcal{W}_1^\infty(\omega, \tau, 0)$ is the solution (9.41) for a random half-space $L \to \infty$.

$$\mathbb{E}\left[|R_\omega^\varepsilon|^2(-L, 0)\right] \xrightarrow{\varepsilon \to 0} \mathbb{P}(N_0 = 0 \mid N_{-L} = 1).$$

This could be used to deduce an explicit integral representation of this limiting moment, in an alternative way to the one used in Section 7.6. Using the analogue of Proposition 7.4 in the strongly heterogeneous white-noise regime (which affects only the definition of $L_{\mathrm{loc}}(\omega)$) we have that

$$\lim_{\varepsilon \to 0} \mathbb{E}\left[|R_\omega^\varepsilon|^2(-L, 0)\right] = \int \mathcal{W}_1(\omega, \tau, -L, 0)d\tau = 1 - \xi_1\left(\frac{L}{L_{\mathrm{loc}}(\omega)}\right), \quad (9.34)$$

where ξ_1 is given by (7.52). We know that

$$\lim_{L \to \infty} \lim_{\varepsilon \to 0} \mathbb{E}\left[|R_\omega^\varepsilon|^2(-L, 0)\right] = 1,$$

which means total reflection by the random half-space, and implies

$$\lim_{L \to \infty} \int \mathcal{W}_1(\omega, \tau, -L, 0)d\tau = 1. \quad (9.35)$$

Finally, we give an alternative representation of the solution $\mathcal{W}_p$ of the transport equations (9.23), which will be used in the next section to derive the asymptoic behavior of $\mathcal{W}_p$ as $L \to \infty$. It is based on a probabilistic representation of the solution to the canonical transport equations (9.27) in terms of the canonical jump Markov process $(\tilde{N}_{\tilde{z}})_{\tilde{z} \geq 0}$ with state space $\mathbb{N}$ and infinitesimal generator $\tilde{\mathcal{L}}$ given by (9.29). The construction of the jump process $(\tilde{N}_{\tilde{z}})_{\tilde{z} \geq 0}$ is as follows. When the jump process reaches the state $n > 0$, a random clock with exponential distribution and parameter $2n^2$ starts running.

When the clock strikes, the process jumps to $n+1$ or $n-1$ with probability $1/2$, with zero an absorbing state. The solution of the canonical transport equations (9.27) has the probabilistic representation

$$\int_{\tilde{\tau}_0}^{\tilde{\tau}_1} \tilde{\mathcal{W}}_p(\tilde{\tau}, \tilde{L}) d\tilde{\tau} = \mathbb{P}\left(\tilde{N}_{\tilde{L}} = 0 \ , \ 2\int_0^{\tilde{L}} \tilde{N}_{\tilde{z}'} d\tilde{z}' \in [\tilde{\tau}_0, \tilde{\tau}_1] \mid \tilde{N}_0 = p \right). \quad (9.36)$$

Using (9.30), the solution $\mathcal{W}_p(\omega, \tau, -L, 0)$ of the original scaled system of transport equations (9.23), has the representation

$$\int_{\tau_0}^{\tau_1} \mathcal{W}_p(\omega, \tau, -L, 0) d\tau$$

$$= \mathbb{P}\left(\tilde{N}_{L/L_{\mathrm{loc}}(\omega)} = 0 \ , \ \frac{2L_{\mathrm{loc}}(\omega)}{\bar{c}} \int_0^{L/L_{\mathrm{loc}}(\omega)} \tilde{N}_{z'} dz' \in [\tau_0, \tau_1] \mid \tilde{N}_0 = p \right). (9.37)$$

This shows that a single jump process (namely, $(\tilde{N}_{\tilde{z}})_{\tilde{z} \geq 0}$ starting from $\tilde{N}_0 = 1$) is sufficient to compute the function $\mathcal{W}_1(\omega, \tau, -L, 0)$ for all ω and τ. This provides the basis for a very efficient Monte Carlo integration method for computing the function $\mathcal{W}_1$.

9.2.3 Explicit Solution for a Random Half-Space

In the limit $L \to \infty$, in which the random slab occupies the full half-space $z \leq 0$, we can compute explicitly the solution of the transport equations (9.23). For this we use the probabilistic representation (9.37) of the solution in terms of the canonical jump Markov process. The jump Markov process $(\tilde{N}_{\tilde{z}})_{\tilde{z} \geq 0}$ behaves like a symmetric random walk on the set of positive integers. It is a well-known result from probability theory that it will eventually reach the state 0, and since 0 is an absorbing state, $\tilde{N}_{\tilde{z}} = 0$ for $\tilde{z}$ large enough. Therefore, the random variable

$$\tilde{T}_\infty = 2\int_0^\infty \tilde{N}_{\tilde{z}} d\tilde{z}$$

is well defined. As a result,

$$\int_{\tilde{\tau}_0}^{\tilde{\tau}_1} \tilde{\mathcal{W}}_p(\tilde{\tau}, \tilde{L}) d\tilde{\tau} \overset{\tilde{L} \to \infty}{\longrightarrow} \mathbb{P}\left(\tilde{T}_\infty \in [\tilde{\tau}_0, \tilde{\tau}_1] \mid \tilde{N}_0 = p \right), \qquad p \geq 0.$$

The probability density function P_p^∞ of the random variable $\tilde{T}_\infty$ (with the initial condition $\tilde{N}_0 = p$) satisfies the system of differential equations

$$\frac{dP_p^\infty}{d\tilde{\tau}} = \frac{p}{2}\left(P_{p+1}^\infty - 2P_p^\infty + P_{p-1}^\infty \right), \qquad p \geq 1,$$

with $P_0^\infty(\tilde{\tau}) = \delta(\tilde{\tau})$, and P_p^∞ does not have a Dirac mass at $\tilde{\tau} = 0$ if $p \geq 1$. This system is obtained by setting $\partial \tilde{\mathcal{W}}_p / \partial \tilde{z} = 0$ in (9.27), since the limiting distribution does not depend on $\tilde{z}$. It is convenient to consider the associated cumulative distribution function

$$H_p^\infty(\tilde{\tau}) = \int_{-\infty}^{\tilde{\tau}} P_p^\infty(\tilde{\tau}')d\tilde{\tau}', \quad p \geq 0,$$

so that the family $(H_p^\infty)_{p\in\mathbb{N}}$ satisfies the same system as $(P_p^\infty)_{p\in\mathbb{N}}$, but with $H_0^\infty(\tau) = \mathbf{1}_{[0,\infty)}(\tau)$ and $H_p^\infty(\tau) = 0$ if $\tau \leq 0$ and $p \geq 1$. By a direct verification, we find that the solution is given by

$$H_p^\infty(\tilde{\tau}) = \left(\frac{\tilde{\tau}}{2+\tilde{\tau}}\right)^p \mathbf{1}_{[0,\infty)}(\tilde{\tau}), \quad p \geq 0.$$

Therefore the solution for the canonical system of transport equations (9.27) has the limit

$$\lim_{\tilde{L}\to\infty} \tilde{\mathcal{W}}_p(\tilde{\tau}, \tilde{L}) = P_p^\infty(\tilde{\tau}), \qquad P_p^\infty(\tilde{\tau}) = \frac{\partial}{\partial\tilde{\tau}}\left[\left(\frac{\tilde{\tau}}{2+\tilde{\tau}}\right)^p \mathbf{1}_{[0,\infty)}(\tilde{\tau})\right]. \quad (9.38)$$

For $p = 0$, we have $P_0^\infty(\tilde{\tau}) = \delta(\tilde{\tau})$, and for $p \geq 1$,

$$P_p^\infty(\tilde{\tau}) = \frac{2p\tilde{\tau}^{p-1}}{(2+\tilde{\tau})^{p+1}}\mathbf{1}_{[0,\infty)}(\tilde{\tau}). \quad (9.39)$$

From (9.30) the solution for the system of transport equations (9.23) has the limit, as $L \to \infty$,

$$\lim_{L\to\infty} \mathcal{W}_p(\omega, \tau, -L, 0) = \frac{\bar{c}}{L_{\mathrm{loc}}(\omega)}P_p^\infty\left(\frac{\bar{c}\tau}{L_{\mathrm{loc}}(\omega)}\right). \quad (9.40)$$

In particular, the function $\mathcal{W}_1$ that appears in the limit expression (9.25) of the autocorrelation function of the reflection coefficient has the limit, as $L \to \infty$,

$$\lim_{L\to\infty} \mathcal{W}_1(\omega, \tau, -L, \theta) = \mathcal{W}_1^\infty(\omega, \tau, 0) = \frac{2\bar{c}/L_{\mathrm{loc}}(\omega)}{(2+\bar{c}\tau/L_{\mathrm{loc}}(\omega))^2}\mathbf{1}_{[0,\infty)}(\tau). \quad (9.41)$$

By integrating the right-hand side in (9.41) with respect to τ, we recover the result (9.35), which implies total reflection by the random half-space.

9.2.4 Multifrequency Moments

In Section 9.2.1 we give a complete description of the moments of the reflection coefficient at two nearby frequencies. This is sufficient for the computation of the time correlation function of the reflected signal, but we need to generalize it to an arbitrary number of frequencies in order to get the complete statistical distribution of the reflected signal. Let us choose n distinct frequencies $(\omega_j)_{1\leq j\leq n}$ and n frequency shifts $(h_j)_{1\leq j\leq n}$. For $(p_j, q_j)_{1\leq j\leq n} \in \mathbb{N}^{2n}$ we introduce the generalized product of reflection coefficients

$$\widetilde{U}_{\tilde{p},\tilde{q}}^{\varepsilon}(\tilde{\omega},\tilde{h},z) = \prod_{j=1}^{n} \left(R_{\omega_j + \varepsilon h_j/2}^{\varepsilon}(-L,z) \right)^{p_j} \left(\overline{R_{\omega_j - \varepsilon h_j/2}^{\varepsilon}(-L,z)} \right)^{q_j}, \qquad (9.42)$$

where we use the multi-index notation

$$\tilde{\omega} = (\omega_j)_{1 \le j \le n}, \qquad \tilde{h} = (h_j)_{1 \le j \le n}, \qquad (\tilde{p},\tilde{q}) = (p_j,q_j)_{1 \le j \le n}.$$

From the Riccati equation (9.10) we can write the system of differential equations satisfied by $\widetilde{U}^{\varepsilon}$,

$$\frac{\partial \widetilde{U}_{\tilde{p},\tilde{q}}^{\varepsilon}}{\partial z} = \sum_{j=1}^{n} \frac{i\omega_j}{\bar{c}} \nu^{\varepsilon}(p_j - q_j) \widetilde{U}_{\tilde{p},\tilde{q}}^{\varepsilon}$$

$$+ \sum_{j=1}^{n} \frac{i\omega_j}{2\bar{c}} \nu^{\varepsilon} e^{\frac{2i\omega_j z}{\bar{c}\varepsilon}} \left(q_j e^{-\frac{ih_j z}{\bar{c}}} \widetilde{U}_{\tilde{p},\tilde{q}-\tilde{e}_j}^{\varepsilon} - p_j e^{\frac{ih_j z}{\bar{c}}} \widetilde{U}_{\tilde{p}+\tilde{e}_j,\tilde{q}}^{\varepsilon} \right)$$

$$+ \sum_{j=1}^{n} \frac{i\omega_j}{2\bar{c}} \nu^{\varepsilon} e^{-\frac{2i\omega_j z}{\bar{c}\varepsilon}} \left(q_j e^{\frac{ih_j z}{\bar{c}}} \widetilde{U}_{\tilde{p},\tilde{q}+\tilde{e}_j}^{\varepsilon} - p_j e^{-\frac{ih_j z}{\bar{c}}} \widetilde{U}_{\tilde{p}-\tilde{e}_j,\tilde{q}}^{\varepsilon} \right),$$

starting from

$$\widetilde{U}_{\tilde{p},\tilde{q}}^{\varepsilon}(\tilde{\omega},\tilde{h},z=-L) = \prod_{j=1}^{n} \mathbf{1}_0(p_j)\mathbf{1}_0(q_j).$$

We have denoted by $\tilde{e}_j$ the vector $(0,\ldots,0,1,0,\ldots,0)$ whose entries are all 0's except the jth entry, which is equal to 1. We apply an n-dimensional Fourier transform with respect to $\tilde{h}$,

$$\widetilde{V}_{\tilde{p},\tilde{q}}^{\varepsilon}(\tilde{\omega},\tilde{\tau},z) = \frac{1}{(2\pi)^n} \int e^{-i\tilde{h}\cdot(\tilde{\tau}-(\tilde{p}+\tilde{q})z/\bar{c})} \widetilde{U}_{\tilde{p},\tilde{q}}^{\varepsilon}(\tilde{\omega},\tilde{h},z)d\tilde{h}, \qquad (9.43)$$

where $\tilde{\tau}$ is a multi-index notation for $(\tau_j)_{1 \le j \le n}$, to obtain

$$\frac{\partial \widetilde{V}_{\tilde{p},\tilde{q}}^{\varepsilon}}{\partial z} = -\sum_{j=1}^{n} \frac{p_j + q_j}{\bar{c}} \frac{\partial \widetilde{V}_{\tilde{p},\tilde{q}}^{\varepsilon}}{\partial \tau_j} + \sum_{j=1}^{n} \frac{i\omega_j}{\bar{c}} \nu^{\varepsilon}(p_j - q_j) \widetilde{V}_{\tilde{p},\tilde{q}}^{\varepsilon}$$

$$+ \sum_{j=1}^{n} \frac{i\omega_j}{2\bar{c}} \nu^{\varepsilon} e^{\frac{2i\omega_j z}{\bar{c}\varepsilon}} \left(q_j \widetilde{V}_{\tilde{p},\tilde{q}-\tilde{e}_j}^{\varepsilon} - p_j \widetilde{V}_{\tilde{p}+\tilde{e}_j,\tilde{q}}^{\varepsilon} \right)$$

$$+ \sum_{j=1}^{n} \frac{i\omega_j}{2\bar{c}} \nu^{\varepsilon} e^{-\frac{2i\omega_j z}{\bar{c}\varepsilon}} \left(q_j \widetilde{V}_{\tilde{p},\tilde{q}+\tilde{e}_j}^{\varepsilon} - p_j \widetilde{V}_{\tilde{p}-\tilde{e}_j,\tilde{q}}^{\varepsilon} \right), \qquad (9.44)$$

starting from

$$\widetilde{V}_{\tilde{p},\tilde{q}}^{\varepsilon}(\tilde{\omega},\tilde{\tau},z=-L) = \prod_{j=1}^{n} \delta(\tau_j)\mathbf{1}_0(p_j)\mathbf{1}_0(q_j).$$

By applying the diffusion approximation as in Section 9.2.1, we find that $\widetilde{V}_{p,q}^{\varepsilon}(\tilde{\omega}, \tilde{\tau}, z)$ converges to the random process $\widetilde{V}_{\tilde{p},\tilde{q}}(\tilde{\omega}, \tilde{\tau}, z)$, the solution of the Itô stochastic differential equations

$$
\begin{aligned}
d\widetilde{V}_{\tilde{p},\tilde{q}} = &-\sum_{j=1}^{n} \frac{p_j + q_j}{\bar{c}} \frac{\partial \widetilde{V}_{\tilde{p},\tilde{q}}}{\partial \tau_j} dz + \sum_{j=1}^{n} \frac{i\sqrt{\gamma}\omega_j}{\bar{c}} (p_j - q_j)\widetilde{V}_{\tilde{p},\tilde{q}} dW_0(z) \\
&+ \sum_{j=1}^{n} \frac{i\sqrt{\gamma}\omega_j}{2\sqrt{2}\bar{c}} \left(q_j \widetilde{V}_{\tilde{p},\tilde{q}-\tilde{e}_j} - p_j \widetilde{V}_{\tilde{p}+\tilde{e}_j,\tilde{q}} + q_j \widetilde{V}_{\tilde{p},\tilde{q}+\tilde{e}_j} - p_j \widetilde{V}_{\tilde{p}-\tilde{e}_j,\tilde{q}} \right) dW_{1,j}(z) \\
&+ \sum_{j=1}^{n} \frac{i\sqrt{\gamma}\omega_j}{2\sqrt{2}\bar{c}} \left(q_j \widetilde{V}_{\tilde{p},\tilde{q}-\tilde{e}_j} - p_j \widetilde{V}_{\tilde{p}+\tilde{e}_j,\tilde{q}} - q_j \widetilde{V}_{\tilde{p},\tilde{q}+\tilde{e}_j} + p_j \widetilde{V}_{\tilde{p}-\tilde{e}_j,\tilde{q}} \right) dW_{2,j}(z) \\
&+ \sum_{j=1}^{n} \frac{\gamma\omega_j^2}{4\bar{c}^2} p_j q_j \left(\widetilde{V}_{\tilde{p}+\tilde{e}_j,\tilde{q}+\tilde{e}_j} + \widetilde{V}_{\tilde{p}-\tilde{e}_j,\tilde{q}-\tilde{e}_j} - 2\widetilde{V}_{\tilde{p},\tilde{q}} \right) dz \\
&- \frac{\gamma}{4\bar{c}^2} \left[\sum_{j=1}^{n} \omega_j^2 (p_j - q_j)^2 + 2 \left(\sum_{j=1}^{n} \omega_j (p_j - q_j) \right)^2 \right] \widetilde{V}_{\tilde{p},\tilde{q}} dz, \quad (9.45)
\end{aligned}
$$

where W_0, $W_{1,j}$, and $W_{2,j}$, $j = 1, \ldots, n$, are $2n + 1$ independent Brownian motions. The initial condition at $z = -L$ is

$$
\widetilde{V}_{\tilde{p},\tilde{q}}(\tilde{\omega}, \tilde{\tau}, z = -L) = \prod_{j=1}^{n} \delta(\tau_j) \mathbf{1}_0(p_j) \mathbf{1}_0(q_j).
$$

This result can be obtained by inspection from (9.44) by replacing the integrals of $\nu^{\varepsilon}(z)$, $\nu^{\varepsilon}(z) \cos(2\omega_j z/(\bar{c}\varepsilon))$, and $\nu^{\varepsilon}(z) \sin(2\omega_j z/(\bar{c}\varepsilon))$, $j = 1, \ldots, n$, by the independent Brownian motions W_0, $W_{1,j}$, and $W_{2,j}$, $j = 1, \ldots, n$. The last two lines in (9.45) are the Itô–Stratonovich corrections (6.100).

We can thus write the following closed-form system for the moments:

$$
\begin{aligned}
\frac{d\mathbb{E}[\widetilde{V}_{\tilde{p},\tilde{q}}]}{dz} = &-\sum_{j=1}^{n} \frac{p_j + q_j}{\bar{c}} \frac{\partial \mathbb{E}[\widetilde{V}_{\tilde{p},\tilde{q}}]}{\partial \tau_j} \\
&+ \sum_{j=1}^{n} \frac{\gamma\omega_j^2}{4\bar{c}^2} p_j q_j \left(\mathbb{E}[\widetilde{V}_{\tilde{p}+\tilde{e}_j,\tilde{q}+\tilde{e}_j}] + \mathbb{E}[\widetilde{V}_{\tilde{p}-\tilde{e}_j,\tilde{q}-\tilde{e}_j}] - 2\mathbb{E}[\widetilde{V}_{\tilde{p},\tilde{q}}] \right) \\
&- \frac{\gamma}{4\bar{c}^2} \left[\sum_{j=1}^{n} \omega_j^2 (p_j - q_j)^2 + 2 \left(\sum_{j=1}^{n} \omega_j (p_j - q_j) \right)^2 \right] \mathbb{E}[\widetilde{V}_{\tilde{p},\tilde{q}}]. \quad (9.46)
\end{aligned}
$$

This system has special structure. For a fixed integer n_0, the subfamily of moments $\left\{ \mathbb{E}[\widetilde{V}_{\tilde{p},\tilde{q}}], (\tilde{p}, \tilde{q}) \in \mathbb{N}^{2n}, \sum_{j=1}^{n} |p_j - q_j| = n_0 \right\}$ satisfies a closed subsystem of transport equations. The main difference between these sub-systems is that for $n_0 = 0$, the initial conditions for the subfamily are not zero, while for $n_0 \neq 0$, the initial condition is zero. As a result, the solution of the subsystem is zero for all $n_0 \neq 0$. This implies the following.

First, if $\tilde{p} \neq \tilde{q}$, that is if there is a $j \in \{1, \ldots, n\}$ such that $p_j \neq q_j$, then the generalized moment of the product of reflection coefficients (9.42) converges to 0,

$$\mathbb{E}\left[\widetilde{U}_{\tilde{p},\tilde{q}}^{\varepsilon}(\tilde{\omega}, \tilde{h}, z)\right] \overset{\varepsilon \to 0}{\longrightarrow} 0, \tag{9.47}$$

and second, if $\tilde{p} = \tilde{q}$, that is, if $p_j = q_j$ for all j, then the moment of the process $\widetilde{V}_{\tilde{p},\tilde{p}}^{\varepsilon}$ defined by (9.43) converges to

$$\widetilde{\mathcal{W}}_{\tilde{p}}(\tilde{\omega}, \tilde{\tau}, -L, z) = \prod_{j=1}^{n} \mathcal{W}_{p_j}(\omega_j, \tau_j, -L, z). \tag{9.48}$$

Here $\mathcal{W}_p$ is the solution of the system of transport equations (9.23). As a result,

$$\mathbb{E}\left[\widetilde{U}_{\tilde{p},\tilde{p}}^{\varepsilon}(\tilde{\omega}, \tilde{h}, 0)\right] \overset{\varepsilon \to 0}{\longrightarrow} \int e^{i\tilde{h}\cdot\tilde{\tau}} \widetilde{\mathcal{W}}_{\tilde{p}}(\tilde{\omega}, \tilde{\tau}, -L, 0) d\tilde{\tau}$$

$$= \prod_{j=1}^{n} \int e^{ih_j \tau_j} \mathcal{W}_{p_j}(\omega_j, \tau_j, -L, 0) d\tau_j. \tag{9.49}$$

From (9.47) and (9.49) we have the following proposition.

Proposition 9.2. *The expectation of the product of $2n$ reflection coefficients*

$$\mathbb{E}\left[\prod_{j=1}^{n} R_{\omega_j + \varepsilon h_j/2}^{\varepsilon}(-L, 0) \overline{R_{\omega_j - \varepsilon h_j/2}^{\varepsilon}(-L, 0)}\right],$$

where n is a positive integer, $(\omega_j)_{1 \leq j \leq n} \in \mathbb{R}^n$ are all distinct, and $(h_j)_{1 \leq j \leq n} \subset \mathbb{R}^n$, converges as $\varepsilon \to 0$ to the limit

$$\prod_{j=1}^{n} \int e^{ih_j \tau_j} \mathcal{W}_1(\omega_j, \tau_j, -L, 0) d\tau_j,$$

where $\mathcal{W}_1$ is the solution of the system of transport equations (9.23).

If there is one or several unmatched frequencies in the product of reflection coefficients, then the limit of the moment is zero.

We note that in the limit $\varepsilon \to 0$, the processes $(R_\omega^\varepsilon, R_{\omega'}^\varepsilon)$ are not independent when $\omega \neq \omega'$. They share the randomness of the common Brownian motion W_0 in (9.45). However, they behave as if they were independent as far as moments of (9.42) are concerned, because from (9.22) and (9.47), and from (9.26) and (9.49), we see by direct computation that for any $(p_j, q_j)_{1 \leq j \leq n}$,

$$\lim_{\varepsilon \to 0} \mathbb{E}\left[\prod_{j=1}^{n} U_{p_j,q_j}^{\varepsilon}(\omega_j, h_j, 0)\right] = \prod_{j=1}^{n} \lim_{\varepsilon \to 0} \mathbb{E}\left[U_{p_j,q_j}^{\varepsilon}(\omega_j, h_j, 0)\right].$$

In particular, choosing $p_j = 1$, $q_j = 0$, and $h_j = 0$, we see that for any set of distinct frequencies $(\omega_j)_{j=1,\ldots,n}$,

$$\lim_{\varepsilon \to 0} \mathbb{E}\left[\prod_{j=1}^{n} R_{\omega_j}^{\varepsilon}(-L, 0)\right] = 0. \tag{9.50}$$

9.3 Statistics of the Reflected Wave in the Time Domain

9.3.1 Mean Amplitude

By (9.13) the mean amplitude of the reflected wave is

$$\mathbb{E}[A^{\varepsilon}(t, 0)] = \frac{1}{2\pi\sqrt{\varepsilon}} \int \mathbb{E}[R_{\omega}^{\varepsilon}(-L, 0)]\hat{f}(\omega)e^{-i\frac{\omega t}{\varepsilon}} d\omega. \tag{9.51}$$

In Chapter 8 we have studied this quantity for an incident pulse with a fixed amplitude, that is, without the factor $1/\sqrt{\varepsilon}$. In that case, it is sufficient to use (9.21), $\mathbb{E}[R_{\omega}^{\varepsilon}(-L, 0)] \to 0$ as $\varepsilon \to 0$, to obtain the convergence of the mean amplitude to zero. We know from the asymptotic analysis of Chapter 6 that $\mathbb{E}[R_{\omega}^{\varepsilon}(-L, 0)]$ converges to 0 with an error of order ε, which neutralizes the singular factor $1/\sqrt{\varepsilon}$, and thus we get the expected result

$$\mathbb{E}[A^{\varepsilon}(t, 0)] \xrightarrow{\varepsilon \to 0} 0. \tag{9.52}$$

9.3.2 Mean Intensity

We consider the representation (9.14) of the mean intensity:

$$\mathbb{E}[A^{\varepsilon}(t, 0)^2] = \frac{1}{4\pi^2} \int \int \mathbb{E}[U_{11}^{\varepsilon}(\omega, h, 0)]\hat{f}(\omega + \varepsilon h/2)\overline{\hat{f}(\omega - \varepsilon h/2)}e^{-iht} d\omega \, dh.$$

From (9.25) we know the limit of the expectation that appears in this integral, so that we can write

$$\mathbb{E}[A^{\varepsilon}(t, 0)^2] \xrightarrow{\varepsilon \to 0} I(t),$$

with

$$I(t) = \frac{1}{4\pi^2} \int \int \int \mathcal{W}_1(\omega, \tau, -L, 0)|\hat{f}(\omega)|^2 e^{ih(\tau - t)} dh \, d\tau \, d\omega$$

$$= \frac{1}{4\pi^2} \int \int \mathcal{W}_1(\omega, \tau, -L, 0)|\hat{f}(\omega)|^2 2\pi\delta(\tau - t) d\tau \, d\omega$$

$$= \frac{1}{2\pi} \int \mathcal{W}_1(\omega, t, -L, 0)|\hat{f}(\omega)|^2 d\omega. \tag{9.53}$$

In Figure 9.4 we show simulations of the reflected signal amplitude and intensity in the numerical setup described in Section 8.2.3. The slow power law decay of the mean reflected intensity is indicative of intense delay spread in the reflected signal, due to random scattering.

Integrating (9.53) with respect to t we obtain the total reflected energy. Using (9.34), we have the explicit expression

$$\int I(t)dt = \frac{1}{2\pi} \int \left[1 - \xi_1 \left(\frac{L}{L_{\text{loc}}(\omega)} \right) \right] |\hat{f}(\omega)|^2 d\omega,$$

where ξ_1 is given by (7.52).

For L large enough, the reflected intensity at a given time t does not depend on L. This is a simple consequence of the hyperbolicity of the acoustic wave equation with a bounded speed of propagation. This has also been pointed out in Section 9.2.2, where we have shown with the probabilistic representation of $\mathcal{W}_1(\omega, \tau, -L, 0)$ that it does not depend on L for $L \geq \bar{c}\tau/2$. In particular, the transmitted intensity (9.53) does not depend on L for L large enough, and therefore it is equal to its limit as $L \to \infty$. In the case of the random half-space analyzed in Section 9.2.3 we have the explicit formula (9.41) for $\mathcal{W}_1$, leading to

$$I^\infty(t) = \frac{1}{2\pi} \int \frac{2\bar{c}/L_{\text{loc}}(\omega)}{(2 + \bar{c}t/L_{\text{loc}}(\omega))^2} |\hat{f}(\omega)|^2 d\omega.$$

As noted in Section 9.2.3, the total reflected energy equals the total incident energy,

$$\int I^\infty(t)dt = \frac{1}{2\pi} \int |\hat{f}(\omega)|^2 d\omega = \int f(t)^2 dt,$$

which confirms that the wave has been completely reflected by the random medium, as predicted by the localization theory in Chapter 7.

When the incident signal is a narrowband pulse with carrier frequency ω_0 and energy $E_0 = \int f(t)^2 dt$, the mean reflected intensity is approximately

$$I^\infty(t) = \frac{2E_0\bar{c}/L_{\text{loc}}(\omega_0)}{(2 + \bar{c}t/L_{\text{loc}}(\omega_0))^2} = \frac{E_0/t_0}{(1 + t/t_0)^2}, \tag{9.54}$$

where $t_0 = 2L_{\text{loc}}(\omega_0)/\bar{c}$. This slow power law decay as t^{-2} is typical of one-dimensional random media that produce reflections that continue for a long time. Half the reflected energy is captured in the time interval $[0, t_0]$. The rough picture is that the wave penetrates into the medium up to the distance $L_{\text{loc}}(\omega_0)$, and then it is scattered back, which takes $2L_{\text{loc}}(\omega_0)/\bar{c}$ time.

9.3.3 Autocorrelation and Time-Domain Localization

We now consider the local time autocorrelation function of the reflected signal at a fixed time t with lag εs, on the scale of the incident pulse, which is defined by

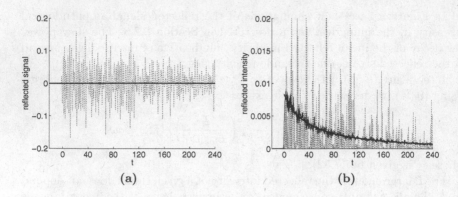

Fig. 9.4. Plot (a): Reflected signal for one realization of the random medium. Plot (b): Reflected intensity for one realization of the random medium (dotted line), mean intensity averaged over 10^3 realizations (thin solid line), and theoretical expected reflected intensity (thick solid line) given by (9.53). The numerical setup described in Section 8.2.3 is used.

$$c_t^\varepsilon(s) = \mathbb{E}[A^\varepsilon(t,0)A^\varepsilon(t+\varepsilon s,0)].$$

Using the integral representation (9.12), we have

$$c_t^\varepsilon(s) = \frac{1}{4\pi^2} \int\int \mathbb{E}[U_{11}^\varepsilon(\omega,h,0)]\hat{f}(\omega+\varepsilon h/2)\overline{\hat{f}(\omega-\varepsilon h/2)}e^{-iht+i\omega s+i\varepsilon hs}d\omega\,dh.$$

Taking the limit $\varepsilon \to 0$, using the finite energy of the pulse and Proposition 9.1, we have that

$$c_t^\varepsilon(s) \xrightarrow{\varepsilon\to 0} c_t(s),$$

where c_t is given by

$$c_t(s) = \frac{1}{4\pi^2} \int\int\int \mathcal{W}_1(\omega,\tau,-L,0)|\hat{f}(\omega)|^2 e^{ih(\tau-t)}e^{i\omega s}d\tau\,d\omega\,dh$$

$$= \frac{1}{2\pi} \int \mathcal{W}_1(\omega,t,-L,0)|\hat{f}(\omega)|^2 e^{i\omega s}d\omega. \tag{9.55}$$

We see therefore that the local power spectral density of the reflected wave around time t is $\mathcal{W}_1(\omega,t,-L,0)|\hat{f}(\omega)|^2$.

In the case of a random half-space, $\mathcal{W}_1$ is given by (9.41), and so

$$\mathcal{W}_1^\infty(\omega,t,0) = \frac{2\bar{c}/L_{\mathrm{loc}}(\omega)}{(2+\bar{c}t/L_{\mathrm{loc}}(\omega))^2}\mathbf{1}_{[0,\infty)}(t).$$

For a fixed time t the maximum of this quantity over ω is attained at $\omega^*(t)$, where

$$t = \frac{2\bar{c}}{L_{\mathrm{loc}}(\omega^*(t))}, \tag{9.56}$$

or

$$\omega^*(t) = \sqrt{\frac{2\bar{c}t}{\gamma}}.$$

We interpret this as follows. Assuming that $|\hat{f}(\omega)|$ is flat over its bandwidth, then the maximum of the local power spectral density of the reflected signal at time t is at $\omega^*(t)$, which is defined by (9.56). This is the frequency for which waves travel to a distance equal to the localization length $L_{\text{loc}}(\omega^*)$ and back. This provides a time-domain interpretation of the localization length as the distance from which the most scattered energy is carried by the reflected waves.

9.3.4 Gaussian Statistics

The goal of this section is to show that the sequence of processes

$$(A^\varepsilon(t + \varepsilon s))_{-\infty < s < \infty},$$

with t fixed, converges as $\varepsilon \to 0$ in distribution to a Gaussian process. We will prove this by showing that for any smooth test function $g(s)$, the sequence of random variables

$$A^\varepsilon_{t,g} = \int A^\varepsilon(t + \varepsilon s) g(s) \, ds$$

converges in distribution to a Gaussian random variable as $\varepsilon \to 0$. This will be done by computing the limiting moments of $A^\varepsilon_{t,g}$.

We first substitute the integral representation of $A(t + \varepsilon s)$ in the definition of $A^\varepsilon_{t,g}$

$$A^\varepsilon_{t,g} = \frac{1}{2\pi\sqrt{\varepsilon}} \int \int R^\varepsilon_\omega(-L, 0) \hat{f}(\omega) g(s) e^{-i\omega s} e^{-i\frac{\omega t}{\varepsilon}} \, ds \, d\omega$$

$$= \frac{1}{2\pi\sqrt{\varepsilon}} \int R^\varepsilon_\omega(-L, 0) \hat{F}(\omega) e^{-i\frac{\omega t}{\varepsilon}} \, d\omega,$$

with the notation

$$\hat{F}(\omega) = \hat{f}(\omega) \hat{g}(\omega).$$

The functions f and g are real-valued, so that $\hat{F}(-\omega) = \overline{\hat{F}(\omega)}$. From the Riccati equation (9.10), we also have $R^\varepsilon_{-\omega} = \overline{R^\varepsilon_\omega}$. As a consequence we can write

$$A^\varepsilon_{t,g} = \frac{1}{2\pi\sqrt{\varepsilon}} \left(\int_{-\infty}^0 R^\varepsilon_\omega(-L, 0) \hat{F}(\omega) e^{-i\frac{\omega t}{\varepsilon}} \, d\omega + \int_0^\infty R^\varepsilon_\omega(-L, 0) \hat{F}(\omega) e^{-i\frac{\omega t}{\varepsilon}} \, d\omega \right)$$

$$= \frac{1}{2\pi\sqrt{\varepsilon}} \left(\int_0^\infty R^\varepsilon_{-\omega}(-L, 0) \hat{F}(-\omega) e^{i\frac{\omega t}{\varepsilon}} \, d\omega + \int_0^\infty R^\varepsilon_\omega(-L, 0) \hat{F}(\omega) e^{-i\frac{\omega t}{\varepsilon}} \, d\omega \right)$$

$$= \frac{1}{2\pi\sqrt{\varepsilon}} \left(\int_0^\infty \overline{R^\varepsilon_\omega(-L, 0)} \overline{\hat{F}(\omega)} e^{i\frac{\omega t}{\varepsilon}} \, d\omega + \int_0^\infty R^\varepsilon_\omega(-L, 0) \hat{F}(\omega) e^{-i\frac{\omega t}{\varepsilon}} \, d\omega \right).$$

We use this decomposition because it involves in a symmetric way the reflection coefficient and its complex conjugate. As a result, the nth moment can be expanded as

$$\mathbb{E}\left[(A^{\varepsilon}_{t,g})^n\right] = \sum_{p=0}^{n} \binom{n}{p} M^{\varepsilon}_{p,n-p}, \tag{9.57}$$

where

$$M^{\varepsilon}_{p,q} = \frac{1}{(2\pi)^{p+q}\varepsilon^{(p+q)/2}} \int_{\omega_j>0} \int_{\omega'_k>0} \prod_{j=1}^{p} \hat{F}(\omega_j) \prod_{k=1}^{q} \overline{\hat{F}(\omega'_k)} e^{i\frac{t}{\varepsilon}(\sum_k \omega'_k - \sum_j \omega_j)}$$

$$\times \mathbb{E}\left[\prod_{j=1}^{p} R^{\varepsilon}_{\omega_j}(-L,0) \prod_{k=1}^{q} \overline{R^{\varepsilon}_{\omega'_k}(-L,0)}\right] \prod_{j=1}^{p} d\omega_j \prod_{k=1}^{q} d\omega'_k. \tag{9.58}$$

We now compute the limit of $M^{\varepsilon}_{p,q}$ as $\varepsilon \to 0$.

First we consider the case $p \neq q$. From (9.47) we know that the limit of the generalized moment that appears in the expression of $M^{\varepsilon}_{p,q}$ is zero. However, this argument is not sufficient because of the presence of the singular factor $1/\varepsilon^{(p+q)/2}$. With a corrector argument similar to those used in Chapter 6, but more involved, we get the convergence of the mean amplitude to zero. We refer to [32] for a detailed analysis. The result is that if $p \neq q$,

$$M^{\varepsilon}_{p,q} \xrightarrow{\varepsilon \to 0} 0. \tag{9.59}$$

We now address the case $p = q$. The integrand in (9.58) is symmetric with respect to the two sets of frequencies $(\omega_j)_{1\leq j\leq p}$, $(\omega'_k)_{1\leq k\leq p}$, so we can write

$$M^{\varepsilon}_{p,p} = \frac{p!^2}{(2\pi)^{2p}\varepsilon^p} \int_{0<\omega_1<\cdots<\omega_p} \int_{0<\omega'_1<\cdots<\omega'_p} \prod_{j=1}^{p} \hat{F}(\omega_j)\overline{\hat{F}(\omega'_j)} e^{i\frac{t}{\varepsilon}\sum_j(\omega'_j-\omega_j)}$$

$$\times \mathbb{E}\left[\prod_{j=1}^{p} R^{\varepsilon}_{\omega_j}(-L,0)\overline{R^{\varepsilon}_{\omega'_j}(-L,0)}\right] \prod_{j=1}^{p} d\omega_j \, d\omega'_j.$$

The limit of the expectation is nonzero only if ω'_j is close to ω_j, as shown by (9.47). By applying the change of variables $\omega'_j = \omega_j - \varepsilon h_j$, we have

$$M^{\varepsilon}_{p,p} = \frac{p!^2}{(2\pi)^{2p}} \int_{0<\omega_1<\cdots<\omega_p} \int_{h_1,\ldots,h_p} \prod_{j=1}^{p} \hat{F}(\omega_j)\overline{\hat{F}(\omega_j - \varepsilon h_j)} e^{-it\sum_j h_j}$$

$$\times \mathbb{E}\left[\prod_{j=1}^{p} R^{\varepsilon}_{\omega_j}(-L,0)\overline{R^{\varepsilon}_{\omega_j-\varepsilon h_j}(-L,0)}\right] \prod_{j=1}^{p} d\omega_j \, dh_j.$$

We shift the variables ω_j to identify the generalized moment of the reflection coefficient studied in Section 9.2.4,

$$M_{p,p}^{\varepsilon} = \frac{p!^2}{(2\pi)^{2p}} \int_{0<\omega_1<\cdots<\omega_p} \int_{h_1,\ldots,h_p} \prod_{j=1}^{p} \hat{F}(\omega_j + \varepsilon h_j/2)\overline{\hat{F}(\omega_j - \varepsilon h_j/2)}e^{-it\sum_j h_j}$$

$$\times \mathbb{E}\left[\prod_{j=1}^{p} R_{\omega_j+\varepsilon h_j/2}^{\varepsilon}(-L,0)\overline{R_{\omega_j-\varepsilon h_j/2}^{\varepsilon}(-L,0)}\right] \prod_{j=1}^{p} d\omega_j\, dh_j.$$

We can now take the limit $\varepsilon \to 0$ and use (9.49):

$$M_{p,p}^{\varepsilon} \xrightarrow{\varepsilon\to 0} \frac{p!^2}{(2\pi)^{2p}} \int_{0<\omega_1<\cdots<\omega_p} \int_{h_1,\cdots,h_p} \prod_{j=1}^{p} |\hat{F}(\omega_j)|^2 e^{-it\sum_j h_j}$$

$$\times \prod_{j=1}^{p} \left[\int \mathcal{W}_1(\omega_j,\tau_j,-L,0)e^{ih_j\tau_j}d\tau_j\right] \prod_{j=1}^{p} d\omega_j\, dh_j.$$

The multiple integral is symmetric with respect to the set of frequencies $(\omega_j)_{1\le j\le p}$, so we can write

$$M_{p,p}^{\varepsilon} \xrightarrow{\varepsilon\to 0} \frac{p!}{(2\pi)^{2p}} \int_{\omega_1>0,\ldots,\omega_p>0} \int_{h_1,\ldots,h_p} \int_{\tau_1,\ldots,\tau_p} \prod_{j=1}^{p} |\hat{F}(\omega_j)|^2 e^{-it\sum_j h_j}$$

$$\times \mathcal{W}_1(\omega_j,\tau_j,-L,0)e^{ih_j\tau_j}d\tau_j\, d\omega_j\, dh_j,$$

which can be factored as

$$M_{p,p}^{\varepsilon} \xrightarrow{\varepsilon\to 0} \frac{p!}{2^p}(\mu_{t,g})^p, \tag{9.60}$$

where

$$\mu_{t,g} = \frac{1}{2\pi^2} \int_{\omega>0} \int \int |\hat{F}(\omega)|^2 \mathcal{W}_1(\omega,\tau,-L,0)e^{ih(\tau-t)}dh\, d\tau\, d\omega$$

$$= \frac{1}{\pi} \int_{\omega>0} |\hat{F}(\omega)|^2 \mathcal{W}_1(\omega,t,-L,0)d\omega$$

$$= \frac{1}{2\pi} \int |\hat{F}(\omega)|^2 \mathcal{W}_1(\omega,t,-L,0)d\omega. \tag{9.61}$$

Substituting (9.59) and (9.60) into the expression (9.57), we have that for any integer p,

$$\mathbb{E}\left[(A_{t,g}^{\varepsilon})^{2p+1}\right] \xrightarrow{\varepsilon\to 0} 0, \tag{9.62}$$

$$\mathbb{E}\left[(A_{t,g}^{\varepsilon})^{2p}\right] \xrightarrow{\varepsilon\to 0} \binom{2p}{p} \frac{p!}{2^p}(\mu_{t,g})^p. \tag{9.63}$$

These are the moments of a zero-mean Gaussian random variable with variance $\mu_{t,g}$. Thus, $A_{t,g}^{\varepsilon}$ converges in distribution to this random variable. This is true for any test function g. As a result, $(A^{\varepsilon}(t + \varepsilon s))_{\infty<s<\infty}$ converges for each t fixed to a Gaussian process $(\mathcal{A}_t(s))_{-\infty < s < \infty}$ that has mean zero and autocorrelation function

$$
\begin{aligned}
c_t(s) &= \mathbb{E}[\mathcal{A}_t(s')\mathcal{A}_t(s'+s)] \\
&= \frac{1}{2\pi} \int \mathcal{W}_1(\omega, t, -L, 0)|\hat{f}(\omega)|^2 e^{i\omega s} d\omega.
\end{aligned}
\tag{9.64}
$$

Note that for each t the limiting process is stationary, since the autocorrelation function does not depend on s'. We state this result in the following proposition.

Proposition 9.3. *The reflected wave around some time t, on the scale ε,*

$$
A^\varepsilon(t + \varepsilon s, 0),
$$

converges as $\varepsilon \to 0$ as a process in s to $(\mathcal{A}_t(s))_{-\infty < s < \infty}$, which is a stationary Gaussian process with mean zero and autocorrelation function

$$
\mathbb{E}[\mathcal{A}_t(s')\mathcal{A}_t(s'+s)] = \frac{1}{2\pi} \int \mathcal{W}_1(\omega, t, -L, 0)|\hat{f}(\omega)|^2 e^{i\omega s} d\omega.
$$

Here $\mathcal{W}_1$ is the solution of the system of transport equations (9.23).

9.4 The Transmitted Wave

As with the reflected signal (9.12), the integral expression of the transmitted wave involves the transmission coefficient $T_\omega^\varepsilon(-L, 0)$:

$$
\begin{aligned}
B^\varepsilon(t, -L) &= b\left(\frac{t - L/\bar{c}}{\varepsilon}, -L\right) \\
&= \frac{1}{2\pi} \int \hat{b}^\varepsilon(\omega, -L) e^{-i\omega \frac{t-L/\bar{c}}{\varepsilon}} d\omega \\
&= \frac{1}{2\pi\sqrt{\varepsilon}} \int T_\omega^\varepsilon(-L, 0)\hat{f}(\omega) e^{-i\omega \frac{t-L/\bar{c}}{\varepsilon}} d\omega.
\end{aligned}
\tag{9.65}
$$

The statistical distribution of the transmitted wave can therefore be obtained from the joint distribution of the transmission coefficients at several frequencies.

9.4.1 Autocorrelation Function of the Transmission Coefficient

We study here the autocorrelation function of the transmission coefficient at two nearby frequencies. The strategy follows the same lines as in Section 9.2.1. We first define a new family of processes indexed by $p, q \in \mathbb{N}$,

$$
U_{p,q}^{(T),\varepsilon}(\omega, h, z) = U_{p,q}^\varepsilon(\omega, h, z) T_{\omega+\varepsilon h/2}^\varepsilon(-L, z)\overline{T_{\omega-\varepsilon h/2}^\varepsilon(-L, z)},
\tag{9.66}
$$

where $U_{p,q}^\varepsilon$ is the product of reflection coefficients defined by (9.15). Using the Riccati equation (9.10) and the differential equation (9.11) satisfied by the transmission coefficient, we see that the family $(U_{p,q}^{(T),\varepsilon})_{p,q\in\mathbb{N}}$ satisfies the closed system

$$\frac{\partial U_{p,q}^{(T),\varepsilon}}{\partial z} = \frac{i\omega}{\bar{c}}\nu^\varepsilon(p-q)U_{p,q}^{(T),\varepsilon} + \frac{i\omega}{2\bar{c}}\nu^\varepsilon e^{\frac{2i\omega z}{\bar{c}\varepsilon}}\left(qe^{-\frac{ihz}{\bar{c}}}U_{p,q-1}^{(T),\varepsilon} - (p+1)e^{\frac{ihz}{\bar{c}}}U_{p+1,q}^{(T),\varepsilon}\right)$$
$$+\frac{i\omega}{2\bar{c}}\nu^\varepsilon e^{-\frac{2i\omega z}{\varepsilon}}\left((q+1)e^{\frac{ihz}{\bar{c}}}U_{p,q+1}^{(T),\varepsilon} - pe^{\frac{-ihz}{\bar{c}}}U_{p-1,q}^{(T),\varepsilon}\right),$$

starting from

$$U_{p,q}^{(T),\varepsilon}(\omega,h,z=-L) = \mathbf{1}_0(p)\mathbf{1}_0(q).$$

Taking a shifted and scaled Fourier transform with respect to h,

$$V_{p,q}^{(T),\varepsilon}(\omega,\tau,z) = \frac{1}{2\pi}\int e^{-ih(\tau-(p+q)z/\bar{c})}U_{n,p}^{(T),\varepsilon}(\omega,h,z)dh,$$

we get

$$\frac{\partial V_{p,q}^{(T),\varepsilon}}{\partial z} = -\frac{p+q}{\bar{c}}\frac{\partial V_{p,q}^{(T),\varepsilon}}{\partial \tau} + \frac{i\omega}{\bar{c}}\nu^\varepsilon(p-q)V_{p,q}^{(T),\varepsilon}$$
$$+\frac{i\omega}{2\bar{c}}\nu^\varepsilon e^{\frac{2i\omega z}{\bar{c}\varepsilon}}\left(qV_{p,q-1}^{(T),\varepsilon} - (p+1)V_{p+1,q}^{(T),\varepsilon}\right)$$
$$+\frac{i\omega}{2\bar{c}}\nu^\varepsilon e^{-\frac{2i\omega z}{\varepsilon}}\left((q+1)V_{p,q+1}^{(T),\varepsilon} - pV_{p-1,q}^{(T),\varepsilon}\right),$$

starting from

$$V_{p,q}^{(T),\varepsilon}(\omega,\tau,z=-L) = \delta(\tau)\mathbf{1}_0(p)\mathbf{1}_0(q).$$

As in Section 9.2.1, we can use the limit theorem of Section 6.7.3 to show that the process $(V_{p,q}^{(T),\varepsilon})_{p,q\in\mathbb{N}}$ converges as $\varepsilon \to 0$ to a diffusion process. In particular, the expectations $\mathbb{E}[V_{p,p}^{(T),\varepsilon}(\omega,\tau,z)]$, $p \in \mathbb{N}$, converge to $\mathcal{W}_p^{(T)}(\omega,\tau,-L,z)$, $p \in \mathbb{N}$, which obey the closed system of transport equations

$$\frac{\partial \mathcal{W}_p^{(T)}}{\partial z} + \frac{2p}{\bar{c}}\frac{\partial \mathcal{W}_p^{(T)}}{\partial \tau} = (\mathcal{L}_\omega^{(T)}\mathcal{W}^{(T)})_p, \qquad z \geq -L, \qquad (9.67)$$

starting from

$$\mathcal{W}_p^{(T)}(\omega,\tau,-L,z=-L) = \delta(\tau)\mathbf{1}_0(p).$$

The operator $\mathcal{L}_\omega^{(T)}$ is defined by

$$(\mathcal{L}_\omega^{(T)}\phi)_p = \frac{1}{L_{\mathrm{loc}}(\omega)}\left((p+1)^2\phi_{p+1} + p^2\phi_{p-1} - ((p+1)^2 + p^2)\phi_p\right), (9.68)$$

with $L_{\mathrm{loc}}(\omega)$ given by (7.81).

By taking an inverse Fourier transform with respect to τ, we get the limit of the autocorrelation function of the transmission coefficient as stated in the following proposition.

Proposition 9.4. *The expectation of the product of two transmission coefficients at two nearby frequencies has a limit as $\varepsilon \to 0$:*

$$\mathbb{E}\left[T^{\varepsilon}_{\omega+\varepsilon h/2}(-L,0)\overline{T^{\varepsilon}_{\omega-\varepsilon h/2}(-L,0)}\right] \overset{\varepsilon\to 0}{\longrightarrow} \int \mathcal{W}_0^{(T)}(\omega,\tau,-L,0)e^{ih\tau}d\tau, \quad (9.69)$$

where $\mathcal{W}_0^{(T)}$ is the solution of the system of transport equations (9.67).

Note that $\tau \mapsto \mathcal{W}_p^{(T)}(\omega,\tau,-L,0)$ is characterized by only two parameters: the travel time $L_{\mathrm{loc}}(\omega)/\bar{c}$ and the ratio $L/L_{\mathrm{loc}}(\omega)$. Indeed, if we introduce the solutions $\tilde{\mathcal{W}}_p^{(T)}(\tilde{\tau},\tilde{z})$ of the canonical system of transport equations

$$\frac{\partial \tilde{\mathcal{W}}_p^{(T)}}{\partial \tilde{z}} + 2p\frac{\partial \tilde{\mathcal{W}}_p^{(T)}}{-\partial \tilde{\tau}} = (\tilde{\mathcal{L}}^{(T)}\tilde{\mathcal{W}})_p, \quad \tilde{z} \geq 0, \quad (9.70)$$

$$(\tilde{\mathcal{L}}^{(T)}\tilde{\phi})_p = (p+1)^2\tilde{\phi}_{p+1} + p^2\tilde{\phi}_{p-1} - (p^2 + (p+1)^2)\tilde{\phi}_p; \quad (9.71)$$

starting from

$$\tilde{\mathcal{W}}_p^{(T)}(\tilde{\tau},\tilde{z}=0) = \delta(\tilde{\tau})\mathbf{1}_0(p),$$

then we simply have

$$\mathcal{W}_p^{(T)}(\omega,\tau,-L,0) = \frac{\bar{c}}{L_{\mathrm{loc}}(\omega)}\tilde{\mathcal{W}}_p^{(T)}\left(\frac{\bar{c}\tau}{L_{\mathrm{loc}}(\omega)}, \frac{L}{L_{\mathrm{loc}}(\omega)}\right).$$

9.4.2 Probabilistic Representation of the Transport Equations

We can interpret the system of transport equations (9.67) in terms of a jump Markov process as in Section 9.2.2. Let us introduce the process $(N_z^{(T)})_{z\geq-L}$ with state space $\mathbb{N}$ and infinitesimal generator $\mathcal{L}_\omega^{(T)}$ defined by (9.68). The explicit description of the jump process is as follows. When the jump process reaches the state $n \in \mathbb{N}$, a random clock with exponential distribution and parameter $[n^2 + (n+1)^2]/L_{\mathrm{loc}}(\omega)$ starts running. When the clock strikes, the process jumps to $n+1$ with probability $(n+1)^2/[n^2 + (n+1)^2]$ and to $n-1$ with probability $n^2/[n^2+(n+1)^2]$. There is no absorbing state. Note that $\mathcal{L}_\omega^{(T)}$ is the adjoint of the generator $\mathcal{L}_\omega$ of the process $(N_z)_{z\geq-L}$ given by (9.24) in Section 9.2.2:

$$\sum_p \phi_p(\mathcal{L}_\omega^{(T)}\psi)_p = \sum_p (\mathcal{L}_\omega\phi)_p\psi_p.$$

By comparing the corresponding Kolmogorov equations, we see that the distribution of $(N_z^{(T)})_{z\geq-L}$ is that of the time-reversed process of $(N_z)_{z\geq-L}$ (which has nothing to do with time reversal!). Proceeding as in Section 9.2.2, we obtain the probabilistic representation

$$\int_{\tau_0}^{\tau_1} \mathcal{W}_0^{(T)}(\omega,d\tau,-L,0) = \mathbb{P}\left(\frac{2}{\bar{c}}\int_{-L}^0 N_z^{(T)}dz \in [\tau_0,\tau_1], \ N_0^{(T)} = 0 \mid N_{-L}^{(T)} = 0\right).$$
$$(9.72)$$

Note that the Markov process is nonnegative-valued, which means that $\mathcal{W}_0^{(T)}$ must vanish for negative τ. Two types of paths of the Markov process contribute to the probability (9.72): those that stay at state zero during the interval $z \in [-L, 0]$, and those that have at least one jump.

(1) The contribution of the paths of the first kind is singular and gives rise to a Dirac mass at $\tau = 0$ with the probability

$$p_{\omega,d} = \mathbb{P}\left(N_z^{(T)} = 0 \ \forall z \in [-L, 0] \mid N_{-L}^{(T)} = 0 \right)$$
$$= \mathbb{P}\left(\text{ no jump before } 0 \mid N_{-L}^{(T)} = 0 \right).$$

The first time when the process jumps has an exponential distribution with parameter $1/L_{\text{loc}}(\omega)$, so that

$$p_{\omega,d} = \exp\left(-\frac{L}{L_{\text{loc}}(\omega)} \right).$$

(2) The contribution of the paths of the second kind is continuous. This result is obtained by conditioning the right-hand side of (9.72) over the first jump time Z_1^J of the process $(N_z^{(T)})_{z \geq -L}$:

$$\int_0^{\tau_1} \mathcal{W}_0^{(T)}(\omega, d\tau, -L, 0) = \int_{-L}^\infty Q(dz_1) = \int_0^\infty Q(dz_1) + \int_{-L}^0 Q(dz_1),$$

with

$$Q(dz_1)$$
$$= \mathbb{P}\left(\frac{2}{\bar{c}} \int_{-L}^0 N_z^{(T)} dz \in [0, \tau_1] \ , \ Z_1^J \in [z_1, z_1 + dz_1) \ , \ N_0^{(T)} = 0 \mid N_{-L}^{(T)} = 0 \right).$$

The first term of the right-hand side is the contribution of the paths of the first kind (i.e., the first jump occurs after 0):

$$\int_0^\infty Q(dz_1) = p_{\omega,d}.$$

If $z_1 \in (-L, 0)$, then we can apply the Markov property:

$$Q(dz_1) = q(z_1, \tau_1) \mathbb{P}\left(Z_1^J \in [z_1, z_1 + dz_1) \mid N_{-L}^{(T)} = 0 \right),$$
$$q(z_1, \tau_1) = \mathbb{P}\left(\frac{2}{\bar{c}} \int_{z_1}^0 N_z^{(T)} dz \in [0, \tau_1] \ , \ N_0^{(T)} = 0 \mid N_{z_1}^{(T)} = 1 \right).$$

The random variable Z_1^J has an exponential distribution with parameter $1/L_{\text{loc}}(\omega)$, so that

$$Q(dz_1) = q(z_1, \tau_1) \frac{1}{L_{\text{loc}}(\omega)} e^{-\frac{L+z_1}{L_{\text{loc}}(\omega)}} dz_1.$$

By the homogeneity of the random process $N^{(T)}$, we can write

$$q(z_1, \tau_1) = \mathbb{P}\left(\frac{2}{\bar{c}} \int_{-L}^{-z_1 - L} N_z^{(T)} dz \in [0, \tau_1] , \ N_{-z_1 - L}^{(T)} = 0 \mid N_{-L}^{(T)} = 1\right)$$

$$= \int_0^{\tau_1} \mathcal{W}_1^{(T)}(\omega, d\tau, -L, -z_1 - L).$$

As a consequence, $\mathcal{W}_0^{(T)}$ is not a density with respect the Lebesgue measure over $\mathbb{R}^+$. It consists of the sum of a Dirac mass at 0 and a density:

$$\mathcal{W}_0^{(T)}(\omega, d\tau, -L, 0) = p_{\omega, d}\delta(d\tau) + \mathcal{W}_{0,c}^{(T)}(\omega, d\tau, -L, 0). \tag{9.73}$$

The absolutely continuous part is given by

$$\mathcal{W}_{0,c}^{(T)}(\omega, d\tau, -L, 0)$$

$$= \frac{1}{L_{\text{loc}}(\omega)} \int_{-L}^0 \mathcal{W}_1^{(T)}(\omega, d\tau, -L, -z_1 - L) e^{-\frac{L + z_1}{L_{\text{loc}}(\omega)}} dz_1. \tag{9.74}$$

It does not seem possible to derive a closed-form expression for the density $\mathcal{W}_{0,c}^{(T)}$. We can either derive expansions, or perform numerical simulations based on Monte Carlo simulations of the random jump process $(N_z^{(T)})_{-L \leq z \leq 0}$.

- We can expand $\mathcal{W}_{0,c}^{(T)}$ for small τ. Indeed, if $\bar{c}\tau / L_{\text{loc}}(\omega) \ll 1$, then only the paths that jump very quickly from 1 to 0, and then stay at 0, contribute to the value of $\mathcal{W}_1^{(T)}$:

$$\mathcal{W}_1^{(T)}(\omega, d\tau, -L, -z_1 - L)$$

$$= \mathbb{P}\left(\frac{2}{\bar{c}} \int_{-L}^{-L - z_1} N_z^{(T)} dz \in [\tau, \tau + d\tau] , \ N_{-z_1 - L}^{(T)} = 0 \mid N_{-L}^{(T)} = 1\right)$$

$$\sim \mathbb{P}\left(Z_1^J \in [-L + \bar{c}\tau/2, -L + \bar{c}(\tau + d\tau)/2),\right.$$

$$\left.\text{first jump to 0, stay at } 0 \mid N_{-L}^{(T)} = 1\right)$$

$$= \frac{5}{L_{\text{loc}}(\omega)} \exp\left(-\frac{5\bar{c}\tau}{2L_{\text{loc}}(\omega)}\right) \frac{\bar{c}d\tau}{2} \times \frac{1}{5} \times \exp\left(\frac{2z_1 + \bar{c}\tau}{2L_{\text{loc}}(\omega)}\right)$$

$$= \frac{\bar{c}}{2L_{\text{loc}}(\omega)} \exp\left(\frac{z_1 - 2\bar{c}\tau}{L_{\text{loc}}(\omega)}\right) d\tau,$$

so that

$$\mathcal{W}_{0,c}^{(T)}(\omega, d\tau, -L, 0) \overset{\bar{c}\tau \ll L_{\text{loc}}(\omega)}{\sim} \frac{\bar{c}L}{2L_{\text{loc}}(\omega)^2} \exp\left(-\frac{L}{L_{\text{loc}}(\omega)}\right) d\tau. \tag{9.75}$$

- We can expand $\mathcal{W}_{0,c}^{(T)}$ for small L. Indeed, if $L/L_{\text{loc}}(\omega) \ll 1$ (in practice, $\leq 1/2$), then only the paths that jump very quickly from 1 to 0, and then stay at 0, contribute to the value $\mathcal{W}_1^{(T)}$. Using the same method as above,

$$\mathcal{W}_{0,c}^{(T)}(\omega, d\tau, -L, 0)$$

$$\underset{L \ll L_{\text{loc}}(\omega)}{\sim} \frac{\bar{c}}{2L_{\text{loc}}(\omega)^2} \left(L - \frac{\bar{c}\tau}{2} \right) \exp\left(-\frac{L + 2\bar{c}\tau}{L_{\text{loc}}(\omega)} \right) \mathbf{1}_{(0, 2L/\bar{c})}(\tau) d\tau. \quad (9.76)$$

- We can perform Monte Carlo simulations to compute the density of the absolutely continuous measure $\mathcal{W}_{0,c}^{(T)}(\omega, d\tau, -L, 0)$. We plot in Figure 9.5 the densities $\tau \mapsto \mathcal{W}_{0,c}^{(T)}(\omega, d\tau, -L, 0)$ for different values of the ratio $L/L_{\text{loc}}(\omega)$.

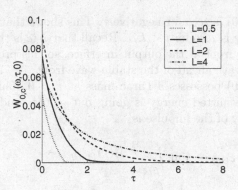

Fig. 9.5. Profiles of the function $\tau \mapsto \mathcal{W}_{0,c}^{(T)}(\omega, \tau, 0)$, which is the density of the absolutely continuous part of the solution $\mathcal{W}_0^{(T)}(\omega, d\tau, -L, 0)$ of the system of transport equations (9.67), for different values for the width of the slab L. Here $\bar{c} = 1$, $L_{\text{loc}}(\omega) = 2$.

9.4.3 Statistics of the Transmitted Wave in the Time Domain

We consider the transmitted signal in a time window of size ε centered at time t:

$$B^\varepsilon(t + \varepsilon s, -L) = \frac{1}{2\pi\sqrt{\varepsilon}} \int T_\omega^\varepsilon(-L, 0)\hat{f}(\omega)e^{-i\frac{\omega(t-L/\bar{c})}{\varepsilon}}e^{-i\omega s}d\omega. \quad (9.77)$$

If $t = L/\bar{c}$, then we know that the transmitted wave contains a coherent part. In our framework, the amplitude of this coherent part is of order $1/\sqrt{\varepsilon}$, and the shape of this coherent wave has been extensively studied in Chapter 8. If $t \neq L/\bar{c}$, then there is no coherent wave in the observed signal.

We can consider the mean intensity of the transmitted wave $\mathbb{E}[B(t, -L)^2]$. Using the integral representation (9.77),

$$\mathbb{E}[B^\varepsilon(t, -L)^2] = \frac{1}{4\pi^2} \int\int \mathbb{E}\left[U_{00}^{(T),\varepsilon}(\omega, h, 0) \right]$$

$$\times \hat{f}(\omega + \varepsilon h/2)\overline{\hat{f}(\omega - \varepsilon h/2)}e^{-ih(t-L/\bar{c})}d\omega\,dh,$$

and taking the limit $\varepsilon \to 0$ yields

$$\mathbb{E}[B^\varepsilon(t, -L)^2] \stackrel{\varepsilon \to 0}{\longrightarrow} I^{(T)}(t), \tag{9.78}$$

with

$$I^{(T)}(t) = \frac{1}{2\pi} \int \mathcal{W}_0^{(T)}(\omega, t - L/\bar{c}, -L, 0)|\hat{f}(\omega)|^2 d\omega. \tag{9.79}$$

We can discuss the qualitative properties of the mean transmitted intensity by considering the form (9.73) of $\mathcal{W}_0^{(T)}$:

1. $\mathcal{W}_0^{(T)}(\omega, \tau, -L, 0)$ is zero for negative τ. This shows that the mean transmitted intensity is zero for $t < L/\bar{c}$. Recall that $L/\bar{c}$ is the arrival time of the stable wave front at the output interface, so this proves that there is no incoherent wave ahead of the stable wave front.

2. $\mathcal{W}_0^{(T)}(\omega, \tau, -L, 0)$ possesses a Dirac mass at $\tau = 0$. This shows that an impulse of transmitted energy is going out of the random slab at time $L/\bar{c}$. The energy of the impulse is

$$
\begin{aligned}
\mathcal{E}_{\mathrm{coh}} &:= \lim_{\delta \to 0} \int_{L/\bar{c}-\delta}^{L/\bar{c}+\delta} I^{(T)}(t)dt \\
&= \frac{1}{2\pi} \int p_{\omega,d}|\hat{f}(\omega)|^2 d\omega \\
&= \frac{1}{2\pi} \int \exp\left(-\frac{L}{L_{\mathrm{loc}}(\omega)}\right) |\hat{f}(\omega)|^2 d\omega.
\end{aligned} \tag{9.80}
$$

This energy impulse actually corresponds to the transmission of the stable wave front, which is consistent with the expression (8.48) of the energy of the stable wave front.

3. $\mathcal{W}_0^{(T)}(\omega, \tau, -L, 0)$ is positive-valued for $\tau > 0$. This shows that we can observe the transmission of incoherent waves following the wave front. The energy of the transmitted incoherent wave fluctuations is

$$
\begin{aligned}
\mathcal{E}_{\mathrm{inc}} &:= \lim_{\delta \to 0} \int_{L/\bar{c}+\delta}^{\infty} I^{(T)}(t)dt \\
&= \frac{1}{2\pi} \int \int \mathcal{W}_{0,c}(\omega, t, -L, 0)dt|\hat{f}(\omega)|^2 d\omega \\
&= \frac{1}{2\pi} \int \left[\xi_1\left(\frac{L}{L_{\mathrm{loc}}(\omega)}\right) - \exp\left(-\frac{L}{L_{\mathrm{loc}}(\omega)}\right) \right] |\hat{f}(\omega)|^2 d\omega , \tag{9.81}
\end{aligned}
$$

where ξ_1 is defined by (7.52). The last identity comes from the fact that the total transmitted energy is known (Proposition 7.5) as well as the energy of the transmitted wave front. As shown by Figure 9.6, for $L/L_{\mathrm{loc}}(\omega) > 2.04$, the energy of the transmitted incoherent wave fluctuations is larger than the energy of the wave front. These incoherent waves require a more detailed analysis.

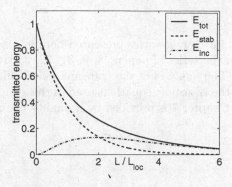

Fig. 9.6. Transmitted energy density at frequency ω as a function of $L/L_{\mathrm{loc}}(\omega)$. The solid line stands for the total transmitted energy, the dashed line for the energy of the transmitted stable wave front, and the dot-dashed line for the incoherent transmitted energy.

Let us consider a time $t > L/\bar{c}$. We would like to study the process $(B(t + \varepsilon s, -L))_{-\infty<s<\infty}$. We first consider the time-correlation function

$$c_t^{(T),\varepsilon}(s) = \mathbb{E}[B^\varepsilon(t,-L)B^\varepsilon(t+\varepsilon s,-L)].$$

Using the integral representation (9.77), we have

$$c_t^{(T),\varepsilon}(s) = \frac{1}{4\pi^2} \int \int \mathbb{E}\left[U_{00}^{(T),\varepsilon}(\omega,h,0)\right]$$
$$\times \hat{f}(\omega+\varepsilon h/2)\overline{\hat{f}(\omega-\varepsilon h/2)}e^{-ih(t-L/\bar{c})+i\omega s+i\varepsilon h s}\,d\omega\,dh.$$

Taking the limit $\varepsilon \to 0$ yields

$$c_t^{(T),\varepsilon}(s) \xrightarrow{\varepsilon\to 0} c_t^{(T)}(s), \tag{9.82}$$

with

$$c_t^{(T)}(s) = \frac{1}{2\pi} \int \mathcal{W}_0^{(T)}(\omega,t-L/\bar{c},-L,0)|\hat{f}(\omega)|^2 e^{i\omega s}\,d\omega. \tag{9.83}$$

Computing higher-order moments as in Section 9.3.4, we eventually get the following proposition.

Proposition 9.5. *The transmitted wave fluctuations have Gaussian statistics in the limit* $\varepsilon \to 0$. *More precisely, for any* $t > L/\bar{c}$, *the process* $(B^\varepsilon(t + \varepsilon s, -L))_{-\infty<s<\infty}$ *converges to a zero-mean stationary Gaussian process with the autocorrelation function* $c_t^{(T)}$ *given by (9.83).*

Notes

The statistics of the incoherent waves presented here in a self-contained way have been derived in the series of papers [8, 9, 31, 32, 33, 104, 107, 133, 137, 152, 153, 170, 169]. The analysis of the transmission coefficient is new. The connection between the transport equations used in this chapter and the polar coordinates used in Chapter 7 is provided by a duality introduced in [137].

10

Time Reversal in Reflection and Spectral Estimation

In this chapter we introduce the concept of time reversal of waves. We first consider the case of time reversal in reflection, in which a source emits a pulse at one end of a one-dimensional slab, and a time-reversal mirror placed at the same location records the reflected signal. The mirror then reemits a part of the recorded signal trace in the reverse direction of time, so that what is recorded last is sent first (last-in-first-out at the mirror). This is in contrast to a standard mirror, which corresponds to first-in-first-out. This basic time reversal setup is illustrated in Figure 10.3. The remarkable properties of time reversal in random media are (i) the **refocusing** (or recompression) of the wave field at a given deterministic time (Section 10.1.2) (ii) the **statistical stability** of the refocused pulse (Section 10.1.3). We will see that the degree or quality of refocusing and stability depends on how much of the reflected signal is recorded.

In this chapter we consider the case in which the background medium is constant with matching conditions at the ends of the random slab. The reflected wave has no coherent part, as described in Chapter 8. It is illustrated in Figure 10.1. The precise description of the statistics of the reflected wave is given in Section 9.3.4. Here we perform time reversal, and we show that, surprisingly, refocusing takes place as seen in Figure 10.2. The refocused pulse is statistically stable, that is, it is independent of the particular realization of the medium, and can be written as a convolution of the initial pulse with a refocusing kernel as shown in (10.6). This kernel is a smoothed version of the local covariance of the reflected signal, and it can be computed explicitly in the uniform-background case.

In more general cases in which the background parameters are variable, refocusing still takes place and the refocusing kernels contain information on the background parameters. This can, in fact, be exploited to estimate these background parameters. In Chapter 11 we will give applications to the detection and imaging of weak reflectors (jump in speed with no contrast of impedance) or dissipative regions. In Chapter 17 we will look at the inverse problem in the case that the background parameters are slowly varying. In all these problems it is imprortant to obtain precise estimates of the local covariance of the reflected signal. In Section 10.2 we compare the estimates obtained using time reversal with those using cross-correlations of the reflected signal. We show that in the presence of measurement noise, the time-reversal method can improve the signal-to-noise ratios of these estimates. Finally, in Section 10.3 we propose a method for adapting the wavelength of the pulse for probing a particular depth in the medium.

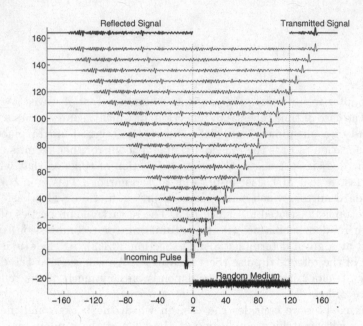

Fig. 10.1. Propagation of a pulse through a slab of random medium $(0, L)$. A right-going wave is incoming from the left. The pulse shape is the second derivative of a Gaussian. Snapshots of the wave profile (here the pressure) at increasing times are plotted from bottom to top. The reflected and transmitted signals at the terminal time of the numerical simulation are plotted at the top. The numerical setup described in Section 8.2.3 is used. The bottom line is a realization of the random medium (here the bulk modulus), but it is not the one used in the simulations (it takes only two values, which would give a black band in the plot).

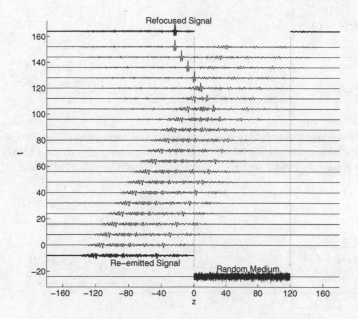

Fig. 10.2. We use the same random medium as in Figure 10.1 and send back the time-reversed reflected signal. Snapshots of the wave profile (here the pressure) at increasing times are plotted from bottom to top. We can see the refocused pulse that emerges from the random medium. In this time-reversal experiment we have recorded the signal up to time 130 shown in Figure 10.1, and we observe the refocusing at the surface at the exact time 130. After the refocusing time, the refocused pulse keeps traveling to the left.

10.1 Time Reversal in Reflection

10.1.1 Time-Reversal Setup

We again consider a random slab $(-L, 0)$ embedded in a homogeneous medium with no background discontinuities. In the strongly heterogeneous white-noise regime, a pulse of the form $f(t/\varepsilon)$ incoming from the right homogeneous half-space is scattered by the random slab. We have seen in (9.12) that the reflected wave $A^\varepsilon(t, 0)$ has the form:

$$A^\varepsilon(t, 0) = \frac{1}{2\pi} \int R_\omega^\varepsilon(-L, 0)\hat{f}(\omega)e^{-\frac{i\omega t}{\varepsilon}} d\omega, \qquad (10.1)$$

where $R_\omega^\varepsilon(-L, 0)$ is the reflection coefficient defined by (9.9). The first step in the time-reversal consists in recording the reflected signal at $z = 0$. It turns out that as $\varepsilon \to 0$, the interesting asymptotic regime arises when we record

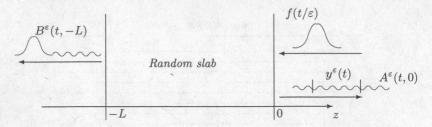

(a) A left-going pulse $f(t/\varepsilon)$ is on impinging the random slab $(-L,0)$ and it generates a reflected signal $A^\varepsilon(t,0)$. The time-reversal-mirror (TRM), used in a passive mode, records a segment $y^\varepsilon(t)$ of the reflected signal.

(b) The TRM is used as an active device that sends back in the medium the signal $y^\varepsilon(t_1 - t)$. We observe the new reflected signal $A^\varepsilon_{\text{new}}(t,0)$.

Fig. 10.3. Setup for a time reversal in reflection (TRR) experiment.

the signal up to a large time of order one, which we denote by t_1 (with $t_1 > 0$). A segment of the recorded signal with support of order one is clipped using a cutoff function $G(t)$. We denote the recorded part of the wave by y^ε so that

$$y^\varepsilon(t) = A^\varepsilon(t,0)G(t).$$

We then time-reverse this segment of signal about t_1 and send it back into the same medium as shown in Figure 10.3. This means that we have a new scattering problem defined by the same acoustic equations, but with the new incoming signal

$$f^\varepsilon_{\text{new}}(t) = y^\varepsilon(t_1 - t) = A^\varepsilon(t_1 - t, L)G(t_1 - t),$$

which corresponds to a left-going wave incoming from the right homogeneous half-space. Since we are dealing with real-valued signals, we can write

$$A^\varepsilon(t,0) = \overline{A^\varepsilon(t,0)} = \frac{1}{2\pi} \int \overline{R^\varepsilon_\omega(-L,0)} \; \overline{\hat{f}(\omega)} e^{\frac{i\omega t}{\varepsilon}} \, d\omega \,,$$

so that the scaled Fourier transform of the new incoming signal has the form

$$\hat{f}_{\text{new}}^{\varepsilon}(\omega) = \int e^{\frac{i\omega t}{\varepsilon}} A^{\varepsilon}(t_1 - t, 0) G(t_1 - t) \, dt$$

$$= \varepsilon \int e^{i\omega s} A^{\varepsilon}(t_1 - \varepsilon s, 0) G(t_1 - \varepsilon s) \, ds$$

$$= \varepsilon \int e^{i\omega s} \left\{ \frac{1}{2\pi} \int e^{-i\omega' s} \, \overline{R_{\omega'}^{\varepsilon}(-L, 0)} \, \overline{\hat{f}(\omega')} e^{\frac{i\omega' t_1}{\varepsilon}} \, d\omega' \right\} G(t_1 - \varepsilon s) \, ds$$

$$= \frac{\varepsilon}{2\pi} \int \overline{R_{\omega'}^{\varepsilon}(-L, 0)} \, \overline{\hat{f}(\omega')} \left\{ \int e^{i(\omega' - \omega)(-s)} \, G(t_1 - \varepsilon s) \, ds \right\} e^{\frac{i\omega' t_1}{\varepsilon}} \, d\omega'$$

$$= \frac{1}{2\pi} \int \overline{R_{\omega'}^{\varepsilon}(-L, 0)} \, \overline{\hat{f}(\omega')} \, \hat{G}\left(\frac{\omega - \omega'}{\varepsilon}\right) e^{\frac{i\omega t_1}{\varepsilon}} \, d\omega' \, .$$

The new incoming signal is scattered by the random slab and gives rise to a reflected wave $A_{\text{new}}^{\varepsilon}(t, 0)$ at $z = 0$ and a transmitted wave $B_{\text{new}}^{\varepsilon}(t, -L)$ at $z = -L$. The reflected signal observed in the time domain around the observation time t_{obs} on the scale ε is given by

$$S_L^{\varepsilon}(t_{\text{obs}} + \varepsilon s) := A_{\text{new}}^{\varepsilon}(t_{\text{obs}} + \varepsilon s, 0) = \frac{1}{2\pi\varepsilon} \int e^{-i\omega(s + \frac{t_{\text{obs}}}{\varepsilon})} R_{\omega}^{\varepsilon}(-L, 0) \hat{f}_{\text{new}}^{\varepsilon}(\omega) \, d\omega \, .$$

Substituting the expression of $\hat{f}_{\text{new}}^{\varepsilon}$ into this equation gives the integral representation of the reflected signal

$$S_L^{\varepsilon}(t_{\text{obs}} + \varepsilon s) = \frac{1}{(2\pi)^2 \varepsilon} \int \int e^{-i\omega_1 s} e^{i\frac{\omega_1(t_1 - t_{\text{obs}})}{\varepsilon}} \, \overline{\hat{f}(\omega_2)} \, \hat{G}\left(\frac{\omega_1 - \omega_2}{\varepsilon}\right)$$

$$\times \overline{R_{\omega_2}^{\varepsilon}(-L, 0)} R_{\omega_1}^{\varepsilon}(-L, 0) \, d\omega_1 \, d\omega_2 \, .$$

Motivated by the scaled argument in $\hat{G}$ we use the change of variables $\omega_1 = \omega + \varepsilon h/2$, $\omega_2 = \omega - \varepsilon h/2$ and get

$$S_L^{\varepsilon}(t_{\text{obs}} + \varepsilon s) = \frac{1}{(2\pi)^2} \int \int e^{-i\omega s} e^{i\frac{\omega(t_1 - t_{\text{obs}})}{\varepsilon}} e^{ih(t_1 - t_{\text{obs}})/2 - i\varepsilon hs/2} \, \overline{\hat{f}(\omega - \varepsilon h/2)}$$

$$\times \hat{G}(h) \overline{R_{\omega - \varepsilon h/2}^{\varepsilon}(-L, 0)} R_{\omega + \varepsilon h/2}^{\varepsilon}(-L, 0) \, dh \, d\omega \, . \qquad (10.2)$$

We will analyze the behavior of this reflected signal in the limit $\varepsilon \to 0$.

10.1.2 Time-Reversal Refocusing

We first observe that the signal (10.2), recorded at $z = 0$, vanishes in the limit $\varepsilon \to 0$ if the time of observation t_{obs} is not the time of recording t_1. Indeed, the rapid phase $\exp(i\omega(t_1 - t_{\text{obs}})/\varepsilon)$ averages out the integral except when $t_{\text{obs}} = t_1$. This means that

refocusing can be observed only at the time $t_{\text{obs}} = t_1$.

In other words, an observer located at $z = 0$ detects no coherent signal at any time different from t_1. The observed small incoherent wave fluctuations vanish in the limit $\varepsilon \to 0$. This is what is called **time-reversal refocusing**, and the precise description of the refocused pulse observed at time t_1 is carried out in the next section. The refocused pulse at time $t_{\mathrm{obs}} = t_1$ has the form

$$S_L^\varepsilon(t_1 + \varepsilon s) = \frac{1}{(2\pi)^2} \int \int e^{-i\omega s - i\varepsilon h s/2} \overline{\hat{f}(\omega - \varepsilon h/2)} \, \hat{G}(h)$$
$$\times R_{\omega + \varepsilon h/2}^\varepsilon(-L, 0) \overline{R_{\omega - \varepsilon h/2}^\varepsilon(-L, 0)} \, dh \, d\omega. \qquad (10.3)$$

Note that the product of reflection coefficients that appears in this integral has been analyzed extensively in Chapter 9.

10.1.3 The Limiting Refocused Pulse

The uniform boundedness of the reflection coefficient, which follows from the conservation of energy as given in (7.11), implies that the finite-dimensional distributions of the process $S_L^\varepsilon(t_1 + \varepsilon \cdot)$ will be characterized by the moments

$$\mathbb{E}[S_L^\varepsilon(t_1 + \varepsilon s_1)^{p_1} \cdots S_L^\varepsilon(t_1 + \varepsilon s_k)^{p_k}] \qquad (10.4)$$

for all real number in the range $s_1 < \cdots < s_k$ and all integer $p_1, \ldots, p_k$.

First Moment

We start by considering the first moment. Using the representation (10.3), the expected value of $S_L^\varepsilon(t_1 + \varepsilon s)$ is

$$\mathbb{E}[S_L^\varepsilon(t_1 + \varepsilon s)] = \frac{1}{(2\pi)^2} \int \int e^{-i\omega s} e^{-i\varepsilon h s/2} \overline{\hat{f}(\omega - \varepsilon h/2)} \, \hat{G}(h)$$
$$\times \mathbb{E}\left[R_{\omega + \varepsilon h/2}^\varepsilon(-L, 0) \overline{R_{\omega - \varepsilon h/2}^\varepsilon(-L, 0)} \right] dh \, d\omega.$$

Taking the limit $\varepsilon \to 0$ and applying Proposition 9.1 gives

$$\mathbb{E}[S_L^\varepsilon(t_1 + \varepsilon s)] \xrightarrow{\varepsilon \to 0} \frac{1}{(2\pi)^2} \int \int e^{-i\omega s} \overline{\hat{f}(\omega)} \hat{G}(h) \left[\int e^{ih\tau} \mathcal{W}_1(\omega, \tau, -L, 0) \, d\tau \right] dh \, d\omega$$
$$= \frac{1}{(2\pi)^2} \int \int e^{-i\omega s} \overline{\hat{f}(\omega)} \left[\int \overline{\hat{G}(h)} e^{ih\tau} dh \right] \mathcal{W}_1(\omega, \tau, -L, 0) \, d\tau \, d\omega$$
$$= \frac{1}{2\pi} \int \int e^{-i\omega s} \overline{\hat{f}(\omega)} G(\tau) \mathcal{W}_1(\omega, \tau, -L, 0) \, d\tau \, d\omega,$$

where the quantity $\mathcal{W}_1(\omega, \tau, -L, 0)$ is obtained by solving the system of transport equations (9.23). We have also used the fact that G is real-valued.

Higher Order Moments

Let us now consider the general moments (10.4). Using the representation (10.3) for each factor $S_L^\varepsilon(t_1 + \varepsilon s_j)$, these moments can be written as multiple integrals over $p = \sum_{j=1}^k p_j$ frequencies:

$$
\mathbb{E}\left[\prod_{j=1}^k S_L^\varepsilon(t_1 + \varepsilon s_j)^{p_j}\right]
$$
$$
= \frac{1}{(2\pi)^{2p}} \int \cdots \int \prod_{\substack{1 \le j \le k \\ 1 \le l \le p_j}} \overline{\hat{f}(\omega_{j,l})} e^{-i\omega_{j,l}s_j} e^{-i\varepsilon h_{j,l}s_j/2} \widehat{G}(h_{j,l})
$$
$$
\times \, \mathbb{E}\left[\prod_{\substack{1 \le j \le k \\ 1 \le l \le p_j}} R^\varepsilon_{\omega_{j,l}+\varepsilon h_{j,l}/2}(-L,0)\overline{R^\varepsilon_{\omega_{j,l}-\varepsilon h_{j,l}/2}(-L,0)}\right] \prod_{\substack{1 \le j \le k \\ 1 \le l \le p_j}} d\omega_{j,l} \, dh_{j,l} \, .
$$

The important quantity is the expectation of the product of reflection coefficients, whose limit as $\varepsilon \to 0$ is given by Proposition 9.2. As a result, taking the limit $\varepsilon \to 0$ gives

$$
\mathbb{E}\left[\prod_{j=1}^k S_L^\varepsilon(t_1 + \varepsilon s_j)^{p_j}\right] \xrightarrow{\varepsilon \to 0} \frac{1}{(2\pi)^p} \int \cdots \int \prod_{\substack{1 \le j \le k \\ 1 \le l \le p_j}} \mathcal{W}_1(\omega_{j,l}, \tau_{j,l}, -L, 0)
$$
$$
\prod_{\substack{1 \le j \le k \\ 1 \le l \le p_j}} \overline{\hat{f}(\omega_{j,l})} e^{-i\omega_{j,l}s_j} G(\tau_{j,l}) \, d\omega_{j,l} \, d\tau_{j,l}
$$
$$
= \prod_{1 \le j \le k} \left(\frac{1}{2\pi} \int \mathcal{W}_1(\omega, \tau, -L, 0)\overline{\hat{f}(\omega)} e^{-i\omega s_j} G(\tau) d\omega \, d\tau\right)^{p_j} .
$$

This shows that the expectation of a product of terms $S_\varepsilon^L(t_1 + \varepsilon s)$ converges to the product of the limits of the expectations:

$$
\lim_{\varepsilon \to 0} \mathbb{E}\left[\prod_{j=1}^k S_L^\varepsilon(t_1 + \varepsilon s_j)^{p_j}\right] = \prod_{j=1}^k \lim_{\varepsilon \to 0} \mathbb{E}\left[S_L^\varepsilon(t_1 + \varepsilon s_j)^{p_j}\right] .
$$

This result is in dramatic contrast to the statistical description of the reflected wave before time reversal in terms of a Gaussian process (see Proposition 9.3). We have therefore shown that the finite-dimensional distributions of $(S_\varepsilon^L(t_1 + \varepsilon s))_{s \in (-\infty, \infty)}$ converge to those of the deterministic function

$$
\frac{1}{2\pi} \int \mathcal{W}_1(\omega, \tau, -L, 0)\overline{\hat{f}(\omega)} e^{-i\omega s} G(\tau) \, d\omega \, d\tau \, .
$$

Tightness

We have characterized the limiting refocused pulse in terms of its finite-dimensional time distributions. In fact, a tightness argument shows that this limit holds in the sense of the convergence in distribution for continuous processes. This is done by showing that the sequence of processes $S_L^\varepsilon(t_1 + \varepsilon\cdot)$, $\varepsilon > 0$, is precompact in the space of continuous functions (see [114]). On the one hand, the conservation of energy relation yields that $|R_\omega^\varepsilon| \leq 1$ and $S_L^\varepsilon(t_1 + \varepsilon s)$ is uniformly bounded by

$$|S_L^\varepsilon(t_1 + \varepsilon s)| \leq \frac{1}{(2\pi)^2} \int |\hat{f}(\omega)|\, d\omega \times \int |\hat{G}(h)|\, dh. \tag{10.5}$$

On the other hand, the modulus of continuity

$$M^\varepsilon(\delta) = \sup_{|s_1 - s_2| \leq \delta} |S_L^\varepsilon(t_1 + \varepsilon s_1) - S_L^\varepsilon(t_1 + \varepsilon s_2)|$$

is bounded by

$$M^\varepsilon(\delta) \leq \frac{1}{(2\pi)^2} \int \sup_{|s_1 - s_2| \leq \delta} |1 - \exp(i\omega(s_1 - s_2))||\hat{f}(\omega)|d\omega \times \int |\hat{G}(h)|\, dh,$$

which goes to zero as δ goes to zero uniformly with respect to ε. As a result, the refocused pulse $((S_L^\varepsilon(t_1 + \varepsilon s))_{-\infty<s<\infty})_{\varepsilon>0}$ is a tight (i.e., weakly compact) family in the space of continuous trajectories equipped with the supremum norm.

Convergence of the Refocused Pulse

We have just shown the tightness of the process $(S_L^\varepsilon(t_1 + \varepsilon s))_{s\in(-\infty,\infty)}$ as well as the convergence of its finite-dimensional distributions. Accordingly, we have shown that this process converges in probability as $\varepsilon \to 0$ to the deterministic function

$$S_L(s) = \frac{1}{2\pi} \int \Lambda_{\mathrm{TRR}}^L(\omega, \tau)\overline{\hat{f}(\omega)}e^{-i\omega s}G(\tau)\, d\omega\, d\tau,$$

where $\Lambda_{\mathrm{TRR}}^L(\omega, \tau) = \mathcal{W}_1(\omega, \tau, -L, 0)$ is given by (9.23) and TRR stands for "time reversal in reflection." We summarize this result in the following proposition.

Proposition 10.1. *The refocused signal $(S_L^\varepsilon(t_1 + \varepsilon s))_{s\in(-\infty,\infty)}$ converges in probability as $\varepsilon \to 0$ to the **deterministic pulse shape***

$$S_L(s) = (f(-\,\cdot)\star K_{\mathrm{TRR}}(\cdot))\,(s), \tag{10.6}$$

*where the Fourier transform of the **refocusing kernel** K_{TRR} is given by*

$$\hat{K}_{\mathrm{TRR}}(\omega) = \int G(\tau) \Lambda_{\mathrm{TRR}}^{L}(\omega,\tau)\,d\tau, \tag{10.7}$$

and the refocusing density $\Lambda_{\mathrm{TRR}}^{L}(\omega,\tau) = \mathcal{W}_{1}(\omega,\tau,-L,0)$ is given by the system (9.23).

If the medium is homogeneous, that is, $\gamma = 0$, then the refocusing kernel is zero. Indeed, in this case nothing is recorded by the TRM, since the initial pulse simply travels to the left without scattering. If the medium is random, $\gamma > 0$, then we get the striking result that we observe a refocused pulse whose shape does not depend on the particular realization of the medium, but only on its statistical distribution through the parameter γ. This is the **statistical stability** property of the refocused pulse due to the **self-averaging** property. This remarkable property is clearly seen in numerical simulations, as shown in Figure 10.4. In the next paragraph we examine a particular case in which an explicit formula can be derived for the refocusing kernel.

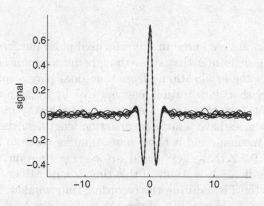

Fig. 10.4. We plot the refocused pulses generated by 10 independent numerical simulations of time reversal (we follow the same procedure as in Figures 10.1-10.2). The initial pulse is the second derivative of a Gaussian (with maximum normalized to one). As predicted by the theory, the refocused pulse does not depend on the realization of the medium, in contrast to the small-amplitude random wave fluctuations before and after the refocusing time. The refocused pulse shape is a filtered version of the initial pulse shape as described by (10.6) (thick dashed line).

The Refocusing Kernel for a Half-Space

We consider the case of a random half-space, that is, $L \to \infty$. We have computed explicitly the solution for the system of transport equations in this case (see (9.40)). We thus get a closed-form expression for the refocusing local spectral density $\Lambda_{\mathrm{TRR}}^{\infty}$ in this case

$$\Lambda_{\mathrm{TRR}}^{\infty}(\omega, \tau) = \frac{8\gamma\omega^2/\bar{c}}{(8 + \gamma\omega^2\tau/\bar{c})^2} = \frac{2\bar{c}\tau/L_{\mathrm{loc}}(\omega)}{(2 + \bar{c}\tau/L_{\mathrm{loc}}(\omega))^2}, \qquad (10.8)$$

where $L_{\mathrm{loc}} = 4\bar{c}^2/(\gamma\omega^2)$ is the localization length defined by (7.81). If we also assume that $G(t) = \mathbf{1}_{[0,t_1]}(t)$, then by computing the integral in (10.7) we find that the refocusing kernel is

$$\hat{K}_{\mathrm{TRR}}(\omega) = \frac{\gamma\omega^2 t_1/\bar{c}}{8 + \gamma\omega^2 t_1/\bar{c}} = \frac{\bar{c}t_1/L_{\mathrm{loc}}(\omega)}{2 + \bar{c}t_1/L_{\mathrm{loc}}(\omega)}.$$

Note that if we assume that we record everything at the mirror ($t_1 = \infty$ and $G \equiv 1$), then $\hat{K}_{\mathrm{TRR}}(\omega) = 1$. This is of course expected: the pulse has been completely scattered back by the random half-space due to localization, as seen in Chapter 7. We have sent back everything that has been recorded, so we get a perfect refocusing as a result of the time-reversibility of the wave equation.

If $t_1 < \infty$, then the kernel $\hat{K}_{\mathrm{TRR}}$ has the form of a high-pass filter with cutoff frequency

$$\omega_c^2 = \frac{8\bar{c}}{\gamma t_1}.$$

Frequencies above ω_c are recovered in the refocused pulse but frequencies below ω_c are lost. The reason is that even though the medium is completely reflecting because of the localization effect, time does play a role. High frequencies have a very short localization length, given by $L_{\mathrm{loc}}(\omega) = 4\bar{c}^2/(\gamma\omega^2)$, as shown in Chapter 7, so that they are scattered back very quickly by the medium. Low frequencies have a large localization length, so they can penetrate deep into the medium, and it takes more time for them to be reflected. We saw in Section 9.3.2 that they spend an average time on the order of $2L_{\mathrm{loc}}(\omega)/\bar{c}$ in the medium. As a result, if this time is larger than t_1, then they are not recorded by the TRM during the recording time window. The relation $2L_{\mathrm{loc}}(\omega)/\bar{c} \leq t_1$ gives the bandwidth of the refocusing kernel $|\omega| \leq \omega_c$.

Numerical Experiments

We consider the situation described in Figure 10.1. We record the reflected signal plotted at the top of the figure, time reverse it, and send it back into the same medium. Figure 10.2 shows the dynamics of this new wave. A refocused pulse can be seen to emerge from the random slab at the predicted time. The shape of this pulse results from the convolution of the original pulse shape with the refocusing kernel, which acts as a high-pass filter. The statistical stability of the refocused pulse derived in this section is illustrated in the numerical experiments shown in Figure 10.4.

10.1.4 Time-Reversal Mirror Versus Standard Mirror

We comment further on the role played by reversing time before sending back the recorded wave into the medium. If a **standard mirror** is used, then the recorded wave is simply reflected, and the integral representation (10.3) of the reflected signal observed at $t_{\text{obs}} = t_1$ has the form

$$\tilde{S}_L^\varepsilon(t_1 + \varepsilon s) = \frac{1}{(2\pi)^2} \int \int e^{-i\omega s - i\varepsilon h s/2} \hat{f}(\omega - \varepsilon h/2) \, \hat{G}(h)$$
$$\times R_{\omega + \varepsilon h/2}^\varepsilon(-L, 0) R_{\omega - \varepsilon h/2}^\varepsilon(-L, 0) \, dh \, d\omega \,.$$

The main difference compared with (10.3) is the presence of the product of two reflection coefficients without conjugation. The integral representation of the second moment $\mathbb{E}\left[|\tilde{S}_L^\varepsilon(t_1 + \varepsilon s)|^2\right]$ involves the following expectation of the product of four reflection coefficients:

$$\mathbb{E}\left[R_{\omega + \varepsilon h/2}^\varepsilon(-L, 0) R_{\omega - \varepsilon h/2}^\varepsilon(-L, 0) \overline{R_{\omega' + \varepsilon h'/2}^\varepsilon(-L, 0)} \; \overline{R_{\omega' - \varepsilon h'/2}^\varepsilon(-L, 0)}\right] \,.$$

Using (9.47) this expectation goes to zero as soon as $\omega \neq \omega'$. Substituting this result into the integral representation of the second moment of $\tilde{S}_L^\varepsilon$ and using the dominated convergence theorem, we obtain

$$\mathbb{E}\left[|\tilde{S}_L^\varepsilon(t_1 + \varepsilon s)|^2\right] \xrightarrow{\varepsilon \to 0} 0 \,,$$

which proves that there is indeed **no refocusing** with a standard mirror.

10.2 Time Reversal Versus Cross Correlations

In this section we shall discuss the advantage of using a time-reversal method for probing the medium. In the context of a homogeneous background, the information about the medium is contained in the function $\Lambda(\omega, t) = \mathcal{W}_1(\omega, t, 0)$. We shall see in Section 17.2 that this still holds when the medium has also a slowly varying background.

As shown in Section 9.3.3, in the regime in which the input pulse is of the form $f(t/\varepsilon)/\sqrt{\varepsilon}$, the function $\Lambda(\omega, t)$ is proportional to the local power spectral density of the reflected wave around time t. The power spectral density is given explicitly by $\Lambda(\omega, t)|\hat{f}(\omega)|^2$.

As shown in this chapter, in the regime in which the input pulse is of the form $f(t/\varepsilon)$, the function $\Lambda(\omega, t)$ is proportional to the refocusing density of the refocused pulse after a time-reversal operation. More precisely, the refocused pulse shape is the Fourier transform of $[\int \Lambda(\omega, t)G(t)dt]\hat{f}(\omega)$ with G being the time-reversal cutoff function.

Thus, the reflected wave and the refocused pulse contain information about the density Λ. But this information is not encoded in the same way, and the purpose of this section is to show that when time reversal is physically feasible, then it is easier and more efficient to extract the information from the refocused pulse than from the reflected signal. Before discussing this important issue, we comment on the statistics of the reflected wave.

10.2.1 The Empirical Correlation Function

We consider the strongly heterogeneous white-noise regime and we assume that an incoming pulse of the form $f(t/\varepsilon)$ is impinging on the random slab. The reflected signal is recorded at the surface and is denoted by $A^\varepsilon(t)$. In this section we compute the statistical distribution of the empirical correlation function of the reflected signal $A^\varepsilon(t)$ in the limit $\varepsilon \to 0$. This empirical correlation function is defined as the time average

$$C_{t_0,t_1}^\varepsilon(s) = \int_{t_0}^{t_1} A^\varepsilon(t + \varepsilon s) A^\varepsilon(t)\, dt. \tag{10.9}$$

The first moment of the correlation function is given by

$$\mathbb{E}[C_{t_0,t_1}^\varepsilon(s)] = \int_{t_0}^{t_1} c_t^\varepsilon(s)\, dt,$$

where the time correlation function $c_t^\varepsilon(s) = \mathbb{E}[A^\varepsilon(t) A^\varepsilon(t + \varepsilon s)]$ has been analyzed in Section 9.3.3 in the regime in which the input pulse is of the form $f(t/\varepsilon)/\sqrt{\varepsilon}$. Indeed, it is necessary to send a pulse with an energy of order one in order to get an incoherent reflected wave with an amplitude of order one. In that scaling, we have shown in Section 9.3.3 that the time correlation function converges as $\varepsilon \to 0$ to

$$c_t(s) = \frac{1}{2\pi} \int \Lambda(\omega, t) |\hat{f}(\omega)|^2 e^{i\omega s}\, d\omega.$$

Using this result and the linearity of the acoustic wave equations, we therefore obtain that in the present scaling (an input pulse of the form $f(t/\varepsilon)$), we have $c_t^\varepsilon/\varepsilon \to c_t$ as $\varepsilon \to 0$. Note that it is necessary to amplify the recorded reflected signal by a factor $1/\sqrt{\varepsilon}$ to get a limit of order one. Thus, we get the limit of the first moment of $C_{t_0,t_1}^\varepsilon(s)$:

$$\lim_{\varepsilon \to 0} \frac{1}{\varepsilon} \mathbb{E}[C_{t_0,t_1}^\varepsilon(s)] = \int_{t_0}^{t_1} c_t(s)\, dt. \tag{10.10}$$

We now consider the second moment of C_{t_0,t_1}^ε:

$$\mathbb{E}[C_{t_0,t_1}^\varepsilon(s)^2] = \int_{t_0}^{t_1} \int_{t_0}^{t_1} \mathbb{E}\left[A^\varepsilon(t + \varepsilon s) A^\varepsilon(t) A^\varepsilon(t' + \varepsilon s) A^\varepsilon(t')\right] dt'\, dt.$$

By the result in Section 9.3.4, the processes $A^\varepsilon(t + \varepsilon\cdot)/\sqrt{\varepsilon}$ and $A^\varepsilon(t' + \varepsilon\cdot)/\sqrt{\varepsilon}$ become independent as $\varepsilon \to 0$ as soon as $t \neq t'$. As a consequence,

$$\frac{1}{\varepsilon^2} \mathbb{E}\left[A^\varepsilon(t + \varepsilon s) A^\varepsilon(t) A^\varepsilon(t' + \varepsilon s) A^\varepsilon(t')\right] \xrightarrow{\varepsilon \to 0} c_t(s) c_{t'}(s),$$

which in turn implies

$$\lim_{\varepsilon \to 0} \frac{1}{\varepsilon^2} \mathbb{E}[C^\varepsilon_{t_0,t_1}(s)^2] = \int_{t_0}^{t_1} \int_{t_0}^{t_1} c_t(s)c_{t'}(s)dt'dt = \left(\int_{t_0}^{t_1} c_t(s)\,dt\right)^2$$

$$= \left(\lim_{\varepsilon \to 0} \frac{1}{\varepsilon} \mathbb{E}[C^\varepsilon_{t_0,t_1}(s)]\right)^2,$$

and establishes the convergence in probability of $\varepsilon^{-1}C^\varepsilon_{t_0,t_1}$ to its limiting expectation (10.10).

10.2.2 Measuring the Spectral Density

We first consider the situation in which time reversal can be done physically. A short pulse is sent into the medium, with a typical wavelength that is large compared to the inhomogeneities of the medium, but small compared to the depth we wish to probe. We record the reflected signal $A^\varepsilon(t)$, and we then perform a series of time-reversal experiments. Let Δt denote a time increment on the macrosopic order-one scale. For a fixed integer n we send back the truncated time-reversed signal $G_{[n\Delta t,(n+1)\Delta t]}(-t)A^\varepsilon(-t)$. This gives rise to a refocused signal, which we denote by $S^\varepsilon_n(t)$. The theory developed in this chapter shows that, if the scales are well separated, then the Fourier transform $\hat{S}^\varepsilon_n(\omega)$ of $S^\varepsilon_n(t)$ can be approximated by

$$\hat{S}^\varepsilon_n(\omega) \sim \overline{\hat{f}(\omega)} \int_{n\Delta t}^{(n+1)\Delta t} \Lambda(\omega, \tau)\,d\tau.$$

By repeating the experiment for $n = 0, \ldots, N-1$, we obtain an estimate of the refocusing density $\Lambda(\omega, \tau)$ with a time resolution of Δt over the time window $[0, N\Delta t]$ and the frequency window corresponding to the support of the spectrum of $\hat{f}$. This method gives a robust and stable estimate of the refocusing density by a series of measures performed with a single realization of the medium.

If time reversal cannot be done physically, then an alternative method is to estimating Λ directly from the reflected signal $A^\varepsilon(t)$. We choose some Δt, and for a fixed integer n we compute the empirical correlation function

$$C^\varepsilon_n(s) = \int_{n\Delta t}^{(n+1)\Delta t} A^\varepsilon(t)A^\varepsilon(t+s)\,dt.$$

The result presented in the previous section shows that, if the scales are well separated, then the Fourier transform $\hat{C}^\varepsilon_n(\omega)$ of $C^\varepsilon_n(s)$ is close to

$$\hat{C}^\varepsilon_n(\omega) \sim \varepsilon|\hat{f}(\omega)|^2 \int_{n\Delta t}^{(n+1)\Delta t} \Lambda(\omega, \tau)\,d\tau.$$

By repeating the experiment for $n = 0, \ldots, N-1$, we obtain an estimate of the spectral density $\Lambda(\omega, \tau)$ with a time resolution of Δt.

Therefore, it seems that we can extract the spectral density Λ without doing time reversal. However, there is an important difference between the two methods. As shown in the previous section, we need to amplify the recorded signal by a large factor when using the reflected signal, while the refocused pulse has the same amplitude as the input pulse. Accordingly, if we take into account **additive external noise** and/or the minimal detection level of our recording system, the amplification process is likely to lead to a very poor **signal-to-noise ratio (SNR)** of the empirical correlation function.

In fact, in the case of the reflected signal, the information about the spectral density is distributed over a large time window, and we have to process a signal of small amplitude over this large time window to extract the information. In contrast, in the case of the time-reversed refocused pulse, the information about the spectral density is contained in a very small time window (with a width comparable to the original pulse width) and it appears in the form of a large-amplitude signal (i.e., with the same amplitude as the original pulse). Time reversal compresses the information available in the reflected wave, and this recompression is performed by the medium itself, so it does not induce any error. In the next section, we propose a quantitative analysis that confirms the qualitative picture described in this section.

10.2.3 Signal-to-Noise Ratio Comparison

In this section we compute and compare the signal-to-noise ratios for the time-reversal method and the correlation method for estimating the power spectral density Λ.

Spectral SNR Using Time Reversal

We assume that time reversal can be done physically, meaning that we can implement the time-reversal steps involving measurement, time reversal, and reemission. We consider the procedure for determining the power spectral density from the refocused signal that we have described in the previous section. We probe the medium with a short pulse of the form $f(t/\varepsilon)$ and denote by $A^\varepsilon(t, 0)$ the reflected signal at the surface. We assume that the measurement of the reflected signal is not perfect and we model the recorded signal with an additive noise component as

$$y^\varepsilon(t) = \left[A^\varepsilon(t, 0) + \frac{\sigma}{\varepsilon^{p/2}} \eta \left(\frac{t}{\varepsilon^p} \right) \right] G(t),$$

where η is a centered stationary random process satisfying strong mixing conditions. We may consider, for instance, a Gaussian process with an integrable autocorrelation function. The scale ε^p characterizes the correlation length of the additive noise and we assume that it is smaller than the typical wavelength of the pulse, that is, $p > 1$.

Note that the process $\frac{1}{\varepsilon^{p/2}}\eta\left(\frac{t}{\varepsilon^p}\right)$ behaves like a **white noise** in the limit $\varepsilon \to 0$. The parameter σ characterizes the amplitude of the additive white noise.

The cutoff function G has its support included in $[0, t_1]$. The new incoming signal that is sent back into the medium is $f_{\text{new}}^\varepsilon(t) = y^\varepsilon(t_1 - t)$. This generates a new reflected signal, and according to the theory developed in this chapter, we know that refocusing occurs around time t_1. The refocused signal is

$$S_L^{\varepsilon,\sigma}(t_1 + \varepsilon s) = S_L^\varepsilon(t_1 + \varepsilon s) + \frac{\sigma}{\varepsilon^{p/2}} S_{L,1}^\varepsilon(t_1 + \varepsilon s) + \frac{\sigma}{\varepsilon^{p/2}} S_{L,2}^\varepsilon(t_1 + \varepsilon s), \quad (10.11)$$

where S_L^ε is the refocused signal (10.3) in absence of additive external noise, $S_{L,1}^\varepsilon$ is the reflected signal corresponding to the noise η,

$$S_{L,1}^\varepsilon(t_1 + \varepsilon s) = \frac{1}{2\pi\varepsilon}\int R_\omega^\varepsilon(-L, 0)\hat{\eta}^\varepsilon(\omega)e^{-\frac{i\omega t_1}{\varepsilon} - i\omega s}\, d\omega\,,$$

$$\hat{\eta}^\varepsilon(\omega) = \int G(t_1 - t)\eta\left(\frac{t_1 - t}{\varepsilon^p}\right)e^{\frac{i\omega t}{\varepsilon}}\, dt = \int G(t)\eta\left(\frac{t}{\varepsilon^p}\right)e^{-\frac{i\omega(t - t_1)}{\varepsilon}}\, dt\,,$$

and $S_{L,2}^\varepsilon$ corresponds to a new additive noise in the measurement of the refocused pulse:

$$S_{L,2}^\varepsilon(t) = \eta'\left(\frac{t}{\varepsilon^p}\right).$$

Here we assume that η and η' are independent and identically distributed.

We want to process the observed refocused signal in order to recover the power spectral density Λ. According to the previous section, we compute the Fourier transform of the noisy refocused signal. In fact, we restrict the integral to some interval $(-T, T)$. Otherwise, we integrate not only the refocused pulse, but the whole incoherent wave. Therefore, we define

$$\hat{S}_L^{\varepsilon,\sigma}(\omega) = \int_{-T}^T S_L^{\varepsilon,\sigma}(t_1 + \varepsilon s)e^{i\omega s}\, ds\,. \quad (10.12)$$

This spectral quantity can be written as a sum of three terms by (10.11). We know that the first term,

$$\hat{S}_L^\varepsilon(\omega) = \int_{-T}^T S_L^\varepsilon(t_1 + \varepsilon s)e^{i\omega s}\, ds\,,$$

has a well-defined limit as $\varepsilon \to 0$:

$$\hat{S}_L^\varepsilon(\omega) \xrightarrow{\varepsilon \to 0} \hat{S}_L(\omega) = \int_{-T}^T S_L(s)e^{i\omega s}\, ds\,,$$

where S_L is the deterministic function given by (10.6). Thus we obtain

$$\hat{S}_L(\omega) = \frac{T}{\pi} \int \hat{K}(\omega')\overline{\hat{f}(\omega')}\mathrm{sinc}[T(\omega - \omega')]\, d\omega', \qquad (10.13)$$

where $\hat{K}(\omega) = \int G(\tau)\Lambda(\omega, \tau)d\tau$ and the sinc function is defined by

$$\mathrm{sinc}(x) = \frac{\sin(x)}{x}. \qquad (10.14)$$

For T large enough, the limit $\hat{S}_L(\omega)$ is equal to $\overline{\hat{f}(\omega)} \int G(\tau)\Lambda(\omega, \tau)\, d\tau$. In fact, using the perturbed-test-function method, we can show that the mean-square error

$$\mathbb{E}\left[|\hat{S}_L^\varepsilon(\omega) - \hat{S}_L(\omega)|^2\right]$$

is of order ε. Therefore, in the absence of additive external noise $\sigma = 0$, we can recover the density Λ. In the presence of noise $\sigma > 0$, the contributions of the last two components in (10.11) should be estimated in order to ensure that the experimental measure gives the expected result.

- Influence of the noise η. We compute

$$\hat{S}_{L,1}^\varepsilon(\omega) := \int_{-T}^{T} S_{L,1}^\varepsilon(t_1 + \varepsilon s)e^{i\omega s}\, ds$$

$$= \frac{1}{2\pi\varepsilon} \int R_{\omega'}^\varepsilon(-L, 0)\hat{\eta}^\varepsilon(\omega')e^{-\frac{i\omega' t_1}{\varepsilon}} \int_{-T}^{T} e^{i(\omega - \omega')s}\, ds\, d\omega'$$

$$= \frac{T}{\pi\varepsilon} \int R_{\omega'}^\varepsilon(-L, 0)\mathrm{sinc}[T(\omega - \omega')] \int G(t)\eta\left(\frac{t}{\varepsilon^p}\right)e^{-\frac{i\omega' t}{\varepsilon}}\, dt\, d\omega'.$$

This quantity has mean zero, and its variance is

$$\mathbb{E}\left[|\hat{S}_{L,1}^\varepsilon(\omega)|^2\right] = \frac{T^2}{\pi^2\varepsilon^2} \int d\omega_1 \int d\omega_2 \int dt_1 \int dt_2\, \mathbb{E}[R_{\omega_1}^\varepsilon(-L, 0)\overline{R_{\omega_2}^\varepsilon(-L, 0)}]$$

$$\times \mathrm{sinc}[T(\omega - \omega_1)]\mathrm{sinc}[T(\omega - \omega_2)]G(t_1)G(t_2)\phi\left(\frac{t_1 - t_2}{\varepsilon^p}\right)e^{\frac{i\omega_2 t_2}{\varepsilon} - \frac{i\omega_1 t_1}{\varepsilon}},$$

where

$$\phi(t) = \mathbb{E}[\eta(t')\eta(t + t')].$$

By the change of variables $\omega_1 = \omega_0 + \varepsilon h/2$, $\omega_2 = \omega_0 - \varepsilon h/2$, $t_1 = t + \varepsilon^p u/2$, $t_2 = t - \varepsilon^p u/2$, we obtain

$$\mathbb{E}\left[|\hat{S}_{L,1}^\varepsilon(\omega)|^2\right]$$

$$= \frac{T^2\varepsilon^{p-1}}{\pi^2} \int d\omega \int dh \int dt \int du\, \mathbb{E}[R_{\omega_0 + \varepsilon h/2}^\varepsilon(-L, 0)\overline{R_{\omega_0 - \varepsilon h/2}^\varepsilon(-L, 0)}]$$

$$\times\ \mathrm{sinc}[T(\omega - \omega_0 - \varepsilon h/2)]\mathrm{sinc}[T(\omega - \omega_0 + \varepsilon h/2)]$$

$$\times\ G(t + \varepsilon^p u/2)G(t - \varepsilon^p u/2)\phi(u)e^{-iht - i\omega_0 u\varepsilon^{p-1}}.$$

Taking the limit $\varepsilon \to 0$ yields

$$\frac{1}{\varepsilon^{p-1}}\mathbb{E}\left[|\hat{S}^{\varepsilon}_{L,1}(\omega)|^2\right] \qquad\qquad (10.15)$$

$$\xrightarrow{\varepsilon \to 0} \frac{T^2}{\pi^2}\left(\int G(t)^2 e^{-iht} U_{\omega_0,h}\mathrm{sinc}^2[T(\omega - \omega_0)]\, dt\, dh\, d\omega_0\right)\left(\int \phi(u)du\right)$$

where $U_{\omega_0,h}$ is the limit as $\varepsilon \to 0$ of $\mathbb{E}[R^{\varepsilon}_{\omega_0+\varepsilon h/2}(-L,0)\overline{R^{\varepsilon}_{\omega_0-\varepsilon h/2}(-L,0)}]$, which is given by (9.25):

$$U_{\omega_0,h} = \int \Lambda(\omega_0,\tau)e^{ih\tau}\, d\tau.$$

By substituting this expression in (10.15) and integrating with respect to h, we obtain

$$\frac{1}{\varepsilon^{p-1}}\mathbb{E}\left[|\hat{S}^{\varepsilon}_{L,1}(\omega)|^2\right]$$

$$\xrightarrow{\varepsilon \to 0} \frac{2T^2}{\pi}\left(\int G(t)^2 \Lambda(\omega_0,t)\mathrm{sinc}^2[T(\omega-\omega_0)]\, dt\, d\omega_0\right)\left(\int \phi(u)\, du\right).$$

- Influence of the noise η'. We consider the term

$$\hat{S}^{\varepsilon}_{L,2}(\omega) := \int_{-T}^{T} S^{\varepsilon}_{L,2}(t_1 + \varepsilon s)e^{i\omega s}\, ds$$

$$= \int_{-T}^{T} \eta'\left(\frac{t_1}{\varepsilon^p} + \frac{s}{\varepsilon^{p-1}}\right)e^{i\omega s}\, ds,$$

whose expectation is zero, and we compute its variance:

$$\mathbb{E}\left[|\hat{S}^{\varepsilon}_{L,2}(\omega)|^2\right] = \int_{-T}^{T}\int_{-T}^{T} \phi\left(\frac{s_1 - s_2}{\varepsilon^{p-1}}\right)ds_1\, ds_2.$$

The limit is easily obtained:

$$\frac{1}{\varepsilon^{p-1}}\mathbb{E}\left[|\hat{S}^{\varepsilon}_{L,2}(\omega)|^2\right] \xrightarrow{\varepsilon \to 0} 2T\int \phi(u)\, du.$$

- **Conclusion.** The signal-to-noise ratio (SNR) of the spectral estimation using time reversal is defined by

$$\mathrm{SNR}_{\mathrm{TR}} = \frac{|\hat{S}^{\sigma}_L(\omega)|^2}{\mathbb{E}[|\hat{S}^{\varepsilon,\sigma}_L(\omega) - \hat{S}^{\sigma}_L(\omega)|^2]},$$

where $\hat{S}^{\varepsilon,\sigma}_L(\omega)$ is the measured noisy quantity given by (10.12) and $\hat{S}_L(\omega)$ is the spectral quantity (10.13) that we want to estimate. By collecting the previous estimates we obtain the order of magnitude of the denominator,

$$\mathbb{E}[|\hat{S}_L^{\varepsilon,\sigma}(\omega) - \hat{S}_L^{\sigma}(\omega)|^2] \sim \varepsilon + \frac{\sigma^2}{\varepsilon^p} \times \varepsilon^{p-1} + \frac{\sigma^2}{\varepsilon^p} \times \varepsilon^{p-1} \sim \varepsilon + \frac{\sigma^2}{\varepsilon},$$

with $p > 1$. As a result, the SNR of the spectral estimation using time reversal has the order of magnitude

$$\mathrm{SNR_{TR}} \sim \min\left(\frac{1}{\varepsilon}, \frac{\varepsilon}{\sigma^2}\right). \tag{10.16}$$

Spectral SNR Using the Reflected Signal

We consider in this section the efficiency of the estimation of the power spectral density Λ using the method based on computing the cross-correlation of the noisy reflected signal. We consider as above that the measurement of the reflected signal is not perfect and we model these imperfections by an additive noise with short correlation length ε^p with $p > 1$. Thus, the reflected signal has the form

$$y^{\varepsilon}(t) = A^{\varepsilon}(t,0) + \frac{\sigma}{\varepsilon^{p/2}}\eta\left(\frac{t}{\varepsilon^p}\right).$$

Therefore, the scaled empirical correlation function is given by

$$C_{t_0,t_1}^{\varepsilon,\sigma}(s) = \frac{1}{\varepsilon}\int_{t_0}^{t_1} y^{\varepsilon}(t)y^{\varepsilon}(t + \varepsilon s)dt \tag{10.17}$$

$$= \frac{1}{\varepsilon}C_{t_0,t_1}^{\varepsilon}(s) + \frac{\sigma}{\varepsilon^{p/2+1}}C_{t_0,t_1}^{\varepsilon,1}(s) + \frac{\sigma}{\varepsilon^{p/2+1}}C_{t_0,t_1}^{\varepsilon,2}(s) + \frac{\sigma^2}{\varepsilon^{p+1}}C_{t_0,t_1}^{\varepsilon,3}(s),$$

where $C_{t_0,t_1}^{\varepsilon}(s)$ is the noiseless empirical correlation function given by (10.9), and

$$C_{t_0,t_1}^{\varepsilon,1}(s) = \int_{t_0}^{t_1} A^{\varepsilon}(t)\eta\left(\frac{t + \varepsilon s}{\varepsilon^p}\right)dt,$$

$$C_{t_0,t_1}^{\varepsilon,2}(s) = \int_{t_0}^{t_1} A^{\varepsilon}(t + \varepsilon s)\eta\left(\frac{t}{\varepsilon^p}\right)dt,$$

$$C_{t_0,t_1}^{\varepsilon,3}(s) = \int_{t_0}^{t_1} \eta\left(\frac{t}{\varepsilon^p}\right)\eta\left(\frac{t + \varepsilon s}{\varepsilon^p}\right)dt.$$

We wish to process the observed correlation function in order to recover the power spectral density Λ. According to the previous section, we compute the Fourier transform of the noisy empirical correlation function $C_{t_0,t_1}^{\varepsilon,\sigma}$. Actually, we restrict the integral to some interval $(-T, T)$ and define

$$\hat{C}_{t_0,t_1}^{\varepsilon,\sigma}(\omega) = \int_{-T}^{T} C_{t_0,t_1}^{\varepsilon,\sigma}(s)e^{i\omega s}ds. \tag{10.18}$$

This spectral quantity can be written as the sum of four terms by (10.17). We first consider the contribution of the noiseless empirical correlation function:

$$\hat{C}^{\varepsilon}_{t_0,t_1}(\omega) = \int_{-T}^{T} C^{\varepsilon}_{t_0,t_1}(s)e^{i\omega s}\,ds\,.$$

We have

$$\frac{1}{\varepsilon}\hat{C}^{\varepsilon}_{t_0,t_1}(\omega) \xrightarrow{\varepsilon \to 0} \hat{C}_{t_0,t_1}(\omega) = \int_{t_0}^{t_1}\int_{-T}^{T} c_t(s)e^{i\omega s}\,ds\,dt\,,$$

where $c_t(s)$ is the limit of $\mathbb{E}[A^{\varepsilon}(t,0)A^{\varepsilon}(t+\varepsilon s)]/\varepsilon$ given by (9.64). Thus

$$\hat{C}_{t_0,t_1}(\omega) = \frac{T}{\pi}\int_{t_0}^{t_1}\int \Lambda(\omega',t)|\hat{f}(\omega')|^2 \mathrm{sinc}[T(\omega-\omega')]\,d\omega'\,dt\,, \qquad (10.19)$$

which is equal to $|\hat{f}(\omega)|^2\int_{t_0}^{t_1}\Lambda(\omega,t)\,dt$ for T large enough. Using the perturbed-test-function method we find that

$$\mathbb{E}\left[\left|\frac{1}{\varepsilon}\hat{C}^{\varepsilon}_{t_0,t_1}(\omega) - \hat{C}_{t_0,t_1}(\omega)\right|^2\right]$$

is of order ε.

- Influence of the noise term $C^{\varepsilon,1}_{t_0,t_1}$. The quantity

$$\hat{C}^{\varepsilon,1}_{t_0,t_1}(\omega) = \int_{-T}^{T} C^{\varepsilon,1}_{t_0,t_1}(s)e^{i\omega s}\,ds$$

$$= \int_{t_0}^{t_1}\int_{-T}^{T} A^{\varepsilon}(t)\eta\left(\frac{t+\varepsilon s}{\varepsilon^p}\right)e^{i\omega s}\,ds\,dt\,,$$

has mean zero and variance

$$\mathbb{E}\left[|\hat{C}^{c,1}_{t_0,t_1}(\omega)|^2\right] = \int_{t_0}^{t_1}\int_{t_0}^{t_1}\int_{-T}^{T}\int_{-T}^{T}\mathbb{E}[A^c(t)A^c(t')]\phi\left(\frac{t-t'+\varepsilon(s-s')}{\varepsilon^p}\right)$$
$$\times e^{i\omega(s-s')}\,ds\,ds'\,dt\,dt'\,.$$

We perform the change of variables $(s,s',t') \mapsto (u,v,w)$ with $u = s - s'$, $v = (s+s')/2$, and $w = (t-t')/\varepsilon$, and we integrate with respect to v to obtain

$$\mathbb{E}\left[|\hat{C}^{\varepsilon,1}_{t_0,t_1}(\omega)|^2\right] = \varepsilon\int_{t_0}^{t_1}\int_{(t_0-t)/\varepsilon}^{(t_1-t)/\varepsilon}\mathbb{E}[A^{\varepsilon}(t)A^{\varepsilon}(t+\varepsilon w)]$$
$$\times \int_{-2T}^{2T}(2T-|u|)\phi\left(\frac{u-w}{\varepsilon^{p-1}}\right)e^{i\omega u}\,dw\,du\,dt\,.$$

By the change of variables $(u,w) \mapsto (v,s)$ with $v = (u-\theta)/\varepsilon^{p-1}$ and $s = w$, we get

$$\mathbb{E}\left[|\hat{C}^{\varepsilon,1}_{t_0,t_1}(\omega)|^2\right] = \varepsilon^p \int_{t_0}^{t_1} \int_{(t_0-t)/\varepsilon}^{(t_1-t)/\varepsilon} \mathbb{E}[A^\varepsilon(t)A^\varepsilon(t+\varepsilon s)]e^{i\omega s}$$

$$\times \int_{(-2T-s)/\varepsilon^{p-1}}^{(2T-s)/\varepsilon^{p-1}} (2T-|s+\varepsilon^{p-1}v|)\phi(v)e^{i\omega\varepsilon^{p-1}v}\,dv\,ds\,dt\,.$$

As $\varepsilon \to 0$, the integral with respect to v converges to $(2T-|s|)\int \phi(v)dv$ if $|s| < 2T$, and to zero otherwise. We also know that $\mathbb{E}[A^\varepsilon(t)A^\varepsilon(t+\varepsilon s)]/\varepsilon$ converges to $c_t(s)$. We therefore get the limit

$$c\frac{1}{\varepsilon^{p+1}}\mathbb{E}\left[|\hat{C}^{\varepsilon,1}_{t_0,t_1}(\omega)|^2\right]$$

$$\xrightarrow{\varepsilon\to 0} 2T\left[\int_{t_0}^{t_1}\int_{-2T}^{2T} c_t(s)e^{i\omega s}\left(1-\frac{|s|}{2T}\right)ds\,dt\right]\left(\int \phi(v)dv\right).$$

In the large-T limit the above right-hand side is asymptotically equivalent to

$$2T\left(\int_{t_0}^{t_1} \Lambda(\omega,t)|\hat{f}(\omega)|^2 dt\right)\left(\int \phi(v)dv\right).$$

- Influence of the noisy term $C^{\varepsilon,2}_{t_0,t_1}$. The computation is similar to the one for $C^{\varepsilon,1}_{t_0,t_1}$, and it gives a term of the same order in ε.
- Influence of the noisy term $C^{\varepsilon,3}_{t_0,t_1}$. We consider the random variable

$$\hat{C}^{\varepsilon,3}_{t_0,t_1}(\omega) := \int_{-T}^{T} C^{\varepsilon,3}_{t_0,t_1}(s)e^{i\omega s}ds$$

$$= \int_{t_0}^{t_1}\int_{-T}^{T}\eta\left(\frac{t}{\varepsilon^p}\right)\eta\left(\frac{t+\varepsilon s}{\varepsilon^p}\right)e^{i\omega s}\,ds\,dt\,.$$

Its expectation is given by

$$\mathbb{E}\left[\hat{C}^{\varepsilon,3}_{t_0,t_1}(\omega)\right] = \int_{t_0}^{t_1}\int_{-T}^{T}\phi\left(\frac{s}{\varepsilon^{p-1}}\right)e^{i\omega s}\,ds\,dt\,.$$

Taking the limit $\varepsilon \to 0$ we obtain

$$\frac{1}{\varepsilon^{p-1}}\mathbb{E}\left[\hat{C}^{\varepsilon,3}_{t_0,t_1}(\omega)\right] \xrightarrow{\varepsilon\to 0} (t_1-t_0)\int \phi(u)\,du\,. \tag{10.20}$$

The second moment of $\hat{C}^{\varepsilon,3}_{t_0,t_1}(\omega)$ has the form

$$\mathbb{E}\left[|\hat{C}^{\varepsilon,3}_{t_0,t_1}(\omega)|^2\right] = \int_{t_0}^{t_1}\int_{t_0}^{t_1}\int_{-T}^{T}\int_{-T}^{T} e^{i\omega(s-s')}$$

$$\times\mathbb{E}\left[\eta\left(\frac{t}{\varepsilon^p}\right)\eta\left(\frac{t+\varepsilon s}{\varepsilon^p}\right)\eta\left(\frac{t'}{\varepsilon^p}\right)\eta\left(\frac{t'+\varepsilon s'}{\varepsilon^p}\right)\right]ds\,ds'\,dt\,dt'\,.$$

Assuming that the process η is Gaussian, the fourth-order moment can be written as a sum of products of second moments and we get that the second moment of $\hat{C}^{\varepsilon,3}_{t_0,t_1}(\omega)$ is given by

$$\mathbb{E}\left[|\hat{C}^{\varepsilon,3}_{t_0,t_1}(\omega)|^2\right] = \int_{t_0}^{t_1}\int_{t_0}^{t_1}\int_{-T}^{T}\int_{-T}^{T}\left[\phi\left(\frac{t-t'-\varepsilon s'}{\varepsilon^p}\right)\phi\left(\frac{t-t'+\varepsilon s}{\varepsilon^p}\right)\right.$$
$$+\left.\phi\left(\frac{t-t'}{\varepsilon^p}\right)\phi\left(\frac{t-t'+\varepsilon(s-s')}{\varepsilon^p}\right)+\phi\left(\frac{s}{\varepsilon^{p-1}}\right)\phi\left(\frac{s'}{\varepsilon^{p-1}}\right)\right]$$
$$\times\, e^{i\omega(s-s')}\,ds\,ds'\,dt\,dt'.$$

Among the three terms in the brackets the third one dominates, and we get the limit

$$\frac{1}{\varepsilon^{2p-2}}\mathbb{E}\left[|\hat{C}^{\varepsilon,3}_{t_0,t_1}(\omega)|^2\right] \xrightarrow{\varepsilon\to 0} (t_1-t_0)^2\left(\int\phi(u)du\right)^2.$$

This shows that $\varepsilon^{1-p}\hat{C}^{\varepsilon,3}_{t_0,t_1}(\omega)$ converges to the positive deterministic quantity (10.20) as $\varepsilon \to 0$. This term is therefore responsible for a bias in the spectral estimation.

- **Conclusion.** The signal-to-noise ratio (SNR) of the spectral estimation using the reflected signal to compute the cross-correlation is defined by

$$\text{SNR}_{\text{Cor}} = \frac{|\hat{C}_{t_0,t_1}(\omega)|^2}{\mathbb{E}[|\hat{C}^{\varepsilon,\sigma}_{t_0,t_1}(\omega) - \hat{C}_{t_0,t_1}(\omega)|^2]},$$

where $\hat{C}^{\varepsilon,\sigma}_{t_0,t_1}(\omega)$ is the measured noisy quantity given by (10.18) and $\hat{C}_{t_0,t_1}(\omega)$ is the spectral quantity (10.19) that we want to estimate. By collecting the previous estimates we obtain the order of magnitude

$$\mathbb{E}[|\hat{C}^{\varepsilon,\sigma}_{t_0,t_1}(\omega) - \hat{C}_{t_0,t_1}(\omega)|^2] \sim \varepsilon + \frac{\sigma^2}{\varepsilon^{p+2}}\times\varepsilon^{p+1} + \frac{\sigma^2}{\varepsilon^{p+2}}\times\varepsilon^{p+1} + \frac{\sigma^2}{\varepsilon^{p+1}}\times\varepsilon^{p-1}$$
$$\sim \varepsilon + \frac{\sigma^2}{\varepsilon} + \frac{\sigma^2}{\varepsilon^2} \sim \varepsilon + \frac{\sigma^2}{\varepsilon^2}.$$

As a result, the SNR of the spectral estimation using cross-correlation has the order of magnitude

$$\text{SNR}_{\text{Cor}} \sim \min\left(\frac{1}{\varepsilon}, \frac{\varepsilon^2}{\sigma^2}\right). \tag{10.21}$$

Discussion

- If $\sigma \leq \varepsilon^{3/2}$, then the dominant terms in the *SNRs* for the time-reversal method (10.16) and for the cross-correlation method (10.21) are the first ones, which are of order ε^{-1}. As a result, the SNRs of the two methods are equivalent and given by

$$\text{SNR}_{\text{TR}} \sim \text{SNR}_{\text{Cor}} \sim \frac{1}{\varepsilon}.$$

Note that in this regime the SNRs are very high, and therefore the additive external noise plays no significant role.

- If $\varepsilon^{3/2} < \sigma < \varepsilon$, then the dominant term in the SNR for the cross-correlation method (10.21) is the second one, while the dominant term in the SNR for the time-reversal method (10.16) is the first one. Thus, the SNRs of the two methods are different, and the SNR of the time-reversal method is higher:

$$\text{SNR}_{\text{TR}} \sim \frac{1}{\varepsilon} \sim \frac{\sigma^2}{\varepsilon^3} \times \frac{\varepsilon^2}{\sigma^2} \sim \frac{\sigma^2}{\varepsilon^3} \times \text{SNR}_{\text{Cor}}.$$

Note that in this regime, the SNRs are high, but the SNR of the time-reversal method is higher.

- If $\varepsilon \le \sigma$, then the dominant terms in the SNRs for the cross-correlation method (10.21) and for the time-reversal method (10.16) are the second ones. Thus, the SNRs of the two methods are different, and the SNR of the time-reversal method is higher:

$$\text{SNR}_{\text{TR}} \sim \frac{\varepsilon}{\sigma^2} \sim \frac{1}{\varepsilon} \times \frac{\varepsilon^2}{\sigma^2} \sim \frac{1}{\varepsilon} \times \text{SNR}_{\text{Cor}}.$$

Note that in this regime, the SNR of the cross-correlation method is small, while the SNR of the time-reversal method is still high as long as $\sigma > \varepsilon^{1/2}$. For instance, in the critical case in which $\sigma \sim \varepsilon$, then $\text{SNR}_{\text{Cor}} \sim 1$, while the SNR for the time-reversal method is of order $\varepsilon^{-1/2}$, which is much larger than one. Surprisingly in this case the correlation method fails, while the time-reversal method enables us to estimate the local power spectral density Λ.

Therefore, we can conclude that as long as the correlation method gives accurate results, it is not crucial to use a time-reversal method, since the two methods have then the same SNR. However, when the correlation method fails ($\text{SNR}_{\text{Cor}} \le 1$), then it turns out that the time reversal-method gives much better results independently of the value of p (in the range $p > 1$). Table 10.1 gives the comparative values of the SNRs for the two methods at different values for the noise amplitude σ.

10.3 Calibrating the Initial Pulse

There is one more issue that we now discuss. In most of the book we consider the strongly heterogeneous white-noise regime in which the correlation length

σ	$\mathrm{SNR_{TR}}$	$\mathrm{SNR_{Cor}}$
ε^2	ε^{-1}	ε^{-1}
$\varepsilon^{3/2}$	ε^{-1}	ε^{-1}
ε	ε^{-1}	1
$\varepsilon^{1/2}$	1	ε^2
1	ε^1	ε^4

Table 10.1. Order of magnitudes of the SNRs for different values of the noise amplitude σ. Note that $\mathrm{SNR_{Cor}} \sim 1$, corresponding to $\sigma \sim \varepsilon$, is the critical case where the correlation method starts failing, while the time-reversal method is still performing with a high SNR.

of the medium is much smaller than the wavelength, which is much smaller than the size of the slab. The size L of the slab, or in the context of imaging the depth one wishes to probe, and the correlation length of the medium l are fixed, in the sense that they are determined by the medium, even if we do not know the correlation length exactly. The regime we consider is based on the separation of these two scales $l \ll L$, and on the use of a pulse whose typical wavelength λ is in between these scales. The question of how one "tunes" λ in the case that one does not know l exactly is an important question.

A first method is based on the following observations. First, note that, in our framework, the correlation length of the medium is defined by the parameter γ. If we send a narrowband pulse with carrier frequency ω_0, then the mean fraction of reflected intensity at time t by a random half-space follows from (9.54) and is given by the local power spectral density

$$J_{\omega_0}(t) = \frac{8\gamma\omega_0^2/\bar{c}}{(8 + \gamma\omega_0^2 t/\bar{c})^2} = \frac{2\bar{c}t/L_{\mathrm{loc}}(\omega)}{(2 + \bar{c}t/L_{\mathrm{loc}}(\omega))^2}.$$

For a fixed time t, this function gives a maximum for the frequency $\omega_0 = 2\sqrt{2}\sqrt{\bar{c}}/\sqrt{\gamma t}$. As discussed in Section 9.3.2, this frequency corresponds to a localization length that is of order $\bar{c}t/2$. Furthermore, the time-averaged reflected intensity

$$I^\varepsilon = \int_{t-\Delta t}^{t+\Delta t} A^\varepsilon(t')^2 dt',$$

is a self-averaging quantity, as shown in Section 10.2.1, so that it attains its maximum at ω_0.

These arguments suggest a method to tune the carrier frequency of a pulse to probe a given depth L_0. It consists in emitting a series of pulses with different carrier frequencies $(\omega_j)_{j=1,\dots,N}$ and with energies $(E_j)_{j=1,\dots,N}$, and to compute for each incoming pulse the relative time averaged reflected intensity of the corresponding reflected wave A_j^ε:

$$I_j^\varepsilon = \frac{1}{E_j} \int_{t_0-\Delta t}^{t_0+\Delta t} A_j^\varepsilon(t)^2\, dt,$$

with $t_0 = 2L_0/\bar{c}$. The maximum of the values I_j^e corresponds to the frequency ω_j whose localization length is the closest to L_0, which means that this pulse penetrates to this depth and is then scattered back.

The method we have just described is sufficient to calibrate a pulse, since we wish to find only an order of magnitude, and not a precise value. In the presence of measurement noise, a time-reversal method may be desirable in order to reduce the SNR as seen in the previous section. This method would consist in emitting a series of pulses at different carrier frequencies, and then performing time reversal on each reflected signal. We would then find the pulse maximizing the refocused energy.

Notes

The reflected signal and its spectral content have been studied in the regime of separation of scales in [8], [9], [31], [32], [33]. Refocusing and self-averaging for time reversal in reflection in the one-dimensional case was derived in 1997 by Clouet and Fouque in the article [40]. An iterative time-reversal method to estimate higher moments is also presented in that reference. A generalization to the weakly heterogeneous regime is given in [158]. Numerical experiments similar to those presented in Section 10.1.3 were first carried out in collaboration with André Nachbin [65, 59]. The discussion on the advantage of using time reversal over a direct processing of the reflected signal is presented in Section 10.2 for the first time.

11

Applications to Detection

In this chapter we discuss how the tools that we have developed for describing the local power spectrum of the reflected signal can be used to detect changes in the medium that do not create coherent or stable reflected pulses. In Section 8.3 we have characterized the wave front reflections generated by a strong interface corresponding to a discontinuity in the average impedance $\bar{\zeta}$. Here we consider three cases in which the average impedance remains constant, so that no reflected coherent wave front is created.

In the first case, in Section 11.1, we consider a discontinuity in the background velocity $\bar{c}$. The detection problem consists in identifying both the depth and the strength of the interface or **weak reflector**.

In the second case, Section 11.2, we show that a similar analysis enables us to detect an interface when the parameter γ, depending on the statistical properties of the medium, is changing while the background medium $(\bar{\zeta}, \bar{c})$ is unchanged.

In the third case, after having introduced **dissipation** in the model in Section 11.3, we analyze in Section 11.4 a medium that has an embedded layer that has an anomalously large dissipation. Our main objective is to identify the location of the layer, and we discuss in particular the situation in which the layer is very thin, in which case the detection formulas simplify.

In all three cases, the approach we use to identify the interfaces consists in scrutinizing the form of the local power spectrum or covariance of the reflections. The local power spectrum has a discontinuity at a time corresponding to the two-way travel time to the depth of the interface, and this fact can be used for detection. Using the local spectral density of the reflections to solve the inverse problem of recovering the slowly varying background parameters is carried out in [8] and [133] and is briefly presented at the end of Section 17.2.3. Here we use the **time-reversal** procedure to identify the spectrum through its relation to the shape of the refocused pulse. As discussed in the previous chapter, this approach may be advantageous when there is measurement noise.

11.1 Detection of a Weak Reflector

We consider time reversal in reflection (TRR) as shown in Figure 10.3, where the random slab is made up of two different media as in Figure 11.1 and described below:

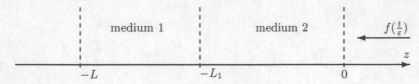

Fig. 11.1. A pulse is incoming from the right homogeneous half-space and impinges on a stack of two heteregeneous slabs with different averaged parameters but the same impedance.

$$\frac{1}{K(z)} = \begin{cases} \frac{1}{\bar{K}_1} & \text{for } z \in (-\infty, -L), \\ \frac{1}{\bar{K}_1}\left(1 + \nu(z/\varepsilon^2)\right) & \text{for } z \in [-L, -L_1), \\ \frac{1}{\bar{K}_2}\left(1 + \nu(z/\varepsilon^2)\right) & \text{for } z \in [-L_1, 0), \\ \frac{1}{\bar{K}_2} & \text{for } z \in [0, \infty), \end{cases}$$

$$\rho(z) = \begin{cases} \bar{\rho}_1 & \text{for } z \in (-\infty, -L_1), \\ \bar{\rho}_2 & \text{for } z \in [-L_1, \infty). \end{cases}$$

The reflection coefficient of the interface $z = -L_1$ is given by

$$R_I = \frac{\bar{\zeta}_2 - \bar{\zeta}_1}{\bar{\zeta}_1 + \bar{\zeta}_2},$$

where $\bar{\zeta}_j = \sqrt{\bar{K}_j \bar{\rho}_j}$, $j = 1, 2$, are the impedances of the two effective background media separated by the interface at $z = -L_1$.

We consider the case in which this reflection is weak, which we simply model here by assuming $\bar{\zeta}_1 = \bar{\zeta}_2$, or equivalently $R_I = 0$. To simplify the presentation, we have assumed matched medium boundary conditions at both ends of the random slab, and that only the bulk modulus is randomly fluctuating.

We introduce the piecewise constant coefficient

$$\bar{c}(z) = \begin{cases} \bar{c}_1 & \text{for } z \in [-L, -L_1), \\ \bar{c}_2 & \text{for } z \in [-L_1, 0], \end{cases} \tag{11.1}$$

which is the local average sound speed, and the (negative) effective travel time

$$\vartheta(z) = \int_0^z \frac{1}{\bar{c}(z')}dz' = \begin{cases} -\dfrac{L_1}{\bar{c}_2} + \dfrac{z + L_1}{\bar{c}_1} & \text{for } z \in [-L, -L_1), \\ \dfrac{z}{\bar{c}_2} & \text{for } z \in [-L_1, 0]. \end{cases} \tag{11.2}$$

We choose to define a negative travel time (since $z < 0$) so that its derivative is exactly the reciprocal of the local sound speed.

The right- and left-going modes are defined by

$$\hat{p}^\varepsilon(\omega, z) = \frac{\sqrt{\bar{\zeta}}}{2} \left(\hat{a}^\varepsilon(\omega, z) e^{\frac{i\omega\vartheta(z)}{\varepsilon}} - \hat{b}^\varepsilon(\omega, z) e^{-\frac{i\omega\vartheta(z)}{\varepsilon}} \right),$$

$$\hat{u}^\varepsilon(\omega, z) = \frac{1}{2\sqrt{\bar{\zeta}}} \left(\hat{a}^\varepsilon(\omega, z) e^{\frac{i\omega\vartheta(z)}{\varepsilon}} + \hat{b}^\varepsilon(\omega, z) e^{-\frac{i\omega\vartheta(z)}{\varepsilon}} \right),$$

where $\bar{\zeta} = \bar{\zeta}_1 = \bar{\zeta}_2$, the constant effective impedance.

For $z \in [-L, -L_1) \cup (-L_1, 0)$, the modes satisfy the coupled system of ordinary differential equations:

$$\frac{d}{dz} \begin{bmatrix} \hat{a}^\varepsilon \\ \hat{b}^\varepsilon \end{bmatrix} = \frac{1}{\varepsilon} \mathbf{H}_\omega \left(\bar{c}(z), \frac{\vartheta(z)}{\varepsilon}, \nu\left(\frac{z}{\varepsilon^2}\right) \right) \begin{bmatrix} \hat{a}^\varepsilon \\ \hat{b}^\varepsilon \end{bmatrix},$$

$$\mathbf{H}_\omega(\bar{c}, \vartheta, \nu) = \frac{i\omega}{2\bar{c}} \nu \begin{bmatrix} 1 & -e^{-2i\omega\vartheta} \\ e^{2i\omega\vartheta} & -1 \end{bmatrix}.$$

They also satisfy the following boundary conditions at the ends of the slab $[-L, 0]$:

$$\hat{a}^\varepsilon(\omega, -L) = 0, \qquad \hat{b}^\varepsilon(\omega, 0) = \hat{f}(\omega).$$

At the interface $z = -L_1$ the continuity of the pressure and velocity fields and the absence of contrast of impedance imply

$$\hat{a}^\varepsilon(\omega, (-L_1)^-) = \hat{a}^\varepsilon(\omega, (-L_1)^+), \qquad \hat{b}^\varepsilon(\omega, (-L_1)^-) = \hat{b}^\varepsilon(\omega, (-L_1)^+). \quad (11.3)$$

For $-L \le z_0 \le z \le 0$, we introduce the propagator matrix $\mathbf{P}_\omega^\varepsilon(z_0, z)$, which satisfies the equation

$$\frac{d}{dz} \mathbf{P}_\omega^\varepsilon(z_0, z) = \frac{1}{\varepsilon} \mathbf{H}_\omega \left(\bar{c}(z), \frac{\vartheta(z)}{\varepsilon}, \nu\left(\frac{z}{\varepsilon^2}\right) \right) \mathbf{P}_\omega^\varepsilon(z_0, z), \quad (11.4)$$

with the initial condition $\mathbf{P}_\omega^\varepsilon(z_0, z = z_0) = \mathbf{I}$. With this definition, we have

$$\mathbf{P}_\omega^\varepsilon(-L, 0) \begin{bmatrix} 0 \\ \hat{b}^\varepsilon(\omega, -L) \end{bmatrix} = \begin{bmatrix} \hat{a}^\varepsilon(\omega, 0) \\ \hat{f}(\omega) \end{bmatrix},$$

obtained using the relation

$$\mathbf{P}_\omega^\varepsilon(-L, 0) = \mathbf{P}_\omega^\varepsilon(-L_1, 0) \mathbf{P}_\omega^\varepsilon(-L, -L_1)$$

and the continuity conditions (11.3) at $z = -L_1$. As in the previous chapter, the trace of $\mathbf{H}_\omega$ is zero, and $\mathbf{P}_\omega^\varepsilon$ can be written as

$$\mathbf{P}_\omega^\varepsilon(-L, z) = \begin{bmatrix} \alpha_\omega^\varepsilon(-L, z) & \overline{\beta_\omega^\varepsilon(-L, z)} \\ \beta_\omega^\varepsilon(-L, z) & \overline{\alpha_\omega^\varepsilon(-L, z)} \end{bmatrix},$$

with $|\alpha_\omega^\varepsilon(-L, z)|^2 - |\beta_\omega^\varepsilon(-L, z)|^2 = 1$. We can then define the reflection coefficient by

$$R_\omega^\varepsilon(-L, z) = \frac{\overline{\beta_\omega^\varepsilon(-L, z)}}{\alpha_\omega^\varepsilon(-L, z)},$$

so that the reflected right-going wave is $\hat{a}^\varepsilon(\omega, 0) = R_\omega^\varepsilon(-L, 0)\hat{f}(\omega)$. From (11.4) we deduce that $R_\omega^\varepsilon(-L, z)$ satisfies in $-L \le z \le 0$ the Riccati equation

$$\frac{dR_\omega^\varepsilon}{dz} = -\frac{i\omega}{2\bar{c}(z)\varepsilon}\nu\left(\frac{z}{\varepsilon^2}\right)\left(e^{\frac{-2i\omega\vartheta(z)}{\varepsilon}} - 2R_\omega^\varepsilon + (R_\omega^\varepsilon)^2 e^{\frac{2i\omega\vartheta(z)}{\varepsilon}}\right),$$

with the initial condition $R_\omega^\varepsilon(-L, z = -L) = 0$.

At this point the analysis of time-reversal refocusing follows the lines of Section 10.1. The refocused pulse given in (10.3) converges to the deterministic pulse given by (10.6). The refocusing kernel K_{TRR} is given by (10.7), where the refocusing density $\Lambda_{\text{TRR}}^L(\omega, \tau) = \mathcal{W}_1(\omega, \tau, -L, 0)$ is the solution of the system of transport equations (9.23) with the only difference that the constant speed $\bar{c}$ is replaced by the piecewise-constant speed $\bar{c}(z)$ defined in (11.1).

We can now make use of the probabilistic representation of the solution of the transport equations presented in detail in Section 9.2.2. Since, by hyperbolicity, the refocused pulse is not affected by L for L large enough, we simplify the problem by letting L go to infinity, and we denote the density $\Lambda_{\text{TRR}}^L(\omega, \tau)$ by $\Lambda_{\text{TRR}}(\omega, \tau)$. We use the results of Section 9.2.3 to solve explicitly the transport equations from $-\infty$ to $-L_1$ in medium 1 with constant speed $\bar{c}_1$. Denoting the solution by $\mathcal{W}_p^{(\bar{c}_1)}$, we deduce from (9.40) that

$$\mathcal{W}_p^{(\bar{c}_1)}(\omega, \tau) = \begin{cases} \delta(\tau) & \text{if } p = 0, \\ \dfrac{8p\gamma\omega^2}{\bar{c}_1} \dfrac{(\gamma\omega^2\tau/\bar{c}_1)^{p-1}}{(8 + \gamma\omega^2\tau/\bar{c}_1)^{p+1}} \mathbf{1}_{[0,\infty)}(\tau) & \text{otherwise}. \end{cases} \tag{11.5}$$

We next solve the transport equations from $-L_1$ to 0, in medium 2 with constant speed $\bar{c}_2$, and with the initial condition $\mathcal{W}_p^{(\bar{c}_1)}(\omega, \tau, -L_1)$ given above. The probabilistic representation (9.32) obtained in Section 9.2.2 gives

$$\Lambda_{\text{TRR}}(\omega, \tau) = \mathbb{E}\left[\mathcal{W}_{N_0^{(\bar{c}_2)}}^{(\bar{c}_1)}\left(\omega, \tau - \frac{2}{\bar{c}_2}\int_{-L_1}^0 N_s^{(\bar{c}_2)}ds\right) \mid N_{-L_1}^{(\bar{c}_2)} = 1\right].$$

Here $N_z^{(\bar{c}_2)}$ jumps by ± 1 with probability $1/2$ and with intensity $n^2\gamma\omega^2/(2\bar{c}_2^2)$, n denoting the value of the process and 0 being an absorbing state.

Denoting by $\mathcal{W}^{(\bar{c}_2)}$ the expression (11.5) with $\bar{c}_2$ instead of $\bar{c}_1$, we write $\mathcal{W}^{(\bar{c}_1)} = \mathcal{W}^{(\bar{c}_2)} + (\mathcal{W}^{(\bar{c}_1)} - \mathcal{W}^{(\bar{c}_2)})$, so that we obtain

$$\Lambda_{\text{TRR}}(\omega, \tau) = \Lambda_{\text{TRR}}^{(\bar{c}_2)}(\omega, \tau) \tag{11.6}$$

$$+ \mathbb{E}\left[\left(\mathcal{W}_{N_0^{(\bar{c}_2)}}^{(\bar{c}_1)} - \mathcal{W}_{N_0^{(\bar{c}_2)}}^{(\bar{c}_2)}\right)\left(\omega, \tau - \frac{2}{\bar{c}_2}\int_{-L_1}^0 N_s^{(\bar{c}_2)}ds\right) \mid N_{-L_1}^{(\bar{c}_2)} = 1\right].$$

The first term is the stationary solution in a half-space with constant speed $\bar{c}_2$, given explicitly by (9.41),

$$\Lambda_{\mathrm{TRR}}^{(\bar{c}_2)}(\omega,\tau) = \frac{8\gamma\omega^2/\bar{c}_2}{\left(8 + \gamma\omega^2\tau/\bar{c}_2\right)^2} \mathbf{1}_{[0,\infty)}(\tau).$$

Formula (11.6) is very convenient for computing the density $\Lambda_{\mathrm{TRR}}(\omega,\tau)$ by Monte Carlo simulations as illustrated in Figures 11.2 and 11.3. In the first case (Figure 11.2) the speed increases at the reflector, producing a negative jump in the density at $\tau = 2L/\bar{c}_2 = 2$. In the second case (Figure 11.3) the speed decreases at the reflector, producing a positive jump in the density at $\tau = 2L/\bar{c}_2 = 2$. We can give the following heuristic interpretation of these jumps. Let us consider the case (Figure 11.2) in which the sound speed increases at the reflector position $-L_1$. As soon as the wave front passes through $-L_1$, it speeds up, and its time profile becomes elongated and smoother relative to the scale of the random inhomogeneities. As a result, the backscattering is reduced and the density has a negative jump. However, scattering by a random half-space results in total reflection. This implies that the power spectral density integrated over all times is equal to one, with or without the weak reflector. That is why we observe in Figures 11.2b and 11.3b an inversion of the behavior of the density for large τ that compensates for the jump at $\tau = 2L/\bar{c}_2$. In fact, as $\tau \to \infty$, we get from (11.6) the asymptotic behavior

$$\Lambda_{\mathrm{TRR}}(\omega,\tau) \simeq \frac{8\bar{c}_1}{\gamma\omega^2\tau^2},$$

while we have, in the case of a stationary random half-space without reflector,

$$\Lambda_{\mathrm{TRR}}^{(\bar{c}_2)}(\omega,\tau) \sim \frac{8\bar{c}_2}{\gamma\omega^2\tau^2}.$$

The trajectories of $N_z^{(\bar{c}_2)}$ reaching 0 do not contribute to the expectation in (11.6), since $\mathcal{W}_0^{(\bar{c}_1)}(\omega,0) - \mathcal{W}_0^{(\bar{c}_2)}(\omega,0) = 0$, and they have in fact been taken into account in the first term $\Lambda_{\mathrm{TRR}}^{(\bar{c}_2)}(\omega,\tau)$. For the trajectories not reaching 0, the integral $\int_{-L_1}^{0} N_s^{(\bar{c}_2)}ds$ is at least equal to L_1, and because of the restriction on the support of the $\mathcal{W}$'s, $\Lambda_{\mathrm{TRR}}(\omega,\tau)$ will differ from $\Lambda_{\mathrm{TRR}}^{(\bar{c}_2)}(\omega,\tau)$ only for $\tau > 2L_1/\bar{c}_2$, which corresponds to the travel time (back and forth) from the surface $z = 0$ to the reflector. In fact, at $\tau = 2L_1/\bar{c}_2$ there is a jump in $\Lambda_{\mathrm{TRR}}(\omega,\tau)$ due to the contribution of the set of trajectories that stay at $N_z = 1$ for all $z \in [-L_1,0]$. The other trajectories will produce an integral $\int_{-L_1}^{0} N_s^{(\bar{c}_2)}ds$ strictly larger than $2L_1/\bar{c}_2$ and therefore will contribute only for $\tau > 2L_1/\bar{c}_2$. The size of this jump can then easily be computed from formula (11.6):

$$\mathbb{E}\left[\left(\mathcal{W}_1^{(\bar{c}_1)} - \mathcal{W}_1^{(\bar{c}_2)}\right)(\omega,0)\mathbf{1}_{\{N_z^{(\bar{c}_2)}=1,z\in[-L_1,0]\}} \mid N_{-L_1}^{(\bar{c}_2)} = 1\right]$$

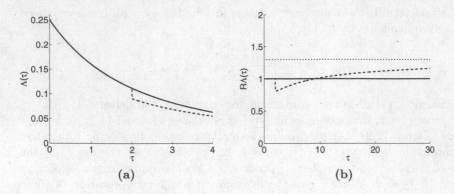

Fig. 11.2. Plot (a): Densities $\tau \mapsto \Lambda_{\mathrm{TRR}}(\omega, \tau)$ (dashed line) and $\tau \mapsto \Lambda_{\mathrm{TRR}}^{(\bar{c}_2)}(\omega, \tau)$ (solid line). Plot (b): Ratio of the densities $\tau \mapsto R\Lambda(\tau) := \Lambda_{\mathrm{TRR}}(\omega, \tau)/\Lambda_{\mathrm{TRR}}^{(\bar{c}_2)}(\omega, \tau)$. Here we assume $\gamma\omega^2 = 2$, $L_1 = 1$, $\bar{c}_1 = 1.3$, and $\bar{c}_2 = 1$. The dotted line stands for the asymptotic value $\bar{c}_1/\bar{c}_2$.

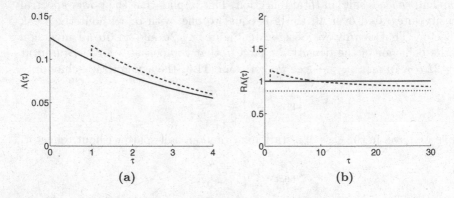

Fig. 11.3. Plot (a): Densities $\tau \mapsto \Lambda_{\mathrm{TRR}}(\omega, \tau)$ (dashed line) and $\tau \mapsto \Lambda_{\mathrm{TRR}}^{(\bar{c}_2)}(\omega, \tau)$ (solid line). Plot (b): Ratio of the densities $\tau \mapsto R\Lambda(\tau) := \Lambda_{\mathrm{TRR}}(\omega, \tau)/\Lambda_{\mathrm{TRR}}^{(\bar{c}_2)}(\omega, \tau)$. Here we assume $\gamma\omega^2 = 2$, $L_1 = 1$, $\bar{c}_1 = 1.7$, and $\bar{c}_2 = 2$.

$$
= \left(\mathcal{W}_1^{(\bar{c}_1)} - \mathcal{W}_1^{(\bar{c}_2)}\right)(\omega, 0) \mathbb{P}\left[N_z^{(\bar{c}_2)} = 1, z \in [-L_1, 0] \mid N_{-L_1}^{(\bar{c}_2)} = 1\right]
$$
$$
= \frac{\gamma\omega^2}{8}\left(\frac{1}{\bar{c}_1} - \frac{1}{\bar{c}_2}\right) e^{-\frac{\gamma\omega^2 L_1}{2\bar{c}_2^2}}. \tag{11.7}
$$

Detection of a weak reflector, assuming that $\bar{c}_2$ is known at the surface, starts by doing time reversal physically, then retrieving the density $\Lambda_{\mathrm{TRR}}(\omega, \tau)$ from the deterministic refocused pulse given in (10.6, 10.7), and next comparing it with $\Lambda_{\mathrm{TRR}}^{(\bar{c}_2)}(\omega, \tau)$. An observed jump at a time τ^* indicates the presence of a reflector at depth $L_1 = \tau^* \bar{c}_2/2$. Moreover, the speed $\bar{c}_1$ can be retrieved from the size of this jump using formula (11.7).

The generalization to multiple weak reflectors is indeed possible. The locations of the jumps in the density $\Lambda_{\mathrm{TRR}}(\omega, \tau)$ correspond to (two-way) travel times to the physical locations of the jumps in speed. However, the formulas for retrieving the speeds are not explicit for a given frequency.

In a typical inverse problem setting, the procedure described in this section is applied for multiple frequencies, and estimates of the locations and magnitudes of the velocity jumps are obtained with a least-squares method.

11.2 Detection of an Interface Between Media

We now consider the case of two media as in Figure 11.1 with the same background parameters $(\bar{K}, \bar{\rho})$, or equivalently $(\bar{\zeta}, \bar{c})$, and with fluctuation processes that have different statistical parameters. The model has the form

$$\frac{1}{K(z)} = \begin{cases} \frac{1}{\bar{K}} & \text{for } z \in (-\infty, -L), \\ \frac{1}{\bar{K}}\left(1 + \nu_1(z/\varepsilon^2)\right) & \text{for } z \in [-L, -L_1), \\ \frac{1}{\bar{K}}\left(1 + \nu_2(z/\varepsilon^2)\right) & \text{for } z \in [-L_1, 0), \\ \frac{1}{\bar{K}} & \text{for } z \in [0, \infty), \end{cases}$$

$$\rho(z) = \bar{\rho} \text{ for } z \in (-\infty, +\infty).$$

To simplify the presentation, we have assumed matched medium boundary conditions at both ends of the random slab, and that only the bulk modulus is randomly fluctuating. The fluctuation processes ν_1 and ν_2 are assumed to be independent, stationary, centered, and ergodic. We denote by γ_j the respective integrated autocorrelations

$$\gamma_j = \int_{-\infty}^{\infty} \mathbb{E}[\nu_j(0)\nu_j(s)]\, ds, \quad j = 1, 2.$$

The interface $z = -L_1$ does not generate a coherent reflection, since the average impedance is constant. Proceeding along the lines of the previous section we find that the reflection coefficient $R_\omega^\varepsilon(-L, 0)$ satisfies in $-L \le z \le 0$ the Riccati equation

$$\frac{dR_\omega^\varepsilon}{dz} = -\frac{i\omega}{2\bar{c}\varepsilon}\nu\left(z, \frac{z}{\varepsilon^2}\right)\left(e^{\frac{-2i\omega z}{\bar{c}\varepsilon}} - 2R_\omega^\varepsilon + (R_\omega^\varepsilon)^2 e^{\frac{2i\omega z}{\bar{c}\varepsilon}}\right),$$

with the initial condition $R_\omega^\varepsilon(-L, -L) = 0$, and with the fluctuation process $\nu(z, \cdot)$ defined by

$$\nu(z, \cdot) = \begin{cases} \nu_1(\cdot) & \text{for } z \in [-L, -L_1), \\ \nu_2(\cdot) & \text{for } z \in [-L_1, 0]. \end{cases} \tag{11.8}$$

At this point, the analysis of time-reversal refocusing follows again the lines of Section 10.1. The refocused pulse given in (10.3) converges to the deterministic pulse given by (10.6). The refocusing kernel K_{TRR} is given by

(10.7), where the density $\Lambda_{\text{TRR}}^L(\omega, \tau) = \mathcal{W}_1(\omega, \tau, 0)$ is obtained by solving the system of transport equations (9.23) with the only difference that the constant coefficient γ is replaced by the piecewise-constant coefficient $\gamma(z)$ defined by

$$\gamma(z) = \begin{cases} \gamma_1 & \text{for } z \in [-L, -L_1), \\ \gamma_2 & \text{for } z \in [-L_1, 0]. \end{cases} \tag{11.9}$$

As in the previous section we let L go to infinity, and we denote the density $\Lambda_{\text{TRR}}^L(\omega, \tau)$ by $\Lambda_{\text{TRR}}(\omega, \tau)$. We solve explicitly the transport equations from $-\infty$ to $-L_1$ in medium 1 with constant $\gamma = \gamma_1$. Denoting the solution by $\mathcal{W}_p^{(\gamma_1)}$, we deduce from (9.40) that

$$\mathcal{W}_p^{(\gamma_1)}(\omega, \tau) = \begin{cases} \delta(\tau) & \text{if } p = 0, \\ \dfrac{8p\gamma_1\omega^2}{\bar{c}} \dfrac{(\gamma_1\omega^2\tau/\bar{c})^{p-1}}{(8 + \gamma_1\omega^2\tau/\bar{c})^{p+1}} \mathbf{1}_{[0,\infty)}(\tau) & \text{otherwise}. \end{cases} \tag{11.10}$$

We solve next the transport equations from $-L_1$ to 0, in medium 2 with constant $\gamma = \gamma_2$, and with the initial condition $\mathcal{W}_p^{(\gamma_1)}(\omega, \tau, -L_1)$ given above. The probabilistic representation of Section 9.2.2 gives

$$\Lambda_{\text{TRR}}(\omega, \tau) = \mathbb{E}\left[\mathcal{W}_{N_0^{(\gamma_2)}}^{(\gamma_1)} \left(\omega, \tau - \frac{2}{\bar{c}} \int_{-L_1}^0 N_s^{(\gamma_2)} ds \right) \mid N_{-L_1}^{(\gamma_2)} = 1 \right],$$

where $N_z^{(\gamma_2)}$ jumps by ± 1 with probability $1/2$ and with intensity $n^2\gamma_2\omega^2/(2\bar{c}^2)$. Here n denotes the state of the process and $n = 0$ is an absorbing state.

Writing $\mathcal{W}^{(\gamma_1)} = \mathcal{W}^{(\gamma_2)} + (\mathcal{W}^{(\gamma_1)} - \mathcal{W}^{(\gamma_2)})$ we obtain

$$\Lambda_{\text{TRR}}(\omega, \tau) = \Lambda_{\text{TRR}}^{(\gamma_2)}(\omega, \tau) \tag{11.11}$$

$$+ \mathbb{E}\left[\left(\mathcal{W}_{N_0^{(\gamma_2)}}^{(\gamma_1)} - \mathcal{W}_{N_0^{(\gamma_2)}}^{(\gamma_2)} \right) \left(\omega, \tau - \frac{2}{\bar{c}} \int_{-L_1}^0 N_s^{(\gamma_2)} ds \right) \mid N_{-L_1}^{(\gamma_2)} = 1 \right],$$

with

$$\Lambda_{\text{TRR}}^{(\gamma_2)}(\omega, \tau) = \frac{8\gamma_2\omega^2/\bar{c}}{(8 + \gamma_2\omega^2\tau/\bar{c})^2} \mathbf{1}_{[0,\infty)}(\tau),$$

and the difference $\left(\mathcal{W}_p^{(\gamma_1)} - \mathcal{W}_p^{(\gamma_2)} \right)$ given explicitly by (11.10) and a similar formula with γ_1 replaced by γ_2.

Using the trajectory analysis of the previous section we find that

$$\Lambda_{\text{TRR}}(\omega, \tau) = \Lambda_{\text{TRR}}^{(\gamma_2)}(\omega, \tau) \quad \text{for} \quad \tau < 2L_1/\bar{c},$$

and that there is a jump at $\tau = 2L_1/\bar{c}$ of size

$$\mathbb{E}\left[\left(\mathcal{W}_1^{(\gamma_1)} - \mathcal{W}_1^{(\gamma_2)} \right)(\omega, 0) \mathbf{1}_{\{N_z^{(\gamma_2)} = 1, z \in [-L_1, 0]\}} \mid N_{-L_1}^{(\gamma_2)} = 1 \right]$$

$$= \left(\mathcal{W}_1^{(\gamma_1)} - \mathcal{W}_1^{(\gamma_2)} \right)(\omega, 0) \mathbb{P}\left[N_z^{(\gamma_2)} = 1, z \in [-L_1, 0] \mid N_{-L_1}^{(\gamma_2)} = 1 \right]$$

$$= \frac{\omega^2}{8\bar{c}} (\gamma_1 - \gamma_2) e^{-\frac{\gamma_2\omega^2 L_1}{2\bar{c}^2}}.$$

Assuming that $\bar{c}$ and γ_2 are known, these two facts can be used to detect, locate, and characterize a sudden change in the fluctuations statistics, including the case of a region without fluctuations, corresponding to $\gamma_1 = 0$.

Comparing the results of this section (jump in γ) with the results of the previous section (jump in $\bar{c}$), with constant impedance in both cases, it is clear that only the ratio $\gamma/\bar{c}$ matters.

A consequence of this remark is that if both the background speed $\bar{c}$ and the statistics γ are changing at an interface $-L_1$, with no jump in the impedance, and if these two coefficients are known in medium 2 before the interface, then the weak reflector is detected as an observed jump in the density $\Lambda_{TRR}(\omega, \tau)$ at τ^*. The depth L_1 is recovered through the formula $L_1 = \tau^* \bar{c}_2/2$. However, only the ratio $\gamma_1/\bar{c}_1$ can be recovered from the expression for the jump size

$$\frac{\omega^2}{8}\left(\frac{\gamma_1}{\bar{c}_1} - \frac{\gamma_2}{\bar{c}_2}\right) e^{-\frac{\gamma_2 \omega^2 L_1}{2\bar{c}_2^2}} = \frac{\omega^2}{8}\left(\frac{\gamma_1}{\bar{c}_1} - \frac{\gamma_2}{\bar{c}_2}\right) e^{-\frac{\gamma_2 \omega^2 \tau^*}{4\bar{c}_2}}.$$

11.3 Waves in One-Dimensional Dissipative Random Media

In this section we generalize the model considered so far for acoustic waves by introducing absorption, which is modeled by a linear dissipative term in the acoustic equations. We show that, up to some technical changes, the statistical properties of transmitted and reflected waves are again described by a system of transport equations in the regime of scale separation. We also show that time reversal is still efficient in recompressing coherent and incoherent waves despite the loss of energy due to absorption.

11.3.1 The Acoustic Model with Random Dissipation

We consider the acoustic wave equations with dissipation

$$\rho \frac{\partial u^\varepsilon}{\partial t} + \frac{\partial p^\varepsilon}{\partial z} + \sigma u^\varepsilon = 0, \qquad (11.12)$$

$$\frac{1}{K} \frac{\partial p^\varepsilon}{\partial t} + \frac{\partial u^\varepsilon}{\partial z} = 0, \qquad (11.13)$$

where p^ε is the pressure, u^ε is the velocity, σ is the dissipation of the medium, ρ is the density, and K is the bulk modulus. The fluctuations of the medium parameters are described by

$$\frac{1}{K} = \begin{cases} \frac{1}{K}\left(1 + \nu(z/\varepsilon^2)\right) & \text{if } z \in [-L, 0] \\ \frac{1}{K} & \text{if } z \in (-\infty, -L) \cup (0, \infty), \end{cases} \qquad (11.14)$$

$$\rho = \bar{\rho} \text{ for all } z \qquad (11.15)$$

as previously, and

$$\sigma = \begin{cases} \sigma(z, z/\varepsilon^2) & \text{if } z \in [-L, 0], \\ 0 & \text{if } z \in (-\infty, -L) \cup (0, \infty), \end{cases} \tag{11.16}$$

where for any $z \in [-L, 0]$, $\zeta \mapsto \sigma(z, \zeta)$ is a nonnegative-valued stationary mixing process with mean $\bar{\sigma}(z) = \mathbb{E}[\sigma(z, \zeta)]$. This allows us to consider cases in which the dissipation background $\bar{\sigma}(z)$ is not uniform. In particular, it includes the following types of media.

1. The dissipation coefficient has a stationary statistical distribution:

$$\sigma\left(z, \frac{z}{\varepsilon^2}\right) = \sigma_0\left(\frac{z}{\varepsilon^2}\right),$$

where σ_0 is a stationary ergodic random process taking nonnegative values with mean $\bar{\sigma}_0 = \mathbb{E}[\sigma_0(\zeta)]$.

2. The dissipation coefficient is different inside some embedded layer $[z_0, z_1]$, $-L < z_0 < z_1 < 0$,

$$\sigma\left(z, \frac{z}{\varepsilon^2}\right) = \begin{cases} \sigma_1\left(\frac{z}{\varepsilon^2}\right) & \text{if } z_0 < z < z_1, \\ \sigma_0\left(\frac{z}{\varepsilon^2}\right) & \text{if } -L \le z \le z_0 \text{ and } z_1 \le z \le 0, \end{cases}$$

where σ_0 and σ_1 are two stationary ergodic random processes taking nonnegative values. We denote by $\bar{\sigma}_j$ their respective means

$$\bar{\sigma}_j = \mathbb{E}[\sigma_j(\zeta)],$$

and we assume that $\bar{\sigma}_1 \ne \bar{\sigma}_0$. The goal of the following section is to detect the layer $[z_0, z_1]$.

We consider scattering by the finite slab $(-L, 0)$, where a left-going pulse is impinging on the random slab.

11.3.2 Propagator Formulation

In this section we first express the scattering problem as a two-point boundary value problem in the frequency domain, and then rewrite it as an initial value problem in terms of the propagator. This analysis follows the lines carried out in the previous chapters. We consider the random acoustic equation (11.12–11.13) and take the scaled-time Fourier transform so that the system reduces to a system of ordinary differential equations:

$$\frac{d\hat{p}^\varepsilon}{dz} - \frac{i\omega\bar{\rho}}{\varepsilon}\hat{u}^\varepsilon + \sigma\left(z, \frac{z}{\varepsilon^2}\right)\hat{u}^\varepsilon = 0, \tag{11.17}$$

$$\frac{d\hat{u}^\varepsilon}{dz} - \frac{i\omega}{\bar{K}\varepsilon}\left(1 + \nu\left(\frac{z}{\varepsilon^2}\right)\right)\hat{p}^\varepsilon = 0. \tag{11.18}$$

As in the case without dissipation, we decompose the wave into right-going modes $\hat{a}^\varepsilon$ and left-going modes $\hat{b}^\varepsilon$:

$$\hat{a}^\varepsilon(\omega, z) = \left(\bar{\zeta}^{1/2}\hat{u}^\varepsilon(\omega, z) + \bar{\zeta}^{-1/2}\hat{p}^\varepsilon(\omega, z)\right) e^{\frac{-i\omega z}{\bar{c}\varepsilon}},$$

$$b^\varepsilon(\omega, z) = \left(\bar{\zeta}^{1/2}\hat{u}^\varepsilon(\omega, z) - \bar{\zeta}^{-1/2}\hat{p}^\varepsilon(\omega, z)\right) e^{\frac{i\omega z}{\bar{c}\varepsilon}}.$$

The modes satisfy the linear system

$$\frac{d}{dz}\begin{bmatrix}\hat{a}^\varepsilon \\ \hat{b}^\varepsilon\end{bmatrix} = \mathbf{H}_\omega\left(\frac{z}{\varepsilon}, \nu^\varepsilon(z), \sigma^\varepsilon(z)\right)\begin{bmatrix}\hat{a}^\varepsilon \\ \hat{b}^\varepsilon\end{bmatrix}, \tag{11.19}$$

where the complex 2×2 matrix $\mathbf{H}_\omega$ is given by

$$\mathbf{H}_\omega(z, \nu, \sigma) = \frac{i\omega}{2\bar{c}}\nu\begin{bmatrix}1 & -e^{-2i\omega z/\bar{c}} \\ e^{2i\omega z/\bar{c}} & -1\end{bmatrix} + \frac{\sigma}{2\zeta}\begin{bmatrix}-1 & -e^{-2i\omega z/\bar{c}} \\ e^{2i\omega z/\bar{c}} & 1\end{bmatrix}, \tag{11.20}$$

using the notation

$$\nu^\varepsilon(z) = \frac{1}{\varepsilon}\nu\left(\frac{z}{\varepsilon^2}\right), \qquad \sigma^\varepsilon(z) = \sigma\left(z, \frac{z}{\varepsilon^2}\right). \tag{11.21}$$

The boundary conditions correspond again to a left-going wave of the form $f(t/\varepsilon)$ incoming from the right:

$$\hat{b}^\varepsilon(\omega, z = 0) = \hat{f}(\omega), \quad \hat{a}^\varepsilon(\omega, z = -L) = 0.$$

We introduce the propagator $\mathbf{P}_\omega^\varepsilon(-L, z)$, which is a complex 2×2 matrix, the solution of

$$\frac{d}{dz}\mathbf{P}_\omega^\varepsilon(\omega, -L, z) = \mathbf{H}_\omega\left(\frac{z}{\varepsilon}, \nu^\varepsilon(z), \sigma^\varepsilon(z)\right)\mathbf{P}_\omega^\varepsilon(-L, z), \quad \mathbf{P}_\omega^\varepsilon(-L, z = -L) = \mathbf{I},$$

such that

$$\mathbf{P}_\omega^\varepsilon(-L, z)\begin{bmatrix}\hat{a}^\varepsilon(\omega, -L) \\ \hat{b}^\varepsilon(\omega, -L)\end{bmatrix} = \begin{bmatrix}\hat{a}^\varepsilon(\omega, z) \\ \hat{b}^\varepsilon(\omega, z)\end{bmatrix}.$$

The main difference with the nondissipative case is that, if (α, β) is a solution of (11.19), then $(\bar{\beta}, \bar{\alpha})$ is not a solution due to the dissipation term in (11.20). The propagator matrix $\mathbf{P}_\omega^\varepsilon$ has the form

$$\mathbf{P}_\omega^\varepsilon(-L, z) = \begin{bmatrix}\alpha_{\omega,1}^\varepsilon(-L, z) & \alpha_{\omega,2}^\varepsilon(-L, z) \\ \beta_{\omega,1}^\varepsilon(-L, z) & \beta_{\omega,2}^\varepsilon(-L, z)\end{bmatrix},$$

where $(\alpha_{\omega,1}^\varepsilon, \beta_{\omega,1}^\varepsilon)^T$ and $(\alpha_{\omega,2}^\varepsilon, \beta_{\omega,2}^\varepsilon)^T$, are solutions of equation (11.19) with, respectively, the initial conditions

$$\alpha_{\omega,1}^\varepsilon(-L, z = -L) = 1, \quad \beta_{\omega,1}^\varepsilon(-L, z = -L) = 0, \tag{11.22}$$

$$\alpha_{\omega,2}^\varepsilon(-L, z = -L) = 0, \quad \beta_{\omega,2}^\varepsilon(-L, z = -L) = 1. \tag{11.23}$$

We define the transmission and reflection coefficients $T_\omega^\varepsilon(-L, z)$ and $R_\omega^\varepsilon(-L, z)$ as before:

$$\mathbf{P}_\omega^\varepsilon(-L, z) \begin{bmatrix} 0 \\ T_\omega^\varepsilon(-L, z) \end{bmatrix} = \begin{bmatrix} R_\omega^\varepsilon(-L, z) \\ 1 \end{bmatrix}. \tag{11.24}$$

In terms of the propagator entries they are given by

$$R_\omega^\varepsilon(-L, z) = \frac{\alpha_{\omega,2}^\varepsilon(-L, z)}{\beta_{\omega,2}^\varepsilon(-L, z)}, \quad T_\omega^\varepsilon(-L, z) = \frac{1}{\beta_{\omega,2}^\varepsilon(-L, z)},$$

and satisfy the closed-form nonlinear differential system

$$\frac{dR_\omega^\varepsilon}{dz} = -\frac{i\omega}{2\bar{c}} \nu^\varepsilon(z) \left(e^{-2i\omega z/(\bar{c}\varepsilon)} - 2R_\omega^\varepsilon + (R_\omega^\varepsilon)^2 e^{2i\omega z/(\bar{c}\varepsilon)} \right)$$
$$- \frac{\sigma^\varepsilon(z)}{2\zeta} \left(e^{-2i\omega z/(\bar{c}\varepsilon)} + 2R_\omega^\varepsilon + (R_\omega^\varepsilon)^2 e^{2i\omega z/(\bar{c}\varepsilon)} \right), \tag{11.25}$$

$$\frac{dT_\omega^\varepsilon}{dz} = \frac{i\omega}{2\bar{c}} \nu^\varepsilon(z) \left(1 - R_\omega^\varepsilon e^{2i\omega z/(\bar{c}\varepsilon)} \right) T_\omega^\varepsilon$$
$$- \frac{\sigma^\varepsilon(z)}{2\zeta} \left(1 + R_\omega^\varepsilon e^{2i\omega z/(\bar{c}\varepsilon)} \right) T_\omega^\varepsilon, \tag{11.26}$$

with the initial conditions

$$R_\omega^\varepsilon(-L, z = -L) = 0, \quad T_\omega^\varepsilon(-L, z = -L) = 1,$$

at $z = -L$. As seen in the previous chapters, the transmitted and reflected waves admit the following integral representations:

$$B^\varepsilon(t, z = -L) = \frac{1}{2\pi} \int e^{i\frac{\omega}{\varepsilon}(\frac{L}{\bar{c}} - t)} \hat{f}(\omega) T_\omega^\varepsilon(-L, 0) \, d\omega, \tag{11.27}$$

$$A^\varepsilon(t, z = 0) = \frac{1}{2\pi} \int e^{-i\frac{\omega t}{\varepsilon}} \hat{f}(\omega) R_\omega^\varepsilon(-L, 0) \, d\omega, \tag{11.28}$$

with the transmission and reflection coefficients satisfying (11.25–11.26).

The solution vectors $(\alpha_{\omega,j}^\varepsilon, \beta_{\omega,j}^\varepsilon)$ satisfy

$$\frac{d}{dz} \left(|\alpha_{\omega,j}^\varepsilon|^2 - |\beta_{\omega,j}^\varepsilon|^2 \right) = -\frac{\sigma^\varepsilon(z)}{\zeta} \left| \alpha_{\omega,j}^\varepsilon e^{i\omega z/(\bar{c}\varepsilon)} + \beta_{\omega,j}^\varepsilon e^{-i\omega z/(\bar{c}\varepsilon)} \right|^2 \leq 0,$$

and thus

$$|\alpha_{\omega,2}^\varepsilon(-L, z)|^2 + 1 \leq |\beta_{\omega,2}^\varepsilon(-L, z)|^2.$$

This implies the energy-dissipation relation

$$|R_\omega^\varepsilon|^2 + |T_\omega^\varepsilon|^2 \leq 1, \tag{11.29}$$

and in turn the uniform boundedness of the transmission and reflection coefficients.

11.3.3 Transmitted Wave Front

Before considering time reversal we give an integral representation for the transmitted stable wave front observed at $z = -L$ around the expected arrival time $L/\bar{c}$. By (11.27), the transmitted wave front observed on the time scale of the initial pulse is given by

$$B^\varepsilon(L/\bar{c} + \varepsilon s, z = -L) = \frac{1}{2\pi} \int e^{-i\omega s} \hat{f}(\omega) T_\omega^\varepsilon(-L, 0) \, d\omega. \qquad (11.30)$$

The results obtained in Chapter 8 can be easily extended. The process $(B^\varepsilon(L/\bar{c} + \varepsilon s, z = -L))_{-\infty < s < \infty}$ converges in the space of the continuous functions to

$$b_L(s) = \frac{1}{2\pi} \int \hat{f}(\omega) \exp\left(i\omega(s - \frac{\sqrt{\gamma}}{2\bar{c}} W_0(L)) - \frac{\gamma\omega^2}{8\bar{c}^2} L - \int_{-L}^0 \frac{\bar{\sigma}(z)}{2\bar{\zeta}} dz \right) d\omega,$$

where $W_0(L)$ is a standard Brownian motion and γ is the integrated covariance of the process ν. Using convolution operators, the transmitted front can be written in a simpler form:

$$b_L(s) = G_{\text{att}} \times (K * f)\left(s - \frac{\sqrt{\gamma}}{2\bar{c}} W_0(L) \right), \qquad (11.31)$$

which means that a random centering appears with the Brownian motion $W_0(L)$, while the pulse shape spreads in a deterministic way through the convolution by the Gaussian kernel K whose Fourier transform is

$$\hat{K}(\omega) = \exp\left(-\frac{\gamma\omega^2 L}{8\bar{c}^2} \right).$$

Dissipation acts, as expected, as an attenuation factor

$$G_{\text{att}} = \exp\left(-\int_{-L}^0 \frac{\bar{\sigma}(z)}{2\bar{\zeta}} dz \right).$$

Only the mean dissipation $\bar{\sigma}(z)$ appears in the attenuation factor, as can be seen by a simple averaging argument. Thus, as far as the stable wave front is concerned, the random effects and the dissipative effects simply add.

11.3.4 The Refocused Pulse for Time Reversal in Reflection

In this section we consider the time-reversal setup described in Section 10.1, that is, the time reversal of the reflected wave. The analysis of the new reflected signal follows the same lines, and we get the same integral representation (10.3):

$$S_L^\varepsilon(t_1 + \varepsilon s) = \frac{1}{(2\pi)^2} \int \int e^{-i\omega s} e^{-i\varepsilon h s/2} \overline{\hat{f}(\omega - \varepsilon h/2)} \hat{G}(h)$$

$$\times R_{\omega+\varepsilon h/2}^\varepsilon(-L,0) \overline{R_{\omega-\varepsilon h/2}^\varepsilon(-L,0)}\, dh\, d\omega\,. \qquad (11.32)$$

Here G is the recording cutoff function. However, the changes in the Riccati equation (11.25) modify the joint distribution of the reflection coefficients at different frequencies.

Frequency Autocorrelation Function of the Reflection Coefficient

The representation (11.32) shows that the statistical distribution of the refocused pulse depends on the frequency autocorrelation function of the reflection coefficient. We proceed as in Chapter 9 and for $p, q \in \mathbb{N}$ introduce

$$U_{p,q}^\varepsilon(\omega, h, z) = \left(R_{\omega+\varepsilon h/2}^\varepsilon(-L, z) \right)^p \left(\overline{R_{\omega-\varepsilon h/2}^\varepsilon(-L, z)} \right)^q.$$

Using the Riccati equation (11.25) satisfied by R_ω^ε, we have that

$$\frac{\partial U_{p,q}^\varepsilon}{\partial z} = \frac{i\omega}{\bar{c}} \nu^\varepsilon (p-q) U_{p,q}^\varepsilon + \frac{i\omega}{2\bar{c}} \nu^\varepsilon e^{\frac{2i\omega z}{\bar{c}\varepsilon}} \left(q e^{-\frac{ihz}{\bar{c}}} U_{p,q-1}^\varepsilon - p e^{\frac{ihz}{\bar{c}}} U_{p+1,q}^\varepsilon \right)$$

$$+ \frac{i\omega}{2\bar{c}} \nu^\varepsilon e^{-\frac{2i\omega z}{\varepsilon}} \left(q e^{\frac{ihz}{\bar{c}}} U_{p,q+1}^\varepsilon - p e^{\frac{-ihz}{\bar{c}}} U_{p-1,q}^\varepsilon \right)$$

$$- (p+q) \frac{\sigma^\varepsilon}{\zeta} U_{p,q}^\varepsilon - e^{\frac{2i\omega z}{\bar{c}\varepsilon}} \frac{\sigma^\varepsilon}{2\bar{\zeta}} \left(p e^{\frac{ihz}{\bar{c}}} U_{p+1,q}^\varepsilon + q e^{-\frac{ihz}{\bar{c}}} U_{p,q-1}^\varepsilon \right)$$

$$- e^{-\frac{2i\omega z}{\bar{c}\varepsilon}} \frac{\sigma^\varepsilon}{2\bar{\zeta}} \left(q e^{\frac{ihz}{\bar{c}}} U_{p,q+1}^\varepsilon + p e^{-\frac{ihz}{\bar{c}}} U_{p-1,q}^\varepsilon \right),$$

starting from $U_{p,q}^\varepsilon(\omega, h, z = -L) = \mathbf{1}_0(p)\mathbf{1}_0(q)$. Taking a shifted scaled Fourier transform with respect to h,

$$V_{p,q}^\varepsilon(\omega, \tau, z) = \frac{1}{2\pi} \int e^{ih(\tau - (q+p)z/\bar{c})} U_{p,q}^\varepsilon(\omega, h, z)\, dh\,, \qquad (11.33)$$

we get

$$\frac{\partial V_{q,p}^\varepsilon}{\partial z} = \frac{i\omega}{\bar{c}} \nu^\varepsilon (p-q) V_{p,q}^\varepsilon - (p+q) \frac{\sigma^\varepsilon}{\zeta} V_{p,q}^\varepsilon$$

$$+ \frac{i\omega}{2\bar{c}} \nu^\varepsilon e^{\frac{2i\omega z}{\bar{c}\varepsilon}} \left(q V_{p,q-1}^\varepsilon - p V_{p+1,q}^\varepsilon \right) + \frac{i\omega}{2\bar{c}} \nu^\varepsilon e^{-\frac{2i\omega z}{\varepsilon}} \left(q V_{p,q+1}^\varepsilon - p V_{p-1,q}^\varepsilon \right)$$

$$- e^{\frac{2i\omega z}{\bar{c}\varepsilon}} \frac{\sigma^\varepsilon}{2\bar{\zeta}} \left(p V_{p+1,q}^\varepsilon + q V_{p,q-1}^\varepsilon \right) - e^{-\frac{2i\omega z}{\bar{c}\varepsilon}} \frac{\sigma^\varepsilon}{2\bar{\zeta}} \left(q V_{p,q+1}^\varepsilon + p V_{p-1,q}^\varepsilon \right),$$

starting from $V_{p,q}^\varepsilon(\omega, \tau, z = -L) = \delta(\tau)\mathbf{1}_0(p)\mathbf{1}_0(q)$. As in Section 9.2.1, we can use the limit theorem of Section 6.7.3 to show that the process $V_{p,q}^\varepsilon$ converges as $\varepsilon \to 0$ to a diffusion process. In particular, the expectations $\mathbb{E}[V_{p,p}^\varepsilon(\omega, \tau, z)]$, $p \in \mathbb{N}$, converge to $\mathcal{W}_p(\omega, \tau, -L, z)$, which obey the closed system of transport equations

$$\frac{\partial \mathcal{W}_p}{\partial z} + \frac{2p}{\bar{c}}\frac{\partial \mathcal{W}_p}{\partial \tau} = (\mathcal{L}_\omega \mathcal{W})_p - \frac{2p\bar{\sigma}(z)}{\bar{\zeta}}\mathcal{W}_p, \tag{11.34}$$

$$(\mathcal{L}_\omega \phi)_p = \frac{p^2}{L_{loc}(\omega)}\left(\phi_{p+1} + \phi_{p-1} - 2\phi_p\right), \tag{11.35}$$

starting from $\mathcal{W}_p(\omega, \tau, -L, z = -L) = \delta(\tau)\mathbf{1}_0(p)$. Here $L_{loc}(\omega) = (4\bar{c}^2)/(\gamma\omega^2)$ is the localization length introduced in (7.81). We then get the limit of the autocorrelation function of the reflection coefficient:

$$\mathbb{E}\left[R^\varepsilon_{\omega+\varepsilon h/2}(-L,0)\overline{R^\varepsilon_{\omega-\varepsilon h/2}(-L,0)}\right] \xrightarrow{\varepsilon\to 0} \int \mathcal{W}_1(\omega,\tau,-L,0)e^{ih\tau}\,d\tau, \tag{11.36}$$

where $\mathcal{W}_1(\omega,\tau,-L,0)$ is obtained by solving the system of transport equations (11.34).

Convergence of the Refocused Pulse

This section is devoted to the analysis of the convergence of the refocused pulse shape $S_L^\varepsilon(t_1 + \varepsilon\cdot)$ to a deterministic shape as $\varepsilon \to 0$.

The proof of the tightness of the process $S_L^\varepsilon(t_1+\varepsilon\cdot)$ is exactly the same as in the nondissipative case, since we use only the fact that $|R_\omega^\varepsilon| \le 1$. The uniform boundedness (10.5) also implies that the finite-dimensional distributions of the process $S_L^\varepsilon(t_1 + \varepsilon\cdot)$ are characterized by the moments

$$\mathbb{E}[S_L^\varepsilon(t_1 + \varepsilon s_1)^{p_1}\cdots S_L^\varepsilon(t_1 + \varepsilon s_k)^{p_k}] \tag{11.37}$$

for all real numbers $\tau_1 < \cdots < s_k$ and all integers $p_1, \ldots, p_k$.

Let us first consider the first moment. Taking the expectation of the representation (11.32) and using (11.36) yields

$$\mathbb{E}[S_L^\varepsilon(t_1 + \varepsilon s)] \xrightarrow{\varepsilon\to 0} \frac{1}{(2\pi)^2}\int\int\int e^{-i\omega s}e^{ih\tau}\overline{\hat{f}(\omega)}\hat{G}(h)\mathcal{W}_1(\omega,\tau,-L,0)\,dh\,d\tau\,d\omega$$

$$= \frac{1}{2\pi}\int\int e^{-i\omega s}\overline{\hat{f}(\omega)}G(\tau)\mathcal{W}_1(\omega,\tau,-L,0)\,d\tau\,d\omega.$$

The computation of the general moment (11.37) follows the same lines as in Section 10.1, and we eventually get the same result: The refocused signal $(S_L^\varepsilon(t_1 + \varepsilon s))_{-\infty<s<\infty}$ converges in probability as $\varepsilon \to 0$ to

$$S_L(s) = \frac{1}{2\pi}\int \Lambda_{\mathrm{TRR}}^L(\omega,\tau)\overline{\hat{f}(\omega)}e^{-i\omega s}G(\tau)\,d\omega\,d\tau,$$

where $\Lambda_{\mathrm{TRR}}^L(\omega,\tau) = \mathcal{W}_1(\omega,\tau,-L,0)$ is the density given by the system (11.34). We can also write this as

$$S_L(s) = (f(-\cdot) * K_{\mathrm{TRR}}(\cdot))(s),$$

where the Fourier transform of the refocusing kernel K_{TRR} is given by

$$\hat{K}_{\text{TRR}}(\omega) = \int G(\tau)\Lambda^L_{\text{TRR}}(\omega,\tau)\,d\tau\,.$$

We can give a probabilistic representation to the density Λ^L_{TRR}, and more generally of the solution of the transport equations (11.34), in terms of the jump Markov process introduced in Section 9.2.2. By means of the Feynman–Kac formula (see Section 6.6.5), we get

$$\int_{\tau_0}^{\tau_1} \mathcal{W}_p(\omega,\tau,-L,0)d\tau = \mathbb{E}\left[\mathbf{1}_{E_L}\exp\left(-\frac{2}{\zeta}\int_{-L}^{0}\bar{\sigma}(-L-s)N_s ds\right)\mid N_{-L}=p\right],$$

$$E_L = \left\{\frac{2}{\bar{c}}\int_{-L}^{0} N_s ds \in [\tau_0,\tau_1]\,,\ N_0=0\right\}. \qquad (11.38)$$

This representation is very useful in deriving qualitative and quantitative properties of the refocusing kernel, as we will see in the next section.

In the particular case of a **constant mean dissipation**, $\bar{\sigma}(z) = \bar{\sigma}_0$, the probabilistic representation of the densities $\mathcal{W}_p$ is

$$\mathcal{W}_p(\omega,\tau,-L,0) = \mathbb{E}\left[\delta\left(\tau-\frac{2}{\bar{c}}\int_{-L}^{0} N_s ds\right)\mathbf{1}_0(N_0)\right.$$
$$\left.\times\exp\left(-\frac{2\bar{\sigma}_0}{\zeta}\int_{-L}^{0} N_s ds\right)\mid N_{-L}=p\right].$$

We note that the argument of the exponential is deterministic because the value of the integral $\int_{-L}^{0} N_s ds$ is constrained to be equal to $\bar{c}\tau/2$. As a result,

$$\mathcal{W}_p(\omega,\tau,-L,0) = \mathcal{W}_p^{(0)}(\omega,\tau,-L,0)e^{-\bar{c}\bar{\sigma}_0\tau/\zeta}, \qquad (11.39)$$

where $\mathcal{W}_p^{(0)}$ is the solution of the transport equations (9.23) in the absence of dissipation. We can also verify directly that (11.39) is indeed a solution of the transport equations (11.34) with $\bar{\sigma}(z) = \bar{\sigma}_0$.

11.4 Application to the Detection of a Dissipative Layer

In this section we show that time reversal can be used as an efficient and statistically stable method to image a dissipative layer embedded in a randomly scattering medium. We consider the same configuration as the one analyzed in the previous section. We compute explicitly the refocusing kernel and show that it contains information about the presence of an embedded dissipative layer. Finally, we show how this information can be extracted. In practical situations the refocusing kernels are estimated from measured reflected signals. We comment on the implementation of such a procedure in Section 10.2.2.

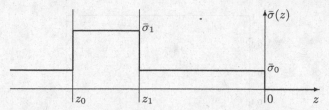

Fig. 11.4. Profile of the mean dissipation.

11.4.1 Constant Mean Dissipation

We consider the case of a **half-space** and we assume that there is no embedded layer. The mean dissipation is constant $\bar{\sigma}(z) \equiv \bar{\sigma}_0$. We compute explicitly the solution for the system of transport equations using (11.39) and (9.40):

$$\mathcal{W}_p(\omega, \tau) = \frac{\partial}{\partial \tau} \left[\left(\frac{\gamma \omega^2 \tau / \bar{c}}{8 + \gamma \omega^2 \tau / \bar{c}} \right)^p \mathbf{1}_{[0,\infty)}(\tau) \right] e^{-\bar{c}\bar{\sigma}_0 \tau / \bar{\zeta}}. \tag{11.40}$$

We thus get a closed-form expression for the density $\Lambda_{\mathrm{TRR}}(\omega, \tau) = \mathcal{W}_1(\omega, \tau)$ in the case of a constant mean dissipation, which we denote by

$$\Lambda_0(\omega, \tau) = \frac{8\gamma \omega^2 / \bar{c}}{\left(8 + \gamma \omega^2 \tau / \bar{c}\right)^2} e^{-\bar{c}\bar{\sigma}_0 \tau / \bar{\zeta}} \mathbf{1}_{[0,\infty)}(\tau). \tag{11.41}$$

Assuming that we record all the reflected signal at the mirror, so that $G \equiv 1$, then we have an explicit formula for the refocusing kernel

$$\hat{K}_{\mathrm{TRR}}(\omega) = 1 - \frac{8\bar{c}^2 \bar{\sigma}_0}{\bar{\zeta} \gamma \omega^2} \exp\left(\frac{8\bar{c}^2 \bar{\sigma}_0}{\bar{\zeta} \gamma \omega^2} \right) E_i \left(\frac{8\bar{c}^2 \bar{\sigma}_0}{\bar{\zeta} \gamma \omega^2} \right),$$

where E_i is the exponential integral function

$$E_i(x) = \int_1^\infty \frac{\exp(-xt)}{t} dt.$$

This refocusing kernel is plotted in Figure 11.5. The fact that it behaves like a high-pass filter can be explained as follows: the pulse is reflected back by the random half-space because of wave localization, it partly dissipates, and the time-reversal mirror sends back everything that is recorded. High frequencies (above the cutoff frequency $\omega_c = 4\bar{c}\sqrt{\bar{\sigma}_0/(\bar{\zeta}\gamma)}$) are well recovered, because they do not penetrate far and they spend a short time in the random dissipative half-space. In contrast, low frequencies are highly dissipated, because they penetrate deeper and spend a longer time in the medium before being scattered back.

11.4.2 Thin Dissipative Layer

We consider here a configuration in which a thin layer with mean dissipation $\bar{\sigma}_1$ located in $[z_0, z_1]$ is embedded into a half plane with mean dissipation

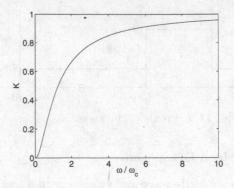

Fig. 11.5. Refocusing kernel in presence of dissipation and in case of complete recording $G_{t_1} \equiv 1$. The cutoff frequency is $\omega_c = 4\bar{c}\sqrt{\bar{\sigma}_0/(\bar{\zeta}\gamma)}$.

$\bar{\sigma}_0$ (see Figure 11.4). The layer is **thin** in the sense that $z_0 - z_1 \to 0$, but its dissipation coefficient is high, so that $\bar{\sigma}_1(z_0 - z_1) \to \lambda$. We discuss the domain of validity of this limit in the next section. First, we solve the system of transport equations (11.34) from $-\infty$ to z_0, so that we get the stationary solution (11.40):

$$\mathcal{W}_p^{(z_0)}(\omega, \tau) = \frac{\partial}{\partial \tau} \left[\left(\frac{\gamma \omega^2 \tau/\bar{c}}{8 + \gamma \omega^2 \tau/\bar{c}} \right)^p \mathbf{1}_{[0,\infty)}(\tau) \right] e^{-\bar{c}\bar{\sigma}_0 \tau/\bar{\zeta}} .$$

Second, we solve the system across the layer from z_0 to z_1. The layer is very thin, so that the Markov process does not jump. As a result, we get

$$\mathcal{W}_p^{(z_1)}(\omega, \tau) = \mathcal{W}_p^{(z_0)}(\omega, \tau) e^{-2\lambda p/\bar{\zeta}}$$
$$= \frac{\partial}{\partial \tau} \left[\left(\frac{\gamma \omega^2 \tau/\bar{c}}{8 + \gamma \omega^2 \tau/\bar{c}} \right)^p \mathbf{1}_{[0,\infty)}(\tau) \right] e^{-\bar{c}\bar{\sigma}_0 \tau/\bar{\zeta} - 2\lambda p/\bar{\zeta}} .$$

Third, we solve the system from z_1 to 0. Using the probabilistic representation of the solution of the system transport equations with an arbitrary initial condition (see Section 9.2.2), we find that the density $\Lambda_{\mathrm{TRR}}(\omega, \tau)$ is given by

$$\Lambda(\omega, \tau) = \mathbb{E}\left[\mathcal{W}_{N_0}^{(z_1)}\left(\omega, \tau - \frac{2}{\bar{c}} \int_{z_1}^0 N_s ds \right) \mid N_{z_1} = 1 \right] .$$

This density would be equal to the stationary solution (11.41) if the multiplicative factor $\exp(-2\lambda N_0/\bar{\zeta})$ were absent. We then expand this factor as $1 - \left(1 - \exp(-2\lambda N_0/\bar{\zeta}) \right)$, so that we obtain

$$\Lambda(\omega, \tau) = \Lambda_0(\omega, \tau) - \mathbb{E}\left[\tilde{\mathcal{W}}_{N_0}^{(0)}\left(\omega, \tau - \frac{2}{\bar{c}} \int_{z_1}^0 N_s ds \right) \right.$$
$$\left. \times \left(1 - e^{-2\lambda N_0/\bar{\zeta}} \right) \mid N_{z_1} = 1 \right] e^{-\bar{c}\bar{\sigma}_0 \tau/\bar{\zeta}} , \quad (11.42)$$

where

$$
\tilde{W}_p^{(0)}(\omega,\tau) = \begin{cases} \delta(\tau) & \text{if } p = 0, \\ \dfrac{8p\gamma\omega^2}{\bar{c}}\dfrac{(\gamma\omega^2\tau/\bar{c})^{p-1}}{(8+\gamma\omega^2\tau/\bar{c})^{p+1}}\mathbf{1}_{[0,\infty)}(\tau) & \text{otherwise}. \end{cases} \tag{11.43}
$$

Next we analyze the expectation in the right-hand side of (11.42). If $N_0 = 0$, then the second term inside the expectation is zero. If $N_0 \geq 1$, then $N_z \geq 1$ for all $z \in [z_1, 0]$ because 0 is an absorbing state. This means that only the paths of the process that never reach zero can contribute to the value of the expectation. These paths satisfy $\int_{z_1}^0 N_s ds \geq |z_1|$. We also know that $\tilde{W}_p^{(0)}(\omega,\tau)$ is zero for $\tau < 0$. This shows that the first term inside the expectation is zero for any $\tau < 2|z_1|/\bar{c}$. Accordingly, the density Λ is indistinguishable from the density Λ_0 corresponding to a constant mean dissipation for any $\tau \leq 2|z_1|/\bar{c}$. When τ crosses this critical value corresponding to a round trip from the surface to the dissipative layer, a density jump occurs. Indeed, a set of paths suddenly contributes to the expectation in the right-hand side of (11.42). This is the set of paths where no jump occurs (i.e., $N_z = 1$ for all $z \in [z_1, 0]$). The density then jumps from

$$
\Lambda\left(\omega,\tau = (2|z_1|/\bar{c})^-\right) = \Lambda_0\left(\omega, 2|z_1|/\bar{c}\right)
$$

to

$$
\Lambda\left(\omega,\tau = (2|z_1|/\bar{c})^+\right) = \Lambda_0(\omega, 2|z_1|/\bar{c}) - \tilde{W}_1^{(0)}(\omega, 0)\left(1 - e^{-2\lambda/\bar{\zeta}}\right)
$$
$$
\times \mathbb{P}\left(\text{no jump before } 0 \mid N_{z_1} = 1\right) e^{-\bar{c}\bar{\sigma}_0\tau/\bar{\zeta}}
$$
$$
= \Lambda_0(\omega, 2|z_1|/\bar{c})\left[1 - \Delta\Lambda\right],
$$

where the relative amplitude of the jump is

$$
\Delta\Lambda = \left(1 - e^{-2\lambda/\bar{\zeta}}\right)\left(1 + \frac{\gamma\omega^2}{4\bar{c}^2}|z_1|\right)^2 \exp\left(-\frac{\gamma\omega^2|z_1|}{2\bar{c}^2}\right). \tag{11.44}
$$

Summary. *In order to detect the depth and the dissipation coefficient of the layer from a measured power spectral density Λ, we plot the ratio of the measured density Λ over the density Λ_0 with constant mean dissipation $\bar{\sigma}_0$. This ratio is one up to $\tau = 2|z_1|/\bar{c}$ and has a jump at $\tau = 2|z_1|/\bar{c}$, which enables us to recover the depth z_1. The amplitude of the jump is given by (11.44), which allows us to recover the dissipation strength λ.*

We have carried out Monte Carlo simulations for the jump Markov process N to compute Λ from the expression (11.42). The results for a particular set of parameters are plotted in Figure 11.6, where the jump in the density can be seen clearly.

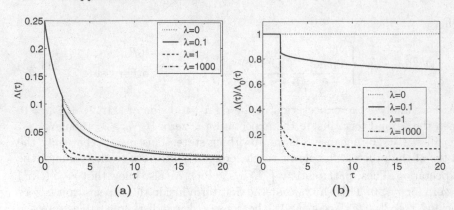

Fig. 11.6. Plot (a): Density $\tau \mapsto \Lambda(\omega, \tau)$. Plot (b): Ratio of the densities $\tau \mapsto \Lambda(\omega, \tau)/\Lambda_0(\omega, \tau)$. Here we assume $\gamma\omega^2/\bar{c}^2 = 2$, $z_0 = 1$, $\bar{c} = 1$, and $\bar{\zeta} = 1$.

11.4.3 Thick Dissipative Layer

We revisit the previous configuration without assuming that the layer is thin. Accordingly, we consider a configuration in which a layer with mean dissipation $\bar{\sigma}_1$ lying in $[z_0, z_1]$ is embedded into a half-space with mean dissipation $\bar{\sigma}_0$. We proceed as above and find that

$$
\Lambda(\omega, \tau) = \mathbb{E}\left\{ \mathcal{W}_{N_0}^{(z_0)}\left(\omega, \tau - \frac{2}{\bar{c}}\int_{z_0}^{0} N_s ds\right) \right.
$$

$$
\left. \times \exp\left(2\frac{\bar{\sigma}_0 - \bar{\sigma}_1}{\bar{\zeta}}\int_{z_0 - z_1}^{0} N_s ds\right) \mid N_{z_0} = 1 \right\} e^{-\frac{\bar{c}\bar{\sigma}_0\tau}{\bar{\zeta}}}
$$

$$
= \Lambda_0(\omega, \tau) - \mathbb{E}\left\{ \tilde{\mathcal{W}}_{N_0}^{(0)}\left(\omega, \tau - \frac{2}{\bar{c}}\int_{z_0}^{0} N_s ds\right) \right.
$$

$$
\left. \times \left[1 - \exp\left(2\frac{\bar{\sigma}_0 - \bar{\sigma}_1}{\bar{\zeta}}\int_{z_0 - z_1}^{0} N_s ds\right)\right] \mid N_{z_0} = 1 \right\} e^{-\frac{\bar{c}\bar{\sigma}_0\tau}{\bar{\zeta}}}. \quad (11.45)
$$

If $N_{z_0 - z_1} = 0$, then $N_z = 0$ for all $z \in [z_0 - z_1, 0]$ and the second term inside the expectation is zero. If $N_{z_0 - z_1} \geq 1$, then $N_z \geq 1$ for all $z \in [z_0, z_0 - z_1]$, so that $2\int_{z_0}^{0} N_s ds \geq 2|z_1|$. The fact that $\tilde{\mathcal{W}}_p^{(0)}(\omega, \tau)$ is zero for $\tau < 0$ then shows that the first term inside the expectation is zero if $\tau \leq 2|z_1|/\bar{c}$. Accordingly, $\Lambda(\omega, \tau) = \Lambda_0(\omega, \tau)$ for $\tau \leq 2|z_1|/\bar{c}$.

The derivative of the density jumps at $\tau = 2|z_1|/\bar{c}$. Indeed, for τ just above $2|z_1|/\bar{c}$ a path contributes to the value of the expectation that appears in the right-hand side of (11.45), namely the path where $N_z = 1$ for all $z \in [z_0, z_0 - z_1]$, and a jump from state 1 to state 0 occurs at $z_0 + \bar{c}\tau/2$. Accordingly, the derivative $\Lambda' = \partial\Lambda/\partial\tau$ goes from

$$
\Lambda'\left(\omega, \tau = (2|z_1|/\bar{c})^-\right) = \Lambda_0'\left(\omega, 2|z_1|/\bar{c}\right)
$$

$$= -\left(\frac{2\gamma\omega^2}{\bar{c}}\right)\frac{\gamma\omega^2/\bar{c} + \bar{c}\bar{\sigma}_0(4 + \gamma\omega^2|z_1|/\bar{c}^2)/\zeta}{(4 + \gamma\omega^2|z_1|/\bar{c}^2)^3}e^{-2\bar{\sigma}_0|z_1|/\bar{\zeta}}$$

to

$$\Lambda'(\omega, \tau = (2|z_1|/\bar{c})^+) = \Lambda'_0(\omega, 2|z_1|/\bar{c}) - \frac{\bar{\sigma}_1 - \bar{\sigma}_0}{8\bar{\zeta}}\gamma\omega_-^2 e^{-\gamma\omega^2|z_1|/(2\bar{c}^2)}e^{-2\bar{\sigma}_0|z_1|}$$

$$= \Lambda'_0(\omega, 2|z_1|/\bar{c})\left[1 + \Delta\Lambda'\right],$$

where the relative amplitude of the jump is given by

$$\Delta\Lambda' = \frac{\bar{\sigma}_1 - \bar{\sigma}_0}{16\bar{\zeta}}\frac{(4 + \gamma\omega^2|z_1|/\bar{c}^2)^3}{\gamma\omega^2/\bar{c}^2 + \bar{\sigma}_0(4 + \gamma\omega^2|z_1|/\bar{c}^2)/\bar{\zeta}}\exp\left(-\frac{\gamma\omega^2|z_1|}{2\bar{c}^2}\right). \quad (11.46)$$

There is a second jump in the derivative of Λ at $\tau = 2|z_0|/\bar{c}$. Indeed, the mechanism described just above fails precisely when τ becomes larger than $\tau = 2|z_0|/\bar{c}$, because the jump of the Markov process at $z_0 + \bar{c}\tau/2 > 0$ has no influence.

Summary. *In order to detect the depth, the thickness, and the dissipation coefficient of the layer from a measured power spectral density Λ, we plot the ratio of the measured density Λ over Λ_0. The ratio is one up to $\tau = 2|z_1|/\bar{c}$. The position of the first jump in the derivative of the density is $2|z_1|/\bar{c}$. The position of the second jump is $2|z_0|/\bar{c}$. The amplitude of the first jump is given by (11.46), which allows us to recover $\bar{\sigma}_1$ assuming $\bar{\sigma}_0$ is known.*

We have carried out Monte Carlo simulations for the jump Markov process N to compute Λ from the expression (11.45). The results for a particular set of parameters are plotted in Figure 11.7, where the first jump of the derivative density can be seen clearly. It may be more difficult to detect the second jump if the layer is thick. The case in which the thickness of the layer is small, $z_1 - z_0 = 0.1$, is very similar to the approximation of a thin layer with $\lambda = 0.1$ presented in Figure 11.6.

The thin-layer approximation developed in Section 11.4.2 can now be discussed more quantitatively. The interpretation in terms of the jump Markov process is helpful for this discussion. Considering expression (11.45), it can be seen that the approximation is valid if the event "the process jumps between $z_0 - z_1$ and 0" is negligible. The brackets [.] in the right-hand side of (11.45) then simplify to $[1 - \exp(2(\bar{\sigma}_0 - \bar{\sigma}_1)(z_1 - z_0)N_0/\bar{\zeta})]$, and we recover precisely (11.42). This event is negligible if $\gamma\omega^2|z_1 - z_0|/\bar{c}^2 \ll 1$, and this condition is the criterion for the validity of the thin-layer approximation.

Concluding remarks In Section 10.2.2 we discused two different approaches for estimating the power spectral density Λ. They give robust and stable estimates of the density by measurents from a single realization of the medium. The results obtained in this section, tell us that the detection of an anomalous layer can be done by looking for a discontinuity in Λ or in its τ-derivative.

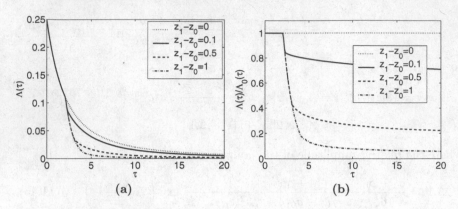

Fig. 11.7. Plot (a): Density $\tau \mapsto \Lambda(\omega, \tau)$. Plot (b): Ratio of the densities $\tau \mapsto \Lambda(\omega, \tau)/\Lambda_0(\omega, \tau)$. Here we assume $\gamma\omega^2/\bar{c}^2 = 2$, $\bar{c} = 1$, $\bar{\zeta} = 1$, $z_1 = 1$, $\bar{\sigma}_0 = 0$, $\bar{\sigma}_1 = 1$, and the thickness of the layer $z_1 - z_0$ goes from zero (absence of dissipative layer) to one.

Notes

The results of this chapter on dissipation and detection of a dissipative layer were derived in 2004 by Fouque–Garnier–Nachbin–Sølna in [63]. The results on detection of weak reflectors were derived in [66]. The general inverse problem, which consists in recovering the slowly varying background parameters from the reflections, is treated in [8] and [133].

12

Time Reversal in Transmission

In this chapter we consider time reversal in transmission where a source emits a pulse at one end of a one-dimensional slab, and a time reversal mirror located at the other end of the slab records the transmitted signal. The mirror then reemits a segment of the recorded signal in the reverse direction of time so that what is recorded last is sent first (last-in-first-out at the mirror). This is in contrast with a standard mirror which corresponds to first-in-first-out. This time-reversal setup is illustrated in Figure 12.1. In Section 12.1 we consider the case in which the mirror records only the stable front of the transmitted signal, which has been studied in detail in Chapter 8. The main effects of randomness are the spreading of the pulse and a random shift in the arrival time. We will see that time reversal does not compensate for this spreading and that the signal spreads even further while propagating back to the original source point. In contrast, we will see in Section 12.2 that, when the fluctuations following the stable wave front are recorded, then the quality of the refocusing is greatly improved. This can be explained in terms of the frequency content of these fluctuations and the localization theory presented in Chapter 7. The time-reversal results derived here will be presented in the framework of separation of scales introduced in the previous chapters, namely in the strongly heterogeneous white-noise regime. In particular, the analysis relies on the precise asymptotics of the moments of the transmission coefficients derived in Chapter 7 and in Chapter 9. Applications to communications using the results presented in this chapter will be addressed in the next chapter. Time reversal in reflection in the case of a one-dimensional random medium was studied in Chapter 10. Other time-reversal situations, including three-dimensional randomly layered media, will be presented in the subsequent chapters.

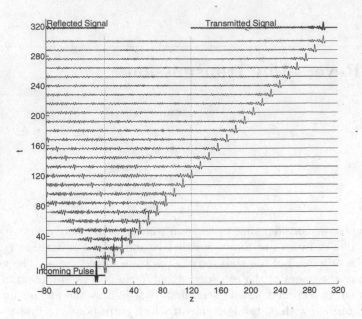

Fig. 12.1. Propagation of a pulse through a slab of random medium $(0, L)$. This figure is similar to Figure 10.1, but we now record a long piece of the transmitted wave.

12.1 Time Reversal of the Stable Front

In Chapter 8 we described the shape of the wave front generated by a pulse propagating through a one-dimensional random slab in the strongly heterogeneous white-noise regime. We found that when this wave front is observed at its random arrival time it can be expressed as a convolution of the transmitted front in the homogeneous case with a deterministic kernel that depends on the statistics of the medium fluctuations. In this section we discuss the time reversal of the transmitted stable front as illustrated in Figure 12.2. We will see that the effect of randomness is simply a spreading of the back-propagated pulse corresponding to twice the travel distance. The effect of time reversal is to cancel the random component in the arrival time of the transmitted pulse. In other words, time reversal does not play a significant role when only the wave front is used. In Section 12.2 we will consider the more interesting case in which a part of the incoherent coda wave is also time reversed.

12.1.1 Time-Reversal Experiment

We consider a random slab $(-L, 0)$ embedded in a homogeneous medium with matched medium boundary conditions. A short pulse of the form $f(t/\varepsilon)$ incoming from the right homogeneous half-space is scattered by the random slab. As seen in Section 9.4, the transmitted wave admits the integral representation

$$B^\varepsilon(t, -L) = \frac{1}{2\pi} \int e^{-i\omega \frac{t - L/\bar{c}}{\varepsilon}} T_\omega^\varepsilon(-L, 0) \hat{f}(\omega) \, d\omega, \qquad (12.1)$$

where $T_\omega^\varepsilon(-L, 0)$ is the transmission coefficient defined in (9.9). The first step in time reversal of the wave front consists in recording the transmitted signal at the end of the slab $z = -L$ during a small time interval of duration of order ε centered at time t_1. More precisely, a segment of order ε around t_1 of the transmitted signal is "clipped" using a cutoff function G, with t_1 to be chosen later. We denote the recorded part of the wave by y^ε, so that

$$y^\varepsilon(t) = B^\varepsilon(t_1 + t, -L) G\left(\frac{t}{\varepsilon}\right).$$

We then time reverse this segment of signal and send it back into the same medium. This means that we need to consider a new problem defined by the same acoustic equations, but with the new incoming signal given by

$$f_{\text{new}}^\varepsilon(t) = y^\varepsilon(-t) = B^\varepsilon(t_1 - t, -L) G\left(-\frac{t}{\varepsilon}\right).$$

It is a right-going wave incoming from the left homogeneous half-space. Using the fact that we are dealing with real-valued signals, we can write

$$B^\varepsilon(t_1 - t, -L) = \overline{B^\varepsilon(t_1 - t, -L)} = \frac{1}{2\pi} \int \overline{T_\omega^\varepsilon(-L, 0)} \overline{\hat{f}(\omega)} e^{i\omega \frac{t_1 - t - L/\bar{c}}{\varepsilon}} \, d\omega,$$

so that the scaled Fourier transform of the new incoming signal is of the form

$$\hat{f}_{\text{new}}^\varepsilon(\omega) = \int e^{i \frac{\omega t}{\varepsilon}} f_{\text{new}}^\varepsilon(t) \, dt = \varepsilon \int e^{i\omega s} B^\varepsilon(t_1 - \varepsilon s, -L) G(-s) \, ds$$

$$= \varepsilon \int e^{i\omega s} \left\{ \frac{1}{2\pi} \int e^{-i\omega' s} \overline{T_{\omega'}^\varepsilon(-L, 0)} \overline{\hat{f}(\omega')} e^{i\omega' \frac{t_1 - L/\bar{c}}{\varepsilon}} \, d\omega' \right\} G(-s) \, ds$$

$$= \frac{\varepsilon}{2\pi} \int \overline{T_{\omega'}^\varepsilon(-L, 0)} \overline{\hat{f}(\omega')} \left\{ \int e^{i(\omega' - \omega)(-s)} G(-s) \, ds \right\} e^{i\omega' \frac{t_1 - L/\bar{c}}{\varepsilon}} \, d\omega'$$

$$= \frac{\varepsilon}{2\pi} \int \overline{T_{\omega'}^\varepsilon(-L, 0)} \overline{\hat{f}(\omega')} \hat{G}(\omega - \omega') e^{i\omega' \frac{t_1 - L/\bar{c}}{\varepsilon}} \, d\omega'.$$

The new incoming signal is scattered by the random slab and gives rise to a reflected wave $B_{\text{new}}^\varepsilon(t, -L)$ at $z = -L$ and a transmitted wave $A_{\text{new}}^\varepsilon(t, 0)$ at $z = 0$ that is the original source point. The transmitted signal observed in the time domain around the observation time t_{obs} reads

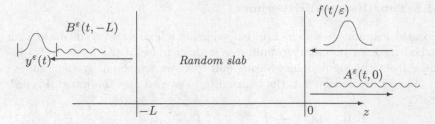

(a) A left-going pulse $f(t/\varepsilon)$ is impinging the random slab $(-L, 0)$, and it generates a transmitted signal $B^\varepsilon(t, -L)$. The TRM, used in a passive mode, records the wave front $y^\varepsilon(t)$.

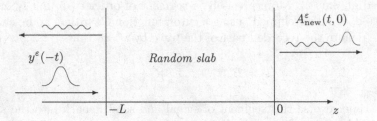

(b) The TRM is used as an active device and it sends back into the medium the signal $y^\varepsilon(-t)$. We observe the new transmitted signal $A_{\text{new}}^\varepsilon(t, 0)$.

Fig. 12.2. Experimental setup for time reversal of the stable wave front.

$$S_L^\varepsilon(t_{\text{obs}} + \varepsilon s) = A_{\text{new}}^\varepsilon(t_{\text{obs}} + \varepsilon s, 0)$$
$$= \frac{1}{2\pi\varepsilon} \int e^{-i\omega(s + \frac{t_{\text{obs}} - L/\bar{c}}{\varepsilon})} \hat{f}_{\text{new}}^\varepsilon(\omega) \widetilde{T}_\omega^\varepsilon(-L, 0) \, d\omega, \quad (12.2)$$

where $\widetilde{T}_\omega^\varepsilon$ (resp. $\widetilde{R}_\omega^\varepsilon$) is defined as the **adjoint** transmission (resp. reflection) coefficient for the experiment corresponding to a right-going input wave incoming from the left (see Figure 12.4 for the definition of the adjoint reflection and transmission coefficients, and compare with Figure 12.3 for the definition of the standard reflection and transmission coefficients). Using the propagator matrix (9.7) we find that the adjoint reflection and transmission coefficient satisfy

$$\mathbf{P}_\omega^\varepsilon(-L, 0) \begin{bmatrix} 1 \\ \widetilde{R}_\omega^\varepsilon(-L, 0) \end{bmatrix} = \begin{bmatrix} \widetilde{T}_\omega^\varepsilon(-L, 0) \\ 0 \end{bmatrix}.$$

The adjoint coefficients are given in terms of the entries $\alpha_\omega^\varepsilon$ and β_ω^ε of the propagator matrix (9.5) by

$$\widetilde{R}_\omega^\varepsilon(-L, 0) = -\frac{\beta_\omega^\varepsilon(-L, 0)}{\alpha_\omega^\varepsilon(-L, 0)}, \qquad \widetilde{T}_\omega^\varepsilon(-L, 0) = \frac{1}{\alpha_\omega^\varepsilon(-L, 0)}. \quad (12.3)$$

By comparing with the corresponding expressions (9.9) for the standard reflection and transmission coefficients, one gets that the two transmission coefficients are equal:

$$\tilde{T}_\omega^\varepsilon(-L,0) = \frac{1}{\alpha_\omega^\varepsilon(-L,0)} = T_\omega^\varepsilon(-L,0)\,.$$

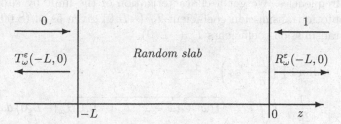

Fig. 12.3. Reflection and transmission coefficients.

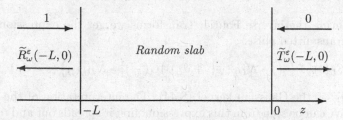

Fig. 12.4. Adjoint reflection and transmission coefficients.

Substituting the expression of $\hat{f}_{\text{new}}^\varepsilon$ into the representation (12.2) of $S_L^\varepsilon(t_{\text{obs}} + \varepsilon s)$ yields the integral representation of the transmitted signal

$$S_L^\varepsilon(t_{\text{obs}} + \varepsilon s) = \frac{1}{(2\pi)^2} \int\!\!\int e^{-i\omega s} e^{-i\frac{\omega}{\varepsilon}(t_{\text{obs}} - \frac{L}{\bar{c}})} e^{i\frac{\omega'}{\varepsilon}(t_1 - \frac{L}{\bar{c}})} \overline{\hat{f}(\omega')} \hat{G}(\omega - \omega')$$

$$\times \overline{T_{\omega'}^\varepsilon(-L,0)} T_\omega^\varepsilon(-L,0)\,d\omega'\,d\omega\,. \qquad (12.4)$$

12.1.2 The Refocused Pulse

The integral representation (12.4) holds for any value of ε, and we now study the behavior of the transmitted signal in the limit $\varepsilon \to 0$. By choosing

$$t_1 = t_{\text{obs}} = L/\bar{c}\,,$$

the rapid phases in (12.4) cancel. This particular choice corresponds to (1) a recording-time window centered at the expected arrival time of the pulse front during the first part of the experiment, (2) an observation-time window centered at the expected arrival time of the new transmitted signal after reemission. We thus expect to observe a stable front. From the mathematical point of view (12.4) has the same structure as (8.35), which means that we deal with a problem similar to the one addressed in Chapter 8. The statistical distribution of the transmitted wave is characterized by its moments, which involve the moments of the transmission coefficients and their conjugates at different frequencies. We get a characterization of the limit by substituting the asymptotic transmission coefficient $\tilde{T}_{\omega'}(-L,0)$ given as in (8.60) for the random transmission coefficients $T_{\omega'}^{\varepsilon}(-L,0)$:

$$
\begin{aligned}
S_L(s) &= \lim_{\varepsilon \to 0} S_L^{\varepsilon}\left(\frac{L}{\bar{c}} + \varepsilon s\right) \\
&= \frac{1}{(2\pi)^2} \int\int e^{-i\omega s}\,\overline{\hat{f}(\omega')\hat{G}(\omega - \omega')\tilde{T}_{\omega'}(-L,0)}\,\tilde{T}_{\omega}(-L,0)\,d\omega'\,d\omega \\
&= \frac{1}{(2\pi)^2} \int\int e^{-i\omega s}\,\overline{\hat{f}(\omega')\hat{G}(\omega - \omega')}e^{-i\omega'\frac{\sqrt{\gamma}}{2\bar{c}}W(L)}e^{-\omega'^2\frac{\gamma L}{4\bar{c}^2}} \\
&\qquad\qquad \times e^{i\omega\frac{\sqrt{\gamma}}{2\bar{c}}W(L)}e^{-\omega^2\frac{\gamma L}{8\bar{c}^2}}\,d\omega'\,d\omega\,.
\end{aligned}
$$

By computing the inverse Fourier transforms we get the expression for the limiting transmitted pulse:

$$
S_L(s) = \left\{\left[\left(f * \mathcal{N}_{D_L}(-(\cdot + \theta_L))\right)G(-\cdot)\right] * \mathcal{N}_{D_L}(\cdot)\right\}(s - \theta_L)\,,
$$

where $\mathcal{N}_{D_L}$ is the Gaussian kernel (8.44). The interpretation of the-time reversed wave can be seen from this expression: first it spreads out and randomly shifts; then it is multiplied by G; then it is time-reversed; and finally it once again spreads out and randomly shifts. Note that the second shift tends to compensate for the first one, but this compensation can be blurred out by the cutoff function G. To see this we rewrite the expression of the limiting transmitted pulse as follows:

$$
S_L(s) = \left\{\left[(f * \mathcal{N}_{D_L}(\cdot))G(\cdot + \theta_L)\right] * \mathcal{N}_{D_L}(\cdot)\right\}(-s)\,. \tag{12.5}
$$

If we take $G = 1$, then this expression can be reduced to

$$
S_L(s) = f * \mathcal{N}_{D_{2L}}(-s)\,, \tag{12.6}
$$

which shows that the time-reversal operation has completely compensated for the random time shift, but not for the deterministic spreading. We shall see in the next sections that time reversal can also compensate for the spreading by recording a significant part of the incoherent coda wave.

If G is such that $(f * \mathcal{N}_{D_L}(\cdot))G(\cdot + \theta_L) = f * \mathcal{N}_{D_L}(\cdot)$, which means that we have recorded the whole pulse front, then the result is equivalent to having $G = 1$. However, θ_L is a Gaussian random variable, which implies in particular that its support is not bounded. Thus, we cannot guarantee that a given compactly supported time-window function G will record the stable part of the transmitted signal, since this signal can be delayed by an uncontrolled random time.

12.2 Time Reversal with Coda Waves

12.2.1 Time-Reversal Experiment

The time-reversal procedure consists in recording the transmitted signal at $z = -L$ over a time interval $[L/\bar{c} + t_0, L/\bar{c} + t_1]$ as shown in Figure 12.5. A piece of the recorded signal is cut using a cutoff function $t \mapsto G(t - L/\bar{c})$, where the support of G is included in $[t_0, t_1]$:

$$y^\varepsilon(t) = B^\varepsilon(L/\bar{c} + t, z = -L)G(t).$$

In contrast to the procedure used in the previous section, here we cut a

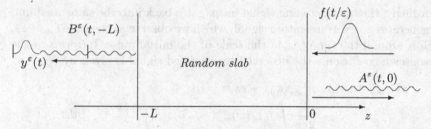

(a) A left-going pulse $f(t/\varepsilon)$ is impinging the random slab $(-L, 0)$ and it generates a transmitted signal $B^\varepsilon(t, -L)$. The TRM, used in a passive mode, records a segment $y^\varepsilon(t)$ of the transmitted signal.

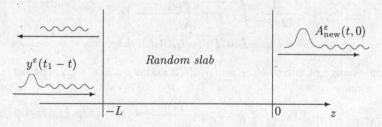

(b) The TRM is used as an active device and it sends back into the medium the signal $y^\varepsilon(t_1 - t)$. We observe the new transmitted signal $A^\varepsilon_{\text{new}}(t, 0)$.

Fig. 12.5. Setup for a time reversal in transmission (TRT) experiment with coda waves.

segment of size one. We then time-reverse that segment of signal and send it back into the same medium. This means that we have a new problem with a new source term located at $z = -L$ and given by

$$f_{\text{new}}^\varepsilon(t) = y^\varepsilon(t_1 - t) = B^\varepsilon(L/\bar c + t_1 - t, z = -L)G(t_1 - t).$$

The scaled Fourier transform of this source term is

$$\hat f_{\text{new}}^\varepsilon(\omega) = \int e^{i\frac{\omega t}{\varepsilon}} f_{\text{new}}^\varepsilon(t)\, dt$$

$$= \varepsilon \int e^{i\omega s} B^\varepsilon(L/\bar c + t_1 - \varepsilon s, z = -L)G(t_1 - \varepsilon s)\, ds$$

$$= \varepsilon \int e^{i\omega s} \left\{ \frac{1}{2\pi} \int e^{-i\omega' s} \overline{T_{\omega'}^\varepsilon(-L,0)}\,\overline{\hat f(\omega')} e^{i\omega'\frac{t_1}{\varepsilon}}\, d\omega' \right\} G(t_1 - \varepsilon s)\, ds$$

$$= \frac{\varepsilon}{2\pi} \int \overline{T_{\omega'}^\varepsilon(-L,0)}\,\overline{\hat f(\omega')} \left\{ \int e^{i(\omega'-\omega)(-s)}G(t_1 - \varepsilon s)\, ds \right\} e^{i\omega'\frac{t_1}{\varepsilon}}\, d\omega'$$

$$= \frac{1}{2\pi} \int \overline{T_{\omega'}^\varepsilon(-L,0)}\overline{\hat f(\omega')}\hat G\left(\frac{\omega - \omega'}{\varepsilon}\right) e^{i\omega\frac{t_1}{\varepsilon}}\, d\omega'.$$

Accordingly, the new incoming signal propagates back into the same medium, and generates a new transmitted signal, which we observe at the time $t_{\text{obs}}+\varepsilon s$, which is around the time t_{obs} in the scale of the initial pulse. In terms of the transmission coefficients the observed transmitted signal is given by

$$S_L^\varepsilon(t_{\text{obs}} + \varepsilon s) = A_{\text{new}}^\varepsilon(t_{\text{obs}} + \varepsilon s, z = 0)$$

$$= \frac{1}{2\pi\varepsilon} \int \hat f_{\text{new}}^\varepsilon(\omega) T_\omega^\varepsilon(-L,0) e^{-i\omega\left(s + \frac{t_{\text{obs}} - L/\bar c}{\varepsilon}\right)}\, d\omega.$$

Substituting the expression of $\hat f_{\text{new}}^\varepsilon$ into this equation yields the following representation of the new transmitted signal:

$$S_L^\varepsilon(t_{\text{obs}} + \varepsilon s) = \frac{1}{(2\pi)^2\varepsilon} \int \overline{\hat f(\omega_2)}\hat G\left(\frac{\omega_1 - \omega_2}{\varepsilon}\right) e^{-i\omega_1\left(s + \frac{t_{\text{obs}} - L/\bar c - t_1}{\varepsilon}\right)}$$

$$\times T_{\omega_1}^\varepsilon(-L,0)\overline{T_{\omega_2}^\varepsilon(-L,0)}\, d\omega_1\, d\omega_2.$$

After the change of variable $\omega_1 = \omega + \varepsilon h/2$ and $\omega_2 = \omega - \varepsilon h/2$, the representation becomes

$$S_L^\varepsilon(t_{\text{obs}} + \varepsilon s) = \frac{1}{(2\pi)^2} \int e^{-i\omega s} e^{i\frac{\omega(t_1 + L/\bar c - t_{\text{obs}})}{\varepsilon}} e^{ih(t_1 + L/\bar c - t_{\text{obs}})/2 - i\varepsilon hs/2}$$

$$\times \overline{\hat f(\omega - \varepsilon h/2)}\hat G(h) T_{\omega+\varepsilon h/2}^\varepsilon(-L,0)\overline{T_{\omega-\varepsilon h/2}^\varepsilon(-L,0)}\, dh\, d\omega. \qquad (12.7)$$

We can now carry out the precise analysis of the transmitted wave. Due to the presence of the fast phase $\exp(i\omega(t_1 + L/\bar c - t_{\text{obs}})/\varepsilon)$ in (12.7), it is easily seen that at the observation point $z = 0$,

refocusing takes place only if $t_{\text{obs}} = L/\bar c + t_1,$

and in a window around this observation time the refocused signal becomes

$$S_L^\varepsilon(L/\bar{c} + t_1 + \varepsilon s) = \frac{1}{(2\pi)^2} \int e^{-i\omega s} e^{-i\varepsilon h s/2} \overline{\hat{f}(\omega - \varepsilon h/2) \hat{G}(h)}$$
$$\times T_{\omega + \varepsilon h/2}^\varepsilon(-L, 0) \overline{T_{\omega - \varepsilon h/2}^\varepsilon(-L, 0)}\, dh\, d\omega\,. \quad (12.8)$$

The product of transmission coefficients has been analyzed in detail in Section 9.4. Using the asymptotic analysis of their moments we can characterize the limit of the refocused pulse $S_L^\varepsilon(L/\bar{c} + t_1 + \varepsilon s)$ as a continuous process with respect to the time variable s. The situation is similar to time reversal in reflection studied in Chapter 10, with transmission coefficients replacing reflection coefficients. A detailed proof of the convergence, including tightness, was given in the context of time reversal in reflection in Section 10.1.3. The proof of the convergence for time reversal in transmission follows exactly the same lines. Here we give directly the limiting refocused pulse for time reversal in transmission with coda waves.

12.2.2 Decomposition of the Refocusing Kernel

Proposition 12.1. *The refocused signal* $(S_L^\varepsilon(L/\bar{c} + t_1 + \varepsilon s))_{s\in(-\infty,\infty)}$ *converges in probability as $\varepsilon \to 0$ to the deterministic pulse*

$$S_L(s) = (f(-\,\cdot\,) * K_{\mathrm{TRT}}(\cdot))\,(s)\,, \quad (12.9)$$

where the Fourier transform of the refocusing kernel K_{TRT} is given by

$$\hat{K}_{\mathrm{TRT}}(\omega) = \int G(\tau) \Lambda_{\mathrm{TRT}}(\omega, d\tau)\,. \quad (12.10)$$

The refocusing spectral measure $\Lambda_{\mathrm{TRT}}(\omega, d\tau) = \mathcal{W}_0^{(T)}(\omega, d\tau, -L, 0)$ is given by the system of transport equations (9.67) corresponding to the transmission problem.

The deterministic nature of the refocused pulse, meaning that it is independent of the particular realization of the medium, is referred to as **statistical stability** or the **self-averaging** property. This property is important in many applications to communications or detection where the aim is to construct stable estimators of quantities of interest.

If we assume that the medium is homogeneous ($\gamma = 0$), then $\Lambda_{\mathrm{TRT}}(\omega, d\tau) = \delta_0(d\tau)$, so that

$$\hat{K}_{\mathrm{TRT}}(\omega) = G(0)\,,$$

and therefore the refocused pulse has exactly the same shape as the input one:

$$S_L(s) = G(0)f(-s)\,.$$

Indeed, as expected, in the homogeneous medium the pulse f travels through the medium at constant speed $\bar{c}$ without deformation. As a result, we record it perfectly if the recording-time window captures it, that is to say, if $t_0 < 0 < t_1$. After reemission by the TRM, the new pulse travels without deformation once again, and we get the original pulse shape.

Let us now consider the effect of randomness and assume that $\gamma > 0$. Considering the form (9.73) of $\mathcal{W}_0^{(T)}$ and its decomposition into singular and continuous parts, we can divide the refocusing kernel into two kernels corresponding respectively to the *stable* and the *incoherent* components of the recorded signal that participate in the refocusing:

$$\hat{K}_{\mathrm{TRT}}(\omega) = \hat{K}_{(\mathrm{TRT,stab})}(\omega) + \hat{K}_{(\mathrm{TRT,inc})}(\omega), \tag{12.11}$$

with

$$\hat{K}_{(\mathrm{TRT,stab})}(\omega) = G(0) \exp\left(-\frac{L}{L_{\mathrm{loc}}(\omega)}\right), \tag{12.12}$$

$$\hat{K}_{(\mathrm{TRT,inc})}(\omega) = \int G(\tau) \mathcal{W}_{0,c}^{(T)}(\omega, d\tau, -L, 0), \tag{12.13}$$

where the localization length $L_{\mathrm{loc}}(\omega) = 4\bar{c}^2/(\gamma\omega^2)$ was defined by (7.81). The refocused pulse is therefore

$$S_L(s) = \left(f(-\cdot) * K_{(\mathrm{TRT,stab})}(\cdot)\right)(s) + \left(f(-\cdot) * K_{(\mathrm{TRT,inc})}(\cdot)\right)(s).$$

The refocusing kernel $K_{(\mathrm{TRT,stab})}$ is the contribution to time reversal of the stable wave front. It results from the double action of the deterministic spreading kernel on the pulse front in forward and backward directions. This is actually exactly the kernel that we found when we considered the time reversal of the stable front in Section 12.1 (see (12.6)). Of course, this contribution completely vanishes if we do not record the pulse front (i.e., if $G(0) = 0$).

The kernel $K_{(\mathrm{TRT,inc})}$ is the contribution to time reversal of the transmitted incoherent coda waves. Its form is rather complicated. In order to discuss its qualitative properties, we consider two particular cases: a short recording-time window, or a very long recording-time window.

12.2.3 Midband Filtering by the Medium

Let us assume that we record a small segment of the transmitted wave, in the sense that the cutoff function G has its support in $[t_0, t_1]$ such that $t_0 < 0 < t_1$, and $\gamma\omega_0^2 t_1/\bar{c} \ll 1$, with ω_0 the typical frequency of the input pulse. This can also be written $\bar{c}t_1 \ll L_{\mathrm{loc}}(\omega_0)$, where $L_{\mathrm{loc}}(\omega_0)$ is the localization length introduced in (7.81). We then use the expansion (9.75),

$$\mathcal{W}_0^{(T)}(\omega, d\tau, -L, 0) \approx \frac{\bar{c}L}{2L_{\mathrm{loc}}^2(\omega)} \exp\left(-\frac{L}{L_{\mathrm{loc}}(\omega)}\right) d\tau,$$

to deduce that

$$\hat{K}_{(\text{TRT,inc})}(\omega) \approx \frac{\bar{c}L}{2L_{\text{loc}}^2(\omega)} \exp\left(-\frac{L}{L_{\text{loc}}(\omega)}\right) \times \int_0^\infty G(t)\,dt\,.$$

As a result, the kernel $K_{(\text{TRT,inc})}$ is a midband filter. It retains only the frequencies around ω_c, where

$$\omega_c^2 = \frac{8\bar{c}^2}{\gamma L}$$

corresponds to the maximum of $\hat{K}_{(\text{TRT,inc})}(\omega)$. The localization length associated with this frequency is $L_{\text{loc}}(\omega_c) = L/2$. Thus the frequencies around ω_c can probe the random slab and experience scattering, but they are not yet completely reflected by the strong localization effect. As a result, these incoherent waves can be recorded by the TRM at the output of the slab, and they can contribute to the refocused pulse when they are sent back.

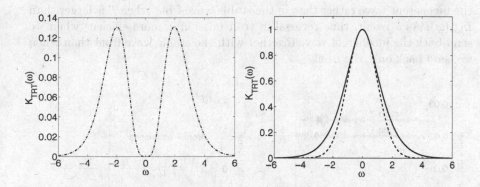

Fig. 12.6. Refocusing kernels for time reversal in transmission. The dashed line in the left plot shows a midband filter that corresponds to the case in which only the incoherent coda wave is sent back. The dotted line in the right plot shows a low-pass filter that corresponds to the case in which only the stable transmitted wave front is sent back, while the solid line corresponds to the case in which everything that is recorded in transmission is sent back. Note the difference in the vertical scaling in the two figures.

12.2.4 Low-Pass Filtering

Let us now address the ideal case in which we record completely the transmitted signal and send back everything that is recorded. This ideal case is obtained by setting $G \equiv 1$. We then find that the refocusing kernel is

$$\hat{K}_{\text{TRT}}(\omega) = \int \Lambda_{\text{TRT}}(\omega, d\tau)\,.$$

By Proposition 9.4 this is the limit of the mean power transmission coefficient:

$$\hat{K}_{\mathrm{TRT}}(\omega) = \lim_{\varepsilon \to 0} \mathbb{E}[|T_\omega^\varepsilon(-L, 0)|^2],$$

which has been computed in Chapter 7 and is given explicitly by Proposition 7.4. Thus we have

$$\hat{K}_{\mathrm{TRT}}(\omega) = \xi_1 \left(\frac{L}{L_{\mathrm{loc}}(\omega)} \right),$$

where ξ_1 is given by (7.52). In Figure 12.6 we compare this kernel with the kernel $\hat{K}_{(\mathrm{TRT,stab})}(\omega)$ given in (12.12) obtained when only the stable transmitted wave front is sent back. The kernel $\hat{K}_{\mathrm{TRT}}(\omega)$ is a low-pass filter whose cutoff frequency is higher than the cutoff frequency of $\hat{K}_{(\mathrm{TRT,stab})}(\omega)$, which is because we send back the whole transmitted wave. We have seen in Section 8.2.2 that the most important part of the transmitted energy is contained in the incoherent wave rather than in the stable wave front when L is larger than $L_{\mathrm{loc}}(\omega)$. As a result, time reversal in transmission is more efficient when we send back the incoherent wave together with the stable wave front than when we send back only the front.

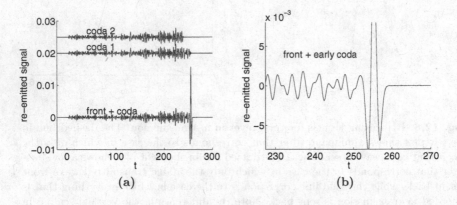

(a)   (b)

Fig. 12.7. Plot (a): Four time-reversal experiments are carried out with the transmitted signal shown in Figure 12.1, depending on the reemitted segments of the signal: (i) only the wave front is sent back; (ii) the stable wave front plus a piece of the coda is sent back (front+coda); (iii) the stable wave front is removed from the reemitted signal (coda 1); (iv) the stable wave front plus the early coda is removed (coda 2). The last three reemitted signals are plotted. Plot (b): Zoom on the stable wave front and the early coda.

12.3 Discussion and Numerical Simulations

We consider the numerical setup described in Section 8.2.3. We record the transmitted signal shown in Figure 12.1. We can then perform a series of time-reversal experiments, by sending back different segments of the time-reversed recorded signal. Two of these experiments are plotted in Figures 12.8–12.9. The different segments are plotted in Figure 12.7a, and the corresponding refocused pulses are plotted in Figure 12.10.

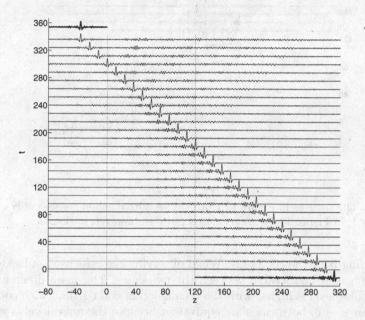

Fig. 12.8. We use the same random medium as in Figure 12.1 and send back the time-reversed transmitted signal. Snapshots of the wave profile at increasing times are plotted from bottom to top. We can see the refocused pulse that emerges from the random medium. Here the reemitted signal consists of the stable wave front plus a piece of the coda.

Comparison of the refocused pulses in Figures 12.10a (reemission of the stable front only) and 12.10b (reemission of the stable front and the coda) shows clearly the refocusing improvement that results when the incoherent coda is sent back with the stable front. The time-reversal process recompresses the incoherent wave fluctuations with the stable front, which enhances the refocusing properties, as predicted by the theory developed in this chapter.

Comparison of the refocused pulses in Figures 12.10c (reemission of coda 1) and 12.10d (reemission of coda 2) shows that refocusing can occur even

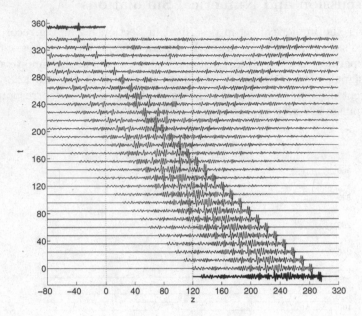

Fig. 12.9. The same as in Figure 12.1 but the reemitted signal consists only of the coda (coda 1, in Figure 12.7a). A refocused pulse is, however, noticeable.

when sending back the coda only. When observing the refocused pulses, only minor changes are noticeable between the two cases. The main difference is that the noise level is higher with the reemission of coda 1 than with coda 2. In the limit $\varepsilon \to 0$, both cases are equivalent because the noise level is going to zero, but the numerical simulations at finite ε give these fine details, which can be explained as follows.

When performing the time-reversal experiment with coda 1, we do not send back the wave front, but we send back the wave fluctuations that were just behind it and that were generated during the forward propagation of the wave front (see Figure 12.7). The back-propagation of these wave fluctuations gives rise to the relatively large noise in the vicinity of the refocused pulse seen in pictures (b) and (c) at negative t. This noise can actually be approximated by a convolved version of the early part of the coda 1 signal seen in Figure 12.7. This noise is large compared to the residual noise in the refocused signal generated by the back-propagation of the time-reversed incoherent waves, which can be seen in picture (c) at positive t and in picture (d) at positive t and negative $t \in (-15, 0)$.

When reemitting coda 2 (picture (d)), the wave fluctuations just behind the transmitted wave front over a time interval of duration 15 are not sent back, so that the large noise in the refocused signal obtained in the case of coda 1 is not present in the vicinity $t \in (-15, 0)$ of the refocused pulse, but can be detected in the region $t \leq -15$. However, the most important part of the wave fluctuations are sent back, so the refocusing of these wave fluctuations gives rise to a refocused pulse that is very similar to the one obtained with coda 1. As a result, the signal-to-noise ratio with the time reversal of coda 2 appears to be higher in the vicinity of the refocused pulse than with coda 1.

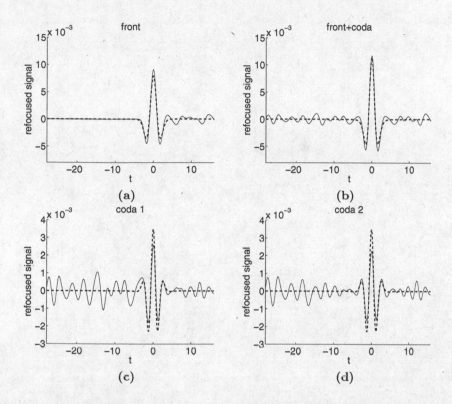

Fig. 12.10. Refocused pulses after time reversal of different segments of the transmitted signal. In thick dashed lines are plotted the theoretical refocused pulses, obtained by convolutions of the original pulse shape with the theoretical refocusing kernels. In plot (a), the theoretical refocusing kernel is (12.12). In plot (b), the refocusing kernel is (12.11). In plots (c) and (d), the refocusing kernel is (12.13).

Notes

Time reversal of stable fronts is studied in [67] by Fouque and Sølna in the context of three-dimensional randomly layered media as introduced in Chapter

14. Time reversal in transmission of incoherent waves is presented for the first time in Section 12.2.

13

Application to Communications

In this chapter we present an application of time reversal in transmission to communications through a one-dimensional channel with random fluctuations. We consider the communications scheme in which the transmitter emits a train of pulses that encodes a binary message. The pulse train is transmitted through a random channel and recorded by the receiver. We assume that the signal-to-noise ratio at the receiver is large (infinite), in the sense that there is no additive noise, such as electronic noise, in the signal recorded by the receiver. For very simple receivers we derive quantitative estimates for the intersymbol interference (ISI), which affects the decoding of the received signal. By very simple receivers we mean here ones for which no advanced signal processing method, such as equalization filtering, is applied to decode the received signal. The received signal is decoded by simply tapping it at multiples of the intersymbol time. We show that the signal-to-interference ratio (SIR), which is a quantity that measures intersymbol interference, is self-averaging in one-dimensional randomly layered media, in the strongly heterogeneous white-noise regime. The computable form of SIR in this asymptotic limit allows us to compare the advantages, if any, when using time reversal at the transmitter as opposed to not using it. It is expected that time reversal will decrease ISI, and hence increase SIR, because of refocusing of pulses, as we saw in the previous chapter. The analysis of this chapter shows, however, that this is not always the case.

The contents of this chapter are not essential for the remainder of the book. However, the reader interested in communications problems can find an interesting application of time reversal here, and an example of how the computations carried out in the previous chapters can be applied to such problems.

13.1 Review of Basic Communications Schemes

13.1.1 Nyquist Pulse

A function $f_0(t)$ is a Nyquist pulse if it satisfies the Nyquist condition for zero intersymbol interference (ISI):

$$f_0(kT_b) = \begin{cases} 1 & \text{if } k = 0, \\ 0 & \text{if } k \in \mathbb{Z} \backslash \{0\}. \end{cases}$$

The *sinc function*

$$f_0(t) = \frac{\sin(\pi t/T_b)}{\pi t/T_b}$$

is the most commonly used Nyquist pulse. In the frequency domain it has the form

$$\hat{f}_0(\omega) = \int f(t)e^{i\omega t}\, dt = \begin{cases} T_b & \text{if } |\omega| \leq \pi/T_b, \\ 0 & \text{if } |\omega| > \pi/T_b, \end{cases}$$

and we denote by $B = 2\pi/T_b$ the **bandwidth**. The pulse function f_0 is used to encode the elementary bits of information for the transmission of a binary message, as we now explain.

Let us consider a communications channel connecting a transmitter and a receiver. If the binary message is the sequence $(\delta_k)_{k=1,\dots,N}$, $\delta_k \in \{0,1\}$, then the transmitter encodes the message in the form of a train of Nyquist pulses

$$S(t) = \sum_{k=1}^{N} \delta_k f_0(t - kT_b). \tag{13.1}$$

Other encoding methods (for instance, $\delta_k \in \{-1,1\}$) are also possible, and the analysis of this chapter can be extended to these situations. The transmitter sends the signal S whose Fourier transform is given by

$$\hat{S}(\omega) = \hat{f}_0(\omega) \sum_{k=1}^{N} \delta_k e^{i\omega kT_b}. \tag{13.2}$$

The signal received by the receiver is $S_{\text{tr}} = K * S$, where K is the **impulse response** of the channel. Here we have assumed zero additive noise at the receiver, that is, infinite signal-to-noise ratio. In the frequency domain

$$\hat{S}_{\text{tr}}(\omega) = \hat{K}(\omega)\hat{S}(\omega),$$

where $\hat{K}$ is the **transfer function** of the channel, that is the Fourier transform of the impulse response, and $\hat{S}$ is the Fourier transform of the signal given in (13.1).

If the transfer function of the channel is frequency-flat as in a homogeneous medium, then, after removing the propagation time delay, the transfer function of the channel is identically one, and the received pulse train S_{tr} is exactly S, given by (13.1). As a consequence the message can be read by **tapping** at $t = kT_b$, that is, by evaluating $S_{\text{tr}}(kT_b) = \delta_k$ for all $k = 1, \dots, N$.

13.1.2 Signal-to-Interference Ratio

If the transmission channel is perturbed, then it degrades the zeros of the Nyquist pulse, producing intersymbol interference. This interference can be quantified with the **signal-to-interference ratio** defined by

$$\text{SIR} = \frac{f_{\text{tr}}^2(0)}{\sum_{k \neq 0} f_{\text{tr}}^2(kT_b)}, \tag{13.3}$$

where f_{tr} is the transmitted pulse and pulse shaping is done with the Nyquist pulse f_0:

$$\hat{f}_{\text{tr}}(\omega) = \hat{K}(\omega)\hat{f}_0(\omega). \tag{13.4}$$

Note that high SIR means low interference between bits, with a zero interference producing an infinite SIR.

The Poisson summation formula

$$\sum_{n=-\infty}^{\infty} \delta(t - n) = \sum_{k=-\infty}^{\infty} \exp(2i\pi kt) \tag{13.5}$$

is useful for expressing the denominator of the SIR in terms of a continuous integral. Indeed, for any t_0 we have

$$\sum_{k=-\infty}^{\infty} f_{\text{tr}}\left(t_0 + \frac{2\pi k}{B}\right)^2 = \sum_k \int f_{\text{tr}}(t_0 + t)^2 \delta\left(t - \frac{2\pi k}{B}\right) dt$$

$$= \frac{B}{2\pi} \sum_n \int f_{\text{tr}}(t_0 + t)^2 e^{inBt} \, dt$$

$$= \frac{B}{(2\pi)^3} \sum_n \int\int \hat{K}(\omega_1)\overline{\hat{K}(\omega_2)}\hat{f}_0(\omega_1)\overline{\hat{f}_0(\omega_2)}$$

$$\times e^{i(\omega_2 - \omega_1)t_0} \int e^{i(\omega_2 - \omega_1 + nB)t} \, dt \, d\omega_1 \, d\omega_2.$$

Introducing the change of variables $\omega_1 = \omega - h/2$, $\omega_2 = \omega + h/2$, and integrating with respect to t gives the Dirac distribution $\delta(h - nB)$. We therefore have

$$\sum_{k=-\infty}^{\infty} f_{\text{tr}}\left(t_0 + \frac{2\pi k}{B}\right)^2 = \frac{B}{(2\pi)^2} \sum_n \int \hat{f}_0\left(\omega - \frac{nB}{2}\right)\overline{\hat{f}_0\left(\omega + \frac{nB}{2}\right)}$$

$$\times e^{inBt_0}\hat{K}\left(\omega - \frac{nB}{2}\right)\overline{\hat{K}\left(\omega + \frac{nB}{2}\right)} \, d\omega.$$

Using the fact that the support of $\hat{f}_0$ is included in $[-B/2, B/2]$, the sum over n is reduced to the term $n = 0$:

$$\sum_{k=-\infty}^{\infty} f_{\text{tr}}\left(t_0 + \frac{2\pi k}{B}\right)^2 = \frac{B}{(2\pi)^2} \int |\hat{f}_0(\omega)|^2 |\hat{K}(\omega)|^2 \, d\omega. \tag{13.6}$$

By the Parseval equality the right-hand side is simply equal to the total energy $\int f_{\text{tr}}^2(t)\, dt$, up to the multiplicative factor $B/(2\pi)$. The fact that the discrete sum is equal to the continuous integral can be viewed as a manifestation of Shannon's sampling theorem for band-limited signals. We thus get two equivalent expressions for SIR:

$$\text{SIR} = \frac{f_{\text{tr}}(0)^2}{\frac{B}{2\pi} \int f_{\text{tr}}^2(t)\, dt - f_{\text{tr}}(0)^2} \tag{13.7}$$

$$= \frac{\left| \int \hat{f}_0(\omega)\hat{K}(\omega)\, d\omega \right|^2}{B \int |\hat{f}_0(\omega)|^2 |\hat{K}(\omega)|^2\, d\omega - \left| \int \hat{f}_0(\omega)\hat{K}(\omega)\, d\omega \right|^2}. \tag{13.8}$$

13.1.3 Modulated Nyquist Pulse

The frequency band of the transmission channel is usually of the form $[\omega_0 - B/2, \omega_0 + B/2]$, with $B < \omega_0$ or even $B \ll \omega_0$. In such a case the signal $S(t)$ given by (13.1) is modulated at the frequency ω_0 before transmission, and the emitted signal is

$$S_{\text{em}}(t) = \frac{1}{2} S(t) e^{-i\omega_0 t} + c.c.,$$

where $c.c.$ stands for complex conjugate. The Fourier transform of S_{em},

$$\hat{S}_{\text{em}}(\omega) = \frac{1}{2}\hat{S}(\omega - \omega_0) + \frac{1}{2}\hat{S}(\omega + \omega_0),$$

has support contained in the frequency band of the transmission channel. The signal received by the intended receiver is $K * S_{\text{em}}$, which gives in the Fourier domain

$$\widehat{K * S_{\text{em}}}(\omega) = \frac{1}{2}\left[\hat{K}(\omega)\hat{S}(\omega - \omega_0) + \hat{K}(\omega)\hat{S}(\omega + \omega_0) \right].$$

This signal is shifted in baseband by demodulation at the receiver, which means that a modulation is applied to the signal

$$S_{\text{tr}}(t) = K * S_{\text{em}}(t) e^{i\omega_0 t} + c.c.,$$

which gives in the Fourier domain

$$\hat{S}_{\text{tr}}(\omega) = \widehat{K * S_{\text{em}}}(\omega + \omega_0) + \widehat{K * S_{\text{em}}}(\omega - \omega_0).$$

The application of a low-pass filter gives

$$\hat{S}_{\text{tr}}(\omega) = \frac{1}{2}[\hat{K}(\omega + \omega_0) + \hat{K}(\omega - \omega_0)]\hat{S}(\omega)$$

$$= \frac{1}{2}[\hat{K}(\omega_0 + \omega) + \overline{\hat{K}(\omega_0 - \omega)}]\hat{S}(\omega)$$

$$= \frac{1}{2}\hat{K}(\omega_0 + \omega)\hat{S}(\omega) + \frac{1}{2}\overline{\hat{K}(\omega_0 - \omega)\hat{S}(-\omega)},$$

where $\hat{S}$ is the Fourier transform of the signal given by (13.2). In the time domain,

$$S_{\text{tr}}(t) = \frac{1}{4\pi} \int \hat{K}(\omega_0 + \omega)\hat{S}(\omega)e^{-i\omega t}\, d\omega + c.c.$$

We can now revisit the results obtained in Section 13.1.2. Using once again the Poisson formula, we obtain instead of (13.6),

$$\sum_{k=-\infty}^{\infty} f_{\text{tr}}\left(t_0 + \frac{2\pi k}{B}\right)^2 = \frac{B}{(2\pi)^2} \int |\hat{f}_0(\omega)|^2 \mathcal{K}_{\omega_0}(\omega)\, d\omega,$$

where

$$\mathcal{K}_{\omega_0}(\omega) = \frac{|\hat{K}(\omega_0 + \omega)|^2 + \text{Re}(\hat{K}(\omega_0 + \omega)\hat{K}(\omega_0 - \omega))}{2},$$

and the SIR can be written in the form

$$\text{SIR} = \frac{\text{Re}\left(\int \hat{f}_0(\omega)\hat{K}(\omega_0 + \omega)\, d\omega\right)^2}{B\int |\hat{f}_0(\omega)|^2 \mathcal{K}_{\omega_0}(\omega)\, d\omega - \text{Re}\left(\int \hat{f}_0(\omega)\hat{K}(\omega_0 + \omega)\, d\omega\right)^2}. \tag{13.9}$$

13.2 Communications in Random Media Using Nyquist Pulses

We consider now a one-dimensional random medium. We want to transmit a sequence of bits $(\delta_k)_{k=1,\ldots,N}$ from a transmitter A to a receiver B with scalar waves governed by the equation

$$\frac{1}{c^2(z)} \frac{\partial^2 u}{\partial t^2} - \frac{\partial^2 u}{\partial z^2} = 0,$$

which is the acoustic wave equation with constant density. More general wave equations, in particular ones that contain dispersion, will be considered in Chapter 18.

In the standard **direct-communications** scheme described in Figure 13.1, the signal

$$S_{\text{dir}}(t) = \sum_{k=1}^{N} \delta_k f_0(t - kT_b)$$

is sent by the transmitter through the medium from A to B, which is the channel. The transmitted signal at the receiver B, denoted by $S_{(\text{tr,dir})}$, has the following Fourier transform:

$$\hat{S}_{(\text{tr,dir})}(\omega) = \hat{K}_{\text{dir}}(\omega)\hat{f}_0(\omega) \sum_{k=1}^{N} \delta_k e^{ikT_b\omega},$$

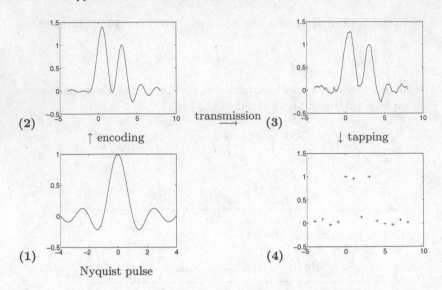

Fig. 13.1. Transmission of the binary message 1101 by a direct-communications scheme. The left part describes the operations performed by the transmitter and the right part describes the operations performed by the receiver. A Nyquist pulse f_0 is used (1). A train of three Nyquist pulses encodes the message 1101 (2). The pulse train is sent by the transmitter and a blurred pulse train is received by the receiver (3). The receiver taps the pulse train to decode the binary message (4).

where $\hat{K}_{\mathrm{dir}}$ is the transfer function of the medium (the channel).

In **time-reversal communications** two stages are required, as shown in Figure 13.2. In the first stage, the Nyquist pulse $f_0(t)$ is sent by the intended receiver B. The intended transmitter A records a segment of the transmitted signal, and then time-reverses it to obtain the new signal $f_1(t)$, which is used to encode the bit sequence in the time-reversal scheme. Thus, in the second stage, the transmitter A sends the signal

$$S_{\mathrm{TR}}(t) = \sum_{k=1}^{N} \delta_k f_1(t - kT_b)$$

through the channel. The transmitted signal at the receiver B, denoted by $S_{(\mathrm{tr},\mathrm{TR})}(t)$, has the Fourier transform

$$\hat{S}_{(\mathrm{tr},\mathrm{TR})}(\omega) = \hat{K}_{\mathrm{TRT}}(\omega)\hat{f}_0(\omega) \sum_{k=1}^{N} \delta_k e^{ikT_b\omega},$$

where $\hat{K}_{\mathrm{TRT}}$ is the time-reversal transfer function.

The goal now is to compare the SIRs, defined by (13.3), for the two communications methods, which are direct transmission using f_0 and time reversal using f_1.

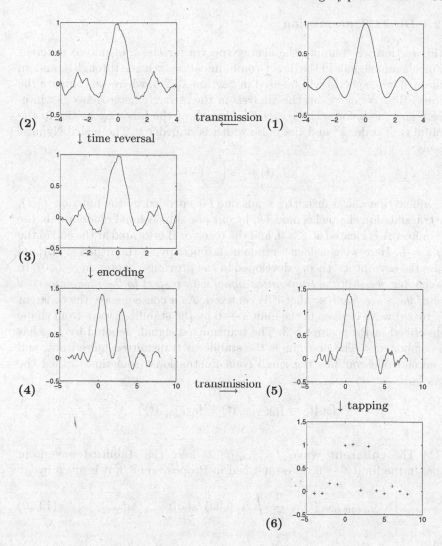

Fig. 13.2. Transmission of the binary message 1101 with the TR-communications scheme. The left part describes the operations performed by the transmitter and the right part describes the operations performed by the receiver. A Nyquist pulse f_0 is emitted by the intended receiver (1). The intended transmitter records a blurred Nyquist pulse (2) and time-reverses it in its memory (3). This pulse f_1 is used by the transmitter to generate a train of three pulses encoding the binary message 1101 (4). The receiver receives a blurred pulse train (5) and taps it to decode the binary message (6).

13.2.1 Direct Transmission

In this section we compute the energy spectra for the transmitted coherent and incoherent signals in the direct-communications scheme through a random medium. These spectra will be used in Section 13.2.3, where we compare the various SIRs. We carry out the analysis in the strongly heterogeneous white-noise regime as before. The correlation length of the inhomogeneities in the medium is of order ε^2 and the pulse width is of order ε. The scaled Nyquist pulse is

$$f_0^\varepsilon(t) = f_0\left(\frac{t}{\varepsilon}\right).$$

We assume that the transmitter sends one bit encoded by the function $f_0^\varepsilon(t)$. The transmission channel is modeled by our one-dimensional random slab, the transmitter A is located at $z = 0$, and the receiver B is located at the end of the slab $z = L$. Here we consider a random slab occupying the region $(0, L)$ and we use the asymptotic theory developed in the previous chapter on $(-L, 0)$. In this chapter, *we shift the transmitted signal with respect to the random arrival time of the wave front*, so that it is centered. As a consequence, the coherent (i.e., mean) wave is given in the limit $\varepsilon \to 0$ by the stabilized wave-front shape as described in Proposition 8.3. The transmitted signal, denoted by f_{dir}, has two components. The first one is the stabilized transmitted pulse front, and the second one contains the small coda fluctuations (sometimes called the "grass"):

$$f_{\mathrm{dir}}(t) = f_{(\mathrm{dir,coh})}(t) + f_{(\mathrm{dir,inc})}(t).$$

(1) The **coherent wave** $f_{(dir,\mathrm{coh})}(t)$ is here the stabilized wave-front shape. In the limit $\varepsilon \to 0$, as established in Proposition 8.3, it is given by

$$f_{(\mathrm{dir,coh})}(t) = \frac{1}{2\pi}\int \hat{K}_{\mathrm{coh}}(\omega)\hat{f}_0(\omega)e^{-i\omega t/\varepsilon}\,d\omega, \tag{13.10}$$

where

$$\hat{K}_{\mathrm{coh}}(\omega) = \exp\left(-\frac{\gamma\omega^2 L}{8\bar{c}^2}\right). \tag{13.11}$$

We write here this kernel in the form

$$\hat{K}_{\mathrm{coh}}(\omega) = \xi_0\left(\frac{L}{L_{\mathrm{loc}}(\omega)}\right),$$

where $L_{\mathrm{loc}}(\omega) = 4\bar{c}^2/(\gamma\omega^2)$ is the localization length given by (7.81) and ξ_0 is the function

$$\xi_0(l) = \exp\left(-\frac{l}{2}\right). \tag{13.12}$$

This stable front has support of order ε and an amplitude of order one, that is, of the same order as the input Nyquist pulse. The total energy of the wave front is given by (9.80):

$$\int f_{(\text{dir,coh})}^2(t)\, dt = \frac{\varepsilon}{2\pi} \int \xi_0^2 \left(\frac{L}{L_{\text{loc}}(\omega)} \right) |\hat{f}_0(\omega)|^2 \, d\omega .$$

(2) The **incoherent wave** fluctuations $f_{(\text{dir,inc})}(t)$ analyzed in Section 9.4.3 are, in the limit $\varepsilon \to 0$, a zero-mean Gaussian process whose mean intensity is given by

$$\mathbb{E}[f_{(\text{dir,inc})}^2(t)] = \frac{\varepsilon}{2\pi} \int \mathcal{W}_{0,c}^{(T)}(\omega, t) |\hat{f}_0(\omega)|^2 \, d\omega ,$$

where $\mathcal{W}_{0,c}^{(T)}$ is the continuous part of the solution of the system of transport equations in (9.74). The incoherent wave has support of order one and amplitude of order $\sqrt{\varepsilon}$. The mean energy of the incoherent wave is given by (9.81):

$$\int \mathbb{E}[f_{(\text{dir,inc})}^2(t)] \, dt = \frac{\varepsilon}{2\pi} \int (\xi_1 - \xi_0^2) \left(\frac{L}{L_{\text{loc}}(\omega)} \right) |\hat{f}_0(\omega)|^2 \, d\omega ,$$

where the function ξ_1 is defined by (7.52). As shown in Section 7.2.2, the intensity of the transmitted wave is random. However, the total transmitted energy

$$\int f_{\text{dir}}^2(t) \, dt = \frac{\varepsilon}{2\pi} \int |\hat{f}_0(\omega)|^2 |T_\omega^\varepsilon|^2 \, d\omega$$

is a self-averaging quantity, in the sense that it converges as $\varepsilon \to 0$ in mean square and in probability to the deterministic energy

$$\int f_{\text{dir}}^2(t) \, dt = \frac{\varepsilon}{2\pi} \int |\hat{f}_0(\omega)|^2 \xi_1 \left(\frac{L}{L_{\text{loc}}(\omega)} \right) d\omega . \tag{13.13}$$

13.2.2 Communications Using Time Reversal

In this section we compute the energy spectra for the transmitted coherent and incoherent signals using the time-reversal-communications scheme through a random medium. As before in previous chapters, we denote by $G(t)$ the recording-time-window function used by the transmitter. The pulse obtained at the receiver, when one bit encoded by f_1 is sent by the transmitter, is again centered in time with respect to its random arrival time. The received signal, denoted by $f_{\text{TR}}(t)$, has two components. The first one is the stabilized refocused signal, and the second one contains the small wave fluctuations (grass):

$$f_{\text{TR}}(t) = f_{(\text{TR,coh})}(t) + f_{(\text{TR,inc})}(t) .$$

(1) The **coherent wave** $f_{(\mathrm{TR,coh})}(t)$, in the limit $\varepsilon \to 0$, is given in (12.9):

$$f_{(\mathrm{TR,coh})}(t) = \frac{1}{2\pi} \int \hat{K}_{\mathrm{TRT}}(\omega)\overline{\hat{f}_0(\omega)}e^{-i\omega t/\varepsilon}\, d\omega\,,$$

with

$$\hat{K}_{\mathrm{TRT}}(\omega) = \int G(\tau)\Lambda_{\mathrm{TRT}}(\omega, d\tau)\,,$$

where Λ_{TRT} is given by (12.10). The coherent wave has support of order ε and amplitude of order one, that is, of the same order as the input pulse. The total energy of the coherent wave is

$$\int f_{(\mathrm{TR,coh})}^2(t)\, dt = \frac{\varepsilon}{2\pi} \int \hat{K}_{\mathrm{TRT}}^2(\omega)|\hat{f}_0(\omega)|^2\, d\omega\,. \tag{13.14}$$

In the case $G = 1$, in which the transmitter uses the full signal trace of the wave first sent by the receiver, we have

$$\hat{K}_{\mathrm{TRT}}(\omega) = \xi_1\left(\frac{L}{L_{\mathrm{loc}}(\omega)}\right)\,, \tag{13.15}$$

where ξ_1 is given by (7.52), so that

$$f_{(\mathrm{TR,coh})}(t) = \frac{1}{2\pi} \int \xi_1\left(\frac{L}{L_{\mathrm{loc}}(\omega)}\right)\overline{\hat{f}_0(\omega)}e^{-i\omega t/\varepsilon}\, d\omega\,. \tag{13.16}$$

We note here that the filtering kernel in the time-reversal case, $\xi_1(L/L_{\mathrm{loc}}(\omega))$, coincides with the kernel that describes the power spectrum in the direct-transmission case, as shown in (13.13).

(2) The small **incoherent** wave fluctuations $f_{(\mathrm{TR,inc})}(t)$ admit the following integral representation obtained using (12.8) and subtracting the coherent component described above:

$$f_{(\mathrm{TR,inc})}(t) = \frac{1}{(2\pi)^2} \int e^{-i\omega t/\varepsilon}e^{-iht/2}\overline{\hat{f}_0(\omega - \varepsilon h/2)}\,\hat{G}(h)T_{\omega+\varepsilon h/2}^{\varepsilon}\overline{T_{\omega-\varepsilon h/2}^{\varepsilon}}\, dh\, d\omega$$
$$-f_{(\mathrm{TR,coh})}(t)\,.$$

Note that the second term on the right-hand side is the expected value of the first term in the limit $\varepsilon \to 0$. Therefore, the mean energy of the incoherent wave, to leading order in ε, is given by

$$\int \mathbb{E}[f_{(\mathrm{TR,inc})}^2(t)]\, dt = \varepsilon\left(\frac{1}{(2\pi)^3} \int \overline{\hat{G}(h)}\hat{G}(h')U(\omega, h, h')\, dh\, dh'\, |\hat{f}_0(\omega)|^2\, d\omega\right.$$
$$\left. - \int f_{(\mathrm{TR,coh})}^2(t)\, dt\right),$$

where

$$U(\omega, h, h') = \lim_{\varepsilon \to 0} \mathbb{E}\left[|T_\omega^\varepsilon|^2 T_{\omega-\varepsilon h'}^\varepsilon \overline{T_{\omega-\varepsilon h}^\varepsilon}\right].$$

Here the number of frequencies has been reduced by one because of the Dirac distribution that results from the integration with respect to time.

In the case $G = 1$ we have $\hat{G}(h) = 2\pi\delta(h)$ and the mean incoherent energy is

$$\int \mathbb{E}[f_{(\mathrm{TR,inc})}^2(t)]\, dt = \frac{\varepsilon}{2\pi} \int \left(\mathbb{E}[\tau_\omega(L)^2] - \mathbb{E}[\tau_\omega(L)]^2\right) |\hat{f}_0(\omega)|^2\, d\omega\,,$$

where $\tau_\omega(L)$ is the limit in distribution of $|T_\omega^\varepsilon|^2$, described in Proposition 7.3. The moments of $\tau_\omega(L)$ are given by (7.49), so that we have

$$\int \mathbb{E}[f_{(\mathrm{TR,inc})}^2(t)]\, dt = \frac{\varepsilon}{2\pi} \int (\xi_2 - \xi_1^2)\left(\frac{L}{L_{\mathrm{loc}}(\omega)}\right) |\hat{f}_0(\omega)|^2\, d\omega\,,$$

where the functions ξ_n are defined by (7.50).

The energy of the transmitted signal is a self-averaging quantity. This can be established easily in the case $G = 1$, since the expression of the total energy can then be reduced to

$$\int f_{\mathrm{TR}}^2(t)\, dt = \frac{\varepsilon}{2\pi} \int |\hat{f}_0(\omega)|^2 |T_\omega^\varepsilon|^4\, d\omega\,.$$

By the same arguments as those used in Section 7.2.2, we can show that the second moment of the energy converges to the square of the first moment. This is due to the decorrelation property of the transmission coefficient in the frequency domain. As a result, the total energy converges as $\varepsilon \to 0$ in mean square and in probability to the deterministic energy

$$\int f_{\mathrm{TR}}^2(t)\, dt = \frac{\varepsilon}{2\pi} \int |\hat{f}_0(\omega)|^2 \xi_2\left(\frac{L}{L_{\mathrm{loc}}(\omega)}\right) d\omega\,. \tag{13.17}$$

13.2.3 SIRs for Coherent Pulses

In this section we compute the signal-to-interference ratios that are associated with the coherent parts of the transmitted signals, in the direct- and the time-reversal-communications schemes. In the scaled regime the SIR is given by

$$\mathrm{SIR} = \frac{f_{\mathrm{tr}}^2(0)}{I}\,, \qquad I = \sum_{k \neq 0} f_{\mathrm{tr}}^2(2\varepsilon\pi k/B)\,. \tag{13.18}$$

The interference term I can be decomposed into two terms:

$$I = I_{\mathrm{coh}} + I_{\mathrm{inc}}\,.$$

Here I_{coh} is the contribution of the coherent wave given by

$$I_{\mathrm{coh}} = \sum_{k \neq 0} f_{\mathrm{coh}}^2(2\varepsilon\pi k/B)\,,$$

while I_{inc} is the contribution of the incoherent wave. The interference term is deterministic in the limit $\varepsilon \to 0$, as we will see below. In this section we neglect the small incoherent wave fluctuations, with amplitude of order $\sqrt{\varepsilon}$, as noted in the previous section, in the transmitted signals f_{tr}. This corresponds to considering that the sum in the expression (13.18) for I is restricted to integers k of order less than ε^{-1}. Denoting the resulting SIR by SIR_{coh} we get

$$\text{SIR}_{\text{coh}} = \frac{f_{\text{coh}}^2(0)}{I_{\text{coh}}}, \qquad I_{\text{coh}} = \lim_{N \to \infty} \lim_{\varepsilon \to 0} \sum_{k \neq 0, |k| \leq N} f_{\text{tr}}^2(2\varepsilon\pi k/B).$$

Direct Transmission

In this section we compute SIR_{coh} in the direct-transmission scheme, and we denote it by $\text{SIR}_{(\text{dir,coh})}$. It is given by

$$\text{SIR}_{(\text{dir,coh})} = \frac{f_{(\text{dir,coh})}^2(0)}{I_{\text{dir,coh}}}, \qquad I_{\text{dir,coh}} = \sum_{k \neq 0} f_{(\text{dir,coh})}^2(2\varepsilon\pi k/B),$$

where the coherent wave is given by (13.10),

$$f_{(\text{dir,coh})}(2\varepsilon\pi k/B) = \frac{1}{2\pi} \int \hat{K}_{\text{coh}}(\omega) \hat{f}_0(\omega) e^{-i\omega 2\pi k/B} \, d\omega.$$

The kernel $\hat{K}_{\text{coh}}(\omega)$ has the form

$$\hat{K}_{\text{coh}}(\omega) = \exp\left(-\frac{L}{2L_{\text{loc}}^B}\left(\frac{2\omega}{B}\right)^2\right),$$

where

$$L_{\text{loc}}^B = L_{\text{loc}}(B/2) = \frac{16\bar{c}^2}{\gamma B^2}$$

is the localization length corresponding to the frequency $B/2$, as introduced in (7.73). As a result, we have

$$\text{SIR}_{(\text{dir,coh})} = \frac{\left[\int_0^1 \xi_0\left(\frac{L}{L_{\text{loc}}^B}x^2\right)\hat{F}_0(x)\,dx\right]^2}{\int_0^1 \xi_0^2\left(\frac{L}{L_{\text{loc}}^B}x^2\right)\hat{F}_0^2(x)\,dx - \left[\int_0^1 \xi_0\left(\frac{L}{L_{\text{loc}}^B}x^2\right)\hat{F}_0(x)\,dx\right]^2}, \tag{13.19}$$

with $\xi_0(l) = \exp(-l/2)$, and $\hat{F}_0$ is the normalized Fourier transform of the pulse:

$$\hat{F}_0(x) = \frac{B}{2}\hat{f}_0\left(\frac{B}{2}x\right). \tag{13.20}$$

The formula (13.19) shows that the SIR is a function of L/L_{loc}^B only.

Time-Reversal Communications

If we assume that all of the transmitted signal in the first stage is recorded by the transmitter A, then the time-reversal kernel is given by $\hat{K}_{TR}(\omega) = \xi_1(L/L_{loc}(\omega))$. The TR kernel has the form

$$\hat{K}_{TR}(\omega) = \xi_1 \left(\frac{L}{L_{loc}^B} \left(\frac{2\omega}{B} \right)^2 \right).$$

As a result, the SIR can be written as a function of L/L_{loc}^B only:

$$\text{SIR}_{(TR,coh)} = \frac{\left[\int_0^1 \xi_1 \left(\frac{L}{L_{loc}^B} x^2 \right) \hat{F}_0(x)\, dx \right]^2}{\int_0^1 \xi_1^2 \left(\frac{L}{L_{loc}^B} x^2 \right) \hat{F}_0^2(x)\, dx - \left[\int_0^1 \xi_1 \left(\frac{L}{L_{loc}^B} x^2 \right) \hat{F}_0(x)\, dx \right]^2}. \tag{13.21}$$

Comparing this with (13.19) shows that the direct-transmission system is *more efficient* than the time-reversal-communications system, in disordered media. Figure 13.3b plots the ratio of the two SIRs.

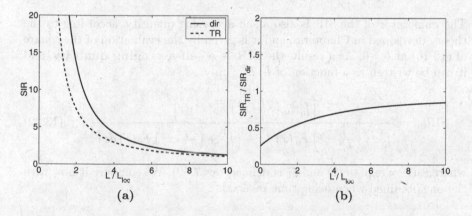

Fig. 13.3. SIRs (a) and ratio of the SIRs (b) for the two transmission systems when only the coherent pulses are taken into account. Here the Nyquist pulse is a sinc, meaning that $\hat{F}_0(x) = \pi \mathbf{1}_{[-1,1]}(x)$.

13.2.4 Influence of the Incoherent Waves

In this section we take into account the contribution to the interference term I of the small incoherent wave fluctuations. Now we consider the sum in the expression (13.18) of I extending to all integers k, including those of order ε^{-1}.

Direct Transmission

In the limit $\varepsilon \to 0$ the contribution of the incoherent wave to the numerator of the expression for SIR is negligible. However, they should be taken into account for the computation of the denominator (the interference term). As shown in Section 13.1.2, we have

$$\sum_{k \neq 0} f_{\mathrm{dir}}^2 \left(\frac{2\varepsilon \pi k}{B} \right) = \frac{B}{(2\pi)^2} \int |\hat{f}_0(\omega)|^2 |T_\omega^\varepsilon|^2 \, d\omega - f_{\mathrm{dir}}^2(0) .$$

As a consequence, the denominator of the SIR is a self-averaging quantity in the limit $\varepsilon \to 0$ that is given by

$$
\begin{aligned}
I_{\mathrm{dir}} &= \frac{B}{(2\pi)^2} \int \mathbb{E}[\tau_\omega(L)] |\hat{f}_0(\omega)|^2 \, d\omega - \frac{1}{(2\pi)^2} \left| \int \hat{K}_{\mathrm{coh}}(\omega) \hat{f}_0(\omega) \, d\omega \right|^2 \\
&= \frac{B}{(2\pi)^2} \int \xi_1 \left(\frac{L}{L_{\mathrm{loc}}(\omega)} \right) |\hat{f}_0(\omega)|^2 \, d\omega \\
&\quad - \frac{1}{(2\pi)^2} \left| \int \xi_0 \left(\frac{L}{L_{\mathrm{loc}}(\omega)} \right) \hat{f}_0(\omega) \, d\omega \right|^2 .
\end{aligned}
$$

The numerator of the SIR is also a self-averaging quantity, according to the theory developed in Chapter 8, and it is given by the evaluation of the square of (13.10) at $t = 0$. As a result, the **SIR is a self-averaging quantity**, and it can be written as a function of L/L_{loc}^B only:

$$\mathrm{SIR}_{\mathrm{dir}} = \frac{\left[\int_0^1 \xi_0 \left(\frac{L}{L_{\mathrm{loc}}^B} x^2 \right) \hat{F}_0(x) \, dx \right]^2}{\int_0^1 \xi_1 \left(\frac{L}{L_{\mathrm{loc}}^B} x^2 \right) \hat{F}_0^2(x) \, dx - \left[\int_0^1 \xi_0 \left(\frac{L}{L_{\mathrm{loc}}^B} x^2 \right) \hat{F}_0(x) \, dx \right]^2} , \tag{13.22}$$

where $\xi_0(l) = \exp(-l/2)$ and ξ_1 is defined by (7.52). We compare it now with the one obtained when using time reversal.

Time-Reversal Communications

We assume again that $G = 1$. In the limit $\varepsilon \to 0$ the contribution of the incoherent wave in the numerator of the expression of SIR is negligible. However, they should be taken into account for the computation of the denominator (the interference term). We have

$$\sum_{k \neq 0} f_{\mathrm{TR}}^2 \left(\frac{2\varepsilon \pi k}{B} \right) = \frac{B}{(2\pi)^2} \int |\hat{f}_0(\omega)|^2 |T_\omega^\varepsilon|^4 \, d\omega - f_{\mathrm{TR}}^2(0) .$$

As a consequence, the denominator of the SIR is a self-averaging quantity in the limit $\varepsilon \to 0$ and it is given by

$$I_{\mathrm{TR}} = \frac{B}{(2\pi)^2} \int \mathbb{E}[\tau_\omega(L)^2] |\hat{f}_0(\omega)|^2 \, d\omega - \frac{1}{(2\pi)^2} \left| \int \hat{K}_{\mathrm{TR}}(\omega) \, \hat{f}_0(\omega) \, d\omega \right|^2$$

$$= \frac{B}{(2\pi)^2} \int \xi_2 \left(\frac{L}{L_{\mathrm{loc}}(\omega)} \right) |\hat{f}_0(\omega)|^2 \, d\omega$$

$$- \frac{1}{(2\pi)^2} \left| \int \xi_1 \left(\frac{L}{L_{\mathrm{loc}}(\omega)} \right) \hat{f}_0(\omega) \, d\omega \right|^2 .$$

The SIR is therefore a self-averaging quantity, and it can be written as a function of L/L_{loc}^B only:

$$\mathrm{SIR}_{\mathrm{TR}} = \frac{\left[\int_0^1 \xi_1 \left(\frac{L}{L_{\mathrm{loc}}^B} x^2 \right) \hat{F}_0(x) \, dx \right]^2}{\int_0^1 \xi_2 \left(\frac{L}{L_{\mathrm{loc}}^B} x^2 \right) \hat{F}_0^2(x) \, dx - \left[\int_0^1 \xi_1 \left(\frac{L}{L_{\mathrm{loc}}^B} x^2 \right) \hat{F}_0(x) \, dx \right]^2} . \tag{13.23}$$

Here the ξ_n's are defined by (7.50). Comparing with (13.22) we see that, also when all contributions to interferences are included, the direct-communications system is *more efficient* than the TR-communications system, for disordered media. The SIRs are plotted in Figure 13.4 for a more quantitative comparison. Note that the SIR is smaller than 1 when L/L_{loc} is larger than 8.

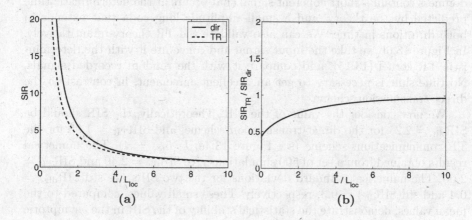

Fig. 13.4. SIRs (a) and ratio of the SIRs (b) for the two communications schemes when the incoherent wave fluctuations are taken into account. Here the Nyquist pulse is a sinc, meaning $\hat{F}_0(x) = \pi \mathbf{1}_{[-1,1]}(x)$.

13.2.5 Numerical Simulations

In this subsection we discuss the results of numerical simulations using the acoustic equations in a random medium in order to validate our theoretical predictions. We consider a medium with a piecewise-constant bulk modulus.

This means that the bulk modulus is constant in each elementary layer of thickness δz. In each layer, ν takes a value that is equal to $\pm\sigma_\kappa$ with probability $1/2$. The other parameters in the numerical simulations are: $B = 2$, $L = 250$, $\delta z = 0.1$, $\sigma_\kappa = 0.8$, $\bar{\rho} = 1$, $\bar{K} = 1$, and thus $\bar{c} = 1$. The random slab is a stack of 2500 layers. With these parameters we have $L_{\text{loc}}(B/2) = 62.5$ and $L/L_{\text{loc}} = 4$. The input pulse is a sinc function.

In Figure 13.5a we compare the signals recorded at the end of the slab in the absence of randomness and with two realizations of the random medium for the direct-transmission scheme. The recorded signals in the presence of randomness contain a short stable front that is randomly shifted, and a small-amplitude long coda. We can thus validate the theory: the shapes of the recorded signals are clearly deterministic because they do not depend on the particular realization of the medium, while the coda is changing when the medium is changed. We can also validate the theory quantitatively. In Figure 13.5b we take the input signal and convolute it with the deterministic kernel (13.11), and compare it with the random recorded signals time-shifted to allow for better comparison. The agreement is excellent.

In Figure 13.6a we compare the signals recorded at the end of the slab in the absence of randomness and with two realizations of the random medium, for the time-reversal scheme. The recorded signals in the presence of randomness contain a short coherent signal that occurs at the deterministic time predicted by the theory, and a small-amplitude long coda that extends in both directions in time. We can also validate the TR theory quantitatively. In Figure 13.6b we take the input signal and convolute it with the deterministic TR kernel (13.15), and compare it with the random recorded signals. No time shift is necessary to get an excellent agreement, in contrast to the direct-transmission scheme.

We now consider the value of the SIR. Theoretically, the SIR should be $\text{SIR}_{\text{dir}} = 2.25$ for the direct-transmission scheme, and $\text{SIR}_{\text{TR}} = 1.95$ for the TR-communications scheme (see Figure 13.4a, $L/L_{\text{loc}} = 4$). The numerical results obtained from a set of 500 simulations give $\text{SIR}_{\text{dir}} = 2.40$ and $\text{SIR}_{\text{TR}} = 2.00$. The numerical standard deviations for the two SIRs are $\text{std}(\text{SIR}_{\text{dir}}) = 0.4$ and $\text{std}(\text{SIR}_{\text{dir}}) = 0.3$, respectively. These small values, compared to the mean values, demonstrate the statistical stability of the SIR in the asymptotic regime that we are considering.

13.3 Communications in Random Media Using Modulated Nyquist Pulses

In this section we revisit the results obtained above by considering modulated Nyquist pulses. The modulation changes the picture, as can be seen by examining the expression (13.9) for the SIR. The term $\hat{K}(\omega_0 + \omega)\hat{K}(\omega_0 - \omega)$ that appears in the integral of the interference term is equal to $|\hat{K}(\omega)|^2$ if $\omega_0 = 0$, but it can contain rapid phases as soon as $\omega_0 \neq 0$.

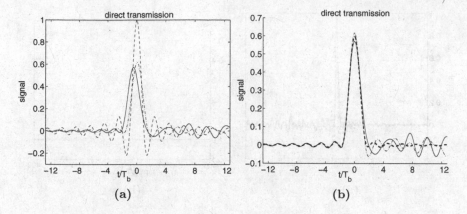

Fig. 13.5. Signals recorded at the end of the slab for the direct-transmission scheme $(T_b = 2\pi/B)$. Plot (a): Comparison between the signal in homogeneous medium (dashed lines) and the signals obtained from simulations with two different realizations of the random medium (solid and dot-dashed lines). Plot (b): Comparison between the theoretical pulse front predicted by the theory (thick dashed lines) and the signals obtained from simulations (and time-shifted by the user to remove the random time shifts). The pulse front is predicted very well.

13.3.1 SIRs of Modulated Nyquist Pulses

The analysis follows the same lines as the one in Section 13.2. It is based on formula (13.9). In the direct-transmission system we remove the random

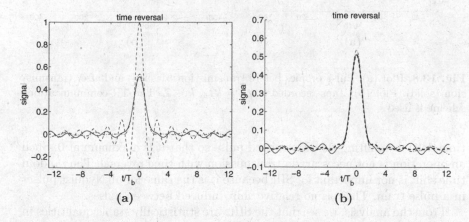

Fig. 13.6. Signals recorded at the end of the slab for the time-reversal scheme. Plot (a): Comparison between the signal in a homogeneous medium (dashed lines) and the signals obtained from simulations with two different realizations of the random medium (solid and dot-dashed lines). Plot (b): Comparison between the theoretical refocused pulse predicted by the TR theory (thick dashed lines) and the signals obtained from simulations (no time shift is performed).

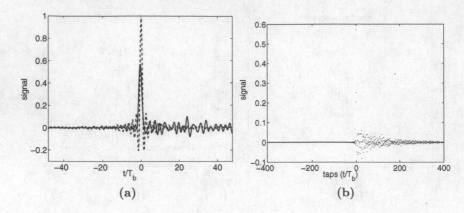

Fig. 13.7. Plot (a): Pulse profiles before transmission (dashed) and after transmission (solid). Plot (b): Taps recorded at $t_k = kT_b$, $k \in \mathbb{Z}$. The direct-transmission scheme is used.

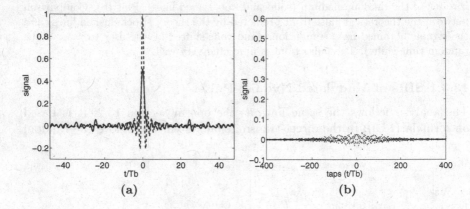

Fig. 13.8. Plot (a): Pulse profiles before transmission (dashed) and after transmission (solid). Plot (b): Taps recorded at $t_k = kT_b$, $k \in \mathbb{Z}$. The TR-communications scheme is used.

time delay by shifting the transmitted pulse so that it is maximum at 0. Such an operation is not necessary in transmission with time reversal. The random time shift is not important for SIR because it is the same for all Nyquist pulses in a pulse train. There is no relative shift induced between pulses.

From the analysis, we see that the SIRs are statistically stable quantities in the sense that they do not depend on the particular realization of the random medium, but only on its statistical properties. Explicit expressions for the SIRs can be obtained, and they can be cast in a form that depends only on $L/L_{\mathrm{loc}}(\omega_0)$ and B/ω_0. The SIRs for direct transmission and for time-reversal transmission are given by the two following expressions:

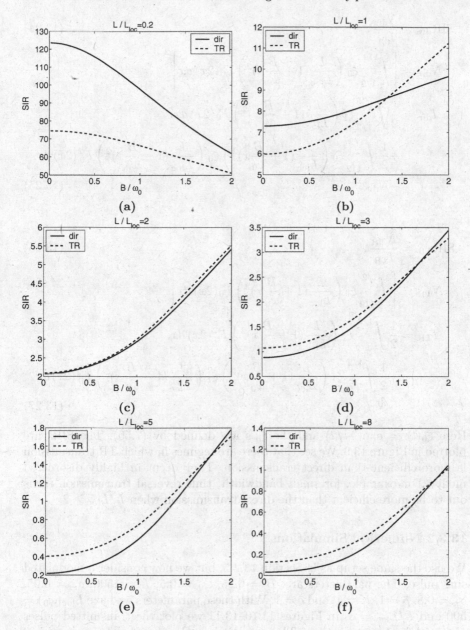

Fig. 13.9. SIRs for direct- and TR-transmission schemes with modulated Nyquist pulses versus the bandwidth. Here $L_{loc} = L_{loc}(\omega_0)$.

$$\mathrm{SIR_{dir}} = \frac{N_{\mathrm{dir}}}{I_{\mathrm{dir}}}, \tag{13.24}$$

$$N_{\mathrm{dir}} = \left[\int_{-1/2}^{1/2} \xi_0 \left(\frac{L}{L_{\mathrm{loc}}} (1 + \frac{B}{\omega_0} x)^2 \right) \hat{F}_0(2x)\, dx \right]^2,$$

$$I_{\mathrm{dir}} = \frac{1}{2} \int_{-1/2}^{1/2} \xi_1 \left(\frac{L}{L_{\mathrm{loc}}} (1 + \frac{B}{\omega_0} x)^2 \right) \hat{F}_0(2x)\, dx$$

$$+ \frac{1}{2} \int_{-1/2}^{1/2} \xi_0 \left(\frac{L}{L_{\mathrm{loc}}} (1 + \frac{B}{\omega_0} x)^2 \right) \xi_0 \left(\frac{L}{L_{\mathrm{loc}}} (1 - \frac{B}{\omega_0} x)^2 \right) \hat{F}_0^2(2x)\, dx$$

$$- N_{\mathrm{dir}}, \tag{13.25}$$

$$\mathrm{SIR_{TR}} = \frac{N_{\mathrm{TR}}}{I_{\mathrm{TR}}}, \tag{13.26}$$

$$N_{\mathrm{TR}} = \left[\int_{-1/2}^{1/2} \xi_1 \left(\frac{L}{L_{\mathrm{loc}}} (1 + \frac{B}{\omega_0} x)^2 \right) \hat{F}_0(2x)\, dx \right]^2,$$

$$I_{\mathrm{TR}} = \frac{1}{2} \int_{-1/2}^{1/2} \xi_2 \left(\frac{L}{L_{\mathrm{loc}}} (1 + \frac{B}{\omega_0} x)^2 \right) \hat{F}_0^2(2x)\, dx$$

$$+ \frac{1}{2} \int_{-1/2}^{1/2} \xi_1 \left(\frac{L}{L_{\mathrm{loc}}} (1 + \frac{B}{\omega_0} x)^2 \right) \xi_1 \left(\frac{L}{L_{\mathrm{loc}}} (1 - \frac{B}{\omega_0} x)^2 \right) \hat{F}_0^2(2x)\, dx$$

$$- N_{\mathrm{TR}}. \tag{13.27}$$

Here $\xi_0(l) = \exp(-l/2)$ and the ξ_n's are defined by (7.50). The SIRs are plotted in Figure 13.9. We see that there are regimes in which TR transmission is more efficient than direct transmission. They occur in highly disordered media. For example, for small bandwidth, time-reversal transmission turns out to be more efficient than the direct transmission when $L/L_{\mathrm{loc}} \geq 2$.

13.3.2 Numerical Simulations

We use the same setup as in Section 13.2.5, but we now consider a modulated sinc pulse. The parameters are: $B = 0.25$, $\omega_0 = 0.5$, $L = 1500$, $\delta z = 0.05$, $\sigma_\kappa = 0.8$, $\bar{\rho} = 1$, $\bar{K} = 1$, and $\bar{c} = 1$. With these parameters we have $L_{\mathrm{loc}}(\omega_0) = 500$ and $L/L_{\mathrm{loc}} = 3$. In Figures 13.10–13.11 we plot the transmitted pulses obtained for both direct and TR transmission. Comparison of the numerical pulse fronts and the theoretical predictions shows once again good agreement and statistical stability.

Theoretically, the SIR should be $\mathrm{SIR_{dir}} = 1.0$ for direct transmission and $\mathrm{SIR_{TR}} = 1.2$ for the TR transmission (see Figure 13.9d, $L/L_{\mathrm{loc}} = 3$, $B/\omega_0 = 0.5$). The numerical results obtained with 500 realizations give $\mathrm{SIR_{dir}} = 1.1$ and $\mathrm{SIR_{TR}} = 1.3$. The numerical standard deviations for the two SIRs are std(SIR_{dir}) = 0.25 and std(SIR_{TR}) = 0.2, respectively. The small values of the standard deviations compared to their respective mean values come from the statistical stability of the SIR. Note that in the considered configuration the SIR is higher for TR transmission than for direct transmission. This can also be seen in the plot of the taps of the transmitted signals (Figure 13.12) where the level of fluctuations of the TR refocused pulse is smaller than that of the directly transmitted pulse.

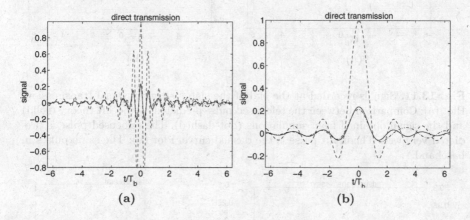

Fig. 13.10. Signals recorded at the end of the slab for direct transmission ($T_b = 2\pi/B$). Plot (a): Comparison between the pulse front predicted by the theory (solid) and the signal obtained from simulations, with the random time shift removed (dot-dashed). The pulse front is predicted very well. The input pulse is the dashed curve. Plot (b): The same pulses in baseband.

13.3.3 Discussion

The differences seen in SIR in media with weak and strong disorder can be explained as follows. Time-reversal transmission has a disadvantage because the random channel is traversed twice, so that more energy is scattered. However, it has an advantage because it recompresses the incoherent waves. Depending on the strength of the disorder, the gain may or may not dominate the loss.

Communications systems in ultra-wideband channels with time reversal, equalization, and finite SNR are considered in [162]. Spatial focusing for time reversal in communications applications is considered in [127]. An important reason for using time reversal is to have transmission with **low probability of intercept**, by using a cutoff function that excludes the coherent part [99]. If a tap is placed in the channel at C, between A and B, then it will be very difficult to detect a coherent signal.

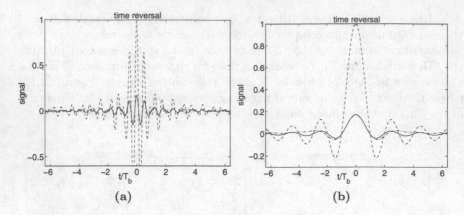

Fig. 13.11. Signals recorded at the end of the slab for time-reversal transmission. Plot (a): Comparison between the refocused pulse predicted by the TR theory (solid) and the signal obtained from simulations (dot-dashed). The refocused pulse is predicted very well. The input pulse is the dashed curve. Plot (b): The same pulses in baseband.

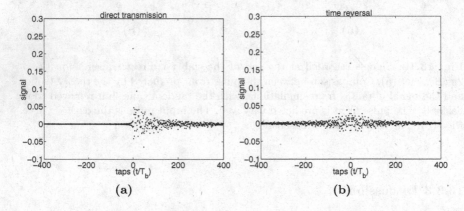

Fig. 13.12. Taps at $t = kT_b$, $k \in \mathbb{Z}$, of the recorded signals in baseband for both methods.

Notes

The application to communications discussed in this chapter is presented for the first time in this book. We refer to [141] for a systematic presentation of broadband communications systems.

Scattering by a Three-Dimensional Randomly Layered Medium

In this chapter we extend the theory presented in the previous chapters that deal with one-dimensional waves to the case with waves generated by a point source and scattered by a three-dimensional randomly layered slab. The layered medium varies only with respect to one coordinate, which we choose to be the z-coordinate in our notation. In this chapter we consider layered media that vary only in the slab section $z \in (-L, 0)$. We do not limit ourselves to piecewise constant media, but rather model the z-variation in the same way as in the one-dimensional case through coefficients depending on z.

We pay special attention to the modeling of a point source emitting a short pulse in Section 14.1. The techniques that we present here involve a decomposition of the wave field into **plane-wave modes** that propagate as one-dimensional waves. Each mode is characterized by a propagator that has a similar form as before, and these different propagators are statistically coupled through the random medium variations. We consider a regime of separation of scales similar to the one studied in the one-dimensional case, that is, the strongly heterogeneous white-noise regime, in which the correlation length of the medium is smaller than the wavelength, which is smaller than the distance of propagation or the size of the slab. The **stationary-phase method**, introduced in this chapter and combined with diffusion approximation results, plays a crucial role in the asymptotic analysis of wave propagation in randomly layered media. The techniques of decomposing the wave into modes and using stationary phase will also be important in Chapters 15 and 16, where we discuss various three-dimensional time-reversal experiments. In Section 14.2 we extend the stabilization of the wave front of Chapter 8 to the three-dimensional setting. In Section 14.3 we study the incoherent waves reflected by the random slab. This study generalizes the results obtained in Chapter 9 in the one-dimensional case.

14.1 Acoustic Waves in Three Dimensions

We consider linear acoustic waves propagating in three spatial dimensions:

$$\rho \frac{\partial \mathbf{u}}{\partial t} + \nabla p = \mathbf{F}, \tag{14.1}$$

$$\frac{1}{K} \frac{\partial p}{\partial t} + \nabla \cdot \mathbf{u} = 0, \tag{14.2}$$

where p is the pressure, $\mathbf{u}$ is the velocity, ρ is the density of the medium, and K the bulk modulus. We will write the three-dimensional spatial variable as $(\mathbf{x}, z) = (x, y, z)$ and we will refer to z as the vertical variable and to $\mathbf{x}$ as the horizontal variable. The velocity field has three components as well, which we denote by $\mathbf{u} = (\mathbf{v}, u) = (v_1, v_2, u)$. The source is modeled by the forcing term

$$\mathbf{F}(t, \mathbf{x}, z) = \mathbf{f}(t, \mathbf{x})\delta(z - z_s).$$

14.1.1 Homogenization Regime

We first consider the regime in which the wave propagates over a distance that is of the same order as the typical wavelength of the source, while the density and the bulk modulus are randomly varying on a fine scale along the z-coordinate:

$$K(\mathbf{x}, z) = K(z/\varepsilon), \qquad \rho(\mathbf{x}, z) = \rho(z/\varepsilon),$$

where ρ and K are ergodic random processes, which we assume are bounded and bounded away from zero. In this section we revisit the homogenization regime studied in the one-dimensional case in Chapter 4. The natural step now is to carry out a Fourier transform with respect to the horizontal variables $\mathbf{x}$. It is actually convenient to carry out a joint Fourier transform in time and space in order to obtain equations in the Fourier domain having the same form as those in the one-dimensional case. This joint transform enables us to decompose the waves into right- and left-going wave modes in the z-direction:

$$\hat{\mathbf{u}}(\omega, \boldsymbol{\kappa}, z) = \int \int e^{i\omega(t - \boldsymbol{\kappa} \cdot \mathbf{x})} \mathbf{u}(t, \mathbf{x}, z) \, dt \, d\mathbf{x},$$

$$\hat{p}(\omega, \boldsymbol{\kappa}, z) = \int \int e^{i\omega(t - \boldsymbol{\kappa} \cdot \mathbf{x})} p(t, \mathbf{x}, z) \, dt \, d\mathbf{x},$$

where $\boldsymbol{\kappa}$ denotes the two-dimensional slowness vector. The inverse transform is given by

$$p(t, \mathbf{x}, z) = \frac{1}{(2\pi)^3} \int \int e^{-i\omega(t - \boldsymbol{\kappa} \cdot \mathbf{x})} \hat{p}(\omega, \boldsymbol{\kappa}, z)\omega^2 \, d\omega \, d\boldsymbol{\kappa},$$

with a similar formula for the velocity. Observe the presence of the factor ω^2 in this Fourier inverse due to our specific choice of Fourier variables and the number of transverse dimensions. The acoustic equations in the Fourier domain read

$$-i\omega\rho\left(\frac{z}{\varepsilon}\right)\hat{\mathbf{v}} + i\omega\boldsymbol{\kappa}\hat{p} = \hat{\mathbf{f}}_{\mathbf{x}}(\omega, \boldsymbol{\kappa})\delta(z - z_s), \qquad (14.3)$$

$$-i\omega\rho\left(\frac{z}{\varepsilon}\right)\hat{u} + \frac{\partial\hat{p}}{\partial z} = \hat{f}_z(\omega, \boldsymbol{\kappa})\delta(z - z_s),$$

$$-\frac{i\omega}{K(z/\varepsilon)}\hat{p} + i\omega\boldsymbol{\kappa}\cdot\hat{\mathbf{v}} + \frac{\partial\hat{u}}{\partial z} = 0.$$

Here $\hat{\mathbf{v}}$ and $\hat{u}$ denote respectively the Fourier transform of the horizontal velocity field $\mathbf{v}$ and that of the vertical velocity field u. By eliminating $\hat{\mathbf{v}}$ we deduce that $(\hat{u}, \hat{p})$ satisfy the following closed system for $z \neq z_s$:

$$\frac{d\hat{p}}{dz} = i\omega\rho\left(\frac{z}{\varepsilon}\right)\hat{u}, \qquad (14.4)$$

$$\frac{d\hat{u}}{dz} = i\omega\left(\frac{1}{K(z/\varepsilon)} - \frac{\kappa^2}{\rho(z/\varepsilon)}\right)\hat{p}, \qquad (14.5)$$

where $\kappa = |\boldsymbol{\kappa}|$ is the slowness of the mode. Thus we have reduced the acoustic equations to a family of random ordinary differential equations for the modes. By applying the averaging theorem presented in Section 4.5.2 in the limit $\varepsilon \to 0$, we get the effective system

$$\frac{d\hat{p}}{dz} = i\omega\bar{\rho}\hat{u}, \qquad (14.6)$$

$$\frac{d\hat{u}}{dz} = i\omega\left(\frac{1}{\overline{K}} - \frac{\kappa^2}{\widetilde{\rho}}\right)\hat{p}, \qquad (14.7)$$

where $\bar{\rho} = \mathbb{E}[\rho]$ is the mean density, $\widetilde{\rho} = (\mathbb{E}[\rho^{-1}])^{-1}$ is the harmonic mean density, and $\overline{K} = (\mathbb{E}[K^{-1}])^{-1}$ is the harmonic mean bulk modulus. Starting from the following subsection, we will consider the case in which there is no fluctuation in the density, so that $\bar{\rho} = \widetilde{\rho}$. We will return to the case with random fluctuations in the density in Section 17.3, where the difference between $\bar{\rho}$ and $\widetilde{\rho}$ will play a role. The system (14.6–14.7) can be reduced to the following second-order homogeneous differential equation for $\hat{u}$:

$$\frac{d^2\hat{u}}{dz^2} + \omega^2 D(\kappa)\hat{u} = 0,$$

where we have defined

$$D(\kappa) = \frac{\bar{\rho}}{\overline{K}} - \frac{\kappa^2\bar{\rho}}{\widetilde{\rho}}.$$

Note that $\hat{p}$ satisfies the same equation. The form of the solution depends on the sign of $D(\kappa)$. If κ is smaller than $\kappa_{\max}$ defined by

$$\kappa_{\max} = \sqrt{\frac{\widetilde{\rho}}{\overline{K}}}, \qquad (14.8)$$

then $D(\kappa) > 0$ and the general solution is a superposition of **propagating modes**

$$\exp\left[\pm i\omega\sqrt{D(\kappa)}(z - z_s)\right].$$

If κ is larger than $\kappa_{\max}$, then $D(\kappa) < 0$ and the solution is an **evanescent mode**,

$$\exp\left[\pm\omega\sqrt{-D(\kappa)}(z - z_s)\right],$$

with the sign chosen so as to select the exponentially decaying solution. In the next section we describe the diffusion approximation regime in which the wave propagates over distances much larger than the wavelength. Therefore, in that context, the evanescent modes will not play any role, and only the propagating modes will contribute to the quantities of interest.

14.1.2 The Diffusion Approximation Regime

We shall now discuss the strongly heterogeneous white-noise regime, in which the correlation length of the medium $\sim \varepsilon^2$ is much smaller than the typical wavelength $\sim \varepsilon$, which is itself much smaller than the propagation distance ~ 1. For simplicity we start by considering the case in which the density is constant and the bulk modulus is randomly varying. As one can see from (14.3), this permits a simple representation of the transverse velocity field $\mathbf{v}^\varepsilon$ in terms of the pressure field p^ε, and subsequently a reduction of the four-dimensional system in $(u^\varepsilon, \mathbf{v}^\varepsilon, p^\varepsilon)$ to a two-dimensional system in $(u^\varepsilon, p^\varepsilon)$. The generalization to the case with a randomly varying density is presented in Section 17.3.

From the results of the previous section we know that the reciprocal of the bulk modulus is homogenized, and therefore we model fluctuations around this quantity. The bulk modulus is z-dependent in the slab $(-L, 0)$ and constant outside:

$$\frac{1}{K(\mathbf{x}, z)} = \frac{1}{K(z)} = \begin{cases} \frac{1}{\bar{K}}\left(1 + \nu(z/\varepsilon^2)\right) & \text{for } z \in [-L, 0], \\ \frac{1}{\bar{K}} & \text{for } z \in (-\infty, -L) \cup (0, \infty), \end{cases}$$

$$\rho(\mathbf{x}, z) = \bar{\rho} \text{ for all } (\mathbf{x}, z).$$

In this chapter we assume a **matched medium** at both ends of the slab, so that the homogenized coefficients $\bar{\rho}$ and $1/\bar{K}$ are constant in the full space. The nonmatched case will be treated in Section 17.1.

The source term has the form

$$\mathbf{F}^\varepsilon(t, \mathbf{x}, z) = \varepsilon^q \begin{bmatrix} \mathbf{f_x} \\ f_z \end{bmatrix} \left(\frac{t}{\varepsilon}, \frac{\mathbf{x}}{\varepsilon}\right) \delta(z - z_s), \tag{14.9}$$

where the multplicative factor ε^q gives the amplitude scaling and the exponent q will be specified later. The time duration of the source is short, of order ε; the source is located to the right of the random slab at $z = z_s$ with $z_s > 0$;

and its horizontal support, centered at the origin, is also on the scale ε. We denote by $\mathbf{f_x}$ the transverse components of the source and by f_z its vertical component. This setup is illustrated in Figure 14.1.

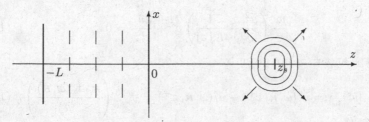

Fig. 14.1. Initial setup. The random slab occupies the region $z \in [-L, 0]$. The source is located at $(\mathbf{0}, z_s)$ with $z_s > 0$.

14.1.3 Plane-Wave Fourier Transform

We carry out a joint Fourier transform in time and transverse spatial coordinates. The goal is to obtain a decomposition of the waves into right- and left-going wave modes in the z-direction. As in the previous chapters we look at the wave on the ε time scale, which is equivalent to using high frequencies ω/ε. We therefore introduce the **specific Fourier transform** of the pressure

$$\hat{p}^{\varepsilon}(\omega, \boldsymbol{\kappa}, z) = \int \int e^{i\frac{\omega}{\varepsilon}(t - \boldsymbol{\kappa} \cdot \mathbf{x})} p^{\varepsilon}(t, \mathbf{x}, z)\, dt\, d\mathbf{x},$$

with a similar formula for the vertical velocity $\hat{u}^{\varepsilon}$ and the horizontal velocity $\hat{\mathbf{v}}^{\varepsilon}$. The inverse transform is given by

$$p^{\varepsilon}(t, \mathbf{x}, z) = \frac{1}{(2\pi\varepsilon)^3} \int \int e^{-i\frac{\omega}{\varepsilon}(t - \boldsymbol{\kappa} \cdot \mathbf{x})} \hat{p}^{\varepsilon}(\omega, \boldsymbol{\kappa}, z)\omega^2\, d\omega\, d\boldsymbol{\kappa}, \quad (14.10)$$

with again a similar formula for the velocity fields. Taking the specific Fourier transform gives that $\hat{\mathbf{u}}^{\varepsilon} = (\hat{\mathbf{v}}^{\varepsilon}, \hat{u}^{\varepsilon})$ and $\hat{p}^{\varepsilon}$ satisfy the system

$$-\bar{\rho}\frac{i\omega}{\varepsilon}\hat{\mathbf{v}}^{\varepsilon} + \frac{i\omega}{\varepsilon}\boldsymbol{\kappa}\hat{p}^{\varepsilon} = \varepsilon^{q+3}\hat{\mathbf{f}}_{\mathbf{x}}(\omega, \boldsymbol{\kappa})\delta(z - z_s), \quad (14.11)$$

$$-\bar{\rho}\frac{i\omega}{\varepsilon}\hat{u}^{\varepsilon} + \frac{d\hat{p}^{\varepsilon}}{dz} = \varepsilon^{q+3}\hat{f}_z(\omega, \boldsymbol{\kappa})\delta(z - z_s), \quad (14.12)$$

$$-\frac{1}{K(z)}\frac{i\omega}{\varepsilon}\hat{p}^{\varepsilon} + \frac{i\omega}{\varepsilon}\boldsymbol{\kappa} \cdot \hat{\mathbf{v}}^{\varepsilon} + \frac{d\hat{u}^{\varepsilon}}{dz} = 0, \quad (14.13)$$

where $\hat{f}$ denotes the unscaled specific Fourier transform:

$$\hat{f}(\omega, \boldsymbol{\kappa}) = \int \int f(t, \mathbf{x})e^{i\omega(t - \boldsymbol{\kappa} \cdot \mathbf{x})}\, dt\, d\mathbf{x}. \quad (14.14)$$

By eliminating $\hat{v}^\varepsilon$ we deduce that $(\hat{u}^\varepsilon, \hat{p}^\varepsilon)$ satisfy the following closed system for $z \neq z_s$:

$$-\frac{i\omega}{\varepsilon}\bar{\rho}\hat{u}^\varepsilon + \frac{d\hat{p}^\varepsilon}{dz} = 0\,, \tag{14.15}$$

$$\frac{i\omega}{\varepsilon}\left(\frac{\kappa^2}{\bar{\rho}} - \frac{1}{K(z)}\right)\hat{p}^\varepsilon + \frac{d\hat{u}^\varepsilon}{dz} = 0\,. \tag{14.16}$$

The jumps at $z = z_s$ are given by

$$[\hat{u}^\varepsilon]_{z_s} := \hat{u}^\varepsilon(\omega, \boldsymbol{\kappa}, z_s^+) - \hat{u}^\varepsilon(\omega, \boldsymbol{\kappa}, z_s^-) = \varepsilon^{q+3}\left(\frac{\boldsymbol{\kappa}\cdot\hat{\mathbf{f}}_{\mathbf{x}}(\omega, \boldsymbol{\kappa})}{\bar{\rho}}\right)\,, \tag{14.17}$$

$$[\hat{p}^\varepsilon]_{z_s} := \hat{p}^\varepsilon(\omega, \boldsymbol{\kappa}, z_s^+) - \hat{p}^\varepsilon(\omega, \boldsymbol{\kappa}, z_s^-) = \varepsilon^{q+3}\left(\hat{f}_z(\omega, \boldsymbol{\kappa})\right)\,. \tag{14.18}$$

We now define the effective **mode speed** for the propagating modes $\kappa < \kappa_{\max}$:

$$\bar{c}(\kappa) = \frac{\bar{c}}{\sqrt{1 - \kappa^2\bar{c}^2}}\,, \tag{14.19}$$

where $\kappa_{\max} = \sqrt{\bar{K}/\bar{\rho}} = 1/\bar{c}$ has been introduced in (14.8). Similarly we define the effective **mode acoustic impedance** and the mode-dependent effective bulk modulus by

$$\bar{\zeta}(\kappa) = \bar{\rho}\bar{c}(\kappa)\,, \qquad \bar{K}(\kappa) = \bar{\rho}\bar{c}(\kappa)^2\,. \tag{14.20}$$

With these definitions the equations for $\hat{p}^\varepsilon$ and $\hat{u}^\varepsilon$ take the form

$$-\frac{i\omega}{\varepsilon}\bar{\rho}\hat{u}^\varepsilon + \frac{d\hat{p}^\varepsilon}{dz} = 0\,, \tag{14.21}$$

$$-\frac{i\omega}{\varepsilon}\frac{1}{\bar{K}(\kappa)}\left(1 + \frac{\bar{c}(\kappa)^2}{\bar{c}^2}\nu\left(\frac{z}{\varepsilon^2}\right)\right)\hat{p}^\varepsilon + \frac{d\hat{u}^\varepsilon}{dz} = 0\,. \tag{14.22}$$

These equations show that, mode by mode, the problem is a one-dimensional wave-propagation problem with the mode-dependent medium fluctuations defined by

$$\nu_\kappa(z) = \frac{\bar{c}(\kappa)^2}{\bar{c}^2}\nu(z)\,. \tag{14.23}$$

14.1.4 One-Dimensional Mode Problems

By analogy with (5.20) in the one-dimensional case, we decompose the wavefield into right- ($\check{a}^\varepsilon = \check{a}^\varepsilon(\omega, \boldsymbol{\kappa}, z)$) and left-going ($\check{b}^\varepsilon = \check{b}^\varepsilon(\omega, \boldsymbol{\kappa}, z)$) waves with respect to the z-direction by setting

$$\hat{p}^\varepsilon(\omega, \boldsymbol{\kappa}, z) = \frac{\sqrt{\bar{\zeta}(\kappa)}}{2}\left(\check{a}^\varepsilon(\omega, \boldsymbol{\kappa}, z)e^{\frac{i\omega z}{\varepsilon\bar{c}(\kappa)}} - \check{b}^\varepsilon(\omega, \boldsymbol{\kappa}, z)e^{-\frac{i\omega z}{\varepsilon\bar{c}(\kappa)}}\right)\,, \tag{14.24}$$

$$\hat{u}^\varepsilon(\omega, \boldsymbol{\kappa}, z) = \frac{1}{2\sqrt{\bar{\zeta}(\kappa)}}\left(\check{a}^\varepsilon(\omega, \boldsymbol{\kappa}, z)e^{\frac{i\omega z}{\varepsilon\bar{c}(\kappa)}} + \check{b}^\varepsilon(\omega, \boldsymbol{\kappa}, z)e^{-\frac{i\omega z}{\varepsilon\bar{c}(\kappa)}}\right)\,. \tag{14.25}$$

Substituting these expressions into (14.21–14.22) establishes the system satisfied by the modes $(\breve{a}^{\varepsilon}, \breve{b}^{\varepsilon})$:

$$\frac{d\breve{a}^{\varepsilon}}{dz} = \frac{i\omega}{2\bar{c}(\kappa)\varepsilon}\nu_{\kappa}\left(\frac{z}{\varepsilon^2}\right)\left(\breve{a}^{\varepsilon} - e^{\frac{-2i\omega z}{\bar{c}(\kappa)\varepsilon}}\breve{b}^{\varepsilon}\right), \qquad (14.26)$$

$$\frac{d\breve{b}^{\varepsilon}}{dz} = \frac{i\omega}{2\bar{c}(\kappa)\varepsilon}\nu_{\kappa}\left(\frac{z}{\varepsilon^2}\right)\left(e^{\frac{2i\omega z}{\bar{c}(\kappa)\varepsilon}}\breve{a}^{\varepsilon} - \breve{b}^{\varepsilon}\right). \qquad (14.27)$$

Using the definitions (14.24) and (14.25) of $\breve{a}^{\varepsilon}$ and $\breve{b}^{\varepsilon}$ and the expressions (14.17) and (14.18) for the jumps in $\hat{u}^{\varepsilon}$ and $\hat{p}^{\varepsilon}$, we deduce the jumps at $z = z_s$ for the modes $\breve{a}^{\varepsilon}$ and $\breve{b}^{\varepsilon}$:

$$\left[\breve{a}^{\varepsilon}\right]_{z_s} = \varepsilon^{q+3}\left(\frac{\sqrt{\zeta(\kappa)}}{\bar{\rho}}\boldsymbol{\kappa}\cdot\hat{\mathbf{f}}_{\mathbf{x}}(\omega,\boldsymbol{\kappa}) + \frac{1}{\sqrt{\zeta(\kappa)}}\hat{f}_z(\omega,\boldsymbol{\kappa})\right)e^{\frac{-i\omega z_s}{\varepsilon\bar{c}(\kappa)}}, \quad (14.28)$$

$$\left[\breve{b}^{\varepsilon}\right]_{z_s} = \varepsilon^{q+3}\left(\frac{\sqrt{\zeta(\kappa)}}{\bar{\rho}}\boldsymbol{\kappa}\cdot\hat{\mathbf{f}}_{\mathbf{x}}(\omega,\boldsymbol{\kappa}) - \frac{1}{\sqrt{\zeta(\kappa)}}\hat{f}_z(\omega,\boldsymbol{\kappa})\right)e^{\frac{i\omega z_s}{\varepsilon\bar{c}(\kappa)}}. \quad (14.29)$$

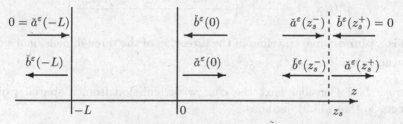

Fig. 14.2. Boundary conditions for the modes $\breve{a}^{\varepsilon}$ and $\breve{b}^{\varepsilon}$.

The system for $\breve{a}^{\varepsilon}$ and $\breve{b}^{\varepsilon}$ is associated with the boundary conditions at $z = z_s$ and $z = -L$ that are shown in Figure 14.2. We assume that no energy is coming from $+\infty$ and $-\infty$, so that we get the radiation conditions

$$\breve{a}^{\varepsilon}(\omega,\boldsymbol{\kappa},-L) = 0, \qquad \breve{b}^{\varepsilon}(\omega,\boldsymbol{\kappa},z_s^+) = 0.$$

The jump condition (14.29) then gives $\breve{b}^{\varepsilon}(\omega,\boldsymbol{\kappa},z_s^-) = \varepsilon^{q+3}\breve{S}(\omega,\boldsymbol{\kappa})$, where

$$\breve{S}(\omega,\boldsymbol{\kappa}) = \left(-\frac{\sqrt{\zeta(\kappa)}}{\bar{\rho}}\boldsymbol{\kappa}\cdot\hat{\mathbf{f}}_{\mathbf{x}}(\omega,\boldsymbol{\kappa}) + \frac{1}{\sqrt{\zeta(\kappa)}}\hat{f}_z(\omega,\boldsymbol{\kappa})\right)e^{\frac{i\omega z_s}{\varepsilon\bar{c}(\kappa)}}. \quad (14.30)$$

The boundary conditions for the right- and left-going wave components entering the slab are therefore

$$\breve{a}^{\varepsilon}(\omega,\boldsymbol{\kappa},-L) = 0, \qquad\qquad (14.31)$$

$$\breve{b}^{\varepsilon}(\omega,\boldsymbol{\kappa},0) = \varepsilon^{q+3}\breve{S}(\omega,\boldsymbol{\kappa}). \qquad\qquad (14.32)$$

The particular form of $\check{S}$ depends on the choice of the physical source. Next we give two simple examples with particular choices of the source term in the case of a homogeneous medium so that $\nu = 0$.

Example 14.1. Assume that medium is homogeneous, $\nu \equiv 0$, and that the source emits in the z-direction and has the following form:

$$\hat{\mathbf{f}}_{\mathbf{x}}(\omega, \boldsymbol{\kappa}) = \mathbf{0}, \qquad \hat{f}_z(\omega, \boldsymbol{\kappa}) = \hat{f}(\omega)\delta(\boldsymbol{\kappa} - \boldsymbol{\kappa}_0),$$

which corresponds to

$$\mathbf{f}_{\mathbf{x}}(t, \mathbf{x}) = \mathbf{0}, \qquad f_z(t, \mathbf{x}) = -\frac{1}{4\pi^2}f''(t - \boldsymbol{\kappa}_0 \cdot \mathbf{x}),$$

using the unscaled specific Fourier transform (14.14). In this case we have

$$\check{S}(\omega, \boldsymbol{\kappa}) = \frac{1}{\sqrt{\zeta(\boldsymbol{\kappa}_0)}}\hat{f}_z(\omega)e^{\frac{i\omega z_s}{\varepsilon\bar{c}(\boldsymbol{\kappa}_0)}}\delta(\boldsymbol{\kappa} - \boldsymbol{\kappa}_0).$$

Let us also assume that $q = 2$, so that for $z < z_s$, we have

$$p^\varepsilon(t, \mathbf{x}, z) = \frac{1}{8\pi^2}f''\left(\frac{t - \boldsymbol{\kappa}_0 \cdot \mathbf{x} + (z - z_s)/\bar{c}(\boldsymbol{\kappa}_0)}{\varepsilon}\right).$$

This is a **plane-wave** traveling in the direction of the three-dimensional slowness vector $\mathbf{s} = \left(\boldsymbol{\kappa}_0, -\sqrt{1 - \kappa_0^2\bar{c}^2}/\bar{c}\right)$. $\square$

Example 14.2. Consider next the case with emission from a spatial **point source**, so that (14.9) becomes

$$\mathbf{F}^\varepsilon(t, \mathbf{x}, z) = \varepsilon^q \begin{bmatrix} \mathbf{f}_{\mathbf{x}} \\ f_z \end{bmatrix}\left(\frac{t}{\varepsilon}\right)\delta\left(\frac{\mathbf{x} - \mathbf{x}_s}{\varepsilon}\right)\delta(z - z_s). \qquad (14.33)$$

Then we find a special case of (14.30),

$$\check{S}(\omega, \boldsymbol{\kappa}) = \left(-\frac{\sqrt{\zeta(\boldsymbol{\kappa})}}{\bar{\rho}}\boldsymbol{\kappa} \cdot \hat{\mathbf{f}}_{\mathbf{x}}(\omega) + \frac{1}{\sqrt{\zeta(\boldsymbol{\kappa})}}\hat{f}_z(\omega)\right)e^{i\frac{\omega}{\varepsilon}(z_s/\bar{c}(\boldsymbol{\kappa}) - \boldsymbol{\kappa}\cdot\mathbf{x}_s)}, \qquad (14.34)$$

and the pressure field in the deterministic case is then obtained by applying the specific inverse Fourier transform (14.10) to $\hat{p}^\varepsilon$ given by (14.24) with the boundary conditions for $\check{a}^\varepsilon$ and $\check{b}^\varepsilon$ given in (14.31–14.32). In the next section we introduce the stationary-phase method, which will enable us to get a simple expression for this transmitted pressure. $\square$

Henceforth we assume that the source is located at the surface, so that

$$z_s = 0,$$

$$\check{S}(\omega, \boldsymbol{\kappa}) = -\frac{\sqrt{\zeta(\boldsymbol{\kappa})}}{\bar{\rho}}\boldsymbol{\kappa} \cdot \hat{\mathbf{f}}_{\mathbf{x}}(\omega, \boldsymbol{\kappa}) + \frac{1}{\sqrt{\zeta(\boldsymbol{\kappa})}}\hat{f}_z(\omega, \boldsymbol{\kappa}).$$

Each mode can now be analyzed as in Chapter 8. We begin by introducing the propagator $\mathbf{P}^{\varepsilon}_{(\omega,\kappa)}(-L,0)$ associated with mode κ and frequency ω. Observe that the propagator depends on the mode κ only through the slowness κ. It satisfies an equation that is analogous to (8.37):

$$\frac{d}{dz}\mathbf{P}^{\varepsilon}_{(\omega,\kappa)}(-L,z) = \frac{1}{\varepsilon}\mathbf{H}_{(\omega,\kappa)}\left(\frac{z}{\varepsilon},\nu_\kappa\left(\frac{z}{\varepsilon^2}\right)\right)\mathbf{P}^{\varepsilon}_{(\omega,\kappa)}(-L,z),\quad (14.35)$$

with the initial conditions $\mathbf{P}^{\varepsilon}_{(\omega,\kappa)}(-L,z=-L) = \mathbf{I}$. This ordinary differential equation is derived exactly as in the previous chapters by replacing $\bar{c}$ with $\bar{c}(\kappa)$ and ν with ν_κ defined in (14.19) and (14.23). In this case the matrix $\mathbf{H}_{(\omega,\kappa)}$ becomes

$$\mathbf{H}_{(\omega,\kappa)}(z,\nu) = \frac{i\omega\nu}{2\bar{c}(\kappa)}\begin{bmatrix} 1 & -e^{\frac{-2i\omega z}{\bar{c}(\kappa)}} \\ e^{\frac{2i\omega z}{\bar{c}(\kappa)}} & -1 \end{bmatrix}. \quad (14.36)$$

As in the one-dimensional case discussed in Chapter 5, the propagator takes the form

$$\mathbf{P}^{\varepsilon}_{(\omega,\kappa)}(-L,0) = \begin{bmatrix} \alpha^{\varepsilon}_{(\omega,\kappa)}(-L,0) & \overline{\beta^{\varepsilon}_{(\omega,\kappa)}(-L,0)} \\ \beta^{\varepsilon}_{(\omega,\kappa)}(-L,0) & \overline{\alpha^{\varepsilon}_{(\omega,\kappa)}(-L,0)} \end{bmatrix},$$

where the coefficients depend on the frequency ω and the slowness modulus κ. The trace of $\mathbf{H}_{(\omega,\kappa)}$ is zero, and it follows again that

$$|\alpha^{\varepsilon}_{\omega,\kappa}|^2 - |\beta^{\varepsilon}_{\omega,\kappa}|^2 = 1. \quad (14.37)$$

The transmitted left-going wave is defined in terms of the harmonic amplitude $\check{b}^{\varepsilon}$, which can be found from the relations

$$\mathbf{P}^{\varepsilon}_{(\omega,\kappa)}(-L,0)\begin{bmatrix} 0 \\ \check{b}^{\varepsilon}(\omega,\kappa,-L) \end{bmatrix} = \begin{bmatrix} \check{a}^{\varepsilon}(\omega,\kappa,0^-) \\ \varepsilon^{q+3}\check{S}(\omega,\kappa) \end{bmatrix}, \quad (14.38)$$

where we have used the boundary conditions (14.31) and (14.32) with $z_s = 0$. Consequently, we have

$$\check{b}^{\varepsilon}(\omega,\kappa,-L) = \varepsilon^{q+3}\check{S}(\omega,\kappa)T^{\varepsilon}_{(\omega,\kappa)}(-L,0), \quad (14.39)$$

where we have defined the **transmission coefficient**

$$T^{\varepsilon}_{(\omega,\kappa)}(-L,0) = \frac{1}{\alpha^{\varepsilon}_{(\omega,\kappa)}(-L,0)}. \quad (14.40)$$

Using (14.37) we obtain that the transmission coefficient $T^{\varepsilon}_{(\omega,\kappa)}(-L,0)$ is uniformly bounded by one. From the same result we also get the important mode-energy-conservation relation

$$|\check{a}^{\varepsilon}(\omega,\kappa,0^-)|^2 + |\check{b}^{\varepsilon}(\omega,\kappa,-L)|^2 = |\varepsilon^{q+3}\check{S}(\omega,\kappa)|^2. \quad (14.41)$$

14.1.5 Transmitted-Pressure Integral Representation

In this section we derive an integral representation for the transmitted pressure pulse $p^\varepsilon(t, \mathbf{x}, -L)$. Substituting the expressions (14.31) for $\breve{a}^\varepsilon(\omega, \boldsymbol{\kappa}, -L)$ and (14.39) for $\breve{b}^\varepsilon(\omega, \boldsymbol{\kappa}, -L)$ into (14.24), we obtain

$$\hat{p}^\varepsilon(\omega, \boldsymbol{\kappa}, -L) = -\frac{\varepsilon^{q+3}\breve{S}(\omega, \boldsymbol{\kappa})\sqrt{\zeta(\boldsymbol{\kappa})}}{2} T^\varepsilon_{(\omega,\boldsymbol{\kappa})}(-L, 0) e^{\frac{i\omega L}{\varepsilon \bar{c}(\boldsymbol{\kappa})}}.$$

By the inverse Fourier transform (14.10) it follows that

$$p^\varepsilon(t_0 + \varepsilon s, \mathbf{x}, -L) = -\frac{1}{(2\pi\varepsilon)^3} \int\int e^{-i\frac{\omega}{\varepsilon}(t_0 + \varepsilon s - \boldsymbol{\kappa}\cdot\mathbf{x} - L/\bar{c}(\boldsymbol{\kappa}))} T^\varepsilon_{(\omega,\boldsymbol{\kappa})}(-L, 0)$$

$$\times \left\{ \frac{\varepsilon^{q+3}\breve{S}(\omega, \boldsymbol{\kappa})\sqrt{\zeta(\boldsymbol{\kappa})}}{2} \right\} \omega^2 \, d\omega \, d\boldsymbol{\kappa}, \qquad (14.42)$$

where the transmitted pressure is observed at the time $t = t_0 + \varepsilon s$, that is, in a time window centered at t_0 and on the scale ε of the initial source pulse, with s being the time variable in this "magnified" window.

14.2 The Transmitted Wave Front

We have seen that in the one-dimensional case, the front of the transmitted wave retains its shape up to a random time shift and a convolution with a Gaussian density given in Proposition 8.3. Our objective is to extend this theory to the three-dimensional case in which the front is now a surface propagating from the point source. We review here the theory that describes this front. The main result is presented in Proposition 14.3, Section 14.2.3. Convergence of finite-dimensional distributions for the wave front is derived in Section 14.2.1 using a moment argument introduced in Section 5.2.3 and discussed in Chapter 8 in the one-dimensional case.

14.2.1 Characterization of Moments

From the integral expression (14.42) we see that the transmission coefficients $T^\varepsilon_{(\omega,\boldsymbol{\kappa})}(-L, 0)$ determine the transmitted wave field. From the energy conservation (14.41) it follows that the moduli of these coefficients are bounded by one. It is important to note that the distribution of the wave in *time* and *space* depends on the *joint* distribution of the transformed wave over all frequencies ω and horizontal wave vectors $\boldsymbol{\kappa}$. We next illustrate that knowledge of the joint moments of the transmitted wave for all *finite* combinations of different frequencies and wave vectors is enough to characterize the distribution of the transmitted wave in time and space. This follows from the fact that the expectations in (14.43) below have arguments that involve a finite number of frequencies and wave vectors. A convenient way to characterize the finite-dimensional distributions of the scalar wave is to compute the joint moments of order $m_1, \ldots, m_n$,

$$\mathbb{E}[p^\varepsilon(t_{0,1} + \varepsilon s_1, \mathbf{x}_1, -L)^{m_1} \cdots p^\varepsilon(t_{0,n} + \varepsilon s_n, \mathbf{x}_n, -L)^{m_n}],$$

which, using (14.42), can be written in an integral form with respect to the variables $\omega_{j,l}, \boldsymbol{\kappa}_{j,l}, 1 \le l \le n, 1 \le j \le m_l$:

$$\frac{(-1)^m}{(2\pi\varepsilon)^{(3m)}} \int \cdots \int e^{-i\sum \omega_{j,l} s_l} e^{i\sum \frac{\omega_{j,l}\phi_{j,l}}{\varepsilon}} \mathbb{E}\left[\prod T^\varepsilon_{(\omega_{j,l}, \boldsymbol{\kappa}_{j,l})}(-L, 0)\right]$$

$$\times \left(\frac{\varepsilon^{(q+3)m} \prod \check{S}(\omega_{j,l}, \boldsymbol{\kappa}_{j,l})\sqrt{\check{\zeta}(\kappa_{j,l})}}{2^m}\right) \prod \omega_{j,l}^2 \, d\omega_{j,l} \, d\boldsymbol{\kappa}_{j,l}, \qquad (14.43)$$

where we have defined

$$m = \sum_{l=1}^{n} m_l,$$

$$\phi_{j,l} = \phi(t_{0,l}, \boldsymbol{\kappa}_{j,l}, \mathbf{x}_l) = -t_{0,l} + \boldsymbol{\kappa}_{j,l} \cdot \mathbf{x}_l + L/\bar{c}(\kappa_{j,l}),$$

and the sum in the exponent and the products are taken over all the distinct frequencies and wave vectors, that is, over l and j, such that

$$1 \le l \le n, \quad 1 \le j \le m_l.$$

Therefore, we are led to study the joint distribution of the transmission coefficients for a finite number of frequencies and wave vectors. We now relabel these by $(\omega_1, \boldsymbol{\kappa}_1), \ldots, (\omega_m, \boldsymbol{\kappa}_m)$. First, consider the situation with the phase $\phi_{j,l} = 0$. Then, if we could obtain the limits

$$\lim_{\varepsilon \to 0} \mathbb{E}\left[T^\varepsilon_{(\omega_1, \boldsymbol{\kappa}_1)}(-L, 0) \cdots T^\varepsilon_{(\omega_m, \boldsymbol{\kappa}_m)}(-L, 0)\right] \qquad (14.44)$$

of all these finite-dimensional problems, we would have characterized all the finite-dimensional distributions of the transmitted wave front in space and time. The argument presented in Section 8.2.5 in the one-dimensional case can be directly applied to the present situation. The limits (14.44) are given by

$$\mathbb{E}\left[\tilde{T}_{(\omega_1, \boldsymbol{\kappa}_1)}(-L, 0) \cdots \tilde{T}_{(\omega_m, \boldsymbol{\kappa}_m)}(-L, 0)\right], \qquad (14.45)$$

where the coefficients $\tilde{T}_{(\omega, \kappa)}(-L, z)$ are solutions of the system of stochastic differential equations

$$d\tilde{T}_{(\omega, \kappa)} = -\omega^2 \frac{\gamma_\kappa}{4\bar{c}(\kappa)^2} \tilde{T}_{(\omega, \kappa)} dz + i\omega \frac{\sqrt{\gamma_\kappa}}{2\bar{c}(\kappa)} \tilde{T}_{(\omega, \kappa)} dW_0(L + z), \quad (14.46)$$

driven by a *single* standard Brownian motion W_0. The initial condition is given at $-L$ by $\tilde{T}_{(\omega,\kappa)}(-L, z = -L) = 1$. The modified correlation length coefficient γ_κ is given by

$$\gamma_\kappa = \int_{-\infty}^{\infty} \mathbb{E}[\nu_\kappa(0)\nu_\kappa(z)]\,dz = \frac{\bar{c}(\kappa)^4}{\bar{c}^4}\gamma, \qquad (14.47)$$

where the important physical parameter γ is discussed in detail in Section 6.3.6. In the example discussed there, $F_2(Y(z))$ corresponds to $\nu(z)$ in the model we analyze here. As in the one-dimensional case in (8.60), an application of Itô's formula shows that equation (14.46) admits the following explicit solution:

$$\tilde{T}_{(\omega,\kappa)}(-L, 0) = \exp\left(i\omega\frac{\sqrt{\gamma_\kappa}}{2\bar{c}(\kappa)}W_0(L) - \omega^2\frac{\gamma_\kappa}{8\bar{c}(\kappa)^2}L\right). \qquad (14.48)$$

Therefore, if we substitute $\tilde{T}$ for T^ε in (14.43), we obtain a characterization of the distribution for the wave front through its moments. This substitution leads to the correct asymptotic limit expression for the wave front also in the case with a fast phase, that is, when $\phi_{j,l}$ is nonzero. The small-ε limit for the front is then obtained via a subsequent **stationary-phase** argument. We denote by $\tilde{p}$ the limit for the pressure that follows from the subsequent stationary phase evaluation, so that

$$\tilde{p}(s, \mathbf{x}, -L) := (\text{sp}) \lim_{\varepsilon \to 0} p^\varepsilon(t_0 + \varepsilon s, \mathbf{x}, -L)$$

$$= (\text{sp}) \lim_{\varepsilon \to 0} \frac{-\varepsilon^q}{(2\pi)^3} \int\int e^{-i\omega s} e^{i\frac{\omega\phi(t_0,\boldsymbol{\kappa},\mathbf{x})}{\varepsilon}} \tilde{T}_{(\omega,\kappa)}(-L, 0)$$

$$\times \left\{\frac{\check{S}(\omega, \boldsymbol{\kappa})\sqrt{\check{\zeta}(\kappa)}}{2}\right\} \omega^2\,d\omega\,d\boldsymbol{\kappa}, \qquad (14.49)$$

where the phase ϕ is given by

$$\phi(t_0, \boldsymbol{\kappa}, \mathbf{x}) = -t_0 + \boldsymbol{\kappa} \cdot \mathbf{x} + L/\bar{c}(\kappa).$$

14.2.2 Stationary-Phase Point

In this section we calculate the stationary-phase point that will give the expression for $\tilde{p}(s, \mathbf{x}, -L)$. The stationary-phase method is briefly reviewed in Appendix 14.4. It follows from the scaling of the two-dimensional stationary-phase result and the expression (14.49) that the transmitted pressure pulse is of order one if q in (14.49) takes the value -1, and we now make this choice:

$$q = -1.$$

We find from (14.77) and (14.19) that the main contribution to the integral expression (14.49) occurs at the stationary-phase point that solves

$$\nabla_{\boldsymbol{\kappa}}\phi = \mathbf{0}.$$

Using the notation $\mathbf{x} = (x_1, x_2)$, $\boldsymbol{\kappa} = (\kappa_1, \kappa_2)$, we have

$$\nabla_{\boldsymbol{\kappa}}\phi = \begin{bmatrix} \dfrac{\partial\phi}{\partial\kappa_1} \\ \dfrac{\partial\phi}{\partial\kappa_2} \end{bmatrix} = \begin{bmatrix} x_1 + \dfrac{\partial}{\partial\kappa_1}\left\{\dfrac{L}{\bar{c}(\kappa)}\right\} \\ x_2 + \dfrac{\partial}{\partial\kappa_2}\left\{\dfrac{L}{\bar{c}(\kappa)}\right\} \end{bmatrix} = \begin{bmatrix} x_1 - L\kappa_1\bar{c}(\kappa) \\ x_2 - L\kappa_2\bar{c}(\kappa) \end{bmatrix},$$

and the Hessian (used in the asymptotic computation of the integral) is

$$\mathbf{H}_{\boldsymbol{\kappa}}(\phi) = \begin{bmatrix} \dfrac{\partial^2\phi}{\partial\kappa_1^2} & \dfrac{\partial^2\phi}{\partial\kappa_1\partial\kappa_2} \\ \dfrac{\partial^2\phi}{\partial\kappa_1\partial\kappa_2} & \dfrac{\partial^2\phi}{\partial\kappa_2^2} \end{bmatrix} = -\dfrac{\bar{c}(\kappa)^3 L}{\bar{c}^2} \begin{bmatrix} 1 - \bar{c}^2\kappa_2^2 & \bar{c}^2\kappa_1\kappa_2 \\ \bar{c}^2\kappa_1\kappa_2 & 1 - \bar{c}^2\kappa_1^2 \end{bmatrix}.$$

Therefore, the stationary horizontal slowness vector $\boldsymbol{\kappa}_{\mathrm{sp}}$ solves

$$\boldsymbol{\kappa}_{\mathrm{sp}} = \frac{\mathbf{x}}{\bar{c}(\boldsymbol{\kappa}_{\mathrm{sp}})L} = \frac{\mathbf{x}\sqrt{1 - \bar{c}^2\kappa_{\mathrm{sp}}^2}}{\bar{c}L}, \tag{14.50}$$

where we again used (14.19) and $\kappa_{\mathrm{sp}}^2 = |\boldsymbol{\kappa}_{\mathrm{sp}}|^2 = \kappa_{\mathrm{sp},1}^2 + \kappa_{\mathrm{sp},2}^2$. Solving (14.50) for κ_{sp}^2 we obtain

$$\bar{c}^2\kappa_{\mathrm{sp}}^2 = \frac{|\mathbf{x}|^2}{|\mathbf{x}|^2 + L^2}, \qquad \bar{c}(\boldsymbol{\kappa}_{\mathrm{sp}})^2 = \bar{c}^2\frac{|\mathbf{x}|^2 + L^2}{L^2}.$$

We now substitute this explicit expression for $\bar{c}^2\kappa_{\mathrm{sp}}^2$ into (14.50) to obtain the stationary-phase point

$$\boldsymbol{\kappa}_{\mathrm{sp}}(\mathbf{x}) = \frac{\mathbf{x}}{\bar{c}\sqrt{|\mathbf{x}|^2 + L^2}}.$$

This value for the slowness vector corresponds to a plane-wave mode that is traveling in the direction $(\mathbf{x}, -L)$, that is, the direction of the vector from the source to the **point of observation** $(\mathbf{x}, -L)$, as shown in Figure 14.3.

Next, we substitute κ_{sp}^2 in (14.47) and (14.20) to obtain

$$\gamma_{\boldsymbol{\kappa}_{\mathrm{sp}}} = \left(1 + \frac{|\mathbf{x}|^2}{L^2}\right)^2\gamma, \qquad \bar{\zeta}(\boldsymbol{\kappa}_{\mathrm{sp}}) = \bar{\zeta}\sqrt{1 + \frac{|\mathbf{x}|^2}{L^2}}.$$

Finally, the value of the phase at the stationary point is given by

$$\phi(t_0, \boldsymbol{\kappa}_{\mathrm{sp}}, \mathbf{x}) = -t_0 + \frac{\sqrt{|\mathbf{x}|^2 + L^2}}{\bar{c}}.$$

By Proposition 14.4, the value of the integral (14.49) goes to zero as $\varepsilon \to 0$ if $\phi(t_0, \boldsymbol{\kappa}_{\mathrm{sp}}, \mathbf{x}) \neq 0$. So we choose t_0 to cancel it:

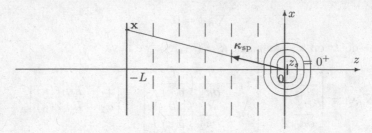

Fig. 14.3. In this figure we show the source position at 0, the observation point at $(\mathbf{x}, -L)$, and the stationary slowness vector κ_{sp}.

$$t_0 = \frac{\sqrt{|\mathbf{x}|^2 + L^2}}{\bar{c}}.$$

This corresponds to choosing t_0 to be the travel time from the source point at the origin to the point of observation $(\mathbf{x}, -L)$ under the constant effective medium sound speed $\bar{c}$. Upon substitution in (14.48) we finally obtain

$$\tilde{T}_{(\omega, \kappa_{\mathrm{sp}})}(-L, 0) = \exp\left(i\omega \frac{\sqrt{\gamma}}{2\bar{c}} \sqrt{1 + \frac{|\mathbf{x}|^2}{L^2}}\, W_0(L) - \omega^2 \frac{\gamma}{8\bar{c}^2}\left(1 + \frac{|\mathbf{x}|^2}{L^2}\right) L \right).$$

14.2.3 Characterization of the Transmitted Wave Front

We have used the diffusion-approximation limit to obtain a joint description of the plane-wave modes, and we next derive a simple explicit formula for the limit of the transmitted wave using the stationary-phase method based on the stationary point calculated above. Applying Proposition 14.4 given in Appendix 14.4 to (14.49), we obtain

$$\tilde{p}(s, \mathbf{x}, -L) = \frac{\bar{\zeta}^{1/2} L^{1/2}}{8\pi^2 \bar{c}(|\mathbf{x}|^2 + L^2)^{3/4}}$$

$$\times \int e^{-i\omega s} e^{i\omega \frac{\sqrt{\gamma}}{2\bar{c}} \sqrt{1 + \frac{|\mathbf{x}|^2}{L^2}}\, W_0(L)} e^{-\omega^2 \frac{\gamma}{8\bar{c}^2}\left(1 + \frac{|\mathbf{x}|^2}{L^2}\right) L}\, i\omega \check{S}(\omega, \kappa_{\mathrm{sp}})\, d\omega.$$

The corresponding limit expression for the transmitted pressure front $\tilde{p}_0$ in a constant medium is obtained by evaluating the above expression for $\gamma = 0$; we obtain

$$\tilde{p}_0(s, \mathbf{x}, -L) = \frac{\bar{\zeta}^{1/2} L^{1/2}}{8\pi^2 \bar{c}(|\mathbf{x}|^2 + L^2)^{3/4}} \int e^{-i\omega s} i\omega \check{S}(\omega, \kappa_{\mathrm{sp}})\, d\omega.$$

Let us consider the case with a spatial point source located at the origin:

$$\mathbf{F}^\varepsilon(t, \mathbf{x}, z) = \frac{1}{\varepsilon}\begin{bmatrix} \mathbf{f_x} \\ f_z \end{bmatrix}\left(\frac{t}{\varepsilon}\right) \delta\left(\frac{\mathbf{x}}{\varepsilon}\right) \delta(z) = \varepsilon \begin{bmatrix} \mathbf{f_x} \\ f_z \end{bmatrix}\left(\frac{t}{\varepsilon}\right) \delta(\mathbf{x}) \delta(z).$$

The expression (14.30) of the functional $\check{S}$ at the stationary slowness vector $\boldsymbol{\kappa}_{\rm sp}$ can be simplified:

$$\check{S}(\omega, \boldsymbol{\kappa}_{\rm sp}) = \frac{1}{\zeta^{1/2}(L^2 + |\mathbf{x}|^2)^{1/4}L^{1/2}} \left(-\mathbf{x} \cdot \hat{\mathbf{f}}_{\mathbf{x}}(\omega) + L\hat{f}_z(\omega) \right).$$

As a result, the limiting pressure field also reads

$$\tilde{p}(s, \mathbf{x}, -L) = -\frac{1}{8\pi^2 \bar{c}(|\mathbf{x}|^2 + L^2)}$$

$$\times \int e^{-i\omega s} e^{i\omega \frac{\sqrt{\gamma}}{2\bar{c}} \sqrt{1 + \frac{|\mathbf{x}|^2}{L^2}} W_0(L)} e^{-\omega^2 \frac{\gamma}{8\bar{c}^2} \left(1 + \frac{|\mathbf{x}|^2}{L^2} \right) L} i\omega \left(\mathbf{x} \cdot \hat{\mathbf{f}}_{\mathbf{x}}(\omega) - L\hat{f}_z(\omega) \right) d\omega,$$

and in the homogeneous medium we get simply

$$\tilde{p}_0(s, \mathbf{x}, -L) = \frac{1}{4\pi \bar{c}(|\mathbf{x}|^2 + L^2)} \left(\mathbf{x} \cdot \mathbf{f}'_{\mathbf{x}}(s) - Lf'_z(s) \right). \tag{14.51}$$

We now summarize the result and state the following proposition. It describes the transmitted pressure pulse through the random medium as a simple modification of the pressure pulse obtained through the homogenized medium.

Proposition 14.3. *In probability distribution the following characterization of the transmitted wave process holds:*

$$\lim_{\varepsilon \to 0} p^\varepsilon \left(\frac{\sqrt{|\mathbf{x}|^2 + L^2}}{\bar{c}} + \varepsilon s, \mathbf{x}, -L \right) = \tilde{p}(s, \mathbf{x}, -L),$$

where

$$\tilde{p}(s, \mathbf{x}, -L) = \left[\mathcal{N}_{D_{(L,\mathbf{x})}} * \tilde{p}_0(\cdot, \mathbf{x}, -L) \right] \left(s - \Theta_{(L,\mathbf{x})} \right), \tag{14.52}$$

and we set

$$D^2_{(L,\mathbf{x})} = \frac{\gamma}{4\bar{c}^2} \left(1 + \frac{|\mathbf{x}|^2}{L^2} \right) L, \tag{14.53}$$

$$\Theta_{(L,\mathbf{x})} = \frac{\sqrt{\gamma \left(1 + \frac{|\mathbf{x}|^2}{L^2} \right)}}{2\bar{c}} W_0(L), \tag{14.54}$$

$$\mathcal{N}_D(s) = \frac{1}{\sqrt{2\pi}D} e^{-s^2/2D^2}.$$

Note therefore that the shape of the wave front is given by the *deterministic* quantity

$$\mathcal{N}_{D_{(L,\mathbf{x})}} * \tilde{p}_0(\cdot, \mathbf{x}, -L),$$

which is the convolution of the homogeneous front with the Gaussian density $\mathcal{N}_{D_{(L,\mathbf{x})}}$. This is often referred to as **stabilization** of the front. The pulse shape corresponds to a "diffusion" in time or a smearing of the transmitted wave process via a convolution with the Gaussian density. The random variable $\Theta_{(L,\mathbf{x})}$ corresponds to a "travel time" shift of the type we discussed in Section 8.2.7: it is proportional to $W_0(L)$, which is a Gaussian random variable with mean zero and variance L. Note that the variance of the random time shift $\Theta_{(L,\mathbf{x})}$ is proportional to the travel distance $\sqrt{L^2 + |\mathbf{x}|^2}$. The relation between the stable front and the coherent wave given by $\mathbb{E}[\tilde{p}(s, \mathbf{x}, -L)]$ is as in the one-dimensional case; namely, the coherent wave is given by the double convolution of the homogeneous front with the Gaussian density $\mathcal{N}_{D_{(L,\mathbf{x})}}$, that is,

$$\mathbb{E}[\tilde{p}(s, \mathbf{x}, -L)] = \left[\mathcal{N}_{2D_{(L,\mathbf{x})}} * \tilde{p}_0(\cdot, \mathbf{x}, -L)\right](s) \ .$$

This completes the description of the transmitted stable front, and we now turn our attention to the study of the incoherent waves starting with the reflected pressure.

14.3 The Mean Reflected Intensity Generated by a Point Source

We consider the case of a point source located at the surface at position $(\mathbf{0}, 0)$, emitting a short pulse at time 0. Our goal here is to analyze the reflected pressure field. The point source generates the source term

$$\mathbf{F}^\varepsilon(t, \mathbf{x}, z) = \varepsilon^{1/2} f\left(\frac{t}{\varepsilon}\right) \delta(\mathbf{x})\delta(z) \begin{bmatrix} \mathbf{0} \\ 1 \end{bmatrix}, \tag{14.55}$$

which imposes the boundary conditions

$$\check{b}^\varepsilon(\omega, \boldsymbol{\kappa}, 0^-) = \varepsilon^{3/2} \check{S}(\omega, \boldsymbol{\kappa}), \qquad \check{S}(\omega, \boldsymbol{\kappa}) = \frac{1}{\sqrt{\zeta(\boldsymbol{\kappa})}} \hat{f}(\omega) \ .$$

The amplitude of the source has been scaled ($q = 1/2$ in (14.9)) so that the mean reflected intensity will be of order one.

14.3.1 Reflected-Pressure Integral Representation

We observe the pressure field at the surface $z = 0$, at the location $\mathbf{x} \neq \mathbf{0}$, and at time $t > 0$. Its Fourier representation is given by (14.10), and by (14.24), its Fourier transform is

$$\hat{p}^\varepsilon(\omega, \boldsymbol{\kappa}, 0^+) = \frac{\sqrt{\zeta(\boldsymbol{\kappa})}}{2} \check{a}^\varepsilon(\omega, \boldsymbol{\kappa}, 0^+) \ ,$$

where we have taken into account the boundary conditions $\check{b}^\varepsilon(\omega, \boldsymbol{\kappa}, 0^+) = 0$. The right-going modes at $z = 0^+$ and at $z = 0^-$ are related by the jump condition (14.28),

$$\breve{a}^{\varepsilon}(\omega, \boldsymbol{\kappa}, 0^+) = \breve{a}^{\varepsilon}(\omega, \boldsymbol{\kappa}, 0^-) + \frac{\varepsilon^{3/2}}{\sqrt{\zeta(\kappa)}} \hat{f}(\omega),$$

and the right-going mode at $z = 0^-$ is obtained from the propagator relation (14.38):

$$\breve{a}^{\varepsilon}(\omega, \boldsymbol{\kappa}, 0^-) = R^{\varepsilon}_{(\omega, \kappa)}(-L, 0) \frac{\varepsilon^{3/2}}{\sqrt{\zeta(\kappa)}} \hat{f}(\omega),$$

where we have introduced the reflection coefficient defined by

$$R^{\varepsilon}_{(\omega, \kappa)}(-L, z) = \frac{\overline{\beta^{\varepsilon}_{(\omega, \kappa)}(-L, z)}}{\alpha^{\varepsilon}_{(\omega, \kappa)}(-L, z)}. \tag{14.56}$$

As a result, the integral representation of the reflected pressure is

$$p^{\varepsilon}(t, \mathbf{x}, 0^+) \tag{14.57}$$

$$= \frac{1}{16\pi^3 \varepsilon^{3/2}} \int e^{-i\frac{\omega}{\varepsilon}(t - \boldsymbol{\kappa} \cdot \mathbf{x})} [1 + R^{\varepsilon}_{(\omega, \kappa)}(-L, 0)] \hat{f}(\omega) \omega^2 \, d\omega \, d\boldsymbol{\kappa}.$$

This field has two components:

- The first component is the direct emission from the source. It is taken into account by the term 1 in the square brackets of the right-hand side of (14.57), and it can be computed easily:

$$p^{\varepsilon}_{dir}(t, \mathbf{x}, 0^+) = \frac{1}{16\pi^3 \varepsilon^{3/2}} \int e^{-i\frac{\omega}{\varepsilon}(t - \boldsymbol{\kappa} \cdot \mathbf{x})} \hat{f}(\omega) \omega^2 \, d\omega \, d\boldsymbol{\kappa} = \frac{\varepsilon^{1/2}}{2} f\left(\frac{t}{\varepsilon}\right) \delta(\mathbf{x}).$$

  Note that this component is concentrated at $\mathbf{x} = \mathbf{0}$ and around the time $t = 0$.

- The second component is the reflection of the source signal by the random medium. It is taken into account by the term $R^{\varepsilon}_{(\omega, \kappa)}(-L, 0)$, and we shall see that it can be detected only after the arrival time $t > 0$ of the direct component.

From now on, we observe the pulse at a time $t > 0$, meaning that the wave due to the direct emission from the source has vanished. As we shall see in the next section (equation (14.69) with $p = 1$ and $q = 0$), the expectation of the reflection coefficient converges to 0 as $\varepsilon \to 0$, so that the mean value of the reflected field is asymptotically 0, meaning that there is no coherent signal in the reflected wave. In order to analyze this incoherent wave field, we compute its second moment.

14.3.2 Autocorrelation Function of the Reflection Coefficient at Two Nearby Slownesses and Frequencies

We consider the mean reflected intensity given by

$$\mathbb{E}[p^\varepsilon(t,\mathbf{x},0)^2] = \frac{1}{256\pi^6\varepsilon^3} \int\int e^{-i\frac{\omega-\omega'}{\varepsilon}t + i\frac{\omega\boldsymbol{\kappa}-\omega'\boldsymbol{\kappa}'}{\varepsilon}\cdot\mathbf{x}}$$

$$\times\, \mathbb{E}\left[R^\varepsilon_{(\omega,\kappa)}(-L,0)\overline{R^\varepsilon_{(\omega',\kappa')}}(-L,0) \right] \hat{f}(\omega)\overline{\hat{f}(\omega')}\omega^2\omega'^2\, d\omega\, d\boldsymbol{\kappa}\, d\omega'\, d\boldsymbol{\kappa}' \,. \quad (14.58)$$

As can be seen from this integral representation, the autocorrelation function of the reflection coefficient plays a crucial role. The analysis is very similar to the one-dimensional case because we essentially deal with plane-waves. We now study the reflection and transmission coefficients $R^\varepsilon_{(\omega,\kappa)}(z_0,z)$ and $T^\varepsilon_{(\omega,\kappa)}(z_0,z)$ defined for a random slab occupying the region $[z_0,z]$. Using the representation (14.56) of the reflection coefficient and the equations satisfied by $(\alpha^\varepsilon_{(\omega,\kappa)}, \beta^\varepsilon_{(\omega,\kappa)})$, we can deduce a closed nonlinear differential system satisfied by the reflection and transmission coefficients:

$$\frac{dR^\varepsilon_{(\omega,\kappa)}}{dz} = \frac{-i\omega}{2\bar{c}(\kappa)\varepsilon}\nu_\kappa\left(\frac{z}{\varepsilon^2}\right)\left(e^{\frac{-2i\omega z}{\bar{c}(\kappa)\varepsilon}} - 2R^\varepsilon_{(\omega,\kappa)} + (R^\varepsilon_{(\omega,\kappa)})^2 e^{\frac{2i\omega z}{\bar{c}(\kappa)\varepsilon}}\right), \quad (14.59)$$

$$\frac{dT^\varepsilon_{(\omega,\kappa)}}{dz} = \frac{i\omega}{2\bar{c}(\kappa)\varepsilon}\nu_\kappa\left(\frac{z}{\varepsilon^2}\right)\left(1 - R^\varepsilon_{(\omega,\kappa)} e^{\frac{2i\omega z}{\bar{c}(\kappa)\varepsilon}}\right) T^\varepsilon_{(\omega,\kappa)}\,. \quad (14.60)$$

The initial conditions for these nonlinear differential equations are, at $z = z_0$,

$$R^\varepsilon_{(\omega,\kappa)}(z_0, z = z_0) = 0\,, \qquad T^\varepsilon_{(\omega,\kappa)}(z_0, z = z_0) = 1\,.$$

We then introduce the family of products of reflection coefficients

$$U^\varepsilon_{p,q}(\omega,\kappa,h,\lambda,z_0,z) = \left(R^\varepsilon_{(\omega+\varepsilon h/2,\kappa+\varepsilon\lambda/2)}(z_0,z)\right)^p \left(\overline{R^\varepsilon_{(\omega-\varepsilon h/2,\kappa-\varepsilon\lambda/2)}(z_0,z)}\right)^q\,.$$

From the Riccati equation (14.59) and the expansion

$$\frac{2(\omega+\varepsilon h/2)}{\bar{c}(\kappa+\varepsilon\lambda/2)} = \frac{2\omega}{\bar{c}(\kappa)} + \varepsilon\left(\frac{h}{\bar{c}(\kappa)} - \omega\lambda\bar{c}(\kappa)\kappa\right) + O(\varepsilon^2)\,,$$

we get

$$\frac{\partial U^\varepsilon_{p,q}}{\partial z} = \frac{i\omega}{\bar{c}(\kappa)}\nu^\varepsilon_\kappa(p-q)U^\varepsilon_{p,q}$$

$$+ \frac{i\omega}{2\bar{c}(\kappa)}\nu^\varepsilon_\kappa e^{\frac{2i\omega z}{\bar{c}(\kappa)\varepsilon}}\left(qe^{-\frac{ihz}{\bar{c}(\kappa)}+i\omega\lambda\bar{c}(\kappa)\kappa z}U^\varepsilon_{p,q-1} - pe^{\frac{ihz}{\bar{c}(\kappa)}-i\omega\lambda\bar{c}(\kappa)\kappa z}U^\varepsilon_{p+1,q}\right)$$

$$+ \frac{i\omega}{2\bar{c}(\kappa)}\nu^\varepsilon_\kappa e^{-\frac{2i\omega z}{\bar{c}(\kappa)\varepsilon}}\left(qe^{\frac{ihz}{\bar{c}(\kappa)}-i\omega\lambda\bar{c}(\kappa)\kappa z}U^\varepsilon_{p,q+1} - pe^{-\frac{ihz}{\bar{c}(\kappa)}+i\omega\lambda\bar{c}(\kappa)\kappa z}U^\varepsilon_{p-1,q}\right)\,,$$

starting from $U^\varepsilon_{p,q}(\omega,\kappa,h,\lambda,z_0,z=z_0) = \mathbf{1}_0(p)\mathbf{1}_0(q)$. Here we have set $\nu^\varepsilon_\kappa(z) = \nu_\kappa(z/\varepsilon^2)/\varepsilon$ and we have not written terms of order ε that will vanish in the limit $\varepsilon \to 0$. We consider the associated family of Fourier transforms

$$V^\varepsilon_{p,q}(\omega,\kappa,\tau,\chi,z_0,z) = \frac{\bar{c}^2\omega}{4\pi^2\bar{c}(\kappa)^2}\int\int e^{-ih(\tau\bar{c}^2/\bar{c}(\kappa)^2 - (p+q)z/\bar{c}(\kappa))}$$

$$\times\, e^{i\omega\lambda(\chi-(p+q)z\bar{c}(\kappa)\kappa)}U^\varepsilon_{p,q}(\omega,\kappa,h,\lambda,z_0,z)\, dh\, d\lambda\,. \quad (14.61)$$

The form of this transform ensures that the variable τ can be interpreted as a travel time, as we shall see below. The family $V_{p,q}^\varepsilon$ satisfies

$$\frac{\partial V_{p,q}^\varepsilon}{\partial z} = -(p+q)\frac{\bar{c}(\kappa)}{\bar{c}^2}\frac{\partial V_{p,q}^\varepsilon}{\partial \tau} - \bar{c}(\kappa)\kappa(p+q)\frac{\partial V_{p,q}^\varepsilon}{\partial \chi} + \frac{i\omega}{\bar{c}(\kappa)}\nu_\kappa^\varepsilon(p-q)V_{p,q}^\varepsilon$$

$$+\frac{i\omega}{2\bar{c}(\kappa)}\nu_\kappa^\varepsilon e^{\frac{2i\omega z}{\bar{c}(\kappa)\varepsilon}}\left(qV_{p,q-1}^\varepsilon - pV_{p+1,q}^\varepsilon\right) + \frac{i\omega}{2\bar{c}(\kappa)}\nu_\kappa^\varepsilon e^{-\frac{2i\omega z}{\bar{c}(\kappa)\varepsilon}}\left(qV_{p,q+1}^\varepsilon - pV_{p-1,q}^\varepsilon\right) .$$

We now apply the limit theorem of Section 6.7.3, in the same way as in Section 9.2.1. This establishes that the process $(V_{p,q}^\varepsilon)_{p,q\in\mathbb{N}}$ converges in distribution as $\varepsilon \to 0$ to a diffusion process. In particular, the moments converge,

$$\mathbb{E}[V_{p,q}^\varepsilon(\omega,\kappa,\tau,\chi,z_0,z)] \stackrel{\varepsilon\to 0}{\longrightarrow} \mathcal{V}_{p,q}(\omega,\kappa,\tau,\chi,z_0,z) ,$$

where the family $(\mathcal{V}_{p,q})_{p,q\in\mathbb{N}}$ satisfies the system of transport equations

$$\frac{\partial \mathcal{V}_{p,q}}{\partial z} + (p+q)\frac{\bar{c}(\kappa)}{\bar{c}^2}\frac{\partial \mathcal{V}_{p,q}}{\partial \tau} + \bar{c}(\kappa)\kappa(p+q)\frac{\partial \mathcal{V}_{p,q}}{\partial \chi}$$

$$= -\frac{3(p-q)^2}{L_{\mathrm{loc}}(\omega,\kappa)}\mathcal{V}_{p,q} + \frac{pq}{L_{\mathrm{loc}}(\omega,\kappa)}\left(\mathcal{V}_{p+1,q+1} + \mathcal{V}_{p-1,q-1} - 2\mathcal{V}_{p,q}\right) ,$$

starting from $\mathcal{V}_{p,q}(\omega,\kappa,\tau,\chi,z_0,z=z_0) = \mathbf{1}_0(p)\mathbf{1}_0(q)\delta(\tau)\delta(\chi)$. The mode-dependent localization length $L_{\mathrm{loc}}(\omega,\kappa)$ is defined by

$$L_{\mathrm{loc}}(\omega,\kappa) = \frac{4\bar{c}^4}{\gamma\bar{c}(\kappa)^2\omega^2} , \qquad \gamma = \int_{-\infty}^{\infty}\mathbb{E}[\nu(0)\nu(z)]\,dz . \qquad (14.62)$$

Note that the localization length for the (ω,κ) mode can be written as

$$L_{\mathrm{loc}}(\omega,\kappa) = \frac{\bar{c}^2}{\bar{c}(\kappa)^2}L_{\mathrm{loc}}(\omega) ,$$

where $L_{\mathrm{loc}}(\omega)$ is the one-dimensional localization length defined in (7.81). For any integer p_0, the subfamily $(\mathcal{V}_{p,p+p_0})_{p\in\mathbb{N}}$ satisfies a closed subsystem. This subsystem of transport equations has zero initial conditions if $p_0 \neq 0$, which shows that

$$\mathcal{V}_{p,q}(\omega,\kappa,\tau,\chi,z_0,z) = 0 \text{ if } p \neq q . \qquad (14.63)$$

Furthermore, for $p = q$, $\mathcal{V}_p := \mathcal{V}_{p,p}$ satisfies the system

$$\frac{\partial \mathcal{V}_p}{\partial z} + 2p\frac{\bar{c}(\kappa)}{\bar{c}^2}\frac{\partial \mathcal{V}_p}{\partial \tau} + 2p\bar{c}(\kappa)\kappa\frac{\partial \mathcal{V}_p}{\partial \chi} = \frac{p^2}{L_{\mathrm{loc}}(\omega,\kappa)}\left(\mathcal{V}_{p+1} + \mathcal{V}_{p-1} - 2\mathcal{V}_p\right) ,$$

$$\mathcal{V}_p(\omega,\kappa,\tau,\chi,z_0,z=z_0) = \mathbf{1}_0(p)\delta(\tau)\delta(\chi) , \qquad (14.64)$$

and it has a probabilistic interpretation. In order to describe this interpretation we introduce the jump Markov process $(N_z)_{z\geq z_0}$ with state space $\mathbb{N}$ and infinitesimal generator

$$\mathcal{L}_{(\omega,\kappa)}\phi(N) = \frac{N^2}{L_{\mathrm{loc}}(\omega,\kappa)}\left(\phi(N+1) + \phi(N-1) - 2\phi(N)\right).$$

The solution $\mathcal{V}_p$ can be written as the expectation of a functional of this jump process:

$$\mathcal{V}_p(\omega,\kappa,\tau,\chi,z_0,z) = \mathbb{E}\left[\mathbf{1}_0(N_z)\delta\left(\tau - 2\frac{\bar{c}(\kappa)}{\bar{c}^2}\int_{z_0}^{z}N_s ds\right)\right.$$
$$\left.\times\delta\left(\chi - 2\bar{c}(\kappa)\kappa\int_{z_0}^{z}N_s ds\right)\mid N_{z_0} = p\right].$$

We can combine the two Dirac distributions to obtain

$$\mathcal{V}_p(\omega,\kappa,\tau,\chi,z_0,z) = \mathcal{W}_p(\omega,\kappa,\tau,z_0,z)\delta\left(\chi - \bar{c}^2\kappa\tau\right), \tag{14.65}$$

where $\mathcal{W}_p$ is defined by

$$\mathcal{W}_p(\omega,\kappa,\tau,z_0,z) = \mathbb{E}\left[\mathbf{1}_0(N_z)\delta\left(\tau - 2\frac{\bar{c}(\kappa)}{\bar{c}^2}\int_{z_0}^{z}N_s ds\right)\mid N_{z_0} = p\right]. \tag{14.66}$$

The simplification (14.65) can also be seen directly from the transport equations (14.64). The functions $\mathcal{W}_p$ can be characterized as the solutions of the system of transport equations

$$\frac{\partial\mathcal{W}_p}{\partial z} + 2p\frac{\bar{c}(\kappa)}{\bar{c}^2}\frac{\partial\mathcal{W}_p}{\partial\tau} = \frac{p^2}{L_{\mathrm{loc}}(\omega,\kappa)}\left(\mathcal{W}_{p+1} + \mathcal{W}_{p-1} - 2\mathcal{W}_p\right), \tag{14.67}$$

$$\mathcal{W}_p(\omega,\kappa,\tau,z_0,z = z_0) = \mathbf{1}_0(p)\delta(\tau). \tag{14.68}$$

We therefore recover the same system of transport equations as in the one-dimensional case. Taking the inverse Fourier transform of (14.61), we get that for $z_0 \leq z$,

$$\mathbb{E}\left[\left(R^{\varepsilon}_{(\omega+\varepsilon h/2,\kappa+\varepsilon\lambda/2)}(z_0,z)\right)^p \left(\overline{R^{\varepsilon}_{(\omega-\varepsilon h/2,\kappa-\varepsilon\lambda/2)}(z_0,z)}\right)^q\right] \tag{14.69}$$

$$\xrightarrow{\varepsilon\to 0} \int\mathcal{W}_p(\omega,\kappa,\tau,z_0,z)e^{i\tau[h\bar{c}^2/\bar{c}(\kappa)^2 - \omega\lambda\bar{c}^2\kappa]}d\tau \times e^{2ipz[-h/\bar{c}(\kappa)+\omega\lambda\bar{c}(\kappa)\kappa]},$$

if $q = p$ and

$$\mathbb{E}\left[\left(R^{\varepsilon}_{(\omega+\varepsilon h/2,\kappa+\varepsilon\lambda/2)}(z_0,z)\right)^p \left(\overline{R^{\varepsilon}_{(\omega-\varepsilon h/2,\kappa-\varepsilon\lambda/2)}(z_0,z)}\right)^q\right] \xrightarrow{\varepsilon\to 0} 0,$$

otherwise.

The case with a semi-infinite slab ($z_0 \to -\infty$) leads to explicit formulas for the autocorrelation function of the reflection coefficient. Applying the same method as in Section 9.2.3, we get that the function $\mathcal{W}_p$ converges as $z_0 \to -\infty$ to the limit

$$\mathcal{W}_p(\omega,\kappa,\tau,z_0,z) \xrightarrow{z_0\to -\infty} P_p^{\infty}\left(\frac{\bar{c}^2\tau}{\bar{c}(\kappa)L_{\mathrm{loc}}(\omega,\kappa)}\right)\frac{\bar{c}^2}{\bar{c}(\kappa)L_{\mathrm{loc}}(\omega,\kappa)}, \tag{14.70}$$

where P_p^∞ is given by (9.39). By the same arguments as those used in Section 9.2.2, we can actually claim that the limit is reached,

$$
\mathcal{W}_p(\omega, \kappa, \tau, z_0, z) = P_p^\infty \left(\frac{\bar{c}^2 \tau}{\bar{c}(\kappa) L_{\text{loc}}(\omega, \kappa)} \right) \frac{\bar{c}^2}{\bar{c}(\kappa) L_{\text{loc}}(\omega, \kappa)} ,
$$

as soon as

$$
\tau \leq \frac{2\bar{c}(\kappa)}{\bar{c}^2}(z - z_0) . \tag{14.71}
$$

Finally, setting $U_j^\varepsilon = R(\omega_j + \frac{\varepsilon h_j}{2}, \mu_j + \frac{\varepsilon \lambda_j}{2}, 0)\overline{R}(\omega_j - \frac{\varepsilon h_j}{2}, \mu_j - \frac{\varepsilon \lambda_j}{2}, 0)$, we can show in the same way as in Section 9.2.4 that for two distinct frequencies $\omega_1 \neq \omega_2$ or for two distinct slownesses $\mu_1 \neq \mu_2$ one has

$$
|\mathbb{E}\left[U_1^\varepsilon U_2^\varepsilon\right] - \mathbb{E}\left[U_1^\varepsilon\right]\mathbb{E}\left[U_2^\varepsilon\right]| \xrightarrow{\varepsilon \to 0} 0 . \tag{14.72}
$$

This decorrelation property will be used in the next chapter to deduce the self-averaging property of the time-reversed refocused pulse.

14.3.3 Asymptotics of the Mean Intensity

We consider the integral representation of the mean reflected intensity (14.58). We parameterize the observation point $\mathbf{x} = (x, 0)$, $x > 0$, and we use polar coordinates for $\boldsymbol{\kappa}$ and $\boldsymbol{\kappa}'$, that is, we write $\boldsymbol{\kappa} = \mu \mathbf{e}_\theta$ and $\boldsymbol{\kappa}' = \mu' \mathbf{e}_{\theta'}$, where $\mathbf{e}_\theta$ is the unit vector $(\cos\theta, \sin\theta)$. The slowness moduli μ and μ' take values between 0 and $1/\bar{c}$, and the slowness angles θ and θ' take values in $[0, 2\pi]$. We obtain

$$
\mathbb{E}[p^\varepsilon(t, \mathbf{x}, 0^+)^2] = \frac{1}{256\pi^6 \varepsilon^3} \int\int e^{-i\frac{\omega - \omega'}{\varepsilon} t + i\frac{\omega x \mu \cos\theta - \omega' x \mu' \cos\theta'}{\varepsilon}}
$$
$$
\times \mathbb{E}\left[R_{(\omega, \mu)}^\varepsilon \overline{R_{(\omega', \mu')}^\varepsilon} \right] \hat{f}(\omega)\hat{f}(\omega') \omega^2 \mu \omega'^2 \mu' \, d\theta \, d\theta' \, d\mu \, d\omega' \, d\mu' .
$$

In Section 14.3.2 it is shown that the standard reflection coefficients are correlated only if the frequencies and the slowness moduli are close to each other at order ε. We therefore perform the change of variables $\omega' = \omega - \varepsilon h$ and $\mu' = \mu - \varepsilon\lambda$:

$$
\mathbb{E}[p^\varepsilon(t, \mathbf{x}, 0^+)^2] = \frac{1}{256\pi^6 \varepsilon} \int\int e^{-iht + i(h\mu + \omega\lambda)x \cos\theta'} e^{i\frac{\omega\mu x}{\varepsilon}[\cos\theta - \cos\theta']}
$$
$$
\times \mathbb{E}\left[R_{(\omega, \mu)}^\varepsilon \overline{R_{(\omega - \varepsilon h, \mu - \varepsilon\lambda)}^\varepsilon} \right] |\hat{f}(\omega)|^2 \omega^4 \mu^2 \, d\theta \, d\theta' \, dh \, d\lambda \, d\omega \, d\mu .
$$

The fast phase can be used in a stationary-phase argument, which gives two contributions. The first one is concentrated on $\theta' = \theta = 0$, the second one is concentrated on $\theta' = \theta = \pi$:

$$
\mathbb{E}[p^\varepsilon(t, \mathbf{x}, 0^+)^2] = \sum_{q \in \{-1, 1\}} \frac{1}{128\pi^5 x} \int\int e^{-iht + iq(h\mu + \omega\lambda)x}
$$
$$
\times \mathbb{E}\left[R_{(\omega, \mu)}^\varepsilon \overline{R_{(\omega - \varepsilon h, \mu - \varepsilon\lambda)}^\varepsilon} \right] |\hat{f}(\omega)|^2 \omega^3 \mu \, dh \, d\lambda \, d\omega \, d\mu . \tag{14.73}
$$

Substituting the limit (14.69) into (14.73), one obtains

$$
\lim_{\varepsilon \to 0} \mathbb{E}[p^\varepsilon(t, \mathbf{x}, 0^+)^2] = \sum_{q \in \{-1,1\}} \frac{1}{128\pi^5 x} \int \int e^{-iht + iq(h\mu + \omega\lambda)x}
$$
$$
\times \, \mathcal{W}_1\left(\omega, \mu, \tau, -L, 0\right) e^{i\tau[h\bar{c}^2/\bar{c}(\mu)^2 - \omega\lambda\bar{c}^2\mu]} |\hat{f}(\omega)|^2 \omega^3 \mu \, d\tau \, dh \, d\lambda \, d\omega \, d\mu.
$$

Integrating with respect to h gives the Dirac mass factor $\delta(t - \tau\bar{c}^2/\bar{c}(\mu)^2 - q\mu x)$. We then integrate with respect to λ, so that we obtain the Dirac mass factor $\delta(\omega q x - \omega\tau\bar{c}^2\mu)$. Since the support of $\mathcal{W}_1$ is restricted to positive τ and $x > 0$, only the integral with $q = 1$ contributes to the value of the mean intensity. This gives

$$
\lim_{\varepsilon \to 0} \mathbb{E}[p^\varepsilon(t, \mathbf{x}, 0^+)^2] = \frac{1}{32\pi^3 x} \int \int \mathcal{W}_1\left(\omega, \mu, \tau, -L, 0\right) \delta\left(\frac{\bar{c}^2}{\bar{c}(\mu)^2}\tau - t + \mu x\right)
$$
$$
\times \, \delta\left(x - \bar{c}^2\mu\tau\right) |\hat{f}(\omega)|^2 \omega^2 \mu \, d\tau \, d\omega \, d\mu.
$$

The second Dirac mass

$$
\delta\left(x - \bar{c}^2\mu\tau\right) = \frac{1}{\bar{c}^2\tau} \delta\left(\mu - \frac{x}{\bar{c}^2\tau}\right)
$$

concentrates the integral with respect to μ on the value $x/(\bar{c}^2\tau)$. For this value of μ, the first Dirac mass becomes $\delta(\tau - t)$, so that the integral with respect to μ is actually concentrated on the value $x/(\bar{c}^2 t)$. If this value does not belong to the interval $[0, 1/\bar{c}]$, which is the support of the integral in μ, then the mean intensity is zero. In other words, the mean intensity is zero if $x > \bar{c}t$. This is not surprising, since $\bar{c}t$ is the distance traveled at time t by the wave emitted from O at time 0. If $x < \bar{c}t$, then the mean reflected intensity is not zero and is given by

$$
\lim_{\varepsilon \to 0} \mathbb{E}[p^\varepsilon(t, \mathbf{x}, 0^+)^2] = \frac{1}{32\pi^3 \bar{c}^4 t^2} \int \mathcal{W}_1\left(\omega, \kappa_{x,t}, t, -L, 0\right) |\hat{f}(\omega)|^2 \omega^2 \, d\omega,
$$

(14.74)

where

$$
\kappa_{x,t} = \frac{x}{\bar{c}^2 t}.
$$

Note that for a given observation point $\mathbf{x} = (x, 0)$, $x > 0$, and a given travel time $t > x/\bar{c}$, in the limit $\varepsilon \to 0$, only the waves with the slowness vector $\boldsymbol{\kappa} = (\kappa_{x,t}, 0)$ contribute to the incoherent intensity given in formula (14.74). Schematically these waves experience multiple scattering by the medium and can be represented as paths from the origin to the observation point $\mathbf{x}$, as we now explain. The total length of these paths is $\bar{c}t$. They are continuous, piecewise linear, and their angles with respect to the surface have cosine equal to $x/(\bar{c}t) = \bar{c}\kappa_{x,t}$. Two such paths are shown in Figure 14.4. This path-angle interpretation is valid only for layered media, where scattering does not modify the slowness vector, but only couples right-going ($\check{a}^\varepsilon(\omega, \boldsymbol{\kappa})$) and left-going ($\check{b}^\varepsilon(\omega, \boldsymbol{\kappa})$) modes with the same frequency and slowness vector.

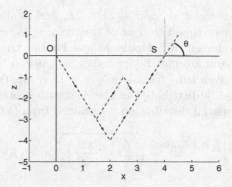

Fig. 14.4. Two typical "paths" participating in the incoherent wave intensity observed at position $\mathbf{x} = (x,0)$, $x = 4$, and at time $t = 4\sqrt{5} \approx 8.94$. The mean velocity is $\bar{c} = 1$. All paths arrive with the same angle θ whose cosine is $\cos(\theta) = x/(\bar{c}t) = 1/\sqrt{5}$. Here we have $\theta \approx 1.11$.

In the asymptotic L large, we have

$$\mathcal{W}_1(\omega, \kappa_{x,t}, t, -L, 0) \overset{L\to\infty}{\longrightarrow} \frac{\bar{c}^2}{\bar{c}(\kappa_{x,t}) L_{\mathrm{loc}}(\omega, \kappa_{x,t})} P_1^\infty \left(\frac{\bar{c}^2 t}{\bar{c}(\kappa_{x,t}) L_{\mathrm{loc}}(\omega, \kappa_{x,t})} \right),$$

where P_1^∞ is given by (9.39) and $L_{\mathrm{loc}}(\omega, \kappa)$ is the mode-dependent localization length defined by (14.62). In fact, this limit is reached as soon as the condition (14.71) is fulfilled:

$$t \le \frac{2\bar{c}(\kappa_{x,t})}{\bar{c}^2} L,$$

which can also be written

$$\sqrt{\bar{c}^2 t^2 - x^2} < 2L.$$

As a result, if the condition $\bar{c}t < 2L$ is satisfied, then we have for all $x < \bar{c}t$,

$$\mathcal{W}_1(\omega, \kappa_{x,t}, t, -L, 0) = \frac{\bar{c}}{L_{\mathrm{loc}}(\omega)} \frac{\bar{c}t}{\sqrt{\bar{c}^2 t^2 - x^2}} P_1^\infty \left(\frac{\bar{c}t}{L_{\mathrm{loc}}(\omega)} \frac{\bar{c}t}{\sqrt{\bar{c}^2 t^2 - x^2}} \right),$$

where $L_{\mathrm{loc}}(\omega) = 4\bar{c}^2/\gamma\omega^2$ is the one-dimensional localization length defined by (7.81). Finally, this gives an explicit representation of the mean reflected intensity in the limit $\varepsilon \to 0$:

$$\lim_{\varepsilon\to 0} \mathbb{E}[p^\varepsilon(t, \mathbf{x}, 0^+)^2] = \frac{1}{32\pi^3 \bar{c}^4 t^2}$$

$$\times \begin{cases} \int \left[\dfrac{\frac{\bar{c}}{2L_{\mathrm{loc}}(\omega)} \frac{\bar{c}t}{\sqrt{\bar{c}^2 t^2 - x^2}}}{\left(1 + \frac{\bar{c}t}{2L_{\mathrm{loc}}(\omega)} \frac{\bar{c}t}{\sqrt{\bar{c}^2 t^2 - x^2}}\right)^2} \right] |\hat{f}(\omega)|^2 \omega^2 \, d\omega & \text{if } t > x/\bar{c}, \\[1em] 0 & \text{if } t < x/\bar{c}. \end{cases} \quad (14.75)$$

This representation is valid as soon as $\bar{c}t < 2L$, and it allows us to discuss quantitatively the space-time behavior of the mean reflected intensity at the surface. Let us consider an initial pulse whose Fourier transform is concentrated in a narrow band around the frequency ω_0. As can be seen in Figures 14.5 and 14.6, at a given time t such that $\bar{c}t \geq L_{\text{loc}}(\omega_0)$, the mean reflected intensity is continuously distributed in the disk with center at 0 and radius $\bar{c}t$. The asymptotic spatial distribution is obtained from (14.75):

$$\lim_{\varepsilon \to 0} \mathbb{E}[p^\varepsilon(t, \mathbf{x}, 0^+)^2] \overset{\bar{c}t \gg L_{\text{loc}}(\omega_0)}{\approx} \sqrt{\left(1 - \frac{|\mathbf{x}|^2}{\bar{c}^2 t^2}\right)_+} \times \lim_{\varepsilon \to 0} \mathbb{E}[p^\varepsilon(t, \mathbf{0}, 0^+)^2],$$

where $(\cdot)_+ = \max(\cdot, 0)$, and the time decay rate of the central intensity is

$$\lim_{\varepsilon \to 0} \mathbb{E}[p^\varepsilon(t, \mathbf{0}, 0^+)^2] \overset{\bar{c}t \gg L_{\text{loc}}(\omega_0)}{\approx} \frac{\omega_0^2 L_{\text{loc}}(\omega_0)}{8\pi^2 \bar{c}^5 t^4} \times \frac{1}{2\pi} \int |\hat{f}(\omega)|^2 \, d\omega$$

$$= \frac{1}{2\pi^2 \gamma \bar{c}^3 t^4} \times \frac{1}{2\pi} \int |\hat{f}(\omega)|^2 \, d\omega, \quad (14.76)$$

which is independent of the carrier frequency. To sum up, for long times, the mean reflected intensity has a self-similar spatial distribution in the disk with center at 0 and radius $\bar{c}t$, and it is given by

$$\lim_{\varepsilon \to 0} \mathbb{E}[p^\varepsilon(t, \mathbf{x}, 0^+)^2] \overset{\bar{c}t \gg L_{\text{loc}}(\omega_0)}{\approx} \frac{1}{2\pi^2 \gamma \bar{c}^3 t^4} \sqrt{\left(1 - \frac{|\mathbf{x}|^2}{\bar{c}^2 t^2}\right)_+} \times \frac{1}{2\pi} \int |\hat{f}(\omega)|^2 \, d\omega.$$

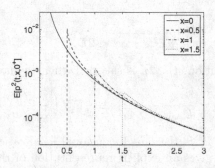

Fig. 14.5. Mean reflected intensity as a function of time $t \mapsto \mathbb{E}[p^\varepsilon(t, \mathbf{x} = (x, 0), 0^+)^2]$ in the limit $\varepsilon \to 0$ as given by (14.75). Here the initial pulse is the second derivative of a Gaussian, with Fourier transform $\hat{f}(\omega) = \omega^2 \exp(-\omega^2/5)$. The mean velocity is $\bar{c} = 1$ and the integrated covariance $\gamma = 4$.

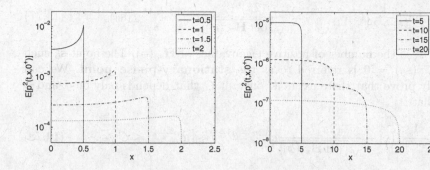

Fig. 14.6. Mean reflected intensity as a function of space $x \mapsto \mathbb{E}[p^\varepsilon(t, \mathbf{x} = (x, 0), 0^+)^2]$ in the limit $\varepsilon \to 0$. The configuration is the one described in Figure 14.5.

14.4 Appendix: Stationary-Phase Method

The One-Dimensional Case

Let ϕ and f be two smooth real-valued functions defined on the interval $(-a, a)$ with a positive. Assume that $\phi'(s)$ vanishes only at $s_0 \in (-a, a)$ and that $\phi''(s_0) \neq 0$. The integral

$$I(\varepsilon) = \int_{-a}^{a} e^{i\frac{\phi(s)}{\varepsilon}} f(s)\, ds$$

can be approximated as $\varepsilon \to 0$ by

$$\lim_{\varepsilon \to \infty} \frac{1}{\sqrt{\varepsilon}} I(\varepsilon) e^{-i\frac{\phi(s_0)}{\varepsilon}} = \frac{\sqrt{2\pi}}{\sqrt{|\phi''(s_0)|}} e^{in^* \frac{\pi}{4}} f(s_0),$$

where $n^* = \mathrm{sgn}(\phi''(s_0))$. This result also holds true if $a = \infty$, $\phi \in C^\infty(\mathbb{R}, \mathbb{R})$, and f belongs to the Schwartz class of infinitely smooth and rapidly decaying functions.

The n-Dimensional Case

The stationary-phase theorem can be generalized to n-dimensional integrals. Let n be a positive integer, and O an open subset of $\mathbb{R}^n$. Let ϕ and f be two smooth functions from O to $\mathbb{R}$. Assume that $\nabla \phi(\mathbf{s})$ vanishes only at $\mathbf{s}_0 \in O$ and that the determinant of the Hessian $\mathbf{H}_{\mathbf{s}_0}(\phi)$ of ϕ at $\mathbf{s}_0$ is nonzero. The integral

$$I(\varepsilon) = \int_O e^{i\frac{\phi(\mathbf{s})}{\varepsilon}} f(\mathbf{s})\, d^n\mathbf{s}$$

can be approximated as $\varepsilon \to 0$ by

$$\lim_{\varepsilon \to \infty} \frac{1}{\varepsilon^{n/2}} I(\varepsilon) e^{-i\frac{\phi(\mathbf{s}_0)}{\varepsilon}} = \frac{(2\pi)^{n/2}}{\sqrt{|\det \mathbf{H}_{\mathbf{s}_0}(\phi)|}} e^{i(2n^*-n)\frac{\pi}{4}} f(\mathbf{s}_0), \tag{14.77}$$

where n^* is the number of positive eigenvalues of $\mathbf{H}_{\mathbf{s}_0}(\phi)$. The point $\mathbf{s}_0$ such that $\nabla\phi(\mathbf{s}_0) = 0$ is referred to as the **stationary-phase point**. We can actually prove that there exists a constant C that depends only on f and ϕ such that

$$\left| \frac{1}{\varepsilon^{n/2}} I(\varepsilon) - \frac{(2\pi)^{n/2}}{\sqrt{|\det \mathbf{H}_{\mathbf{s}_0}(\phi)|}} e^{i(2n^*-n)\frac{\pi}{4}} f(\mathbf{s}_0) e^{i\frac{\phi(\mathbf{s}_0)}{\varepsilon}} \right| \leq C\sqrt{\varepsilon}. \tag{14.78}$$

A Degenerate Case

The typical configuration that is encountered in this book is actually degenerate. The result that we need is summarized in the following proposition.

Proposition 14.4. *For any $\varepsilon > 0$, let us consider the integral*

$$I(\varepsilon) = \int_{\mathbb{R}} \int_{\mathbb{R}^n} e^{i\frac{\omega\phi(\mathbf{s})}{\varepsilon}} f(\omega, \mathbf{s})\omega^{n/2} \, d^n\mathbf{s} \, d\omega,$$

where $\phi \in C^\infty(\mathbb{R}^n, \mathbb{R})$ and f belongs to the Schwartz class of infinitely smooth and rapidly decaying functions. We assume that $\nabla\phi(\mathbf{s})$ vanishes only at $\mathbf{s}_0 \in \mathbb{R}^n$ and that the determinant of the Hessian $\mathbf{H}_{\mathbf{s}_0}(\phi)$ of ϕ at $\mathbf{s}_0$ is nonzero. There are two cases:

1. If $\phi(\mathbf{s}_0) \neq 0$, then

$$\lim_{\varepsilon \to 0} \frac{1}{\varepsilon^{n/2}} I(\varepsilon) = 0. \tag{14.79}$$

2. If $\phi(\mathbf{s}_0) = 0$, then

$$\lim_{\varepsilon \to 0} \frac{1}{\varepsilon^{n/2}} I(\varepsilon) = \frac{(2\pi)^{n/2}}{\sqrt{|\det \mathbf{H}_{\mathbf{s}_0}(\phi)|}} e^{i(2n^*-n)\frac{\pi}{4}} \int f(\omega, \mathbf{s}_0) \, d\omega, \tag{14.80}$$

where n^* is the number of positive eigenvalues of $\mathbf{H}_{\mathbf{s}_0}(\phi)$.

The proof of this proposition is based on the estimate (14.78) applied with a fixed ω. Setting

$$I(\varepsilon, \omega) = \int_{\mathbb{R}^n} e^{i\frac{\omega\phi(\mathbf{s})}{\varepsilon}} f(\omega, \mathbf{s})\omega^{n/2} \, d^n\mathbf{s},$$

we obtain that there exists $C(\omega)$ such that

$$\left| \frac{1}{\varepsilon^{n/2}} I(\varepsilon, \omega) - \frac{(2\pi)^{n/2}}{\sqrt{|\det \mathbf{H}_{\mathbf{s}_0}(\phi)|}} e^{i(2n^*-n)\frac{\pi}{4}} e^{i\frac{\omega\phi(\mathbf{s}_0)}{\varepsilon}} f(\omega, \mathbf{s}_0) \right| \leq C(\omega)\sqrt{\varepsilon}.$$

The factor $\omega^{n/2}$ in the integral $I(\varepsilon, \omega)$ is important, since the phase scales with ω/ε. Furthermore, the smooth and bounded properties of f ensure that

$C(\omega)$ is integrable. We can now integrate with respect to ω. If $\phi(\mathbf{s}_0) \neq 0$, then the integral

$$\int e^{i\frac{\omega\phi(\mathbf{s}_0)}{\varepsilon}} f(\omega, \mathbf{s}_0) \, d\omega$$

goes to zero as $\varepsilon \to 0$ by the Riemann–Lebesgue lemma, which yields (14.79). If $\phi(\mathbf{s}_0) = 0$, then we immediately get (14.80).

Notes

Most of the material presented in Section 14.1 is from the review paper [8]. The formula (14.52) describing the shape of the front pulse has been derived in the three-dimensional case in [38] and in the case with additional horizontal weak fluctuations in [109, 160]. This problem was also addressed in [46, 118]. An introduction to the stationary-phase method can be found in the book by Bleisten and Handelsman [15]. The result needed in the book is summarized in Appendix 14.4. One of the main areas of applications of the theory of waves propagating in disordered media is imaging of embedded active sources or passive scatterers. Coherent interferometry in finely layered random media, exploiting the stable structure of the front derived in this chapter, is presented in [18].

15

Time Reversal in a Three-Dimensional Layered Medium

In this chapter we discuss the basic properties of time reversal in three-dimensional randomly layered media, and we compare the fundamental diffraction limit phenomenon in the homogeneous and random cases. An **active source** located inside the medium emits a pulse that is recorded on a time-reversal mirror. The wave is sent back into the medium, either numerically with the knowledge of the medium, or physically into the real medium. The goal of this chapter is to give a precise description of the refocusing of the pulse. In our regime of separation of scales we show that the pulse refocuses at the original location of the source and at a critical time. In fact, time-reversal refocusing contains information about the source that cannot be obtained by a direct arrival-time analysis. The resolution at the source may be enhanced by the randomness in the medium. This phenomenon is sometimes referred to as a **superresolution** effect.

15.1 The Embedded-Source Problem

We consider our familiar strongly heterogeneous white-noise regime, where the typical wavelength is large relative to the correlation length of the medium, but short relative to the depth of the source. In this chapter, we first study wave propagation in a three-dimensional layered medium in the case that the source is inside the medium. We then consider time reversal of the wave emanating from the source and analyze the refocusing properties of the wave field. Our analysis shows how the time-reversal technique can be useful for source estimation in the context of randomly layered media. We consider linear acoustic waves propagating in three spatial dimensions:

$$\rho \frac{\partial \mathbf{u}^\varepsilon}{\partial t} + \nabla p^\varepsilon = \mathbf{F}^\varepsilon , \tag{15.1}$$

$$\frac{1}{K} \frac{\partial p^\varepsilon}{\partial t} + \nabla \cdot \mathbf{u}^\varepsilon = 0 , \tag{15.2}$$

where p^ε is the pressure, $\mathbf{u}^\varepsilon$ is the velocity, ρ is the density of the medium, and K the bulk modulus. The forcing term $\mathbf{F}^\varepsilon$ is due to the source. We consider the case with a constant density (for simplicity) and a randomly fluctuating bulk modulus that is z-dependent only in the slab $(-L, 0)$. Note that by hyperbolicity of the system of governing equations we can choose L large enough so that the termination of the slab does not affect the wave field at the surface $z = 0$ over the time period that we consider. In view of the homogenization results presented in Section 4.4, we consider a medium with random fluctuations centered at the homogenized quantities. We therefore discuss the model

$$\rho \equiv \bar{\rho}, \tag{15.3}$$

$$\frac{1}{K} = \begin{cases} \dfrac{1}{\overline{K}}\left(1 + \nu\left(\dfrac{z}{\varepsilon^2}\right)\right) & \text{if } z \in [-L, 0], \\ \dfrac{1}{\overline{K}} & \text{if } z \in (-\infty, -L) \cup (0, \infty), \end{cases} \tag{15.4}$$

where ν is a zero-mean mixing process and ε^2 is a small dimensionless parameter that characterizes the ratio between the correlation length of the medium and the typical depth of the source. A point source located at $(\mathbf{x}_s, z_s)$, $z_s \leq 0$, generates a forcing term $\mathbf{F}^\varepsilon$ at time t_s that is given by

$$\mathbf{F}^\varepsilon(t, x, y, z) = \begin{bmatrix} \mathbf{f_x} \\ f_z \end{bmatrix}\left(\frac{t - t_s}{\varepsilon}\right)\delta\left(\frac{\mathbf{x} - \mathbf{x}_s}{\varepsilon}\right)\delta(z - z_s)$$

$$= \varepsilon^2\begin{bmatrix} \mathbf{f_x} \\ f_z \end{bmatrix}\left(\frac{t - t_s}{\varepsilon}\right)\delta(\mathbf{x} - \mathbf{x}_s)\,\delta(z - z_s). \tag{15.5}$$

Note that the time duration of the source is short and scaled by ε, which is large compared to the correlation length of the medium; which is $\mathcal{O}(\varepsilon^2)$. In our time-reversal setup we place a time-reversal mirror of spatial size $\mathcal{O}(1)$ at the origin. Our setup is illustrated in Figure 15.1.

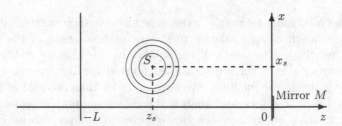

Fig. 15.1. Emission from a point source.

In Section 15.2 we derive an integral representation for the time-reversed wave field. The integral representation is obtained by taking a Fourier transform in the time and horizontal space coordinates. This reduces the problem to a family of one-dimensional problems that can be analyzed by decomposing the wave field into right- and left-going waves as in the previous chapter. In Section 15.3 we consider the case of a homogeneous medium and we compute in particular the size of the refocused spot size (Rayleigh resolution formula). In Section 15.4 we analyze time reversal with a source embedded into a randomly layered medium with a mirror at the surface. We carry out a careful stationary-phase analysis, which is combined with diffusion approximation results in the limit of small ε. This gives a limit for the time-reversed wave field, which reveals a refocusing of the pulse at a critical time and at the original source location. The focal-spot size is explicitly computed in some particular configurations and compared with the results obtained in the homogeneous-medium case, which exhibits the so-called superresolution effect. The theoretical formulas are based on a description of the moments of the mode-dependent reflection and transmission coefficients presented in Chapter 14, and which will be extended in Appendix 15.6.

15.2 Time Reversal with Embedded Source

15.2.1 Emission from a Point Source

In the scaling that we consider, the typical wavelength of the source is small, $\mathcal{O}(\varepsilon)$, and, as in the previous chapter, we use the following specific Fourier transform and its inverse with respect to time and the transverse direction:

$$\hat{p}^\varepsilon(\omega, \boldsymbol{\kappa}, z) = \int \int p^\varepsilon(t, \mathbf{x}, z) e^{\frac{i\omega}{\varepsilon}(t - \boldsymbol{\kappa} \cdot \mathbf{x})} \, dt \, d\mathbf{x},$$

$$p^\varepsilon(t, \mathbf{x}, z) = \frac{1}{(2\pi\varepsilon)^3} \int \int \hat{p}^\varepsilon(\omega, \boldsymbol{\kappa}, z) e^{-\frac{i\omega}{\varepsilon}(t - \boldsymbol{\kappa} \cdot \mathbf{x})} \omega^2 \, d\omega \, d\boldsymbol{\kappa},$$

where $\mathbf{x} = (x, y)$ stands for the transverse spatial variables. Taking the scaled Fourier transform in (15.1) and (15.2) and using (15.5) shows that $\hat{\mathbf{u}}^\varepsilon = (\hat{\mathbf{v}}^\varepsilon, \hat{u}^\varepsilon)$ and $\hat{p}^\varepsilon$ satisfy the system

$$-\bar{\rho}\frac{i\omega}{\varepsilon}\hat{\mathbf{v}}^\varepsilon + \frac{i\omega}{\varepsilon}\boldsymbol{\kappa}\hat{p}^\varepsilon = \varepsilon^3 \hat{\mathbf{f}}_\mathbf{x}(\omega) e^{\frac{i\omega}{\varepsilon}(t_s - \boldsymbol{\kappa} \cdot \mathbf{x}_s)} \delta(z - z_s), \quad (15.6)$$

$$-\bar{\rho}\frac{i\omega}{\varepsilon}\hat{u}^\varepsilon + \frac{\partial \hat{p}^\varepsilon}{\partial z} = \varepsilon^3 \hat{f}_z(\omega) e^{\frac{i\omega}{\varepsilon}(t_s - \boldsymbol{\kappa} \cdot \mathbf{x}_s)} \delta(z - z_s), \quad (15.7)$$

$$-\frac{1}{K(z)}\frac{i\omega}{\varepsilon}\hat{p}^\varepsilon + \frac{i\omega}{\varepsilon}\boldsymbol{\kappa} \cdot \hat{\mathbf{v}}^\varepsilon + \frac{\partial \hat{u}}{\partial z} = 0, \quad (15.8)$$

where $\hat{f}$ is the ordinary unscaled Fourier transform of the pulse profile

$$\hat{f}(\omega) = \int f(t) e^{i\omega t} \, dt, \qquad f(t) = \frac{1}{2\pi} \int \hat{f}(\omega) e^{-i\omega t} \, d\omega.$$

By eliminating $\hat{\mathbf{v}}^\varepsilon$, we deduce that $(\hat{u}^\varepsilon, \hat{p}^\varepsilon)$ satisfy the following closed system for $-L < z < z_s$ and $z_s < z < 0$:

$$\frac{\partial \hat{u}^\varepsilon}{\partial z} + \frac{i\omega}{\varepsilon}\left(-\frac{1}{K(z)} + \frac{|\kappa|^2}{\bar{\rho}}\right)\hat{p}^\varepsilon = 0, \tag{15.9}$$

$$\frac{\partial \hat{p}^\varepsilon}{\partial z} - \frac{i\omega}{\varepsilon}\bar{\rho}\hat{u}^\varepsilon = 0, \tag{15.10}$$

with the jumps at $z = z_s$ given by

$$[\hat{u}^\varepsilon]_{z_s} = \varepsilon^3\frac{\kappa \cdot \hat{\mathbf{f}}_\mathbf{x}(\omega)}{\bar{\rho}}e^{\frac{i\omega}{\varepsilon}(t_s - \kappa \cdot \mathbf{x}_s)}, \tag{15.11}$$

$$[\hat{p}^\varepsilon]_{z_s} = \varepsilon^3\hat{f}_z(\omega)e^{\frac{i\omega}{\varepsilon}(t_s - \kappa \cdot \mathbf{x}_s)}. \tag{15.12}$$

We introduce the right- and left-propagating wave modes $\check{a}^\varepsilon$ and $\check{b}^\varepsilon$, which are defined as in (14.24) and (14.25) by

$$\hat{p}^\varepsilon(\omega, \kappa, z) = \frac{\sqrt{\bar{\zeta}(\kappa)}}{2}\left(\check{a}^\varepsilon(\omega, \kappa, z)e^{\frac{i\omega z}{\varepsilon\bar{c}(\kappa)}} - \check{b}^\varepsilon(\omega, \kappa, z)e^{-\frac{i\omega z}{\varepsilon\bar{c}(\kappa)}}\right), \tag{15.13}$$

$$\hat{u}^\varepsilon(\omega, \kappa, z) = \frac{1}{2\sqrt{\bar{\zeta}(\kappa)}}\left(\check{a}^\varepsilon(\omega, \kappa, z)e^{\frac{i\omega z}{\varepsilon\bar{c}(\kappa)}} + \check{b}^\varepsilon(\omega, \kappa, z)e^{-\frac{i\omega z}{\varepsilon\bar{c}(\kappa)}}\right), \tag{15.14}$$

where $\kappa = |\kappa|$, $\bar{c} = \sqrt{\bar{K}/\bar{\rho}}$ is the effective speed, $\bar{c}(\kappa)$ is the effective mode-dependent vertical velocity (14.19), and $\bar{\zeta}(\kappa)$ is the mode-dependent acoustic impedance (14.20), which are recalled here:

$$\bar{c}(\kappa) = \frac{\bar{c}}{\sqrt{1 - \bar{c}^2\kappa^2}}, \qquad \bar{\zeta}(\kappa) = \bar{\rho}\bar{c}(\kappa). \tag{15.15}$$

We consider only propagating modes and ignore evanescent modes, meaning that $|\kappa| < \bar{c}^{-1}$ as discussed in Chapter 14. The system for $\check{a}^\varepsilon$ and $\check{b}^\varepsilon$ can be written in the form

$$\frac{\partial}{\partial z}\begin{bmatrix}\check{a}^\varepsilon \\ \check{b}^\varepsilon\end{bmatrix} = \frac{1}{\varepsilon}\mathbf{H}_{(\omega,\kappa)}\left(\frac{z}{\varepsilon}, \nu_\kappa\left(\frac{z}{\varepsilon^2}\right)\right)\begin{bmatrix}\check{a}^\varepsilon \\ \check{b}^\varepsilon\end{bmatrix}, \tag{15.16}$$

where the complex 2×2 matrix $\mathbf{H}_{(\omega,\kappa)}$ is given by (14.36)

$$\mathbf{H}_{(\omega,\kappa)}(z, \nu) = \frac{i\omega\nu}{2\bar{c}(\kappa)}\begin{bmatrix}1 & -e^{-\frac{2i\omega z}{\bar{c}(\kappa)}} \\ e^{\frac{2i\omega z}{\bar{c}(\kappa)}} & -1\end{bmatrix}. \tag{15.17}$$

The mode-dependent random process ν_κ is defined by (14.23):

$$\nu_\kappa(z) = \frac{\bar{c}(\kappa)^2}{\bar{c}^2}\nu(z) = \frac{1}{1 - \bar{c}^2\kappa^2}\nu(z).$$

Using the definitions (15.13) and (15.14) and the expressions (15.11) and (15.12) for the jumps in $\hat{u}^\varepsilon$ and $\hat{p}^\varepsilon$, we deduce the jumps at $z = z_s$ for the modes $\check{a}^\varepsilon$ and $\check{b}^\varepsilon$:

$$\left[\breve{a}^{\varepsilon}\right]_{z_s} = \varepsilon^3 e^{\frac{i\omega}{\varepsilon}(t_s - \boldsymbol{\kappa} \cdot \mathbf{x}_s - z_s/\bar{c}(\kappa))} S_a(\omega, \boldsymbol{\kappa}), \tag{15.18}$$

$$\left[\breve{b}^{\varepsilon}\right]_{z_s} = \varepsilon^3 e^{\frac{i\omega}{\varepsilon}(t_s - \boldsymbol{\kappa} \cdot \mathbf{x}_s + z_s/\bar{c}(\kappa))} S_b(\omega, \boldsymbol{\kappa}), \tag{15.19}$$

with the source contributions given by

$$S_a(\omega, \boldsymbol{\kappa}) = \frac{\sqrt{\zeta(\kappa)}}{\bar{\rho}} \boldsymbol{\kappa} \cdot \hat{\mathbf{f}}_{\mathbf{x}}(\omega) + \frac{1}{\sqrt{\zeta(\kappa)}} \hat{f}_z(\omega), \tag{15.20}$$

$$S_b(\omega, \boldsymbol{\kappa}) = \frac{\sqrt{\zeta(\kappa)}}{\bar{\rho}} \boldsymbol{\kappa} \cdot \hat{\mathbf{f}}_{\mathbf{x}}(\omega) - \frac{1}{\sqrt{\zeta(\kappa)}} \hat{f}_z(\omega). \tag{15.21}$$

The system for $\breve{a}^{\varepsilon}$ and $\breve{b}^{\varepsilon}$ is supplemented with the boundary conditions at $z = 0$ and $z = -L$ that are shown in Figure 15.2. We assume that no energy is coming from $+\infty$ or $-\infty$, so that we get the conditions

$$\breve{a}^{\varepsilon}(\omega, \boldsymbol{\kappa}, -L) = 0, \quad \breve{b}^{\varepsilon}(\omega, \boldsymbol{\kappa}, 0) = 0.$$

The quantity of interest is the wave field at the surface, which is completely characterized by $\breve{a}^{\varepsilon}(\omega, \boldsymbol{\kappa}, 0)$, since $\breve{b}^{\varepsilon}(\omega, \boldsymbol{\kappa}, 0) = 0$.

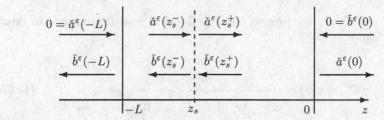

Fig. 15.2. Boundary conditions at $z = -L$ and $z = 0$ corresponding to the emission from the point source located at depth z_s.

We transform this boundary value problem into an initial value problem by introducing the propagator $\mathbf{P}^{\varepsilon}_{(\omega,\kappa)}(z_0, z)$, $-L \leq z_0 \leq z \leq 0$, which is a family of complex 2×2 matrices that solve

$$\frac{\partial \mathbf{P}^{\varepsilon}_{(\omega,\kappa)}}{\partial z} = \frac{1}{\varepsilon} \mathbf{H}_{(\omega,\kappa)}\left(\frac{z}{\varepsilon}, \nu_\kappa\left(\frac{z}{\varepsilon^2}\right)\right) \mathbf{P}^{\varepsilon}_{(\omega,\kappa)}, \quad \mathbf{P}^{\varepsilon}_{(\omega,\kappa)}(z_0, z = z_0) = \mathbf{I}.$$

Using the particular form of the matrix $\mathbf{H}_{(\omega,\kappa)}$, one can show that the propagator matrix can be written as

$$\mathbf{P}^{\varepsilon}_{(\omega,\kappa)}(z_0, z) = \begin{bmatrix} \alpha^{\varepsilon}_{(\omega,\kappa)} & \overline{\beta^{\varepsilon}_{(\omega,\kappa)}} \\ \beta^{\varepsilon}_{(\omega,\kappa)} & \overline{\alpha^{\varepsilon}_{(\omega,\kappa)}} \end{bmatrix}(z_0, z),$$

where the column vector $(\alpha^{\varepsilon}_{(\omega,\kappa)}, \beta^{\varepsilon}_{(\omega,\kappa)})^T$ solves (15.16) with the initial conditions

$$\alpha^{\varepsilon}_{(\omega,\kappa)}(z_0, z = z_0) = 1\,, \qquad \beta^{\varepsilon}_{(\omega,\kappa)}(z_0, z = z_0) = 0\,. \qquad (15.22)$$

The boundary conditions at $z = -L$ and $z = 0$ and the jump conditions (15.18) and (15.19) imply that

$$\begin{bmatrix} \check{a}^{\varepsilon}(z_s^-) \\ \check{b}^{\varepsilon}(z_s^-) \end{bmatrix} = \mathbf{P}^{\varepsilon}_{(\omega,\kappa)}(-L, z_s) \begin{bmatrix} 0 \\ \check{b}^{\varepsilon}(-L) \end{bmatrix}\,, \qquad (15.23)$$

$$\begin{bmatrix} \check{a}^{\varepsilon}(0) \\ 0 \end{bmatrix} = \mathbf{P}^{\varepsilon}_{(\omega,\kappa)}(z_s, 0) \begin{bmatrix} \check{a}^{\varepsilon}(z_s^+) \\ \check{b}^{\varepsilon}(z_s^+) \end{bmatrix} = \mathbf{P}^{\varepsilon}_{(\omega,\kappa)}(z_s, 0) \begin{bmatrix} \check{a}^{\varepsilon}(z_s^-) + [\check{a}^{\varepsilon}]_{z_s} \\ \check{b}^{\varepsilon}(z_s^-) + [\check{b}^{\varepsilon}]_{z_s} \end{bmatrix}\,, (15.24)$$

where for notational simplicity, we do not display the arguments (ω, κ) in $\check{a}^{\varepsilon}$ and $\check{b}^{\varepsilon}$. Inverting $\mathbf{P}^{\varepsilon}_{(\omega,\kappa)}(z_s, 0)$ in (15.24) and substituting (15.23) in (15.24) gives the following system:

$$\begin{bmatrix} \overline{\beta^{\varepsilon}_{(\omega,\kappa)}(-L, z_s)}\check{b}^{\varepsilon}(-L) + [\check{a}^{\varepsilon}]_{z_s} \\ \alpha^{\varepsilon}_{(\omega,\kappa)}(-L, z_s)\check{b}^{\varepsilon}(-L) + [\check{b}^{\varepsilon}]_{z_s} \end{bmatrix} = \begin{bmatrix} \overline{\alpha^{\varepsilon}_{(\omega,\kappa)}(z_s, 0)}\check{a}^{\varepsilon}(0) \\ -\beta^{\varepsilon}_{(\omega,\kappa)}(z_s, 0)\check{a}^{\varepsilon}(0) \end{bmatrix}\,.$$

Solving these equations for $\check{a}^{\varepsilon}(0)$ gives

$$\check{a}^{\varepsilon}(\omega, \kappa, 0) = T^{\varepsilon}_g(\omega, \kappa, z_s) [\check{a}^{\varepsilon}]_{z_s} - R^{\varepsilon}_g(\omega, \kappa, z_s) [\check{b}^{\varepsilon}]_{z_s}\,,$$

where R^{ε}_g and T^{ε}_g are the **generalized reflection and transmission coefficients** defined by

$$R^{\varepsilon}_g(\omega, \kappa, z) = \frac{\dfrac{\overline{\beta^{\varepsilon}_{(\omega,\kappa)}(-L, z)}}{\alpha^{\varepsilon}_{(\omega,\kappa)}(-L, z)}}{\alpha^{\varepsilon}_{(\omega,\kappa)}(z, 0) + \beta^{\varepsilon}_{(\omega,\kappa)}(z, 0)\dfrac{\overline{\beta^{\varepsilon}_{(\omega,\kappa)}(-L, z)}}{\alpha^{\varepsilon}_{(\omega,\kappa)}(-L, z)}}\,, \qquad (15.25)$$

$$T^{\varepsilon}_g(\omega, \kappa, z) = \frac{1}{\alpha^{\varepsilon}_{(\omega,\kappa)}(z, 0) + \beta^{\varepsilon}_{(\omega,\kappa)}(z, 0)\dfrac{\overline{\beta^{\varepsilon}_{(\omega,\kappa)}(-L, z)}}{\alpha^{\varepsilon}_{(\omega,\kappa)}(-L, z)}}\,, \qquad (15.26)$$

which are evaluated at a general depth z in the slab $(-L, 0)$. Using the expressions (15.18) and (15.19) for the jumps gives the following explicit formula for $\check{a}^{\varepsilon}(0)$:

$$\check{a}^{\varepsilon}(\omega, \kappa, 0) = \varepsilon^3 e^{\frac{i\omega}{\varepsilon}(t_s - \kappa \cdot \mathbf{x}_s)} \left[e^{-\frac{i\omega z_s}{\varepsilon \bar{c}(\kappa)}} T^{\varepsilon}_g(\omega, \kappa, z_s) S_a(\omega, \kappa) \right.$$
$$\left. - e^{\frac{i\omega z_s}{\varepsilon \bar{c}(\kappa)}} R^{\varepsilon}_g(\omega, \kappa, z_s) S_b(\omega, \kappa) \right]\,. \qquad (15.27)$$

The coefficients R^{ε}_g and T^{ε}_g are generalized versions of the reflection and transmission coefficients used in the previous chapters, as we explain now. This will enable us to give an interpretation of the content of formula (15.27) at the end of this section. The transmission and reflection coefficients $T^{\varepsilon}_{(\omega,\kappa)}(-L, z)$ and $R^{\varepsilon}_{(\omega,\kappa)}(-L, z)$ for a slab $[-L, z]$ (see Figure 15.3 and Section 9.1) are given in terms of $\alpha^{\varepsilon}_{(\omega,\kappa)}$ and $\beta^{\varepsilon}_{(\omega,\kappa)}$ by

$$R^\varepsilon_{(\omega,\kappa)}(-L,z) = \frac{\overline{\beta^\varepsilon_{(\omega,\kappa)}(-L,z)}}{\alpha^\varepsilon_{(\omega,\kappa)}(-L,z)}, \qquad T^\varepsilon_{(\omega,\kappa)}(-L,z) = \frac{1}{\alpha^\varepsilon_{(\omega,\kappa)}(-L,z)}. \quad (15.28)$$

We also introduce $\widetilde{R}^\varepsilon_{(\omega,\kappa)}$ and $\widetilde{T}^\varepsilon_{(\omega,\kappa)}$ defined as the **adjoint** reflection and transmission coefficients for the experiment corresponding to a right-going input wave incoming from the left (see Figure 15.4). They were introduced in the one-dimensional case in Section 12.1.1. They are given in terms of $\alpha^\varepsilon_{(\omega,\kappa)}$ and $\beta^\varepsilon_{(\omega,\kappa)}$ by

$$\widetilde{R}^\varepsilon_{(\omega,\kappa)}(z,0) = -\frac{\beta^\varepsilon_{(\omega,\kappa)}(z,0)}{\alpha^\varepsilon_{(\omega,\kappa)}(z,0)}, \qquad \widetilde{T}^\varepsilon_{(\omega,\kappa)}(z,0) = \frac{1}{\alpha^\varepsilon_{(\omega,\kappa)}(z,0)}. \quad (15.29)$$

Note that the two transmission coefficients are equal: $\widetilde{T}^\varepsilon_{(\omega,\kappa)}(z,0) = T^\varepsilon_{(\omega,\kappa)}(z,0)$.

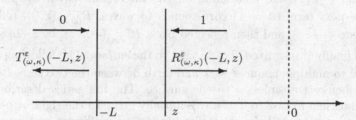

Fig. 15.3. Reflection and transmission coefficients.

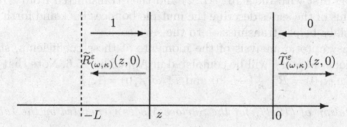

Fig. 15.4. Adjoint reflection and transmission coefficients.

We can express the generalized coefficients R^ε_g and T^ε_g in terms of the usual reflection and transmission coefficients $R^\varepsilon_{(\omega,\kappa)}$ and $T^\varepsilon_{(\omega,\kappa)}$ and the adjoint coefficients $\widetilde{R}^\varepsilon_{(\omega,\kappa)}$ and $\widetilde{T}^\varepsilon_{(\omega,\kappa)}$:

$$R^\varepsilon_g(\omega,\kappa,z) = \frac{\widetilde{T}^\varepsilon_{(\omega,\kappa)}(z,0)R^\varepsilon_{(\omega,\kappa)}(-L,z)}{1 - \widetilde{R}^\varepsilon_{(\omega,\kappa)}(z,0)R^\varepsilon_{(\omega,\kappa)}(-L,z)}, \quad (15.30)$$

$$T^\varepsilon_g(\omega,\kappa,z) = \frac{\widetilde{T}^\varepsilon_{(\omega,\kappa)}(z,0)}{1 - \widetilde{R}^\varepsilon_{(\omega,\kappa)}(z,0)R^\varepsilon_{(\omega,\kappa)}(-L,z)}. \quad (15.31)$$

Interpretation of the Generalized Coefficients

Keeping in mind the interpretation of the usual reflection and transmission coefficients by a random slab (see Section 5.1.6), we now explain the content of the generalized coefficients. We first consider the generalized transmission coefficient T_g^ε. We can rewrite it as a series

$$T_g^\varepsilon(\omega, \kappa, z) = \widetilde{T}_{(\omega,\kappa)}^\varepsilon(z, 0) \sum_{n=0}^{\infty} \left[\widetilde{R}_{(\omega,\kappa)}^\varepsilon(z, 0) R_{(\omega,\kappa)}^\varepsilon(-L, z) \right]^n, \qquad (15.32)$$

where we use the fact that $|\widetilde{R}_{(\omega,\kappa)}^\varepsilon(z,0)| < 1$ and $|R_{(\omega,\kappa)}^\varepsilon(-L,z)| < 1$, as follows from the energy conservation relation (7.11). The first term $\widetilde{T}_{(\omega,\kappa)}^\varepsilon(z,0)$ corresponds to waves transmitted from z to 0 without interacting with the left medium $(-L, z)$ and containing all the interactions with the medium $(z, 0)$. The next term $(n = 1)$ corresponds to waves, $\widetilde{R}_{(\omega,\kappa)}^\varepsilon(z, 0)$, reflected by $(z, 0)$ into $(-L, z)$, and then reflected back, $R_{(\omega,\kappa)}^\varepsilon(-L, z)$, by $(-L, z)$ into $(z, 0)$, and finally transmitted, $\widetilde{T}_{(\omega,\kappa)}^\varepsilon(z, 0)$, to the surface. The following terms correspond to multiple bounces back and forth between the two slabs $(-L, z)$ and $(z, 0)$ before transmission to the surface. The full series describes the total transmission from z to 0 of waves initially going to the right at point z. Similarly, we can expand the generalized reflection coefficient and we obtain a series for the total transmission from z to 0 of waves initially going to the left at point z. The first term $R_{(\omega,\kappa)}^\varepsilon(-L, z)\widetilde{T}_{(\omega,\kappa)}^\varepsilon(z, 0)$ corresponds to waves interacting first with the slab $(-L, z)$ and then transmitted from z to 0. The other terms of the series describe the mutiple bounces back and forth between the two slabs before transmission to the surface.

   The asymptotic analysis of the moments of these coefficients, started in the previous chapters, will be completed in Appendix 15.6. Note that if $z = 0$, then $R_g^\varepsilon(\omega, \kappa, 0) = R_{(\omega,\kappa)}^\varepsilon(-L, 0)$ and $T_g^\varepsilon(\omega, \kappa, 0) = 1$.

Interpretation of (15.27) for the Surface Modes Generated by the Internal Source

There are two contributions in (15.27): $T_g^\varepsilon S_a$ and $R_g^\varepsilon S_b$. The first one corresponds to the waves generated by the source, initially propagating to the right, and transmitted at the surface by the generalized transmission coefficient T_g^ε. The second term contains the waves generated by the source, initially propagating to the left, and sent back by the medium to the surface by the generalized reflection coefficient R_g^ε.

Integral Representation of the Field at the Surface

We denote the wave at the surface $z = 0$ by $(\mathbf{u}_s^\varepsilon, p_s^\varepsilon)$. By taking an inverse Fourier transform we obtain

$$p_s^\varepsilon(t,\mathbf{x}) = \frac{1}{(2\pi\varepsilon)^3} \int \frac{\sqrt{\zeta(\kappa)}}{2} \breve{a}^\varepsilon(\omega,\boldsymbol{\kappa},0) e^{-\frac{i\omega}{\varepsilon}(t-\boldsymbol{\kappa}\cdot\mathbf{x})} \omega^2 \, d\omega \, d\boldsymbol{\kappa} , \quad (15.33)$$

$$u_s^\varepsilon(t,\mathbf{x}) = \frac{1}{(2\pi\varepsilon)^3} \int \frac{1}{2\sqrt{\zeta(\kappa)}} \breve{a}^\varepsilon(\omega,\boldsymbol{\kappa},0) e^{-\frac{i\omega}{\varepsilon}(t-\boldsymbol{\kappa}\cdot\mathbf{x})} \omega^2 \, d\omega \, d\boldsymbol{\kappa} , \quad (15.34)$$

$$\mathbf{v}_s^\varepsilon(t,\mathbf{x}) = \frac{1}{(2\pi\varepsilon)^3} \int \frac{\sqrt{\zeta(\kappa)}}{2\bar{\rho}} \boldsymbol{\kappa}\breve{a}^\varepsilon(\omega,\boldsymbol{\kappa},0) e^{-\frac{i\omega}{\varepsilon}(t-\boldsymbol{\kappa}\cdot\mathbf{x})} \omega^2 \, d\omega \, d\boldsymbol{\kappa} , \quad (15.35)$$

where $\mathbf{u}_s^\varepsilon = (\mathbf{v}_s^\varepsilon, u_s^\varepsilon)$. These signals contain a stable wave front of duration $\mathcal{O}(\varepsilon)$ corresponding to the duration of the source and a long noisy coda part that is caused by the multiple scattering by the layers. These coda waves are part of, and play a crucial role in, the time-reversal procedure that we describe next.

15.2.2 Recording, Time Reversal, and Reemission

The first step of the time-reversal procedure consists in recording the velocity signal and/or the pressure signal at the surface $z = 0$ on the mirror $M = \{(\mathbf{x}, z), \mathbf{x} \in D, z = 0\}$ during some time interval centered at $t = 0$. The shape of the mirror whose center is located at the point $\mathbf{0}$ is given by $D \subset \mathbb{R}^2$. It turns out that as $\varepsilon \to 0$, the interesting asymptotic regime arises when we record the signal during a long time interval whose duration is of order one. We consider here the situation in which only the velocity is recorded. The case in which the pressure signal is recorded will be discussed in Section 15.5.5.

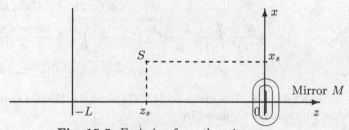

Fig. 15.5. Emission from the mirror.

In the second step of the time-reversal procedure a piece of the recorded signal is clipped using a cutoff function $t \mapsto G_1(t)$, where the support of G_1 is included in $[-t_1/2, t_1/2]$. We denote the recorded part of the wave by $\mathbf{u}_{\text{rec}}^\varepsilon$, so that

$$\mathbf{u}_{\text{rec}}^\varepsilon(t,\mathbf{x}) = \mathbf{u}_s^\varepsilon(t,\mathbf{x})G_1(t)G_2(\mathbf{x}) , \quad (15.36)$$

where G_2 is the spatial cutoff function introduced by the mirror,

$$G_2(\mathbf{x}) = \mathbf{1}_D(\mathbf{x}),$$

and D is the shape of the mirror. We could alternatively choose more-general spatial cutoff functions G_2 with integrability conditions (for instance $L^1 \cap L^2$). We then time-reverse this piece of the signal and send it back into the same medium, as illustrated in Figure 15.5. This means that we consider a new problem defined by the acoustic equations (15.1–15.2) with the source term

$$\mathbf{F}^\varepsilon_{\mathrm{TR}}(t, \mathbf{x}, z) = \bar{\rho}\bar{c}\,\mathbf{u}^\varepsilon_{\mathrm{rec}}(-t, \mathbf{x})\delta(z)\,, \qquad (15.37)$$

where the subscript TR stands for "time reversal" and the factor $\bar{\rho}\bar{c}$ has been added to restore the physical dimension of the expression. Note that by linearity of the problem this factor plays no role in the analysis. In terms of right- and left-going wave modes, the problem is defined by the linear system (15.16) for $-L \leq z < 0$, with the boundary conditions

$$\breve{b}^\varepsilon_{\mathrm{TR}}(\omega, \boldsymbol{\kappa}, 0^+) = 0\,, \qquad \breve{a}^\varepsilon_{\mathrm{TR}}(\omega, \boldsymbol{\kappa}, -L) = 0\,;$$

see Figure 15.6, and the jump condition

$$[\breve{b}^\varepsilon_{\mathrm{TR}}]_0 = \frac{\sqrt{\zeta(\kappa)}}{\bar{\rho}}\,\boldsymbol{\kappa} \cdot \hat{\mathbf{F}}^\varepsilon_{\mathrm{TR},\mathbf{x}}(\omega, \boldsymbol{\kappa}) - \frac{1}{\sqrt{\zeta(\kappa)}}\hat{F}^\varepsilon_{\mathrm{TR},z}(\omega, \boldsymbol{\kappa})\,,$$

where, using successively (15.37), (15.36), (15.35), and (15.34), we have

$$\hat{\mathbf{F}}^\varepsilon_{\mathrm{TR},\mathbf{x}}(\omega, \boldsymbol{\kappa}) = \frac{\bar{\rho}\bar{c}}{(2\pi\varepsilon)^3}\int \frac{\sqrt{\zeta(\kappa')}}{2\bar{\rho}}\boldsymbol{\kappa}'\overline{\hat{G}_1\left(\frac{\omega - \omega'}{\varepsilon}\right)}\hat{G}_2\left(\frac{\omega\boldsymbol{\kappa} + \omega'\boldsymbol{\kappa}'}{\varepsilon}\right)$$
$$\times \overline{\breve{a}^\varepsilon(\omega', \boldsymbol{\kappa}', 0)}\omega'^2\,d\omega'\,d\boldsymbol{\kappa}'\,,$$

$$\hat{F}^\varepsilon_{\mathrm{TR},z}(\omega, \boldsymbol{\kappa}) = \frac{\bar{\rho}\bar{c}}{(2\pi\varepsilon)^3}\int \frac{1}{2\sqrt{\zeta(\kappa')}}\overline{\hat{G}_1\left(\frac{\omega - \omega'}{\varepsilon}\right)}\hat{G}_2\left(\frac{\omega\boldsymbol{\kappa} + \omega'\boldsymbol{\kappa}'}{\varepsilon}\right)$$
$$\times \overline{\breve{a}^\varepsilon(\omega', \boldsymbol{\kappa}', 0)}\omega'^2\,d\omega'\,d\boldsymbol{\kappa}'\,.$$

The quantity $\breve{a}^\varepsilon(\omega, \boldsymbol{\kappa}, 0)$ is given by (15.27), and the Fourier transforms of the window functions are defined by

$$\hat{G}_1(\omega) = \int G_1(t)e^{i\omega t}\,dt\,, \qquad \hat{G}_2(\mathbf{k}) = \int G_2(\mathbf{x})e^{-i\mathbf{k}\cdot\mathbf{x}}\,d\mathbf{x}\,.$$

This configuration is now described by the system (15.16) for $-L \leq z < 0$. From $\breve{b}^\varepsilon_{\mathrm{TR}}(0^-) = -[\breve{b}^\varepsilon_{\mathrm{TR}}]_0$ and the expressions for $\hat{\mathbf{F}}^\varepsilon_{\mathrm{TR},\mathbf{x}}$ and $\hat{F}^\varepsilon_{\mathrm{TR},z}$ given above we deduce the following boundary condition at $z = 0^-$

$$\breve{b}^\varepsilon_{\mathrm{TR}}(\omega, \boldsymbol{\kappa}, 0^-) = \frac{1}{(2\pi\varepsilon)^3}\int \frac{H_0(\boldsymbol{\kappa}, \boldsymbol{\kappa}')}{2}\overline{\hat{G}_1\left(\frac{\omega - \omega'}{\varepsilon}\right)}\hat{G}_2\left(\frac{\omega\boldsymbol{\kappa} + \omega'\boldsymbol{\kappa}'}{\varepsilon}\right)$$
$$\times \overline{\breve{a}^\varepsilon(\omega', \boldsymbol{\kappa}', 0)}\omega'^2\,d\omega'\,d\boldsymbol{\kappa}'\,, \qquad (15.38)$$

where

$$H_0(\boldsymbol{\kappa}, \boldsymbol{\kappa}') = \frac{\bar{\rho}\bar{c}}{\sqrt{\bar{\zeta}(\kappa)\bar{\zeta}(\kappa')}} - \frac{\bar{c}\sqrt{\bar{\zeta}(\kappa)\bar{\zeta}(\kappa')}}{\bar{\rho}}\boldsymbol{\kappa} \cdot \boldsymbol{\kappa}'. \qquad (15.39)$$

The boundary condition at $z = -L$ is simply the radiation condition

$$\breve{a}_{\mathrm{TR}}^{\varepsilon}(\omega, \boldsymbol{\kappa}, -L) = 0 . \qquad (15.40)$$

The Fourier transforms of the pressure and the velocity are now given by (15.13) and (15.14), where $\breve{a}^{\varepsilon}$ and $\breve{b}^{\varepsilon}$ are replaced by $\breve{a}_{\mathrm{TR}}^{\varepsilon}$ and $\breve{b}_{\mathrm{TR}}^{\varepsilon}$.

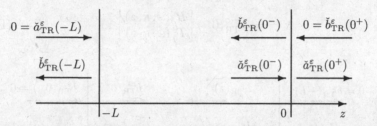

Fig. 15.6. Boundary conditions at $z = -L$ and $z = 0$ corresponding to the emission from the mirror located at $z = 0$.

15.2.3 The Time-Reversed Wave Field

The new incoming signal propagates into the same medium and produces the time-reversed wave field. Here we derive an exact integral representation for this wave field, which will be exploited in the following sections to analyze the refocusing properties of the time-reversed field.

The radiation condition (15.40) at $z = -L$ implies that

$$\begin{bmatrix} \breve{a}_{\mathrm{TR}}^{\varepsilon}(0^-) \\ \breve{b}_{\mathrm{TR}}^{\varepsilon}(0^-) \end{bmatrix} = \mathbf{P}_{(\omega,\kappa)}^{\varepsilon}(-L,0) \begin{bmatrix} 0 \\ \breve{b}_{\mathrm{TR}}^{\varepsilon}(-L) \end{bmatrix} = \breve{b}_{\mathrm{TR}}^{\varepsilon}(-L) \begin{bmatrix} \beta_{(\omega,\kappa)}^{\varepsilon}(-L,0) \\ \alpha_{(\omega,\kappa)}^{\varepsilon}(-L,0) \end{bmatrix} .$$

Solving this equation for $\breve{b}_{\mathrm{TR}}^{\varepsilon}(-L)$ gives

$$\breve{b}_{\mathrm{TR}}^{\varepsilon}(-L) = \frac{\breve{b}_{\mathrm{TR}}^{\varepsilon}(0^-)}{\alpha_{(\omega,\kappa)}^{\varepsilon}(-L,0)} .$$

Computing the $(2,2)$-entry of the decomposition,

$$\mathbf{P}_{(\omega,\kappa)}^{\varepsilon}(-L,0) = \mathbf{P}_{(\omega,\kappa)}^{\varepsilon}(z,0)\mathbf{P}_{(\omega,\kappa)}^{\varepsilon}(-L,z) ,$$

shows, using the definitions of R_g^{ε} and T_g^{ε} in (15.25) and (15.26), that

$$\overline{\alpha_{(\omega,\kappa)}^{\varepsilon}(-L,0)} = \frac{\overline{\alpha_{(\omega,\kappa)}^{\varepsilon}(-L,z)}}{T_g^{\varepsilon}(\omega,\kappa,z)} = \frac{\overline{\beta_{(\omega,\kappa)}^{\varepsilon}(-L,z)}}{R_g^{\varepsilon}(\omega,\kappa,z)} ,$$

and from this we obtain

$$\begin{bmatrix} \check{a}^\varepsilon_{\mathrm{TR}}(z) \\ \check{b}^\varepsilon_{\mathrm{TR}}(z) \end{bmatrix} = \mathbf{P}^\varepsilon_{(\omega,\kappa)}(-L,z) \begin{bmatrix} 0 \\ \check{b}^\varepsilon_{\mathrm{TR}}(-L) \end{bmatrix}$$

$$= \check{b}^\varepsilon_{\mathrm{TR}}(-L) \begin{bmatrix} \beta^\varepsilon_{(\omega,\kappa)}(-L,z) \\ \alpha^\varepsilon_{(\omega,\kappa)}(-L,z) \end{bmatrix}$$

$$= \frac{\check{b}^\varepsilon_{\mathrm{TR}}(0^-)}{\alpha^\varepsilon_{(\omega,\kappa)}(-L,0)} \begin{bmatrix} \beta^\varepsilon_{(\omega,\kappa)}(-L,z) \\ \alpha^\varepsilon_{(\omega,\kappa)}(-L,z) \end{bmatrix}$$

$$= \check{b}^\varepsilon_{\mathrm{TR}}(0^-) \begin{bmatrix} R^\varepsilon_g(\omega,\kappa,z) \\ T^\varepsilon_g(\omega,\kappa,z) \end{bmatrix}. \qquad (15.41)$$

$$
\begin{array}{c}
0 = \check{a}^\varepsilon_{\mathrm{TR}}(-L) \longrightarrow \quad \check{a}^\varepsilon_{\mathrm{TR}}(z) \longrightarrow \qquad \check{b}^\varepsilon_{\mathrm{TR}}(0^-) \longleftarrow \quad \check{b}^\varepsilon_{\mathrm{TR}}(0^+) = 0 \\[2mm]
\check{b}^\varepsilon_{\mathrm{TR}}(-L) \longleftarrow \quad \check{b}^\varepsilon_{\mathrm{TR}}(z) \longleftarrow \qquad \check{a}^\varepsilon_{\mathrm{TR}}(0^-) \longrightarrow \quad \check{a}^\varepsilon_{\mathrm{TR}}(0^+) \\[2mm]
\hline
\qquad -L \qquad z \qquad\qquad\qquad 0
\end{array}
$$

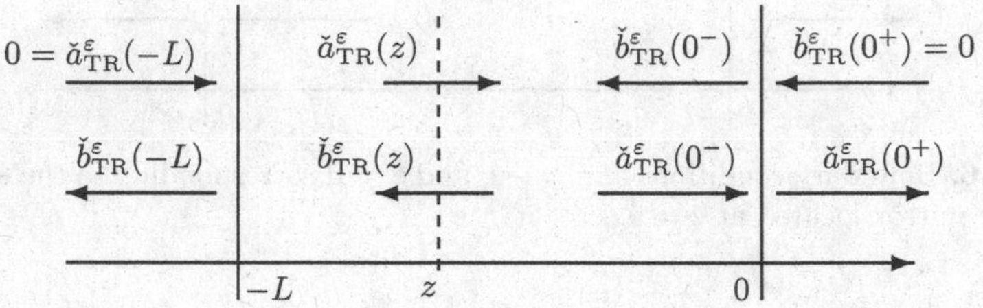

Fig. 15.7. Boundary conditions at $z = -L$ and $z = 0$ corresponding to reemission from the TR-mirror located on the plane $z = 0$. The field is observed at the point $z \in (-L, 0)$.

Using (15.13), (15.14), and (15.41) and applying an inverse Fourier transform, we find that the wave for $-L \le z < 0$ is given by

$$p^\varepsilon_{\mathrm{TR}}(t, \mathbf{x}, z) = \frac{1}{(2\pi\varepsilon)^3} \int \frac{\sqrt{\zeta(\kappa_1)}}{2} \check{b}^\varepsilon_{\mathrm{TR}}(\omega_1, \kappa_1, 0^-) \Big[R^\varepsilon_g(\omega_1, \kappa_1, z) e^{\frac{i\omega_1 z}{\varepsilon \bar{c}(\kappa_1)}}$$

$$- T^\varepsilon_g(\omega_1, \kappa_1, z) e^{-\frac{i\omega_1 z}{\varepsilon \bar{c}(\kappa_1)}} \Big] e^{-\frac{i\omega_1}{\varepsilon}(t - \kappa_1 \cdot \mathbf{x})} \omega_1^2 \, d\omega_1 \, d\kappa_1, \qquad (15.42)$$

$$u^\varepsilon_{\mathrm{TR}}(t, \mathbf{x}, z) = \frac{1}{(2\pi\varepsilon)^3} \int \frac{1}{2\sqrt{\zeta(\kappa_1)}} \check{b}^\varepsilon_{\mathrm{TR}}(\omega_1, \kappa_1, 0^-) \Big[R^\varepsilon_g(\omega_1, \kappa_1, z) e^{\frac{i\omega_1 z}{\varepsilon \bar{c}(\kappa_1)}}$$

$$+ T^\varepsilon_g(\omega_1, \kappa_1, z) e^{-\frac{i\omega_1 z}{\varepsilon \bar{c}(\kappa_1)}} \Big] e^{-\frac{i\omega_1}{\varepsilon}(t - \kappa_1 \cdot \mathbf{x})} \omega_1^2 \, d\omega_1 \, d\kappa_1. \qquad (15.43)$$

Substituting the expression (15.38) for $\check{b}^\varepsilon_{\mathrm{TR}}(\omega, \kappa, 0^-)$ into the equation for $u^\varepsilon_{\mathrm{TR}}$ yields the following representation for the vertical velocity:

$$u^\varepsilon_{\mathrm{TR}}(t, \mathbf{x}, z) = \frac{1}{(2\pi)^6 \varepsilon^3} \int \int \frac{H_0(\kappa_1, \kappa_2)}{4\sqrt{\zeta(\kappa_1)}} \overline{\hat{G}_1 \left(\frac{\omega_1 - \omega_2}{\varepsilon} \right)} \hat{G}_2 \left(\frac{\omega_1 \kappa_1 + \omega_2 \kappa_2}{\varepsilon} \right)$$

$$\times e^{i\left(\frac{-(\omega_2 t_s + \omega_1 t) + (\omega_2 \kappa_2 \cdot \mathbf{x}_s + \omega_1 \kappa_1 \cdot \mathbf{x})}{\varepsilon} \right)} \Big[\sum_{j=1}^{4} P^\varepsilon_j \Big] \omega_1^2 \omega_2^2 \, d\omega_1 \, d\kappa_1 \, d\omega_2 \, d\kappa_2, \quad (15.44)$$

where we define the P_j^ε's by

$$P_1^\varepsilon = -e^{i\left(-\frac{\omega_2 z_s}{\varepsilon \bar{c}(\kappa_2)} + \frac{\omega_1 z}{\varepsilon \bar{c}(\kappa_1)}\right)} R_g^\varepsilon(\omega_1, \kappa_1, z)\overline{R_g^\varepsilon(\omega_2, \kappa_2, z_s)}\,\overline{S_b(\omega_2, \kappa_2)}\,,$$

$$P_2^\varepsilon = e^{i\left(\frac{\omega_2 z_s}{\varepsilon \bar{c}(\kappa_2)} + \frac{\omega_1 z}{\varepsilon \bar{c}(\kappa_1)}\right)} R_g^\varepsilon(\omega_1, \kappa_1, z)\overline{T_g^\varepsilon(\omega_2, \kappa_2, z_s)}\,S_a(\omega_2, \kappa_2)\,,$$

$$P_3^\varepsilon = e^{i\left(\frac{\omega_2 z_s}{\varepsilon \bar{c}(\kappa_2)} - \frac{\omega_1 z}{\varepsilon \bar{c}(\kappa_1)}\right)} T_g^\varepsilon(\omega_1, \kappa_1, z)\overline{T_g^\varepsilon(\omega_2, \kappa_2, z_s)}\,S_a(\omega_2, \kappa_2)\,,$$

$$P_4^\varepsilon = -e^{i\left(-\frac{\omega_2 z_s}{\varepsilon \bar{c}(\kappa_2)} - \frac{\omega_1 z}{\varepsilon \bar{c}(\kappa_1)}\right)} T_g^\varepsilon(\omega_1, \kappa_1, z)\overline{R_g^\varepsilon(\omega_2, \kappa_2, z_s)}\,\overline{S_b(\omega_2, \kappa_2)}\,.$$

Motivated by the presence of the terms $\hat{G}_1\left(\frac{\omega_1 - \omega_2}{\varepsilon}\right)$ and $\hat{G}_2\left(\frac{\omega_1 \kappa_1 + \omega_2 \kappa_2}{\varepsilon}\right)$, we carry out the change of variables $\omega_1 = \omega + \varepsilon h/2, \omega_2 = \omega - \varepsilon h/2, \kappa_1 = \kappa + \varepsilon \lambda/2$, $\kappa_2 = -\kappa + \varepsilon \lambda/2$, which to leading order gives

$$u_{\mathrm{TR}}^\varepsilon(t, \mathbf{x}, z) = \frac{1}{(2\pi)^6} \int \int \frac{H_0(\kappa, -\kappa)}{4\sqrt{\zeta(\kappa)}} \overline{\hat{G}_1(h)} \hat{G}_2\left(h\kappa + \omega\lambda\right) \left[\sum_{j=1}^4 \tilde{P}_j^\varepsilon\right]$$

$$\times e^{\frac{i\omega}{\varepsilon}(-(t_s+t)+\kappa\cdot(\mathbf{x}-\mathbf{x}_s))} e^{\frac{ih}{2}(t_s - t + \kappa\cdot(\mathbf{x}+\mathbf{x}_s)) + \frac{i\omega}{2}\lambda\cdot(\mathbf{x}+\mathbf{x}_s)} \omega^4 d\omega\, dh\, d\kappa\, d\lambda\,, \quad (15.45)$$

with

$$\tilde{P}_1^\varepsilon = -e^{\frac{-i\omega}{\varepsilon}\left(\frac{z_s - z}{\bar{c}(\kappa)}\right)} e^{\frac{ih}{2}\frac{z_s + z}{\bar{c}(\kappa)} - \frac{i\omega}{2}\lambda\cdot\kappa\bar{c}(\kappa)(z_s + z)} R_g^\varepsilon \overline{R_g^\varepsilon}\,\overline{S_b(\omega, -\kappa)}\,,$$

$$\tilde{P}_2^\varepsilon = e^{\frac{i\omega'}{\varepsilon}\left(\frac{z_s + z}{\bar{c}(\kappa)}\right)} e^{-\frac{ih}{2}\frac{z_s - z}{\bar{c}(\kappa)} + \frac{i\omega}{2}\lambda\cdot\kappa\bar{c}(\kappa)(z_s - z)} R_g^\varepsilon \overline{T_g^\varepsilon}\,\overline{S_a(\omega, -\kappa)}\,,$$

$$\tilde{P}_3^\varepsilon = e^{\frac{i\omega}{\varepsilon}\left(\frac{z_s - z}{\bar{c}(\kappa)}\right)} e^{-\frac{ih}{2}\frac{z_s + z}{\bar{c}(\kappa)} + \frac{i\omega}{2}\lambda\cdot\kappa\bar{c}(\kappa)(z_s + z)} T_g^\varepsilon \overline{T_g^\varepsilon}\,\overline{S_a(\omega, -\kappa)}\,,$$

$$\tilde{P}_4^\varepsilon = -e^{-\frac{i\omega}{\varepsilon}\left(\frac{z_s + z}{\bar{c}(\kappa)}\right)} e^{\frac{ih}{2}\frac{z_s - z}{\bar{c}(\kappa)} - \frac{i\omega}{2}\lambda\cdot\kappa\bar{c}(\kappa)(z_s - z)} T_g^\varepsilon \overline{R_g^\varepsilon}\,\overline{S_b(\omega, -\kappa)}\,,$$

where the complex-conjugated coefficients are evaluated at $(\omega - \varepsilon h/2, -\kappa + \varepsilon\lambda/2, z_s)$, the other coefficients are evaluated at $(\omega + \varepsilon h/2, \kappa + \varepsilon\lambda/2, z)$, and we have used that $\nabla_\kappa \bar{c}(\kappa) = \bar{c}^3(\kappa)\kappa$ from the definition (15.15). Note that using the definition (15.39), we have

$$H_0(\kappa, -\kappa) = \frac{\bar{c}(\kappa)}{\bar{c}}\,.$$

In the following sections we study the asymptotic behavior of the time-reversed wave field $u_{\mathrm{TR}}^\varepsilon$ in the limit $\varepsilon \to 0$.

15.3 Homogeneous Medium

We first examine the **deterministic** case with $\nu \equiv 0$, which corresponds to a source embedded into a homogeneous medium at the depth $z_s < 0$.

15.3.1 The Field Recorded at the Surface

The recorded wave field at a point $M = (\mathbf{x}, 0)$ at the surface is described in terms of (15.27) with $R_g^\varepsilon = 0$ and $T_g^\varepsilon = 1$, which gives

$$\breve{a}^\varepsilon(\omega, \boldsymbol{\kappa}, 0) = \varepsilon^3 e^{\frac{i\omega}{\varepsilon}(t_s - \boldsymbol{\kappa} \cdot \mathbf{x}_s - z_s/\bar{c}(\kappa))} S_a(\omega, \boldsymbol{\kappa}).$$

The signal that can be recorded at the surface is a spherical wave. Using the Fourier representation (15.34) of the pressure field at the surface and the definition (15.20) of the source contribution S_a, we find that the pressure signal is given by

$$p_s^\varepsilon(t, \mathbf{x}) = \frac{1}{2(2\pi)^3} \int \left(\frac{\bar{\zeta}(\kappa)\boldsymbol{\kappa}}{\bar{\rho}} \cdot \hat{\mathbf{f}}_\mathbf{x}(\omega) + \hat{f}_z(\omega) \right) e^{\frac{i\omega\phi(\boldsymbol{\kappa})}{\varepsilon}} \omega^2 \, d\omega \, d\boldsymbol{\kappa}, \qquad (15.46)$$

where the phase term ϕ is

$$\phi(\boldsymbol{\kappa}) = t_s - t + \boldsymbol{\kappa} \cdot (\mathbf{x} - \mathbf{x}_s) - \frac{z_s}{\bar{c}(\kappa)}.$$

As $\varepsilon \to 0$ the asymptotic behavior of the integral (15.46) is governed by its fast phase ϕ/ε. We apply the **stationary-phase method** to the integral (see Appendix 14.4). The partial derivatives of the phase with respect to the phase variables $\boldsymbol{\kappa}$ are given by

$$\nabla_{\boldsymbol{\kappa}} \phi = \mathbf{x} - \mathbf{x}_s + \boldsymbol{\kappa} \bar{c}(\kappa) z_s.$$

There exists a unique stationary point given by $\boldsymbol{\kappa} = \boldsymbol{\kappa}_{\mathrm{sp}}$, with

$$\boldsymbol{\kappa}_{\mathrm{sp}} = \frac{1}{\bar{c}} \frac{\mathbf{x} - \mathbf{x}_s}{\sqrt{z_s^2 + |\mathbf{x} - \mathbf{x}_s|^2}} = \frac{1}{\bar{c}} \frac{\mathbf{x} - \mathbf{x}_s}{SM}.$$

By Proposition 14.4, the limit value of the integral (15.46) is of order $o(\varepsilon)$ if $\phi(\boldsymbol{\kappa}_{\mathrm{sp}}) \neq 0$ and it is of leading order ε only if $\phi(\boldsymbol{\kappa}_{\mathrm{sp}}) = 0$, that is, if the observation time t satisfies $t_s - t + SM/\bar{c} = 0$, where SM is the distance from the source to the mirror. Thus, we evaluate the pressure field $p_s^\varepsilon(t, \mathbf{x})$ at the time $t = t_s + SM/\bar{c} + \varepsilon T$, and a direct application of the stationary-phase result in Appendix 14.4 gives the approximation

$$p_s^\varepsilon \left(t_s + \frac{SM}{\bar{c}} + \varepsilon T, \mathbf{x} \right) \approx \frac{\varepsilon}{4\pi \bar{c} SM^2} \mathbf{SM} \cdot \begin{bmatrix} \mathbf{f}_\mathbf{x}' \\ f_z' \end{bmatrix} (T),$$

with $\mathbf{SM}$ being the vector from the source to the mirror. Similarly, from (15.34) and (15.35), we deduce that the three-dimensional velocity field is approximated by

$$\mathbf{u}_s^\varepsilon \left(t_s + \frac{SM}{\bar{c}} + \varepsilon T, \mathbf{x} \right) \approx \varepsilon \frac{\mathbf{SM}}{4\pi \bar{\rho} \bar{c}^2 SM^3} \mathbf{SM} \cdot \begin{bmatrix} \mathbf{f}_\mathbf{x}' \\ f_z' \end{bmatrix} (T). \qquad (15.47)$$

Note that the amplitudes of these waves are small, of order ε, and they have a short duration of order ε.

15.3.2 The Time-Reversed Field

In the homogenous case the time-reversed wave field is described by (15.45) with $R_g^\varepsilon = 0$ and $T_g^\varepsilon = 1$. In this case only the $\tilde{P}_3^\varepsilon$ term contributes to $u_{\mathrm{TR}}^\varepsilon$ which reduces to

$$u_{\mathrm{TR}}^\varepsilon(t, \mathbf{x}, z) = \frac{1}{(2\pi)^6} \int \int \frac{\bar{c}(\kappa)}{4\bar{c}\sqrt{\zeta(\kappa)}} \overline{S_a(\omega, -\kappa)} \hat{G}_1(h) \hat{G}_2(h\kappa + \omega\lambda) \qquad (15.48)$$

$$\times e^{\frac{i\omega\phi(\kappa)}{\varepsilon}} e^{\frac{ih}{2}\left(t_s - t + \kappa\cdot(\mathbf{x}+\mathbf{x}_s) - \frac{z_s+z}{\bar{c}(\kappa)}\right) + \frac{i\omega}{2}\left(\lambda\cdot(\mathbf{x}+\mathbf{x}_s) + \lambda\cdot\kappa\bar{c}(\kappa)(z_s+z)\right)} \omega^4 d\omega\, dh\, d\kappa\, d\lambda \,,$$

where the rapid phase is

$$\phi(\kappa) = -(t_s + t) + \kappa\cdot(\mathbf{x} - \mathbf{x}_s) + \frac{z_s - z}{\bar{c}(\kappa)} \,. \qquad (15.49)$$

We again apply the stationary-phase method and observe first that

$$\nabla_\kappa \phi = \mathbf{x} - \mathbf{x}_s - \kappa\bar{c}(\kappa)(z_s - z) \,.$$

There are three cases depending on the observation time t and the observation point $M = (\mathbf{x}, z)$. The first case corresponds to observing the wave front before it refocuses at the original source point, whereas the second case corresponds to an observation point on the wave front after it has refocused, and we show below that these two contributions are small, of order ε. The third case corresponds to observing the wave front at the original source point and at the critical time, and the refocused wave is of order one.

- If $t < -t_s$, then there exists one stationary point given by $\kappa = \kappa_{\mathrm{sp}}$, with

$$\kappa_{\mathrm{sp}} = -\frac{1}{\bar{c}} \frac{\mathbf{x} - \mathbf{x}_s}{\sqrt{|z - z_s|^2 + |\mathbf{x} - \mathbf{x}_s|^2}} = -\frac{1}{\bar{c}} \frac{\mathbf{x} - \mathbf{x}_s}{SM} \,.$$

By Proposition 14.4, the value of the integral (15.48) is of order ε if the observation point M satisfies $z > z_s$ and $SM = \bar{c}|t + t_s|$. Otherwise, the value of the integral is $o(\varepsilon)$.

- If $t > -t_s$, then there exists one stationary point given by $\kappa = \kappa_{\mathrm{sp}}$, with

$$\kappa_{\mathrm{sp}} = \frac{1}{\bar{c}} \frac{\mathbf{x} - \mathbf{x}_s}{\sqrt{|z - z_s|^2 + |\mathbf{x} - \mathbf{x}_s|^2}} = \frac{1}{\bar{c}} \frac{\mathbf{x} - \mathbf{x}_s}{SM} \,.$$

By Proposition 14.4, the value of the integral (15.48) is of order ε if the observation point M satisfies $z < z_s$ and $SM = \bar{c}(t + t_s)$. Otherwise, the value of the integral is $o(\varepsilon)$.

- If $t = -t_s$, the critical time, there exists one stationary map if and only if $\mathbf{x} = \mathbf{x}_s$ and $z = z_s$, and the map is globally stationary, meaning that the rapid phase vanishes identically, which in turn implies that the value of the integral (15.48) is of order one.

We thus consider an observation time t close to $-t_s$ and an observation point close to $(\mathbf{x}_s, z_s)$. This is done using the following parameterization:

$$t = -t_s + \varepsilon T, \quad \mathbf{x} = \mathbf{x}_s + \varepsilon \mathbf{X}, \quad z = z_s + \varepsilon Z. \tag{15.50}$$

The integral representation of the refocused velocity field becomes

$$u_{\mathrm{TR}}^{\varepsilon}(t, \mathbf{x}, z) = \frac{1}{(2\pi)^6} \int \int \frac{\bar{c}(\kappa)}{4\bar{c}\sqrt{\zeta(\kappa)}} \overline{S_a(\omega, -\kappa)} \widehat{G}_1(h) \widehat{G}_2(h\kappa + \omega\lambda)$$

$$\times e^{i\omega\left(-T + \kappa\cdot\mathbf{X} - \frac{Z}{\bar{c}(\kappa)}\right)} e^{ih\left(t_s + \kappa\cdot\mathbf{x}_s - \frac{z_s}{\bar{c}(\kappa)}\right) + i\omega\left(\lambda\cdot\mathbf{x}_s + \lambda\cdot\kappa\bar{c}(\kappa)z_s\right)} \omega^4 d\omega\, dh\, d\kappa\, d\lambda$$

$$= \frac{1}{(2\pi)^6} \int \int \frac{\bar{c}(\kappa)}{4\bar{c}\sqrt{\zeta(\kappa)}} \widehat{G}_1(h) \widehat{G}_2(h\kappa + \omega\lambda) \tag{15.51}$$

$$\times \left(-\frac{\sqrt{\zeta(\kappa)}}{\bar{\rho}} \kappa \cdot \overline{\widehat{\mathbf{f}}_{\mathbf{x}}(\omega)} + \frac{1}{\sqrt{\zeta(\kappa)}} \overline{\widehat{f}_z(\omega)} \right)$$

$$\times e^{i\omega\left(-T + \kappa\cdot\mathbf{X} - \frac{Z}{\bar{c}(\kappa)}\right)} e^{ih\left(t_s + \kappa\cdot\mathbf{x}_s - \frac{z_s}{\bar{c}(\kappa)}\right) + i\omega\left(\lambda\cdot\mathbf{x}_s + \lambda\cdot\kappa\bar{c}(\kappa)z_s\right)} \omega^4 d\omega\, dh\, d\kappa\, d\lambda,$$

where we have used (15.20). We next apply the change of variables $\lambda \mapsto \mathbf{k} = \omega\lambda + h\kappa$ so that

$$u_{\mathrm{TR}}^{\varepsilon}(t, \mathbf{x}, z) = \frac{1}{(2\pi)^6} \int \int \frac{\bar{c}(\kappa)}{4\bar{c}\sqrt{\zeta(\kappa)}} \overline{S_a(\omega, -\kappa)} \widehat{G}_1(h) \widehat{G}_2(\mathbf{k})$$

$$\times e^{i\omega\left(-T + \kappa\cdot\mathbf{X} - \frac{Z}{\bar{c}(\kappa)}\right)} e^{ih\left(t_s - \frac{\bar{c}(\kappa)z_s}{\bar{c}^2}\right) + i\mathbf{k}\cdot(\mathbf{x}_s + \kappa\bar{c}(\kappa)z_s)} \omega^2 d\omega\, dh\, d\kappa\, d\mathbf{k},$$

and we integrate with respect to $\mathbf{k}$ and h to obtain that in the limit $\varepsilon \to 0$, the vertical velocity field is given by

$$u_{\mathrm{TR}}(t, \mathbf{x}, z) = \frac{1}{(2\pi)^3} \int \int \frac{\bar{c}(\kappa)}{4\bar{c}\sqrt{\zeta(\kappa)}} G_1\left(t_s - z_s\frac{\bar{c}(\kappa)}{\bar{c}^2}\right) G_2\left(\mathbf{x}_s + \kappa\bar{c}(\kappa)z_s\right)$$

$$\times \overline{S_a(\omega, -\kappa)} e^{i\omega\left(-T + \kappa\cdot\mathbf{X} - \frac{Z}{\bar{c}(\kappa)}\right)} \omega^2 d\omega\, d\kappa. \tag{15.52}$$

Qualitatively, the mirror emits a converging spherical wave whose amplitude is of order ε and whose support at time $t < -t_s$ lies in the upper part of the sphere with center S and radius $\bar{c}|t + t_s|$. At the critical time $-t_s$ the refocusing occurs at the original source location S. The refocused pulse has an amplitude of order one and it is given by (15.52). After the time $-t_s$ a diverging spherical wave is going downward from the point S (see Figure 15.8). We can be more precise and give quantitative information on the **focal spot**.

In order to simplify the formula and clarify the roles of the different quantities in the refocusing, we assume that the source is concentrated in frequency in a narrow band around a large carrier frequency ω_0, and is located relatively far from the mirror. These assumptions translate into

$$\mathbf{f}(t) = \mathbf{f}_0\left(\frac{t}{T_w}\right) e^{-i\omega_0 t} + c.c., \tag{15.53}$$

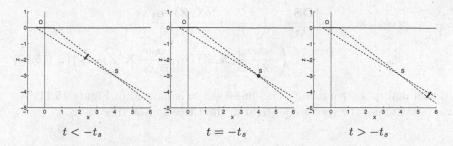

$$t < -t_s \qquad\qquad t = -t_s \qquad\qquad t > -t_s$$

Fig. 15.8. Pulse refocusing in a homogeneous medium. The focal-spot size depends on the numerical aperture, which measures the angular diversity of the refocused waves that participate in the refocusing.

with $\omega_0 T_w \gg 1$, and $|z_s| \gg a$, where T_w is the initial pulse width, a is the diameter of the mirror, and "c.c." stands for the complex-conjugated quantity. The function $\mathbf{f}_0$ is the envelope of the pulse profile, whose support is normalized to be of order one. The spatial cutoff function determining the mirror has the form

$$G_2(\mathbf{x}) = g_2\left(\frac{\mathbf{x}}{a}\right), \tag{15.54}$$

where g_2 is the normalized cutoff function. We choose the (x,y)-axes so that the horizontal position of the source is $\mathbf{x}_s = (x_s, 0)$ with $x_s \geq 0$. Then, we carry out the changes of variables $\boldsymbol{\kappa} \mapsto \mathbf{y} = \mathbf{x}_s + \boldsymbol{\kappa}\bar{c}(\kappa)z_s$ and $\omega \mapsto \tilde{\omega} = T_w(\omega - \omega_0)$, and we integrate with respect to $\mathbf{y}$ and $\tilde{\omega}$, so that we get

$$u_{\mathrm{TR}}(t,\mathbf{x},z) = \frac{a^2 z_s \omega_0^2}{16\pi^2 \bar{c}^3 \bar{\rho} OS^4}\mathbf{OS}\cdot\mathbf{f}_0\left(-\frac{T}{T_w} + \frac{(\mathbf{X},Z)\cdot\mathbf{OS}}{\bar{c}OST_w}\right)e^{i\omega_0\left(-T+\frac{(\mathbf{X},Z)\cdot\mathbf{OS}}{\bar{c}OS}\right)}$$

$$\times G_1\left(t_s + \frac{OS}{\bar{c}}\right)\hat{g}_2\left(\frac{\omega_0 a|z_s|}{\bar{c}OS^3}(x_s Z - z_s X), \frac{\omega_0 a}{\bar{c}OS}Y\right) + c.c., \tag{15.55}$$

where $\mathbf{OS}$ is the vector from the origin to the source. Note that the vertical refocused velocity field is nonzero only if the signal originating from the source has been recorded by the mirror, meaning that $t_s + OS/\bar{c}$ lies in the support of G_1. The focal shape depends on the Fourier transform of the mirror shape and the initial pulse profile. We introduce the new orthonormal frame $(\mathbf{e}_1, \mathbf{e}_2, \mathbf{e}_3)$ defined by

$$\mathbf{e}_1 = \frac{1}{OS}\begin{bmatrix} -z_s \\ 0 \\ x_s \end{bmatrix}, \qquad \mathbf{e}_2 = \begin{bmatrix} 0 \\ 1 \\ 0 \end{bmatrix}, \qquad \mathbf{e}_3 = \frac{1}{OS}\begin{bmatrix} x_s \\ 0 \\ z_s \end{bmatrix},$$

and described in Figure 15.9. If we take out the rapid oscillatory term at frequency ω_0, the pulse shape envelope in the frame $(\mathbf{e}_1, \mathbf{e}_2, \mathbf{e}_3)$ is given by

$$|u_{\mathrm{TR}}(t,\mathbf{x},z)| = \left|\frac{\mathbf{OS}}{OS} \cdot \mathbf{f}_0\left(-\frac{T}{T_w} + \frac{(\mathbf{X},Z)\cdot\mathbf{e}_3}{\bar{c}T_w}\right)\right|$$

$$\times \left|\hat{g}_2\left(\frac{\omega_0 a|z_s|}{\bar{c}OS^2}(\mathbf{X},Z)\cdot\mathbf{e}_1, \frac{\omega_0 a}{\bar{c}OS}(\mathbf{X},Z)\cdot\mathbf{e}_2\right)\right|, \quad (15.56)$$

up to a multiplicative factor. This pulse shape is plotted in Figure 15.11. We can now discuss the spot-shape radii in the three spatial directions.

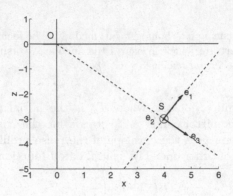

Fig. 15.9. Frame adapted to the refocusing.

- In the $\mathbf{e}_3$-direction (along the vector $\mathbf{OS}$), the focal spot size is determined by the initial pulse width T_w and it is given by

$$R_3 = \bar{c}T_w.$$

- In the other two directions, the focal-spot size is determined by the mirror size a. In the $\mathbf{e}_2$-direction, the focal spot has the size

$$R_2 = \frac{\lambda_0 OS}{a},$$

where $\lambda_0 = 2\pi\bar{c}/\omega_0$ is the carrier wavelength of the pulse. In the $\mathbf{e}_1$-direction, the focal spot has the size

$$R_1 = \frac{\lambda_0 OS^2}{a|z_s|} = \frac{\lambda_0 OS}{a_1} \text{ with } a_1 = a\frac{|z_s|}{OS}.$$

These formulas are consistent with the standard **Rayleigh resolution** formula, which claims that the focal spot of a beam with carrier wavelength λ_0 focused with a system of effective size a_s from a distance OS is of order $\lambda_0 OS/a_s$ [20]. Here the effective size is a in the $\mathbf{e}_2$-direction and a_1 in the $\mathbf{e}_1$-direction. We can explain the differences in the effective sizes, which involve the differences in the focal-spot radii, by looking at the numerical apertures of the waves that participate in the refocusing (see Figure 15.10). The spot sizes in the directions $\mathbf{e}_j$ are given by $R_j = \lambda_0/\Delta\phi_j$, $j = 1, 2$, where

$$\Delta\phi_1 = \arctan\left(\frac{x_s + \frac{a}{2}}{|z_s|}\right) - \arctan\left(\frac{x_s - \frac{a}{2}}{|z_s|}\right) \overset{|z_s| \gg a}{\approx} \frac{|z_s| a}{OS^2}, \quad (15.57)$$

$$\Delta\phi_2 = \arctan\left(\frac{a}{2OS}\right) - \arctan\left(-\frac{a}{2OS}\right) \overset{|z_s| \gg a}{\approx} \frac{a}{OS}. \quad (15.58)$$

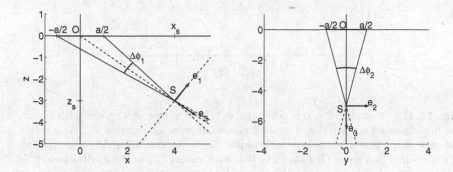

Fig. 15.10. Cones of aperture. The left plot shows the $(\mathbf{e}_1, \mathbf{e}_3)$-section of the cone of aperture. The right plot shows the $(\mathbf{e}_2, \mathbf{e}_3)$-section.

15.4 Complete Description of the Time-Reversed Field in a Random Medium

Now we analyze the case with a random medium. We first consider the P_3^ε term in the expression (15.45) for the vertical velocity. As shown in Appendix 15.6, the generalized transmission coefficient depends only on the modulus of the wave vector, so we can write

$$u_{\mathrm{TR}}^{(\varepsilon,3)}(t, \mathbf{x}, z) = \frac{1}{(2\pi)^6} \int \int \frac{\bar{c}(\kappa)}{4\bar{c}\sqrt{\zeta(\kappa)}} \overline{S_a(\omega, -\boldsymbol{\kappa})\hat{G}_1(h)} \hat{G}_2(h\boldsymbol{\kappa} + \omega\boldsymbol{\lambda}) \quad (15.59)$$

$$\times T_g^\varepsilon\left(\omega - \frac{\varepsilon h}{2}, |-\boldsymbol{\kappa} + \frac{\varepsilon\boldsymbol{\lambda}}{2}|, z_s\right) T_g^\varepsilon\left(\omega + \frac{\varepsilon h}{2}, |\boldsymbol{\kappa} + \frac{\varepsilon\boldsymbol{\lambda}}{2}|, z\right)$$

$$\times e^{\frac{i\omega\phi}{\varepsilon}} e^{\frac{ih}{2}\left(t_s - t + \boldsymbol{\kappa}\cdot(\mathbf{x}+\mathbf{x}_s) - \frac{z_s + z}{\bar{c}(\kappa)}\right) + \frac{i\omega}{2}\left(\boldsymbol{\lambda}\cdot(\mathbf{x}+\mathbf{x}_s) + \boldsymbol{\lambda}\cdot\boldsymbol{\kappa}\bar{c}(\kappa)(z_s + z)\right)} \omega^4 \, d\omega \, dh \, d\boldsymbol{\kappa} \, d\boldsymbol{\lambda},$$

where the rapid phase ϕ is given by (15.49). As $\varepsilon \to 0$ the asymptotic behavior of this integral is governed by its fast phase and by the product of the two transmission coefficients that contains the effect of randomness. We first apply

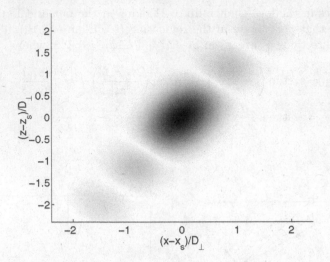

Fig. 15.11. Pulse shape in the plane $(\mathbf{e}_1, \mathbf{e}_3)$. The mirror is a square with size a, the initial source is vertical and emits a Gaussian pulse of duration T_w. As a result the spot shape is a sinc function with radius $D_\perp = \lambda_0/\Delta\phi_1$ with $\Delta\phi_1 = a|z_s|/OS^2$ in the $\mathbf{e}_1$-direction and a Gaussian function with radius $\bar{c}T_w$ in the $\mathbf{e}_3$-direction. It is also a sinc function with radius $\lambda_0/\Delta\phi_2$ with $\Delta\phi_2 = a/OS$ in the $\mathbf{e}_2$-direction (not shown here).

the stationary-phase method, and we will deal with the random part of the integral in a second step. The rapid phase is the same as in the homogeneous case, and we find that the globally stationary point is determined by $t+t_s = 0$, $\mathbf{x} = \mathbf{x}_s$, and $z = z_s$. We then consider an observation time t close to $-t_s$ and an observation point close to $(\mathbf{x}_s, z_s)$ according to the parameterization (15.50). The integral representation of the refocused velocity field is then

$$u_{\mathrm{TR}}^{(\varepsilon,3)}(t,\mathbf{x},z) = \frac{1}{(2\pi)^6} \int \int \frac{\bar{c}(\kappa)}{4\bar{c}\sqrt{\zeta(\kappa)}} \overline{S_a(\omega,-\kappa)\hat{G}_1(h)}\hat{G}_2(h\kappa + \omega\lambda) \quad (15.60)$$

$$\times \overline{T_g^\varepsilon\left(\omega - \frac{\varepsilon h}{2}, |-\kappa + \frac{\varepsilon\lambda}{2}|, z_s\right)}T_g^\varepsilon\left(\omega + \frac{\varepsilon h}{2}, |\kappa + \frac{\varepsilon\lambda}{2}|, z_s + \varepsilon Z\right)$$

$$\times e^{i\omega\left(-T+\kappa\cdot\mathbf{X}-\frac{Z}{\bar{c}(\kappa)}\right)}e^{ih\left(t_s+\kappa\cdot\mathbf{x}_s-\frac{z_s}{\bar{c}(\kappa)}\right)+i\omega\left(\lambda\cdot\mathbf{x}_s+\lambda\cdot\kappa\bar{c}(\kappa)z_s\right)}\omega^4 \, d\omega \, dh \, d\kappa \, d\lambda.$$

The effect of the randomness is contained in the product of the generalized transmission coefficients. In the next section we exploit the asymptotic analysis of the autocorrelation function studied in Appendix 15.6 to deduce the refocusing properties of the pulse.

15.4.1 Expectation of the Refocused Pulse

The expectation of $u_{\mathrm{TR}}^{(\varepsilon,3)}(t,\mathbf{x},z)$ given in (15.60) involves the first moment of the product of generalized transmission coefficients. In Appendix 15.6 it is shown that the generalized transmission coefficients at two frequencies and slowness vectors are correlated only if the frequencies and the moduli of the slowness vectors are close to each other at order ε. In (15.88) it is shown that for any l and any h, we have

$$\mathbb{E}\left[T_g^\varepsilon\left(\omega - \frac{\varepsilon h}{2}, \kappa - \frac{\varepsilon l}{2}, z_s\right)T_g^\varepsilon\left(\omega + \frac{\varepsilon h}{2}, \kappa + \frac{\varepsilon l}{2}, z_s + \varepsilon Z\right)\right] \quad (15.61)$$

$$\stackrel{\varepsilon \to 0}{\longrightarrow} \int e^{i[h\bar{c}^2/\bar{c}(\kappa)^2 - \omega l \bar{c}^2 \kappa]\tau} \mathcal{W}_g^{(T)}(\omega,\kappa,\tau)\, d\tau\,,$$

where

$$\mathcal{W}_g^{(T)}(\omega,\kappa,\tau) = \sum_{n=0}^\infty \left[\mathcal{W}_n(\omega,\kappa,\cdot,-L,z_s) * \mathcal{W}_n^{(T)}(\omega,\kappa,\cdot,z_s,0)\right](\tau). \quad (15.62)$$

Using

$$\left|\kappa + \frac{\varepsilon\lambda}{2}\right| = \kappa + \frac{\varepsilon l}{2}, \qquad l = \frac{\kappa \cdot \lambda}{\kappa} + \mathcal{O}(\varepsilon)\,,$$

we get after substitution of (15.61) into (15.60) that in the limit $\varepsilon \to 0$, the expectation $\mathbb{E}\left[u_{\mathrm{TR}}^{(\varepsilon,3)}(t,\mathbf{x},z)\right]$ converges to

$$u_{\mathrm{TR}}^{(3)}(t,\mathbf{x},z) = \frac{1}{(2\pi)^6}\int\int \frac{\bar{c}(\kappa)}{4\bar{c}\sqrt{\zeta(\kappa)}}\overline{S_a(\omega,-\kappa)}\hat{G}_1(h)\hat{G}_2\left(h\kappa + \omega\lambda\right) \quad (15.63)$$

$$\times \int e^{i\left[h\bar{c}^2/\bar{c}(\kappa)^2 - \omega\left(\frac{\kappa\cdot\lambda}{\kappa}\right)\bar{c}^2\kappa\right]\tau}\mathcal{W}_g^{(T)}(\omega,\kappa,\tau)\, d\tau$$

$$\times e^{i\omega(-T+\kappa\cdot\mathbf{X} - \frac{Z}{\bar{c}(\kappa)})} e^{ih\left(t_s + \kappa\cdot\mathbf{x}_s - \frac{z_s}{\bar{c}(\kappa)}\right) + i\omega\left(\lambda\cdot\mathbf{x}_s + \lambda\cdot\kappa\bar{c}(\kappa)z_s\right)}\omega^4\, d\omega\, dh\, d\kappa\, d\lambda\,.$$

Performing the change of variables $\lambda \mapsto \mathbf{k} = \omega\lambda + h\kappa$ and grouping the exponential terms together gives

$$u_{\mathrm{TR}}^{(3)}(t,\mathbf{x},z) = \frac{1}{(2\pi)^6}\int\int \frac{\bar{c}(\kappa)}{4\bar{c}\sqrt{\zeta(\kappa)}}\overline{S_a(\omega,-\kappa)}\hat{G}_1(h)\hat{G}_2(\mathbf{k})\,\mathcal{W}_g^{(T)}(\omega,\kappa,\tau)$$

$$\times e^{i\omega(-T+\kappa\cdot\mathbf{X}-\frac{Z}{\bar{c}(\kappa)})} e^{ih\left(t_s - \frac{\bar{c}(\kappa)z_s}{\bar{c}^2} + \tau\right) + i\mathbf{k}\cdot\left(\mathbf{x}_s + \kappa\bar{c}(\kappa)z_s - \bar{c}^2\kappa\tau\right)}\omega^2\, d\omega\, dh\, d\kappa\, d\mathbf{k}\, d\tau\,.$$

We finally integrate with respect to $\mathbf{k}$ and h to obtain

$$u_{\mathrm{TR}}^{(3)}(t,\mathbf{x},z) = \frac{1}{(2\pi)^3}\int\int \frac{\bar{c}(\kappa)}{4\bar{c}\sqrt{\zeta(\kappa)}}\overline{S_a(\omega,-\kappa)}G_1\left(t_s - z_s\frac{\bar{c}(\kappa)}{\bar{c}^2} + \tau\right) \quad (15.64)$$

$$\times G_2\left(\mathbf{x}_s + \kappa\bar{c}(\kappa)z_s - \bar{c}^2\kappa\tau\right)\mathcal{W}_g^{(T)}(\omega,\kappa,\tau)e^{i\omega\left(-T+\kappa\cdot\mathbf{X}-\frac{Z}{\bar{c}(\kappa)}\right)}\omega^2\, d\omega\, d\kappa\, d\tau\,.$$

We can deal with the other three components by the same method. The result (15.90) obtained in Appendix 15.6.3 shows that the limit as $\varepsilon \to 0$ of the cross moments of R_g^ε and T_g^ε are zero, so that the contributions of the P_2^ε and P_4^ε terms vanish in the limit $\varepsilon \to 0$. Therefore we only need to study the P_1^ε term. We obtain that in the limit $\varepsilon \to 0$, the expectation $\mathbb{E}\left[u_{\mathrm{TR}}^{(\varepsilon,1)}(t, \mathbf{x}, z)\right]$ converges to

$$u_{\mathrm{TR}}^{(1)}(t, \mathbf{x}, z) = \frac{-1}{(2\pi)^3} \int \int \frac{\bar{c}(\kappa)}{4\bar{c}\sqrt{\zeta(\kappa)}} \overline{S_b(\omega, -\kappa)} G_1\left(t_s - z_s \frac{\bar{c}(\kappa)}{\bar{c}^2} + \tau\right) \quad (15.65)$$

$$\times G_2\left(\mathbf{x}_s + \kappa \bar{c}(\kappa)z_s - \bar{c}^2\kappa\tau\right) \mathcal{W}_g^{(R)}(\omega, \kappa, \tau) e^{i\omega\left(-T + \kappa \cdot \mathbf{X} + \frac{Z}{\bar{c}(\kappa)}\right)} \omega^2 \, d\omega \, d\kappa \, d\tau \,,$$

where

$$\mathcal{W}_g^{(R)}(\omega, \kappa, \tau) = \sum_{n=0}^{\infty} \left[\mathcal{W}_{n+1}(\omega, \kappa, \cdot, -L, z_s) * \mathcal{W}_n^{(T)}(\omega, \kappa, \cdot, z_s, 0)\right](\tau). \quad (15.66)$$

15.4.2 Refocusing of the Pulse

The main results of this section are the refocusing of the pulse and its self-averaging property. These results are precisely stated in the following theorem, which gives the refocusing property and the convergence of the refocused pulse to a deterministic shape concentrated at $-t_s$ in time and at the original source location $(\mathbf{x}_s, z_s)$ in space.

Theorem 15.1.
(a) For any $T_0 > 0$, $R_0 > 0$, $Z_0 > 0$, $\delta > 0$, and $(t_0, \mathbf{x}_0, z_0) \neq (-t_s, \mathbf{x}_s, z_s)$, we have

$$\mathbb{P}\left(\sup_{|t-t_0|\leq\varepsilon T_0, |\mathbf{x}-\mathbf{x}_s|\leq\varepsilon R_0, |z-z_0|\leq\varepsilon Z_0} |u_{\mathrm{TR}}^\varepsilon(t, \mathbf{x}, z)| > \delta\right) \xrightarrow{\varepsilon \to 0} 0.$$

(b) For any $T_0 > 0$, $R_0 > 0$, $Z_0 > 0$, and $\delta > 0$, we have

$$\mathbb{P}\left(\sup_{|T|\leq T_0, |\mathbf{X}|\leq R_0, |Z|\leq Z_0} |u_{\mathrm{TR}}^\varepsilon(-t_s + \varepsilon T, \mathbf{x}_s + \varepsilon\mathbf{X}, z_s + \varepsilon Z)\right.$$

$$\left. - U_{\mathrm{TR}}(T, \mathbf{X}, Z)| > \delta\right) \xrightarrow{\varepsilon \to 0} 0,$$

where U_{TR} is the deterministic pulse shape

$$U_{\mathrm{TR}}(T, \mathbf{X}, Z) = \frac{1}{(2\pi)^3} \int K^+(\omega, \kappa) \left[\bar{c}(\kappa)\kappa \cdot \overline{\hat{\mathbf{f}}_\mathbf{x}(\omega)} + \overline{\hat{f}_z(\omega)}\right]$$

$$\times e^{i\omega\left(-T + \kappa \cdot \mathbf{X} + \frac{Z}{\bar{c}(\kappa)}\right)} \omega^2 \, d\omega \, d\kappa$$

$$+ \frac{1}{(2\pi)^3} \int K^-(\omega, \kappa) \left[-\bar{c}(\kappa)\kappa \cdot \overline{\hat{\mathbf{f}}_\mathbf{x}(\omega)} + \overline{\hat{f}_z(\omega)}\right]$$

$$\times e^{i\omega\left(-T + \kappa \cdot \mathbf{X} - \frac{Z}{\bar{c}(\kappa)}\right)} \omega^2 \, d\omega \, d\kappa \,, \quad (15.67)$$

and the refocusing kernels are given by

$$K^+(\omega, \boldsymbol{\kappa}) = \frac{1}{4\bar{c}\bar{\rho}} \int G_1\left(t_s - z_s\frac{\bar{c}(\kappa)}{\bar{c}^2} + \tau\right) G_2\left(\mathbf{x}_s + \boldsymbol{\kappa}\bar{c}(\kappa)z_s - \bar{c}^2\boldsymbol{\kappa}\tau\right)$$

$$\times \, \mathcal{W}_g^{(R)}\left(\omega, \boldsymbol{\kappa}, \tau\right) \, d\tau\,, \tag{15.68}$$

$$K^-(\omega, \boldsymbol{\kappa}) = \frac{1}{4\bar{c}\bar{\rho}} \int G_1\left(t_s - z_s\frac{\bar{c}(\kappa)}{\bar{c}^2} + \tau\right) G_2\left(\mathbf{x}_s + \boldsymbol{\kappa}\bar{c}(\kappa)z_s - \bar{c}^2\boldsymbol{\kappa}\tau\right)$$

$$\times \, \mathcal{W}_g^{(T)}\left(\omega, \boldsymbol{\kappa}, \tau\right) \, d\tau\,. \tag{15.69}$$

The picture is qualitatively the same for the time-reversed transverse velocity and pressure fields. The precise expressions for the refocused fields are the following:

$$P_{\mathrm{TR}}(T, \mathbf{X}, Z) = \frac{\bar{\rho}}{(2\pi)^3} \int K^+(\omega, \boldsymbol{\kappa})\bar{c}(\kappa)\left[\bar{c}(\kappa)\boldsymbol{\kappa}\cdot\widehat{\overline{\mathbf{f}_\mathbf{x}}}(\omega) + \widehat{\overline{f_z}}(\omega)\right]$$

$$\times \, e^{i\omega\left(-T+\boldsymbol{\kappa}\cdot\mathbf{X}+\frac{Z}{\bar{c}(\kappa)}\right)}\omega^2 \, d\omega \, d\boldsymbol{\kappa}$$

$$+ \frac{\bar{\rho}}{(2\pi)^3} \int K^-(\omega, \boldsymbol{\kappa})\bar{c}(\kappa)\left[\bar{c}(\kappa)\boldsymbol{\kappa}\cdot\widehat{\overline{\mathbf{f}_\mathbf{x}}}(\omega) - \widehat{\overline{f_z}}(\omega)\right]$$

$$\times \, e^{i\omega\left(-T+\boldsymbol{\kappa}\cdot\mathbf{X}-\frac{Z}{\bar{c}(\kappa)}\right)}\omega^2 \, d\omega \, d\boldsymbol{\kappa}\,,$$

$$\mathbf{V}_{\mathrm{TR}}(T, \mathbf{X}, Z) = \frac{1}{(2\pi)^3} \int K^+(\omega, \boldsymbol{\kappa})\bar{c}(\kappa)\boldsymbol{\kappa}\left[\bar{c}(\kappa)\boldsymbol{\kappa}\cdot\widehat{\overline{\mathbf{f}_\mathbf{x}}}(\omega) + \widehat{\overline{f_z}}(\omega)\right]$$

$$\times \, e^{i\omega\left(-T+\boldsymbol{\kappa}\cdot\mathbf{X}+\frac{Z}{\bar{c}(\kappa)}\right)}\omega^2 \, d\omega \, d\boldsymbol{\kappa}$$

$$+ \frac{1}{(2\pi)^3} \int K^-(\omega, \boldsymbol{\kappa})\bar{c}(\kappa)\boldsymbol{\kappa}\left[\bar{c}(\kappa)\boldsymbol{\kappa}\cdot\widehat{\overline{\mathbf{f}_\mathbf{x}}}(\omega) - \widehat{\overline{f_z}}(\omega)\right]$$

$$\times \, e^{i\omega\left(-T+\boldsymbol{\kappa}\cdot\mathbf{X}-\frac{Z}{\bar{c}(\kappa)}\right)}\omega^2 \, d\omega \, d\boldsymbol{\kappa}\,.$$

The proof of the theorem is a generalization of the arguments described in Chapter 9 and goes along the following main steps.

- We first consider the expected value of $u_{\mathrm{TR}}^\varepsilon$. By the result of Section 15.4.1 we find that this expectation converges to the limiting value given in the theorem.
- We then consider the variance of $u_{\mathrm{TR}}^\varepsilon$. We write the second moment as a multiple integral involving the product of four generalized reflection or transmission coefficients at four different frequencies as in (14.72). Using the decorrelation property of these coefficients we deduce that the variance goes to zero. Note that an integral over frequency (ensured by the time-domain nature of time reversal) is needed for the stabilization or the self-averaging of the refocused pulse.

15.5 Refocusing Properties in a Random Medium

15.5.1 The Case $|z_s| \ll L_{\mathrm{loc}}$

The complete expressions of the refocusing kernels are complicated and do not allow a simple discussion. In order to be more quantitative we consider the case

$$|z_s| \ll L_{\mathrm{loc}}(\omega_0) \,,$$

where $L_{\mathrm{loc}}(\omega_0) = 4\bar{c}^2/(\gamma \omega_0^2)$ is the localization length associated with the carrier wavelength ω_0. This means that we address the case of a source whose depth is small enough so that we can detect a stable wave front at the surface. We also assume that $L \gg L_{\mathrm{loc}}$, which means that we deal with a random half-space, and, using (14.70), we deduce

$$\mathcal{W}_0(\omega, \kappa, \tau, -L, z_s) = \delta(\tau) \,,$$

$$\mathcal{W}_1(\omega, \kappa, \tau, -L, z_s) = \frac{\bar{c}(\kappa)}{2 L_{\mathrm{loc}}(\omega)} \frac{1}{\left(1 + \frac{\bar{c}(\kappa)\tau}{2 L_{\mathrm{loc}}(\omega)}\right)^2} \mathbf{1}_{[0,\infty)}(\tau) \,,$$

$$\mathcal{W}_2(\omega, \kappa, \tau, -L, z_s) = \frac{\bar{c}(\kappa)^2 \tau}{2 L_{\mathrm{loc}}(\omega)^2} \frac{1}{\left(1 + \frac{\bar{c}(\kappa)\tau}{2 L_{\mathrm{loc}}(\omega)}\right)^3} \mathbf{1}_{[0,\infty)}(\tau) \,.$$

Since $|z_s| \ll L_{\mathrm{loc}}$, we can expand $\mathcal{W}_p^{(T)}$ with respect to $|z_s|/L_{\mathrm{loc}}$. Using the probabilistic interpretation (15.84) and with the same method as that used to derive (9.76), we obtain that only $\mathcal{W}_0^{(T)}$ and $\mathcal{W}_1^{(T)}$ give contributions of order one and of order $|z_s|/L_{\mathrm{loc}}$. They are given by

$$\mathcal{W}_0^{(T)}(\omega, \kappa, \tau, z_s, 0) = \exp\left(-\frac{\bar{c}(\kappa)^2}{\bar{c}^2} \frac{|z_s|}{L_{\mathrm{loc}}}\right) \delta(\tau) + \mathcal{W}_{0,c}^{(T)}(\omega, \kappa, \tau, z_s, 0)$$

$$\approx \exp\left(-\frac{\bar{c}(\kappa)^2}{\bar{c}^2} \frac{|z_s|}{L_{\mathrm{loc}}}\right) \delta(\tau) \,,$$

$$\mathcal{W}_1^{(T)}(\omega, \kappa, \tau, z_s, 0) \approx \frac{\bar{c}(\kappa)}{2 L_{\mathrm{loc}}} \mathbf{1}_{[0, 2|z_s|\bar{c}(\kappa)/\bar{c}^2]}(\tau) \,.$$

As a result, we obtain

$$\mathcal{W}_g^{(R)}(\omega, \kappa, \tau) = \frac{\bar{c}(\kappa)}{2 L_{\mathrm{loc}}} \frac{1}{\left(1 + \frac{\bar{c}(\kappa)\tau}{2 L_{\mathrm{loc}}}\right)^2} - \frac{|z_s|}{L_{\mathrm{loc}}} \frac{\bar{c}(\kappa)^3}{2\bar{c}^2 L_{\mathrm{loc}}} \frac{1 - \frac{\bar{c}(\kappa)\tau}{2 L_{\mathrm{loc}}}}{\left(1 + \frac{\bar{c}(\kappa)\tau}{2 L_{\mathrm{loc}}}\right)^3} \,, \tag{15.70}$$

$$\mathcal{W}_g^{(T)}(\omega, \kappa, \tau) = \exp\left(-\frac{\bar{c}(\kappa)^2}{\bar{c}^2} \frac{|z_s|}{L_{\mathrm{loc}}}\right) \delta(\tau) + \frac{|z_s|}{L_{\mathrm{loc}}} \frac{\bar{c}(\kappa)^3}{2\bar{c}^2 L_{\mathrm{loc}}} \frac{1}{\left(1 + \frac{\bar{c}(\kappa)\tau}{2 L_{\mathrm{loc}}}\right)^2} \,, \tag{15.71}$$

to leading order. The first term in the right-hand of (15.71) corresponds to the contribution of the stable wave front. The other terms are the contributions of the incoherent waves. Note that the first term in the right-hand side of (15.70) has a mass of the same order $O(1)$ with respect to $|z_s|/L_{\text{loc}}$ as the contribution of the stable front in (15.71).

We parameterize the horizontal position of the source by $\mathbf{x}_s = (x_s, 0)$, $x_s > 0$, and we assume that the spatial cutoff function associated with the mirror has the form (15.54): $G_2(\mathbf{x}) = g_2(\mathbf{x}/a)$, where a is the diameter of the mirror and g_2 is the normalized cutoff function. In order to simplify the forthcoming expressions we also assume that $\mathbf{f}$ is given by (15.53), where ω_0 is the high-carrier frequency and T_w is the time-pulse width, with $\omega_0 T_w \gg 1$.

15.5.2 Time Reversal of the Front

Let us assume that we record only the front. This means that the support of the function G_1 is of the form $[T_0 - \Delta_T, T_0 + \Delta_T]$ with $\varepsilon \ll \Delta_T \ll 1$, and

$$T_0 = \frac{OS}{\bar{c}} + t_s$$

corresponds to the arrival time at the mirror of the wave front emitted by the source. Due to the small support of G_1, only the components in $\delta(\tau)$ of the densities $\mathcal{W}_g^{(R)}$ and $\mathcal{W}_g^{(T)}$ contribute to the refocusing kernels $K^{\pm}(\omega, \boldsymbol{\kappa})$. As seen above, $\mathcal{W}_g^{(R)}$ has no Dirac contribution, while $\mathcal{W}_g^{(T)}$ has one, with the weight $\exp[-\bar{c}(\kappa)^2|z_s|/(\bar{c}^2 L_{\text{loc}})]$. We then get the same expression for the refocused pulse shape as in the homogeneous case, up to the exponential damping term. More precisely, the envelope of the focal spot is no longer given by (15.56), but by

$$|U_{\text{TR}}(T, \mathbf{X}, Z)| \approx \exp\left(-\frac{OS^2}{|z_s|L_{\text{loc}}}\right) \left|\frac{\mathbf{OS}}{OS} \cdot \mathbf{f}_0\left(-\frac{T}{T_w} + \frac{(\mathbf{X}, Z) \cdot \mathbf{e}_3}{\bar{c} T_w}\right)\right|$$

$$\times \left|\hat{g}_2\left(\frac{\omega_0 \Delta\phi_1}{\bar{c}}(\mathbf{X}, Z) \cdot \mathbf{e}_1, \frac{\omega_0 \Delta\phi_2}{\bar{c}}(\mathbf{X}, Z) \cdot \mathbf{e}_2\right)\right|, \quad (15.72)$$

where the $\Delta\phi_j$'s are defined by (15.57–15.58) and $L_{\text{loc}} = L_{\text{loc}}(\omega_0) = 4\bar{c}^2/(\gamma\omega_0^2)$. This shows that the focal-spot shape is the same as in the homogeneous case. The only difference is a slight reduction of the amplitude due to the decay of the energy of the stable front.

In the next sections, we assume that we record some part of the long incoherent waves. This means that $G_1(t) = \mathbf{1}_{[T_1, T_2]}(t)$, where $T_0 < T_1 < T_2$.

15.5.3 Time Reversal of the Incoherent Waves with Offset

We assume that the mirror is not exactly above the source and denote by $x_s > 0$ the offset. The expression (15.67) for the refocused field U_{TR} can be significantly simplified if we assume that the following four hypotheses are satisfied:

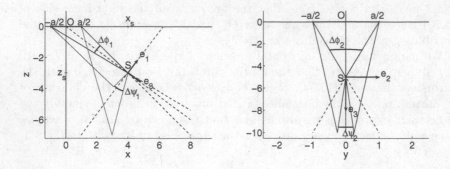

Fig. 15.12. Cones of aperture $(\Delta\phi_1, \Delta\phi_2)$ and $(\Delta\psi_1, \Delta\psi_2)$ generated respectively by the stable wave front and by the incoherent scattered waves $((\mathbf{e}_1, \mathbf{e}_3)$-section on the left and $(\mathbf{e}_2, \mathbf{e}_3)$-section on the right).

H_1. The timewindow cutoff function is of the form $G_1(\tau) = \mathbf{1}_{[T_1, T_2]}(\tau)$ with $T_2 > T_1 > T_0$, where $T_0 = t_s + OS/\bar{c}$ is the arrival time of the front. This means that we do not capture the coherent front.

H_2. The spatial shape of the time-reversal mirror is of the form $G_2(\mathbf{x}) = g_2(\mathbf{x}/a)$, the reference frame is oriented such that the source location is $\mathbf{x}_s = (x_s, 0)$, and the offset x_s is much larger than the mirror diameter a. This means that we consider a narrow-aperture situation.

H_3. The source emits a pulse with a carrier frequency ω_0 that is much larger than the bandwidth $1/T_w$ (equal to the inverse of the pulse width). This allows us to derive a simplified high-frequency expression for the refocused field.

H_4. The random slab is a semi-infinite random half-space and the source depth is much smaller than the localization length $L_{\text{loc}} = 4\bar{c}^2/(\gamma\omega_0^2)$, with ω_0 the carrier frequency. This allows us to obtain explicit approximations for the kernel densities $\mathcal{W}_g$, which in turn will allow us to discuss quantitatively the properties of the refocusing.

We introduce the angles $0 < \theta_1 < \theta_2 < \pi/2$:

$$\theta_j = \arccos\left(\frac{x_s}{\bar{c}(T_j - t_s)}\right). \tag{15.73}$$

As discussed below and shown in Figure 15.13, this defines a "temporal" cone delimited by the angle of the shortest ray θ_1 and the angle of the longest ray θ_2 recorded by the center point of the mirror. After some algebra detailed in Appendix 15.8 based on the set of hypotheses $H_1 - -H_4$, we obtain that the refocused focal spot is given by

$$U_{\mathrm{TR}}(T, \mathbf{X}, Z) = \frac{a^2 \omega_0^2}{32\pi^2 \bar{\rho}\bar{c}^3 x_s L_{\mathrm{loc}}} \int_{\cos(\theta_2)}^{\cos(\theta_1)} \frac{1}{\left(1 + \frac{x_s}{2L_{\mathrm{loc}}\eta\sqrt{1-\eta^2}}\right)^2} \frac{1}{\sqrt{1-\eta^2}}$$

$$\times \hat{g}_2 \left(\frac{\omega_0 a\eta}{\bar{c}x_s}\left(X - \frac{\eta}{\sqrt{1-\eta^2}}Z\right), \frac{\omega_0 a\eta}{\bar{c}x_s}Y\right) e^{i\frac{\omega_0}{\bar{c}}\left(\eta X + Z\sqrt{1-\eta^2} - T\right)}$$

$$\times \left[f_{0z} + \frac{\eta}{\sqrt{1-\eta^2}}f_{0x}\right]\left(-\frac{T}{T_w} + \frac{X\eta + Z\sqrt{1-\eta^2}}{\bar{c}T_w}\right) d\eta + c.c.. \quad (15.74)$$

The y-component f_{0y} of the source does not appear in this leading-order expression, because it has not been recorded by the mirror due to the fact that the source location has a zero y-coordinate. As pointed out above, this expression holds true if $|z_s| \ll L_{\mathrm{loc}}$ and $x_s > 0$. If we moreover assume that x_s is of the same order as $|z_s|$, so that $x_s \ll L_{\mathrm{loc}}$, then the first factor in the integral simplifies and becomes one. Let us write $\theta_1 = \bar{\theta} - \Delta\theta/2$ and $\theta_2 = \bar{\theta} + \Delta\theta/2$, with

$$\bar{\theta} = \frac{\theta_1 + \theta_2}{2} = \frac{1}{2}\left[\arccos\left(\frac{x_s}{\bar{c}(T_2 - t_s)}\right) + \arccos\left(\frac{x_s}{\bar{c}(T_1 - t_s)}\right)\right]. \quad (15.75)$$

We now introduce the new orthonormal frame $(\mathbf{w}_1, \mathbf{w}_2, \mathbf{w}_3)$ defined by

$$\mathbf{w}_1 = \begin{bmatrix} -\sin\bar{\theta} \\ 0 \\ \cos\bar{\theta} \end{bmatrix}, \quad \mathbf{w}_2 = \begin{bmatrix} 0 \\ 1 \\ 0 \end{bmatrix}, \quad \mathbf{w}_3 = \begin{bmatrix} \cos\bar{\theta} \\ 0 \\ \sin\bar{\theta} \end{bmatrix}, \quad (15.76)$$

and described in Figure 15.14. In this frame, $\mathbf{w}_3$ is the main direction of arrival of the time-reversed incoherent waves that have interacted with the medium below z_s and that participate in the refocusing at the original source location. Note the striking difference with the direction of arrival $\mathbf{e}_3$ of the time-reversed front (compare the two pictures in Figure 15.14). The range of $\bar{\theta}$ is $(\theta_0, \pi/2)$ with $\theta_0 = \arccos(x_s/OS)$: $\bar{\theta}$ is close to θ_0 when $T_0 < T_1 < T_2 \searrow T_0$, corresponding to the recording and reemission of a short piece of incoherent waves near the front. The angle $\bar{\theta}$ is close to $\pi/2$ when $T_2 > T_1 \nearrow \infty$, corresponding to a piece of incoherent waves far from the front.

We can simplify further the expression (15.74) by assuming that $\Delta\theta$ is relatively small, so that we can expand the arguments of the integral in $\eta \in (\cos(\bar{\theta} - \Delta\theta/2), \cos(\bar{\theta} + \Delta\theta/2))$ around the central value $\cos(\bar{\theta})$. We then obtain that in the frame $(\mathbf{w}_1, \mathbf{w}_2, \mathbf{w}_3)$, the envelope of the refocused field (15.74) can be approximated as

$$|U_{\mathrm{TR}}(T, \mathbf{X}, Z)| \approx \left|\hat{g}_2\left(-\frac{\omega_0 \Delta\psi_1}{\bar{c}}(\mathbf{X}, Z)\cdot\mathbf{w}_1, \frac{\omega_0 \Delta\psi_2}{\bar{c}}(\mathbf{X}, Z)\cdot\mathbf{w}_2\right)\right|$$

$$\times \left|\mathrm{sinc}\left(\frac{\omega_0 \Delta\theta}{2\bar{c}}(\mathbf{X}, Z)\cdot\mathbf{w}_1\right)\right| \left|\mathbf{w}_3 \cdot \mathbf{f}_0\left(-\frac{T}{T_w} + \frac{(\mathbf{X}, Z).\mathbf{w}_3}{\bar{c}T_w}\right)\right|, \quad (15.77)$$

up to a multiplicative factor. Here we have introduced the angles

$$\Delta\psi_1 = \frac{a}{x_s \tan\bar\theta}, \qquad \Delta\psi_2 = \frac{a\cos\bar\theta}{x_s}. \tag{15.78}$$

1. In the $\mathbf{w}_3$-direction, the focal-spot size is governed by the initial pulse width T_w and it is given by $\bar c T_w$.
2. In the $\mathbf{w}_2$-direction, the focal spot is determined by the mirror size a and it has the size $\lambda_0/\Delta\psi_2$, where $\Delta\psi_2 = a(\cos\bar\theta)/x_s$ ($\Delta\psi_2$ is represented in the right picture of Figure 15.12). This size is consistent with the Rayleigh formula. The random medium cannot help refocusing in this transverse direction because no scattering occurs in this direction. This is connected to the layered structure of the medium and should not happen in a three-dimensional random medium with isotropic medium fluctuations.
3. In the $\mathbf{w}_1$-direction, the focal spot radius depends on the two angular cones $\Delta\theta$ and $\Delta\psi_1$, which can be chosen independently:

- The angle $\Delta\theta$ is determined by the recording time interval $[T_1, T_2]$:

$$\Delta\theta = \theta_2 - \theta_1 = \arccos\left(\frac{x_s}{\bar c(T_2 - t_s)}\right) - \arccos\left(\frac{x_s}{\bar c(T_1 - t_s)}\right). \tag{15.79}$$

  This **temporal cone** is delimited by the angle of the shortest ray θ_1 and the angle of the longest ray θ_2 recorded by the center point of the mirror (see the right picture of Figure 15.14).

- The angle $\Delta\psi_1$ is determined by the spatial size of the mirror. This **geometrical cone** is delimited by the two rays with the same travel time $x_s/[\bar c \cos(\bar\theta)]$ originating from the two ends of the mirror, at $x = -a/2$ and $x = a/2$. These rays are plotted in the left picture of Figure 15.12. The value of the angle $\Delta\psi_1$ is computed by trigonometry. The crossover of these two rays is responsible for the minus sign in the first argument of $\hat g_2$ in (15.77).

Two cases should then be distinguished to determine the size of the focal spot size in the $\mathbf{w}_1$-direction.

(1) If $\Delta\theta \ll \Delta\psi_1 \ll 1$, which is the case in particular if only a very short piece of coda is recorded, then (15.74) can be simplified to

$$|U_{\mathrm{TR}}(T, \mathbf{X}, Z)| \approx \left|\hat g_2\left(-\frac{\omega_0\Delta\psi_1}{\bar c}(\mathbf{X}, Z)\cdot\mathbf{w}_1, \frac{\omega_0\Delta\psi_2}{\bar c}(\mathbf{X}, Z)\cdot\mathbf{w}_2\right)\right|$$
$$\times \left|\mathbf{w}_3\cdot\mathbf{f}_0\left(-\frac{T}{T_w} + \frac{(\mathbf{X}, Z)\cdot\mathbf{w}_3}{\bar c T_w}\right)\right|. \tag{15.80}$$

In this case, the angular diversity of the refocused waves mainly originates from the *numerical aperture of the mirror*, and we get a formula in qualitative agreement with the Rayleigh resolution formula. More quantitatively, if we record a piece of the coda just after the front, meaning that T_1, T_2 are close to T_0, then $\bar{\theta} = \arccos(x_s/OS)$ and the focal spot size is $\lambda_0|z_s|/a$. Recall that the focal spot size generated by the front is given by $\lambda_0 OS^2/(a|z_s|)$. This shows that the random medium enables **correction for the offset** x_s.

(2) If $\Delta\psi_1 \ll \Delta\theta \ll 1$, then (15.74) simplifies to

$$|U_{\mathrm{TR}}(T, \mathbf{X}, Z)| \approx \left| \hat{g}_2 \left(0, \frac{\omega_0 \Delta\psi_2}{\bar{c}} (\mathbf{X}, Z) \cdot \mathbf{w}_2 \right) \right| \left| \mathrm{sinc} \left(\frac{\omega_0 \Delta\theta}{2\bar{c}} (\mathbf{X}, Z) \cdot \mathbf{w}_1 \right) \right|$$

$$\times \left| \mathbf{w}_3 \cdot \mathbf{f}_0 \left(-\frac{T}{T_w} + \frac{(\mathbf{X}, Z) \cdot \mathbf{w}_3}{\bar{c} T_w} \right) \right|. \tag{15.81}$$

In this case, the angular diversity of the refocused waves mainly originates from the *temporal refocusing cone* $\Delta\theta$ described above. Therefore the focal spot size is $\lambda_0/\Delta\theta$ as soon as $\Delta\theta > a/OS$. This size is determined by the angular diversity of the refocused incoherent waves and it is much smaller than the prediction of the Rayleigh resolution formula. As we show below, (15.81) can be extended to $\Delta\theta \approx 1$. The focal-spot size is then of order the **diffraction limit** if $\Delta\theta \approx 1$, that is, if $T_2 - T_1 \geq \bar{c}/x_s$. This means that

recording a long coda allows us to enhance dramatically the **effective aperture** *of the mirror thanks to the multiple scattering in the random medium below the source.*

The spot shape is plotted in Figure 15.15. Qualitatively, if we increase the recording time window by taking a larger T_2 and a smaller T_1, then we get a wider and wider numerical aperture for the virtual mirror located below S. The largest numerical aperture is $[\theta_0, \pi/2]$, where $\theta_0 = \arccos(x_s/OS)$. It is obtained for T_1 close to T_0 and $T_2 - T_0 \gg \bar{c}/x_s$.

(3) In the case $\Delta\psi_1 \ll \Delta\theta \approx 1$, we can reconsider the expression (15.74) and focus our attention on the spatial profile in the $\mathbf{w}_1$-direction. We obtain that for any r,

$$|U_{\mathrm{TR}}(T = 0, (\mathbf{X}, Z) = r\mathbf{w}_1)| \approx \left| \int_{-1/2}^{1/2} \cos \left(\frac{\omega_0 r}{\bar{c}} \cos(\Delta\theta\xi) \right) d\xi \right|,$$

which shows that the size of the focal spot is of the order of the carrier wavelength λ_0 when $\Delta\theta \approx 1$. If $\Delta\theta$ could become close to 2π, then the focal spot would converge to the Bessel function $J_0(\omega_0 r/\bar{c})$ and its radius would be equal to $0.383\lambda_0$.

Note that in the regime $|z_s| \ll L_{\mathrm{loc}}$ studied in this section the incoherent waves scattered by the random slab $(z_s, 0)$ do not contribute to the refocusing at the original source location.

Let us finally comment on the case $|z_s| \geq L_{\mathrm{loc}}$. If z_s is large and becomes comparable to the localization length, then the refocusing can become even better, since incoherent waves scattered in the slab $(z_s, 0)$ can also contribute

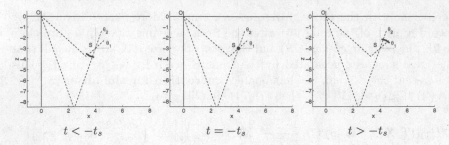

$$t < -t_s \qquad\qquad t = -t_s \qquad\qquad t > -t_s$$

Fig. 15.13. Pulse refocusing when the incoherent waves are recorded. The two typical paths shown correspond to the shortest and the longest ones recorded, or equivalently sent back by the mirror, and producing the angles θ_1 and θ_2, respectively. This path-angle interpretation is valid in the layered case considered here as discussed in Section 14.3, Figure 14.4.

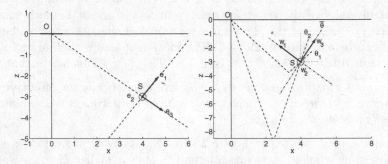

Fig. 15.14. Frame adapted to the refocusing of the front wave (left picture) and frame adapted to the refocusing of the incoherent waves (right picture).

to the refocusing of the pulse (see Figure 15.16). However, it is more cumbersome to write explicit formulas because no analytical expressions are available for the refocusing densities. In the best conditions (where the recording time window is infinite) the numerical aperture is generated by two cones $(\theta_0, \pi/2)$ and $(-\pi/2, -\theta_0)$.

15.5.4 Time Reversal of the Incoherent Waves Without Offset

We now discuss the case in which the mirror is located just above the source: $x_s = 0$. If we assume the set of hypotheses H_1 to H_4 stated at the beginning of Section 15.5.3 (except the requirement that $x_s \gg a$), then we can apply the method described in Appendix 15.8 to simplify the expression (15.67) of the refocused field U_{TR}. We obtain, to leading order in $|z_s|/L_{loc}$,

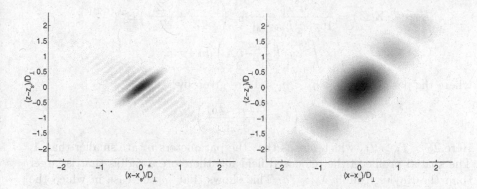

Fig. 15.15. Left picture: Pulse shape generated by the refocusing of the incoherent waves in the plane $(\mathbf{w}_1, \mathbf{w}_3)$. The mirror is square with size a and the initial pulse profile is Gaussian with duration T_w. Here $a = 0.33$, $z_s = -3$, $x_s = 4$ $(OS = 5)$, $t_s = 0$ $(\bar{c}T_0 = 5)$, $\bar{c}T_1 = 6$, $\bar{c}T_2 = 8$, so that $\tan\bar{\theta} = 1.38$, $\Delta\phi_1 = 0.04$, $\Delta\psi_1 = 0.06$, and $\Delta\theta = 0.2$. As a result, the spot shape is a sinc function with radius $D_\perp/5$, where $D_\perp = \lambda_0 x_s \tan\bar{\theta}/a$ in the $\mathbf{w}_1$-direction and a Gaussian function with radius $\bar{c}T_w$ in the $\mathbf{w}_3$-direction. The spot shape in the $\mathbf{w}_2$-direction is a sinc function with radius $\lambda_0/\Delta\psi_2$, where $\Delta\psi_2 = a\cos\bar{\theta}/x_s$. For comparison we show the focal spot in the homogeneous case in the right picture.

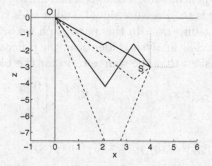

Fig. 15.16. Different contributions of the multiply scattered waves to the aperture enhancement in the case $|z_s| > L_{\mathrm{loc}}$. The dashed lines stand for two typical paths that participate in the lower aperture cone $(\theta_0, \pi/2)$. The solid lines stand for two typical paths that participate in the upper aperture cone $(-\pi/2, \theta_0)$.

$$U_{\mathrm{TR}}(T, \mathbf{X}, Z) = \frac{a^2 \omega_0^2}{16\pi \bar\rho \bar c^3 L_{\mathrm{loc}} |z_s|} \hat f_{0z} \left(-\frac{T}{T_w} + \frac{Z}{\bar c T_w} \right) e^{i\omega_0 (-T + \frac{Z}{\bar c})}$$

$$\times \int_{\eta_2}^{\eta_1} \hat g_2 \left(\frac{\omega_0 a}{\bar c |z_s|} \eta \mathbf{X} \right) d\eta + c.c.,$$

where the parameters $0 < \eta_2 < \eta_1 < 1$ are given by

$$\eta_j = \left(1 + \frac{\bar c (T_j - T_0)}{|z_s|} \right)^{-1}.$$

Here $T_2 > T_1 > T_0$, which means that the parameters η_j are smaller than 1. The spatial profile of the refocused field has therefore a radius that is larger than the Rayleigh limit $\lambda_0 |z_s|/a$. This shows that in the case in which the mirror is located just above the source, the refocusing of the incoherent waves cannot involve superresolution, because the angular diversity of the refocused waves is very small (the geometry is as in the right picture of Figure 15.12 with the cone of aperture $\Delta \psi_2 = \eta_1 a / |z_s|$).

15.5.5 Record of the Pressure Signal

In the previous sections we have assumed that the time-reversal mirror records the three components of the velocity signal, and uses this signal as a new three-dimensional source for the back-propagation. Practically, with the usual experimental constraints in acoustics, a time-reversal mirror consists of an array of transducers; it records the pressure signal, and it uses this signal as a new source emitting in the z-direction. In this paragraph we briefly revisit the results in this framework. The analysis is identical; the difference is that instead of (15.37), we now consider that the source term for the back propagation is

$$\mathbf{F}_{\mathrm{TR}}^\varepsilon (t, \mathbf{x}, z) = p_{\mathrm{rec}}^\varepsilon (-t, \mathbf{x}) \delta(z) \begin{bmatrix} \mathbf{0} \\ 1 \end{bmatrix},$$

with

$$p_{\mathrm{rec}}^\varepsilon (t, \mathbf{x}) = p_s^\varepsilon (t, \mathbf{x}) G_1(t) G_2(\mathbf{x}).$$

The results are then qualitatively unchanged, but the refocusing kernels given by (15.68, 15.69) should be multiplied by the factor $\bar c / \bar c(\kappa)$.

15.6 Appendix A: Moments of the Reflection and Transmission Coefficients

15.6.1 Autocorrelation Function of the Transmission Coefficient at Two Nearby Slownesses and Frequencies

Cross-moments of transmission and reflection coefficients are required in Section 15.4. The analysis of moments of reflection coefficients has been performed in Section 14.3.2. We perform the same analysis for the transmission coefficient, and we get that if $q = p$:

$$\mathbb{E}\left[\left(R^{\varepsilon}_{(\omega+\varepsilon h/2,\kappa+\varepsilon\lambda/2)}(z_0,z)\right)^p T^{\varepsilon}_{(\omega+\varepsilon h/2,\kappa+\varepsilon\lambda/2)}(z_0,z)\right.$$

$$\left. \times \left(\overline{R^{\varepsilon}_{(\omega-\varepsilon h/2,\kappa-\varepsilon\lambda/2)}(z_0,z)}\right)^q \overline{T^{\varepsilon}_{(\omega-\varepsilon h/2,\kappa-\varepsilon\lambda/2)}(z_0,z)}\right] \tag{15.82}$$

$$\xrightarrow{\varepsilon\to 0} \int \mathcal{W}_p^{(T)}(\omega,\kappa,\tau,z_0,z)e^{i\tau[h\bar{c}^2/\bar{c}(\kappa)^2-\omega\lambda\bar{c}^2\kappa]}\, d\tau \times e^{2ipz[-h/\bar{c}(\kappa)+\omega\lambda\bar{c}(\kappa)]}\,,$$

and otherwise

$$\mathbb{E}\left[\left(R^{\varepsilon}_{(\omega+\varepsilon h/2,\kappa+\varepsilon\lambda/2)}(z_0,z)\right)^p T^{\varepsilon}_{(\omega+\varepsilon h/2,\kappa+\varepsilon\lambda/2)}(z_0,z)\right.$$

$$\left. \times \left(\overline{R^{\varepsilon}_{(\omega-\varepsilon h/2,\kappa-\varepsilon\lambda/2)}(z_0,z)}\right)^q \overline{T^{\varepsilon}_{(\omega-\varepsilon h/2,\kappa-\varepsilon\lambda/2)}(z_0,z)}\right] \xrightarrow{\varepsilon\to 0} 0,\ (15.83)$$

where the quantity $\mathcal{W}_p^{(T)}$ is obtained through the following system of transport equations:

$$\frac{\partial \mathcal{W}_p^{(T)}}{\partial z} + 2p\frac{\bar{c}(\kappa)}{\bar{c}^2}\frac{\partial \mathcal{W}_p^{(T)}}{\partial \tau}$$

$$= \frac{1}{L_{\text{loc}}(\omega,\kappa)}\left((p+1)^2\mathcal{W}_{p+1}^{(T)} + p^2\mathcal{W}_{p-1}^{(T)} - (p^2+(p+1)^2)\mathcal{W}_p^{(T)}\right),$$

$$\mathcal{W}_p^{(T)}(\omega,\kappa,\tau,z_0,z=z_0) = \mathbf{1}_0(p)\delta(\tau)\,.$$

The solution $\mathcal{W}_p^{(T)}$ has the probabilistic interpretation

$$\mathcal{W}_p^{(T)}(\omega,\kappa,\tau,z_0,z) = \mathbb{E}\left[\delta\left(\tau - 2\frac{\bar{c}(\kappa)}{\bar{c}^2}\int_{z_0}^z N_u^{(T)}\,du\right)\mathbf{1}_0(N_z^{(T)})\mid N_{z_0}^{(T)} = p\right] \tag{15.84}$$

in terms of the jump Markov processes $N^{(T)}$ with infinitesimal generator given by

$$\mathcal{L}^{(T)}\phi(N) = \frac{1}{L_{\text{loc}}(\omega,\kappa)}\ \left[(N+1)^2\phi(N+1) + N^2\phi(N-1)\right.$$

$$\left. -((N+1)^2 + N^2)\phi(N)\right],$$

where $L_{\text{loc}}(\omega,\kappa)$ is defined by (14.62).

15.6.2 Shift Properties

Straightforward manipulations based on shifts of the governing equations show that

$$\left(R^{\varepsilon}_{(\omega,\kappa)}(z,z_1), T^{\varepsilon}_{(\omega,\kappa)}(z,z_1), \nu_{\kappa}^{\varepsilon}(z)\right)_{z_0\le z\le z_1}$$

and

$$\left(R^{\varepsilon}_{(\omega,\kappa)}(z-z_1,0)e^{\frac{2i\omega z_1}{\bar{c}(\kappa)\varepsilon}}, T^{\varepsilon}_{(\omega,\kappa)}(z-z_1,0), \nu_{\kappa}^{\varepsilon}(z-z_1)\right)_{z_0\le z\le z_1}$$

have the same distribution. Similarly,

$$\left(R^\varepsilon_{(\omega,\kappa)}(z,0), T^\varepsilon_{(\omega,\kappa)}(z,0), \nu^\varepsilon_\kappa(z)\right)_{z_0 \leq z \leq 0}$$

and

$$\left(\widetilde{R}^\varepsilon_{(\omega,\kappa)}(z,0)e^{\frac{-2i\omega z_0}{\bar{c}(\kappa)\varepsilon}}, \widetilde{T}^\varepsilon_{(\omega,\kappa)}(z,0), \nu^\varepsilon_\kappa(z_0 - z)\right)_{z_0 \leq z \leq 0}$$

have the same distribution. These shift properties are used in the next section, devoted to the analysis of the generalized transmission and reflection coefficients.

15.6.3 Generalized Coefficients

As shown by the integral representation (15.45), the cross-correlation of the generalized reflection coefficients plays an important role. We define

$$U^\varepsilon_g = \mathbb{E}\left[R^\varepsilon_g\left(\omega + \frac{\varepsilon h}{2}, \mu + \frac{\varepsilon \lambda}{2}, z_s + \varepsilon \mathcal{Z}\right)\overline{R^\varepsilon_g\left(\omega - \frac{\varepsilon h}{2}, \mu - \frac{\varepsilon \lambda}{2}, z_s\right)}\right].$$

(15.85)

Using the representation (15.30) of the generalized coefficient R^ε_g in terms of the usual reflection and transmission coefficients (with moduli less than 1), we obtain that

$$U^\varepsilon_g = \sum_{n,m=0}^{\infty} \mathbb{E}\left[\overline{R^\varepsilon}^{m+1} R^{\varepsilon n+1} \widetilde{R}^{\varepsilon^n}\widetilde{T}^\varepsilon \overline{\widetilde{R}^\varepsilon}^m\overline{\widetilde{T}^\varepsilon}\right],$$

where $\overline{R^\varepsilon}^{m+1}$ is evaluated at $(\omega - \varepsilon h/2, \mu - \varepsilon \lambda/2, -L, z_s)$, $R^{\varepsilon n+1}$ is evaluated at $(\omega + \varepsilon h/2, \mu + \varepsilon \lambda/2, -L, z_s + \varepsilon \mathcal{Z})$, $\widetilde{R}^{\varepsilon^n}\widetilde{T}^\varepsilon$ is evaluated at $(\omega + \varepsilon h/2, \mu + \varepsilon \lambda/2, z_s + \varepsilon \mathcal{Z}, 0)$, and $\overline{\widetilde{R}^\varepsilon}^m \overline{\widetilde{T}^\varepsilon}$ is evaluated at $(\omega - \varepsilon h/2, \mu - \varepsilon \lambda/2, z_s, 0)$. As $\varepsilon \to 0$ the propagators between $-L$ and z_s and between z_s and 0 become independent. By continuity with respect to z of the limits of the moments of the reflection and transmission coefficients, the small offset $\varepsilon \mathcal{Z}$ does not play any role in the limit of U^ε_g. Accordingly, we shall obtain the limit of U^ε_g as $\varepsilon \to 0$ by looking at the limits of $\mathbb{E}[\overline{R^\varepsilon}^{m+1} R^{\varepsilon n+1}]$ and $\mathbb{E}[\widetilde{R}^{\varepsilon^n}\widetilde{T}^\varepsilon \overline{\widetilde{R}^\varepsilon}^m \overline{\widetilde{T}^\varepsilon}]$. By using the shift properties of the reflection coefficients described in Appendix 15.6.2 and the expressions of the limit values for moments of reflection and transmission coefficients, we obtain that

$$\mathbb{E}\left[R^{\varepsilon n+1}\overline{R^\varepsilon}^{m+1}\right] \xrightarrow{\varepsilon \to 0} 0 \quad \text{and} \quad \mathbb{E}\left[\widetilde{R}^{\varepsilon^n}\widetilde{T}^\varepsilon \overline{\widetilde{R}^\varepsilon}^m\overline{\widetilde{T}^\varepsilon}\right] \xrightarrow{\varepsilon \to 0} 0$$

if $m \neq n$ and

$$\mathbb{E}\left[R^{\varepsilon n+1}\overline{R^\varepsilon}^{n+1}\right] \xrightarrow{\varepsilon \to 0} e^{2i(n+1)[-h/\bar{c}(\mu)+\lambda\omega\bar{c}(\mu)\mu]z_s}$$

$$\times \int \mathcal{W}_{n+1}(\omega,\mu,\tau,-L,z_s)e^{i\tau[h\bar{c}^2/\bar{c}(\mu)^2-\omega\lambda\bar{c}^2\mu]}\, d\tau,$$

$$\mathbb{E}\left[\widetilde{R}^{\varepsilon^n}\widetilde{T}^\varepsilon \ \overline{\widetilde{R}^\varepsilon}^n\overline{\widetilde{T}^\varepsilon}\right] \overset{\varepsilon\to 0}{\longrightarrow} e^{2in[h/\bar{c}(\mu)-\lambda\omega\bar{c}(\mu)\mu]z_s}$$

$$\times \int \mathcal{W}_n^{(T)}(\omega,\mu,\tau,z_s,0)e^{i\tau[h\bar{c}^2/\bar{c}(\mu)^2-\omega\lambda\bar{c}^2\mu]}\,d\tau,$$

where $\mathcal{W}_n$ and $\mathcal{W}_n^{(T)}$ are described in Section 14.3.2 and Appendix 15.6.1 respectively. We can then deduce that

$$U_g^\varepsilon \overset{\varepsilon\to 0}{\longrightarrow} \sum_{n=0}^\infty e^{2i[-h/\bar{c}(\mu)+\lambda\omega\bar{c}(\mu)\mu]z_s}\int \mathcal{W}_{n+1}(\omega,\mu,\tau)e^{i\tau H}\,d\tau$$

$$\times \int \mathcal{W}_n^{(T)}(\omega,\mu,\tau)e^{i\tau H}\,d\tau$$

$$= e^{2i[-h/\bar{c}(\mu)+\lambda\omega\bar{c}(\mu)\mu]z_s}\sum_{n=0}^\infty \int \mathcal{W}_{n+1}*_\tau \mathcal{W}_n^{(T)}(\omega,\mu,\tau)e^{iH\tau}\,d\tau,\tag{15.86}$$

where $H = h\bar{c}^2/\bar{c}(\mu)^2 - \omega\lambda\bar{c}^2\mu$.

Similarly, if we consider the product of two generalized transmission coefficients,

$$U_g^{(T),\varepsilon} = \mathbb{E}\left[T_g^\varepsilon\left(\omega+\frac{\varepsilon h}{2},\mu+\frac{\varepsilon\lambda}{2},z_s+\varepsilon\mathcal{Z}\right)\overline{T_g^\varepsilon\left(\omega-\frac{\varepsilon h}{2},\mu-\frac{\varepsilon\lambda}{2},z_s\right)}\right],$$
$$\tag{15.87}$$

then, using the expansion

$$U_g^{(T),\varepsilon} = \sum_{n,m=0}^\infty \mathbb{E}\left[\overline{R^\varepsilon}^m R^{\varepsilon n}\ \widetilde{R}^{\varepsilon^n}\widetilde{T}^\varepsilon\ \overline{\widetilde{R}^\varepsilon}^m\overline{\widetilde{T}^\varepsilon}\right],$$

we can show that

$$U_g^{(T),\varepsilon} \overset{\varepsilon\to 0}{\longrightarrow} \sum_{n=0}^\infty \int \mathcal{W}_n(\omega,\mu,\tau)e^{i\tau H}\,d\tau \int \mathcal{W}_n^{(T)}(\omega,\mu,\tau)e^{i\tau H}\,d\tau$$

$$= \sum_{n=0}^\infty \int \mathcal{W}_n *_\tau \mathcal{W}_n^{(T)}(\omega,\mu,\tau)e^{iH\tau}\,d\tau,\tag{15.88}$$

where $H = h\bar{c}^2/\bar{c}(\mu)^2 - \omega\lambda\bar{c}^2\mu$.

The limit of the cross moment

$$U_g^{(TR),\varepsilon} = \mathbb{E}\left[T_g^\varepsilon\left(\omega+\frac{\varepsilon h}{2},\mu+\frac{\varepsilon\lambda}{2},z_s+\varepsilon\mathcal{Z}\right)\overline{R_g^\varepsilon\left(\omega-\frac{\varepsilon h}{2},\mu-\frac{\varepsilon\lambda}{2},z_s\right)}\right]$$
$$\tag{15.89}$$

can be obtained by expanding in series the expressions (15.30–15.31) of R_g^ε and T_g^ε:

$$U_g^{(TR),\varepsilon} = \sum_{n,m=0}^{\infty} \mathbb{E}\left[\overline{R^{\varepsilon}}^m \ R^{\varepsilon n} \ \widetilde{R}^{\varepsilon n} \widetilde{T}^{\varepsilon} \ \overline{\widetilde{R}^{\varepsilon}}^{m+1} \overline{\widetilde{T}^{\varepsilon}} \right]$$

$$\xrightarrow{\varepsilon \to 0} \sum_{n,m=0}^{\infty} \lim_{\varepsilon \to 0} \mathbb{E}\left[\overline{R^{\varepsilon}}^m \ R^{\varepsilon n} \right] \lim_{\varepsilon \to 0} \mathbb{E}\left[\widetilde{R}^{\varepsilon n} \widetilde{T}^{\varepsilon} \ \overline{\widetilde{R}^{\varepsilon}}^{m+1} \overline{\widetilde{T}^{\varepsilon}} \right].$$

For each term with indices (m, n) of this series, one of the two limits is zero, so the global limit is zero. This shows that

$$U_g^{(TR),\varepsilon} \xrightarrow{\varepsilon \to 0} 0, \tag{15.90}$$

and we get the same result if we exchange the roles of R_g^{ε} and T_g^{ε}.

15.7 Appendix B: A Priori Estimates for the Generalized Coefficients

Estimates of moments of the generalized transmission and reflection coefficients are required to establish tightness and convergence results. In this subsection we prove that T_g^{ε} admits moments of any order that are uniformly bounded with respect to ε. This in turn implies that R_g^{ε} admits moments of any order that are uniformly bounded with respect to ε, since the definitions (15.25–15.26) of the generalized coefficients show that

$$R_g^{\varepsilon}(\omega, \kappa, z_s) = T_g^{\varepsilon}(\omega, \kappa, z_s) R_{(\omega,\kappa)}^{\varepsilon}(-L, z_s),$$

and $|R_{(\omega,\kappa)}^{\varepsilon}(-L, z_s)|$ is bounded by 1.

We first give a simple representation of the generalized transmission coefficient in terms of the usual transmission coefficient. From the propagator relation $\mathbf{P}_{(\omega,\kappa)}^{\varepsilon}(-L, 0) = \mathbf{P}_{(\omega,\kappa)}^{\varepsilon}(z_s, 0)\mathbf{P}_{(\omega,\kappa)}^{\varepsilon}(-L, z_s)$, we get

$$\overline{\alpha_{(\omega,\kappa)}^{\varepsilon}(-L, 0)} = \overline{\alpha_{(\omega,\kappa)}^{\varepsilon}(z_s, 0)}\alpha_{(\omega,\kappa)}^{\varepsilon}(-L, z_s) + \overline{\beta_{(\omega,\kappa)}^{\varepsilon}(z_s, 0)}\overline{\beta_{(\omega,\kappa)}^{\varepsilon}(-L, z_s)}.$$

Using the expression (15.28) of the usual transmission coefficient, we have

$$\frac{T_{(\omega,\kappa)}^{\varepsilon}(-L, 0)}{T_{(\omega,\kappa)}^{\varepsilon}(-L, z_s)} = \frac{\overline{\alpha_{(\omega,\kappa)}^{\varepsilon}(-L, z_s)}}{\overline{\alpha_{(\omega,\kappa)}^{\varepsilon}(-L, 0)}}$$

$$= \frac{\overline{\alpha_{(\omega,\kappa)}^{\varepsilon}(-L, z_s)}}{\overline{\alpha_{(\omega,\kappa)}^{\varepsilon}(z_s, 0)}\alpha_{(\omega,\kappa)}^{\varepsilon}(-L, z_s) + \beta_{(\omega,\kappa)}^{\varepsilon}(z_s, 0)\overline{\beta_{(\omega,\kappa)}^{\varepsilon}(-L, z_s)}}.$$

Comparing this identity with the definition (15.26) of the generalized transmission, we obtain

$$T_g^{\varepsilon}(\omega, \kappa, z_s) = \frac{T_{(\omega,\kappa)}^{\varepsilon}(-L, 0)}{T_{(\omega,\kappa)}^{\varepsilon}(-L, z_s)}. \tag{15.91}$$

The reflection and transmission coefficients satisfy the nonlinear differential equations (14.59–14.60):

$$\frac{dR^\varepsilon_{(\omega,\kappa)}}{dz} = -\frac{i\omega}{2\bar{c}(\kappa)\varepsilon}\nu_\kappa\left(\frac{z}{\varepsilon^2}\right)\left(e^{\frac{-2i\omega z}{\bar{c}(\kappa)\varepsilon}} - 2R^\varepsilon_{(\omega,\kappa)} + (R^\varepsilon_{(\omega,\kappa)})^2 e^{\frac{2i\omega z}{\bar{c}(\kappa)\varepsilon}}\right),$$

$$\frac{dT^\varepsilon_{(\omega,\kappa)}}{dz} = \frac{i\omega}{2\bar{c}(\kappa)\varepsilon}\nu_\kappa\left(\frac{z}{\varepsilon^2}\right)\left(1 - R^\varepsilon_{(\omega,\kappa)} e^{\frac{2i\omega z}{\bar{c}(\kappa)\varepsilon}}\right)T^\varepsilon_{(\omega,\kappa)},$$

with the initial conditions at $z = -L$ given by $R^\varepsilon_{(\omega,\kappa)}(-L, z = -L) = 0$ and $T^\varepsilon_{(\omega,\kappa)}(-L, z = -L) = 1$. The equation for $T^\varepsilon_{(\omega,\kappa)}$ can be integrated,

$$T^\varepsilon_{(\omega,\kappa)}(-L, z) = \exp\left[\int_{-L}^z \frac{i\omega}{2\bar{c}(\kappa)\varepsilon}\nu_\kappa\left(\frac{z'}{\varepsilon^2}\right)\left(1 - R^\varepsilon_{(\omega,\kappa)}(-L, z')e^{\frac{2i\omega z'}{\bar{c}(\kappa)\varepsilon}}\right)\right],$$

so that we get from (15.91) the integral representation $|T^\varepsilon_g| = \exp(Y^\varepsilon)$ with

$$Y^\varepsilon = \frac{1}{\varepsilon}\int_{z_s}^0 f\left(R^\varepsilon_1(z), R^\varepsilon_2(z), \frac{z}{\varepsilon}, \nu_\kappa\left(\frac{z}{\varepsilon^2}\right)\right)\, dz,$$

$$f(R_1, R_2, \tau, \nu) = \frac{\omega\nu}{2\bar{c}(\kappa)}\left[R_2\cos(2\omega\tau/\bar{c}(\kappa)) + R_1\sin(2\omega\tau/\bar{c}(\kappa))\right],$$

where $R^\varepsilon_1(z) = \text{Re}(R^\varepsilon_{(\omega,\kappa)}(-L, z))$ and $R^\varepsilon_2(z) = \text{Im}(R^\varepsilon_{(\omega,\kappa)}(-L, z))$. The estimates of the exponential moments of Y^ε are based on the perturbed-test-function method. To simplify the presentation we assume that the process ν is a bounded ergodic Markov process whose generator Q satisfies the Fredholm alternative. Since $\nu \mapsto f(R_1, R_2, \tau, \nu)$ has zero mean with respect to the invariant probability measure of ν for any R_1, R_2, τ, there exists a bounded function f_1 such that $Qf_1 = -f$. Let us introduce the periodic function $\tau(z) = z \bmod \pi\bar{c}(\kappa)/\omega$. The vector $(R^\varepsilon_1(z), R^\varepsilon_2(z), \tau(z/\varepsilon), \nu_\kappa(z/\varepsilon^2))$ is a Markov process with a compact state space and generator

$$\mathcal{L}^\varepsilon = \frac{1}{\varepsilon}F_1(R_1, R_2, \tau, \nu)\frac{\partial}{\partial R_1} + \frac{1}{\varepsilon}F_2(R_1, R_2, \tau, \nu)\frac{\partial}{\partial R_2} + \frac{1}{\varepsilon}\frac{\partial}{\partial\tau} + \frac{1}{\varepsilon^2}Q,$$

where

$$F_1(R_1, R_2, \tau, \nu) = \frac{\omega\nu}{2\bar{c}(\kappa)}\left\{-2R_2 + (R_1^2 - R_2^2 - 1)\sin(2\omega\tau/\bar{c}(\kappa))\right.$$
$$\left. + 2R_1 R_2\cos(2\omega\tau/\bar{c}(\kappa))\right\},$$

$$F_2(R_1, R_2, \tau, \nu) = \frac{\omega\nu}{2\bar{c}(\kappa)}\left\{2R_1 + (R_2^2 - R_1^2 - 1)\cos(2\omega\tau/\bar{c}(\kappa))\right.$$
$$\left. + 2R_1 R_2\sin(2\omega\tau/\bar{c}(\kappa))\right\}.$$

The bounded functions $F_{1,2}$ are obtained from the real and imaginary parts of the right-hand side of (14.59) multiplied by ε. The process

$$M^\varepsilon(z) = \varepsilon f_1\left(R_1^\varepsilon(z), R_2^\varepsilon(z), \tau\left(\frac{z}{\varepsilon}\right), \nu_\kappa\left(\frac{z}{\varepsilon^2}\right)\right)$$
$$- \varepsilon f_1\left(R_1^\varepsilon(z_s), R_2^\varepsilon(z_s), \tau\left(\frac{z_s}{\varepsilon}\right), \nu_\kappa\left(\frac{z_s}{\varepsilon^2}\right)\right)$$
$$- \varepsilon\int_{z_s}^z \mathcal{L}^\varepsilon f_1\left(R_1^\varepsilon(z'), R_2^\varepsilon(z'), \tau\left(\frac{z'}{\varepsilon}\right), \nu_\kappa\left(\frac{z'}{\varepsilon^2}\right)\right)\,dz'$$

is a martingale whose quadratic variation

$$\langle M^\varepsilon\rangle_z = \varepsilon^2\int_{z_s}^z \left(\mathcal{L}^\varepsilon f_1^2 - 2f_1\mathcal{L}^\varepsilon f_1\right)\left(R_1^\varepsilon(z'), R_2^\varepsilon(z'), \tau\left(\frac{z'}{\varepsilon}\right), \nu_\kappa\left(\frac{z'}{\varepsilon^2}\right)\right)\,dz'$$

is uniformly bounded with respect to ε by a constant C_M. By decomposing $\mathcal{L}^\varepsilon f_1$ and using $Qf_1 = -f$, we get

$$\frac{1}{\varepsilon}f = -\varepsilon\mathcal{L}^\varepsilon f_1 + F_1\partial_{R_1}f_1 + F_2\partial_{R_2}f_1 + \partial_\tau f_1\,,$$

so that we can write

$$Y^\varepsilon = M^\varepsilon(0) + N^\varepsilon(0)\,,$$

where the term

$$N^\varepsilon(z) =$$
$$\varepsilon f_1\left(R_1^\varepsilon(z_s), R_2^\varepsilon(z_s), \tau\left(\frac{z_s}{\varepsilon}\right), \nu_\kappa\left(\frac{z_s}{\varepsilon^2}\right)\right) - \varepsilon f_1\left(R_1^\varepsilon(z), R_2^\varepsilon(z), \tau\left(\frac{z}{\varepsilon}\right), \nu_\kappa\left(\frac{z}{\varepsilon^2}\right)\right)$$
$$+ \int_{z_s}^z \left(F_1\frac{\partial}{\partial R_1} + F_2\frac{\partial}{\partial R_2} + \frac{\partial}{\partial\tau}\right)f_1\left(R_1^\varepsilon(z'), R_2^\varepsilon(z'), \tau\left(\frac{z'}{\varepsilon}\right), \nu_\kappa\left(\frac{z'}{\varepsilon^2}\right)\right)\,dz'$$

is uniformly bounded with respect to ε by a constant C_N. As a result,

$$\mathbb{E}[|T_g^\varepsilon|^p] = \mathbb{E}[\exp(pY^\varepsilon)] \le \mathbb{E}[\exp(pM^\varepsilon(0))]\exp(C_N p) \le \exp\left(C_M\frac{p^2}{2} + C_N p\right)\,.$$

The last inequality comes from the identity

$$\mathbb{E}\left[\exp\left(pM^\varepsilon(z) - \frac{p^2}{2}\langle M^\varepsilon\rangle_z\right)\right] = 1\,,$$

which implies $\mathbb{E}[\exp(pM^\varepsilon(z))]\exp(-\frac{p^2}{2}C_M) \le 1$.

15.8 Appendix C: Derivation of (15.74)

The goal of this appendix is to derive the simplified expression (15.74) for the refocused field U_{TR}. This derivation is based on the set of hypotheses H_1 to H_4 stated at the beginning of Section 15.5.3, which can be summarized by

$$x_s \gg a\,, \qquad \omega_0 T_w \gg 1\,, \qquad |z_s| \ll L_{\mathrm{loc}}\,.$$

We start from the expressions (15.68–15.69) of the refocusing kernels. We first perform the change of variables $\tau \mapsto \tilde{\tau}$ with the mapping $\tilde{\tau} = t_s - z_s \bar{c}(\kappa)/\bar{c}^2 + \tau$, and we obtain that

$$K^+(\omega, \kappa) = \frac{1}{4\bar{c}\bar{\rho}} \int G_1(\tilde{\tau})\, G_2\left(\mathbf{x}_s + \kappa(t_s - \tilde{\tau})\bar{c}^2\right) \mathcal{W}_g^{(R)}(\omega, \kappa, \tau(\tilde{\tau}))\, d\tilde{\tau},$$

where $\tau(\tilde{\tau})$ is the inverse mapping $\tau(\tilde{\tau}) = \tilde{\tau} - t_s + z_s \bar{c}(\kappa)/\bar{c}^2$, and $K^-(\omega, \kappa)$ is given by the same expression with $\mathcal{W}_g^{(T)}$ instead of $\mathcal{W}_g^{(R)}$. By taking into account the form of the time-window cutoff function $G_1(\tau) = \mathbf{1}_{[T_1, T_2]}(\tau)$ with $T_2 > T_1 > T_0$, the form of the spatial shape of the time-reversal mirror $G_2(\mathbf{x}) = g_2(\mathbf{x}/a)$, and the fact that the reference frame is oriented such that $\mathbf{x}_s = (x_s, 0)$, we get after the change of variable $\tilde{\tau} \mapsto \eta$ with $\eta = x_s/[\bar{c}(\tilde{\tau} - t_s)]$ that

$$K^+(\omega, \kappa) = \frac{x_s}{4\bar{c}^2\bar{\rho}} \int_{\cos(\theta_2)}^{\cos(\theta_1)} g_2\left(\frac{x_s}{a}\left(1 - \frac{\kappa_1 \bar{c}}{\eta}\right), -\frac{x_s}{a}\frac{\bar{c}\kappa_2}{\eta}\right) \mathcal{W}_g^{(R)}(\omega, \kappa, \tau_\kappa(\eta))\, \frac{d\eta}{\eta^2},$$

where $\cos(\theta_j) = x_s/[\bar{c}(T_j - t_s)]$, $j = 1, 2$, and

$$\tau_\kappa(\eta) = \frac{1}{\bar{c}}\left(\frac{x_s}{\eta} + \frac{\bar{c}(\kappa)z_s}{\bar{c}}\right).$$

We next substitute this expression for $K^+(\omega, \kappa)$ and the corresponding one for $K^-(\omega, \kappa)$ into the integral representation (15.67) of U_{TR}. Then we perform the change of variables $\kappa = (\kappa_1, \kappa_2) \mapsto \tilde{\kappa} = (\tilde{\kappa}_1, \tilde{\kappa}_2)$ with $\bar{c}\kappa_1 = \eta - \tilde{\kappa}_1 \eta a/x_s$ and $\bar{c}\kappa_2 = -\tilde{\kappa}_2 \eta a/x_s$. Using the fact that $x_s \gg a$, we write the following expansion:

$$\frac{1}{\bar{c}(\kappa)} = \frac{1}{\bar{c}}\left[\sqrt{1 - \eta^2} + \frac{\eta^2}{\sqrt{1 - \eta^2}}\tilde{\kappa}_1\frac{a}{x_s} + O\left(\frac{a^2}{x_s^2}\right)\right].$$

We thus obtain

$$U_{\mathrm{TR}}(T, \mathbf{X}, Z) = U_{\mathrm{TR},+} + U_{\mathrm{TR},-},$$

with

$$U_{\mathrm{TR},+} = \frac{1}{(2\pi)^3} \int_{\cos(\theta_2)}^{\cos(\theta_1)} \int \left[\frac{\eta}{\sqrt{1 - \eta^2}}\widehat{f_x}(\omega) + \widehat{f_z}(\omega)\right] \mathcal{W}_g^{(R)}\left(\omega, \frac{\eta}{\bar{c}}, \tau(\eta)\right)$$

$$\times \frac{a^2}{4x_s\bar{c}^4\bar{\rho}} \int g_2(\tilde{\kappa})\, e^{i\omega\left(-T + \frac{\eta}{\bar{c}}X + \frac{\sqrt{1-\eta^2}}{\bar{c}}Z\right)} e^{-i\frac{\omega a\eta}{\bar{c}x_s}\left(\tilde{\kappa}\cdot\mathbf{X} - \tilde{\kappa}_1 Z \frac{\eta}{\sqrt{1-\eta^2}}\right)}\, d\tilde{\kappa}\, \omega^2\, d\omega\, d\eta.$$

Here we have used the notation $\mathbf{X} = (X, Y)$, and $\tau(\eta)$ is defined by

$$\tau(\eta) = \frac{1}{\bar{c}}\left(\frac{x_s}{\eta} + \frac{z_s}{\sqrt{1 - \eta^2}}\right). \tag{15.92}$$

The second term, $U_{\mathrm{TR},-}$, is given by a similar expression with $\mathcal{W}_g^{(T)}$ instead of $\mathcal{W}_g^{(R)}$. The time $\tau(\eta)$ can be interpreted as the relative travel time from the source S to the mirror O for a plane wave with the slowness vector $\boldsymbol{\kappa} = (\eta/\bar{c}, 0)$. The term "relative" is due to the fact that the travel times in the density $\mathcal{W}_g^{(R)}$ are measured relatively to the travel time of the coherent front, so the correct interpretation of $\tau(\eta)$ is a difference between two travel times, the first one corresponding to the length of a path that goes from S to O with rays whose cosines are equal to η, the second one corresponding to the length of the direct path that goes from S to the surface with the ray whose cosine is equal to η, as described in Figure 15.17.

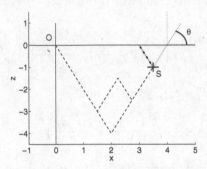

Fig. 15.17. Travel time interpretation of $\tau(\eta)$ given by (15.92). The dashed lines correspond to two paths that go from S to O with rays whose cosines $\cos(\theta)$ are equal to η. These paths have equal lengths l_1. The thick dot-dashed line corresponds to the direct path that goes from S to the surface with the ray whose cosine is equal to η. This path has length l_2. The value $\tau(\eta)$ is the difference between the travel times of these two paths with the velocity $\bar{c}$, that is, $\tau(\eta) = \bar{c}(l_1 - l_2)$.

We now exploit the form $\mathbf{f}(t) = \mathbf{f}_0(t/T_w)e^{-i\omega_0 t} + c.c.$ of the source, and we perform the change of variables $\omega \mapsto \tilde{\omega}$ with $\omega = \omega_0 + \tilde{\omega}/T_w$ and we keep only the leading-order terms in the small parameter $1/(\omega_0 T_w)$:

$$U_{\mathrm{TR},+} = \frac{1}{(2\pi)^3} \int_{\cos(\theta_2)}^{\cos(\theta_1)} \int \left[\frac{\eta}{\sqrt{1-\eta^2}} \overline{\hat{f}_{0x}(\tilde{\omega})} + \overline{\hat{f}_{0z}(\tilde{\omega})} \right] \mathcal{W}_g^{(R)}\left(\omega_0, \frac{\eta}{\bar{c}}, \tau(\eta) \right)$$

$$\times \frac{a^2}{4x_s \bar{c}^4 \bar{\rho}} e^{i\omega_0\left(-T + \frac{\eta}{\bar{c}}X + \frac{\sqrt{1-\eta^2}}{\bar{c}}Z \right)} e^{i\frac{\tilde{\omega}}{T_w}\left(-T + \frac{\eta}{\bar{c}}X + \frac{\sqrt{1-\eta^2}}{\bar{c}}Z \right)}$$

$$\times \int g_2(\tilde{\boldsymbol{\kappa}}) e^{-i\frac{\omega_0 a \eta}{\bar{c}x_s}\left(\tilde{\boldsymbol{\kappa}}\cdot\mathbf{X} - \tilde{\kappa}_1 Z \frac{\eta}{\sqrt{1-\eta^2}} \right)} d\tilde{\boldsymbol{\kappa}}\, \omega_0^2\, d\tilde{\omega}\, d\eta + c.c..$$

By integrating first with respect to $\tilde{\boldsymbol{\kappa}}$, then $\tilde{\omega}$, we obtain

$$
U_{\text{TR},+} = \int_{\cos(\theta_2)}^{\cos(\theta_1)} \left[\frac{\eta}{\sqrt{1-\eta^2}} f_{0x} + f_{0z} \right] \left(-\frac{T}{T_w} + \frac{\eta X + \sqrt{1-\eta^2}Z}{\bar{c}T_w} \right)
$$

$$
\times \, \mathcal{W}_g^{(R)} \left(\omega_0, \frac{\eta}{\bar{c}}, \tau(\eta) \right) \frac{a^2 \omega_0^2}{16\pi^2 x_s \bar{c}^4 \bar{\rho}} e^{i\omega_0 \left(-T + \frac{\eta}{\bar{c}} X + \frac{\sqrt{1-\eta^2}}{\bar{c}} Z \right)}
$$

$$
\times \, \hat{g}_2 \left(\frac{\omega_0 a \eta}{\bar{c} x_s} \left(X - Z \frac{\eta}{\sqrt{1-\eta^2}} \right), \frac{\omega_0 a \eta}{\bar{c} x_s} Y \right) d\eta + c.c.. \tag{15.93}
$$

Finally, we use the hypothesis $|z_s| \ll L_{\text{loc}}$ to simplify further this expression. We consider the approximations (15.70–15.71), and keep only the leading-order terms in $|z_s|/L_{\text{loc}}$. On the one hand, the expression of $U_{\text{TR},-}$ involves the density $\mathcal{W}_g^{(T)}$ evaluated at the positive times $\tau(\eta)$. By (15.71), this term is of order $|z_s|/L_{\text{loc}}$, and thus we get that $U_{\text{TR},-}$ is vanishing. On the other hand, the expression of $U_{\text{TR},+}$ involves the density $\mathcal{W}_g^{(R)}$ evaluated at $\tau(\eta)$, which is of order one by (15.70):

$$
\mathcal{W}_g^{(R)} \left(\omega_0, \frac{\eta}{\bar{c}}, \tau(\eta) \right) = \frac{\bar{c}}{2L_{\text{loc}}\sqrt{1-\eta^2}} \frac{1}{\left(1 + \frac{x_s}{2\sqrt{1-\eta^2}\eta L_{\text{loc}}} \right)^2} + O\left(\frac{|z_s|}{L_{\text{loc}}} \right),
$$

where $L_{\text{loc}} = 4\bar{c}^2/(\gamma\omega_0^2)$. Substituting this expression into (15.93), we finally obtain that the refocused focal spot is given by (15.74).

Notes

Time-reversal ultrasound acoustics has been thoroughly investigated experimentally by M. Fink and his collaborators [55, 57], and by W. Kuperman's group in the context of underwater acoustics [113]. Temporal and spatial refocusing in disordered media has been observed and explained by multipathing effects due to multiple scattering. The spatial refocusing and statistical stability properties have been derived mathematically in the parabolic approximation regime [10, 16, 136, 53], in the radiative transfer regime [11, 52], and in the case of three-dimensional randomly layered media [62] (and in details in this chapter). In general, the statistical stability of the refocusing is ensured by the superposition in the time domain of many approximately uncorrelated frequency components. The superresolution effect is due to multipathing, and in the case of randomly layered media treated in this chapter the paths contributing to the aperture enhancement are clearly identified as paths scattered by the medium below the source.

Application to Echo-Mode Time Reversal

The analysis carried out in the previous chapter can be extended to the case in which a passive scatterer embedded in a random medium is illuminated by a source located at the surface. A time-reversal mirror records the scattered signal, where the scattering results from the interaction of the wave field with the small inhomogeneities and with the embedded scatterer. The mirror sends back the time-reversed signal into the medium. As in the previous chapter, the mirror has a large spatial extent and records a long time segment of the wave field. This experiment gives rise to the richest scattering situation addressed in this book. We will show that the time-reversed wave field focuses tightly on the scatterer. Again we will observe a superresolution effect: The focusing at the scatterer is enhanced by random fluctuations in the medium.

16.1 The Born Approximation for an Embedded Scatterer

The random medium occupying the slab $(-L, 0)$ is again defined by (15.3) and (15.4). Now a point source is located just above the surface at $(0, 0^+)$ and emits at time zero a short pulse. Moreover, a scatterer is buried inside the random medium. We consider acoustic waves described by (15.1) and (15.2) with the external source given as in (15.5),

$$\mathbf{F}^\varepsilon(t, \mathbf{x}, z) = \varepsilon^2 \begin{bmatrix} \mathbf{0} \\ 1 \end{bmatrix} f\left(\frac{t}{\varepsilon}\right) \delta(\mathbf{x}) \delta(z), \qquad (16.1)$$

and scaled so that it will produce a refocused field of order one. Figure 16.1 illustrates the geometry of the configuration: a point source is located at the origin $O = (\mathbf{0}, 0)$, the scatterer position is $S = (\mathbf{x}_s, z_s)$, and the time-reversal mirror M is located in the plane $z = 0$, but not necessarily at the origin O.

We now consider a scatterer embedded at $S = (\mathbf{x}_s, z_s)$. We model this scatterer as a local change in the density of the medium

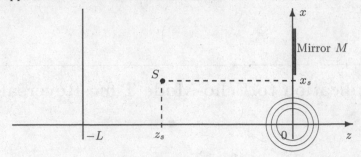

Fig. 16.1. Emission from a point source located at $O = (0,0)$. The scatterer position is $S = (\mathbf{x}_s, z_s)$. The signal is recorded at the mirror M.

$$\rho(\mathbf{x}, z) = \bar{\rho} + \rho_1 \mathbf{1}_B(\mathbf{x}, z),$$

where B is a small domain around S. The system that governs the propagation of the acoustic waves can be written in the form

$$\bar{\rho}\frac{\partial \mathbf{u}^\varepsilon}{\partial t} + \nabla p^\varepsilon = \mathbf{F}^\varepsilon - \rho_1 \mathbf{1}_B(\mathbf{x}, z)\frac{\partial \mathbf{u}^\varepsilon}{\partial t},$$

$$\frac{1}{K(z)}\frac{\partial p^\varepsilon}{\partial t} + \nabla \cdot \mathbf{u}^\varepsilon = 0.$$

We apply the **Born approximation** (or single-scattering approximation) for the modeling of the scattering by the scatterer S [123]: the total field

$$\mathbf{u}^\varepsilon = \mathbf{u}_0^\varepsilon + \mathbf{u}_1^\varepsilon + \mathbf{u}_r^\varepsilon, \qquad p^\varepsilon = p_0^\varepsilon + p_1^\varepsilon + p_r^\varepsilon \tag{16.2}$$

is the superposition of the primary field $(\mathbf{u}_0^\varepsilon, p_0^\varepsilon)$ that solves

$$\bar{\rho}\frac{\partial \mathbf{u}_0^\varepsilon}{\partial t} + \nabla p_0^\varepsilon = \mathbf{F}^\varepsilon,$$

$$\frac{1}{K(z)}\frac{\partial p_0^\varepsilon}{\partial t} + \nabla \cdot \mathbf{u}_0^\varepsilon = 0,$$

and of a secondary field $(\mathbf{u}_1^\varepsilon, p_1^\varepsilon)$ that originates from the emission of a secondary source located at S. The emission of the secondary source is proportional to the primary field at the position of the scatterer:

$$\bar{\rho}\frac{\partial \mathbf{u}_1^\varepsilon}{\partial t} + \nabla p_1^\varepsilon = -\rho_1 \mathbf{1}_B(\mathbf{x}, z)\frac{\partial \mathbf{u}_0^\varepsilon}{\partial t}, \tag{16.3}$$

$$\frac{1}{K(z)}\frac{\partial p_1^\varepsilon}{\partial t} + \nabla \cdot \mathbf{u}_1^\varepsilon = 0. \tag{16.4}$$

In the Born approximation, the reminder $(\mathbf{u}_r^\varepsilon, p_r^\varepsilon)$ is assumed to be negligible (see (16.10) below for conditions ensuring the validity of this approximation). Note that the Born approximation concerns only the interaction of the wave with the scatterer at S. The multiple scattering of the wave with the random medium is taken into account.

16.1.1 Integral Expressions for the Wave Fields

We first consider the primary field $(\mathbf{u}_0^\varepsilon, p_0^\varepsilon)$. The pressure field p_0^ε just below the surface is given by (15.42) with $z = 0$ and $\check{b}^\varepsilon(0^-) = \varepsilon^3 \hat{f}(\omega)/\sqrt{\bar{\zeta}(\kappa)}$ replacing $\check{b}_{TR}^\varepsilon(\omega, \kappa, 0^-)$:

$$p_0^\varepsilon(t, \mathbf{x}, 0^-) = \frac{1}{(2\pi)^3} \int \frac{R_{(\omega,\kappa)}^\varepsilon(-L, 0) - 1}{2} e^{-i\frac{\omega}{\varepsilon}(t - \boldsymbol{\kappa} \cdot \mathbf{x})} \hat{f}(\omega) \omega^2 \, d\omega \, d\boldsymbol{\kappa}.$$

Taking into account the jump condition (15.18), the field just above the surface is

$$p_0^\varepsilon(t, \mathbf{x}) = \frac{1}{(2\pi)^3} \int \frac{R_{(\omega,\kappa)}^\varepsilon(-L, 0) + 1}{2} e^{-i\frac{\omega}{\varepsilon}(t - \boldsymbol{\kappa} \cdot \mathbf{x})} \hat{f}(\omega) \omega^2 \, d\omega \, d\boldsymbol{\kappa}. \tag{16.5}$$

The vertical velocity field is correspondingly given by (15.43):

$$u_0^\varepsilon(t, \mathbf{x}) = \frac{1}{(2\pi)^3} \int \frac{R_{(\omega,\kappa)}^\varepsilon(-L, 0) + 1}{2\bar{\zeta}(\kappa)} e^{-i\frac{\omega}{\varepsilon}(t - \boldsymbol{\kappa} \cdot \mathbf{x})} \hat{f}(\omega) \omega^2 \, d\omega \, d\boldsymbol{\kappa}. \tag{16.6}$$

Similarly, the pressure and vertical velocity fields at the position $(\mathbf{x}_s, z_s)$ inside the medium $(z_s < 0)$ are given by (15.42) and (15.43):

$$p_0^\varepsilon(t, \mathbf{x}_s, z_s) = \frac{1}{(2\pi)^3} \int \left[\frac{R_g^\varepsilon(\omega, \kappa, z_s)}{2} e^{-i\frac{\omega}{\varepsilon}(t - \boldsymbol{\kappa} \cdot \mathbf{x}_s - z_s/\bar{c}(\kappa))} \right. \tag{16.7}$$
$$\left. - \frac{T_g^\varepsilon(\omega, \kappa, z_s)}{2} e^{-i\frac{\omega}{\varepsilon}(t - \boldsymbol{\kappa} \cdot \mathbf{x}_s + z_s/\bar{c}(\kappa))} \right] \hat{f}(\omega) \omega^2 \, d\omega \, d\boldsymbol{\kappa},$$

$$u_0^\varepsilon(t, \mathbf{x}_s, z_s) = \frac{1}{(2\pi)^3} \int \left[\frac{R_g^\varepsilon(\omega, \kappa, z_s)}{2\bar{\zeta}(\kappa)} e^{-i\frac{\omega}{\varepsilon}(t - \boldsymbol{\kappa} \cdot \mathbf{x}_s - z_s/\bar{c}(\kappa))} \right. \tag{16.8}$$
$$\left. + \frac{T_g^\varepsilon(\omega, \kappa, z_s)}{2\bar{\zeta}(\kappa)} e^{-i\frac{\omega}{\varepsilon}(t - \boldsymbol{\kappa} \cdot \mathbf{x}_s + z_s/\bar{c}(\kappa))} \right] \hat{f}(\omega) \omega^2 \, d\omega \, d\boldsymbol{\kappa},$$

$$\mathbf{v}_0^\varepsilon(t, \mathbf{x}_s, z_s) = \frac{1}{(2\pi)^3 \bar{\rho}} \int \boldsymbol{\kappa} \left[\frac{R_g^\varepsilon(\omega, \kappa, z_s)}{2} e^{-i\frac{\omega}{\varepsilon}(t - \boldsymbol{\kappa} \cdot \mathbf{x}_s - z_s/\bar{c}(\kappa))} \right. \tag{16.9}$$
$$\left. - \frac{T_g^\varepsilon(\omega, \kappa, z_s)}{2} e^{-i\frac{\omega}{\varepsilon}(t - \boldsymbol{\kappa} \cdot \mathbf{x}_s + z_s/\bar{c}(\kappa))} \right] \hat{f}(\omega) \omega^2 \, d\omega \, d\boldsymbol{\kappa}.$$

We next consider the secondary field and take a Fourier transform with respect to the time and the transverse spatial variables in (16.3–16.4) to obtain

$$-\bar{\rho}\frac{i\omega}{\varepsilon}\hat{\mathbf{v}}_1^\varepsilon + i\frac{\omega}{\varepsilon}\boldsymbol{\kappa}\hat{p}_1^\varepsilon = \rho_1 \frac{i\omega}{\varepsilon} \int \mathbf{v}_0^\varepsilon(t, \mathbf{x}, z) \mathbf{1}_B(\mathbf{x}, z) e^{i\frac{\omega}{\varepsilon}(t - \boldsymbol{\kappa} \cdot \mathbf{x})} \, dt \, d\mathbf{x},$$

$$-\bar{\rho}\frac{i\omega}{\varepsilon}\hat{u}_1^\varepsilon + \frac{\partial \hat{p}_1^\varepsilon}{\partial z} = \rho_1 \frac{i\omega}{\varepsilon} \int u_0^\varepsilon(t, \mathbf{x}, z) \mathbf{1}_B(\mathbf{x}, z) e^{i\frac{\omega}{\varepsilon}(t - \boldsymbol{\kappa} \cdot \mathbf{x})} \, dt \, d\mathbf{x},$$

$$-\frac{1}{K(z)}\frac{i\omega}{\varepsilon}\hat{p}_1^\varepsilon + i\frac{\omega}{\varepsilon}\boldsymbol{\kappa} \cdot \hat{\mathbf{v}}_1^\varepsilon + \frac{\partial \hat{u}_1^\varepsilon}{\partial z} = 0,$$

where $\mathbf{u}_1^\varepsilon = (\mathbf{v}_1^\varepsilon, u_1^\varepsilon)$ and $\mathbf{u}_0^\varepsilon = (\mathbf{v}_0^\varepsilon, u_0^\varepsilon)$. We assume that the scattering region B is smaller than the wavelength, and we model it by a point scatterer with **scattering volume** $\varepsilon^3 \sigma_s$, with σ_s small, so that $\rho_1 \mathrm{Vol}(B) = \bar\rho \sigma_s \varepsilon^3$ and

$$\rho_1 \mathbf{1}_B(\mathbf{x}, z) = \varepsilon^3 \bar\rho \sigma_s \delta(\mathbf{x} - \mathbf{x}_s) \delta(z - z_s) \,.$$

Moreover, the fact that the parameter σ_s is small ensures the validity of the Born approximation, in the sense that in the expansion (16.2) we have

$$(\mathbf{u}_0^\varepsilon, p_0^\varepsilon) \sim \varepsilon^2 \,, \quad (\mathbf{u}_1^\varepsilon, p_1^\varepsilon) \sim \varepsilon^2 \sigma_s \,, \quad (\mathbf{u}_r^\varepsilon, p_r^\varepsilon) \sim \varepsilon^2 \sigma_s^2 \,. \tag{16.10}$$

This result is not obvious, but it will be explained in the next sections, in particular in Propositions 16.1 and 16.2.

Therefore, the secondary field $(\mathbf{u}_1^\varepsilon, p_1^\varepsilon)$ solves

$$-\bar\rho \frac{i\omega}{\varepsilon} \hat{\mathbf{v}}_1^\varepsilon + i\frac{\omega}{\varepsilon} \boldsymbol{\kappa} \hat{p}_1^\varepsilon = \varepsilon^3 \mathbf{S}_{1,\mathbf{x}}^\varepsilon(\omega) e^{-i\frac{\omega}{\varepsilon} \boldsymbol{\kappa} \cdot \mathbf{x}_s} \delta(z - z_s) \,, \tag{16.11}$$

$$-\bar\rho \frac{i\omega}{\varepsilon} \hat{u}_1^\varepsilon + \frac{\partial \hat{p}_1^\varepsilon}{\partial z} = \varepsilon^3 S_{1,z}^\varepsilon(\omega) e^{-i\frac{\omega}{\varepsilon} \boldsymbol{\kappa} \cdot \mathbf{x}_s} \delta(z - z_s) \,, \tag{16.12}$$

$$-\frac{1}{K(z)} \frac{i\omega}{\varepsilon} \hat{p}_1^\varepsilon + i\frac{\omega}{\varepsilon} \boldsymbol{\kappa} \cdot \hat{\mathbf{v}}_1^\varepsilon + \frac{\partial \hat{u}_1^\varepsilon}{\partial z} = 0 \,,$$

with the secondary source terms given by

$$\mathbf{S}_{1,\mathbf{x}}^\varepsilon(\omega) = \frac{i\sigma_s}{(2\pi)^2} \int \frac{\boldsymbol{\kappa}'}{2} \left[R_g^\varepsilon(\omega, \kappa', z_s) e^{i\frac{\omega}{\varepsilon}(\boldsymbol{\kappa}' \cdot \mathbf{x}_s + z_s / \bar{c}(\kappa'))} \right. \tag{16.13}$$

$$\left. - T_g^\varepsilon(\omega, \kappa', z_s) e^{i\frac{\omega}{\varepsilon}(\boldsymbol{\kappa}' \cdot \mathbf{x}_s - z_s / \bar{c}(\kappa'))} \right] \hat{f}(\omega) \omega^3 \, d\kappa' \,,$$

$$S_{1,z}^\varepsilon(\omega) = \frac{i\sigma_s}{(2\pi)^2} \int \frac{\bar\rho}{2\bar\zeta(\kappa')} \left[R_g^\varepsilon(\omega, \kappa', z_s) e^{i\frac{\omega}{\varepsilon}(\boldsymbol{\kappa}' \cdot \mathbf{x}_s + z_s / \bar{c}(\kappa'))} \right. \tag{16.14}$$

$$\left. + T_g^\varepsilon(\omega, \kappa', z_s) e^{i\frac{\omega}{\varepsilon}(\boldsymbol{\kappa}' \cdot \mathbf{x}_s - z_s / \bar{c}(\kappa'))} \right] \hat{f}(\omega) \omega^3 \, d\kappa' \,.$$

Note that these source terms correspond to the emission from a point source similar to the embedded source problem (15.6–15.8) addressed in Section 15.2.

Let us consider the secondary pressure field p_1^ε at the observation point $M = (\mathbf{x}_m, 0)$ as illustrated in Figure 16.1 and representing a point mirror, say. From the analysis carried out in Section 15.2, we get

$$p_1^\varepsilon(t, \mathbf{x}_m, 0) = \frac{1}{(2\pi\varepsilon)^3} \int \frac{\sqrt{\bar\zeta(\kappa)}}{2} \breve{a}_1^\varepsilon(\omega, \boldsymbol{\kappa}, 0) e^{-i\frac{\omega}{\varepsilon}(t - \boldsymbol{\kappa} \cdot \mathbf{x}_m)} \omega^2 \, d\omega \, d\boldsymbol{\kappa} \,,$$

with

$$\breve{a}_1^\varepsilon(\omega, \boldsymbol{\kappa}, 0) = \varepsilon^3 e^{i\frac{\omega}{\varepsilon}(-\boldsymbol{\kappa} \cdot \mathbf{x}_s)} \left[e^{-i\frac{\omega z_s}{\varepsilon \bar{c}(\kappa)}} T_g^\varepsilon(\omega, \boldsymbol{\kappa}, z_s) S_a^\varepsilon(\omega, \boldsymbol{\kappa}) \right.$$

$$\left. - e^{i\frac{\omega z_s}{\varepsilon \bar{c}(\kappa)}} R_g^\varepsilon(\omega, \boldsymbol{\kappa}, z_s) S_b^\varepsilon(\omega, \boldsymbol{\kappa}) \right] \,, \tag{16.15}$$

$$S_a^\varepsilon(\omega, \boldsymbol{\kappa}) = \frac{\sqrt{\bar{\zeta}(\kappa)}}{\bar{\rho}} \boldsymbol{\kappa} \cdot \mathbf{S}_{1,\mathbf{x}}^\varepsilon(\omega) + \frac{1}{\sqrt{\bar{\zeta}(\kappa)}} S_{1,z}^\varepsilon(\omega) \,,$$

$$S_b^\varepsilon(\omega, \boldsymbol{\kappa}) = \frac{\sqrt{\bar{\zeta}(\kappa)}}{\bar{\rho}} \boldsymbol{\kappa} \cdot \mathbf{S}_{1,\mathbf{x}}^\varepsilon(\omega) - \frac{1}{\sqrt{\bar{\zeta}(\kappa)}} S_{1,z}^\varepsilon(\omega) \,.$$

Thus, p_1^ε consists of four terms. The first one is

$$p_{1,I}^\varepsilon(t_m + \varepsilon\sigma, \mathbf{x}_m, 0) = \frac{\sigma_s}{(2\pi)^5} \frac{i}{4} \int T_g^\varepsilon(\omega, \kappa, z_s) T_g^\varepsilon(\omega, \kappa', z_s)$$

$$\times \left(-\frac{\bar{\zeta}(\kappa)\boldsymbol{\kappa} \cdot \boldsymbol{\kappa}'}{\bar{\rho}} + \frac{\bar{\rho}}{\bar{\zeta}(\kappa')} \right) e^{-i\omega\sigma} e^{i\frac{\omega}{\varepsilon}\phi_I(\boldsymbol{\kappa},\boldsymbol{\kappa}')} \hat{f}(\omega)\omega^5 \, d\omega \, d\boldsymbol{\kappa} \, d\boldsymbol{\kappa}' \,, \quad (16.16)$$

where the rapid phase is

$$\phi_I(\boldsymbol{\kappa}, \boldsymbol{\kappa}') = -t_m + \boldsymbol{\kappa} \cdot \mathbf{x}_m - z_s/\bar{c}(\kappa) - \boldsymbol{\kappa} \cdot \mathbf{x}_s + \boldsymbol{\kappa}' \cdot \mathbf{x}_s - z_s/\bar{c}(\kappa') \,.$$

The three other terms $p_{1,II}^\varepsilon$, $p_{1,III}^\varepsilon$, and $p_{1,IV}^\varepsilon$ have similar expressions but with crossed products $R_g^\varepsilon T_g^\varepsilon$, $T_g^\varepsilon R_g^\varepsilon$, and $R_g^\varepsilon R_g^\varepsilon$, respectively, and rapid phases with different signs in front of $z_s/\bar{c}(\kappa)$ and $z_s/\bar{c}(\kappa')$.

16.2 Asymptotic Theory for the Scattered Field

The field that can be observed at the surface consists of the superposition of the primary field (with subscript 0) and the secondary field (with subscript 1).

16.2.1 The Primary Field

We first look at the primary field in (16.5–16.6). This field has two components:

1. A deterministic component, which can be estimated in the limit $\varepsilon \to 0$ by use of the Weyl representation of a spherical wave [120, Section 3.2.4], which we discuss in Chapter 2.
2. A random component that involves the reflection coefficient $R_{(\omega,\kappa)}^\varepsilon(-L, 0)$.

Concerning the coherent field, we get the following proposition by integrating the expectation of (16.5) and (16.6), which shows that the coherent waves have an amplitude of order ε^2 at the surface.

Proposition 16.1. *Let $t_m \in \mathbb{R}$ be an observation time and $M = (\mathbf{x}_m, 0)$ an observation point at the surface $(\mathbf{x}_m \neq \mathbf{0})$. The rescaled mean signal detected at M converges as*

$$\frac{\mathbb{E}[p_0^\varepsilon(t_m + \varepsilon\sigma, \mathbf{x}_m, 0)]}{\varepsilon^2} \xrightarrow{\varepsilon \to 0} 0, \tag{16.17}$$

$$\frac{\mathbb{E}[\mathbf{v}_0^\varepsilon(t_m + \varepsilon\sigma, \mathbf{x}_m, 0)]}{\varepsilon^2} \xrightarrow{\varepsilon \to 0} \mathbf{0}, \tag{16.18}$$

$$\frac{\mathbb{E}[u_0^\varepsilon(t_m + \varepsilon\sigma, \mathbf{x}_m, 0)]}{\varepsilon^2} \xrightarrow{\varepsilon \to 0} \begin{cases} -\dfrac{1}{4\pi\bar{\rho}\bar{c}|\mathbf{x}_m|^2} f(\sigma) & \text{if } \bar{c}t_m = |\mathbf{x}_m|, \\ 0 & \text{otherwise.} \end{cases} \tag{16.19}$$

The result stated in Proposition 16.1 means that the coherent wave has amplitude of order ε^2 and it is very particular to the vertical source case. For a general source emitting in the three spatial dimensions, we would observe a coherent signal of order ε at the surface.

The particular scaling we used for the source (16.1) means that the incoherent waves observed at the surface have an amplitude of order $\varepsilon^{3/2}$. The evaluation of the order of magnitude of the incoherent primary waves follows from the analysis carried out in Section 14.3, where the point source (14.55) has the amplitude factor $\varepsilon^{1/2}$ and the intensity of the incoherent waves is of order one. Here the source (16.1) has the amplitude factor ε^2, so the incoherent primary waves have amplitude of order $\varepsilon^{3/2}$. By comparing the typical amplitude of the incoherent waves with that of the coherent waves described in the previous proposition, we can then conclude that the primary field observed at the surface is dominated by the incoherent wave fluctuations of order $\varepsilon^{3/2}$.

16.2.2 The Secondary Field

We now address the secondary field and consider first the term $p_{1,I}^\varepsilon$ given by (16.16). It is convenient to use polar coordinates, and we parameterize $\mathbf{x}_s = |\mathbf{x}_s|\mathbf{e}_{\theta_s}$, $\mathbf{x}_m = \mathbf{x}_s + |\mathbf{x}_m - \mathbf{x}_s|\mathbf{e}_{\bar{\theta}}$, $\boldsymbol{\kappa} = \mu\mathbf{e}_\theta$, and $\boldsymbol{\kappa}' = \mu'\mathbf{e}_{\theta'}$. We apply a stationary-phase argument similar to the one used in Chapter 14. We find that there exists a unique stationary point given by

$$\mu_c' = \frac{1}{\bar{c}}\frac{|\mathbf{x}_s|}{\sqrt{|\mathbf{x}_s|^2 + z_s^2}}, \qquad \mu_c = \frac{1}{\bar{c}}\frac{|\mathbf{x}_m - \mathbf{x}_s|}{\sqrt{|\mathbf{x}_m - \mathbf{x}_s|^2 + z_s^2}}, \qquad \theta_c' = \theta_s, \qquad \theta_c = \bar{\theta}.$$

The point $\boldsymbol{\kappa}_c' = (\mu_c', \theta_c')$ corresponds to the direction of the ray going from the source O to the scatterer S, while the point $\boldsymbol{\kappa}_c = (\mu_c, \theta_c)$ corresponds to the direction of the ray going from the source S to the scatterer M. By Proposition 14.4, the limit value of the integral (16.16) is of order $o(\varepsilon^2)$ if $\phi_I(\boldsymbol{\kappa}_c, \boldsymbol{\kappa}_c') \neq 0$, and it is of leading order ε^2 if $\phi_I(\boldsymbol{\kappa}_c, \boldsymbol{\kappa}_c') = 0$, that is, if $t_m = t_c$ with

$$t_c = \frac{1}{\bar{c}}\left(\sqrt{|\mathbf{x}_s|^2 + z_s^2} + \sqrt{|\mathbf{x}_m - \mathbf{x}_s|^2 + z_s^2}\right) = \frac{1}{\bar{c}}\left(|OS| + |SM|\right). \tag{16.20}$$

We then find that to leading order in ε,

$$p^\varepsilon_{1,I}(t_m + \varepsilon\sigma, \mathbf{x}_m, 0) = \frac{\varepsilon^2 \sigma_s}{(2\pi)^3} \frac{i}{4} \frac{|z_s|}{\bar{c}^3} \frac{\mathbf{OS} \cdot \mathbf{SM}}{|OS|^3 |SM|^2} \tag{16.21}$$

$$\times \int T^\varepsilon_g(\omega, \mu_c, z_s) T^\varepsilon_g(\omega, \mu'_c, z_s) e^{-i\omega\sigma} \hat{f}(\omega) \omega^3 \, d\omega \,.$$

We also find that the three other terms $p_{1,II}$, $p_{1,III}$, and $p_{1,IV}$ do not have such a stationary point, so that they bring a contribution to the value of p_1 that is at least of order $\sqrt{\varepsilon}$ lower than $p_{1,I}$.

We finally use the expansion (15.32) of the generalized transmission coefficient and we apply a moment analysis similar as the one carried out in Section 14.2.1 to obtain the statistical limit of the product $T^\varepsilon_g(\omega, \mu_c, z_s) T^\varepsilon_g(\omega, \mu'_c, z_s)$. This allows us to state the following proposition, which describes the structure of the secondary field observed at the surface.

Proposition 16.2.
(a) If $\bar{c}t_m \neq |OS| + |SM|$, then the rescaled pressure field $p^\varepsilon_1(t_m + \varepsilon\cdot, \mathbf{x}_m, 0)/\varepsilon^2$ at the observation point $M = (\mathbf{x}_m, 0)$ converges to 0.
(b) If $\bar{c}t_m = |OS| + |SM|$, then the rescaled pressure field converges in distribution to a random function

$$\frac{p^\varepsilon_1(t_m + \varepsilon\sigma, \mathbf{x}_m, 0)}{\varepsilon^2} \xrightarrow{\varepsilon \to 0} \frac{\sigma_s}{(2\pi)^3} \frac{i}{4} \frac{|z_s|}{\bar{c}^3} \frac{\mathbf{OS} \cdot \mathbf{SM}}{|OS|^3 |SM|^2} \tag{16.22}$$

$$\times \int \hat{f}(\omega) \exp\left[i\omega\,(T_s - \sigma) - \omega^2 \left(\frac{\gamma}{8\bar{c}^2} \frac{|OS|^2 + |SM|^2}{|z_s|} \right) \right] \omega^3 \, d\omega \,,$$

where T_s is a random time delay

$$T_s \overset{\triangle}{=} \frac{1}{\bar{c}} \frac{\sqrt{\gamma}}{2} \left(\frac{|OS| + |SM|}{|z_s|} \right) W_0(z_s), \tag{16.23}$$

and W_0 is a standard Brownian motion.

The result stated in Proposition 16.2 holds true only if M is not on the surface ring with center $\mathbf{x}_s$ and passing through O. In that case, the two random travel times from O to S and from S to M are perfectly correlated, and the global random travel time is $T_s = \frac{1}{\bar{c}}\sqrt{2\gamma}|OS|/|z_s|W_0(z_s)$.

This proposition means that the secondary field at the observation point has a deterministic shape given by the inverse Fourier transform of

$$\frac{\varepsilon^2 \sigma_s}{(2\pi)^2} \frac{i\omega^3}{4} \frac{|z_s|}{\bar{c}^3} \left(\frac{\mathbf{OS} \cdot \mathbf{SM}}{|OS|^3 |SM|^2} \right) \hat{f}(\omega) \exp\left[-\frac{\gamma\omega^2}{8\bar{c}^2} \left(\frac{|OS|^2 + |SM|^2}{|z_s|} \right) \right] \,.$$

This deterministic shape is the convolution of the original pulse shape of the source with a deterministic kernel. The field has also a random center, which is given by

$$T_1 = t_c + \varepsilon T_s = \frac{1}{\bar{c}} \left(|OS| + |MS| \right) \left(1 + \varepsilon \frac{\sqrt{\gamma}}{2} \frac{1}{|z_s|} W_0(z_s) \right) \,. \tag{16.24}$$

Concerning the vertical velocity field, we get a similar result. This shows that the primary and secondary fields have coherent components whose amplitudes are of order ε^2. However, these coherent components are buried in the incoherent primary waves, whose typical amplitude is of order $\varepsilon^{3/2}$. To be complete, we can add that the secondary incoherent field is even smaller, of order $\varepsilon^{5/2}$. However, we will see in the next section that the secondary incoherent waves participate to leading order in the refocusing of the time-reversed wave field at the location of the scatterer S.

16.3 Time Reversal of the Recorded Wave

16.3.1 Integral Representation of the Time-Reversed Field

We assume that we record the velocity signal at the surface, at the mirror M. This second step of the time-reversal procedure is implemented as described in Section 15.2.2, where we discussed the situation with an internal source. The only difference will be that we **amplify** the recorded signal before reemission. In this chapter the internal source has been replaced by the waves being reflected by an internal scatterer. We next show how the superresolution phenomenon observed in the previous chapter generalizes to the current configuration. Recall that the first time-reversal step consists in recording the velocity signal and/or the pressure signal at the surface $z = 0$ on the mirror $M = \{(\mathbf{x}, z), \mathbf{x} \in D, z = 0\}$ during some time interval centered at $t = 0$. The shape of the mirror is given by $D \subset \mathbb{R}^2$. We record the signal during a large time interval whose duration is of order one. We consider again the situation in which only the velocity is recorded.

In the second step of the time-reversal procedure we clip a piece of the recorded signal by a cutoff function $t \mapsto G_1(t)$, where the support of G_1 is included in $[-t_1/2, t_1/2]$, with $t_1 > 0$. The recorded part of the wave is denoted by $\mathbf{u}^\varepsilon_{\text{rec}}$ and is

$$\mathbf{u}^\varepsilon_{\text{rec}}(t, \mathbf{x}) = \mathbf{u}^\varepsilon(t, \mathbf{x}) G_1(t) G_2(\mathbf{x}), \qquad (16.25)$$

with $\mathbf{u}^\varepsilon$ being the total velocity field and where G_2 is the spatial cutoff function introduced by the mirror, whose support is in the domain D. We then time-reverse this piece of the signal and send it back into the same medium as illustrated in Figure 15.5. We therefore consider a new problem defined by the acoustic wave equations (15.1–15.2) with the new source term

$$\mathbf{F}^\varepsilon_{\text{TR}}(t, \mathbf{x}, z) = \frac{\bar{\rho}\bar{c}\,\mathbf{u}^\varepsilon_{\text{rec}}(-t, \mathbf{x})\delta(z)}{\varepsilon}, \qquad (16.26)$$

where the factor $\bar{\rho}\bar{c}$ has been added to restore the physical dimension of the expression. The amplification factor $1/\varepsilon$ has been introduced so that the refocused field on the scatterer will be of order one. Note that the recorded signals have amplitudes of order $\varepsilon^{3/2}$, so the reemitted signals have amplitudes of order $\varepsilon^{1/2}$. Apart from the amplification factor, the mirror is implemented exactly as in Section 15.2.2.

The wave field in the second part of the time-reversal experiment can be decomposed and described by:

(1) *The reemitted primary field*, which is not scattered by S during the back-propagation.

(2) *The reemitted secondary field*. In order to be consistent with the Born approximation, we neglect the scattering of this field at S, since it produces a field of order σ_s^2.

(3) *The new scattered field*, which is the reemitted primary field that is scattered by S during the back-propagation.

As we will show, the component (2) will refocus at the scatterer S and give the leading contribution to the wave field there. We first study the component (2) of $u_{\mathrm{TR}}^\varepsilon$. We will discuss the other wave components (1) and (3) in Section 16.3.4.

We now find by comparing (16.11) and (16.12) with (15.6) and (15.7) respectively and taking the amplification factor into account that the time-reversed signal $u_{\mathrm{TR}}^\varepsilon$ is described by (15.44) under the replacements

$$t_s \mapsto 0, \quad \hat{\mathbf{f}}_{\mathbf{x}}(\omega) \mapsto \mathbf{S}_{1,\mathbf{x}}^\varepsilon/\varepsilon, \quad \hat{f}_z(\omega) \mapsto S_{1,z}^\varepsilon(\omega)/\varepsilon.$$

The time-reversed vertical velocity can therefore be described by

$$u_{\mathrm{TR}}^\varepsilon(t,\mathbf{x},z) = \frac{1}{(2\pi)^6\varepsilon^4} \int\int \frac{H_0(\boldsymbol{\kappa}_1,\boldsymbol{\kappa}_2)}{4\sqrt{\zeta(\boldsymbol{\kappa}_1)}} \overline{\hat{G}_1\left(\frac{\omega_1-\omega_2}{\varepsilon}\right)} \hat{G}_2\left(\frac{\omega_1\boldsymbol{\kappa}_1+\omega_2\boldsymbol{\kappa}_2}{\varepsilon}\right)$$

$$\times e^{i\frac{-\omega_1 t + \omega_2 \boldsymbol{\kappa}_2 \cdot \mathbf{x}_s + \omega_1 \boldsymbol{\kappa}_1 \cdot \mathbf{x}}{\varepsilon}} \left[\sum_{j=1}^{4} P_j^\varepsilon\right] \omega_1^2\omega_2^2 \, d\omega_1 \, d\boldsymbol{\kappa}_1 \, d\omega_2 \, d\boldsymbol{\kappa}_2, \qquad (16.27)$$

where we define the P_j^ε's by

$$P_1^\varepsilon = -e^{i\left(-\frac{\omega_2 z_s}{\varepsilon\bar{c}(\boldsymbol{\kappa}_2)}+\frac{\omega_1 z}{\varepsilon\bar{c}(\boldsymbol{\kappa}_1)}\right)} \overline{R_g^\varepsilon(\omega_2,\boldsymbol{\kappa}_2,z_s)} R_g^\varepsilon(\omega_1,\boldsymbol{\kappa}_1,z) \overline{S_b^\varepsilon(\omega_2,\boldsymbol{\kappa}_2)},$$

$$P_2^\varepsilon = e^{i\left(\frac{\omega_2 z_s}{\varepsilon\bar{c}(\boldsymbol{\kappa}_2)}+\frac{\omega_1 z}{\varepsilon\bar{c}(\boldsymbol{\kappa}_1)}\right)} \overline{T_g^\varepsilon(\omega_2,\boldsymbol{\kappa}_2,z_s)} R_g^\varepsilon(\omega_1,\boldsymbol{\kappa}_1,z) \overline{S_a^\varepsilon(\omega_2,\boldsymbol{\kappa}_2)},$$

$$P_3^\varepsilon = e^{i\left(\frac{\omega_2 z_s}{\varepsilon\bar{c}(\boldsymbol{\kappa}_2)}-\frac{\omega_1 z}{\varepsilon\bar{c}(\boldsymbol{\kappa}_1)}\right)} \overline{T_g^\varepsilon(\omega_2,\boldsymbol{\kappa}_2,z_s)} T_g^\varepsilon(\omega_1,\boldsymbol{\kappa}_1,z) \overline{S_a^\varepsilon(\omega_2,\boldsymbol{\kappa}_2)},$$

$$P_4^\varepsilon = -e^{i\left(-\frac{\omega_2 z_s}{\varepsilon\bar{c}(\boldsymbol{\kappa}_2)}-\frac{\omega_1 z}{\varepsilon\bar{c}(\boldsymbol{\kappa}_1)}\right)} \overline{R_g^\varepsilon(\omega_2,\boldsymbol{\kappa}_2,z_s)} T_g^\varepsilon(\omega_1,\boldsymbol{\kappa}_1,z) \overline{S_b^\varepsilon(\omega_2,\boldsymbol{\kappa}_2)},$$

and where we recall that

$$S_a^\varepsilon(\omega,\boldsymbol{\kappa}) = \frac{\sqrt{\zeta(\boldsymbol{\kappa})}}{\bar{\rho}}\boldsymbol{\kappa}\cdot\mathbf{S}_{1,\mathbf{x}}^\varepsilon(\omega) + \frac{1}{\sqrt{\zeta(\boldsymbol{\kappa})}}S_{1,z}^\varepsilon(\omega), \qquad (16.28)$$

$$S_b^\varepsilon(\omega,\boldsymbol{\kappa}) = \frac{\sqrt{\zeta(\boldsymbol{\kappa})}}{\bar{\rho}}\boldsymbol{\kappa}\cdot\mathbf{S}_{1,\mathbf{x}}^\varepsilon(\omega) - \frac{1}{\sqrt{\zeta(\boldsymbol{\kappa})}}S_{1,z}^\varepsilon(\omega), \qquad (16.29)$$

$$\mathbf{S}_{1,\mathbf{x}}^\varepsilon(\omega) = \frac{i\sigma_s}{(2\pi)^2} \int \frac{\boldsymbol{\kappa}'}{2} \left[R_g^\varepsilon(\omega, \boldsymbol{\kappa}', z_s) e^{i\frac{\omega}{\varepsilon}(\boldsymbol{\kappa}'\cdot\mathbf{x}_s + z_s/\bar{c}(\boldsymbol{\kappa}'))} \right.$$

$$\left. - T_g^\varepsilon(\omega, \boldsymbol{\kappa}', z_s) e^{i\frac{\omega}{\varepsilon}(\boldsymbol{\kappa}'\cdot\mathbf{x}_s - z_s/\bar{c}(\boldsymbol{\kappa}'))} \right] \hat{f}(\omega)\omega^3 \, d\boldsymbol{\kappa}',$$

$$S_{1,z}^\varepsilon(\omega) = \frac{i\sigma_s}{(2\pi)^2} \int \frac{\bar{\rho}}{2\bar{\zeta}(\boldsymbol{\kappa}')} \left[R_g^\varepsilon(\omega, \boldsymbol{\kappa}', z_s) e^{i\frac{\omega}{\varepsilon}(\boldsymbol{\kappa}'\cdot\mathbf{x}_s + z_s/\bar{c}(\boldsymbol{\kappa}'))} \right.$$

$$\left. + T_g^\varepsilon(\omega, \boldsymbol{\kappa}', z_s) e^{i\frac{\omega}{\varepsilon}(\boldsymbol{\kappa}'\cdot\mathbf{x}_s - z_s/\bar{c}(\boldsymbol{\kappa}'))} \right] \hat{f}(\omega)\omega^3 \, d\boldsymbol{\kappa}'.$$

16.3.2 Refocusing in the Homogeneous Case

We first consider the homogeneous medium situation with $T_g^\varepsilon \equiv 1$ and $R_g^\varepsilon \equiv 0$. The refocusing of the primary field has been studied in Chapter 15. Here we concentrate our attention on the refocusing of the secondary field. Then (16.27) gives

$$u_{\mathrm{TR}}^\varepsilon(t,\mathbf{x},z) = \frac{1}{\varepsilon(2\pi)^6} \int \int \frac{H_0(\boldsymbol{\kappa}, -\boldsymbol{\kappa})}{4\sqrt{\bar{\zeta}(\boldsymbol{\kappa})}} \overline{\hat{G}_1(h)} \hat{G}_2 \left(h\boldsymbol{\kappa} + \omega\boldsymbol{\lambda} \right)$$

$$\times e^{\frac{i\omega}{\varepsilon}\left(\frac{z_s - z}{\bar{c}(\boldsymbol{\kappa})}\right)} e^{-\frac{ih}{2}\frac{z_s + z}{\bar{c}(\boldsymbol{\kappa})} + \frac{i\omega}{2}\boldsymbol{\lambda}\cdot\boldsymbol{\kappa}\bar{c}(\boldsymbol{\kappa})(z_s + z)} \frac{(-i\sigma_s)}{(2\pi)^2} \int e^{-i\frac{\omega}{\varepsilon}(\boldsymbol{\kappa}'\cdot\mathbf{x}_s - z_s/\bar{c}(\boldsymbol{\kappa}'))} \overline{\hat{f}(\omega)} \omega^3$$

$$\times \left(\frac{\sqrt{\bar{\zeta}(\boldsymbol{\kappa})}}{2\bar{\rho}} \boldsymbol{\kappa}\cdot\boldsymbol{\kappa}' + \frac{1}{\sqrt{\bar{\zeta}(\boldsymbol{\kappa})}} \left(\frac{\bar{\rho}}{2\bar{\zeta}(\boldsymbol{\kappa}')} \right) \right) e^{i\frac{h}{2}(\boldsymbol{\kappa}'\cdot\mathbf{x}_s - z_s/\bar{c}(\boldsymbol{\kappa}'))} \, d\boldsymbol{\kappa}'$$

$$\times e^{\frac{i\omega}{\varepsilon}(-t + \boldsymbol{\kappa}\cdot(\mathbf{x}-\mathbf{x}_s))} e^{\frac{ih}{2}(-t + \boldsymbol{\kappa}\cdot(\mathbf{x}+\mathbf{x}_s)) + \frac{i\omega}{2}\boldsymbol{\lambda}\cdot(\mathbf{x}+\mathbf{x}_s)} \omega^4 \, d\omega \, dh \, d\boldsymbol{\kappa} \, d\boldsymbol{\lambda}.$$

We now apply the stationary-phase method and find that the fast-phase component involving $\boldsymbol{\kappa}'$ gives the unique stationary-phase slowness vector

$$\boldsymbol{\kappa}_c' = \frac{\mathbf{x}_s}{\bar{c}\sqrt{\mathbf{x}_s^2 + z_s^2}},$$

which leads to the approximation

$$u_{\mathrm{TR}}^\varepsilon(t,\mathbf{x},z) = \frac{1}{(2\pi)^6} \int \int \frac{H_0(\boldsymbol{\kappa}, -\boldsymbol{\kappa})}{4\sqrt{\bar{\zeta}(\boldsymbol{\kappa})}} \overline{\hat{G}_1(h)} \hat{G}_2 \left(h\boldsymbol{\kappa} + \omega\boldsymbol{\lambda} \right) \qquad (16.30)$$

$$\times e^{\frac{i\omega}{\varepsilon}\left(\frac{z_s - z}{\bar{c}(\boldsymbol{\kappa})}\right)} e^{-\frac{ih}{2}\frac{z_s + z}{\bar{c}(\boldsymbol{\kappa})} + \frac{i\omega}{2}\boldsymbol{\lambda}\cdot\boldsymbol{\kappa}\bar{c}(\boldsymbol{\kappa})(z_s + z)}$$

$$\times \left(-\frac{\sqrt{\bar{\zeta}(\boldsymbol{\kappa})}}{\bar{\rho}} \boldsymbol{\kappa}\cdot\overline{\hat{\mathbf{f}}_{2,\mathbf{x}}(\omega)} + \frac{1}{\sqrt{\bar{\zeta}(\boldsymbol{\kappa})}} \overline{\hat{f}_{2,z}(\omega)} \right)$$

$$\times e^{\frac{i\omega}{\varepsilon}(-(OS/\bar{c}+t)+\boldsymbol{\kappa}\cdot(\mathbf{x}-\mathbf{x}_s))} e^{\frac{ih}{2}(OS/\bar{c}-t+\boldsymbol{\kappa}\cdot(\mathbf{x}+\mathbf{x}_s)) + \frac{i\omega}{2}\boldsymbol{\lambda}\cdot(\mathbf{x}+\mathbf{x}_s)} \omega^4 \, d\omega \, dh \, d\boldsymbol{\kappa} \, d\boldsymbol{\lambda},$$

with the travel time from the source to the scatterer being $OS/\bar{c}$ and where

$$\hat{\mathbf{f}}_{2,\mathbf{x}}(\omega) = -\mathbf{x}_s \hat{\mathcal{H}}(\omega), \quad \hat{f}_{2,z}(\omega) = -z_s \hat{\mathcal{H}}(\omega), \quad \hat{\mathcal{H}}(\omega) = \frac{\sigma_s |z_s| \omega^2 \hat{f}(\omega)}{4\pi \bar{c}^2 OS^3}.$$

$$(16.31)$$

The time-reversed field has exactly the same form as in (15.48) upon the replacements

$$t_s \mapsto OS/\bar{c}, \quad \hat{\mathbf{f}}_{\mathbf{x}} \mapsto \hat{\mathbf{f}}_{2,\mathbf{x}}, \quad \hat{f}_z \mapsto \hat{f}_{2,z}. \qquad (16.32)$$

Therefore, we can conclude that the secondary reemitted field will refocus and be of order one at the scatterer and be small elsewhere.

In order to illustrate the refocusing we consider now a configuration corresponding to the one discussed in Section 15.3:

- We assume that the source is concentrated in frequency in a narrow band around a large carrier frequency ω_0. This means that the function f in (16.1), the source pulse profile, has the form

$$f(t) = f_0\left(\frac{t}{T_w}\right) e^{-i\omega_0 t} + c.c., \qquad (16.33)$$

with $\omega_0 T_w \gg 1$, where T_w is the initial pulse width and the function f_0 is the envelope of the pulse profile with a normalized support.
- The mirror is supposed to be located at the same location as the source, with a diameter a that is small compared to the depth of the scatterer $|z_s|$. We introduce the normalized spatial cutoff function g_2 determining the mirror shape:

$$G_2(\mathbf{x}) = g_2\left(\frac{\mathbf{x}}{a}\right).$$

We choose the (x, y)-axes so that the horizontal position of the scatterer is $\mathbf{x}_s = (x_s, 0)$, with $x_s \geq 0$. Introducing the parameterization

$$t = -\frac{OS}{\bar{c}} + \varepsilon T, \quad \mathbf{x} = \mathbf{x}_s + \varepsilon \mathbf{X}, \text{ and } z = z_s + \varepsilon Z, \qquad (16.34)$$

and using the orthonormal basis vectors

$$\mathbf{e}_1 = \frac{1}{OS}\begin{bmatrix} -z_s \\ 0 \\ x_s \end{bmatrix}, \quad \mathbf{e}_2 = \begin{bmatrix} 0 \\ 1 \\ 0 \end{bmatrix}, \quad \mathbf{e}_3 = \frac{1}{OS}\begin{bmatrix} x_s \\ 0 \\ z_s \end{bmatrix},$$

we find by analogy with the result in (15.55) that in the limit $\varepsilon \to 0$, the refocused field is

$$u_{\mathrm{TR}}(t, \mathbf{x}, z) = \left(\frac{\sigma_s a^2 z_s^2 \omega_0^4}{64\pi^3 \bar{c}^5 \bar{\rho} OS^4}\right) f_0\left(-\frac{T}{T_w} + \frac{(\mathbf{X}, Z) \cdot \mathbf{e}_3}{\bar{c} T_w}\right) e^{i\omega_0\left(-T + \frac{(\mathbf{X}, Z) \cdot \mathbf{e}_3}{\bar{c}}\right)}$$

$$\times G_1\left(2\frac{OS}{\bar{c}}\right) \hat{g}_2\left(\frac{\omega_0 a |z_s|}{\bar{c} OS^2}(\mathbf{X}, Z) \cdot \mathbf{e}_1, \frac{\omega_0 a}{\bar{c} OS}(\mathbf{X}, Z) \cdot \mathbf{e}_2\right) + c.c.. \qquad (16.35)$$

This formula shows that we observe a refocused field only if $2OS/\bar{c}$ lies in the support of G_1. Here O is the source location, which emits at time 0 a short pulse, and the position of the mirror. The time $2OS/\bar{c}$ corresponds to a round trip from O to S, and $G_1(2OS/\bar{c}) > 0$ means that the signal emitted by the source at O and scattered by the scatterer at S is recorded by the mirror at O. The envelope of the refocused pulse is, up to a multiplicative factor,

$$|u_{\mathrm{TR}}(t, \mathbf{x}, z)| = \left| f_0 \left(-\frac{T}{T_w} + \frac{(\mathbf{X}, Z) \cdot \mathbf{e}_3}{\bar{c} T_w} \right) \right|$$
$$\times \left| \hat{g}_2 \left(\frac{\omega_0 a |z_s|}{\bar{c} O S^2} (\mathbf{X}, Z) \cdot \mathbf{e}_1, \frac{\omega_0 a}{\bar{c} O S} (\mathbf{X}, Z) \cdot \mathbf{e}_2 \right) \right| . \quad (16.36)$$

Therefore, in the homogeneous medium the refocused field in the case with a scatterer has the same form as in the case with an embedded source (compare (16.36) with (15.56)). The main aspects of the refocusing mechanism are as found in Section 15.3:

- In the $\mathbf{e}_3$-direction (direction from the source to the scatterer), the focal spot size is approximately $\bar{c} T_w$, that is, the envelope pulse width.
- In the two other directions, the focal spot size is determined by the mirror size a. In the $\mathbf{e}_1$- and $\mathbf{e}_2$-directions, the focal spot has approximately the sizes $\lambda_0 O S^2 / (a |z_s|)$ and $\lambda_0 O S / a$ respectively, where $\lambda_0 = 2\pi \bar{c} / \omega_0$ is the carrier wavelength of the pulse. These formulas correspond again to the standard Rayleigh resolution formula.

16.3.3 Refocusing of the Secondary Field in the Random Case

We now consider the situation with a random medium and study carefully the reemitted secondary field. This subsection contains the technical details, while the main result will be summarized in Theorem 16.3 in Section 16.4. By substituting (16.28) and (16.29) in (16.27) we obtain the integral representation of the time-reversed vertical velocity $u_{\mathrm{TR}}^{\varepsilon}$. This integral expression can be split into a sum of eight terms, each of them involving a product of three reflection or transmission coefficients of the form

$$Q_{g,1}^{\varepsilon} \overline{Q_{g,2}^{\varepsilon} Q_{g,3}^{\varepsilon}},$$

with $Q = R$ or T. From the probabilistic point of view, only the terms involving an even number of reflection coefficients give rise to a significant contribution; otherwise, the expectation goes to zero. Thus, only four of the eight components remain, namely those involving

$$(A): \quad T_{g,1}^{\varepsilon} \overline{R_{g,2}^{\varepsilon} R_{g,3}^{\varepsilon}}, \quad (16.37)$$
$$(B): \quad T_{g,1}^{\varepsilon} \overline{T_{g,2}^{\varepsilon} T_{g,3}^{\varepsilon}},$$
$$(C): \quad R_{g,1}^{\varepsilon} \overline{R_{g,2}^{\varepsilon} T_{g,3}^{\varepsilon}},$$
$$(D): \quad R_{g,1}^{\varepsilon} \overline{T_{g,2}^{\varepsilon} R_{g,3}^{\varepsilon}}.$$

These components are schematically described in Figure 16.2. The index 3 stands for the propagation from the source O to the scatterer S. The index 2 stands for the propagation from the scatterer S to the mirror M. The index 1 stands for the back-propagation from M to the scatterer position.

The A Component

One can check that the expectation of the product of coefficients of Configuration A goes to zero as $\varepsilon \to 0$. This is because the two reflection coefficients are complex-conjugated, while only one complex conjugation is necessary and sufficient to cancel the random phase, as follows from the moment analysis. Therefore only three nontrivial components remain.

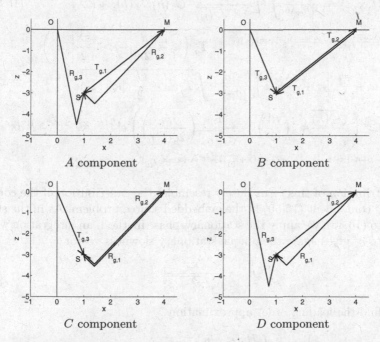

A component   B component

C component   D component

Fig. 16.2. The different wave components. The original source is located at O. The scatterer position is $S = (\mathbf{x}_s, z_s)$. The signal is recorded at the observation point $M = (\mathbf{x}_m, 0)$.

The B Component

Let us write explicitly the B component:

$$
u^\varepsilon_{TR,B}(t, \mathbf{x}, z) = \frac{1}{(2\pi)^6 \varepsilon^4} \int \int \frac{H_0(\boldsymbol{\kappa}_1, \boldsymbol{\kappa}_2)}{4\sqrt{\zeta(\kappa_1)}} \overline{\hat{G}_1\left(\frac{\omega_1 - \omega_2}{\varepsilon}\right)} \hat{G}_2\left(\frac{\omega_1 \boldsymbol{\kappa}_1 + \omega_2 \boldsymbol{\kappa}_2}{\varepsilon}\right)
$$

$$
\times e^{i\left(\frac{-\omega_1 t + \omega_2 \boldsymbol{\kappa}_2 \cdot \mathbf{x}_s + \omega_1 \boldsymbol{\kappa}_1 \cdot \mathbf{x}}{\varepsilon}\right)} e^{i\left(\frac{\omega_2 z_s}{\varepsilon \bar{c}(\kappa_2)} - \frac{\omega_1 z}{\varepsilon \bar{c}(\kappa_1)}\right)} \overline{T^\varepsilon_g(\omega_2, \boldsymbol{\kappa}_2, z_s)} T^\varepsilon_g(\omega_1, \boldsymbol{\kappa}_1, z)
$$

$$
\times \frac{i\sigma_s}{(2\pi)^2} \int \overline{T^\varepsilon_g(\omega_2, \boldsymbol{\kappa}', z_s)} e^{-i\frac{\omega_2}{\varepsilon}(\boldsymbol{\kappa}' \cdot \mathbf{x}_s - z_s / \bar{c}(\kappa'))} \hat{f}(\omega_2) \omega_2^3
$$

$$\times \left(\frac{\sqrt{\zeta(\kappa_2)}}{2\bar{\rho}} \kappa_2 \cdot \kappa' - \frac{1}{\sqrt{\zeta(\kappa_2)}} \left(\frac{\bar{\rho}}{2\bar{\zeta}(\kappa')} \right) \right) d\kappa' \, \omega_1^2 \omega_2^2 \, d\omega_1 \, d\kappa_1 \, d\omega_2 \, d\kappa_2 \,.$$

Again we are motivated by the presence of the terms $\hat{G}_1$ and $\hat{G}_2$ to carry out the change of variables $\omega_1 = \omega + \varepsilon h/2$, $\omega_2 = \omega - \varepsilon h/2$, $\kappa_1 = \kappa + \varepsilon \lambda/2$, $\kappa_2 = -\kappa + \varepsilon \lambda/2$, which to leading order gives

$$u_{TR,B}^{\varepsilon}(t, \mathbf{x}, z) = \frac{1}{\varepsilon(2\pi)^6} \int \int \frac{H_0(\kappa, -\kappa)}{4\sqrt{\zeta(\kappa)}} \overline{\hat{G}_1(h)} \hat{G}_2 \left(h\kappa + \omega\lambda \right) \qquad (16.38)$$

$$\times e^{\frac{i\omega}{\varepsilon}\left(\frac{z_s - z}{\bar{c}(\kappa)} \right)} e^{-\frac{ih}{2} \frac{z_s + z}{\bar{c}(\kappa)} + \frac{i\omega}{2} \lambda \cdot \kappa \bar{c}(\kappa)(z_s + z)} \overline{T_g^{\varepsilon} \left(\omega - \frac{\varepsilon h}{2}, \left| -\kappa + \frac{\varepsilon\lambda}{2} \right|, z_s \right)}$$

$$\times T_g^{\varepsilon} \left(\omega + \frac{\varepsilon h}{2}, \left| \kappa + \frac{\varepsilon\lambda}{2} \right|, z \right) \frac{(-i\sigma_s)}{(2\pi)^2} \int \overline{T_g^{\varepsilon}(\omega - \frac{\varepsilon h}{2}, |\kappa'|, z_s)} e^{-i\frac{\omega}{\varepsilon}(\kappa' \cdot \mathbf{x}_s - \frac{z_s}{\bar{c}(\kappa')})}$$

$$\times \overline{\hat{f}(\omega)} \omega^3 \left(\frac{\sqrt{\zeta(\kappa)}}{\bar{\rho}} \kappa \cdot \left(\frac{\kappa'}{2} \right) + \frac{1}{\sqrt{\zeta(\kappa)}} \left(\frac{\bar{\rho}}{2\bar{\zeta}(\kappa')} \right) \right) e^{i\frac{h}{2}(\kappa' \cdot \mathbf{x}_s - z_s/\bar{c}(\kappa'))} \, d\kappa'$$

$$\times e^{\frac{i\omega}{\varepsilon}(-t + \kappa \cdot (\mathbf{x} - \mathbf{x}_s))} e^{\frac{ih}{2}(-t + \kappa \cdot (\mathbf{x} + \mathbf{x}_s)) + \frac{i\omega}{2} \lambda \cdot (\mathbf{x} + \mathbf{x}_s)} \omega^4 \, d\omega \, dh \, d\kappa \, d\lambda \,.$$

The contribution of this term closely resembles the contribution of the corresponding component (15.59) in the embedded source problem. As in the step leading to (16.30), we apply the stationary-phase method, an integration with respect to κ' gives again the unique stationary slowness vector

$$\kappa_c' = \frac{\mathbf{x}_s}{\bar{c}\sqrt{\mathbf{x}_s^2 + z_s^2}} \,,$$

and we find the leading order approximation

$$u_{TR,B}^{\varepsilon}(t, \mathbf{x}, z) = \frac{1}{(2\pi)^6} \int \int \frac{H_0(\kappa, -\kappa)}{4\sqrt{\zeta(\kappa)}} \overline{\hat{G}_1(h)} \hat{G}_2 \left(h\kappa + \omega\lambda \right)$$

$$\times e^{\frac{i\omega}{\varepsilon}\left(\frac{z_s - z}{\bar{c}(\kappa)} \right)} e^{-\frac{ih}{2} \frac{z_s + z}{\bar{c}(\kappa)} + \frac{i\omega}{2} \lambda \cdot \kappa \bar{c}(\kappa)(z_s + z)} \left(-\frac{\sqrt{\zeta(\kappa)}}{\bar{\rho}} \kappa \cdot \overline{\hat{\mathbf{f}}_{2,\mathbf{x}}(\omega)} + \frac{1}{\sqrt{\zeta(\kappa)}} \overline{\hat{f}_{2,z}(\omega)} \right)$$

$$\times \overline{T_g^{\varepsilon} \left(\omega - \frac{\varepsilon h}{2}, \left| -\kappa + \frac{\varepsilon\lambda}{2} \right|, z_s \right)} T_g^{\varepsilon} \left(\omega + \frac{\varepsilon h}{2}, \left| \kappa + \frac{\varepsilon\lambda}{2} \right|, z \right) \overline{T_g^{\varepsilon} \left(\omega - \frac{\varepsilon h}{2}, \kappa_c', z_s \right)}$$

$$\times e^{\frac{i\omega}{\varepsilon}(-(OS/\bar{c} + t) + \kappa \cdot (\mathbf{x} - \mathbf{x}_s))} e^{\frac{ih}{2}(OS/\bar{c} - t + \kappa \cdot (\mathbf{x} + \mathbf{x}_s)) + \frac{i\omega}{2} \lambda \cdot (\mathbf{x} + \mathbf{x}_s)} \omega^4 \, d\omega \, dh \, d\kappa \, d\lambda \,,$$

where $\hat{\mathbf{f}}_{2,\mathbf{x}}$ and $\hat{f}_{2,z}$ are defined by (16.31). Therefore, the time-reversed field has exactly the same form as the one discussed in (15.59) upon the replacements $t_s \mapsto OS/\bar{c}$, $\hat{\mathbf{f}}_{\mathbf{x}} \mapsto \hat{\mathbf{f}}_{2,\mathbf{x}}$, $\hat{f}_z \mapsto \hat{f}_{2,z}$ and the multiplication by a third transmission coefficient $\overline{T_g^{\varepsilon}(\omega - \varepsilon h/2, \kappa_c', z_s)}$. The rapid phase is now the same as in the homogeneous case and we have a globally stationary point for $t = -OS/\bar{c}$, $\mathbf{x} = \mathbf{x}_s$, and $z = z_s$ corresponding to the vanishing of the rapid phase. We next consider the limit of the product of the three transmission coefficients evaluated at the globally stationary point. The integral with respect to κ averages the product of the first two transmission coefficients so that it can be substituted by its limiting expectation (15.88) as $\varepsilon \to 0$. The third transmission coefficient is taken at the fixed slowness vector κ'_c and it can be substituted by its limiting expression as $\varepsilon \to 0$:

$$T_g^\varepsilon\left(\omega - \frac{\varepsilon h}{2}, \kappa'_c, z_s\right) \overset{\varepsilon \to 0}{\longrightarrow} e^{-i\frac{1}{\sqrt{L_{\text{loc}}(\omega,\kappa'_c)}}W_0(z_s) + \frac{z_s}{2L_{\text{loc}}(\omega,\kappa'_c)}}, \qquad (16.39)$$

where $L_{\text{loc}}(\omega, \kappa'_c) = (4|z_s|\bar{c}^2)/(\omega^2 \gamma OS)$ and W_0 is a standard Brownian motion. Substituting these limits into the integral representation and evaluating in the vicinity of the globally stationary point using the parameterization (16.34) yields the following limit expression for the B component as $\varepsilon \to 0$:

$$u_{TR,B}(t, \mathbf{x}, z) = \frac{1}{(2\pi)^6} \int \int \frac{\bar{c}(\kappa)}{4\bar{c}\sqrt{\zeta(\kappa)}} \hat{G}_1(h)\hat{G}_2(h\kappa + \omega\boldsymbol{\lambda})$$

$$\times \int e^{i[h\bar{c}^2/\bar{c}(\kappa)^2 - \omega\kappa\cdot\boldsymbol{\lambda}\bar{c}^2]\tau}\mathcal{W}_g^{(T)}(\omega, \kappa, \tau)\, d\tau e^{-i\frac{1}{\sqrt{L_{\text{loc}}(\omega,\kappa'_c)}}W_0(z_s) + \frac{z_s}{2L_{\text{loc}}(\omega,\kappa'_c)}}$$

$$\times \left(-\frac{\sqrt{\zeta(\kappa)}}{\bar{\rho}}\kappa \cdot \hat{\mathbf{f}}_{2,\mathbf{x}}(\omega) + \frac{1}{\sqrt{\zeta(\kappa)}}\hat{f}_{2,z}(\omega)\right)$$

$$\times e^{i\omega\left(-T+\kappa\cdot\mathbf{X}-\frac{Z}{\bar{c}(\kappa)}\right)}e^{ih\left(\frac{OS}{\bar{c}}+\kappa\cdot\mathbf{x}_s-\frac{z_s}{\bar{c}(\kappa)}\right)+i\omega\left(\boldsymbol{\lambda}\cdot\mathbf{x}_s+\boldsymbol{\lambda}\cdot\kappa\bar{c}(\kappa)z_s\right)}\omega^4\, d\omega\, dh\, d\kappa\, d\boldsymbol{\lambda}\,.$$

We rewrite this expression as

$$u_{TR,B}(t, \mathbf{x}, z) = \frac{1}{(2\pi)^6} \int \int \frac{\bar{c}(\kappa)}{4\bar{c}\sqrt{\zeta(\kappa)}} \hat{G}_1(h)\hat{G}_2(h\kappa + \omega\boldsymbol{\lambda}) \qquad (16.40)$$

$$\times \int e^{i[h\bar{c}^2/\bar{c}(\kappa)^2 - \omega\kappa\cdot\boldsymbol{\lambda}\bar{c}^2]\tau}\mathcal{W}_g^{(T)}(\omega, \kappa, \tau)\, d\tau$$

$$\times \left(-\frac{\sqrt{\zeta(\kappa)}}{\bar{\rho}}\kappa \cdot \hat{\mathbf{f}}_{3,\mathbf{x}}(\omega) + \frac{1}{\sqrt{\zeta(\kappa)}}\hat{f}_{3,z}(\omega)\right)$$

$$\times e^{i\omega\left(-(T+\chi_c)+\kappa\cdot\mathbf{X}-\frac{Z}{\bar{c}(\kappa)}\right)}e^{ih\left(\frac{OS}{\bar{c}}+\kappa\cdot\mathbf{x}_s-\frac{z_s}{\bar{c}(\kappa)}\right)+i\omega\left(\boldsymbol{\lambda}\cdot\mathbf{x}_s+\boldsymbol{\lambda}\cdot\kappa\bar{c}(\kappa)z_s\right)}\omega^4\, d\omega\, dh\, d\kappa\, d\boldsymbol{\lambda}\,,$$

for

$$\hat{\mathbf{f}}_{3,\mathbf{x}}(\omega) = -\mathbf{x}_s\tilde{\mathcal{H}}(\omega)\,, \quad \hat{f}_{3,z}(\omega) = -z_s\tilde{\mathcal{H}}(\omega)\,, \quad \chi_c = \frac{\gamma OS}{4|z_s|\bar{c}^2}W_0(z_s)\,,$$

$$\tilde{\mathcal{H}}(\omega) = \frac{\sigma_s|z_s|\omega^2}{4\pi\bar{c}^2 OS^3}\hat{f}(\omega)e^{\frac{z_s}{2L_{\text{loc}}(\omega,\kappa'_c)}}\,.$$

The C Component

We next consider the contribution associated with the C terms in (16.37). By repeating the steps used in the derivation of the B component (16.40) we find the following approximation for this term:

$$u^{\varepsilon}_{TR,C}(t,\mathbf{x},z) = \frac{1}{(2\pi)^6} \int\int \frac{\bar{c}(\kappa)}{4\bar{c}\sqrt{\zeta(\kappa)}} \overline{\hat{G}_1(h)}\hat{G}_2(h\kappa+\omega\lambda) \tag{16.41}$$

$$\times \int e^{i[h\bar{c}^2/\bar{c}(\kappa)^2 - \omega\kappa\cdot\lambda\bar{c}^2]\tau} \mathcal{W}^{(R)}_g (\omega,\kappa,\tau)\, d\tau$$

$$\times \left(\frac{\sqrt{\zeta(\kappa)}}{\bar{\rho}}\kappa \cdot \overline{\hat{\mathbf{f}}_{3,\mathbf{x}}(\omega)} + \frac{1}{\sqrt{\zeta(\kappa)}}\overline{\hat{f}_{3,z}(\omega)} \right)$$

$$\times e^{i\omega\left(-(T+\chi_c)+\kappa\cdot\mathbf{X}+\frac{Z}{\bar{c}(\kappa)}\right)} e^{ih\left(\frac{OS}{\bar{c}}+\kappa\cdot\mathbf{x}_s-\frac{z_s}{\bar{c}(\kappa)}\right)+i\omega\left(\lambda\cdot\mathbf{x}_s+\lambda\cdot\kappa\bar{c}(\kappa)z_s\right)}\omega^4\, d\omega\, dh\, d\kappa\, d\lambda.$$

The B and C components are the main contributions to the refocused field, as we discuss explicitly below.

The D Component

We discuss the contribution of the D term in (16.37). We show that this component does not contribute to a refocused wave field. The integral expression for this term can be obtained via the steps leading to (16.38), and we obtain the leading-order expression

$$u^{\varepsilon}_{TR,D}(t,\mathbf{x},z) = \frac{1}{\varepsilon(2\pi)^6} \int\int \frac{H_0(\kappa,-\kappa)}{4\sqrt{\zeta(\kappa)}} \overline{\hat{G}_1(h)}\hat{G}_2(h\kappa+\omega\lambda) \tag{16.42}$$

$$\times e^{\frac{i\omega}{\varepsilon}\left(\frac{z_s+z}{\bar{c}(\kappa)}\right)} e^{-\frac{ih}{2}\frac{z_s-z}{\bar{c}(\kappa)}+\frac{i\omega}{2}\lambda\cdot\kappa\bar{c}(\kappa)(z_s-z)} \overline{T^{\varepsilon}_g \left(\omega - \frac{\varepsilon h}{2}, \left|-\kappa+\frac{\varepsilon\lambda}{2}\right|, z_s \right)}$$

$$\times R^{\varepsilon}_g \left(\omega + \frac{\varepsilon h}{2}, \left|\kappa+\frac{\varepsilon\lambda}{2}\right|, z \right) \frac{(-i\sigma_s)}{(2\pi)^2} \int \overline{R^{\varepsilon}_g \left(\omega - \frac{\varepsilon h}{2}, |\kappa'|, z_s \right)} e^{-i\frac{\omega}{\varepsilon}\left(\kappa'\cdot\mathbf{x}_s+\frac{z_s}{\bar{c}(\kappa')}\right)}$$

$$\times \overline{\hat{f}(\omega)}\omega^3 \left(\frac{\sqrt{\zeta(\kappa)}}{\bar{\rho}}\kappa \cdot \left(\frac{\kappa'}{2}\right) + \frac{1}{\sqrt{\zeta(\kappa)}} \left(\frac{\bar{\rho}}{2\bar{\zeta}(\kappa')}\right) \right) e^{i\frac{h}{2}(\kappa'\cdot\mathbf{x}_s-z_s/\bar{c}(\kappa'))}\, d\kappa'$$

$$\times e^{\frac{i\omega}{\varepsilon}(-t+\kappa\cdot(\mathbf{x}-\mathbf{x}_s))} e^{\frac{ih}{2}(-t+\kappa\cdot(\mathbf{x}+\mathbf{x}_s))+\frac{i\omega}{2}\lambda\cdot(\mathbf{x}+\mathbf{x}_s)}\omega^4\, d\omega\, dh\, d\kappa\, d\lambda.$$

As seen in Chapter 15, the generalized reflection coefficients evaluated at two different slowness vectors are correlated only as long as the moduli of the slowness vectors are within an ε-neighborhood of each other. We therefore carry out the change of variables $\kappa' = (\mu+\varepsilon l)\mathbf{e}_{\theta'}$, $\kappa = \mu\mathbf{e}_\theta$, and write $\mathbf{x} = \mu_x\mathbf{e}_{\theta_x}$, $\mathbf{x}_s = \mu_{x_s}\mathbf{e}_{\theta_{x_s}}$. This gives the following representation for the fast phase in the integral expression for $u_{TR,D}$:

$$\phi(\mu,\theta,\theta') = \frac{z}{\bar{c}(\mu)} - \mu\mu_x\cos(\theta'-\theta_{x_s}) - t + \mu\mu_x\cos(\theta-\theta_x) - \mu\mu_{x_s}\cos(\theta-\theta_{x_s}).$$

It follows that we have a global stationary point with $\phi \equiv 0$ only if

$$z = t = \mu_x = \mu_{x_s} = 0,$$

and only then may $u^{\varepsilon}_{TR,D}$ be of order one. However, this means that the two generalized reflection coefficients in (16.42) are evaluated at the two different depths $z = 0$ and $z = z_s$ and one can show that their expectation then goes to zero and that $u^{\varepsilon}_{TR,D}$ does not contribute to the refocused field.

16.3.4 Contributions of the Other Wave Components

We have just studied the reemission of the secondary field, labeled (2) in Section 16.3.1, into the random medium, and found that it refocuses at the position of the scatterer and that it generates a focal spot whose amplitude is of order one. We now discuss the other wave components, labeled (1) and (3) in Section 16.3.1, which in fact will contribute at a lower order at the scatterer.

The Reemitted Primary Field

We first discuss the role of the reemitted primary field $\mathbf{u}_0$. We assume, however, that we do not reemit the deterministic wave front directly transmitted from the source to the mirror. Reemission and refocusing of the incoherent primary field was discussed in Chapter 15. The source at O is defined as in (15.5) and the analysis in Chapter 15 shows that we will have refocusing at a later time of a strong signal of order ε^{-1} at the original **source** location O. However, the reemitted incoherent primary field will be small at the scatterer position S. It can be shown using the integral representation of this wave component and the moment analysis presented in Chapter 15 that indeed this wave component is small, of order $\varepsilon^{1/2}$, at S, so that it does not generate a significant wave component at the position S of the scatterer.

The New Scattered Field

We finally discuss the **new scattered** field, that is, the wave field generated by reemission from the mirror that is scattered by the embedded scatterer during the back-propagation. This field can again be described in terms of the Born approximation. To be consistent with the Born approximation and consider a field that scales with the scattering volume σ_s, we discuss only the new scattered field generated by the primary field $\mathbf{u}_0^\varepsilon$ (since the component generated by $\mathbf{u}_1^\varepsilon$ will scale with σ_s^2). The reflections of this wave component off the internal scatterer do not create any coherent structure during the back-propagation. Indeed, it can be checked that this wave component involves products of transmission and reflection coefficients whose expectations vanish in the limit $\varepsilon \to 0$.

16.4 Time-Reversal Superresolution with a Passive Scatterer

16.4.1 The Refocused Pulse Shape

The main results of this section are the refocusing of the pulse and its self-averaging property in a randomly corrected time frame. These results are precisely stated in the following theorem, which gives the refocusing property and shows that the wave field converges to a deterministic shape when observed relative to the random time $-OS/\bar{c} + \varepsilon\chi_c$ and at the original source location $(\mathbf{x}_s, z_s)$ in space.

Theorem 16.3.
(a) For any $T_0 > 0$, $R_0 > 0$, $Z_0 > 0$, $\delta > 0$, and $(t_0, \mathbf{x}_0, z_0) \neq (-OS/\bar{c}, \mathbf{x}_s, z_s)$, and $(t_0, \mathbf{x}_0, z_0) \neq (0, \mathbf{0}, 0)$, we have

$$\mathbb{P}\left(\sup_{|t-t_0| \leq \varepsilon T_0, |\mathbf{x}-\mathbf{x}_s| \leq \varepsilon R_0, |z-z_0| \leq \varepsilon Z_0} |u_{\mathrm{TR}}^{\varepsilon}(t, \mathbf{x}, z)| > \delta \right) \xrightarrow{\varepsilon \to 0} 0 .$$

(b) For any $T_0 > 0$, $R_0 > 0$, $Z_0 > 0$, and $\delta > 0$, we have

$$\mathbb{P}\left(\sup_{|T| \leq T_0, |\mathbf{X}| \leq R_0, |Z| \leq Z_0} \left| u_{\mathrm{TR}}^{\varepsilon} \left(-\frac{OS}{\bar{c}} + \varepsilon T, \mathbf{x}_s + \varepsilon \mathbf{X}, z_s + \varepsilon Z \right) \right. \right.$$
$$\left. \left. - U_{\mathrm{TR}}(T + \chi_c, \mathbf{X}, Z) \right| > \delta \right) \xrightarrow{\varepsilon \to 0} 0 ,$$

where U_{TR} is the deterministic pulse shape

$$U_{\mathrm{TR}}(T, \mathbf{X}, Z) = \frac{1}{(2\pi)^3} \int K^+(\omega, \boldsymbol{\kappa}) \left[-\bar{c}(\kappa)\boldsymbol{\kappa} \cdot \mathbf{x}_s - z_s \right]$$
$$\times \overline{\hat{f}(\omega)} e^{i\omega(-T + \boldsymbol{\kappa} \cdot \mathbf{X} + \frac{Z}{\bar{c}(\kappa)})} \omega^2 \, d\omega \, d\boldsymbol{\kappa}$$
$$+ \frac{1}{(2\pi)^3} \int K^-(\omega, \boldsymbol{\kappa}) \left[\bar{c}(\kappa)\boldsymbol{\kappa} \cdot \mathbf{x}_s - z_s \right]$$
$$\times \overline{\hat{f}(\omega)} e^{i\omega(-T + \boldsymbol{\kappa} \cdot \mathbf{X} - \frac{Z}{\bar{c}(\kappa)})} \omega^2 \, d\omega \, d\boldsymbol{\kappa} . \qquad (16.43)$$

The refocusing kernels are given by

$$K^+(\omega, \boldsymbol{\kappa}) = \int G_1 \left(\frac{OS}{\bar{c}} - z_s \frac{\bar{c}(\kappa)}{\bar{c}^2} + \tau \right) G_2 \left(\mathbf{x}_{2,s} + \boldsymbol{\kappa}\bar{c}(\kappa)z_s - \bar{c}^2 \boldsymbol{\kappa}\tau \right)$$
$$\times \mathcal{W}_g^{(R)}(\omega, \kappa, \tau) \, d\tau \left(\frac{\sigma_s |z_s| \omega^2}{16\pi \bar{c}^3 \bar{\rho} OS^3} e^{-\frac{|z_s|}{2L_{\mathrm{loc}}(\omega, \kappa_c')}} \right) , \qquad (16.44)$$

$$K^-(\omega, \boldsymbol{\kappa}) = \int G_1 \left(\frac{OS}{\bar{c}} - z_s \frac{\bar{c}(\kappa)}{\bar{c}^2} + \tau \right) G_2 \left(\mathbf{x}_s + \boldsymbol{\kappa}\bar{c}(\kappa)z_s - \bar{c}^2 \boldsymbol{\kappa}\tau \right)$$
$$\times \mathcal{W}_g^{(T)}(\omega, \kappa, \tau) \, d\tau \left(\frac{\sigma_s |z_s| \omega^2}{16\pi \bar{c}^3 \bar{\rho} OS^3} e^{-\frac{|z_s|}{2L_{\mathrm{loc}}(\omega, \kappa_c')}} \right) , \qquad (16.45)$$

where $L_{\mathrm{loc}}(\omega, \kappa_c') = (\omega^2 \gamma OS)/(4|z_s| \bar{c}^2)$. The random travel time correction χ_c is a zero-mean Gaussian random variable with variance $\gamma OS/(4\bar{c}^2)$.

The picture is qualitatively the same for the time-reversed transverse velocity and pressure fields. The precise expressions for the refocused fields are the following:

$$P_{\text{TR}}(T, \mathbf{X}, Z) = \frac{\bar{\rho}}{(2\pi)^3} \int K^+(\omega, \boldsymbol{\kappa}) \bar{c}(\kappa) \left[\bar{c}(\kappa) \boldsymbol{\kappa} \cdot \overline{\hat{\mathbf{f}}_{3,\mathbf{x}}(\omega)} + \hat{f}_{3,z}(\omega) \right]$$

$$\times e^{i\omega \left(-(T+\chi_c) + \boldsymbol{\kappa} \cdot \mathbf{X} + \frac{Z}{\bar{c}(\kappa)} \right)} \omega^2 \, d\omega \, d\boldsymbol{\kappa}$$

$$+ \frac{\bar{\rho}}{(2\pi)^3} \int K^-(\omega, \boldsymbol{\kappa}) \bar{c}(\kappa) \left[\bar{c}(\kappa) \boldsymbol{\kappa} \cdot \overline{\hat{\mathbf{f}}_{3,\mathbf{x}}(\omega)} - \hat{f}_{3,z}(\omega) \right]$$

$$\times e^{i\omega \left(-(T+\chi_c) + \boldsymbol{\kappa} \cdot \mathbf{X} - \frac{Z}{\bar{c}(\kappa)} \right)} \omega^2 \, d\omega \, d\boldsymbol{\kappa} \, ,$$

$$\mathbf{V}_{\text{TR}}(T, \mathbf{X}, Z) = \frac{1}{(2\pi)^3} \int K^+(\omega, \boldsymbol{\kappa}) \bar{c}(\kappa) \boldsymbol{\kappa} \left[\bar{c}(\kappa) \boldsymbol{\kappa} \cdot \overline{\hat{\mathbf{f}}_{3,\mathbf{x}}(\omega)} + \hat{f}_{3,z}(\omega) \right]$$

$$\times e^{i\omega \left(-(T+\chi_c) + \boldsymbol{\kappa} \cdot \mathbf{X} + \frac{Z}{\bar{c}(\kappa)} \right)} \omega^2 \, d\omega \, d\boldsymbol{\kappa}$$

$$+ \frac{1}{(2\pi)^3} \int K^-(\omega, \boldsymbol{\kappa}) \bar{c}(\kappa) \boldsymbol{\kappa} \left[\bar{c}(\kappa) \boldsymbol{\kappa} \cdot \overline{\hat{\mathbf{f}}_{3,\mathbf{x}}(\omega)} - \hat{f}_{3,z}(\omega) \right]$$

$$\times e^{i\omega \left(-(T+\chi_c) + \boldsymbol{\kappa} \cdot \mathbf{X} - \frac{Z}{\bar{c}(\kappa)} \right)} \omega^2 \, d\omega \, d\boldsymbol{\kappa} \, .$$

The proof of the theorem is again a generalization of the arguments described in Chapter 9 and goes along the following main steps:

- We first consider the expected value of $u^\varepsilon_{\text{TR}}$ correctly centered with respect to the fine-scale random time correction χ_c. By a generalization of the results of Section 15.4.1 we find that this expectation converges to the limiting value given in the theorem.
- We then consider the variance of $u^\varepsilon_{\text{TR}}$ correctly centered with respect to χ_c. We write the second moment as a multiple integral involving the product of four reflection coefficients at four different frequencies as in (14.72). Using the decorrelation property of the reflection coefficients we deduce that the variance goes to zero.
- Note that an integral over frequency (ensured by the time-domain nature of time reversal) is needed for the stabilization or the self-averaging of the refocused pulse.

16.4.2 Superresolution with a Random Medium

If we compare the expression (16.43) of the refocused pulse shape in the embedded-scatterer problem with the equivalent expression (15.67) in the internal-source problem, we find that the principal superresolution effect will be as in the case of an internal source up to a spreading by a Gaussian kernel. Note, however, that the statistically stable refocused wave field is observed at the randomly corrected "fine-scale" time $T + \chi_c$.

We continue the time-reversal example introduced in Section 16.3.2, where we discussed the homogeneous case. We make the same assumptions with the additional assumption that $a \ll |z_s| \ll L_{\text{loc}} \ll L$, where $L_{\text{loc}} = 4\bar{c}^2/(\gamma\omega_0^2)$. Here we make use of the analysis presented in Section 16.4 to characterize the effects of medium heterogeneity. The superresolution phenomenon unraveled in Section 15.4.2 in the context of an internal source carries over to the present configuration with an internal scatterer. In the case of an internal source, the envelope of the refocused field has the form (15.77). In the case of an internal scatterer, we take into account the effect of the modification due to the third transmission coefficient in (16.39), and we find that the envelope of the refocused field is given by

$$|U_{TR}(T, \mathbf{X}, Z)| \approx \left| \hat{g}_2 \left(-\frac{\omega_0 \Delta \psi_1}{\bar{c}}(\mathbf{X}, Z) \cdot \mathbf{w}_1, \frac{\omega_0 \Delta \psi_2}{\bar{c}}(\mathbf{X}, Z) \cdot \mathbf{w}_2 \right) \right|$$

$$\times \left| \mathrm{sinc} \left(\frac{\omega_0 \Delta \theta}{2\bar{c}}(\mathbf{X}, Z) \cdot \mathbf{w}_1 \right) \right| \left| f \left(-\frac{T + \chi_c}{T_w} + \frac{(\mathbf{X}, Z) \cdot \mathbf{w}_3}{\bar{c} T_w} \right) \right|, \quad (16.46)$$

up to a multiplicative factor. This expression is obtained in terms of the parameterization (16.34) and in the frame $(\mathbf{w}_1, \mathbf{w}_2, \mathbf{w}_3)$ defined by (15.76):

$$\mathbf{w}_1 = \begin{bmatrix} -\sin\bar{\theta} \\ 0 \\ \cos\bar{\theta} \end{bmatrix}, \quad \mathbf{w}_2 = \begin{bmatrix} 0 \\ 1 \\ 0 \end{bmatrix}, \quad \mathbf{w}_3 = \begin{bmatrix} \cos\bar{\theta} \\ 0 \\ \sin\bar{\theta} \end{bmatrix}.$$

Here $\bar{\theta}$ and $\Delta\theta$ are defined respectively by (15.75–15.79), $\Delta\psi_1$ and $\Delta\psi_2$ by (15.78). The envelope of the refocused field (16.46) is very similar to the envelope of the refocused field (15.77) obtained in the case of an internal source. The only difference is a random time delay represented by the term χ_c and which originates from the random travel time from the source to the scatterer. There is also an exponential damping factor that is hidden in the multiplicative factor, and which originates from the decay of the stable wave front emitted by the source and received by the scatterer. Thus, in the $\mathbf{w}_1$-direction, the focal spot again has size $\lambda_0/\Delta\theta$ when $\Delta\theta > a/OS$. As described in Section 15.5.3, the condition $\Delta\theta > a/OS$ means that the angular diversity of the refocused wave mainly originates from multiple scattering effects rather than the numerical aperture of the mirror. The focal-spot size $\lambda_0/\Delta\theta$ is then imposed by the angular diversity of the refocused incoherent waves, and it is again much smaller than the prediction of the Rayleigh resolution formula. In the $\mathbf{w}_2$- and $\mathbf{w}_3$-directions, the refocusing radii are as in the case of the internal source and described by (15.81).

Note finally that in the situation described in this chapter, the time-reversal mirror receives a very low amplitude signal, of order $\varepsilon^{3/2}$; it amplifies the received signal by the factor ε^{-1}, so that it sends back an acoustic signal whose amplitude is small, of order $\varepsilon^{1/2}$. This generates a focal spot at the location of the scatterer whose amplitude is of order one. However, the time-reversal mirror could send back a signal with amplitude of order one, which means that it could amplify the received signal by the factor $\varepsilon^{-3/2}$. If the time-reversal experiment is performed with this amplification factor, then a very high amplitude, of order $\varepsilon^{-1/2}$, is obtained at the location of the scatterer.

Notes

In this chapter we have shown that by time-reversing incoherent waves, it is possible to concentrate energy on a passive scatterer buried in a random medium. The amplification of these time-reversed waves is the basic principle of time-reversal ultrasound target destruction, applied, for instance, to kidney stones [56]. The iteration of this procedure can be used to enhance the refocusing at the selected target, as described in [144, 145, 146, 147]. In such experiments, both sides of the medium surrounding the target are accessible, and the addition of a randomly layered slab below the medium (opposite to the source and time-reversal mirror apparatus) can dramatically enhance the refocusing at the target according to the theory developed in this chapter. The randomly layered slab can be produced once for all due to the statistical stability, with a correlation length compatible with the source and the regime of separation of scales studied in this book.

Time-reversal ideas have been used recently to propose a method for imaging a heterogeneous medium by cross-correlating the noisy traces recorded at the surface [166, 149, 157]. The mathematical analysis in the case of randomly layered media is given in [76]. The idea is related to the discussion in Section 10.2 comparing the cross-correlation with time-reversal refocusing.

Other Layered Media

In this chapter we extend the theory of wave propagation and time reversal in random media to more general randomly layered media. We still consider the linear acoustic wave equations in the strongly heterogeneous white-noise regime as in the previous chapters. In Section 17.1 we incorporate in the theory the effects of discontinuities in the effective parameters at the boundaries of the random medium. It is important to take into account these effects in practical situations in which the source or the time-reversal mirror is usually located outside of the material to be probed. In Section 17.2 we consider the case in which the effective medium parameters vary smoothly on the macro-scale. In Section 17.3 we consider the situation in which the density parameter is also randomly varying. The main additional difficulty is that the problem cannot be reduced to a decoupled family of one-dimensional problems. We study the coupled system and we derive the corresponding asymptotics for the stable front and the refocused pulse.

17.1 Nonmatched Effective Medium

In this section we summarize modifications in the theory that follow when the random slab has nonmatched effective parameters. That is, we consider the linear acoustic wave equations (14.1) with the medium parameters

$$\frac{1}{K(\mathbf{x},z)} = \frac{1}{K(z)} = \begin{cases} \frac{1}{K_1} & \text{for } z \in (-\infty, -L), \\ \frac{1}{K}\left(1 + \nu(z/\varepsilon^2)\right) & \text{for } z \in [-L, 0], \\ \frac{1}{K_2} & \text{for } z \in (0, \infty), \end{cases}$$

$$\rho(\mathbf{x},z) = \bar{\rho} \quad \text{for all } (\mathbf{x},z).$$

The fluctuations ν are defined as before, but the wave speeds to the left and to the right of the random slab are in general different from the effective medium wave speed $\bar{c} = \sqrt{\bar{K}/\bar{\rho}}$ in the interior of the slab.

17.1.1 Boundary and Jump Conditions

We introduce the mode- and section-dependent effective speeds

$$\bar{c}(\kappa, z) = \begin{cases} c_1(\kappa) = \dfrac{c_1}{\sqrt{1-c_1^2\kappa^2}} & \text{for } z \in (-\infty, -L), \\[2mm] \bar{c}(\kappa) = \dfrac{\bar{c}}{\sqrt{1-\bar{c}^2\kappa^2}} & \text{for } z \in [-L, 0], \\[2mm] c_2(\kappa) = \dfrac{c_2}{\sqrt{1-c_2^2\kappa^2}} & \text{for } z \in (0, \infty), \end{cases}$$

with

$$c_1 = \sqrt{K_1/\bar{\rho}}, \quad c_2 = \sqrt{K_2/\bar{\rho}}.$$

The corresponding impedances are defined by

$$\bar{\zeta}(\kappa, z) = \bar{\rho}\bar{c}(\kappa, z),$$

and we use the notation $\zeta_1(\kappa)$, $\bar{\zeta}(\kappa)$, and $\zeta_2(\kappa)$ for the mode-dependent impedances respectively to the left, in, and to the right of the random slab. Then, we decompose the wave field as before:

$$\hat{p}^\varepsilon(\omega, \boldsymbol{\kappa}, z) = \frac{\sqrt{\bar{\zeta}(\kappa, z)}}{2} \left(\check{a}^\varepsilon(\omega, \boldsymbol{\kappa}, z)e^{\frac{i\omega z}{\varepsilon\bar{c}(\kappa, z)}} - \check{b}^\varepsilon(\omega, \boldsymbol{\kappa}, z)e^{-\frac{i\omega z}{\varepsilon\bar{c}(\kappa, z)}} \right),$$

$$\hat{u}^\varepsilon(\omega, \boldsymbol{\kappa}, z) = \frac{1}{2\sqrt{\bar{\zeta}(\kappa, z)}} \left(\check{a}^\varepsilon(\omega, \boldsymbol{\kappa}, z)e^{\frac{i\omega z}{\varepsilon\bar{c}(\kappa, z)}} + \check{b}^\varepsilon(\omega, \boldsymbol{\kappa}, z)e^{-\frac{i\omega z}{\varepsilon\bar{c}(\kappa, z)}} \right).$$

The z dependence of the "effective" wave speed now leads to jump conditions for the coefficients at $z = 0$ and $z = -L$. Next we use the continuity conditions on the velocity $\hat{u}^\varepsilon$ and the pressure $\hat{p}^\varepsilon$ to derive the jump conditions. First, we introduce the parameters

$$r_1^{(\pm)}(\kappa) = \frac{1}{2} \left(\sqrt{\bar{\zeta}/\zeta_1(\kappa)} \pm \sqrt{\zeta_1/\bar{\zeta}(\kappa)} \right), \tag{17.1}$$

$$r_2^{(\pm)}(\kappa) = \frac{1}{2} \left(\sqrt{\zeta_2/\bar{\zeta}(\kappa)} \pm \sqrt{\bar{\zeta}/\zeta_2(\kappa)} \right), \tag{17.2}$$

and the matrices

$$\mathbf{J}_{\omega,\kappa}^{\varepsilon,1} = \begin{bmatrix} r_1^+(\kappa)e^{\frac{i\omega L}{\varepsilon}\left(\frac{1}{\bar{c}(\kappa)} - \frac{1}{c_1(\kappa)}\right)} & r_1^-(\kappa)e^{\frac{i\omega L}{\varepsilon}\left(\frac{1}{\bar{c}(\kappa)} + \frac{1}{c_1(\kappa)}\right)} \\[2mm] r_1^-(\kappa)e^{\frac{i\omega L}{\varepsilon}\left(-\frac{1}{\bar{c}(\kappa)} - \frac{1}{c_1(\kappa)}\right)} & r_1^+(\kappa)e^{\frac{i\omega L}{\varepsilon}\left(-\frac{1}{\bar{c}(\kappa)} + \frac{1}{c_1(\kappa)}\right)} \end{bmatrix}, \tag{17.3}$$

$$\mathbf{J}_{\omega,\kappa}^{\varepsilon,2} = \begin{bmatrix} r_2^+(\kappa) & r_2^-(\kappa) \\ r_2^-(\kappa) & r_2^+(\kappa) \end{bmatrix}, \tag{17.4}$$

where we have

$$(r_j^+)^2(\kappa) - (r_j^-)^2(\kappa) = 1, \qquad j = 1, 2.$$

This gives the jump condition at the surface

$$\begin{bmatrix} \check{a}^\varepsilon(z = 0^+) \\ \check{b}^\varepsilon(z = 0^+) \end{bmatrix} = \mathbf{J}^{\varepsilon,2}_{\omega,\kappa} \begin{bmatrix} \check{a}^\varepsilon(z = 0^-) \\ \check{b}^\varepsilon(z = 0^-) \end{bmatrix}, \tag{17.5}$$

with the corresponding relation satisfied at $z = -L$:

$$\begin{bmatrix} \check{a}^\varepsilon(z = (-L)^+) \\ \check{b}^\varepsilon(z = (-L)^+) \end{bmatrix} = \mathbf{J}^{\varepsilon,1}_{\omega,\kappa} \begin{bmatrix} \check{a}^\varepsilon(z = (-L)^-) \\ \check{b}^\varepsilon(z = (-L)^-) \end{bmatrix}. \tag{17.6}$$

The mode-dependent interface reflection coefficients $R_{I,j}(\kappa)$ and transmission coefficients $T_{I,j}(\kappa)$ are defined by

$$T_{I,j}(\kappa) = \frac{1}{r_j^+(\kappa)} = \frac{2\sqrt{\zeta_j \bar{\zeta}(\kappa)}}{\zeta_j(\kappa) + \bar{\zeta}(\kappa)}, \tag{17.7}$$

$$R_{I,j}(\kappa) = \frac{r_j^-(\kappa)}{r_j^+(\kappa)} = (-1)^j \frac{\zeta_j(\kappa) - \bar{\zeta}(\kappa)}{\zeta_j(\kappa) + \bar{\zeta}(\kappa)}, \tag{17.8}$$

for $j \in \{1,2\}$. Here the subscript "I" stands for "Interface." Then we have, for instance, at the surface

$$\begin{bmatrix} \check{a}^\varepsilon(z = 0^+) \\ \check{b}^\varepsilon(z = 0^-) \end{bmatrix} = \begin{bmatrix} T_{I,2} & R_{I,2} \\ -R_{I,2} & T_{I,2} \end{bmatrix} \begin{bmatrix} \check{a}^\varepsilon(z = 0^-) \\ \check{b}^\varepsilon(z = 0^+) \end{bmatrix}, \tag{17.9}$$

with

$$|T_{I,2}|^2 + |R_{I,2}|^2 = 1.$$

In the random slab the propagator $\mathbf{P}^\varepsilon_{(\omega,\kappa)}(-L, z)$ solves the same equation (14.35) as before, and we obtain

$$\begin{bmatrix} \check{a}^\varepsilon(0^+) \\ \check{b}^\varepsilon(0^+) \end{bmatrix} = \mathbf{J}^{\varepsilon,2}_{\omega,\kappa} \, \mathbf{P}^\varepsilon_{(\omega,\kappa)}(-L, 0) \, \mathbf{J}^{\varepsilon,1}_{\omega,\kappa} \begin{bmatrix} \check{a}^\varepsilon((-L)^-) \\ \check{b}^\varepsilon((-L)^-) \end{bmatrix}. \tag{17.10}$$

This is the setup used to generalize the configurations and experiments studied in the previous chapters. In the next sections we give two examples of such generalizations, namely the transmitted cohererent pressure field through a nonmatched random slab and the reflected wave by a nonmatched random half-space.

17.1.2 Transmission of a Pulse through a Nonmatched Random Slab

We consider the situation with a source located in the half-space $z > 0$, and we want to characterize the transmitted pressure field in the half-space $z < -L$. Thus, we introduce the nonmatched transmission and reflection coefficients $\mathcal{T}^\varepsilon_{(\omega,\kappa)}$ and $\mathcal{R}^\varepsilon_{(\omega,\kappa)}$, which solve

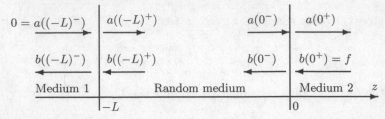

Fig. 17.1. Boundary conditions for the modes in the case of a nonmatched medium.

$$\begin{bmatrix} \mathcal{R}^\varepsilon_{(\omega,\kappa)} \\ 1 \end{bmatrix} = \mathbf{J}^{\varepsilon,2}_{\omega,\kappa} \; \mathbf{P}^\varepsilon_{(\omega,\kappa)}(-L,0) \; \mathbf{J}^{\varepsilon,1}_{\omega,\kappa} \begin{bmatrix} 0 \\ e^{\frac{i\omega L}{\varepsilon}\left(\frac{1}{\bar{c}(\kappa)}-\frac{1}{c_1(\kappa)}\right)} \mathcal{T}^\varepsilon_{(\omega,\kappa)} \end{bmatrix}. \qquad (17.11)$$

Here we choose to add a phase to the generalized transmission coefficient because it simplifies with the phase factor of the jump matrix $\mathbf{J}^{\varepsilon,1}_{\omega,\kappa}$ defined by (17.3). Writing the propagator as

$$\mathbf{P}^\varepsilon_{(\omega,\kappa)} = \begin{bmatrix} \alpha^\varepsilon_{(\omega,\kappa)} & \overline{\beta^\varepsilon_{(\omega,\kappa)}} \\ \beta^\varepsilon_{(\omega,\kappa)} & \overline{\alpha^\varepsilon_{(\omega,\kappa)}} \end{bmatrix},$$

and using the definitions (17.3–17.4) for the jump matrices $\mathbf{J}^{\varepsilon,j}_{\omega,\kappa}$, we obtain

$$\mathcal{T}^\varepsilon_{(\omega,\kappa)} = \frac{1}{\alpha^\varepsilon_{(\omega,\kappa)}\, r_1^+ r_2^+ + \beta^\varepsilon_{(\omega,\kappa)}\, r_1^- r_2^+ e^{\frac{2i\omega L}{\varepsilon\bar{c}(\kappa)}} + \overline{\beta^\varepsilon_{(\omega,\kappa)}}\, r_1^+ r_2^- + \overline{\alpha^\varepsilon_{(\omega,\kappa)}}\, r_1^- r_2^- e^{-\frac{2i\omega L}{\varepsilon\bar{c}(\kappa)}}}.$$

Using the notation (17.7) and (17.8), the definitions

$$T^\varepsilon_{(\omega,\kappa)} = \frac{1}{\alpha^\varepsilon_{(\omega,\kappa)}}, \quad \tilde{R}^\varepsilon_{(\omega,\kappa)} = \frac{-\beta^\varepsilon_{(\omega,\kappa)}}{\alpha^\varepsilon_{(\omega,\kappa)}}, \quad R^\varepsilon_{(\omega,\kappa)} = \frac{\overline{\beta^\varepsilon_{(\omega,\kappa)}}}{\alpha^\varepsilon_{(\omega,\kappa)}},$$

and the relation

$$\frac{\overline{\alpha^\varepsilon_{(\omega,\kappa)}}}{\alpha^\varepsilon_{(\omega,\kappa)}} = \frac{\alpha^\varepsilon_{(\omega,\kappa)}\overline{\alpha^\varepsilon_{(\omega,\kappa)}} - \beta^\varepsilon_{(\omega,\kappa)}\overline{\beta^\varepsilon_{(\omega,\kappa)}}}{\overline{\alpha^\varepsilon_{(\omega,\kappa)}}^2} + \frac{\beta^\varepsilon_{(\omega,\kappa)}\overline{\beta^\varepsilon_{(\omega,\kappa)}}}{\overline{\alpha^\varepsilon_{(\omega,\kappa)}}^2}$$

$$= \frac{1}{\overline{\alpha^\varepsilon_{(\omega,\kappa)}}^2} + \frac{\beta^\varepsilon_{(\omega,\kappa)}\overline{\beta^\varepsilon_{(\omega,\kappa)}}}{\overline{\alpha^\varepsilon_{(\omega,\kappa)}}^2} = (T^\varepsilon_{(\omega,\kappa)})^2 - R^\varepsilon_{(\omega,\kappa)}\tilde{R}^\varepsilon_{(\omega,\kappa)},$$

we find that

$$\mathcal{T}^\varepsilon_{(\omega,\kappa)} = \frac{T_{I,1}T_{I,2}T^\varepsilon_{(\omega,\kappa)}}{1 - \mathcal{U}^\varepsilon_{(\omega,\kappa)}},$$

with

$$\mathcal{U}^\varepsilon_{(\omega,\kappa)} = -R_{I,1}R_{I,2}\left((T^\varepsilon_{(\omega,\kappa)})^2 - R^\varepsilon_{(\omega,\kappa)}\tilde{R}^\varepsilon_{(\omega,\kappa)}\right) e^{\frac{2i\omega L}{\varepsilon\bar{c}(\kappa)}}$$
$$+ R_{I,1}\tilde{R}^\varepsilon_{(\omega,\kappa)} e^{\frac{2i\omega L}{\varepsilon\bar{c}(\kappa)}} - R_{I,2}R^\varepsilon_{(\omega,\kappa)}. \qquad (17.12)$$

We can then expand the nonmatched transmission coefficient in the form

$$T^\varepsilon_{(\omega,\kappa)} = T_{I,1} T_{I,2} T^\varepsilon_{(\omega,\kappa)} \sum_{j=0}^{\infty} [\mathcal{U}^\varepsilon_{(\omega,\kappa)}]^j. \tag{17.13}$$

The first term ($j = 0$) contains the direct arrival of the wave front, and the following terms contain all the bounces back and forth by the interfaces before transmission. The integral representation for the transmitted pressure pulse is

$$p^\varepsilon(t_0 + \varepsilon s, \mathbf{x}, (-L)^-) = \frac{1}{(2\pi\varepsilon)^3} \int \int e^{-i\frac{\omega}{\varepsilon}(t_0 + \varepsilon s - \boldsymbol{\kappa} \cdot \mathbf{x})} \frac{\sqrt{\zeta_1(\kappa)}}{2} e^{\frac{i\omega L}{\varepsilon c_1(\kappa)}}$$

$$\times \left\{ \check{b}^\varepsilon(\omega, \boldsymbol{\kappa}, 0^+) T^\varepsilon_{(\omega,\kappa)} e^{\frac{i\omega L}{\varepsilon}\left(\frac{1}{\bar{c}(\kappa)} - \frac{1}{c_1(\kappa)}\right)} \right\} \omega^2 \, d\omega \, d\boldsymbol{\kappa}. \tag{17.14}$$

where $\check{b}^\varepsilon(\omega, \boldsymbol{\kappa}, 0^+)$ models the incoming wave (see (17.16–17.17) below for a particular case). Note that the two exponential factors $e^{\frac{i\omega L}{\varepsilon c_1(\kappa)}}$ and $e^{\frac{i\omega L}{\varepsilon}\left(\frac{1}{\bar{c}(\kappa)} - \frac{1}{c_1(\kappa)}\right)}$ simplify to $e^{\frac{i\omega L}{\varepsilon \bar{c}(\kappa)}}$.

Deterministic Case

In the deterministic case we have $T^\varepsilon_{(\omega,\kappa)} = 1$, $R^\varepsilon_{(\omega,\kappa)} - R^\varepsilon_{(\omega,\kappa)} = 0$, and we obtain the following expression for the transmitted pressure:

$$p^\varepsilon(t_0 + \varepsilon s, \mathbf{x}, (-L)^-) = \frac{-1}{2(2\pi\varepsilon)^3} \int \int e^{-i\omega s} \sqrt{\zeta_1(\kappa)} \check{b}^\varepsilon(\omega, \boldsymbol{\kappa}, 0^+)$$

$$\times T_{I,1} T_{I,2} \left(\sum_{j=0}^{\infty} e^{\frac{i\omega \phi(j, \boldsymbol{\kappa}, t_0, \mathbf{x}, L)}{\varepsilon}} (-R_{I,1} R_{I,2})^j \right) \omega^2 \, d\omega \, d\boldsymbol{\kappa},$$

with the fast phase factor

$$\phi(j, \boldsymbol{\kappa}, t, \mathbf{x}, L) = -t + \boldsymbol{\kappa} \cdot \mathbf{x} + \frac{(1 + 2j)L}{\bar{c}(\kappa)}.$$

We assume that the source is a point source located at the right of the surface and that it is compactly supported in time on the ε scale and also centered in time. Then the initial coefficient $\check{b}^\varepsilon(\omega, \boldsymbol{\kappa}, 0^+)$ contains no fast-phase term and the stationary phase point associated with the phase $\phi(j, \kappa, t, \mathbf{x}, L)$ is

$$\boldsymbol{\kappa}_{\mathrm{sp},j}(\mathbf{x}) = \frac{\mathbf{x}}{\bar{c}\sqrt{|\mathbf{x}|^2 + ((1+2j)L)^2}},$$

and is computed as in Section 14.2.2. This value for the slowness vector corresponds to a plane-wave mode that is traveling in the direction $(\mathbf{x}, -(1+2j)L)$, that is, the direction of the vector from the source to the virtual point of observation $(\mathbf{x}, -(1+2j)L)$ (see Figure 17.2). Therefore, at the point $(\mathbf{x}, -L)$ we will observe a sequence of pressure pulses $p_{0,j}$ corresponding to the primary arrivals and the higher-order multiples associated with the reflections from the slab interfaces. The arrival time of the jth pulse is

$$t^{(j)} = \frac{\sqrt{|\mathbf{x}|^2 + ((1+2j)L)^2}}{\bar{c}}, \tag{17.15}$$

and we can write

$$p^\varepsilon(t^{(j)} + \varepsilon s, \mathbf{x}, (-L)^-) \overset{\varepsilon \to 0}{\longrightarrow} p_{0,j}(s, \mathbf{x}, -L),$$

where the particular form of $p_{0,j}$ depends on the particular form for the source pulse. We assume that the source term has a scaling as in Section 14.2.2, so that the transmitted pulses have amplitudes of order one. For instance, we can consider the case of a point source located just at the right of the surface at position $(\mathbf{0}, 0^+)$, emitting a short pulse at time 0, and generating the forcing term

$$\mathbf{F}^\varepsilon(t, \mathbf{x}, z) = \varepsilon f\left(\frac{t}{\varepsilon}\right) \delta(\mathbf{x}) \delta(z) \begin{bmatrix} \mathbf{0} \\ 1 \end{bmatrix}. \tag{17.16}$$

We then have

$$\check{b}^\varepsilon(\omega, \boldsymbol{\kappa}, 0^+) = \frac{\varepsilon^2}{\sqrt{\zeta_2(\kappa)}} \hat{f}(\omega) \tag{17.17}$$

and

$$p_{0,j}(s, \mathbf{x}, -L) = -\frac{T_{I,1}T_{I,2}(\kappa_{\mathrm{sp},j})[-R_{I,1}R_{I,2}(\kappa_{\mathrm{sp},j})]^j c_1(\kappa_{\mathrm{sp},j})^{1/2}}{4\pi\bar{c}c_2(\kappa_{\mathrm{sp},j})^{1/2}(2j+1)L} f'(s). \tag{17.18}$$

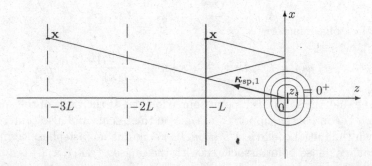

Fig. 17.2. In this figure we show the source position at 0, the observation point at $(\mathbf{x}, -L)$, and the stationary slowness vector for $j = 1$: $\kappa_{\mathrm{sp},1} = \mathbf{x}/[\bar{c}\sqrt{|\mathbf{x}|^2 + (3L)^2}]$.

Random Case

In the random case we proceed as in Section 14.2, and we get a characterization in distribution of the (jth-order mutiple) transmitted pressure pulse process by the replacements

$$T^\varepsilon_{(\omega,\kappa)}(-L,0) \mapsto \widetilde{T}_{(\omega,\kappa)}(-L,0),$$
$$R^\varepsilon_{(\omega,\kappa)}(-L,0) \mapsto 0,$$
$$\check{R}^\varepsilon_{(\omega,\kappa)}(-L,0) \mapsto 0,$$

in the expression (17.13) for the nonmatched transmission coefficient, where

$$\widetilde{T}_{(\omega,\kappa)}(-L,0) = \exp\left(i\omega \frac{\sqrt{\gamma_\kappa}}{2\bar{c}(\kappa)} W_0(L) - \omega^2 \frac{\gamma_\kappa}{8\bar{c}(\kappa)^2} L \right),$$

W_0 is a standard Brownian motion, and

$$\gamma_\kappa = \frac{\bar{c}^4(\kappa)}{\bar{c}^4} \gamma, \qquad \gamma = \int_{-\infty}^{\infty} \mathbb{E}[\nu(0)\nu(z)]dz$$

are obtained as in Section 14.2.1, equation (14.48). Substituting into the integral representation (17.14) of the transmitted pressure field, we obtain that the associated pressure approximation $\widetilde{p}^\varepsilon$ is given by

$$\widetilde{p}^\varepsilon(t_0 + \varepsilon s, \mathbf{x}, (-L)^-) = \frac{-1}{2(2\pi\varepsilon)^3} \int \int e^{-i\omega s} \sqrt{\zeta_1(\kappa)} \check{b}^\varepsilon(\omega, \boldsymbol{\kappa}, 0^+)$$

$$\times T_{I,1} T_{I,2} \widetilde{T}_{(\omega,\kappa)} \left(\sum_{j=0}^{\infty} e^{\frac{i\omega\phi(j,\boldsymbol{\kappa},t_0,\mathbf{x},L)}{\varepsilon}} \left(-R_{I,1} R_{I,2} (\widetilde{T}_{(\omega,\kappa)})^2 \right)^j \right) \omega^2 \, d\omega \, d\boldsymbol{\kappa} .$$

Finally, we get the following version of the stable-front formula in the case of a nonmatched medium:

Proposition 17.1. *In probability distribution the following characterization of the jth multiple of the transmitted wave process holds:*

$$\lim_{\varepsilon \to 0} p^\varepsilon \left(t^{(j)} + \varepsilon s, \mathbf{x}, (-L)^- \right) = \widetilde{p}_j(s, \mathbf{x}, -L),$$

where

$$t^{(j)} = \frac{\sqrt{|\mathbf{x}|^2 + ((1+2j)L)^2}}{\bar{c}},$$

$$\widetilde{p}_j(s, \mathbf{x}, -L) = \left[p_{0,j}(\cdot, \mathbf{x}, -L) * \mathcal{N}_{D_{(L,\mathbf{x},j)}} \right] \left(s - \Theta_{(L,\mathbf{x},j)} \right),$$

and we set, for $j = 1, 2, \ldots,$

$$D^2_{(L,\mathbf{x},j)} = \frac{\gamma}{4\bar{c}^2} \left(1 + \frac{|\mathbf{x}|^2}{((1+2j)L)^2} \right) (1+2j)L \,,$$

$$\Theta_{(L,\mathbf{x},j)} = \frac{\sqrt{\gamma \left(1 + \frac{|\mathbf{x}|^2}{((1+2j)L)^2} \right)}}{2\bar{c}} (1+2j)W_0(L) \,,$$

$$\mathcal{N}_D(s) = \frac{1}{\sqrt{2\pi}D} e^{-s^2/2D^2} \,,$$

with W_0 a standard Brownian motion. Here $p_{0,j}$ is the pulse shape obtained in a homogeneous medium (given by (17.18) if the source is (17.16)).

Note that the variance of the travel-time correction term,

$$\mathbb{E}[(\Theta_{(L,\mathbf{x},j)})^2] = (1+2j)D^2_{(L,\mathbf{x},j)} \,,$$

becomes large relative to the support of the determinsitic spreading for high-order multiples. This follows, since the pulse experiences the same random medium on its successive lags and the travel-time corrections are additive.

17.1.3 Reflection by a Nonmatched Random Half-Space

We consider the same situation as in the previous subsection, and analogously we compute the nonmatched reflection coefficient $\mathcal{R}^\varepsilon_{(\omega,\kappa)}$ from (17.11):

$$\mathcal{R}^\varepsilon_{(\omega,\kappa)} = \frac{R_{I,2} + \mathcal{V}^\varepsilon_{(\omega,\kappa)}}{1 - \mathcal{U}^\varepsilon_{(\omega,\kappa)}} \,, \tag{17.19}$$

where $\mathcal{U}^\varepsilon_{(\omega,\kappa)}$ is given by (17.12) and

$$\mathcal{V}^\varepsilon_{(\omega,\kappa)} = R^\varepsilon_{(\omega,\kappa)} + R_{I,1} \left((T^\varepsilon_{(\omega,\kappa)})^2 - R^\varepsilon_{(\omega,\kappa)}\tilde{R}^\varepsilon_{(\omega,\kappa)} \right) e^{\frac{2i\omega L}{\varepsilon\bar{c}(\kappa)}} - R_{I,1}R_{I,2}\tilde{R}^\varepsilon_{(\omega,\kappa)} e^{\frac{2i\omega L}{\varepsilon\bar{c}(\kappa)}} \,.$$

If we consider an incoming pulse arriving from the right half-space and observe the reflected wave during a finite time interval, then the end of the slab at $z = -L$ plays no role if L is large enough due to the finite speed of the wave propagation. Accordingly, we can restrict ourselves to the case in which the medium is matched at $z = -L$. In other words, we now assume that $R_{I,1} = 0$ ($c_1 = \bar{c}$ and $\zeta_1 = \bar{\zeta}$), and the reflection coefficient takes the simpler form

$$\mathcal{R}^\varepsilon_{(\omega,\kappa)} = \frac{R_{I,2} + R^\varepsilon_{(\omega,\kappa)}}{1 + R_{I,2}R^\varepsilon_{(\omega,\kappa)}} \,,$$

which can be expanded as

$$\mathcal{R}^\varepsilon_{(\omega,\kappa)} = R_{I,2} + \sum_{n=0}^{\infty} (-R_{I,2})^n T^2_{I,2}(R^\varepsilon_{(\omega,\kappa)})^{n+1} \,. \tag{17.20}$$

The first term $R_{I,2}$ represents the wave that has been directly reflected by the interface before entering the slab. The second term $(n = 0)$ is $T_{I,2}^2 R_{(\omega,\kappa)}^\varepsilon$, which represents the wave that goes through the interface, is reflected back by the inhomogeneities of the slab, and goes out to the right half-space through the interface. It is the only term that is present in the matched medium case. The other terms $(n \geq 1)$ represent waves that have been reflected back n times in the random slab by the interface.

The integral representation for the reflected pressure wave is

$$p^\varepsilon(t, \mathbf{x}, 0^+) = \frac{-1}{2(2\pi\varepsilon)^3} \int\int e^{-i\frac{\omega}{\varepsilon}(t-\boldsymbol{\kappa}\cdot\mathbf{x})} \mathcal{R}_{(\omega,\kappa)}^\varepsilon \sqrt{\zeta_2(\kappa)}\,\breve{b}^\varepsilon(\omega, \boldsymbol{\kappa}, 0^+)\omega^2\, d\omega\, d\boldsymbol{\kappa},$$

where $\breve{b}^\varepsilon(\omega, \boldsymbol{\kappa}, 0^+)$ models the incoming wave. We can now carry through the various moment analyses needed in the study of the reflected wave or the time-reversed refocused pulse. These moment analyses involve products of standard reflection coefficients $R_{(\omega,\kappa)}^\varepsilon$ and therefore the solution $\mathcal{W}_p$ of the transport equations (14.67).

For instance, we can consider the case of a point source located just at the right of the surface at position $(\mathbf{0}, 0^+)$, emitting a short pulse at time 0, and compute the mean reflected intensity. The point source generates the source term

$$\mathbf{F}^\varepsilon(t, \mathbf{x}, z) = \varepsilon^{1/2} f\left(\frac{t}{\varepsilon}\right) \delta(\mathbf{x})\delta(z) \begin{bmatrix} \mathbf{0} \\ 1 \end{bmatrix},$$

which imposes the boundary condition

$$\breve{b}^\varepsilon(\omega, \boldsymbol{\kappa}, 0^+) = \frac{\varepsilon^{3/2}}{\sqrt{\zeta_2(\kappa)}} \hat{f}(\omega).$$

We observe the mean reflected intensity at the surface $z = 0$, at the location $\mathbf{x} \neq \mathbf{0}$, and at time $t > 0$:

$$\mathbb{E}[p^\varepsilon(t, \mathbf{x}, 0^+)^2] = \frac{1}{256\pi^6\varepsilon^3} \int\int e^{-i\frac{\omega-\omega'}{\varepsilon}t + i\frac{\omega\boldsymbol{\kappa}-\omega'\boldsymbol{\kappa}'}{\varepsilon}\cdot\mathbf{x}}$$

$$\times\mathbb{E}\left[\mathcal{R}_{(\omega,\kappa)}^\varepsilon \overline{\mathcal{R}_{(\omega',\kappa')}^\varepsilon}\right] \hat{f}(\omega)\overline{\hat{f}(\omega')}\omega^2\kappa\omega'^2\kappa'\, d\omega\, d\boldsymbol{\kappa}\, d\omega'\, d\boldsymbol{\kappa}'.$$

We parameterize $\mathbf{x} = (x, 0)$, $x > 0$, and use polar coordinates for $\boldsymbol{\kappa}$ and $\boldsymbol{\kappa}'$:

$$\mathbb{E}[p^\varepsilon(t, \mathbf{x}, 0^+)^2] = \frac{1}{256\pi^6\varepsilon^3} \int\int e^{-i\frac{\omega-\omega'}{\varepsilon}t + i\frac{\omega x\mu\cos(\theta)-\omega' x\mu'\cos(\theta')}{\varepsilon}}$$

$$\times\mathbb{E}\left[\mathcal{R}_{(\omega,\mu)}^\varepsilon \overline{\mathcal{R}_{(\omega',\mu')}^\varepsilon}\right] \hat{f}(\omega)\overline{\hat{f}(\omega')}\omega^2\mu\omega'^2\mu'\, d\omega\, d\theta\, d\mu\, d\omega'\, d\theta'\, d\mu'. \quad (17.21)$$

We follow the stationary-phase analysis carried out in Section 14.3.3, which addresses the same problem in the matched-medium case. The difference is contained in the expectation of the product of the two generalized reflection coefficients $\mathcal{R}^{\varepsilon}_{(\omega,\mu)}$. From the expansion (17.20) and the limits (14.69), we get that

$$
\mathbb{E}\left[\mathcal{R}^{\varepsilon}_{(\omega,\mu)}\,\overline{\mathcal{R}^{\varepsilon}_{(\omega+\varepsilon h,\mu+\varepsilon\lambda)}}\right]
$$
$$
\xrightarrow{\varepsilon\to 0} R^2_{I,2} + \sum_{n=0}^{\infty} R^{2n}_{I,2} T^4_{I,2} \int \mathcal{W}_{n+1}\left(\omega,\mu,\tau,-L,0\right) e^{i\tau[h\bar{c}^2/\bar{c}(\mu)^2-\omega\lambda\bar{c}^2\mu]} d\tau\,.
$$

By substitution into (17.21) we obtain that in the limit $\varepsilon \to 0$, the mean intensity is zero if $x > \bar{c}t$, and if $x < \bar{c}t$, then it is given by

$$
\lim_{\varepsilon\to 0} \mathbb{E}[p^{\varepsilon}(t,\mathbf{x},0^+)^2] = \sum_{n=0}^{\infty} \frac{R^{2n}_{I,2} T^4_{I,2}}{32\pi^3 \bar{c}^4 t^2} \int \mathcal{W}_{n+1}\left(\omega,\kappa_{x,t},t,-L,0\right) |\hat{f}(\omega)|^2 \omega^2\, d\omega\,,
$$

where

$$
\kappa_{x,t} = \frac{x}{\bar{c}^2 t}\,.
$$

In the asymptotic L large (in fact, as soon as $L \geq \bar{c}t/2$),

$$
\mathcal{W}_p(\omega,\kappa_{x,t},t,-L,0) \xrightarrow{L\to\infty} \frac{\bar{c}^2}{\bar{c}(\kappa_{x,t})L_{\mathrm{loc}}(\omega,\kappa_{x,t})} P^{\infty}_p\left(\frac{\bar{c}^2 t}{\bar{c}(\kappa_{x,t})L_{\mathrm{loc}}(\omega,\kappa_{x,t})}\right)\,,
$$

where P^{∞}_p is given by (9.39) and $L_{\mathrm{loc}}(\omega,\kappa)$ is the localization length defined by (14.62). As a result, if the condition $\bar{c}t < 2L$ is satisfied, then we have for all $x < \bar{c}t$,

$$
\mathcal{W}_{n+1}\left(\omega,\kappa_{x,t},t,-L,0\right) = \frac{\bar{c}}{L_{\mathrm{loc}}(\omega)} \frac{\bar{c}t}{\sqrt{\bar{c}^2 t^2 - x^2}} P^{\infty}_{n+1}\left(\frac{\bar{c}t}{L_{\mathrm{loc}}(\omega)} \frac{\bar{c}t}{\sqrt{\bar{c}^2 t^2 - x^2}}\right)\,,
$$

where $L_{\mathrm{loc}}(\omega) = 4\bar{c}^2/\gamma\omega^2$ is the one-dimensional localization length defined by (7.81). This gives an explicit representation of the mean reflected intensity for $x < \bar{c}t$:

$$
\lim_{\varepsilon\to 0} \mathbb{E}[p^{\varepsilon}(t,\mathbf{x},0^+)^2] = \frac{T^4_{I,2}}{32\pi^3 \bar{c}^4 t^2} \int \frac{\frac{\bar{c}}{2L_{\mathrm{loc}}(\omega)} \frac{\bar{c}t}{\sqrt{\bar{c}^2 t^2 - x^2}}}{\left(1 + \frac{\bar{c}t}{2L_{\mathrm{loc}}(\omega)} \frac{\bar{c}t}{\sqrt{\bar{c}^2 t^2 - x^2}} T^2_{I,2}\right)^2} |\hat{f}(\omega)|^2 \omega^2\, d\omega\,,
$$

where $T_{I,2}$ is given by (17.7) evaluated at $\kappa_{x,t}$. This type of computation can be generalized in an analogous manner in order to treat other situations such as a source emitting from above the surface ($z_s > 0$) or a source embedded in the medium ($z_s < 0$). Similarly, time-reversal refocusing could be computed in the case of a nonmatched medium, as a generalization of the results obtained in Chapters 15 and 16.

17.2 General Background

Up to this point we have dealt with layered media with a random bulk modulus that is rapidly varying around a constant value or a piecewise constant value, as in Sections 8.3 and 11.1. In this section we extend the results to the case in which the bulk modulus is also modulated by a slow background variation described by the deterministic **smooth** function $K_0(z)$. Thus, we model the parameters of the medium as

$$\frac{1}{K(\mathbf{x},z)} = \frac{1}{K(z)} = \begin{cases} \frac{1}{K_0(z)}\left(1+\nu(z/\varepsilon^2)\right) & \text{for } z \in [-L,0], \\ \frac{1}{\bar{K}} & \text{for } z \in (-\infty,-L) \cup (0,\infty), \end{cases}$$

$$\rho(\mathbf{x},z) = \bar{\rho} \text{ for all } (\mathbf{x},z).$$

To simplify, we assume a matched medium at both ends of the slab in the sense that $K_0(-L) = \bar{K} = K_0(0)$ and we address our familiar strongly heterogeneous white-noise regime, where the correlation length of the medium, which is of order ε^2, is much smaller than the typical wavelength of order ε, which is itself much smaller than the propagation distance of order 1. The new aspect of the analysis is primarily that the mode decomposition needs to be adapted to the general background. More precisely, we have to center the modes along the slowly varying characteristics imposed by the slowly varying background. We are then able to define reflection and transmission coefficients in the same way as in the constant-background case. The key ingredient is still the separation-of-scales technique and the associated diffusion-approximation theorem. We can then characterize the limiting statistical distribution of the reflection coefficient in terms of a solution of a system of transport equations that follow the slowly varying characteristics. The application of this result to time reversal is then similar to the constant-background case.

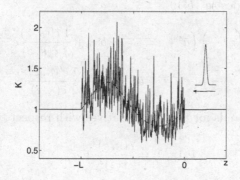

Fig. 17.3. A typical profile (solid line) of the bulk modulus with a slowly varying background (dashed line).

17.2.1 Mode Decomposition

The slowly varying background can be characterized by its local velocity and impedance

$$\bar{c}(z) = \sqrt{K_0(z)/\bar{\rho}}, \qquad \bar{\zeta}(z) = \bar{\rho}\bar{c}(z).$$

We next adapt the sequence of transforms that we have introduced in the constant-background case. The mode velocity and impedance are in this case

$$\bar{c}(\kappa, z) = \frac{\bar{c}(z)}{\sqrt{1 - \kappa^2 \bar{c}^2(z)}}, \qquad \bar{\zeta}(\kappa, z) = \bar{\rho}\bar{c}(\kappa, z). \qquad (17.22)$$

We assume that there are **no turning points**, that is, we consider only modes such that $\kappa\bar{c}(z) < 1$ for all $z \in (-L, 0)$. In order to define the characteristics along which we center the modes, we introduce the travel time

$$\vartheta_\kappa(z) = \int_0^z \frac{dz'}{\bar{c}(\kappa, z')}, \qquad (17.23)$$

which is the **effective time** necessary for a plane wave with slowness κ to reach the depth z (more precisely, this time is $-\vartheta_\kappa$, since z is negative). We decompose the wave field into right- ($\check{a}^\varepsilon = \check{a}^\varepsilon(\omega, \kappa, z)$) and left-going ($\check{b}^\varepsilon = \check{b}^\varepsilon(\omega, \kappa, z)$) waves with respect to the z-direction by setting

$$\hat{p}^\varepsilon(\omega, \kappa, z) = \frac{\sqrt{\bar{\zeta}(\kappa, z)}}{2} \left(\check{a}^\varepsilon(\omega, \kappa, z) e^{\frac{i\omega\vartheta_\kappa(z)}{\varepsilon}} - \check{b}^\varepsilon(\omega, \kappa, z) e^{-\frac{i\omega\vartheta_\kappa(z)}{\varepsilon}} \right), \quad (17.24)$$

$$\hat{u}^\varepsilon(\omega, \kappa, z) = \frac{1}{2\sqrt{\bar{\zeta}(\kappa, z)}} \left(\check{a}^\varepsilon(\omega, \kappa, z) e^{\frac{i\omega\vartheta_\kappa(z)}{\varepsilon}} + \check{b}^\varepsilon(\omega, \kappa, z) e^{-\frac{i\omega\vartheta_\kappa(z)}{\varepsilon}} \right). \quad (17.25)$$

Substituting these expressions into the acoustic equations gives the system satisfied by the modes $(\check{a}^\varepsilon, \check{b}^\varepsilon)$:

$$\frac{d\check{a}^\varepsilon}{dz} = \frac{i\omega}{2\bar{c}(\kappa, z)\varepsilon} \nu_\kappa \left(z, \frac{z}{\varepsilon^2} \right) \left(\check{a}^\varepsilon - e^{\frac{-2i\omega\vartheta_\kappa(z)}{\varepsilon}} \check{b}^\varepsilon \right) + \frac{1}{2} \frac{\bar{c}'(\kappa, z)}{\bar{c}(\kappa, z)} e^{\frac{-2i\omega\vartheta_\kappa(z)}{\varepsilon}} \check{b}^\varepsilon, (17.26)$$

$$\frac{d\check{b}^\varepsilon}{dz} = \frac{i\omega}{2\bar{c}(\kappa, z)\varepsilon} \nu_\kappa \left(z, \frac{z}{\varepsilon^2} \right) \left(e^{\frac{2i\omega\vartheta_\kappa(z)}{\varepsilon}} \check{a}^\varepsilon - \check{b}^\varepsilon \right) + \frac{1}{2} \frac{\bar{c}'(\kappa, z)}{\bar{c}(\kappa, z)} e^{\frac{2i\omega\vartheta_\kappa(z)}{\varepsilon}} \check{a}^\varepsilon, \quad (17.27)$$

where the prime stands for partial derivatives with respect to z, and

$$\nu_\kappa(z, \tilde{z}) = \frac{\bar{c}(\kappa, z)^2}{\bar{c}(z)^2} \nu(\tilde{z}).$$

The system (17.26–17.27) describes the coupling between the right- and left-going modes. This coupling results from the scattering with the rapidly varying fluctuations described by the terms proportional to the random process ν_κ and also from the scattering with the slowly varying background captured by the terms proportional to $\bar{c}'/\bar{c}$. We can use the propagator formulation and introduce the reflection and transmission coefficients in the same way as we did in the constant-background case in Section 14.1. The Riccati equation satisfied by the reflection coefficient is

$$\frac{dR^{\varepsilon}_{(\omega,\kappa)}}{dz} = -\frac{i\omega}{2\bar{c}(\kappa,z)\varepsilon}\nu_{\kappa}\left(z,\frac{z}{\varepsilon^2}\right)\left(e^{\frac{-2i\omega\vartheta_{\kappa}(z)}{\varepsilon}} - 2R^{\varepsilon}_{(\omega,\kappa)} + (R^{\varepsilon}_{(\omega,\kappa)})^2 e^{\frac{2i\omega\vartheta_{\kappa}(z)}{\varepsilon}}\right)$$

$$+ \frac{1}{2}\frac{\bar{c}'(\kappa,z)}{\bar{c}(\kappa,z)}\left(e^{\frac{-2i\omega\vartheta_{\kappa}(z)}{\varepsilon}} - (R^{\varepsilon}_{(\omega,\kappa)})^2 e^{\frac{2i\omega\vartheta_{\kappa}(z)}{\varepsilon}}\right). \qquad (17.28)$$

The terms proportional to $\bar{c}'(\kappa,z)/\bar{c}(\kappa,z)$ in the right-hand side do not play any role in the limit $\varepsilon \to 0$ because they average out due to the fast phases (see Theorem 6.4). Thus the Riccati equation for the reflection coefficient is very similar to the one encountered in the constant-background case.

17.2.2 Transport Equations

As seen in the previous chapters, the autocorrelation function of the reflection coefficient plays a primary role, and its limiting statistical distribution has been found to be characterized by a system of transport equations. It turns out that this is still the case with a general background. We introduce the family of products of reflection coefficients

$$U^{\varepsilon}_{p,q}(\omega,\kappa,h,\lambda,z_0,z) = \left(R^{\varepsilon}_{(\omega+\varepsilon h/2,\kappa+\varepsilon\lambda/2)}(z_0,z)\right)^p \left(\overline{R^{\varepsilon}_{(\omega-\varepsilon h/2,\kappa-\varepsilon\lambda/2)}(z_0,z)}\right)^q.$$

From the Riccati equation (17.28) and the expansion

$$2(\omega + \varepsilon h/2)\vartheta_{\kappa+\varepsilon\lambda/2}(z) = 2\omega\vartheta_{\kappa}(z) + \varepsilon\left(h\vartheta_{\kappa}(z) - \omega\lambda\xi_{\kappa}(z)\right) + O(\varepsilon^2),$$

where

$$\xi_{\kappa}(z) = \int_0^z \kappa\bar{c}(\kappa,z')dz',$$

we get that U^{ε} satisfies to leading order the system

$$\frac{\partial U^{\varepsilon}_{p,q}}{\partial z} = \frac{i\omega}{\bar{c}(\kappa,z)}\nu^{\varepsilon}_{\kappa}(p-q)U^{\varepsilon}_{p,q}$$

$$+ \frac{i\omega}{2\bar{c}(\kappa,z)}\nu^{\varepsilon}_{\kappa}e^{\frac{2i\omega\vartheta_{\kappa}(z)}{\varepsilon}}\left(qe^{-ih\vartheta_{\kappa}(z)+i\omega\lambda\xi_{\kappa}(z)}U^{\varepsilon}_{p,q-1} - pe^{ih\vartheta_{\kappa}(z)-i\omega\lambda\xi_{\kappa}(z)}U^{\varepsilon}_{p+1,q}\right)$$

$$+ \frac{i\omega}{2\bar{c}(\kappa,z)}\nu^{\varepsilon}_{\kappa}e^{-\frac{2i\omega\vartheta_{\kappa}(z)}{\varepsilon}}\left(qe^{ih\vartheta_{\kappa}(z)-i\omega\lambda\xi_{\kappa}(z)}U^{\varepsilon}_{p,q+1} - pe^{-ih\vartheta_{\kappa}(z)+i\omega\lambda\xi_{\kappa}(z)}U^{\varepsilon}_{p-1,q}\right),$$

starting from $U^{\varepsilon}_{p,q}(\omega,\kappa,h,\lambda,z_0,z=z_0) = 1_0(p)1_0(q)$. Here we have set $\nu^{\varepsilon}_{\kappa}(z) = \nu_{\kappa}(z,z/\varepsilon^2)/\varepsilon$, and we have omitted the terms proportional to $\bar{c}'/\bar{c}$ because they average out as $\varepsilon \to 0$. We consider the associated family of Fourier transforms

$$V^{\varepsilon}_{p,q}(\omega,\kappa,\tau,\chi,z_0,z) = \frac{\omega}{4\pi^2}\int\int e^{-ih[\tau-(p+q)\vartheta_{\kappa}(z)]+i\omega\lambda[\chi-(p+q)\xi_{\kappa}(z)]}$$

$$\times U^{\varepsilon}_{p,q}(\omega,\kappa,h,\lambda,z_0,z)\,dh\,d\lambda. \qquad (17.29)$$

By applying the limit theorem of Section 6.7.3, in the same way as in Section 9.2.1, we obtain that the process $(V^\varepsilon_{p,q})_{p,q \in \mathbb{N}}$ converges in distribution as $\varepsilon \to 0$ to a diffusion process. In particular, the moments converge,

$$\mathbb{E}[V^\varepsilon_{p,p}(\omega, \kappa, \tau, \chi, z_0, z)] \xrightarrow{\varepsilon \to 0} \mathcal{V}_p(\omega, \kappa, \tau, \chi, z_0, z),$$

where $\mathcal{V}_p$ satisfies the system of transport equations

$$\frac{\partial \mathcal{V}_p}{\partial z} + \frac{2p}{\bar{c}(\kappa, z)}\frac{\partial \mathcal{V}_p}{\partial \tau} + 2p\bar{c}(\kappa, z)\kappa\frac{\partial \mathcal{V}_p}{\partial \chi} = \frac{p^2}{L_{\mathrm{loc}}(\omega, \kappa, z)}(\mathcal{V}_{p+1} + \mathcal{V}_{p-1} - 2\mathcal{V}_p),$$

$$(17.30)$$

starting from $\mathcal{V}_p(\tau, \omega, \kappa, \chi, z_0, z = z_0) = \mathbf{1}_0(p)\delta(\tau)\delta(\chi)$. The "local" localization length is

$$L_{\mathrm{loc}}(\omega, \kappa, z) = \frac{4\bar{c}(z)^4}{\gamma\bar{c}(\kappa, z)^2\omega^2}, \qquad \gamma = \int_{-\infty}^{\infty} \mathbb{E}[\nu(0)\nu(z)]\,dz.$$

The solution of this system admits a representation in terms of the inhomogeneous jump Markov process $(N_z)_{z \geq z_0}$ with state space $\mathbb{N}$ and infinitesimal generator

$$\mathcal{L}_{(\omega, \kappa, z)}\phi(N) = \frac{N^2}{L_{\mathrm{loc}}(\omega, \kappa, z)}(\phi(N+1) + \phi(N-1) - 2\phi(N)).$$

The solution $\mathcal{V}_p$ can be written as the expectation of a functional of this jump process:

$$\mathcal{V}_p(\omega, \kappa, \tau, \chi, z_0, z) = \mathbb{E}\left[\mathbf{1}_0(N_z)\delta\left(\tau - 2\int_{z_0}^{z}\frac{N_s}{\bar{c}(\kappa, z_0 + z - s)}ds\right)\right.$$
$$\left. \times \delta\left(\chi - 2\int_{z_0}^{z}\kappa\bar{c}(\kappa, z_0 + z - s)N_s ds\right) \mid N_{z_0} = p\right].$$

Note that we cannot combine the two Dirac distributions as we could do in the constant-background case, so we cannot eliminate the χ-variable from the transport equations. Taking the inverse Fourier transform of (17.29), we get that for $z_0 \leq z$,

$$\mathbb{E}\left[\left(R^\varepsilon_{(\omega+\varepsilon h/2, \kappa+\varepsilon\lambda/2)}(z_0, z)\right)^p \left(\overline{R^\varepsilon_{(\omega-\varepsilon h/2, \kappa-\varepsilon\lambda/2)}(z_0, z)}\right)^q\right] \qquad (17.31)$$

$$\xrightarrow{\varepsilon \to 0} \begin{cases} e^{2ip[-h\vartheta_\kappa(z) + \omega\lambda\xi_\kappa(z)]}\displaystyle\int\int \mathcal{V}_p(\omega, \kappa, \tau, \chi, z_0, z)e^{i[\tau h - \omega\lambda\chi]}\,d\tau\,d\chi & \text{if } q = p, \\ 0 & \text{otherwise.} \end{cases}$$

These asymptotic moments, and similar ones for the transmission coefficients, can be used to characterize quantities of interest, such as the transmitted stable front, the mean reflected intensity, the time-reversed refocused pulse, for instance. The main difference with the constant-background case is that the stationary points involved in the computations of integral representations are given implicitly, and do not lead to simple formulas. We give some of these results in the next section.

17.2.3 Applications

Transmitted Stable Front

The results obtained in Chapter 14 concerning the transmitted stable front can be generalized to the case of a general background. At the offset $\mathbf{x}$, we can observe a stable front around the time

$$t_0 = \int_{-L}^{0} \frac{1}{\bar{c}(z)\sqrt{1 - \bar{c}(z)^2 \kappa_0^2}} dz \,.$$

Here $\kappa_0 = |\boldsymbol{\kappa}_0|$ is defined implicitly by

$$\mathbf{x} = \boldsymbol{\kappa}_0 \int_{-L}^{0} \frac{\bar{c}(z)}{\sqrt{1 - \bar{c}(z)^2 |\boldsymbol{\kappa}_0|^2}} dz \,,$$

which is assumed to be unique. For the reader familiar with geometrical optics [20], the stationary slowness vector corresponds to the fastest ray going from the source O to the observation point $(\mathbf{x}, -L)$ in absence of random fluctuations. This ray is uniquely defined if we assume that there are no caustics until depth L. The time t_0 corresponds to the travel time along this ray path.

As in the constant-background case, the transmitted stable front through a random slab is modified in two ways compared to the homogeneous case. First, it experiences a small random time shift, which can be described as a Gaussian random variable with zero mean and variance

$$D^2_{(L,\mathbf{x})} = \frac{\gamma}{4} \int_{-L}^{0} \frac{1}{\bar{c}(z)^2[1 - \bar{c}(z)^2 \kappa_0^2]} dz \,.$$

Second, its shape is the convolution of the homogeneous front with a deterministic Gaussian kernel whose variance is $D^2_{(L,\mathbf{x})}$.

Mean Reflected Intensity

The mean reflected intensity is obtained as in Section 14.3.3. One starts from the integral representation (14.73) and uses the asymptotic expression (17.31) (with $p = q = 1$) for the autocorrelation function of the reflection coefficient. One then obtains

$$\lim_{\varepsilon \to 0} \mathbb{E}[p^\varepsilon(t, \mathbf{x}, 0^+)^2] = \frac{1}{32\pi^3 x} \int \int \mathcal{V}_1(\omega, \mu, t - \mu x, x, -L, 0) |\hat{f}(\omega)|^2 \omega^2 \mu \, d\omega \, d\mu \,,$$

for $\mathbf{x} = (x, 0)$, $x > 0$. The effect of the varying background is contained in the density $\mathcal{V}_1$ given by the system of transport equations (17.30).

Time Reversal

We can also revisit the results obtained in Chapter 15 for time reversal in the presence of a general background. The analysis is similar to the constant-background case. The difference arises when we substitute the limiting expression of the autocorrelation function of the reflection coefficient. We get the same result as stated in Theorem 15.1: refocusing is observed on the original source position, and refocusing occurs exactly at the expected time $-t_s$. The pulse shape can be expressed as a convolution of the original pulse shape with a refocusing kernel that is given in terms of the solution of the system of transport equations (17.30).

Inverse Problem

The inverse problem, which consists in reconstructing the slowly varying background $K_0(z)$ from the observation of the reflected waves can be approached in the following way. The quantity $\mathcal{V}_1(\omega, \kappa, \tau, \chi, -L, 0)$ is estimated from the cross-correlations of the reflected signals, or from the time-reversal refocusing kernels. Then, the coefficient $K_0(z)$ is retrieved by solving the deterministic inverse problem associated with the system of transport equations (17.30). These two steps are delicate and involve sophisticated statistical estimators and numerical procedures beyond the scope of this book. We refer to [8, 133] for more details.

17.3 Medium with Random Density Fluctuations

In this section we discuss the case in which not only the bulk modulus, but also the density fluctuates randomly. We consider again linear acoustic waves propagating in three spatial dimensions:

$$\rho \frac{\partial \mathbf{u}^\varepsilon}{\partial t} + \nabla p = \mathbf{F}^\varepsilon \,, \tag{17.32}$$

$$\frac{1}{K} \frac{\partial p^\varepsilon}{\partial t} + \nabla \cdot \mathbf{u}^\varepsilon = 0 \,, \tag{17.33}$$

with ρ the density of the medium and K the bulk modulus.

As before, we model fluctuations around the reciprocal of the bulk modulus and introduce now also fluctuations in the density as

$$\frac{1}{K(\mathbf{x}, z)} = \frac{1}{K(z)} = \begin{cases} \frac{1}{\overline{K}} \left(1 + \nu(z/\varepsilon^2) \right) & \text{for } z \in [-L, 0] \,, \\ \frac{1}{\overline{K}} & \text{for } z \in (-\infty, -L) \cup (0, \infty) \,, \end{cases}$$

$$\rho(\mathbf{x}, z) = \begin{cases} \overline{\rho} \left(1 + \eta(z/\varepsilon^2) \right) & \text{for } z \in [-L, 0] \,, \\ \overline{\rho} & \text{for } z \in (-\infty, -L) \cup (0, \infty) \,, \end{cases}$$

where the random fluctuations ν and η are ergodic Markov processes. We consider a source located in the right homogeneous half-space at the location $(\mathbf{0}, z_s)$, $z_s > 0$, which imposes the forcing term

$$\mathbf{F}^\varepsilon(t, \mathbf{x}, z) = \varepsilon^q \begin{bmatrix} \mathbf{0} \\ 1 \end{bmatrix} f\left(\frac{t}{\varepsilon}, \frac{\mathbf{x}}{\varepsilon}\right) \delta(z - z_s). \qquad (17.34)$$

Taking the specific Fourier transform gives in this case the system

$$-\bar{\rho}(1+\eta)\frac{i\omega}{\varepsilon}\hat{\mathbf{v}}^\varepsilon + \frac{i\omega}{\varepsilon}\boldsymbol{\kappa}\hat{p}^\varepsilon = \mathbf{0},$$

$$-\bar{\rho}(1+\eta)\frac{i\omega}{\varepsilon}\hat{u}^\varepsilon + \frac{\partial\hat{p}^\varepsilon}{\partial z} = \varepsilon^{q+3}\hat{f}(\omega, \boldsymbol{\kappa})\delta(z - z_s),$$

$$-\frac{1}{\bar{K}}(1+\nu)\frac{i\omega}{\varepsilon}\hat{p}^\varepsilon + \frac{i\omega}{\varepsilon}\boldsymbol{\kappa}\cdot\hat{\mathbf{v}}^\varepsilon + \frac{\partial\hat{u}^\varepsilon}{\partial z} = 0,$$

with $\hat{f}$ being the unscaled specific Fourier transform. From (14.6–14.7) we see that both the mean density $\bar{\rho}$ and the harmonic mean density $\widetilde{\rho}$ defined by

$$\frac{1}{\widetilde{\rho}} = \mathbb{E}\left[\frac{1}{\bar{\rho}(1+\eta)}\right]$$

are important in determining the effective medium. In general, $\widetilde{\rho} \leq \bar{\rho}$ and in the case that $\widetilde{\rho} < \bar{\rho}$ we **do not have matched boundary conditions**, as we discuss next. We eliminate $\hat{\mathbf{v}}^\varepsilon$ to get the system

$$-\frac{i\omega}{\varepsilon}\bar{\rho}(1+\eta)\hat{u}^\varepsilon + \frac{\partial\hat{p}^\varepsilon}{\partial z} = \varepsilon^{q+3}\hat{f}(\omega, \boldsymbol{\kappa})\delta(z - z_s), \qquad (17.35)$$

$$-\frac{i\omega}{\varepsilon}\left(\frac{1}{\bar{K}}(1+\nu) - \frac{\kappa^2}{\widetilde{\rho}}(1+\tilde{\eta})\right)\hat{p}^\varepsilon + \frac{\partial\hat{u}^\varepsilon}{\partial z} = 0, \qquad (17.36)$$

where we have introduced the centered process $\tilde{\eta}$ defined by

$$\frac{1}{\bar{\rho}(1+\eta)} = \frac{1}{\widetilde{\rho}}(1+\tilde{\eta}).$$

We moreover define the mode-dependent wave speed by

$$\bar{c}(\kappa, z) = \begin{cases} \bar{c}(\kappa) = \dfrac{\bar{c}}{\sqrt{1-\kappa^2\bar{c}^2(\bar{\rho}/\widetilde{\rho})}} & \text{for } z \in [-L, 0], \\ c_0(\kappa) = \dfrac{\bar{c}}{\sqrt{1-\bar{c}^2\kappa^2}} & \text{for } z \in (-\infty, -L) \cup (0, \infty), \end{cases} \qquad (17.37)$$

where $\bar{c}^2 = \bar{K}/\bar{\rho}$, and the effective mode-dependent acoustic impedance is given by

$$\bar{\zeta}(\kappa, z) = \begin{cases} \bar{\zeta}(\kappa) = \bar{\rho}\bar{c}(\kappa) & \text{for } z \in [-L, 0], \\ \zeta_0(\kappa) = \bar{\rho}c_0(\kappa) & \text{for } z \in (-\infty, -L) \cup (0, \infty). \end{cases} \qquad (17.38)$$

These definitions allow us to write

$$-\frac{i\omega}{\varepsilon}\bar{\rho}(1+\eta)\hat{u}^{\varepsilon} + \frac{\partial\hat{p}^{\varepsilon}}{\partial z} = \varepsilon^{q+3}\hat{f}(\omega,\boldsymbol{\kappa})\delta(z-z_{s}),   \quad(17.39)$$

$$-\frac{i\omega}{\varepsilon}\frac{1}{\bar{\rho}\bar{c}(\kappa,z)^{2}}(1+\nu_{\kappa})\hat{p}^{\varepsilon} + \frac{\partial\hat{u}^{\varepsilon}}{\partial z} = 0,   \quad(17.40)$$

by defining the mode-dependent medium fluctuations

$$\nu_{\kappa} = \frac{\bar{c}(\kappa)^{2}}{\bar{c}^{2}}(\nu-\tilde{\eta}) + \tilde{\eta}.$$

Thus, we have transformed the problem into a family of one-dimensional mode propagation problems that take the same form as those we have analyzed before. We accordingly decompose the pressure and velocity into right- ($\check{a}^{\varepsilon}$) and left-going ($\check{b}^{\varepsilon}$) modes:

$$\hat{p}^{\varepsilon}(\omega,\boldsymbol{\kappa},z) = \frac{\sqrt{\bar{\zeta}(\kappa,z)}}{2}\left(\check{a}^{\varepsilon}(\omega,\boldsymbol{\kappa},z)e^{\frac{i\omega z}{\varepsilon\bar{c}(\kappa,z)}} - \check{b}^{\varepsilon}(\omega,\boldsymbol{\kappa},z)e^{-\frac{i\omega z}{\varepsilon\bar{c}(\kappa,z)}}\right),   \quad(17.41)$$

$$\hat{u}^{\varepsilon}(\omega,\boldsymbol{\kappa},z) = \frac{1}{2\sqrt{\bar{\zeta}(\kappa,z)}}\left(\check{a}^{\varepsilon}(\omega,\boldsymbol{\kappa},z)e^{\frac{i\omega z}{\varepsilon\bar{c}(\kappa,z)}} + \check{b}^{\varepsilon}(\omega,\boldsymbol{\kappa},z)e^{-\frac{i\omega z}{\varepsilon\bar{c}(\kappa,z)}}\right).   \quad(17.42)$$

In the case with fluctuations in the density, the mode system involves the two processes

$$m_{\kappa} = \nu_{\kappa} + \eta,   \qquad n_{\kappa} = \nu_{\kappa} - \eta,   \quad(17.43)$$

since by substituting (17.41) and (17.42) in (17.39) and (17.40) we get

$$\frac{d}{dz}\begin{bmatrix}\check{a}^{\varepsilon}\\\check{b}^{\varepsilon}\end{bmatrix} = \frac{i\omega}{2\bar{c}(\kappa)\varepsilon}\begin{bmatrix}m_{\kappa}\left(\frac{z}{\varepsilon^{2}}\right) & -n_{\kappa}\left(\frac{z}{\varepsilon^{2}}\right)e^{\frac{-2i\omega z}{\bar{c}(\kappa)\varepsilon}}\\n_{\kappa}\left(\frac{z}{\varepsilon^{2}}\right)e^{\frac{2i\omega z}{\bar{c}(\kappa)\varepsilon}} & -m_{\kappa}\left(\frac{z}{\varepsilon^{2}}\right)\end{bmatrix}\begin{bmatrix}\check{a}^{\varepsilon}\\\check{b}^{\varepsilon}\end{bmatrix},   \quad(17.44)$$

for $z \in (-L,0)$. The continuity conditions for $\hat{p}^{\varepsilon}$ and $\hat{u}^{\varepsilon}$ give the following jump conditions for the amplitudes at the location of the source:

$$[\check{a}^{\varepsilon}]_{z_{s}} = \varepsilon^{q+3}\frac{1}{\sqrt{\zeta_{0}(\kappa)}}\hat{f}(\omega,\boldsymbol{\kappa})e^{\frac{-i\omega z_{s}}{\varepsilon c_{0}(\kappa)}},   \quad(17.45)$$

$$[\check{b}^{\varepsilon}]_{z_{s}} = -\varepsilon^{q+3}\frac{1}{\sqrt{\zeta_{0}(\kappa)}}\hat{f}(\omega,\boldsymbol{\kappa})e^{\frac{i\omega z_{s}}{\varepsilon c_{0}(\kappa)}}.   \quad(17.46)$$

In the following subsection we discuss how the propagator and the distribution of the slab-transmission coefficients are affected by the density variations. Then in Sections 17.3.2 and 17.3.4 respectively we will use these results to analyze how the transmitted and reflected fields are affected.

17.3.1 The Coupled-Propagator White-Noise Model

The propagators now solve the differential equation

$$\frac{d}{dz}\mathbf{P}^{\varepsilon}_{(\omega,\kappa)}(-L,z) = \frac{1}{\varepsilon}\mathbf{H}_{(\omega,\kappa)}\left(\frac{z}{\varepsilon},m_{\kappa}\left(\frac{z}{\varepsilon^{2}}\right),n_{\kappa}\left(\frac{z}{\varepsilon^{2}}\right)\right)\mathbf{P}^{\varepsilon}_{(\omega,\kappa)}(-L,z),   \quad(17.47)$$

with the 2×2 matrix $\mathbf{H}_{(\omega,\kappa)}$ given by

$$\mathbf{H}_{(\omega,\kappa)}(z,m,n) = \frac{i\omega}{2\bar{c}(\kappa)} \begin{bmatrix} m & -ne^{\frac{-2i\omega z}{\bar{c}(\kappa)}} \\ ne^{\frac{2i\omega z}{\bar{c}(\kappa)}} & -m \end{bmatrix},$$

and $\mathbf{P}^{\varepsilon}_{(\omega,\kappa,)}(-L, z = -L) = \mathbf{I}$. The trace of $\mathbf{H}_{(\omega,\kappa)}$ is still zero, and the propagator can be written in the form

$$\mathbf{P}^{\varepsilon}_{(\omega,\kappa)}(-L,z) = \begin{bmatrix} \alpha^{\varepsilon}_{(\omega,\kappa)}(-L,z) & \overline{\beta^{\varepsilon}_{(\omega,\kappa)}(-L,z)} \\ \beta^{\varepsilon}_{(\omega,\kappa)}(-L,z) & \overline{\alpha^{\varepsilon}_{(\omega,\kappa)}(-L,z)} \end{bmatrix}.$$

As in Section 14.2.1, in order to characterize the transmitted pulse, we are led to study the joint distribution of the transmission coefficients

$$T^{\varepsilon}_{(\omega_j,\kappa_j)}(-L,0) = \frac{1}{\alpha^{\varepsilon}_{(\omega_j,\kappa_j)}(-L,0)},$$

for a finite number of frequencies and wave vectors $(\omega_1, \boldsymbol{\kappa}_1), \ldots, (\omega_M, \boldsymbol{\kappa}_M)$. In particular, we need to characterize the limits

$$\lim_{\varepsilon \to 0} \mathbb{E}\left[T^{\varepsilon}_{(\omega_1,\kappa_1)}(-L,0) \cdots T^{\varepsilon}_{(\omega_M,\kappa_M)}(-L,0) \right]. \tag{17.48}$$

The joint statistics of the propagators follow, from the distribution of the multifrequency and multislowness propagator $\mathbf{P}^{\varepsilon}_M(-L,z) = \mathbf{P}^{\varepsilon}_{(\omega_1,\ldots,\kappa_M)}(-L,z)$, which is the $2M \times 2M$ block-diagonal complex matrix

$$\mathbf{P}^{\varepsilon}_M(-L,z) = \begin{bmatrix} \mathbf{P}^{\varepsilon}_{(\omega_1,\kappa_1)}(-L,z) & \cdots & 0 \\ & \cdot & \\ & & \cdot \\ & & & \cdot \\ 0 & \cdots & \mathbf{P}^{\varepsilon}_{(\omega_M,\kappa_M)}(-L,z) \end{bmatrix},$$

where the complex linear system for $\mathbf{P}^{\varepsilon}$ is defined in terms of

$$\mathbf{H}_M(z, m_1, \cdots, n_M) = \begin{bmatrix} \mathbf{H}_{(\omega_1,\kappa_1)}(z, m_1, n_1) & \cdots & 0 \\ & \cdot & \\ & & \cdot \\ 0 & \cdots & \mathbf{H}_{(\omega_M,\kappa_M)}(z, m_M, n_M) \end{bmatrix}.$$

The multifrequency propagator system is

$$\frac{d}{dz}\mathbf{P}^{\varepsilon}_M(-L,z) = \frac{1}{\varepsilon}\mathbf{H}_M\left(\frac{z}{\varepsilon}, m_{\kappa_1}\left(\frac{z}{\varepsilon^2}\right), \ldots, n_{\kappa_M}\left(\frac{z}{\varepsilon^2}\right)\right) \mathbf{P}^{\varepsilon}_M(-L,z), \tag{17.49}$$

with the initial condition $\mathbf{P}^{\varepsilon}_M(-L, z = -L) = \mathbf{I}$, where $\mathbf{I}$ is the identity matrix of dimension $2M$.

We want now to apply the diffusion-approximation result introduced in Section 6.7.3 to obtain the diffusion limit for the multifrequency propagator, similarly as in Section 8.2.4. We first rewrite the differential equations for the 2×2 diagonal entries in (17.49) in the expanded form

$$
\frac{d}{dz}\mathbf{P}^\varepsilon_{(\omega_j,\kappa_j)}(-L,z) = \frac{i\omega_j}{2\bar{c}(\kappa_j)\varepsilon}(\eta + \tilde{\eta})\left(\frac{z}{\varepsilon^2}\right)\begin{bmatrix} 1 & 0 \\ 0 & -1 \end{bmatrix}\mathbf{P}^\varepsilon_{(\omega_j,\kappa_j)}(-L,z) \qquad (17.50)
$$

$$
+ \frac{i\omega_j\bar{c}(\kappa_j)}{2\bar{c}^2\varepsilon}(\nu - \tilde{\eta})\left(\frac{z}{\varepsilon^2}\right)\begin{bmatrix} 1 & 0 \\ 0 & -1 \end{bmatrix}\mathbf{P}^\varepsilon_{(\omega_j,\kappa_j)}(-L,z)
$$

$$
- \frac{\omega_j}{2\bar{c}(\kappa_j)\varepsilon}n_{\kappa_j}\left(\frac{z}{\varepsilon^2}\right)\sin\left(\frac{2\omega_j z}{\bar{c}(\kappa_j)\varepsilon}\right)\begin{bmatrix} 0 & 1 \\ 1 & 0 \end{bmatrix}\mathbf{P}_{(\omega_j,\kappa_j)}(-L,z)
$$

$$
- \frac{i\omega_j}{2\bar{c}(\kappa_j)\varepsilon}n_{\kappa_j}\left(\frac{z}{\varepsilon^2}\right)\cos\left(\frac{2\omega_j z}{\bar{c}(\kappa_j)\varepsilon}\right)\begin{bmatrix} 0 & 1 \\ -1 & 0 \end{bmatrix}\mathbf{P}^\varepsilon_{(\omega_j,\kappa_j)}(-L,z),
$$

where we have used the relation

$$
m_\kappa(z) = (\eta + \tilde{\eta}) + \frac{\bar{c}(\kappa)^2}{\bar{c}^2}(\nu - \tilde{\eta}).
$$

The equation for the limiting propagator can be obtained by "white-noise substitutions," as explained in Section 6.7.3 and as we detail below. These substitutions require that one study the joint asymptotics of the driving processes that appear in the right-hand side of (17.50).

By application of the diffusion approximation Theorem 6.1, we get that the $\mathbb{R}^2$-valued process

$$
\left(\frac{1}{\varepsilon}\int_0^z (\eta + \tilde{\eta})\left(\frac{z'}{\varepsilon^2}\right)dz', \frac{1}{\varepsilon}\int_0^z (\nu - \tilde{\eta})\left(\frac{z'}{\varepsilon^2}\right)dz'\right)
$$

converges in distribution to the diffusion Markov process (X_1, X_2) with the infinitesimal generator

$$
\mathcal{L} = \frac{\gamma_{11}}{2}\frac{\partial^2}{\partial X_1^2} + \frac{\gamma_{22}}{2}\frac{\partial^2}{\partial X_2^2} + \gamma_{12}\frac{\partial^2}{\partial X_1 \partial X_2},
$$

where

$$
\gamma_{11} = \int_{-\infty}^{\infty} \mathbb{E}[(\eta(0) + \tilde{\eta}(0))(\eta(z) + \tilde{\eta}(z))]dz, \qquad (17.51)
$$

$$
\gamma_{12} = \int_{-\infty}^{\infty} \mathbb{E}[(\eta(0) + \tilde{\eta}(0))(\nu(z) - \tilde{\eta}(z))]dz,
$$

$$
\gamma_{22} = \int_{-\infty}^{\infty} \mathbb{E}[(\nu(0) - \tilde{\eta}(0))(\nu(z) - \tilde{\eta}(z))]dz.
$$

The diffusion Markov process (X_1, X_2) can be represented as

$$
X_1(z) = \sqrt{\gamma_{11}}W^{(1)}(z),
$$

$$
X_2(z) = \frac{\gamma_{12}}{\sqrt{\gamma_{11}}}W^{(1)}(z) + \sqrt{\gamma_{22}}\sqrt{1 - \frac{\gamma_{12}^2}{\gamma_{11}\gamma_{22}}}W^{(2)}(z),
$$

where $W^{(1)}$ and $W^{(2)}$ are two independent standard Brownian motions. As a consequence, the following convergence holds true jointly for the set of M slownesses $(\kappa_1, \ldots, \kappa_M)$:

$$\frac{1}{\varepsilon} \int_0^z m_{\kappa_j}\left(\frac{z'}{\varepsilon^2}\right) dz' \xrightarrow{\varepsilon \to 0} \sqrt{\gamma_{1,\kappa_j}} W^{(1)}(z) + \sqrt{\gamma_{2,\kappa_j}} W^{(2)}(z), \qquad (17.52)$$

where

$$\gamma_{1,\kappa} = \gamma_{11}\left(1 + \frac{\gamma_{12}}{\gamma_{11}} \frac{\bar{c}(\kappa)^2}{\bar{c}^2}\right)^2, \qquad \gamma_{2,\kappa} = \gamma_{22}\left(1 - \frac{\gamma_{12}^2}{\gamma_{11}\gamma_{22}}\right) \frac{\bar{c}(\kappa)^4}{\bar{c}^4}. \qquad (17.53)$$

We also have the convergence in distribution of the $\mathbb{R}^{2M}$-valued process

$$\left(\frac{1}{\varepsilon} \int_0^z n_{\kappa_j}\left(\frac{z'}{\varepsilon^2}\right) \sin\left(\frac{2\omega_j z'}{\bar{c}(\kappa_j)\varepsilon}\right) dz', \frac{1}{\varepsilon} \int_0^z n_{\kappa_j}\left(\frac{z'}{\varepsilon^2}\right) \cos\left(\frac{2\omega_j z'}{\bar{c}(\kappa_j)\varepsilon}\right) dz'\right)_{1 \leq j \leq M}$$

$$\xrightarrow{\varepsilon \to 0} \frac{\sqrt{\gamma_{n\kappa_j}}}{\sqrt{2}}\left(W_j(z), \tilde{W}_j(z)\right)_{1 \leq j \leq M}, \qquad (17.54)$$

where $W_j, \tilde{W}_j, j = 1, \ldots, M$ are independent standard Brownian motions and

$$\gamma_{n\kappa} = \int_{-\infty}^{\infty} \mathbb{E}[n_\kappa(0) n_\kappa(z)]\, dz. \qquad (17.55)$$

Here the orthogonality of the Fourier basis leads to the independence of the Brownian motions W_j and $\tilde{W}_j$ with respect to each other, and also with respect to $W^{(1)}$ and $W^{(2)}$. From the expression

$$n_\kappa = (\tilde{\eta} - \eta) + \frac{\bar{c}(\kappa)^2}{\bar{c}^2}(\nu - \tilde{\eta}),$$

we can write

$$\gamma_{n\kappa} = \tilde{\gamma}_{11} + 2\tilde{\gamma}_{12} \frac{\bar{c}(\kappa)^2}{\bar{c}^2} + \gamma_{22} \frac{c(\kappa)^4}{\bar{c}^4}, \qquad (17.56)$$

where we have introduced the coefficients

$$\tilde{\gamma}_{11} = \int_{-\infty}^{\infty} \mathbb{E}[(\eta(0) - \tilde{\eta}(0))(\eta(z) - \tilde{\eta}(z))]\, dz,$$

$$\tilde{\gamma}_{12} = \int_{-\infty}^{\infty} \mathbb{E}[(\eta(0) - \tilde{\eta}(0))(\nu(z) - \tilde{\eta}(z))]\, dz.$$

We now apply the diffusion approximation result stated in Section 6.7.3 to the system of random ordinary differential equations (17.50), and we obtain that the limit multifrequency propagator satisfies the following system of stochastic differential equations written is Stratonovich form:

$$dP_{(\omega_j,\kappa_j)}(-L,z) = \frac{i\omega_j\sqrt{\gamma_{1,\kappa_j}}}{2\bar{c}(\kappa_j)} \begin{bmatrix} 1 & 0 \\ 0 & -1 \end{bmatrix} P_{(\omega_j,\kappa_j)}(-L,z) \circ dW^{(1)}(z)$$

$$+ \frac{i\omega_j\sqrt{\gamma_{2,\kappa_j}}}{2\bar{c}(\kappa_j)} \begin{bmatrix} 1 & 0 \\ 0 & -1 \end{bmatrix} P_{(\omega_j,\kappa_j)}(-L,z) \circ dW^{(2)}(z)$$

$$- \frac{\omega_j\sqrt{\gamma_{n\kappa_j}}}{2\sqrt{2}\bar{c}(\kappa_j)} \begin{bmatrix} 0 & 1 \\ 1 & 0 \end{bmatrix} P_{(\omega_j,\kappa_j)}(-L,z) \circ dW_j(z)$$

$$- \frac{i\omega_j\sqrt{\gamma_{n\kappa_j}}}{2\sqrt{2}\bar{c}(\kappa_j)} \begin{bmatrix} 0 & 1 \\ -1 & 0 \end{bmatrix} P_{(\omega_j,\kappa_j)}(-L,z) \circ d\tilde{W}_j(z). \quad (17.57)$$

This result shows that the 2×2 elementary propagators that are associated with a particular frequency and slowness are coupled through the **two** Brownian motions $W^{(1)}$ and $W^{(2)}$. This reflects the fact that now both the density and the bulk modulus are randomly varying. In the case with a constant density, only one Brownian motion couples the propagators as described in (8.51).

The transmission coefficients are given by

$$T^\varepsilon_{(\omega_j,\kappa_j)}(-L,z) = \frac{1}{\alpha^\varepsilon_{(\omega_j,\kappa_j)}(-L,z)},$$

and we can now use the above diffusion limit to obtain the limits as $\varepsilon \to 0$ of these coefficients, which we denote by $T_{(\omega,\kappa)}$. We follow the same strategy as the one in Section 8.2.5, to show that the transmission coefficients have martingale representations similar to (8.57). From (17.57) the random vector $(T^\varepsilon_{(\omega_1,\kappa_1)}(-L,0),\ldots,T^\varepsilon_{(\omega_M,\kappa_M)}(-L,0))$ converges in distribution as $\varepsilon \to 0$ to the limit $(T_{(\omega_1,\kappa_1)}(-L,0),\ldots,T_{(\omega_M,\kappa_M)}(-L,0))$, where the limit transmission coefficients have the martingale representations

$$T_{(\omega_j,\kappa_j)}(-L,0) = \tilde{M}_{(\omega_j,\kappa_j)}(-L,0)\tilde{T}_{(\omega_j,\kappa_j)}(-L,0), \quad j=1,\ldots,M. \quad (17.58)$$

Here

$$\tilde{T}_{(\omega_j,\kappa_j)}(-L,0) = \exp\left(i\omega_j\frac{\sqrt{\gamma_{1,\kappa_j}}}{2\bar{c}(\kappa_j)}W^{(1)}(L) + i\omega_j\frac{\sqrt{\gamma_{2,\kappa_j}}}{2\bar{c}(\kappa_j)}W^{(2)}(L)\right)$$

$$\times \exp\left(-\omega^2\frac{\gamma_{n\kappa_j}}{8\bar{c}(\kappa_j)^2}L\right), \quad (17.59)$$

and for each j, the process $\tilde{M}_{(\omega_j,\kappa_j)}(-L,0)$ is a complex martingale that depends only on the pair of Brownian motions $(W_j,\tilde{W}_j)$. Therefore, these martingales are independent of each other, and independent of $\tilde{T}_{\omega_j}(-L,0)$, which is a function of the Brownian motions $W^{(1)}$ and $W^{(2)}$ only. The joint limits for the transmission coefficient at frequencies and slownesses $(\omega_1,\kappa_1),\ldots,(\omega_M,\kappa_M)$ will lead to the weak limit for the transmitted front.

17.3.2 The Transmitted Field

In this section we derive an approximation for the transmitted front. Recall that the specific Fourier transform for the pressure and the longitudional velocity component are given by (17.41) and (17.42) with the impedance having a jump at the surface:

$$
\bar{\zeta}(\kappa, z) = \begin{cases} \bar{\zeta}(\kappa) = \dfrac{\bar{\zeta}}{\sqrt{1 - \kappa^2 \bar{c}^2 (\bar{\rho}/\bar{\rho})}} & \text{for } z \in [-L, 0], \\[4mm] \zeta_0(\kappa) = \dfrac{\bar{\zeta}}{\sqrt{1 - \kappa^2 \bar{c}^2}} & \text{for } z \in (-\infty, -L) \cup (0, \infty), \end{cases} \tag{17.60}
$$

with $\bar{\zeta} = \sqrt{\bar{\rho}\bar{K}}$. The continuity condition on the pressure and the velocity imposes a jump condition on the coefficients $\breve{a}^\varepsilon$ and $\breve{b}^\varepsilon$ at the two interfaces $z = 0$ and $z = -L$. Let the mode-dependent surface interface reflection coefficients $R_I(\kappa)$ and transmission coefficients $T_I(\kappa)$ be defined by

$$
R_I(\kappa) = \frac{\zeta_0(\kappa) - \bar{\zeta}(\kappa)}{\zeta_0(\kappa) + \bar{\zeta}(\kappa)}, \qquad T_I(\kappa) = \sqrt{1 - R_I(\kappa)^2} = \frac{2\sqrt{\zeta_0(\kappa)\bar{\zeta}(\kappa)}}{\zeta_0(\kappa) + \bar{\zeta}(\kappa)}, \tag{17.61}
$$

and define also

$$
r^{(\pm)}(\kappa) = \frac{1}{2}\left(\sqrt{\zeta_0/\bar{\zeta}(\kappa)} \pm \sqrt{\bar{\zeta}/\zeta_0(\kappa)} \right),
$$

so that $T_I = 1/r^+$ and $R_I = r^-/r^+$. We observe the transmitted pressure at the time $t = t_0 + \varepsilon s$ at the horizontal offset $\mathbf{x}$. The situation can be actually interpreted as a particular case of the nonmatched medium configuration studied in Section 17.1. The integral representation for the transmitted pressure field is given by (17.14),

$$
p^\varepsilon(t_0 + \varepsilon s, \mathbf{x}, (-L)^-) = \frac{-1}{2(2\pi\varepsilon)^3} \int\int e^{-i\frac{\omega}{\varepsilon}(t_0 + \varepsilon s - \boldsymbol{\kappa}\cdot\mathbf{x})}\sqrt{\zeta_0(\kappa)}e^{-\frac{i\omega L}{\varepsilon\bar{c}(\kappa)}}
$$
$$
\times T^\varepsilon_{(\omega,\kappa)}\breve{b}^\varepsilon(\omega, \boldsymbol{\kappa}, 0^+)\omega^2 \, d\omega \, d\boldsymbol{\kappa}, \tag{17.62}
$$

where $\breve{b}^\varepsilon(\omega, \boldsymbol{\kappa}, 0^+)$ models the incoming wave and is given by (17.46) with the radiation condition $\breve{b}^\varepsilon(\omega, \boldsymbol{\kappa}, z_s^+) = 0$:

$$
\breve{b}^\varepsilon(\omega, \boldsymbol{\kappa}, 0^+) = \varepsilon^{q+3}\frac{1}{\sqrt{\zeta_0(\kappa)}}\hat{f}(\omega, \boldsymbol{\kappa})e^{\frac{i\omega z_s}{\varepsilon c_0(\kappa)}}.
$$

The generalized transmission coefficient is given by the relation (17.11):

$$
\begin{bmatrix} \mathcal{R}^\varepsilon_{(\omega,\kappa)} \\ 1 \end{bmatrix} = \mathbf{J}^{\varepsilon,2}_{\omega,\kappa}\,\mathbf{P}^\varepsilon_{(\omega,\kappa)}(-L,0)\,\mathbf{J}^{\varepsilon,1}_{\omega,\kappa}\begin{bmatrix} 0 \\ e^{\frac{i\omega L}{\varepsilon}\left(\frac{1}{\bar{c}(\kappa)} - \frac{1}{c_0(\kappa)}\right)}T^\varepsilon_{(\omega,\kappa)} \end{bmatrix}, \tag{17.63}
$$

where the interface jump matrices are given by

$$\mathbf{J}_{\omega,\kappa}^{\varepsilon,1} = \begin{bmatrix} r^+(\kappa)e^{\frac{i\omega L}{\varepsilon}\left(\frac{1}{\bar{c}(\kappa)}-\frac{1}{c_0(\kappa)}\right)} & -r^-(\kappa)e^{\frac{i\omega L}{\varepsilon}\left(\frac{1}{\bar{c}(\kappa)}+\frac{1}{c_0(\kappa)}\right)} \\ -r^-(\kappa)e^{\frac{i\omega L}{\varepsilon}\left(-\frac{1}{\bar{c}(\kappa)}-\frac{1}{c_0(\kappa)}\right)} & r^+(\kappa)e^{\frac{i\omega L}{\varepsilon}\left(-\frac{1}{\bar{c}(\kappa)}+\frac{1}{c_0(\kappa)}\right)} \end{bmatrix},$$

$$\mathbf{J}_{\omega,\kappa}^{\varepsilon,2} = \begin{bmatrix} r^+(\kappa) & r^-(\kappa) \\ r^-(\kappa) & r^+(\kappa) \end{bmatrix}.$$

We can expand the generalized transmission coefficient as in (17.13),

$$\mathcal{T}_{(\omega,\kappa)}^{\varepsilon} = T_I^2 T_{(\omega,\kappa)}^{\varepsilon} \sum_{j=0}^{\infty} \left[\mathcal{U}_{(\omega,\kappa)}^{\varepsilon}\right]^j,$$

$$\mathcal{U}_{(\omega,\kappa)}^{\varepsilon} = R_I^2 \left((T_{(\omega,\kappa)}^{\varepsilon})^2 + R_{(\omega,\kappa)}^{\varepsilon}\tilde{R}_{(\omega,\kappa)}^{\varepsilon}\right)e^{\frac{2i\omega L}{\varepsilon\bar{c}(\kappa)}} + R_I \left(\tilde{R}_{(\omega,\kappa)}^{\varepsilon}e^{\frac{2i\omega L}{\varepsilon\bar{c}(\kappa)}} - R_{(\omega,\kappa)}^{\varepsilon}\right),$$

where $R_{(\omega,\kappa)}^{\varepsilon}$ and $T_{(\omega,\kappa)}^{\varepsilon}$ are the usual reflection and transmission coefficients, while $\tilde{R}_{(\omega,\kappa)}^{\varepsilon}$ is the adjoint reflection coefficient as defined in (15.28–15.29):

$$R_{(\omega,\kappa)}^{\varepsilon} = \frac{\overline{\beta_{(\omega,\kappa)}^{\varepsilon}}}{\overline{\alpha_{(\omega,\kappa)}^{\varepsilon}}}, \qquad \tilde{R}_{(\omega,\kappa)}^{\varepsilon} = -\frac{\beta_{(\omega,\kappa)}^{\varepsilon}}{\alpha_{(\omega,\kappa)}^{\varepsilon}}, \qquad T_{(\omega,\kappa)}^{\varepsilon} = \frac{1}{\alpha_{(\omega,\kappa)}^{\varepsilon}}. \qquad (17.64)$$

By the same computation as the one carried out in Section 14.2.1, we get a characterization in distribution of the transmitted pressure pulse field by the replacements

$$T_{(\omega,\kappa)}^{\varepsilon}(-L,0) \mapsto \tilde{T}_{(\omega,\kappa)}, \qquad R_{(\omega,\kappa)}^{\varepsilon}(-L,0) \mapsto 0, \qquad \tilde{R}_{(\omega,\kappa)}^{\varepsilon}(-L,0) \mapsto 0,$$

where

$$\tilde{T}_{(\omega,\kappa)} = \exp\left(i\omega\frac{\sqrt{\gamma_{1,\kappa}}}{2\bar{c}(\kappa)}W^{(1)}(L) + i\omega\frac{\sqrt{\gamma_{2,\kappa}}}{2\bar{c}(\kappa)}W^{(2)}(L) - \omega^2\frac{\gamma_{n\kappa}}{8\bar{c}(\kappa)^2}L\right)$$

$$= \exp\left(i\omega\frac{\sqrt{\gamma_{m\kappa}}}{2\bar{c}(\kappa)}W_0(L) - \omega^2\frac{\gamma_{n\kappa}}{8\bar{c}(\kappa)^2}L\right), \qquad (17.65)$$

$W_0(L)$ is a standard Brownian motion, and the coefficient $\gamma_{m\kappa}$ defined by

$$\gamma_{m\kappa} = \int_{-\infty}^{\infty} \mathbb{E}[m_\kappa(0)m_\kappa(z)]dz \qquad (17.66)$$

depends on the slowness κ and can also be written as

$$\gamma_{m\kappa} = \gamma_{1,\kappa} + \gamma_{2,\kappa} = \gamma_{11} + 2\gamma_{12}\frac{\bar{c}(\kappa)^2}{\bar{c}^2} + \gamma_{22}\frac{\bar{c}(\kappa)^4}{\bar{c}^4}. \qquad (17.67)$$

The other coefficients $\gamma_{1,\kappa}$, $\gamma_{2,\kappa}$, and $\gamma_{n\kappa}$ are given by (17.53) and (17.56). We therefore find that

$$\tilde{p}^{\varepsilon}(t_0 + \varepsilon s, \mathbf{x}, (-L)^-) = \frac{-\varepsilon^q}{2(2\pi)^3} \int \int e^{-i\omega s}\hat{f}(\omega,\boldsymbol{\kappa})$$

$$\times T_I^2\tilde{T}_{(\omega,\kappa)} \sum_{j=0}^{\infty} e^{\frac{i\omega\phi(j,\boldsymbol{\kappa},t_0,\mathbf{x},L)}{\varepsilon}} \left(R_I\tilde{T}_{(\omega,\kappa)}\right)^{2j} \omega^2 \, d\omega \, d\boldsymbol{\kappa},$$

where

$$\phi(j, \boldsymbol{\kappa}, t_0, \mathbf{x}, L) = -t_0 + \boldsymbol{\kappa} \cdot \mathbf{x} + \frac{(1+2j)L}{\bar{c}(\kappa)} + \frac{z_s}{c_0(\kappa)} .$$

We consider now the situation in which the source is located at the surface: $z_s \to 0$, and with $q = -1$, so that the transmitted pulse has amplitude of order one. Then, we can carry out the stationary-phase evaluation as before and find that for each j, the rapid phase $\phi(j, \cdot, t_0, \mathbf{x}, L)$ has one stationary slowness vector given by

$$\boldsymbol{\kappa}_{\mathrm{sp},j}(\mathbf{x}) = \frac{\mathbf{x}}{\bar{c}\sqrt{|\mathbf{x}|^2 \bar{\rho}/\widetilde{\rho} + (1+2j)^2 L^2 (\bar{\rho}/\widetilde{\rho})^2}} .$$

For this stationary point, we have $\phi(j, \boldsymbol{\kappa}_{\mathrm{sp},j}, t_0, \mathbf{x}, L) = 0$ only if the observation time is equal to $t^{(j)}$, given by

$$t^{(j)}(\mathbf{x}) = \frac{(1+2j)L}{\bar{c}} \left(1 + \frac{|\mathbf{x}|^2 \widetilde{\rho}}{(1+2j)^2 L^2 \bar{\rho}} \right)^{1/2} . \tag{17.68}$$

Furthermore, we have

$$\bar{c}(\kappa_{\mathrm{sp},j})^2 = \bar{c}^2 \left(1 + \frac{|\mathbf{x}|^2 \widetilde{\rho}}{(1+2j)^2 L^2 \bar{\rho}} \right) ,$$

$$c_0(\kappa_{\mathrm{sp},j})^2 = \bar{c}^2 \frac{1 + \frac{|\mathbf{x}|^2 \widetilde{\rho}}{(1+2j)^2 L^2 \bar{\rho}}}{1 + \frac{|\mathbf{x}|^2 \widetilde{\rho}}{(1+2j)^2 L^2 \bar{\rho}} \left(1 - \frac{\widetilde{\varrho}}{\varrho} \right)} ,$$

$$R_I(\kappa_{\mathrm{sp},j}) = \frac{\frac{|\mathbf{x}|^2 \widetilde{\rho}}{(1+2j)^2 L^2 \bar{\rho}} \left(1 - \frac{\widetilde{\varrho}}{\varrho} \right)}{2 + \frac{|\mathbf{x}|^2 \widetilde{\rho}}{(1+2j)^2 L^2 \bar{\rho}} \left(1 - \frac{\widetilde{\varrho}}{\varrho} \right)} ,$$

$$T_I(\kappa_{\mathrm{sp},j}) = \frac{2 \left[1 + \frac{|\mathbf{x}|^2 \widetilde{\rho}}{(1+2j)^2 L^2 \bar{\rho}} \left(1 - \frac{\widetilde{\varrho}}{\varrho} \right) \right]^{1/2}}{2 + \frac{|\mathbf{x}|^2 \widetilde{\rho}}{(1+2j)^2 L^2 \bar{\rho}} \left(1 - \frac{\widetilde{\varrho}}{\varrho} \right)} .$$

We thus arrive at the following generalization of the stabilization of the stable-front result in Proposition 14.3.

Proposition 17.2. *In probability distribution the following characterization of the transmitted wave process holds. At the offset $\mathbf{x}$, we can observe a transmitted pulse at time $t^{(j)}$ given by (17.68) for any $j = 0, 1, 2, \ldots$, whose asymptotic form is*

$$\lim_{\varepsilon \to 0} p^\varepsilon \left(t^{(j)} + \varepsilon s, \mathbf{x}, -L \right) = \widetilde{p}_j(s, \mathbf{x}, -L) ,$$

where

$$\widetilde{p}_j(s, \mathbf{x} - L) = [\mathcal{N}_{D_{(L,\mathbf{x},j)}} * \widetilde{p}_{0,j}(\cdot, \mathbf{x}, -L)] \left(s - \Theta_{(L,\mathbf{x},j)} \right) ,$$

and we set

$$D^2_{(L,\mathbf{x},j)} = \frac{1}{4\tilde{c}^2}\left[\tilde{\gamma}_{11}\left(1 + \frac{|\mathbf{x}|^2\tilde{\rho}}{(1+2j)^2 L^2\bar{\rho}}\right)^{-1} + 2\tilde{\gamma}_{12}\right.$$

$$\left.+ \gamma_{22}\left(1 + \frac{|\mathbf{x}|^2\tilde{\rho}}{(1+2j)^2 L^2\bar{\rho}}\right)\right](1+2j)L\,,$$

$$\Theta_{(L,\mathbf{x},j)} = \frac{1}{2\tilde{c}}\left[\gamma_{11}\left(1 + \frac{|\mathbf{x}|^2\tilde{\rho}}{(1+2j)^2 L^2\bar{\rho}}\right)^{-1} + 2\gamma_{12}\right.$$

$$\left.+ \gamma_{22}\left(1 + \frac{|\mathbf{x}|^2\tilde{\rho}}{(1+2j)^2 L^2\bar{\rho}}\right)\right]^{1/2}(1+2j)W_0(L)\,,$$

$$\mathcal{N}_D(s) = \frac{1}{\sqrt{2\pi}D}e^{-s^2/2D^2}\,,$$

with $W_0(L)$ a standard Brownian motion. Here $\tilde{p}_{0,j}$ is the pulse shape

$$\tilde{p}_{0,j}(s,\mathbf{x},-L) = -\left[\frac{R_I(\kappa_{\mathrm{sp},j})^{2j}T_I(\kappa_{\mathrm{sp},j})^2\tilde{c}\tilde{\rho}}{4\pi\tilde{c}(\kappa_{\mathrm{sp},j})^2\bar{\rho}(2j+1)L}\right]f'_j(s)\,,$$

where

$$f_j(t) = \int f(t + \kappa_{\mathrm{sp},j}(\mathbf{x})\cdot\mathbf{y},\mathbf{y})d\mathbf{y}\,.$$

In the case that there are no density fluctuations we have

- $\bar{\rho} = \tilde{\rho}$,
- $\gamma_{11} = \tilde{\gamma}_{11} = \gamma_{12} = \tilde{\gamma}_{12} = 0$,
- $R_I = 0$, $T_I = 1$, so that $\tilde{p}_{0,j} = 0$ if $j \geq 1$ and $\tilde{p}_{0,0}$ is the transmitted pulse shape obtained in the homogeneous medium

$$\tilde{p}_{0,0}(t,\mathbf{x},-L) = -\frac{1}{4\pi\tilde{c}(|\mathbf{x}|^2 + L^2)}f'_0(t)\,,$$

$$f_0(t) = \int f\left(t + \frac{\mathbf{x}\cdot\mathbf{y}}{\tilde{c}\sqrt{|\mathbf{x}|^2 + L^2}},\mathbf{y}\right)d\mathbf{y}\,.$$

We thus recover the result stated in Proposition 14.3.

17.3.3 Transport Equations

As seen in the previous chapters, the limit autocorrelation function of the reflection coefficient determines important physical quantities such as the mean reflected intensity. The asymptotic analysis of the two-frequency correlation function of the reflection coefficient is carried out in this section, and it will be applied to the computation of the mean reflected intensity in Section 17.3.4. From (17.44) we deduce the closed nonlinear differential system satisfied by the reflection and transmission coefficients:

$$\frac{dR_{(\omega,\kappa)}^{\varepsilon}}{dz} = -\frac{i\omega}{2\bar{c}(\kappa)\varepsilon} \left(n_{\kappa}\left(\frac{z}{\varepsilon^2}\right) e^{\frac{-2i\omega z}{\bar{c}(\kappa)\varepsilon}} - m_{\kappa}\left(\frac{z}{\varepsilon^2}\right) 2R_{(\omega,\kappa)}^{\varepsilon} \right.$$

$$\left. + n_{\kappa}\left(\frac{z}{\varepsilon^2}\right) (R_{(\omega,\kappa)}^{\varepsilon})^2 e^{\frac{2i\omega z}{\bar{c}(\kappa)\varepsilon}} \right),$$

$$\frac{dT_{(\omega,\kappa)}^{\varepsilon}}{dz} = -\frac{i\omega}{2\bar{c}(\kappa)\varepsilon} \left(m_{\kappa}\left(\frac{z}{\varepsilon^2}\right) - n_{\kappa}\left(\frac{z}{\varepsilon^2}\right) R_{(\omega,\kappa)}^{\varepsilon} e^{\frac{2i\omega z}{\bar{c}(\kappa)\varepsilon}} \right) T_{(\omega,\kappa)}^{\varepsilon},$$

starting from $R_{(\omega,\kappa)}^{\varepsilon}(z_0, z = z_0) = 0$, $T_{(\omega,\kappa)}^{\varepsilon}(z_0, z = z_0) = 1$. We again introduce the family of products

$$U_{p,q}^{\varepsilon}(\omega, \kappa, h, \lambda, z_0, z) = \left(R_{(\omega+\varepsilon h/2, \kappa+\varepsilon\lambda/2)}^{\varepsilon}(z_0, z) \right)^p \left(\overline{R_{(\omega-\varepsilon h/2, \kappa-\varepsilon\lambda/2)}^{\varepsilon}(z_0, z)} \right)^q,$$

and obtain the following slightly modified system for $U_{(p,q)}^{\varepsilon}$:

$$\frac{dU_{p,q}^{\varepsilon}}{dz} = \frac{i\omega}{\bar{c}(\kappa)} m_{\kappa}^{\varepsilon}(p - q) U_{p,q}^{\varepsilon}$$

$$+ \frac{i\omega}{2\bar{c}(\kappa)} n_{\kappa}^{\varepsilon} e^{\frac{2i\omega z}{\bar{c}(\kappa)\varepsilon}} \left(q e^{\frac{-ihz}{\bar{c}(\kappa)}+i\omega\lambda\bar{c}(\kappa)\kappa z} U_{p,q-1}^{\varepsilon} - p e^{\frac{ihz}{\bar{c}(\kappa)}-i\omega\lambda\bar{c}(\kappa)\kappa z} U_{p+1,q}^{\varepsilon} \right)$$

$$+ \frac{i\omega}{2\bar{c}(\kappa)} n_{\kappa}^{\varepsilon} e^{-\frac{2i\omega z}{\bar{c}(\kappa)\varepsilon}} \left(q e^{\frac{ihz}{\bar{c}(\kappa)}-i\omega\lambda\bar{c}(\kappa)\kappa z} U_{p,q+1}^{\varepsilon} - p e^{\frac{-ihz}{\bar{c}(\kappa)}+i\omega\lambda\bar{c}(\kappa)\kappa z} U_{p-1,q}^{\varepsilon} \right),$$

starting from $U_{p,q}^{\varepsilon}(\omega, \kappa, h, \lambda, z_0, z = z_0) = \mathbf{1}_0(p)\mathbf{1}_0(q)$ and where we again ignore terms of order ε that will vanish in the limit $\varepsilon \to 0$. We have also set $m_{\kappa}^{\varepsilon}(z) = m_{\kappa}(z/\varepsilon^2)/\varepsilon$ and $n_{\kappa}^{\varepsilon}(z) = n_{\kappa}(z/\varepsilon^2)/\varepsilon$. Next we introduce the associated family of Fourier transforms:

$$V_{p,q}^{\varepsilon}(\omega, \kappa, \tau, \chi, z_0, z) = \frac{\omega\psi(\kappa)}{4\pi^2} \int\int e^{-ih(\tau\psi(\kappa)-(p+q)z/\bar{c}(\kappa))}$$

$$\times e^{i\omega\lambda(\chi-(p+q)z\bar{c}(\kappa)\kappa)} U_{p,q}^{\varepsilon}(\omega, \kappa, h, \lambda, z_0, z)\, dh\, d\lambda,$$

where

$$\psi(\kappa) = \frac{1 - \bar{c}^2\kappa^2}{1 - \bar{c}^2\kappa^2(\frac{\bar{\rho}}{\rho} - 1)}.$$

This particular normalization with $\psi(\kappa)$ ensures that the variable τ can be interpreted as a travel time, as we shall see below. The family $(V_{p,q}^{\varepsilon})_{p,q}$ now satisfies

$$\frac{\partial V_{p,q}^{\varepsilon}}{\partial z} = -\frac{p+q}{\psi(\kappa)\bar{c}(\kappa)} \frac{\partial V_{p,q}^{\varepsilon}}{\partial \tau} - \bar{c}(\kappa)\kappa(p+q)\frac{\partial V_{p,q}^{\varepsilon}}{\partial \chi} + \frac{i\omega}{\bar{c}(\kappa)} m_{\kappa}^{\varepsilon}(p-q) V_{p,q}^{\varepsilon}$$

$$+ \frac{i\omega}{2\bar{c}(\kappa)} n_{\kappa}^{\varepsilon} e^{\frac{2i\omega z}{\bar{c}(\kappa)\varepsilon}} \left(q V_{p,q-1}^{\varepsilon} - p V_{p+1,q}^{\varepsilon} \right) + \frac{i\omega}{2\bar{c}(\kappa)} n_{\kappa}^{\varepsilon} e^{-\frac{2i\omega z}{\bar{c}(\kappa)\varepsilon}} \left(q V_{p,q+1}^{\varepsilon} - p V_{p-1,q}^{\varepsilon} \right).$$

By applying the limit theorem of Section 6.7.3, in the same way as in Section 9.2.1, we obtain that the process $(V_{p,q}^{\varepsilon})_{p,q \in \mathbb{N}}$ converges in distribution as $\varepsilon \to 0$ to a diffusion process. In particular, the moments converge to

$$\mathbb{E}[V_{p,q}^{\varepsilon}(\omega, \kappa, \tau, \chi, z_0, z)] \xrightarrow{\varepsilon \to 0} V_{p,q}(\omega, \kappa, \tau, \chi, z_0, z),$$

where the family $(V_{p,q})_{p,q \in \mathbb{N}}$ satisfies the system of transport equations

$$\frac{\partial V_{p,q}}{\partial z} + \frac{p+q}{\psi(\kappa)\bar{c}(\kappa)} \frac{\partial V_{p,q}}{\partial \tau} + \bar{c}(\kappa)\kappa(p+q) \frac{\partial V_{p,q}}{\partial \chi}$$

$$= \frac{pq}{L_{\mathrm{loc}}(\omega, \kappa)} (V_{p+1,q+1} + V_{p-1,q-1} - 2V_{p,q}) - \left(\frac{(p-q)^2}{L_{\mathrm{loc}}(\omega, \kappa)} + \frac{2(p-q)^2}{\tilde{L}_{\mathrm{loc}}(\omega, \kappa)} \right) V_{p,q}$$

starting from $V_{p,q}(\omega, \kappa, \tau, \chi, z_0, z = z_0) = \mathbf{1}_0(p)\mathbf{1}_0(q)\delta(\tau)\delta(\chi)$. Here the mode-dependent localization length is given by

$$L_{\mathrm{loc}}(\omega, \kappa) = \frac{4\bar{c}^2(\kappa)}{\omega^2 \gamma_{n\kappa}}, \qquad \tilde{L}_{\mathrm{loc}}(\omega, \kappa) = \frac{4\bar{c}^2(\kappa)}{\omega^2 \gamma_{m\kappa}}, \qquad (17.69)$$

where $\gamma_{n\kappa}$ and $\gamma_{m\kappa}$ are given by (17.56) and (17.66). As in the case with constant density we have that

$$V_{p,q}(\omega, \kappa, \tau, \chi, z_0, z) = 0 \text{ if } p \neq q, \qquad (17.70)$$

and if we consider $p = q$, then we have

$$V_{p,p}(\omega, \kappa, \tau, \chi, z_0, z) = W_p(\omega, \kappa, \tau, z_0, z)\delta(\chi - \psi(\kappa)\bar{c}(\kappa)^2 \kappa \tau),$$

where W_p satisfies the closed system

$$\frac{\partial W_p}{\partial z} + \frac{2p}{\psi(\kappa)\bar{c}(\kappa)} \frac{\partial W_p}{\partial \tau} = \frac{p^2}{L_{\mathrm{loc}}(\omega, \kappa)} (W_{p+1} + W_{p-1} - 2W_p),$$

starting from $W_p(\omega, \kappa, \tau, z_0, z = z_0) = \mathbf{1}_0(p)\delta(\tau)$. Note that these are the same equations for W_p as the ones we arrived at in Section 14.3.2. However, now the localization length $L_{\mathrm{loc}}(\omega, \kappa)$ is defined differently and is given by (17.69).

17.3.4 Reflection by a Random Half-Space

We consider the case of a point source located at the surface at position $(\mathbf{0}, 0)$, emitting a short pulse at time 0, and compute the mean reflected intensity. The point source generates the source term

$$\mathbf{F}^{\varepsilon}(t, \mathbf{x}, z) = \varepsilon^{1/2} f\left(\frac{t}{\varepsilon}\right) \delta(\mathbf{x})\delta(z) \begin{bmatrix} \mathbf{0} \\ 1 \end{bmatrix},$$

which imposes the boundary condition $\check{b}^{\varepsilon}(\omega, \boldsymbol{\kappa}, 0^+) = \varepsilon^{3/2} \hat{f}(\omega)/\sqrt{\zeta_0(\kappa)}$. The integral representation for the reflected pressure wave is

$$p^{\varepsilon}(t, \mathbf{x}, 0^+) = \frac{1}{2(2\pi)^3 \varepsilon^{3/2}} \int \int e^{-i\frac{\omega}{\varepsilon}(t - \boldsymbol{\kappa} \cdot \mathbf{x})} \mathcal{R}_{(\omega, \kappa)}^{\varepsilon} \hat{f}(\omega)\omega^2 \, d\omega \, d\boldsymbol{\kappa},$$

where $\mathcal{R}^\varepsilon_{(\omega,\kappa)}$ is given by (17.63). In the case of a random half-space, as shown in Section 17.1.3, we can expand the generalized reflection coefficient

$$\mathcal{R}^\varepsilon_{(\omega,\kappa)} = R_I + \sum_{n=0}^{\infty} (-R_I)^n T_I^2 (R^\varepsilon_{(\omega,\kappa)})^{n+1}, \qquad (17.71)$$

where $R_I(\kappa)$ and $T_I(\kappa)$ are given by (17.61). We observe the mean reflected intensity at the surface $z = 0$, at the location $\mathbf{x} \neq \mathbf{0}$, and at time $t > 0$. The computation follows the lines of the one carried out in Section 14.3.3, and we obtain, instead of (14.73), the following expression for the mean intensity:

$$\mathbb{E}[p^\varepsilon(t,\mathbf{x},0^+)^2] = \sum_{q \in \{-1,1\}} \frac{1}{128\pi^5 x} \int\int e^{-iht+iq(h\mu+\omega\lambda)x}$$

$$\times \mathbb{E}\left[\mathcal{R}^\varepsilon_{(\omega,\mu)} \overline{\mathcal{R}^\varepsilon_{(\omega-\varepsilon h,\mu-\varepsilon\lambda)}}\right] |\hat{f}(\omega)|^2 \omega^3 \mu \, d\lambda \, dh \, d\omega \, d\mu. \qquad (17.72)$$

There are two differences with the computation carried out in Section 14.3.3.

- The first difference is the expression for the limit autocorrelation of the generalized reflection coefficient. From the expansion (17.71) we get

$$\mathbb{E}\left[\mathcal{R}^\varepsilon_{(\omega,\mu)} \overline{\mathcal{R}^\varepsilon_{(\omega+\varepsilon h,\mu+\varepsilon\lambda)}}\right]$$

$$\xrightarrow{\varepsilon\to0} R_I^2 + \sum_{n=0}^{\infty} R_I^{2n} T_I^4 \int \mathcal{W}_{n+1}(\omega,\mu,\tau) \, e^{i\tau[h\psi(\mu)-\omega\lambda\bar{c}^2\mu]} \, d\tau,$$

where $\mathcal{W}_{n+1}$ is given by

$$\mathcal{W}_{n+1}(\omega,\mu,\tau) = \frac{\bar{c}(\mu)\psi(\mu)}{L_{\text{loc}}(\omega,\mu)} P^\infty_{n+1}\left(\frac{\bar{c}(\mu)\psi(\mu)\tau}{L_{\text{loc}}(\omega,\mu)}\right),$$

P^∞_{n+1} is given by (9.39), and $L_{\text{loc}}(\omega,\mu)$ is the localization length defined by (17.69). Substituting into (17.72), and integrating in h and λ as in Section 14.3.3, we get

$$\lim_{\varepsilon\to0} \mathbb{E}[p^\varepsilon(t,\mathbf{x},0^+)^2] = \sum_{n=0}^{\infty} \frac{1}{32\pi^3 x} \int\int R_I^{2n} T_I^4 \delta\left(\psi(\mu)\tau - t + \mu x\right)$$

$$\times \delta\left(x - \psi(\mu)\bar{c}(\mu)^2 \mu\tau\right) \mathcal{W}_{n+1}(\omega,\mu,\tau) |\hat{f}(\omega)|^2 \omega^2 \mu \, d\tau \, d\omega \, d\mu.$$

- The integral in τ concentrates on t (which shows that the variable τ in $\mathcal{W}_n$ is indeed a travel time) and the integral in μ concentrates on $\kappa_{x,t}$, which is such that

$$x - \psi(\kappa_{x,t})\bar{c}(\kappa_{x,t})^2 \kappa_{x,t} t = 0,$$

that is to say,

$$\kappa_{x,t} = \frac{\sqrt{1 + 4(\frac{\bar{\rho}}{\rho} - 1)(\frac{x}{\bar{c}t})^2} - 1}{2(\frac{\bar{\rho}}{\rho} - 1)\frac{x}{t}}. \qquad (17.73)$$

The integral in μ runs over $(0, \kappa_{\max})$ with $\kappa_{\max} = \sqrt{\widetilde{\rho}}/(\bar{c}\sqrt{\bar{\rho}})$. This interval determines the slownesses of the propagating modes (see Section 14.1.1, equation (17.37)). The condition to get a nonzero mean intensity is then $\kappa_{x,t} < \kappa_{\max}$, which also reads

$$x < \widetilde{c}t, \qquad \widetilde{c} = \bar{c}\frac{\sqrt{\bar{\rho}}}{\sqrt{\widetilde{\rho}}} = \frac{\sqrt{\overline{K}}}{\sqrt{\widetilde{\rho}}}.$$

This condition is in fact natural, since $\widetilde{c}$ is the horizontal effective velocity given by the homogenization theory described in Section 14.1.1. As a result, in the limit $\varepsilon \to 0$, the mean intensity is zero if $x > \widetilde{c}t$, and if $x < \widetilde{c}t$, then it is given by

$$\lim_{\varepsilon \to 0} \mathbb{E}[p^\varepsilon(t, \mathbf{x}, 0^+)^2] = \sum_{n=0}^{\infty} \frac{R_I^{2n} T_I^4}{32\pi^3 \bar{c}^4 t^2} \times \frac{1}{(\frac{x}{\bar{c}^2 t \kappa_{x,t}})^2 [\frac{2x}{\bar{c}^2 t \kappa_{x,t}} - 1]}$$

$$\times \int \mathcal{W}_{n+1}(\omega, \kappa_{x,t}, t) \, |\hat{f}(\omega)|^2 \omega^2 \, d\omega.$$

Taking into account the explicit expression for $\mathcal{W}_{n+1}$, we get the following expression for the mean reflected intensity for $x < \widetilde{c}t$:

$$\lim_{\varepsilon \to 0} \mathbb{E}[p^\varepsilon(t, \mathbf{x}, 0^+)^2] = \frac{T_I^4(\kappa_{x,t})}{32\pi^3 \bar{c}^4 t^2} \times \frac{1}{(\frac{x}{\bar{c}^2 t \kappa_{x,t}})^2 [\frac{2x}{\bar{c}^2 t \kappa_{x,t}} - 1]}$$

$$\times \int \frac{\frac{\bar{c}(\kappa_{x,t})\psi(\kappa_{x,t})}{2L_{\mathrm{loc}}(\omega, \kappa_{x,t})}}{\left(1 + \frac{\bar{c}(\kappa_{x,t})\psi(\kappa_{x,t})t}{2L_{\mathrm{loc}}(\omega, \kappa_{x,t})} T_I^2(\kappa_{x,t})\right)^2} |\hat{f}(\omega)|^2 \omega^2 \, d\omega,$$

which, compared with the formula (14.75) in the case $\widetilde{\rho} = \bar{\rho}$, summarizes the effect of random fluctuations in the density.

Notes

The nonmatched-medium case and the general-background case addressed in Sections 17.1 and 17.2 have been presented for instance in [8]. In particular, the propagation of the stable front in the presence of a general background is analyzed in detail in [38, 100, 46, 118]. The study of a varying density presented in Section 17.3 is new.

Other Regimes of Propagation

In this chapter we present several generalizations of the theory developed in the previous chapters and we describe the effects on time-reversal properties. Each section is devoted to a particular modification of the theory to accommodate new features of the model. For clarity we choose to present these modifications separately in order to isolate the effects of each new feature of the model. In Section 18.1 we study the weakly heterogeneous regime in which the diffusion-approximation theory applies in randomly layered three-dimensional media. In this realistic regime introduced in Chapter 5, and developed in Chapter 8 for the study of the one-dimensional transmitted wave front, the size of the fluctuations in the medium parameters is small and the pulse is short, with a typical wavelength of the same order as the correlation length of the medium. In Section 18.2 we study the effects of dispersion in the medium. We consider the example of the Boussinesq system describing low-amplitude shallow water waves. We show that time reversal recompresses both the dispersive oscillatory tail of the wave front and the random incoherent waves. In Section 18.3 we incorporate a weak nonlinearity in the equations and we study the combined effects of nonlinearity and randomness on the propagation of the front pulse and its time reversal. In Section 18.4 we address the natural question of the impact of a change of the medium during a time-reversal experiment. We quantify the loss of statistical stability of the refocusing.

18.1 The Weakly Heterogeneous Regime in Randomly Layered Three-Dimensional Media

In Chapters 14–17 we have analyzed wave-propagation phenomena in three-dimensional randomly layered media in a specific regime of separation of scales, the *strongly heterogeneous white-noise* regime, where the correlation length of the medium $\sim \varepsilon^2$ is smaller than the typical propagation distance ~ 1 and the amplitude of the fluctuations of the medium is of order 1. We have seen that for such a configuration, the propagation of a pulse with a typical wavelength of order ε gives rise to a macroscopic interplay that can be analyzed in detail. In this section we show that this regime is not the only one that can be analyzed. We now consider the *weakly heterogeneous* regime in the context of three-dimensional waves. One-dimensional wave propagation was analyzed in this regime in Chapters 7–8. In this case the correlation length of the medium $\sim \varepsilon^2$ is smaller than the propagation distance ~ 1, but the fluctuations of the medium are weak, of order ε. If we send a pulse with a typical wavelength of order ε, then the effect of randomness completely vanishes in the limit $\varepsilon \to 0$, as can be seen from the analysis of the previous chapter, where the integrated covariance γ would vanish (as ε^2). However, if we consider a pulse with a typical wavelength $\sim \varepsilon^2$, that is, of the same order as the correlation length of the medium, then the interaction between the fluctuations of the medium and the wave becomes stronger, and it turns out that this regime leads to a macroscopic interplay and to an effective regime that can be analyzed through the diffusion approximation theory. Thus, we model the medium parameters as

$$\frac{1}{K(\mathbf{x}, z)} = \frac{1}{K(z)} = \begin{cases} \frac{1}{\overline{K}} \left(1 + \varepsilon \nu \left(\frac{z}{\varepsilon^2}\right)\right) & \text{for } z \in [-L, 0], \\ \frac{1}{\overline{K}} & \text{for } z \in (-\infty, -L) \cup (0, \infty), \end{cases}$$

$$\rho(\mathbf{x}, z) = \bar{\rho} \text{ for all } (\mathbf{x}, z).$$

Probing this medium with a pulse with a carrier wavelength of order ε^2 allows us to apply a separation-of-scales technique and to get a remarkable effective solution. The analysis is more or less identical to the one performed in the first regime. From the wave-propagation point of view we use the same mode decomposition. From the probabilistic point of view we apply a modified version of the diffusion-approximation theorem that takes into account the new scales. In this section we shall point out the new features in the weakly heterogeneous regime.

18.1.1 Mode Decomposition

The specific Fourier transform is modified to take into account the scales present in this problem:

$$\hat{p}^\varepsilon(\omega, \boldsymbol{\kappa}, z) = \int p^\varepsilon(t, x, z) e^{i\omega(t - \boldsymbol{\kappa} \cdot \mathbf{x})/\varepsilon^2} dt \, dx,$$

$$p^\varepsilon(t, \mathbf{x}, z) = \frac{1}{(2\pi\varepsilon^2)^3} \int \hat{p}^\varepsilon(t, x, z) e^{-i\omega(t - \boldsymbol{\kappa} \cdot \mathbf{x})/\varepsilon^2} \omega^2 \, d\omega \, d\boldsymbol{\kappa}.$$

We decompose the wave field into right- and left-going waves by setting

$$\hat{p}^\varepsilon(\omega, \boldsymbol{\kappa}, z) = \frac{\sqrt{\zeta(\kappa)}}{2} \left(\breve{a}^\varepsilon(\omega, \boldsymbol{\kappa}, z) e^{\frac{i\omega}{\varepsilon^2 \bar{c}(\kappa)}} - \breve{b}^\varepsilon(\omega, \boldsymbol{\kappa}, z) e^{-\frac{i\omega z}{\varepsilon^2 \bar{c}(\kappa)}} \right), \quad (18.1)$$

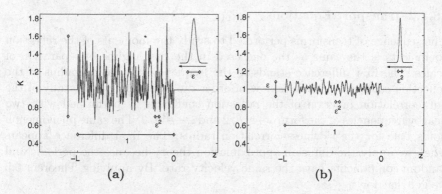

Fig. 18.1. The two regimes of separation of scales. Picture (a): correlation length $(\varepsilon^2) \ll$ wavelength $(\varepsilon) \ll$ propagation distance (1); strong fluctuations (1). Picture (b): correlation length $(\varepsilon^2) \sim$ wavelength $(\varepsilon^2) \ll$ propagation distance (1); weak fluctuations (ε).

$$\hat{u}^\varepsilon(\omega, \boldsymbol{\kappa}, z) = \frac{1}{2\sqrt{\zeta(\kappa)}} \left(\check{a}^\varepsilon(\omega, \boldsymbol{\kappa}, z) e^{\frac{i\omega z}{\varepsilon^2 \bar{c}(\kappa)}} + \check{b}^\varepsilon(\omega, \boldsymbol{\kappa}, z) e^{-\frac{i\omega z}{\varepsilon^2 \bar{c}(\kappa)}} \right), (18.2)$$

where we recall that $\bar{c}(\kappa) = \bar{c}/\sqrt{1 - \bar{c}^2 \kappa^2}$ and $\bar{\zeta}(\kappa) = \bar{\rho}\bar{c}(\kappa)$. Substituting these expressions into the random acoustic wave equations establishes the system satisfied by the modes $(\check{a}^\varepsilon, \check{b}^\varepsilon)$:

$$\frac{d\check{a}^\varepsilon}{dz} = \frac{i\omega}{2\bar{c}(\kappa)\varepsilon} \nu_\kappa \left(\frac{z}{\varepsilon^2} \right) \left(\check{a}^\varepsilon - e^{\frac{-2i\omega z}{\varepsilon^2 \bar{c}(\kappa)}} \check{b}^\varepsilon \right), \tag{18.3}$$

$$\frac{d\check{b}^\varepsilon}{dz} = \frac{i\omega}{2\bar{c}(\kappa)\varepsilon} \nu_\kappa \left(\frac{z}{\varepsilon^2} \right) \left(e^{\frac{2i\omega z}{\varepsilon^2 \bar{c}(\kappa)}} \check{a}^\varepsilon - \check{b}^\varepsilon \right), \tag{18.4}$$

where

$$\nu_\kappa(z) = \frac{\bar{c}(\kappa)^2}{\bar{c}^2} \nu(z).$$

We can again introduce the reflection and transmission coefficients and find that these coefficients satisfy closed-form equations. We write explicitly the Riccati equation satisfied by the reflection coefficient:

$$\frac{dR^\varepsilon_{(\omega,\kappa)}}{dz} = -\frac{i\omega}{2\bar{c}(\kappa)\varepsilon} \nu_\kappa \left(\frac{z}{\varepsilon^2} \right) \left(e^{\frac{-2i\omega z}{\varepsilon^2 \bar{c}(\kappa)}} - 2R^\varepsilon_{(\omega,\kappa)} + (R^\varepsilon_{(\omega,\kappa)})^2 e^{\frac{2i\omega z}{\varepsilon^2 \bar{c}(\kappa)}} \right).$$

The main difference is that the period of the rapid phase is now of the same order as the correlation length of the medium. We therefore expect that the wave will be more sensitive to the fine structure of the fluctuations in the medium parameters.

18.1.2 Transport Equations

The sequence of transforms performed to study the moments of the reflection coefficients is the same as the one used in the first regime of separation of scales. The first difference stands in the frequency-correlation radius of the reflection coefficient, which is now of order ε^2. Accordingly, the study of the autocorrelation function of the reflection coefficient is performed with two nearby frequencies of the form $\omega + \varepsilon^2 h/2$ and $\omega - \varepsilon^2 h/2$. The same phenomenon holds true for the slowness-correlation radius. The final difference appears when we apply the diffusion-approximation theory because the periodic and random components have the same velocity rate. By applying Theorem 6.5 we get that for $z_0 \le z$,

$$\lim_{\varepsilon \to 0} \mathbb{E} \left[\left(R^\varepsilon_{(\omega+\varepsilon^2 h/2, \kappa+\varepsilon^2 \lambda/2)}(z_0, z) \right)^p \left(\overline{R^\varepsilon_{(\omega-\varepsilon^2 h/2, \kappa-\varepsilon^2 \lambda/2)}(z_0, z)} \right)^q \right] \qquad (18.5)$$

$$= \begin{cases} e^{2ipz[-h/\bar{c}(\kappa)+\omega\lambda\bar{c}(\kappa)\kappa]} \displaystyle\int \mathcal{W}_p(\omega, \kappa, \tau, z_0, z) e^{i\tau[h\bar{c}^2/\bar{c}(\kappa)^2 - \omega\lambda\bar{c}^2\kappa]} d\tau & \text{if } q = p, \\ 0 & \text{if } q \ne p, \end{cases}$$

where $\mathcal{W}_p$ is the solution of the system of transport equations

$$\frac{\partial \mathcal{W}_p}{\partial z} + \frac{2p\bar{c}(\kappa)}{\bar{c}^2} \frac{\partial \mathcal{W}_p}{\partial \tau} = \frac{p^2}{L^{(w)}_{\text{loc}}(\omega, \kappa)} \left(\mathcal{W}_{p+1} + \mathcal{W}_{p-1} - 2\mathcal{W}_p \right), \qquad (18.6)$$

starting from $\mathcal{W}_p(\omega, \kappa, \tau, z_0, z = z_0) = \mathbf{1}_0(p)\delta(\tau)$, where

$$L^{(w)}_{\text{loc}}(\omega, \kappa) = \frac{4\bar{c}^4}{\gamma(2\omega/\bar{c}(\kappa))\bar{c}(\kappa)^2\omega^2}, \qquad (18.7)$$

$$\gamma(k) = \int_{-\infty}^{\infty} \mathbb{E}[\nu(0)\nu(z)] \cos(kz) dz. \qquad (18.8)$$

The function $\gamma(k)$ is the power spectral density of the fluctuations of the medium. In the strongly heterogeneous white-noise regime addressed in the previous chapters, the only parameter that remains from the fluctuations of the medium parameters is the correlation length defined as

$$\gamma(0) = \int_{-\infty}^{\infty} \mathbb{E}[\nu(0)\nu(z)] dz.$$

In the regime we consider in this section, the full spectrum of fluctuations of the medium parameters plays a role. Note that by considering the low-frequency limit $\omega \to 0$ in (18.7), we recover the expression of the strongly heterogeneous white-noise regime $L_{\text{loc}} = [4\bar{c}^4]/[\gamma(0)\bar{c}(\kappa)^2\omega^2]$.

18.1.3 Applications

In this subsection, we briefly extend to the weakly heterogeneous regime some of the main results obtained in this book in the strongly heterogeneous white noise regime.

Transmission of the Stable Front

The stabilization of wave front theory discussed in Section 14.2 is still valid qualitatively in the weakly heterogeneous regime. Namely, the transmitted front pulse through a slab of randomly layered medium is modified in two ways compared to the transmitted pulse obtained in a homogeneous medium. This has already been seen in the one-dimensional case in Chapter 8. First, it is randomly time-shifted, and this random time delay can be described in terms of a Brownian motion, and its variance is proportional to the power spectral density $\gamma(0)$ evaluated at zero frequency. Second, it experiences a deterministic shape modification described by the convolution of the homogeneous front with a deterministic kernel. However, this kernel is no longer Gaussian, but it depends in an explicit manner of the power spectral density. More quantitatively, if the source is of the form

$$\mathbf{F}^{\varepsilon}(t, \mathbf{x}, z) = \frac{1}{\varepsilon^2} \begin{bmatrix} \mathbf{f_x} \\ f_z \end{bmatrix} \left(\frac{t}{\varepsilon^2}, \frac{\mathbf{x}}{\varepsilon^2} \right) \delta(z - z_s),$$

then the statement of Proposition 14.3 is modified as follows.

Proposition 18.1. *In probability distribution the following characterization of the transmitted wave process holds:*

$$\lim_{\varepsilon \to 0} p^{\varepsilon} \left(\frac{\sqrt{|\mathbf{x}|^2 + L^2}}{\bar{c}} + \varepsilon^2 s, \mathbf{x}, -L \right) = \tilde{p}(s, \mathbf{x}, -L),$$

where

$$\tilde{p}(s, \mathbf{x}, -L) = \left[\mathcal{K}_{(L,\mathbf{x})} * \tilde{p}_0(\cdot, \mathbf{x}, -L) \right] \left(s - \Theta_{(L,\mathbf{x})} \right).$$

Here

- *The pulse $\tilde{p}_0(\cdot, \mathbf{x}, -L)$ is the front pulse obtained through the homogeneous medium. In the case of a point source located at the origin, it is given by (14.51).*
- *The time shift $\Theta_{(L,\mathbf{x})}$ is Gaussian distributed and can be written as*

$$\Theta_{(L,\mathbf{x})} = \frac{\sqrt{\gamma(0) \left(1 + \frac{|\mathbf{x}|^2}{L^2} \right)}}{2\bar{c}} W_0(L),$$

where W_0 is a standard Brownian motion.
- *The deterministic convolution kernel $\mathcal{K}_{(L,\mathbf{x})}(t)$ is given in the Fourier domain by*

$$\hat{\mathcal{K}}_{(L,\mathbf{x})}(\omega) = \exp \left\{ -\frac{\omega^2 \bar{c}_{(L,\mathbf{x})}^2}{8\bar{c}^4} \left[\gamma \left(\frac{2\omega}{\bar{c}_{(L,\mathbf{x})}} \right) + i\gamma^{(s)} \left(\frac{2\omega}{\bar{c}_{(L,\mathbf{x})}} \right) \right] L \right\},$$

where $\bar{c}_{(L,\mathbf{x})} = \bar{c}\sqrt{1 + \frac{|\mathbf{x}|^2}{L^2}}$ *and*

$$\gamma^{(s)}(k) = 2\int_0^\infty \mathbb{E}[\nu(0)\nu(z)]\sin(kz)dz.$$

Mean Reflected Intensity

We can study the waves reflected by a random slab in the weakly heterogeneous regime by following the same steps as in Section 14.3. Let us consider a point source located at the origin and generating the forcing term

$$\mathbf{F}^\varepsilon(t,\mathbf{x},z) = f\left(\frac{t}{\varepsilon^2}\right)\begin{bmatrix}\mathbf{0}\\1\end{bmatrix}\delta(\mathbf{x})\delta(z).$$

We observe the reflected waves at the surface, at the offset $\mathbf{x} = (x,0)$, $x > 0$, and at time $t > 0$. The mean reflected intensity is zero if $x > \bar{c}t$, and if $x < \bar{c}t$, then it is given by (14.74) with $\mathcal{W}_1$ the solution of the system (18.6). The picture is therefore unchanged.

Time Reversal

We can revisit the results obtained in Chapter 15 in the new regime of separation of scales. We address the case of the exterior point source of the form

$$\mathbf{F}^\varepsilon(t,\mathbf{x},z) = \varepsilon^4 f\left(\frac{t-t_s}{\varepsilon^2}\right)\begin{bmatrix}\mathbf{0}\\1\end{bmatrix}\delta(\mathbf{x}-\mathbf{x}_s)\delta(z).$$

The result is then exactly the same as in the first regime, and we recover the statement of Theorem 15.1 with the refocusing kernel given in terms of the solution $\mathcal{W}_1$ of the system (18.6).

18.2 Dispersive Media

The acoustic-wave equations that we have considered so far form a **hyperbolic system**. It is interesting to address dispersive systems, since many different types of wave-propagation phenomena are actually modeled by dispersive systems. In this section we consider a one-dimensional dispersive random system. We show that the key point is the decomposition of the wave field into suitable right- and left-propagating modes. The analysis of the statistical distribution of the reflection coefficients then follows the same lines as in the hyperbolic case. An interesting feature concerning time reversal is that it can recompress both the incoherent fluctuations promoted by randomness and the **oscillatory tail** caused by dispersion. As a result, the localization of a source can be made with more accuracy in a dispersive context than in a hyperbolic one because the source location is precisely the transition point between the recompression of the dispersive oscillations and the generation of a new oscillatory tail.

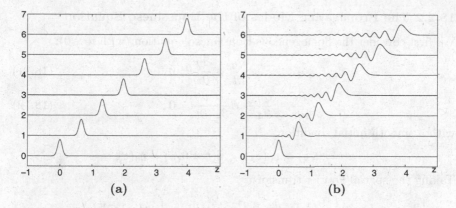

Fig. 18.2. Propagation of a Gaussian pulse in a hyperbolic medium (a) and in a dispersive medium (b). The pulse profiles are plotted at different times. The pulse travels without distortion in the hyperbolic medium. In the dispersive medium an oscillatory tail develops behind the front pulse.

18.2.1 The Terrain-Following Boussinesq Model

We consider the Boussinesq equation that describes the evolution of surface waves in shallow channels [124]:

$$M^\varepsilon(z)\frac{\partial \eta^\varepsilon}{\partial t} + \frac{\partial u^\varepsilon}{\partial z} = 0, \qquad (18.9)$$

$$\frac{\partial u^\varepsilon}{\partial t} + \frac{\partial \eta^\varepsilon}{\partial z} - \beta\frac{\partial^3 u^\varepsilon}{\partial z^2 \partial t} = 0, \qquad (18.10)$$

where η^ε is the wave elevation and u^ε is the depth-averaged velocity, and z and t are the space and time coordinates, respectively. The spatial variations of the coefficient M^ε are imposed by the bottom profile

$$M^\varepsilon(z) = 1 + \varepsilon\nu\left(\frac{z}{\varepsilon^2}\right),$$

where 1 stands for the constant mean depth, and the dimensionless small parameter ε characterizes the amplitude of the relative fluctuations of the bottom and their correlation length. These fluctuations are modeled by the zero-mean stationary random process $\nu(z)$. Note that we consider here the weakly heterogeneous regime. According to the results of Section 18.1, we probe the medium with a pulse whose support is comparable to the correlation length of the random medium, that is, of order ε^2.

The coefficient β measures the dispersion strength. In this section we consider the case in which the dispersion parameter β is of order ε^4: $\beta = \beta_0\varepsilon^4$. The dispersion term involves three derivatives, which shows that dispersive effects for the pulse under consideration show up after a propagation distance of order ε^2. The dispersive effects after a propagation distance of order 1 are therefore very strong.

18.2.2 The Propagating Modes of the Boussinesq Equation

We first consider the homogeneous Boussinesq equation (with $\nu \equiv 0$):

$$\frac{\partial \eta}{\partial t} + \frac{\partial u}{\partial z} = 0, \tag{18.11}$$

$$\frac{\partial u}{\partial t} + \frac{\partial \eta}{\partial z} - \beta \frac{\partial^3 u}{\partial z^2 \partial t} = 0, \tag{18.12}$$

with a smooth initial condition

$$u(t = 0, z) = u_0(z), \quad \eta(t = 0, z) = \eta_0(z).$$

Taking the spatial Fourier transform

$$\check{u}(t, k) = \int u(t, z)e^{-ikz}\, dz, \quad \check{\eta}(t, k) = \int \eta(t, z)e^{-ikz}\, dz,$$

the Boussinesq equation (18.11–18.12) can be reduced to a set of ordinary differential equations:

$$\frac{d\check{\eta}}{dt} = -ik\check{u}, \tag{18.13}$$

$$(1 + \beta k^2)\frac{d\check{u}}{dt} = -ik\check{\eta}. \tag{18.14}$$

Introducing the frequency corresponding to the wave number k through the *dispersion relation*

$$\omega(k) = \frac{k}{\sqrt{1 + \beta k^2}}, \tag{18.15}$$

we get closed-form expressions for the solutions:

$$\check{u}(t, k) = \frac{1}{2}\left(\check{u}_0(k) + \frac{\omega}{k}\check{\eta}_0(k)\right)e^{-i\omega t} + \frac{1}{2}\left(\check{u}_0(k) - \frac{\omega}{k}\check{\eta}_0(k)\right)e^{i\omega t},$$

$$\check{\eta}(t, k) = \frac{1}{2}\left(\frac{k}{\omega}\check{u}_0(k) + \check{\eta}_0(k)\right)e^{-i\omega t} - \frac{1}{2}\left(\frac{k}{\omega}\check{u}_0(k) - \check{\eta}_0(k)\right)e^{i\omega t}.$$

From these expressions we can conclude that any solution can be decomposed as the superposition of left-propagating modes $(u^{(l)}, \eta^{(l)})$ and right-propagating modes $(u^{(r)}, \eta^{(r)})$:

$$u(t, z) = u^{(r)}(t, z) + u^{(l)}(t, z), \qquad \eta(t, z) = \eta^{(r)}(t, z) + \eta^{(l)}(t, z),$$

where

$$u^{(r)}(t, z) = \frac{1}{4\pi}\int\left(\check{u}_0(k) + \frac{\omega}{k}\check{\eta}_0(k)\right)e^{-i\omega(k)t+ikz}\, dk,$$

$$\eta^{(r)}(t, z) = \frac{1}{4\pi}\int\frac{k}{\omega}\left(\check{u}_0(k) + \frac{\omega}{k}\check{\eta}_0(k)\right)e^{-i\omega(k)t+ikz}\, dk,$$

$$u^{(l)}(t, z) = \frac{1}{4\pi}\int\left(\check{u}_0(k) - \frac{\omega}{k}\check{\eta}_0(k)\right)e^{i\omega(k)t+ikz}\, dk,$$

$$\eta^{(l)}(t, z) = -\frac{1}{4\pi}\int\frac{k}{\omega}\left(\check{u}_0(k) - \frac{\omega}{k}\check{\eta}_0(k)\right)e^{i\omega(k)t+ikz}\, dk.$$

This decomposition will be used in the inhomogeneous case in the next section. It plays the same role as the simple decomposition we have used in the previous sections devoted to hyperbolic wave equations. Here the mode decomposition is exact for the dispersive waves under consideration.

18.2.3 Mode Propagation in a Dispersive Random Medium

We consider the problem on the finite slab $-L \le z \le 0$ where boundary conditions are imposed at $-L$ and 0 corresponding to a left-going pulse entering the slab from the right at $z = 0$. We can generalize the standard approach for acoustic equations to the dispersive case using the decomposition introduced in the previous section. We consider the random Boussinesq equation (18.9–18.10) and take the specific time Fourier transform

$$\hat{u}^\varepsilon(\omega, z) = \int u^\varepsilon(t, z) e^{\frac{i\omega t}{\varepsilon^2}}\, dt\,, \quad \hat{\eta}^\varepsilon(\omega, z) = \int \eta^\varepsilon(t, z) e^{\frac{i\omega t}{\varepsilon^2}}\, dt\,,$$

so that the system reduces to a set of ordinary differential equations:

$$\left[1 - \beta_0 \omega^2 \left(1 + \varepsilon \nu \left(\frac{z}{\varepsilon^2} \right) \right) \right] \frac{d\hat{\eta}^\varepsilon}{dz} = \frac{i\omega}{\varepsilon^2} \hat{u}^\varepsilon + \frac{\beta_0 \omega^2}{\varepsilon} \nu' \left(\frac{z}{\varepsilon^2} \right) \hat{\eta}^\varepsilon\,, \quad (18.16)$$

$$\frac{d\hat{u}^\varepsilon}{dz} = \frac{i\omega}{\varepsilon^2} \left(1 + \varepsilon \nu \left(\frac{z}{\varepsilon^2} \right) \right) \hat{\eta}^\varepsilon\,, \quad (18.17)$$

where ν' stands for the spatial derivative of ν. We introduce the scaled wave number k_0 corresponding to the scaled frequency ω:

$$k_0(\omega) = \frac{\omega}{\sqrt{1 - \beta_0 \omega^2}}\,. \quad (18.18)$$

The modes with frequency $\omega^2 > 1/\beta_0$ correspond to evanescent modes that are not taken into account here. Note that the true frequency is ω/ε^2, and we have the identity

$$k \left(\frac{\omega}{\varepsilon^2} \right) = \frac{k_0(\omega)}{\varepsilon^2}\,,$$

where k is the true wave number [the reciprocal of the function defined by (18.15)]. We can decompose the wave into right-going modes $\hat{A}^\varepsilon$ and left-going modes $\hat{B}^\varepsilon$:

$$\hat{A}^\varepsilon(\omega, z) = \hat{\eta}^\varepsilon(\omega, z) + \frac{k_0}{\omega} \hat{u}^\varepsilon(\omega, z)\,,$$

$$\hat{B}^\varepsilon(\omega, z) = \hat{\eta}^\varepsilon(\omega, z) - \frac{k_0}{\omega} \hat{u}^\varepsilon(\omega, z)\,.$$

The modes $(\hat{A}^\varepsilon, \hat{B}^\varepsilon)$ satisfy

$$\frac{d\hat{A}^\varepsilon}{dz} = \frac{ik_0}{\varepsilon^2}\hat{A}^\varepsilon + \frac{ik_0}{2\varepsilon}\nu\left(\frac{z}{\varepsilon^2}\right)(\hat{A}^\varepsilon + \hat{B}^\varepsilon) + \frac{\beta_0 k_0^2}{2\varepsilon}\nu'\left(\frac{z}{\varepsilon^2}\right)(\hat{A}^\varepsilon + \hat{B}^\varepsilon)$$

$$+\frac{i\omega^2}{2k_0\varepsilon^2}\left(\frac{1}{1-\beta_0\omega^2(1+\varepsilon\nu(z/\varepsilon^2))} - \frac{1}{1-\beta_0\omega^2}\right)(\hat{A}^\varepsilon - \hat{B}^\varepsilon)$$

$$+\frac{\beta\omega^2}{2\varepsilon}\nu'(\frac{z}{\varepsilon^2})\left(\frac{1}{1-\beta_0\omega^2(1+\varepsilon\nu(z/\varepsilon^2))} - \frac{1}{1-\beta_0\omega^2}\right)(\hat{A}^\varepsilon + \hat{B}^\varepsilon),$$

$$\frac{d\hat{B}^\varepsilon}{dz} = -\frac{ik_0}{\varepsilon^2}\hat{B}^\varepsilon - \frac{ik_0}{2\varepsilon}\nu\left(\frac{z}{\varepsilon^2}\right)(\hat{A}^\varepsilon + \hat{B}^\varepsilon) + \frac{\beta_0 k_0^2}{2\varepsilon}\nu'\left(\frac{z}{\varepsilon^2}\right)(\hat{A}^\varepsilon + \hat{B}^\varepsilon)$$

$$+\frac{i\omega^2}{2k_0\varepsilon^2}\left(\frac{1}{1-\beta_0\omega^2(1+\varepsilon\nu(z/\varepsilon^2))} - \frac{1}{1-\beta_0\omega^2}\right)(\hat{A}^\varepsilon - \hat{B}^\varepsilon)$$

$$+\frac{\beta\omega^2}{2\varepsilon}\nu'(\frac{z}{\varepsilon^2})\left(\frac{1}{1-\beta_0\omega^2(1+\varepsilon\nu(z/\varepsilon^2))} - \frac{1}{1-\beta_0\omega^2}\right)(\hat{A}^\varepsilon + \hat{B}^\varepsilon).$$

We expand the last terms of the right-hand sides up to $O(\varepsilon^3)$ terms

$$\frac{\omega^2}{1-\beta_0\omega^2(1+\varepsilon\nu(z/\varepsilon^2))} - \frac{\omega^2}{1-\beta_0\omega^2} = \varepsilon\beta_0 k_0^4\nu\left(\frac{z}{\varepsilon^2}\right) + \varepsilon^2\beta_0^2 k_0^6\nu^2\left(\frac{z}{\varepsilon^2}\right) + O(\varepsilon^3).$$
(18.19)

We now look at the waves along the *frequency-dependent modified characteristics* defined by

$$\hat{a}^\varepsilon(\omega,z) = \hat{A}^\varepsilon(\omega,z)\exp\left(-\frac{ik_0 z}{\varepsilon^2}\right)\exp\left[-\frac{\varepsilon\beta_0 k_0^2}{2}\nu\left(\frac{z}{\varepsilon^2}\right) - \frac{\varepsilon^2\beta_0^2 k_0^4}{4}\nu^2\left(\frac{z}{\varepsilon^2}\right)\right],$$

$$\hat{b}^\varepsilon(\omega,z) = \hat{B}^\varepsilon(\omega,z)\exp\left(\frac{ik_0 z}{\varepsilon^2}\right)\exp\left[-\frac{\varepsilon\beta_0 k_0^2}{2}\nu\left(\frac{z}{\varepsilon^2}\right) - \frac{\varepsilon^2\beta_0^2 k_0^4}{4}\nu^2\left(\frac{z}{\varepsilon^2}\right)\right],$$

which satisfy the linear equation

$$\frac{d}{dz}\begin{bmatrix}\hat{a}^\varepsilon \\ \hat{b}^\varepsilon\end{bmatrix}(\omega,z) = \mathbf{H}_\omega^\varepsilon(z)\begin{bmatrix}\hat{a}^\varepsilon \\ \hat{b}^\varepsilon\end{bmatrix}(\omega,z).$$
(18.20)

The complex 2×2 matrix $\mathbf{H}_\omega^\varepsilon$ is given by

$$\mathbf{H}_\omega^\varepsilon(z) = \begin{bmatrix} Q_1^\varepsilon(\omega,z) & Q_2^\varepsilon(\omega,z)e^{-\frac{2ik_0 z}{\varepsilon^2}} \\ Q_2^\varepsilon(\omega,z)e^{\frac{2ik_0 z}{\varepsilon^2}} & Q_1^\varepsilon(\omega,z) \end{bmatrix},$$
(18.21)

with

$$Q_1^\varepsilon(\omega,z) = \frac{ik_0}{2\varepsilon}(1+\beta_0 k_0^2)\nu\left(\frac{z}{\varepsilon^2}\right) + \frac{i\beta_0^2 k_0^5}{2}\nu^2\left(\frac{z}{\varepsilon^2}\right) + O(\varepsilon),$$
(18.22)

$$Q_2^\varepsilon(\omega,z) = \frac{ik_0}{2\varepsilon}(1-\beta_0 k_0^2)\nu\left(\frac{z}{\varepsilon^2}\right) + \frac{\beta k_0^2}{2\varepsilon}\nu'\left(\frac{z}{\varepsilon^2}\right) - \frac{i\beta_0^2 k_0^5}{2}\nu^2\left(\frac{z}{\varepsilon^2}\right)$$

$$+\frac{\beta_0^2 k_0^4}{2}\nu\left(\frac{z}{\varepsilon^2}\right)\nu'\left(\frac{z}{\varepsilon^2}\right) + O(\varepsilon).$$
(18.23)

The small terms of order ε come from the $O(\varepsilon^3)$ term in the expansion (18.19). The reflection and transmission coefficients satisfy the closed-form nonlinear differential system

$$\frac{dR_\omega^\varepsilon}{dz} = 2Q_1^\varepsilon(\omega, z)R_\omega^\varepsilon - e^{-\frac{2ik_0z}{\varepsilon^2}}\overline{Q_2^\varepsilon(\omega, z)}(R_\omega^\varepsilon)^2 + e^{\frac{2ik_0z}{\varepsilon^2}}Q_2^\varepsilon(\omega, z), \quad (18.24)$$

$$\frac{dT_\omega^\varepsilon}{dz} = -T_\omega^\varepsilon\left(e^{-\frac{2ik_0z}{\varepsilon^2}}\overline{Q_2^\varepsilon(\omega, z)}R_\omega^\varepsilon + \overline{Q_1^\varepsilon(\omega, z)}\right). \quad (18.25)$$

These nonlinear equations have the same structure as the Riccati equation encountered in the hyperbolic case and weakly heterogeneous regime. We can therefore carry out a similar analysis.

18.2.4 Transport Equations

The autocorrelation function of the reflection coefficient plays an important role. In particular, it naturally appears in the integral representation of the refocused pulse obtained as a result of a time-reversal experiment. In order to get its limiting distribution as $\varepsilon \to 0$, we introduce for $p, q \in \mathbb{N}$,

$$U_{p,q}^\varepsilon(\omega, h, z_0, z) = \left(R_{\omega+\varepsilon^2 h/2}^\varepsilon(z_0, z)\right)^p \left(\overline{R_{\omega-\varepsilon^2 h/2}^\varepsilon(z_0, z)}\right)^q.$$

Setting

$$k_0'(\omega) = \frac{\partial k_0}{\partial \omega}(\omega) = \frac{1}{(1 - \beta_0\omega^2)^{3/2}} = (1 + \beta_0 k_0^2)^{3/2}, \quad (18.26)$$

and using the Riccati equation (18.24) satisfied by R_ω^ε, we deduce

$$\frac{\partial U_{p,q}^\varepsilon}{\partial z} = 2(p - q)Q_1^\varepsilon U_{p,q}^\varepsilon + Q_2^\varepsilon e^{\frac{-2ik_0(\omega)z}{\varepsilon^2}}\left(pe^{-ik_0'(\omega)hz}U_{p-1,q}^\varepsilon - qe^{ik_0'(\omega)hz}U_{p,q+1}^\varepsilon\right)$$

$$+ \overline{Q_2^\varepsilon}e^{\frac{2ik_0(\omega)z}{\varepsilon^2}}\left(qe^{-ik_0'(\omega)hz}U_{p,q-1}^\varepsilon - pe^{ik_0'(\omega)hz}U_{p+1,q}^\varepsilon\right),$$

starting from $U_{p,q}^\varepsilon(\omega, h, z_0, z = z_0) = \mathbf{1}_0(p)\mathbf{1}_0(q)$. Taking a shifted scaled Fourier transform with respect to h,

$$V_{p,q}^\varepsilon(\omega, \tau, z_0, z) = \frac{1}{2\pi}\int e^{-ih[\tau - k_0'(\omega)(p+q)z]}U_{p,q}^\varepsilon(\omega, h, z)\, dh,$$

we get

$$\frac{\partial V_{p,q}^\varepsilon}{\partial z} = -k_0'(\omega)(p + q)\frac{\partial V_{p,q}^\varepsilon}{\partial \tau} + 2(p - q)Q_1^\varepsilon V_{p,q}^\varepsilon$$

$$+ Q_2^\varepsilon e^{\frac{-2ik_0(\omega)z}{\varepsilon^2}}\left(pV_{p-1,q}^\varepsilon - qV_{p,q+1}^\varepsilon\right) + \overline{Q_2^\varepsilon}e^{\frac{2ik_0(\omega)z}{\varepsilon^2}}\left(qV_{p,q-1}^\varepsilon - pV_{p+1,q}^\varepsilon\right),$$

starting from $V_{p,q}^\varepsilon(\omega, \tau, z_0, z = z_0) = \delta(\tau)\mathbf{1}_0(p)\mathbf{1}_0(q)$. Applying the limit theorem of Section 6.7.3 in the same way as in Section 9.2.1 establishes that the process $V_{p,q}^{\varepsilon}$ converges to a diffusion process as $\varepsilon \to 0$. In particular, the expectations $\mathbb{E}[V_{p,p}^{\varepsilon}(\omega, \tau, z_0, z)]$, $p \in \mathbb{N}$, converge to $\mathcal{W}_p(\omega, \tau, z_0, z)$, which obey the closed system of transport equations

$$\frac{\partial \mathcal{W}_p}{\partial z} + 2k_0'(\omega)p\frac{\partial \mathcal{W}_p}{\partial \tau} = \frac{p^2}{L_{\mathrm{loc}}^{(\beta)}(k_0(\omega))}\left(\mathcal{W}_{p+1} + \mathcal{W}_{p-1} - 2\mathcal{W}_p\right), \qquad (18.27)$$

starting from $\mathcal{W}_p(\omega, \tau, z_0, z = z_0) = \delta(\tau)\mathbf{1}_0(p)$, where

$$L_{\mathrm{loc}}^{(\beta)}(k_0) = \frac{4}{\gamma(2k_0)k_0^2(1 + \beta_0 k_0^2)^2}, \qquad (18.28)$$

and γ is the power spectral density of the random process ν,

$$\gamma(k) = \int_{-\infty}^{\infty} \mathbb{E}[\nu(0)\nu(z)]\cos(kz)\, dz. \qquad (18.29)$$

Note that the limit transport equations (18.27) have the same form as those (9.23) obtained in the nondispersive case. The difference is contained in the expression of the group velocity $1/k_0'(\omega)$ and that of the localization length $L_{\mathrm{loc}}^{(\beta)}(k_0(\omega))$.

18.2.5 Time Reversal

By revisiting the analysis carried out in Section 10.1, we get the integral representation of the refocused pulse at a position close to the surface:

$$S_L^{\varepsilon}(t_1 + \varepsilon^2 s, z = \varepsilon^2\zeta) = \frac{1}{(2\pi)^2}\int\int e^{-i\omega s + ik_0(\omega)\zeta}e^{-i\varepsilon hs/2 + i\varepsilon k_0'(\omega)\zeta}$$
$$\times\hat{f}(\omega - \varepsilon h/2)\hat{G}(h)R_{\omega+\varepsilon h/2}^{\varepsilon}(-L, 0)\overline{R_{\omega-\varepsilon h/2}^{\varepsilon}(-L, 0)}\, dh\, d\omega.$$

The analysis is similar to the hyperbolic case, and we find that the limiting refocused pulse shape at the original source location is

$$S_L(s, \zeta = 0) = (K_{\mathrm{TRR}}(\cdot) * f(-\cdot))(s), \qquad (18.30)$$

where the Fourier transform of the refocusing kernel is

$$\hat{K}_{\mathrm{TRR}}(\omega) = \int G(\tau)\mathcal{W}_1(\omega, \tau, -L, 0)\, d\tau.$$

The analysis of the refocusing kernel can be made simple by considering the case of a random half-space and a cutoff function $G(t) = \mathbf{1}_{[0,t_1]}(t)$. The refocusing kernel is then given by

$$\hat{K}_{\mathrm{TRR}}(\omega) = \frac{\omega^2/\Omega^2(\omega)}{1 + \omega^2/\Omega^2(\omega)} \quad \text{with} \quad \Omega^2(\omega) = \frac{8\left(1 - \beta_0\omega^2\right)^{3/2}}{\gamma(2\omega/\sqrt{1 - \beta_0\omega^2})t_1}.$$

Thus the refocusing kernel has the form of a high-pass filter whose cutoff frequency decays with increasing β_0. This shows that time-reversal focusing in reflection is more efficient in the dispersive case than in the hyperbolic case. This is essentially due to the fact that the effective wave number is larger in the dispersive medium, which enhances the localization effect and the back-scattering promoted by randomness.

When the wave moves away from $z = 0$, its shape is rapidly affected by dispersion, and it is given by

$$S_L(s,\zeta) = \left(K_\zeta^\beta(\cdot) * K_{\mathrm{TRR}}(\cdot) * f(-\cdot) \right)(s - \zeta), \qquad (18.31)$$

where the Fourier transform of the dispersive kernel is

$$\hat{K}_\zeta^\beta(\omega) = e^{i[k_0(\omega)-\omega]\zeta}.$$

In the hyperbolic case $\beta_0 = 0$, the refocused pulse that emerges from the medium is a traveling pulse that propagates to the right without deformation:

$$S_L(s,\zeta) = s_L(s - \zeta, 0).$$

In the dispersive case $\beta_0 > 0$, the wave develops an oscillatory tail as it propagates away from the original source location:

$$S_L(s,\zeta) = K_\zeta^\beta * s_L(s - \zeta, 0).$$

Thus dispersion provides an improvement for source location, in one-dimension, as a consequence of the recompression of the dispersive tail, which in the time-reversal experiment is ahead of the pulse. As soon as the recompressed pulse travels over the original source location, a dispersive oscillatory tail starts to develop behind the pulse. The source location is precisely where there is no dispersive oscillation, either in front or behind the pulse.

18.3 Nonlinear Media

This section is concerned with the study of the deformation of a nonlinear pulse traveling in a random medium. We consider shallow-water waves with a spatially random depth. We demonstrate that in the presence of properly scaled stochastic forcing the solution to the nonlinear conservation law is regularized, leading to a **viscous shock profile**. This enables us to perform time reversal experiments beyond the critical time for shock formation.

We extend the theory developed in this book to nonlinear waves by decomposing the solution of the perturbed system using the Riemann invariants of the unperturbed system. Using a stochastic averaging theorem we show that the right-going Riemann invariant satisfies a viscous Burgers-like equation. The apparent viscosity is a pseudodifferential operator defined in terms of the power spectral density of the random fluctuations.

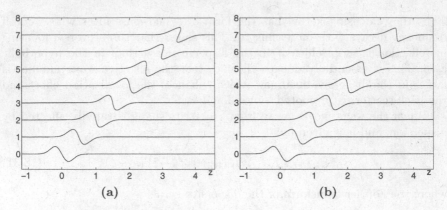

Fig. 18.3. Propagation of a pulse of the form $-t \exp(-t^2)$ in a nonlinear medium without viscosity (a) and with a small viscosity (b). Viscosity reads as a term of the form $\mu \partial_z^2 u$ in the right-hand side of (18.33). In the absence of viscosity a shock occurs at a critical time. The uniqueness of the solution is lost after this time. We plot in picture (a) the solution given by the characteristic method. A small viscosity prevents shock formation. A solution of the inviscid system after the critical time that is more physical than the one given by the characteristic method can be obtained by considering the solution of the viscous system with an evanescent viscosity.

18.3.1 Shallow-Water Waves with Random Depth

The shallow-water equations are given by [45]

$$\frac{\partial \eta}{\partial t} + \frac{\partial(1 + \varepsilon \nu + \alpha \eta)u}{\partial z} = 0, \tag{18.32}$$

$$\frac{\partial u}{\partial t} + \frac{\partial \eta}{\partial z} + \alpha u \frac{\partial u}{\partial z} = 0, \tag{18.33}$$

where η is the free surface elevation and u is the horizontal velocity component. Note that we do not show explicitly the ε-dependence of η and u in order to simplify the notation. We assume that the bottom of the channel is randomly varying in $(0, L)$. More precisely, the fluid body is given by

$$H(t, z) = \begin{cases} 1 + \varepsilon \nu(z) + \alpha \eta(t, z) & \text{if } z \in (0, L), \\ 1 + \alpha \eta(t, z) & \text{if } z \in (-\infty, 0) \cup (L, \infty), \end{cases}$$

where 1 is the normalized mean depth. The parameter α is the ratio of the typical wave amplitude over the mean depth. It governs the strength of the nonlinearity. The parameter ε is the order of magnitude of the fluctuations of the bottom, which are described by the stationary zero-mean random process $\nu(z)$.

We consider the regime in which the amplitude of the fluctuations is small, i.e., ε is small, and the typical wavelength of the initial wave is of the same order of magnitude as the correlation length of the fluctuations of the bottom. As shown in Section 18.1, we should consider a propagation distance of order ε^{-2} to experience a macroscopic effect due to randomness. In this context we must also prescribe the order of magnitude of the nonlinear parameter α. It turns out that the suitable scaling between ε and α to exhibit the interplay between the nonlinear and random effects is

$$\alpha = \varepsilon^2 \alpha_0 , \tag{18.34}$$

where α_0 is the normalized nonlinear parameter, which is a fixed nonnegative number of order one.

The random process ν is assumed to be stationary, to be smooth, and to satisfy the moment conditions $\mathbb{E}[\nu(0)] = 0$, $\mathbb{E}[\nu(0)^2] < \infty$, and $\mathbb{E}[(\partial_z \nu(0))^2] < \infty$. Its autocorrelation function

$$\phi_0(z) = \mathbb{E}[\nu(z_0)\nu(z_0 + z)] \tag{18.35}$$

is also assumed to decay fast enough so that $\int_0^\infty |\phi_0(z)|^{1/2} dz < \infty$.

Let us introduce the "deterministic" local propagation speed

$$c = \sqrt{1 + \alpha\eta} , \tag{18.36}$$

which does not include the term $\varepsilon\nu$, but it is nevertheless random through the term $\alpha\eta$. We can reformulate the above equations in terms of c and u to obtain

$$\frac{\partial c}{\partial t} + \frac{\alpha}{2}c\frac{\partial u}{\partial z} + \alpha u\frac{\partial c}{\partial z} + \frac{\alpha\varepsilon}{2c}\frac{\partial \nu u}{\partial z} = 0 , \tag{18.37}$$

$$\frac{\partial u}{\partial t} + \alpha u\frac{\partial u}{\partial z} + \frac{2c}{\alpha}\frac{\partial c}{\partial z} = 0 . \tag{18.38}$$

We define the Riemann invariants (corresponding to the unperturbed nonlinear hyperbolic system)

$$A(t,z) = \frac{\alpha u + 2c - 2}{\alpha} , \qquad B(t,z) = \frac{\alpha u - 2c + 2}{\alpha} . \tag{18.39}$$

If the bottom is flat $\nu \equiv 0$, then we get back the standard right- and left-going modes (A and B, respectively) of the nonlinear hyperbolic system

$$\frac{\partial A}{\partial t} + c_+\frac{\partial A}{\partial z} = 0 , \qquad \frac{\partial B}{\partial t} - c_-\frac{\partial B}{\partial z} = 0 ,$$

with $c_+ = c + \alpha u = 1 + \alpha(3A + B)/4$ and $c_- = c - \alpha u = 1 - \alpha(A + 3B)/4$. The identities (18.39) can be inverted:

$$u = \frac{A + B}{2} , \qquad c = 1 + \alpha\frac{A - B}{4} .$$

Substituting these expressions into (18.37–18.38), we get the system governing the dynamics of the Riemann invariants in the presence of nonlinearity and randomness:

$$\frac{\partial A}{\partial t} + \left(1 + \alpha \frac{3A + B}{4}\right)\frac{\partial A}{\partial z} = -\frac{\varepsilon}{2}\frac{\partial \nu (A + B)}{\partial z}\frac{1}{1 + \alpha(A - B)/4}, \quad (18.40)$$

$$\frac{\partial B}{\partial t} + \left(-1 + \alpha \frac{A + 3B}{4}\right)\frac{\partial B}{\partial z} = \frac{\varepsilon}{2}\frac{\partial \nu (A + B)}{\partial z}\frac{1}{1 + \alpha(A - B)/4}. \quad (18.41)$$

The system is completed by the initial condition corresponding to a right-going wave incoming from the homogeneous half-space $z < 0$ and impinging the random slab $(0, L)$:

$$A(t, z) = f(t - z), \quad B(t, z) = 0, \quad t < 0,$$

where the function f is compactly supported in $(0, \infty)$.

18.3.2 The Linear Hyperbolic Approximation

If we neglect terms of order α, that is to say if we neglect all nonlinear contributions, then the system for the Riemann invariants can be reduced to

$$\mathbf{Q}(z)\frac{\partial}{\partial z}\begin{bmatrix} A \\ B \end{bmatrix} = \frac{\partial}{\partial t}\begin{bmatrix} A \\ B \end{bmatrix} + \frac{\varepsilon}{2}\nu'(z)\begin{bmatrix} 1 & 1 \\ -1 & -1 \end{bmatrix}\begin{bmatrix} A \\ B \end{bmatrix},$$

where ν' stands for the derivative of ν and the 2×2 matrix $\mathbf{Q}(z)$ is defined by

$$\mathbf{Q}(z) = \frac{1}{2}\begin{bmatrix} -2 - \varepsilon\nu(z) & -\varepsilon\nu(z) \\ \varepsilon\nu(z) & 2 + \varepsilon\nu(z) \end{bmatrix}.$$

This equation can be inverted, which gives

$$\frac{\partial}{\partial z}\begin{bmatrix} A \\ B \end{bmatrix} = \mathbf{Q}^{-1}(z)\frac{\partial}{\partial t}\begin{bmatrix} A \\ B \end{bmatrix} - \frac{\varepsilon}{2}\frac{\nu'(z)}{1 + \varepsilon\nu(z)}\begin{bmatrix} 1 & 1 \\ 1 & 1 \end{bmatrix}\begin{bmatrix} A \\ B \end{bmatrix}, \quad (18.42)$$

where

$$\mathbf{Q}^{-1}(z) = \frac{1}{1 + \varepsilon\nu(z)}\mathbf{Q}(z).$$

The identity (18.42), which holds true up to terms of order $O(\alpha)$, will be used in the forthcoming sections to rewrite the system (18.40–18.41) for the Riemann invariants as a partial differential equation of the form

$$\frac{\partial}{\partial z}\begin{bmatrix} A \\ B \end{bmatrix} = F\left(A, B, A_t, B_t, \nu, \nu'\right),$$

with the same accuracy as the original system.

In the linear hyperbolic approximation, (18.42) can be explicitly analyzed to study the wave dynamics. Indeed, the matrix $\mathbf{Q}^{-1}$ can be diagonalized. The eigenvalues of the matrix $\mathbf{Q}^{-1}(z)$ are $\pm\lambda^{\varepsilon}(z)$, with

$$\lambda^{\varepsilon}(z) = \frac{1}{\sqrt{1+\varepsilon\nu(z)}}. \tag{18.43}$$

We introduce the matrix $\mathbf{U}(z)$,

$$\mathbf{U}(z) = \frac{1}{2}\begin{bmatrix} \lambda^{\varepsilon}(z)^{1/2} + \lambda^{\varepsilon}(z)^{-1/2} & \lambda^{\varepsilon}(z)^{1/2} - \lambda^{\varepsilon}(z)^{-1/2} \\ \lambda^{\varepsilon}(z)^{1/2} - \lambda^{\varepsilon}(z)^{-1/2} & \lambda^{\varepsilon}(z)^{1/2} + \lambda^{\varepsilon}(z)^{-1/2} \end{bmatrix}, \tag{18.44}$$

which is such that

$$\mathbf{U}^{-1}(z)\mathbf{Q}^{-1}(z)\mathbf{U}(z) = \lambda^{\varepsilon}(z)\begin{bmatrix} -1 & 0 \\ 0 & 1 \end{bmatrix}.$$

By introducing

$$\begin{bmatrix} A_1 \\ B_1 \end{bmatrix}(t,z) = \lambda^{\varepsilon}(z)^{-1}\mathbf{U}^{-1}(z)\begin{bmatrix} A \\ B \end{bmatrix}(t,z),$$

the system (18.42) takes the simple form

$$\frac{\partial}{\partial z}\begin{bmatrix} A_1 \\ B_1 \end{bmatrix} = \lambda^{\varepsilon}(z)\begin{bmatrix} -1 & 0 \\ 0 & 1 \end{bmatrix}\frac{\partial}{\partial t}\begin{bmatrix} A_1 \\ B_1 \end{bmatrix} - \frac{\varepsilon}{4}\frac{\nu'(z)}{1+\varepsilon\nu(z)}\begin{bmatrix} 0 & 1 \\ 1 & 0 \end{bmatrix}\begin{bmatrix} A_1 \\ B_1 \end{bmatrix}, \tag{18.45}$$

where we have used the fact that

$$(\mathbf{U}^{-1})'(z)\mathbf{U}(z) = -\frac{\varepsilon\nu'(z)}{4(1+\varepsilon\nu(z))}\begin{bmatrix} 0 & 1 \\ 1 & 0 \end{bmatrix}, \qquad \mathbf{U}^{-1}(z)\begin{bmatrix} 1 & 1 \\ 1 & 1 \end{bmatrix}\mathbf{U}(z) = \begin{bmatrix} 1 & 1 \\ 1 & 1 \end{bmatrix}.$$

Equation (18.45) clearly exhibits the two relevant phenomena. The first term in the right-hand side describes a change of the velocity described by $\lambda^{\varepsilon}(z)$. The second term in the right-hand side describes a coupling between the two modes due to the term $\varepsilon\nu'$.

The system (18.45) has the same form as (4.24) obtained in the context of linear acoustic waves with a centering along local characteristics (Section 4.3.1). In order to complete the analogy between the linear approximation of the random-shallow-water wave equations and the random-acoustic-wave equations studied in the previous chapters, we can consider the twisted modes defined by

$$\begin{bmatrix} A_2 \\ B_2 \end{bmatrix}(t,z) = \mathbf{Q}(z)\begin{bmatrix} A \\ B \end{bmatrix}(t,z).$$

The twisted modes satisfy

$$\frac{\partial}{\partial z}\begin{bmatrix} A_2 \\ B_2 \end{bmatrix} = \mathbf{Q}^{-1}(z)\frac{\partial}{\partial t}\begin{bmatrix} A_2 \\ B_2 \end{bmatrix} = \begin{bmatrix} -1 & 0 \\ 0 & 1 \end{bmatrix}\frac{\partial}{\partial t}\begin{bmatrix} A_2 \\ B_2 \end{bmatrix} + \frac{\nu^{\varepsilon}(z)}{2}\begin{bmatrix} 1 & -1 \\ 1 & -1 \end{bmatrix}\frac{\partial}{\partial t}\begin{bmatrix} A_2 \\ B_2 \end{bmatrix},$$

where

$$\nu^\varepsilon(z) = \frac{\varepsilon\nu(z)}{1+\varepsilon\nu(z)}\,.$$

This system has the same form as (4.28) obtained in the context of linear acoustic waves with a centering along constant characteristics (Section 4.3.2). By taking the Fourier transform with respect to time and considering long propagation distances of the form z/ε^2, the random partial differential equation can be reduced to a set of random ordinary differential equations. We can then perform the same analysis as for the acoustic wave equations with a random bulk modulus in the weakly heterogeneous regime.

18.3.3 The Effective Equation for the Nonlinear Front Pulse

In this section we perform a series of transformations to rewrite the nonlinear evolution equations of the modes (18.40–18.41) by centering along the characteristic of the right-going mode. We will then obtain a system that can be integrated more easily. In a second step we shall apply an averaging theorem to this system in order to establish an effective nonlinear equation for the front pulse. The method follows closely the strategy used in Section 8.1 for the linear acoustic equations.

Our goal is to study the wave propagation for times and distances of order ε^{-2}. Accordingly, we can neglect in (18.40–18.41) the terms of order ε^3. We can also use (18.42), valid up to order ε, to rewrite some z derivatives as time derivatives. This can be done with a sufficient accuracy for the nonlinear terms by taking into account that $\alpha = \varepsilon^2\alpha_0$. As a result, we obtain

$$\frac{\partial}{\partial z}\begin{bmatrix} A \\ B \end{bmatrix} = \mathbf{Q}^{-1}(z)\frac{\partial}{\partial t}\begin{bmatrix} A \\ B \end{bmatrix} - \varepsilon\frac{\nu'}{2(1+\varepsilon\nu)}\begin{bmatrix} 1 & 1 \\ 1 & 1 \end{bmatrix}\begin{bmatrix} A \\ B \end{bmatrix}$$
$$+\varepsilon^2\frac{\alpha_0}{4}\begin{bmatrix} 3A+B & 0 \\ 0 & A+3B \end{bmatrix}\frac{\partial}{\partial t}\begin{bmatrix} A \\ B \end{bmatrix} + O(\varepsilon^3)\,. \qquad (18.46)$$

The random topography affects the propagation of the Riemann invariants by perturbing their characteristics, so that the matrix $\mathbf{Q}^{-1}$ in (18.46) is not the identity matrix. Two main effects can be distinguished: the diagonal terms describe random corrections to the local speed, while the off-diagonal parts describe random coupling. Our first goal is to center the propagation equations along the randomly perturbed characteristics. This can be done by the change of variables proposed in Section 18.3.2: the propagation equation in the eigenbasis of the matrix $\mathbf{Q}^{-1}$ frame exhibits a propagation matrix that is diagonal with z-dependent entries. We push the simplification forward by considering a new spatial variable that is related to the travel time along the characteristics

$$\zeta(z) = \int_0^z \lambda^\varepsilon(s)\,ds\,. \qquad (18.47)$$

We now introduce the modified modes

$$\begin{bmatrix} A_1 \\ B_1 \end{bmatrix}(t,\zeta) = \lambda^\varepsilon(z(\zeta))^{-1}\mathbf{U}^{-1}(z(\zeta))\begin{bmatrix} A \\ B \end{bmatrix}(t,z(\zeta)). \tag{18.48}$$

Note that $\frac{\nu'(z(\zeta))}{\lambda^\varepsilon(z(\zeta))} = \frac{d}{d\zeta}\nu(z(\zeta))$. We still denote this quantity by ν'. Finally we consider the reference frame

$$\tau = t - \zeta, \tag{18.49}$$

which moves with the right-going mode A_1, so that the equation for (A_1, B_1) reads

$$\frac{\partial}{\partial\zeta}\begin{bmatrix} A_1 \\ B_1 \end{bmatrix} = \begin{bmatrix} 0 & 0 \\ 0 & 2 \end{bmatrix}\frac{\partial}{\partial\tau}\begin{bmatrix} A_1 \\ B_1 \end{bmatrix} - \varepsilon\frac{\nu'}{4(1+\varepsilon\nu)}\begin{bmatrix} 0 & 1 \\ 1 & 0 \end{bmatrix}\begin{bmatrix} A_1 \\ B_1 \end{bmatrix}$$

$$+\varepsilon^2\frac{\alpha_0}{4}\begin{bmatrix} 3A_1 + B_1 & 0 \\ 0 & A_1 + 3B_1 \end{bmatrix}\frac{\partial}{\partial\tau}\begin{bmatrix} A_1 \\ B_1 \end{bmatrix} + O(\varepsilon^3). \tag{18.50}$$

The random medium introduces a coupling between the two modes through the term proportional to ν' as a consequence of multiple scattering. The equation for A_1 can be integrated for $\zeta > 0$ as

$$A_1(\tau,\zeta) = \int_0^\zeta S_A(\tau,y)dy + f(\tau), \tag{18.51}$$

$$S_A(\tau,\zeta) = \frac{-\varepsilon\nu'(\zeta)B_1(\tau,\zeta)}{4(1+\varepsilon\nu(\zeta))} + \frac{\varepsilon^2\alpha_0(3A_1 + B_1)}{4}\frac{\partial A_1}{\partial\tau}(\tau,\zeta) + O(\varepsilon^3). \tag{18.52}$$

Recall that we consider a propagation distance ζ of order ε^{-2}. However, we need to be careful to get precise estimates because of the shock-forming nature of the equations. Let $K > 0$ and $T > 0$. We introduce the stopping distance

$$L_K = \inf\left\{ l \geq 0 \text{ s.t. } \sup_{\tau\in[-T,T]}\left(\left|A_1\left(\tau,\frac{l}{\varepsilon^2}\right)\right| + \left|\frac{\partial A_1}{\partial\tau}\left(\tau,\frac{l}{\varepsilon^2}\right)\right|\right) \geq K \right\}. \tag{18.53}$$

As long as $\zeta \leq L_K/\varepsilon^2$, the solutions of the equations (18.50) are well defined. We shall show that B_1 is of order ε, so that S_A is of order ε^2, and the integral in (18.51) will turn out to be of order one. The equation for B_1 can be integrated as

$$B_1(\tau,\zeta) = -\frac{1}{2}\int_{-\infty}^\tau S_B\left(s,\zeta + \frac{\tau - s}{2}\right)ds, \tag{18.54}$$

$$S_B(\tau,\zeta) = \frac{-\varepsilon\nu'(\zeta)A_1(\tau,\zeta)}{4(1+\varepsilon\nu(\zeta))} + \frac{\varepsilon^2\alpha_0(A_1 + 3B_1)}{4}\frac{\partial B_1}{\partial\tau}(\tau,\zeta) + O(\varepsilon^3). \tag{18.55}$$

The integral in (18.54) seems to have an infinite support $(-\infty, \tau)$. However, we are interested in the front pulse, which means that we consider only shifted times τ lying in some interval $[-T, T]$ with a fixed T of order one. On the other hand, the initial conditions impose that A_1 and B_1 are zero for $\tau < 0$ and $\zeta = 0$. The transport equations (18.50) then show that A_1 and B_1 are zero for $\tau < 0$ whatever $\zeta \geq 0$. Thus the integral with respect to s in (18.54) actually goes from 0 to τ. Furthermore, (18.55) shows that S_B is of order ε. This allows us to claim that

$$\sup_{\tau \in [-T, T], \zeta \in [0, L_K/\varepsilon^2]} \left\{ |B_1(\tau, \zeta)| + \left| \frac{\partial B_1}{\partial \tau}(\tau, \zeta) \right| \right\} \leq K\varepsilon. \qquad (18.56)$$

We can now substitute the integral representation (18.54) for B_1 into the one (18.51) for A_1:

$$A_1(\tau, \zeta) = f(\tau) - \frac{\varepsilon^2}{32} \int_0^\zeta \nu'(y) \int_{-\infty}^\tau \nu'\left(y + \frac{\tau - s}{2}\right) A_1\left(s, y + \frac{\tau - s}{2}\right) ds\, dy$$

$$+ \varepsilon^2 \frac{3\alpha_0}{4} \int_0^\zeta A_1 \frac{\partial A_1}{\partial \tau}(\tau, y)\, dy + O(\varepsilon^3(1 + \zeta)). \qquad (18.57)$$

Note that we have eliminated the terms $\varepsilon^2 B_1 \partial_\tau A_1$, $\varepsilon^2 B_1 \partial_\tau B_1$, $\varepsilon^2 A_1 \partial_\tau B_1$, since they are of order ε^3 and are negligible for a propagation distance of order ε^{-2}. We introduce the rescaled right-going mode

$$A_1^\varepsilon(\tau, \zeta) = A_1\left(\tau, \frac{\zeta}{\varepsilon^2}\right),$$

which satisfies

$$A_1^\varepsilon(\tau, \zeta) = f(\tau) + \frac{3\alpha_0}{4} \int_0^\zeta A_1^\varepsilon \frac{\partial A_1^\varepsilon}{\partial \tau}(\tau, y)\, dy$$

$$- \frac{1}{32} \int_0^\zeta \nu'\left(\frac{y}{\varepsilon^2}\right) \int_{-\infty}^\tau \nu'\left(\frac{y}{\varepsilon^2} + \frac{\tau - s}{2}\right) A_1^\varepsilon\left(s, y + \varepsilon^2 \frac{\tau - s}{2}\right) ds\, dy + O(\varepsilon).$$

In a formal way, we can write this equation in the form

$$A_1^\varepsilon(\zeta) = f + \int_0^\zeta G(A_1^\varepsilon(y))\, dy + \int_0^\zeta F\left(\frac{y}{\varepsilon^2}\right) A_1^\varepsilon(y)\, dy, \qquad (18.58)$$

where $F(y)$ is a linear random operator acting on functions $A(\tau)$ as

$$[F(y)A](\tau) = -\frac{1}{32} \nu'(y) \int_{-\infty}^\tau \nu'\left(y + \frac{\tau - s}{2}\right) A(s)\, ds.$$

The random operator $F(y)$ possesses nice ergodic properties inherited through ν'. Thus an averaging over the fast-varying component of (18.58) can be applied as in the appendix of Chapter 8. In the limit $\varepsilon \to 0$, we get that A_1^ε converges to $\tilde{A}_1$, the solution of

$$\tilde{A}_1(\zeta) = f + \int_0^\zeta G(\tilde{A}_1(y))\, dy + \int_0^\zeta \tilde{F} \tilde{A}_1(y)\, dy,$$

where $\tilde{F} = \mathbb{E}[F(y)]$, that is to say,

$$[\tilde{F}A](\tau) = -\frac{1}{32} \int_{-\infty}^{\tau} \mathbb{E}\left[\nu'(y)\nu'\left(y + \frac{\tau - s}{2}\right)\right] A(s)\, ds\,.$$

The integral equation satisfied by the limiting pulse front $\tilde{A}_1$ is explicitly

$$\tilde{A}_1(\tau, \zeta) = f(\tau) + \frac{3\alpha_0}{4} \int_0^{\zeta} \tilde{A}_1 \frac{\partial \tilde{A}_1}{\partial \tau}(\tau, y)\, dy - \frac{1}{16} \int_0^{\zeta} \Lambda \tilde{A}_1(\tau, y)\, dy\,, \quad (18.59)$$

where the operator Λ is

$$\Lambda A(\tau) = \frac{1}{2} \int_0^{\tau} \phi_1\left(\frac{s}{2}\right) A(\tau - s)\, ds = \left[\frac{1}{2}\phi_1\left(\frac{\cdot}{2}\right) \mathbf{1}_{[0,\infty)}(\cdot)\right] * A(\tau)\,,$$

with

$$\phi_1(y) = \mathbb{E}[\nu'(z)\nu'(z + y)]\,.$$

The convergence holds true in the space of the continuous functions $\mathcal{C}([0, \tilde{L}_K] \times [-T, T], \mathbb{R})$, where T and K are arbitrary, and $\tilde{L}_K$ is the deterministic stopping distance defined by (18.53) for the deterministic function $\tilde{A}_1$. In the Fourier domain,

$$\int_{-\infty}^{\infty} \Lambda A(\tau) e^{i\omega\tau}\, d\tau = b_1(2\omega) \int_{-\infty}^{\infty} A(\tau) e^{i\omega\tau}\, d\tau\,, \quad (18.60)$$

$$b_1(\omega) = \int_0^{\infty} \phi_1(\tau) e^{i\omega\tau}\, d\tau\,. \quad (18.61)$$

Note that the integral is going from 0 to ∞, and not from $-\infty$ to ∞. As we shall discuss in the next section, this restriction has consequences in terms of hyperbolicity and dispersion. By use of Fourier analysis, b_1 can be expressed in terms of the autocorrelation function of the random stationary process ν. Let us set

$$b_0(\omega) = \int_0^{\infty} \phi_0(y) e^{i\omega y}\, dy\,, \quad (18.62)$$

where ϕ_0 is the autocorrelation function of ν,

$$\phi_0(y) = \mathbb{E}[\nu(z)\nu(z + y)]\,.$$

The relation (8.29) obtained in Section 8.1.3 holds true:

$$b_1(\omega) = -i\omega\phi_0(0) + \omega^2 b_0(\omega)\,. \quad (18.63)$$

Accordingly, Λ can be decomposed into the sum of a transport term corresponding to the term $-i\omega\phi_0(0)$ in (18.63) and a pseudodifferential operator corresponding to $\omega^2 b_0(\omega)$. In terms of the true mode A, we have to take care of the change of variable $z \mapsto \zeta(z)$. In the macroscopic scales,

$$\zeta\left(\frac{z}{\varepsilon^2}\right) = \frac{z}{\varepsilon^2} - \frac{\varepsilon}{2}\int_0^{\frac{z}{\varepsilon^2}}\nu(x)\,dx + \frac{3\varepsilon^2}{8}\int_0^{\frac{z}{\varepsilon^2}}\nu(x)^2 dx + O(\varepsilon)\,, \tag{18.64}$$

so that it converges as

$$\zeta\left(\frac{z}{\varepsilon^2}\right) - \frac{z}{\varepsilon^2} \xrightarrow{\varepsilon\to 0} \frac{1}{\sqrt{2}}\sqrt{b_0(0)}W_0(z) + \frac{3}{8}\phi_0(0)z\,, \tag{18.65}$$

where $W_0(z)$ is a standard Brownian motion. We can then state the following result.

Let $\tilde{A}_0$ be the solution of

$$\frac{\partial \tilde{A}_0}{\partial z} = \mathcal{L}\tilde{A}_0 + \frac{3\alpha_0}{4}\tilde{A}_0\frac{\partial \tilde{A}_0}{\partial \tau}\,, \tag{18.66}$$

starting from $\tilde{A}_0(\tau,0) = f(\tau)$. We denote by L_{shock} the shock distance of $\tilde{A}_0$. For any $L < L_{\mathrm{shock}}$ and T, the front pulse $A^\varepsilon(\tau,z) := A(\tau + z/\varepsilon^2, z/\varepsilon^2)$, $z \in [0,L]$, $\tau \in [-T,T]$, converges in distribution in the space of continuous functions to $\tilde{A}$ given by

$$\tilde{A}(\tau,z) = \tilde{A}_0\left(\tau - \frac{\sqrt{b_0(0)}}{\sqrt{2}}W_0(z) - \frac{\phi_0(0)}{2}z, z\right)\,. \tag{18.67}$$

The operator $\mathcal{L}$ can be written explicitly in the Fourier domain as

$$\int_{-\infty}^\infty \mathcal{L}A(\tau)e^{i\omega\tau}\,d\tau = -\frac{b_0(2\omega)\omega^2}{4}\int_{-\infty}^\infty A(\tau)e^{i\omega\tau}\,d\tau\,.$$

This result extends the stable-front theory addressed in Chapter 8 to weakly nonlinear waves. As in the linear case, the front-pulse propagation is modified in two ways for the random fluctuations of the medium. First, the front pulse is delayed by a random time shift, described by the Brownian motion $W_0(z)$. Second, it experiences diffusion, dispersion, and attenuation, modeled by the pseudodifferential operator $\mathcal{L}$. The physical description of this operator is given in detail in the next section.

18.3.4 Analysis of the Pseudospectral Operator

In this section we analyze the main properties of the effective equation for the front pulse. The important function that determines the dynamics is the Fourier transform $b_0(\omega)$ of the positive lag part of the autocorrelation function of the random fluctuations of the bottom. The pseudodifferential operator $\mathcal{L}$ models the deterministic pulse spreading imposed by the random fluctuations of the bottom. The effective equation for the front pulse depends both on randomness (through the function b_0) and on nonlinearity (through the parameter α_0). The pseudospectral operator $\mathcal{L}$ can be divided into two parts:

$$\mathcal{L} = \mathcal{L}_r + \mathcal{L}_i \,, \tag{18.68}$$

$$\int_{-\infty}^{\infty} \mathcal{L}_r A(\tau) e^{i\omega\tau} d\tau = -\frac{\gamma(2\omega)\omega^2}{8} \int_{-\infty}^{\infty} A(\tau) e^{i\omega\tau} d\tau \,, \tag{18.69}$$

$$\int_{-\infty}^{\infty} \mathcal{L}_i A(\tau) e^{i\omega\tau} d\tau = -\frac{i\gamma^{(s)}(2\omega)\omega^2}{8} \int_{-\infty}^{\infty} A(\tau) e^{i\omega\tau} d\tau \,, \tag{18.70}$$

where γ and $\gamma^{(s)}$ are respectively twice the real and imaginary parts of b_0,

$$\gamma(\omega) = \int_{-\infty}^{\infty} \mathbb{E}[\nu(0)\nu(z)]\cos(\omega z)\, dz \,, \quad \gamma^{(s)}(\omega) = 2\int_{0}^{\infty} \mathbb{E}[\nu(0)\nu(z)]\sin(\omega z)\, dz \,.$$

In particular, γ is the power spectral density of the random process ν and is nonnegative. Thus $\mathcal{L}_r$ can be interpreted as an effective **diffusion** operator. Moreover, $\mathcal{L}_i$ generates a frequency-dependent phase modulation, and it preserves the wave energy. It can thus be interpreted as an effective **dispersion** operator.

Let us address the case in which the power spectral density of the process ν can be considered as constant over the spectral range of f: $b_0(\omega) \equiv \mu_0 = \gamma(0)/2$. This arises if the typical wavelength of the pulse is larger than the correlation radius of the medium. In this situation the first phase of the effective evolution is governed by the viscous Burgers equation

$$\frac{\partial \tilde{A}_0}{\partial z} = \frac{\mu_0}{4}\frac{\partial^2 \tilde{A}_0}{\partial \tau^2} + \frac{3\alpha_0}{4}\tilde{A}_0\frac{\partial \tilde{A}_0}{\partial \tau} \,. \tag{18.71}$$

However, new frequencies are generated by the nonlinear term, which may fall in the tail of the function b_0. Then the last equation may eventually fail to describe exactly the dynamics of the front pulse, and one must consider the true equation (18.67) with the pseudodifferential operator.

The viscous Burgers equation (18.71) supports self-similar waves or even traveling waves as shown in [167, Chapter 4]. A simple example is the dam-breaking problem, where an initial step propagates into the random medium. The corresponding traveling wave is given in [167, Section 4.3].

18.3.5 Time Reversal

The time-reversal theory for nonlinear acoustic waves has been investigated experimentally by Tanter et al. [164]. They analyzed the nonlinear mechanism for energy transfer to higher harmonic components during forward propagation. The main goal of their experiments was to check for the reversibility of this energy transfer. The acoustic experiments were carried out for a nonlinear sinusoidal wave propagating in a homogeneous medium. The energy reversibility among harmonics was shown to be broken only for propagation longer than the shock-formation distance. The previous analysis allows us to address this issue for a broadband pulse in the presence of randomness.

We have just shown how the random medium regularizes the problem, allowing for propagation beyond the shock distance. This is another example in
which randomness helps in a dramatic fashion. Having prevented a derivative singularity from emerging, the fast transition layer saturates (i.e., the
shock structure forms) and a self-similar traveling wave can emerge from the
balance between nonlinearity and the stochastic forcing. This combined mechanism is enough to prevent the shock from fully developing and therefore it
allows for the propagation beyond the shock distance. Hence time reversal
can be performed beyond the critical time. This situation has been studied
in detail in [61]. We show that time reversal is indeed successful in inverting
the harmonic generation in order to recover the original spectrum and in recompressing the incoherent fluctuations. Indeed, the apparent viscosity does
not remove energy from the system. It only converts coherent wave energy
into incoherent fluctuations, and these incoherent waves can be recompressed
along the back-propagation of the front pulse.

18.4 Time Reversal with Changing Media

This section addresses the impact of a modification of the random medium on
refocusing during a time-reversal experiment. Even in the presence of significant perturbations a coherent refocused pulse is observed. The theory predicts
the level of recompression observed as well as the conditions for the loss of
statistical stabilization. It is shown that the statistical properties of the refocused pulse depend on a simple set of parameters that describe the correlation
degree of the medium. The refocused pulse has in general a random shape that
can be described in terms of a system of stochastic transport equations driven
by a single Brownian motion.

18.4.1 The Experiment

We revisit the time-reversal experiment in reflection that we studied in Section
10.1. We assume here that the fluctuations of the medium may have been
modified between the time windows corresponding to the first and second
parts of the time-reversal experiment. We shall denote by $\nu^{(1)}$, respectively
$\nu^{(2)}$, the random process that represents the fluctuations of the medium during
the first, respectively second, part of the time-reversal experiment. Here $\nu^{(1)}$
and $\nu^{(2)}$ are assumed to have the same statistical distribution, but they are
different realizations and we shall see that the impact of this difference will be
characterized by a correlation degree between the two processes. By revisiting
the analysis carried out in Section 10.1, we get the integral representation of
the refocused pulse

$$S_L^\varepsilon(t_1 + \varepsilon s, z = 0) = \frac{1}{(2\pi)^2} \int \int e^{-i\omega s} e^{-i\varepsilon hs/2} \overline{\hat{f}(\omega - \varepsilon h/2)} \hat{G}(h)$$

$$\times R_{\omega+\varepsilon h/2}^{\varepsilon,2}(-L,0) \overline{R_{\omega-\varepsilon h/2}^{\varepsilon,1}(-L,0)} \, dh \, d\omega, \tag{18.72}$$

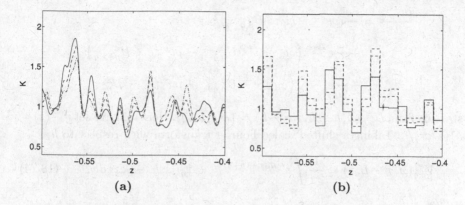

Fig. 18.4. Different realizations of the random bulk modulus. Picture (a) corresponds to a Gaussian continuous model. Picture (b) corresponds to a stepwise constant model. The realizations plotted in dashed (respectively dotted) lines have a correlation degree $\delta = 0.9$ (respectively $\delta = 0.75$) with the realizations plotted in solid lines (see (18.78) for the definition of δ).

where $R_\omega^{\varepsilon,1}$, respectively $R_\omega^{\varepsilon,2}$, stands for the reflection coefficient corresponding to the first, respectively second, realization of the random medium. This section is devoted to the proof of the convergence of the refocused pulse shape to an effective shape as $\varepsilon \to 0$.

18.4.2 Convergence of the Finite-Dimensional Distributions

The uniform boundedness (10.5) still holds true with two different reflection coefficients, since we use only the fact that $|R_\omega^{\varepsilon,j}| \leq 1$. This in turn implies that the finite-dimensional distributions of the process $S_L^\varepsilon(t_1 + \cdot)$ will be characterized by the moments

$$\mathbb{E}[S_L^\varepsilon(t_1 + \varepsilon s_1)^{p_1} \cdots S_L^\varepsilon(t_1 + \varepsilon s_k)^{p_k}], \tag{18.73}$$

for all real numbers $s_1 < \cdots < s_k$ and all integers $p_1, \ldots, p_k$.

First Moment

The statistical distribution of the refocused pulse depends on the frequency autocorrelation function of the reflection coefficient. We extend the approach developed in the previous chapters. It is necessary to consider a family of moments so as to get a closed system of equations. We introduce, for $p, q \in \mathbb{N}$,

$$U_{p,q}^\varepsilon(\omega, h, -L, z) = \left(R_{\omega+\varepsilon h/2}^{\varepsilon,2}(-L, z) \right)^p \left(\overline{R_{\omega-\varepsilon h/2}^{\varepsilon,1}(-L, z)} \right)^q.$$

Using the Riccati equation satisfied by $R_\omega^{\varepsilon,j}$, we deduce

$$\frac{\partial U^\varepsilon_{p,q}}{\partial z} = \frac{i\omega}{\bar{c}}(p\nu^{\varepsilon,2} - q\nu^{\varepsilon,1})U^\varepsilon_{p,q}$$

$$+ \frac{i\omega}{2\bar{c}}e^{\frac{2i\omega z}{\bar{c}\varepsilon}}\left(q\nu^{\varepsilon,1}e^{-\frac{ihz}{\bar{c}}}U^\varepsilon_{p,q-1} - p\nu^{\varepsilon,2}e^{\frac{ihz}{\bar{c}}}U^\varepsilon_{p+1,q}\right)$$

$$+ \frac{i\omega}{2\bar{c}}e^{-\frac{2i\omega z}{\bar{c}\varepsilon}}\left(q\nu^{\varepsilon,1}e^{\frac{ihz}{\bar{c}}}U^\varepsilon_{p,q+1} - p\nu^{\varepsilon,2}e^{-\frac{ihz}{\bar{c}}}U^\varepsilon_{p-1,q}\right),$$

starting from $U^\varepsilon_{p,q}(\omega, h, -L, z = -L) = \mathbf{1}_0(p)\mathbf{1}_0(q)$. Here we have set $\nu^{\varepsilon,j}(z) = \nu^j(z/\varepsilon^2)/\varepsilon$. Taking a shifted scaled Fourier transform with respect to h,

$$V^\varepsilon_{p,q}(\omega, \tau, -L, z) = \frac{1}{2\pi}\int e^{-ih[\tau-(p+q)z/\bar{c}]}U^\varepsilon_{p,q}(\omega, h, -L, z)\,dh\,, \qquad (18.74)$$

we get

$$\frac{\partial V^\varepsilon_{p,q}}{\partial z} = -\frac{p+q}{\bar{c}}\frac{\partial V^\varepsilon_{p,q}}{\partial \tau} + \frac{i\omega}{\bar{c}}(p\nu^{\varepsilon,2} - q\nu^{\varepsilon,1})V^\varepsilon_{p,q}$$

$$+ \frac{i\omega}{2\bar{c}}e^{\frac{2i\omega z}{\bar{c}\varepsilon}}\left(q\nu^{\varepsilon,1}V^\varepsilon_{p,q-1} + p\nu^{\varepsilon,2}V^\varepsilon_{p+1,q}\right)$$

$$+ \frac{i\omega}{2\bar{c}}e^{-\frac{2i\omega z}{\bar{c}\varepsilon}}\left(q\nu^{\varepsilon,1}V^\varepsilon_{p,q+1} - p\nu^{\varepsilon,2}V^\varepsilon_{p-1,q}\right),$$

starting from $V^\varepsilon_{p,q}(\omega, \tau, -L, z = -L) = \delta_0(\tau)\mathbf{1}_0(p)\mathbf{1}_0(q)$. Applying the limit theorem of Section 6.7.3 in the same way as in Section 9.2.1 establishes that the processes $V^\varepsilon_{p,q}$ converge to diffusion processes as $\varepsilon \to 0$. In particular, the expectations $\mathbb{E}[V^\varepsilon_{p,p}(\omega, \tau, -Lz)]$, $p \in \mathbb{N}$, converge to $v_p(\omega, \tau, -L, z)$, which obey the closed system of transport equations

$$\frac{\partial v_p}{\partial z} + \frac{2p}{\bar{c}}\frac{\partial v_p}{\partial \tau} = (\mathcal{L}_\omega v)_p - \frac{\gamma(1-\delta)\omega^2}{\bar{c}^2}p^2 v_p\,, \qquad (18.75)$$

$$(\mathcal{L}_\omega v)_p = \frac{\gamma\delta\omega^2}{4\bar{c}^2}p^2\left(v_{p+1} + v_{p-1} - 2v_p\right) - \frac{\gamma(1-\delta)\omega^2}{2\bar{c}^2}p^2 v_p\,, \qquad (18.76)$$

starting from $v_p(\omega, \tau, -L, z = -L) = \delta_0(\tau)\mathbf{1}_0(p)$, where

$$\gamma = \int_{-\infty}^{\infty}\mathbb{E}[\nu^{(1)}(0)\nu^{(1)}(z)]dz = \int_{-\infty}^{\infty}\mathbb{E}[\nu^{(2)}(0)\nu^{(2)}(z)]\,dz\,, \qquad (18.77)$$

$$\delta = \frac{1}{\gamma}\int_{-\infty}^{\infty}\mathbb{E}[\nu^{(2)}(0)\nu^{(1)}(z)]dz = \frac{1}{\gamma}\int_{-\infty}^{\infty}\mathbb{E}[\nu^{(1)}(0)\nu^{(2)}(z)]\,dz\,. \qquad (18.78)$$

Here γ is the standard integrated autocorrelation of the process ν in absence of time perturbations, and δ characterizes the correlation degree between the processes $\nu^{(1)}$ and $\nu^{(2)}$. If the fluctuations of the medium are the same in the two steps of the experiment ($\nu^{(1)} \equiv \nu^{(2)}$), then $\delta = 1$, and the limit transport equations (18.75) have the same form as the ones obtained in Section 9.2.1. If the fluctuations of the medium are completely uncorrelated in the two steps of the experiment ($\nu^{(1)}$ and $\nu^{(2)}$ independent), then $\delta = 0$, so that $v_p(\omega, \tau, -L, z) = \delta_0(\tau)\mathbf{1}_0(p)$.

We then get the limit of the autocorrelation function of the reflection coefficient:

$$\mathbb{E}\left[R^{\varepsilon,2}_{\omega+\varepsilon h/2}(-L,0)\overline{R^{\varepsilon,1}_{\omega-\varepsilon h/2}(-L,0)}\right] \overset{\varepsilon\to 0}{\longrightarrow} \int v_1(\omega,\tau,-L,0)e^{ih\tau}d\tau.$$

Higher Moments

The convergence of the refocused pulse will be obtained by a moment analysis similar to the one carried out in Section 10.1.3 for the corresponding problem when the medium is not changing. Our objective is then to identify stochastic processes $\mathcal{W}_p$ whose moments

$$\mathbf{E}\left[\prod_{j=1}^{m}\mathcal{W}_{p_j}(\omega_j,\tau_j,-L,0)\right]$$

are the limits of the moments

$$\mathbb{E}\left[\prod_{j=1}^{m}V^{\varepsilon}_{p_j,p_j}(\omega_j,\tau_j,-L,0)\right]$$

for m distinct frequencies $(\omega_j)_{1\le j\le m}$ and any sets $(p_j)_{1\le j\le m}\in \mathbb{N}^m$ and $(\tau_j)_{1\le j\le m}\in\mathbb{R}^m$.

The quantity $v_p(\omega,\tau,-L,0)$ is the limit of $\mathbb{E}[V^{\varepsilon}_{p,p}(\omega,\tau,-L,0)]$ as $\varepsilon\to 0$ and is obtained through the system of transport equations (18.75). Unfortunately, as we shall see below, the limit of $\mathbb{E}[V^{\varepsilon}_{p_1,p_1}(\omega_1,\tau_1,-L,0)V^{\varepsilon}_{p_2,p_2}(\omega_2,\tau_2,-L,0)]$ as $\varepsilon\to 0$ is not equal to $v_{p_1}(\omega_1,\tau_1,-L,0)v_{p_2}(\omega_2,\tau_2,-L,0)$ when $\delta\neq 1$, which shows that the desired process $\mathcal{W}_p$ cannot be the deterministic process v_p.

We now introduce the family of processes $\mathcal{W}_p$ defined as the solutions of the system of **stochastic transport equations**

$$d\mathcal{W}_p + \frac{2p}{\bar{c}}\frac{\partial\mathcal{W}_p}{\partial\tau}dz = \frac{i\omega\sqrt{2\gamma(1-\delta)}}{\bar{c}}p\mathcal{W}_p\,dW_0(z)$$
$$-\frac{\gamma(1-\delta)\omega^2}{\bar{c}^2}p^2\mathcal{W}_p\,dz + (\mathcal{L}_\omega\mathcal{W})_p\,dz, \quad (18.79)$$

driven by a standard Brownian motion $W_0(z)$. In Stratonovich form, this system can be written as

$$d\mathcal{W}_p + \frac{2p}{\bar{c}}\frac{\partial\mathcal{W}_p}{\partial\tau}dz = \frac{i\omega\sqrt{2\gamma(1-\delta)}}{\bar{c}}p\mathcal{W}_p\circ dW_0(z) + (\mathcal{L}_\omega\mathcal{W})_p\,dz.$$

It is straightforward to check by Itô's formula that

$$v_p(\omega,\tau,-L,z) = \mathbf{E}[\mathcal{W}_p(\omega,\tau,-L,z)],$$

where the expectation $\mathbf{E}$ is taken with respect to the distribution of the Brownian motion $W_0(z)$. Substituting this expression into the limit of the expectation of (18.72) yields

$$\mathbb{E}[S_L^\varepsilon(t_1 + \varepsilon s)] \xrightarrow{\varepsilon \to 0} \frac{1}{(2\pi)^2} \int \int \int e^{-i\omega s} e^{ih\tau} \overline{\hat{f}(\omega)\hat{G}(h)} \mathbf{E}\left[\mathcal{W}_1(\omega, \tau, -L, 0)\right] dh\, d\tau\, d\omega$$

$$= \frac{1}{2\pi} \int \int e^{-i\omega s} \overline{\hat{f}(\omega)} G(\tau) \mathbf{E}\left[\mathcal{W}_1(\omega, \tau, -L, 0)\right] d\tau\, d\omega.$$

Let us now consider the general moment (18.73). Using the representation (18.72) for each factor $S_L^\varepsilon(t_1 + \varepsilon s_j)$, these moments can be written as multiple integrals over $m = \sum_{j=1}^k p_j$ frequencies:

$$\mathbb{E}\left[\prod_{j=1}^k S_L^\varepsilon(t_1 + \varepsilon s_j)^{p_j}\right] = \frac{1}{(2\pi)^p} \int \cdots \int \mathbb{E}\left[\prod_{\substack{1 \le j \le k \\ 1 \le l \le p_j}} U_{1,1}^\varepsilon(\omega_{j,l}, h_{j,l}, -L, 0)\right].$$

$$\times \prod_{\substack{1 \le j \le k \\ 1 \le l \le p_j}} \overline{\hat{f}(\omega_{j,l})} e^{-i\omega_{j,l} s_j} e^{-i\varepsilon h_{j,l} s_j / 2} \overline{\hat{G}(h_{j,l})} d\omega_{j,l}\, dh_{j,l}.$$

The important quantity is $\mathbb{E}\left[\prod_{j,l} U_{1,1}^\varepsilon(\omega_{j,l}, h_{j,l}, -L, 0)\right]$. Our problem is now to find the limit, as ε goes to 0, of these moments for m distinct frequencies. This limit will be deduced from the study of the convergence of the distribution of $(U_{p_1,q_1}^\varepsilon(\omega_1, h_1, -L, z), \ldots, U_{p_m,q_m}^\varepsilon(\omega_m, h_m, -L, z))$, which results once again from the application of a diffusion-approximation theorem. Introducing V^ε as in (18.74), it is found that $(V_{p_1,q_1}^\varepsilon(\omega_1, \tau_1, -L, z), \ldots, V_{p_m,q_m}^\varepsilon(\omega_m, \tau_m, -L, z))$ converges as $\varepsilon \to 0$ to a diffusion process. In particular,

$$v_{p_1,\ldots,p_m}(\omega_1, \ldots, \omega_m, \tau_1, \ldots, \tau_m, -L, z) := \lim_{\varepsilon \to 0} \mathbb{E}\left[\prod_j V_{p_j,p_j}^\varepsilon(\omega_j, \tau_j, -L, z)\right]$$

is the solution of

$$\frac{\partial v_{p_1,\ldots,p_m}}{\partial z} + \frac{2}{\bar{c}} \sum_j p_j \frac{\partial v_{p_1,\ldots,p_m}}{\partial \tau_j} = \sum_j \mathcal{L}_{\omega_j} v_{p_1,\ldots,p_m}$$

$$- \frac{2\gamma(1-\delta)}{\bar{c}} \left(\sum_j \omega_j p_j\right)^2 v_{p_1,\ldots,p_m},$$

starting from $v_{p_1,\ldots,p_m}(\omega_1, \ldots, \omega_m, \tau_1, \ldots, \tau_m, -L, z = -L) = \prod_j \delta_0(\tau_j) \mathbf{1}_0(p_j)$. Using the families of processes $\mathcal{W}_p$ introduced in (18.79) defined for every frequency ω with the same Brownian motion $W_0(z)$, a direct calculation using Itô's formula shows that $\mathbf{E}\left[\prod_j \mathcal{W}_{p_j}(\omega_j, \tau_j, -L, z)\right]$ satisfies the above system, so that we have

$$v_{p_1,\ldots,p_m}(\omega_1,\ldots,\omega_m,\tau_1,\ldots,\tau_m,-L,z) = \mathbf{E}\left[\prod_{j=1}^{m} \mathcal{W}_{p_j}(\omega_j,\tau_j,-L,z)\right],$$

and consequently,

$$\mathbf{E}\left[S_L^\varepsilon(t_1+\varepsilon s_1)^{p_1}\cdots S_L^\varepsilon(t_1+\varepsilon s_k)^{p_k}\right]$$

$$\xrightarrow{\varepsilon\to 0} \frac{1}{(2\pi)^m}\int\cdots\int \mathbf{E}\left[\prod_{\substack{1\le j\le k\\ 1\le l\le p_j}} \mathcal{W}_1(\omega_{j,l},\tau_{j,l}-L,0)\right]$$

$$\times \prod_{\substack{1\le j\le k\\ 1\le l\le p_j}} \overline{\hat{f}(\omega_{j,l})}e^{-i\omega_{j,l}s_j}G(\tau_{j,l})\,d\omega_{j,l}\,d\tau_{j,l}$$

$$= \mathbf{E}\left[\prod_{1\le j\le k}\left(\frac{1}{2\pi}\int \mathcal{W}_1(\omega,\tau,-L,0)\overline{\hat{f}(\omega)}e^{-i\omega s_j}G(\tau)\,d\omega\,d\tau\right)^{p_j}\right].$$

This shows the convergence of the finite-dimensional distributions of $(S_L^\varepsilon(t_1 + \varepsilon s))_{s\in(-\infty,\infty)}$ to those of

$$\frac{1}{2\pi}\int \mathcal{W}_1(\omega,\tau,-L,0)\overline{\hat{f}(\omega)}e^{-i\omega s}G(\tau)\,d\omega\,d\tau.$$

18.4.3 Convergence of the Refocused Pulse

The tightness of the process $\left(S_L^\varepsilon(t_1+\varepsilon s, z=0)\right)_{s\in(-\infty,\infty)}$ can be established exactly as in Section 10.1. Together with the convergence of the finite-dimensional distributions that we have just established, this demonstrates the following proposition.

Proposition 18.2. *The refocused signal* $\left(S_L^\varepsilon(t_1+\varepsilon s, z=0)\right)_{s\in(-\infty,\infty)}$ *converges in distribution as* $\varepsilon\to 0$ *to*

$$S_L(s) = \frac{1}{2\pi}\int \mathcal{W}_1(\omega,\tau,-L,0)\overline{\hat{f}(\omega)}e^{-i\omega s}G(\tau)\,d\omega\,d\tau,$$

where $\mathcal{W}_1(\omega,\tau,-L,0)$ *is the* **random** *density that derives from the system (18.79). We can also write*

$$S_L(s) = (f(-\cdot)*K_{\mathrm{TRR}}(\cdot))(s),$$

where the Fourier transform of the random refocusing kernel K_{TRR} *is given by*

$$\hat{K}_{\mathrm{TRR}}(\omega) = \int G(\tau)\mathcal{W}_1(\omega,\tau,-L,0)\,d\tau. \qquad (18.80)$$

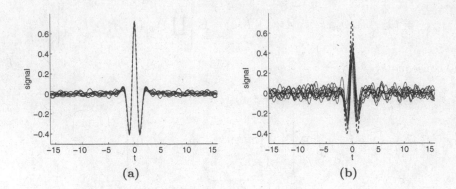

Fig. 18.5. Illustration of the loss of statistical stability when the medium is changing. We plot the refocused pulse shapes for an input pulse that is the second derivative of a Gaussian in the same configuration as in Figures 10.1–10.2. Picture (a) corresponds to the standard time-reversal experiment with the same medium, so that $\delta = 1$, and we present the results of 10 experiments performed with different realizations of the medium with the same statistical distribution (this figure is a copy of Figure 10.4). Picture (b) corresponds to the results of 10 experiments in which the medium is changing during the experiment so that $\delta = 0.75$.

We can give a probabilistic representation of the random density $\mathcal{W}_1$ in terms of a jump Markov process. Let us introduce the process $(N_z)_{z \geq -L}$ with state space $\mathbb{N} \cup \{\diamond\}$ (where $\diamond$ is the cemetery state) and infinitesimal generator $\mathcal{L}_\omega$ given by (18.76). Note that as soon as $\delta < 1$, the jump process can be killed. When the jump process reaches the state $x \in \mathbb{N}^*$, a random clock with exponential distribution and mean $\tau(\omega, x) = 2\bar{c}^2/(x^2 \gamma \omega^2)$ starts running. When the clock strikes, the process is killed and goes to $\diamond$ with probability $p_\diamond = 1 - \delta$; it jumps to $x + 1$ with probability $\delta/2$, and to $x - 1$ with probability $\delta/2$. Finally, 0 is an absorbing state. We can extend the representation proposed in Section 9.2.2 for the case $\delta = 0$ to the system (18.79) by means of a Feynman–Kac formula:

$$
\int_{\tau_0}^{\tau_1} \mathcal{W}_1(\omega, \tau, -L, 0) d\tau = \mathbb{E}\left[\mathbf{1}_{[\tau_0, \tau_1]}\left(\frac{2}{\bar{c}} \int_{-L}^{0} N_z dz \right) \mathbf{1}_0(N_0) \right.
$$

$$
\left. \times \exp\left(i\frac{\sqrt{2\gamma(1-\delta)}\omega}{\bar{c}} \int_{-L}^{0} N_{-L-z} dW_0(z) \right) \mid N_{-L} = 1 \right], \quad (18.81)
$$

where the expectation is taken with respect to the distribution of the jump process, but *not* with respect to the distribution of the Brownian motion W_0. The refocused pulse shape has therefore a random shape in the presence of perturbations of the medium fluctuations. There is no longer statistical stability as there used to be in the standard configurations where the medium is the same throughout the time-reversal experiment (see Figure 18.5).

Notes

The weakly heterogeneous regime addressed in Section 18.1 has been presented, for instance, in [8]. The effects on time reversal are new. The results with dispersion, nonlinearity, and time variations presented in Sections 18.2, 18.3, and 18.4, were derived in 2004, respectively in [60], [61], and [4]. The asymptotic analysis of wave propagation in randomly layered media has been extended to electromagnetic waves in [106], and time reversal in that context is studied in [78]. Among other waves not introduced in this book, elastic waves are of importance in geophysical imaging [151], and their asymptotic analysis has been considered in detail in [108].

19

The Random Schrödinger Model

In this chapter we consider the propagation of linear and nonlinear waves. We focus here for simplicity and pedagogical reasons on the Schrödinger equation, but the forthcoming results can be applied to other situations. The problem for the linear case is very similar to the case of the acoustic waves addressed in Chapter 7 in the frequency domain, but not in the time domain. The results for the transmission problem are stated in the first part of the chapter. The main statement is that, for a given incident wave, the transmission coefficient for a system of finite length decays exponentially with the size of the system. This phenomenon is one of the manifestations of wave localization in one-dimensional random media. We address both time-harmonic and time-dependent problems.

The main aim of the chapter is to discuss the robustness of localization with respect to nonlinearity. More exactly, we want to know how the exponential decay of the transmission can be modified by a nonlinearity. Some nonlinear dispersive systems such as the nonlinear Schrödinger (NLS) equation have special solutions called solitons that can propagate without change in form or diminution of speed in a homogeneous medium. Solitons are therefore candidates for testing the stability of the exponential localization in nonlinear and random media. In the second part of this chapter we study the propagation of a soliton through a slab of a nonlinear and random medium. We use a perturbed version of the inverse transform to exhibit several typical behaviors depending on the amplitude of the incoming soliton.

19.1 Linear Regime

19.1.1 The Linear Schrödinger Equation

Throughout the chapter we consider the Schrödinger equation, which models many important physical phenomena such as, for instance, the dynamics of the state function in quantum mechanics. In a linear and homogeneous medium it reads

$$iu_t + u_{xx} = 0,\tag{19.1}$$

where the partial derivatives are denoted by subscripts. This equation admits elementary solutions of the form

$$u = a \exp i\left(kx - k^2 t\right),$$

where $k \in \mathbb{R}$ is the wave number. The phase of this monochromatic wave can be written as $k(x - kt)$, which shows that the phase velocity is equal to k. The fact that the phase velocity depends on the wave number, in contrast to the standard wave equation, plays an important role in the wave dynamics. Let us now consider the initial value problem that is defined by (19.1) together with an initial condition at $t = 0$: $u(t = 0, x) = u_0(x)$, where $u_0 \in L^2$. A solution procedure for this problem is by Fourier transform. One first applies a direct Fourier transform (DFT) to the initial condition:

$$\hat{u}(0, k) = \frac{1}{2\pi} \int_{-\infty}^{\infty} u(0, x) e^{-ikx} dx.$$

The partial differential equation (19.1) is thus transformed into a set of uncoupled ordinary differential equations:

$$\hat{u}_t = -ik^2 \hat{u} \implies \hat{u}(t, k) = \hat{u}(0, k) e^{-ik^2 t}.\tag{19.2}$$

The solution at any time t can be obtained by applying the inverse Fourier transform (IFT):

$$u(t, x) = \int_{-\infty}^{\infty} \hat{u}(t, k) e^{ikx} dk.$$

Schematically, we have

$$u(0, x) \xrightarrow{\text{DFT}} \hat{u}(0, k)$$

$$(19.1) \downarrow \qquad\qquad \downarrow \text{Explicit and uncoupled evolutions (19.2)}$$

$$u(t, x) \xleftarrow{\text{IFT}} \hat{u}(t, k)$$

This resolution method allows us to describe the main difference between the standard wave equation, which is hyperbolic and supports the propagation of traveling waves, and the Schrödinger equation, which is a dispersive system. As pointed out above, the frequency components of a pulse solution of the Schrödinger equation do not travel with the same velocity, which involves pulse spreading. Using the notation $\hat{f}(k) = \hat{u}(0, k)$, the solution at any time t can be written as

$$u(t, x) = \int_{-\infty}^{\infty} \hat{f}(k) \exp i\left(kx - k^2 t\right) dk.$$

If $\hat{f}(k)$ is smooth enough so that the stationary-phase method can be applied (see Appendix 14.4), then for $t \gg 1$,

$$u(t,x) \sim \sqrt{\frac{\pi}{t}} \hat{f}\left(\frac{x}{2t}\right) \exp i \left(\frac{x^2}{4t} - \frac{\pi}{4}\right),$$

which shows that the amplitude of the wave decays as $1/\sqrt{t}$, while its support increases as t. Furthermore, if the initial condition is such that $\hat{f}$ is concentrated around the carrier wave number k_0, then the wave propagates with the group velocity $2k_0$, as seen in Figure 19.1.

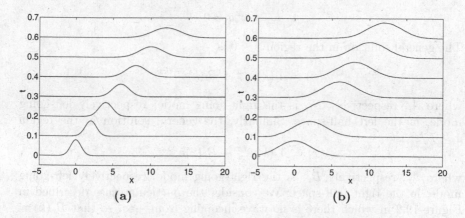

Fig. 19.1. Wave propagation governed by the linear Schrödinger equation (19.1). The spatial profiles $x \mapsto |u(t,x)|$ are plotted at times $t = 0, 0.1, \ldots, 0.6$. The initial condition is $u_0(x) = \exp(-x^2 + 10ix)$ (picture (a)) and $u_0(x) = \exp(-x^2/10 + 10ix)$ (picture (b)). In both cases the carrier wave number is 10, so that the velocity is 20. In picture (a) the initial pulse is short, and dispersion takes place quickly. In picture (b) the initial pulse is broad, and dispersive effects are not yet noticeable.

19.1.2 Transmission of a Monochromatic Wave

This subsection is devoted to the study of the propagation of monochromatic waves. Let $\hat{u}(x)$ be the complex amplitude at $x \in \mathbb{R}$ of a monochromatic wave $u(t,x) = \exp(-ik^2 t)\hat{u}(x)$ traveling in the one-dimensional medium described in Figure 19.2, where a random slab is embedded between two homogeneous half-spaces. We address in this chapter the weakly heterogeneous regime, in which the typical wavelength of the input wave is of the same order as the correlation length of the random potential, the typical amplitude of the fluctuations of the potential is small, of order ε, and the propagation distance is large, of order ε^{-2}.

Inside the slab $[0, L/\varepsilon^2]$ the wave satisfies the inhomogeneous Schrödinger equation $iu_t + u_{xx} = \varepsilon V(x)u$. By Fourier transforming this equation, we get that the field $\hat{u}$ satisfies the Helmholtz-type equation

$$\hat{u}_{xx} + (k^2 - \varepsilon V(x))\hat{u} = 0, \qquad \text{for } x \in [0, L/\varepsilon^2], \qquad (19.3)$$

where V is the realization of a random, stationary, ergodic, and zero-mean process.

The medium is homogeneous outside the slab $[0, L/\varepsilon^2]$ and the wave u obeys the Schrödinger equation $iu_t + u_{xx} = 0$. Accordingly, $\hat{u}$ satisfies in the region $x \leq 0$ and $x \geq L/\varepsilon^2$ the equation

$$\hat{u}_{xx} + k^2\hat{u} = 0.$$

The general solution in the region $x \leq 0$ is

$$\hat{u}(x) = A_l(k)e^{ikx} + B_l(k)e^{-ikx},$$

where A_l, respectively B_l, is the right-going mode, respectively left-going mode, in the left half-space. Similarly, the general solution in the region $x \geq L/\varepsilon^2$ is

$$\hat{u}(x) = A_r(k)e^{ikx} + B_r(k)e^{-ikx},$$

where A_r, respectively B_r, is the right-going mode, respectively left-going mode, in the right half-space. We consider the particular case described in Figure 19.2 in which there is no wave incoming from $+\infty$, so that $B_r(k) = 0$, while a wave with amplitude 1 is coming from $-\infty$, so that $A_l(k) = 1$. Therefore, we have

$$\hat{u}(x) = e^{ikx} + R^\varepsilon(k, L)e^{-ikx}, \quad \text{for} \quad x \leq 0,$$

and

$$\hat{u}(x) = T^\varepsilon(k, L)e^{ikx}, \quad \text{for} \quad x \geq L/\varepsilon^2,$$

where the complex-valued random variables R^ε and T^ε are the reflection and transmission coefficients, respectively. Note that the wave satisfies

$$ik\hat{u} + \hat{u}_x = 2ike^{ikx}, \text{ for } x \leq 0,$$

and

$$ik\hat{u} - \hat{u}_x = 0, \text{ for } x \geq L/\varepsilon^2.$$

The continuity of $\hat{u}$ and $\hat{u}_x$ at $x = 0$ and $x = L/\varepsilon^2$ then implies that the solution $\hat{u}$ of (19.3) also satisfies the two point boundary conditions

$$ik\hat{u} + \hat{u}_x = 2ik \text{ at } x = 0, \quad ik\hat{u} - \hat{u}_x = 0 \text{ at } x = L/\varepsilon^2. \qquad (19.4)$$

The problem turns out to be equivalent to the one studied in Chapter 7, where we addressed the acoustic wave equations. The only difference is that

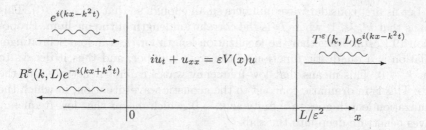

Fig. 19.2. Transmission of a monochromatic wave.

(19.3) appeared in the form $\hat{u}_{xx} + k^2(1 + \varepsilon m(x))\hat{u} = 0$. The two problems are thus completely equivalent if we set $V(x) = -k^2 m(x)$. This can be done for a monochromatic wave involving only one wave number k, but the case of a pulse involving several frequencies is qualitatively different.

The following proposition holds true when the potential V is a stationary process, that has finite moments of all orders and is rapidly mixing. We may assume, for instance, that V is a Markov, stationary, ergodic process on a compact space satisfying the Fredholm alternative (see Chapter 6).

Proposition 19.1. *For ε small enough and fixed, there exists a finite localization length $L_{loc}^{\varepsilon}(k)$ such that with probability one.*

$$\lim_{L \to \infty} \frac{1}{L} \ln |T^{\varepsilon}|^2(k, L) = -\frac{1}{L_{loc}^{\varepsilon}(k)}.$$ (19.5)

This localization length has the limit

$$\lim_{\varepsilon \to 0} \frac{1}{L_{loc}^{\varepsilon}(k)} = \frac{1}{L_{loc}(k)},$$ (19.6)

where $L_{loc}(k)$ is given by

$$\frac{1}{L_{loc}(k)} = \frac{\gamma(2k)}{4k^2}, \qquad \gamma(k) = \int_{-\infty}^{\infty} \mathbb{E}[V(0)V(x)] \cos(kx) \, dx.$$ (19.7)

We give an outline of the proof below. First, we compare this result with the one obtained in the framework of the random acoustic wave equation and stated in Proposition 7.6. The localization length for the acoustic wave equation (normalized so that $\bar{c} = 1$ and $k = \omega$) in the weakly heterogeneous regime can be expanded as

$$\frac{1}{L_{loc}} \bigg|_{\text{acoustic}} = \frac{\gamma(2k)k^2}{4}.$$

Note that the factor k^2 has moved compared to (19.7), because the acoustic model and the Schrödinger model both correspond to Helmholtz-type equations, but the coefficients do not have the same wave-number dependence.

Let us first consider wave numbers small enough so that $\gamma(k) \sim \gamma(0)$. This means that $kl_c \ll 1$, where l_c is the correlation length of the medium. Proposition 19.1 establishes that the localization length for the random Schrödinger equation is a quadratic function of the wave number, and thus it decays to 0 as $k \to 0$. This means that low-frequency waves cannot penetrate into the slab. This is in dramatic contrast to the acoustic wave situation, in which the localization length goes to infinity as $k \to 0$, which means that low-frequency waves penetrate deep into the slab.

The above results hold true if the wave number k satisfies $kl_c \ll 1$, so that it does not lie in the tail of the power spectral density γ. For high wave numbers $kl_c \gg 1$, the dependence of the localization length with respect to the wave number is imposed by the decay of the power spectral density, and thus we may encounter different configurations. If we consider the case of an exponential autocorrelation function $\mathbb{E}[V(0)V(x)] = \sigma^2 \exp(-|x|/l_c)$, then the power spectral density is the Lorentzian

$$\gamma(k) = \frac{2\sigma^2 l_c}{1 + k^2 l_c^2}.$$

As a result, the localization length for the acoustic-wave model saturates to the limit value $2l_c/\sigma^2$ for high wave numbers (see also Figure 19.3a), while the localization length for the Schrödinger model increases as k^4 (meaning that high-frequency waves can penetrate deep into the medium).

If we consider the case of a Gaussian autocorrelation function

$$\mathbb{E}[V(0)V(x)] = \sigma^2 \exp(-x^2/l_c^2),$$

then the power spectral density is the Gaussian

$$\gamma(k) = \sqrt{\pi}\sigma^2 l_c \exp\left(-\frac{k^2 l_c^2}{4}\right).$$

As a consequence, the localization lengths for the acoustic wave model and the Schrödinger model both increase very quickly for high frequencies (see Figure 19.3b).

These two covariance functions describe respectively very rough and very smooth random medium fluctuations. We see that this gives rise to a striking contrast in how far the high frequencies can penetrate into the random medium.

Proof (of Proposition 19.1). We follow closely the strategy developed in Section 7.3 in the framework of the acoustic-wave equation. The study of the exponential behavior of the power-transmission coefficient $|T^\varepsilon|^2$ can be divided into two steps. First, the localization length is shown to be equal to the inverse of the Lyapunov exponent associated with the random oscillator

$$v_{xx} + (k^2 - \varepsilon V(x))v = 0.$$

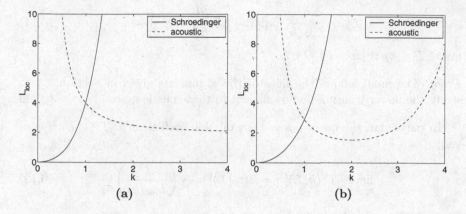

Fig. 19.3. Localization length versus wave number for an exponential autocorrelation function $\mathbb{E}[V(0)V(x)] = \exp(-|x|)$ (picture (a)) and a Gaussian autocorrelation function $\mathbb{E}[V(0)V(x)] = \exp(-x^2)$ (picture (b)).

Second, the expansion of the Lyapunov exponent of the random oscillator is computed.

For consistency, we show how to transform the boundary value problem (19.3–19.4) into an initial value problem similar to (7.3). Inside the perturbed slab we expand $\hat{u}$ in the form

$$\hat{u}\left(k, \frac{x}{\varepsilon^2}\right) = a^\varepsilon(k,x)e^{ik\frac{x}{\varepsilon^2}} + b^\varepsilon(k,x)e^{-ik\frac{x}{\varepsilon^2}}, \tag{19.8}$$

where a^ε and b^ε are respectively the right- and left-going modes defined by

$$a^\varepsilon = \frac{ik\hat{u} + \hat{u}_x}{2ik}e^{-ik\frac{x}{\varepsilon^2}}, \qquad b^\varepsilon = \frac{ik\hat{u} - \hat{u}_x}{2ik}e^{ik\frac{x}{\varepsilon^2}}.$$

The process $(a^\varepsilon, b^\varepsilon)$ is a solution of

$$\frac{d}{dx}\begin{bmatrix} a^\varepsilon \\ b^\varepsilon \end{bmatrix} = \frac{1}{\varepsilon}\mathbf{H}_k\left(\frac{x}{\varepsilon^2}\right)\begin{bmatrix} a^\varepsilon \\ b^\varepsilon \end{bmatrix}, \qquad \mathbf{H}_k(x) = \frac{i}{2k}V(x)\begin{bmatrix} -1 & -e^{-2ikx} \\ e^{2ikx} & 1 \end{bmatrix}. \tag{19.9}$$

The boundary conditions (19.4) read, in terms of a^ε and b^ε,

$$a^\varepsilon(k,0) = 1, \quad b^\varepsilon(k,L) = 0. \tag{19.10}$$

The end of the proof is then identical to that presented in Section 7.3.   □

Proposition 19.1 gives the typical decay of the power-transmission coefficient. The next proposition gives its average decay.

Proposition 19.2. *The power-transmission coefficient* $|T^\varepsilon(k,L)|^2$ *converges in distribution as* $\varepsilon \to 0$ *to the Markov process* $\tau_k(L)$ *whose infinitesimal generator is*

$$\mathcal{L}_k = \frac{1}{L_{\mathrm{loc}}(k)} \left[\tau_k^2 (1 - \tau_k) \frac{\partial^2}{\partial \tau_k^2} - \tau_k^2 \frac{\partial}{\partial \tau_k} \right], \tag{19.11}$$

where $L_{\mathrm{loc}}(k)$ is given by (19.7).

Proof. The proof follows the lines of the arguments given in Section 7.1 to study the power-transmission coefficient in the acoustic case.   □

In particular, the mean-power-transmission coefficient $\mathbb{E}[|T^\varepsilon(k, L)|^2]$ converges as $\varepsilon \to 0$,

$$\lim_{\varepsilon \to 0} \mathbb{E}[|T^\varepsilon(k, L)|^2] = \mathbb{E}[\tau_k(L)] = \xi_1 \left(\frac{L}{L_{\mathrm{loc}}(k)} \right), \tag{19.12}$$

where $\xi_1(l)$ is given by (7.52):

$$\xi_1(l) = \exp\left(-\frac{l}{4}\right) \int_0^\infty e^{-\mu^2 l} \frac{2\pi\mu \sinh(\mu\pi)}{\cosh^2(\mu\pi)} d\mu.$$

The decay of the mean power transmission coefficient is therefore

$$\lim_{L \to \infty} \frac{1}{L} \ln \left(\mathbb{E}[\tau_k(L)] \right) = -\frac{1}{4 L_{\mathrm{loc}}(k)},$$

which shows that the exponential behavior of the expectation of the power-transmission coefficient is very different from its typical behavior. The "right" localization length is intuitively the "typical" one (19.7), in the sense that it is the one that is observed for a "typical" realization of the medium. In fact, the next subsection shows that this holds true only for purely monochromatic waves.

19.1.3 Transmission of a Pulse

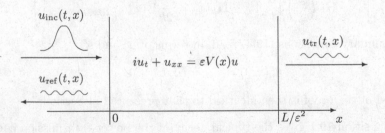

Fig. 19.4. Transmission of a pulse.

We consider a wave incoming from the left:

$$u_{\text{inc}}(t, x) = \frac{1}{2\pi} \int_0^\infty \hat{f}(k) \exp i\left(kx - k^2 t\right) \, dk \,, \quad x \leq 0 \,, \tag{19.13}$$

where $\hat{f} \in L^2$ and has support in $(0, \infty)$. The assumption on the support of $\hat{f}$ is necessary to ensure that we are dealing with a right-going wave (remember that the phase velocity is k). The total field in the region $x \leq 0$ thus consists of the superposition of the incoming wave u_{inc} and the reflected wave:

$$u_{\text{ref}}(t, x) = \frac{1}{2\pi} \int_0^\infty \hat{f}(k) R^\varepsilon(k, L) \exp i\left(-kx - k^2 t\right) \, dk \,, \quad x \leq 0 \,,$$

where $R^\varepsilon(k, L)$ is the reflection coefficient. The field in the region $x \geq L/\varepsilon^2$ consists only of the transmitted wave that is right-going:

$$u_{\text{tr}}(t, x) = \frac{1}{2\pi} \int_0^\infty \hat{f}(k) T^\varepsilon(k, L) \exp i\left(kx - k^2 t\right) \, dk \,, \quad x \geq L/\varepsilon^2 \,, \tag{19.14}$$

where $T^\varepsilon(k, L)$ is the transmission coefficient. Inside the slab the wave has the general form

$$u(t, x) = \frac{1}{2\pi} \int_{-\infty}^\infty \hat{u}(k, x) \exp\left(-ik^2 t\right) \, dk \,, \quad 0 \leq x \leq L/\varepsilon^2 \,.$$

The total transmitted energy is

$$\mathcal{T}^\varepsilon(L) = \frac{1}{2\pi} \int_0^\infty |\hat{f}(k)|^2 |T^\varepsilon(k, L)|^2 \, dk \,,$$

and it converges to its limit expectation.

Proposition 19.3. *The transmitted energy $\mathcal{T}^\varepsilon(L)$ converges in probability to $\mathcal{T}(L)$:*

$$\mathcal{T}(L) = \frac{1}{2\pi} \int_0^\infty |\hat{f}(k)|^2 \xi_1\left(\frac{L}{L_{\text{loc}}(k)}\right) \, dk \,,$$

where $\xi_1(L/L_{\text{loc}}(k))$ is the asymptotic value (19.12) of the expectation of the power-transmission coefficient $|T^\varepsilon(k, L)|^2$.

Proof. The idea is to show that the power-transmission coefficients at two distinct frequencies k_1 and k_2 are asymptotically uncorrelated as $\varepsilon \to 0$. With this result, it is easy to derive the asymptotic behavior of the power-transmission coefficient using the same arguments as those used in Section 7.2.2 in the framework of the acoustic-wave equations.

More precisely, Proposition 19.2 gives the limit value of the expectation of $|T^\varepsilon(k, L)|^2$ for one frequency k, so that

$$\mathbb{E}\left[\mathcal{T}^\varepsilon(L)\right] \xrightarrow{\varepsilon \to 0} \frac{1}{2\pi} \int_0^\infty |\hat{f}(k)|^2 \xi_1\left(\frac{L}{L_{\text{loc}}(k)}\right) \, dk \,.$$

Next one considers the second moment:

$$\mathbb{E}\left[\mathcal{T}^\varepsilon(L)^2\right] = \frac{1}{4\pi^2} \int_0^\infty \int_0^\infty |\hat{f}(k)|^2 |\hat{f}(k')|^2 \mathbb{E}\left[|T^\varepsilon(k,L)|^2 |T^\varepsilon(k',L)|^2\right] dk\, dk'.$$

The computation of this moment requires that one study the two-frequency correlation function $\mathbb{E}\left[|T^\varepsilon(k_1,L)|^2 |T^\varepsilon(k_2,L)|^2\right]$. Using the same arguments as those used in Section 7.2.3 in the framework of the acoustic-wave equations, we obtain that the pair $(|T^\varepsilon(k_1,L)|^2, |T^\varepsilon(k_2,L)|^2)$ converges in distribution to $(\tau_{k_1}(L), \tau_{k_2}(L))$, where the two processes $\tau_{k_1}(L)$ and $\tau_{k_2}(L)$ are two independent Markov processes whose infinitesimal generators are respectively $\mathcal{L}_{k_1}$ and $\mathcal{L}_{k_2}$ defined by (19.11). Consequently,

$$\mathbb{E}\left[\mathcal{T}^\varepsilon(L)^2\right] \overset{\varepsilon \to 0}{\longrightarrow} \left(\frac{1}{2\pi} \int_0^\infty |\hat{f}(k)|^2 \xi_1\left(\frac{L}{L_{\text{loc}}(k)}\right) dk\right)^2,$$

which proves the convergence of $\mathcal{T}^\varepsilon(L)$ to $\mathcal{T}(L)$ in $L^2(\mathbb{P})$. Since the limit value is deterministic, the convergence also holds true in probability.   □

Let us assume that the incoming wave is narrowband, that is, that the spectrum $\hat{f}$ is concentrated around the carrier wave number k_0 with a bandwidth that is smaller than 1, but larger than ε^2. Then $\mathcal{T}(L)$ decays exponentially with the width of the slab as

$$\frac{1}{L} \ln \mathcal{T}(L) \overset{L \gg 1}{\sim} -\frac{1}{4L_{\text{loc}}(k_0)}. \tag{19.15}$$

Note that this is the typical behavior of the *expected* value of the power-transmission coefficient of a monochromatic wave with wave number k_0. This self-averaging property is implied by the asymptotic decorrelation of the power-transmission coefficients at different frequencies.

Remark 19.4. In Chapter 8 we studied the stabilization of the front pulse in the case of the acoustic equations. Such a theory does not exist in the case of the Schrödinger equation, because the front pulse does not exist. The reason is that the system is dispersive, and the computation carried out in Section 19.1.1 shows that after a propagation distance (or time) of order ε^{-2}, the amplitude of a pulse traveling in a homogeneous medium is of order ε.

19.2 Nonlinear Regime

19.2.1 Waves Called Solitons

Solitary Waves in Communications

A solitary wave is a wave that propagates without change of form or diminution of speed. The study of solitary waves began in 1838 with the observation by J. Scott Russel of such a water wave while riding on a horse along a canal. However, no mathematical theory available at the time predicted a solitary wave. The problem was resolved in 1895 by Korteweg and de Vries, who derived an equation (now known as the KdV equation) that governs small shallow-water waves [110]. Boussinesq in 1871 also derived a nonlinear wave equation governing such long waves [22]. Despite this early work, no further application was discovered until the 1960s. In 1967 Gardner, Green, Kruskal, and Miura first discovered an original method for solving KdV by applying an implicit linearization of the equation; the so-called inverse scattering transform [72]. Lax (1968) considerably generalized these ideas [115], and Zakharov and Shabat (1972) showed that the method worked for the nonlinear Schrödinger (NLS) equation [171]:

$$iu_t + u_{xx} + |u|^2 u = 0 .$$

At this time it was known that the NLS equation describes the propagation of short pulses in single-mode optical fibers [125]. Hasegawa (1973) then claimed that the "soliton" was the ideal candidate to be the information bit for the next generation of optical fibers [87].

Indeed, communications in optical fibers [88] consist in sending binary messages at very high rates. A sequence of 0's and 1's can be coded as a train of short pulses, where a 1 is represented by a pulse and a 0 by the absence of a pulse in the corresponding arrival time slot of the train. The success of this method is based on the fact that modern technology has succeeded in producing purified glass fiber with a very low level of attenuation. Unfortunately, another phenomenon appears to be a limitation in the race toward higher and higher transmission rates. Indeed, dispersion makes pulses spread out. However, nonlinear effects such as self-focusing compete with dispersion. The NLS equation, which describes this competition to a good approximation, has a special solution, the so-called soliton, for which the nonlinear effects exactly counterbalance dispersion. It is therefore a good candidate to be the information bit for a new generation of optical fibers [86]. In order to confirm this hope, it is relevant to study the behavior of a soliton when it propagates through weakly perturbed media over very large distances.

To be complete, we must add that soliton-based communications schemes present some serious drawbacks, such as four-wave mixing, which is detrimental for wavelength-division multiplexing. As a result, they have so far never been implemented in real communications systems. Alternative solutions have been explored. A simple solution consists in a direct dispersion compensation for linear pulse propagation by the use of a periodic concatenation of pieces of fibers with opposite signs of dispersion. However, in any realistic optical network it will not be possible to compensate for all the dispersion in each element, so that there will remain some residual dispersion. Furthermore, the amplitude of the signal is bounded from below to keep a reasonable signal-to-noise ratio, so that the nonlinearity should also be taken into account. In [71] it was shown that the pulse propagation in such conditions is described by the NLS equation with a distance-varying dispersion coefficient. As a result, the concept of a dispersion-managed soliton in dispersion-compensated lines was proposed. It combines the advantages of the traditional fundamental soliton of the NLS equation and the dispersion-managed non-return-to-zero signal transmission. Both computational and experimental investigations have shown the existence and the stability of this new type of optical solitary wave.

Solitary Waves in Bose–Einstein Condensates

In 1924, Einstein pointed out that bosons (particles that have integer spin) could "condense" at low temperature in unlimited numbers into a single ground state, because they are not constrained by the Pauli exclusion principle (the Pauli exclusion principle claims that two fermions, particles of half-integer spins, cannot have identical quantum numbers). The result of this "condensation" is a macroscopic quantum state, a Bose–Einstein condensate (BEC), which is a physically intriguing phenomenon. The conditions for achieving a BEC are quite extreme. The participating particles must be identical, and this is a condition that is difficult to achieve for whole atoms. The condition of indistinguishability requires that the de Broglie wavelengths of the particles overlap significantly. This requires extremely low temperatures, so that the de Broglie wavelengths are long enough, but this also requires a rather high particle density to narrow the gaps between the particles.

The experimental realization of BEC in dilute atomic gases [44] founded a rapidly progressing new field of research [42]. This major achievement was permitted by the development of laser cooling techniques, to cool quantum gases to extremely low temperatures, and of magnetic trapping methods, allowing the production of temperatures on the order of the nanokelvin.

The most widely used approach for the description of a quantum degenerate bosonic system is the mean-field Gross–Pitaevskii (GP) theory [84]. In this approach all particles are considered to be in the same quantum state described by the condensate wave function, which evolves in time according to the GP equation

$$i\hbar \frac{\partial \psi}{\partial t} = -\frac{\hbar^2}{2m} \Delta \psi + V(\mathbf{r})\psi + g|\psi|^2 \psi,$$

where m is the mass of a particle and V is the external trapping potential. The nonlinearity parameter is $g = 4\pi\hbar^2 a_s/m$, where a_s is the s-wave scattering length. This scattering length is the interaction range associated with the s-wave scattering between pairs of bosons (the only interaction that is significant at low temperature). The GP equation has a straightforward interpretation: each boson evolves in the external potential V and in the mean-field potential produced by the other bosons. This equation is valid for a large number of atoms at low temperature, in the case that the mean distance between the atoms is larger than a_s. This equation shows that interactions play an important role although the gas is dilute.

In cigar-like trapping potentials, the dynamics in the radial direction are averaged out and the longitudinal profile of the wave function satisfies the one-dimensional GP equation. We can then recover the NLS equation by recasting this equation in dimensionless units. In this framework, a soliton is a perturbation of the density that propagates with constant speed, without deformation. The absence of dispersion, as in nonlinear optics, is due to the compensation between the nonlinear term and the kinetic term in the GP equation.

19.2.2 Dispersion and Nonlinearity

Dispersion in Wave-Propagation Phenomena

A linear dispersive system is any system that admits elementary solutions of the form

$$u = a \exp i \left(kx - \omega t \right) , \tag{19.16}$$

where the frequency ω is a definite real function of the wave number k, and the so-called dispersion relation $\omega(k)$ is determined by the particular system. For instance, the Boussinesq equation studied in Section 18.2 has the dispersion relation $\omega(k) = k/\sqrt{1 + \beta k^2}$. Any general solution can be obtained by superposition of elementary wave trains (19.16) to form Fourier integrals

$$u = \int F(k) \exp i \left(kx - \omega(k)t \right) dk ,$$

where F is chosen to fit the boundary or initial conditions with use of the Fourier inversion theorem. The wave components travel with their own phase velocity $\omega(k)/k$. Dispersion is due to the fact that the dispersion relation is usually not linear. As a consequence, the phase velocity depends on the wave number k. As time evolves, the different component modes disperse and the pulse spreads out. A more quantitative analysis can be easily carried out for the linear Schrödinger equation $iu_t + u_{xx} = 0$, whose dispersion relation is $\omega(k) = k^2$. As shown in Section 19.1.1, the amplitude of the wave decays as $1/\sqrt{t}$, while its support increases as t.

Catastrophic Collapse in Nonlinear Media

The simplest equation describing nonlinear propagation effects is the Burgers equation:

$$u_t + uu_x = 0 . \tag{19.17}$$

This equation can be solved analytically by the standard method of characteristics. If the initial condition $u(t = 0, x) = u_0(x)$ is smooth, then the solution remains smooth until time t_c, and its derivative is explicitly given by

$$u_x(t, x) = \frac{u_{0x}(x)}{1 + u_{0x}(x)t} ,$$

where t_c is defined by $t_c^{-1} := \max_x \{-u_{0x}(x)\}$. At time t_c the solution breaks up (more exactly, a shock develops). An extensive study of this equation can be found, for instance, in [167].

19.2.3 The Nonlinear Schrödinger Equation

We have just seen in the previous section that dispersion tends to spread out a pulse, while certain types of nonlinearity tend to concentrate the wave. It is therefore natural to address the case of dispersive and nonlinear systems and to look for solutions that achieve a balance between dispersion and nonlinearity. We study here the NLS equation, which supports such solutions.

An Introduction to the Inverse Scattering Transform

The goal of this subsection is to present the inverse scattering transform (IST), which transforms the NLS equation into a set of uncoupled ordinary differential equations. As we shall see, the IST can be seen as the analogue of the Fourier transform that achieves the same result for the linear Schrödinger equation. More detail can be found in [119].

The IST aims at studying the solutions of nonlinear partial differential equations of the type $u_t = F(u)$ with rapidly decaying initial conditions. It can be applied in the case that the evolution equation is equivalent to the linear operator relation

$$\frac{\partial \mathbf{L}(u)}{\partial t} + [\mathbf{L}(u), \mathbf{A}(u)] = 0 , \tag{19.18}$$

where $[\mathbf{L}, \mathbf{A}] = \mathbf{L}\mathbf{A} - \mathbf{A}\mathbf{L}$. It is based on the fact that $u(t, .)$ can be characterized by some spectral data of the operator $\mathbf{L}(u(t, .))$. The homogeneous NLS equation

$$iu_t + u_{xx} + 2|u|^2 u = 0 \tag{19.19}$$

can be expressed in the form (19.18) if we set

$$\mathbf{L}(u) = i\mathbf{P}\frac{\partial}{\partial x} + \mathbf{Q}(u) , \text{ with } \mathbf{P} = \begin{bmatrix} 1 & 0 \\ 0 & -1 \end{bmatrix} \text{ and } \mathbf{Q}(u) = \begin{bmatrix} 0 & \overline{u} \\ -u & 0 \end{bmatrix}.$$

The operator $\mathbf{A}$ is of the type $-2i\mathbf{P}\frac{\partial^2}{\partial x^2} + \mathbf{C}(u)$, with $\mathbf{C}(u) \to 0$ when $u \to 0$, $u_x \to 0$. The domain of $\mathbf{L}(u)$ is the space $\mathbb{H}^1(\mathbb{R})$,

$$\mathbb{H}^1(\mathbb{R}) = \left\{ \psi \text{ such that } \psi \in L^2(\mathbb{R}), \psi_x \in L^2(\mathbb{R}) \right\} ,$$

which is a dense subset of the Hilbert space $\mathbb{L}^2(\mathbb{R})$ defined by

$$\mathbb{L}^2(\mathbb{R}) = \left\{ \psi = \psi_1 \mathbf{e_1} + \psi_2 \mathbf{e_2}, \psi_j \in L^2(\mathbb{R}) \right\} , \quad \mathbf{e_1} = \begin{bmatrix} 1 \\ 0 \end{bmatrix}, \quad \mathbf{e_2} = \begin{bmatrix} 0 \\ 1 \end{bmatrix},$$

equipped with the scalar product

$$\langle \psi, \phi \rangle = \int_{-\infty}^{+\infty} [\overline{\psi_1}\phi_1(x) + \overline{\psi_2}\phi_2(x)] \, dx .$$

Operator $\mathbf{L}(0)$

The operator $\mathbf{L}(0)$ is self-adjoint. The real axis constitutes its essential spectrum. The eigenspace associated with the eigenvalue $\lambda \in \mathbb{R}$ has dimension 2 and admits as a basis the couple $\left(\mathbf{e_1}e^{-i\lambda x}, \mathbf{e_2}e^{i\lambda x}\right)$. Moreover, the point spectrum of $\mathbf{L}(0)$ is empty, because the nontrivial solutions of $v_x = i\lambda v$ are not in $L^2(\mathbb{R})$.

Essential Spectrum of the Operator $\mathbf{L}(u(t = t_0, .))$

Let us consider the spectral problem associated with the operator $\mathbf{L}(u) = \mathbf{L}(0) + \mathbf{Q}(u)$:

$$\mathbf{L}(u(t,x))\psi(t,x) = \lambda(t)\psi(t,x)\,, \quad \psi = \psi_1 \mathbf{e_1} + \psi_2 \mathbf{e_2}\,. \tag{19.20}$$

If $u(t = t_0, .) \in L^1(\mathbb{R})$, then $\mathbf{Q}(u)$ is $\mathbf{L}(0)$-compact. As a consequence of the Weyl theorem, the essential spectrum of $\mathbf{L}(u)$ is equal to the real axis. Equation (19.20) actually admits two linearly independent solutions when λ is real. We introduce the so-called **Jost functions** f and g, defined as the eigenfunctions of $\mathbf{L}(u)$ associated with the real eigenvalue λ that satisfy the following boundary conditions:

$$f(x,\lambda) \overset{x \to +\infty}{\longrightarrow} \mathbf{e_2}e^{i\lambda x}\,, \quad g(x,\lambda) \overset{x \to -\infty}{\longrightarrow} \mathbf{e_1}e^{-i\lambda x}\,.$$

If we denote by $\breve{\psi}$ the vector $(\overline{\psi_2}, -\overline{\psi_1})$ associated with a vector ψ solution of (19.20), then $\breve{\psi}$ is a solution of $\mathbf{L}\breve{\psi} = \overline{\lambda}\breve{\psi}$. In the case of a real eigenvalue, ψ and $\breve{\psi}$ are linearly independent and form a basis of the space of the solutions of (19.20). It can then be proved that the Jost functions are related by

$$g(x,\lambda) = a(\lambda)\breve{f}(x,\lambda) + b(\lambda)f(x,\lambda), \tag{19.21}$$
$$f(x,\lambda) = -a(\lambda)\breve{g}(x,\lambda) + \overline{b}(\lambda)g(x,\lambda)\,. \tag{19.22}$$

Substituting the second equality into the first one, we also exhibit the following conservation relation:

$$|a(\lambda)|^2 + |b(\lambda)|^2 = 1\,. \tag{19.23}$$

Using (19.20) we get two additional conservation relations, which concern the norms of the Jost functions f and g:

$$|f_1(x,\lambda)|^2 + |f_2(x,\lambda)|^2 = 1\,, \quad |g_1(x,\lambda)|^2 + |g_2(x,\lambda)|^2 = 1\,.$$

Multiplying (19.21) by the vector $\overline{\breve{f}}$, we get an explicit representation of the coefficient a as the Wronskian of f and g:

$$a(\lambda) = g_1(x,\lambda)f_2(x,\lambda) - g_2(x,\lambda)f_1(x,\lambda)\,. \tag{19.24}$$

We are able to provide a more explicit representation of the Jost functions f and g. Writing $\tilde{f}_1(x,\lambda) = e^{i\lambda x}f_1(x,\lambda)$ and $\tilde{f}_2(x,\lambda) = e^{-i\lambda x}f_2(x,\lambda)$, we find from (19.20) that $\tilde{f}$ satisfies a system of integral equations. Furthermore, $\tilde{f}_1$ can be eliminated from this system by substitution, so that we get a closed equation for $\tilde{f}_2$, whose solution is

$$\tilde{f}_2(x,\lambda) = 1 + \int_x^\infty dy M(y,x,\lambda)\left(1 + \int_y^\infty dz M(z,x,\lambda)\,(\ldots)\right),$$

where $M(y,x,\lambda) = -\overline{u(y)}\int_x^y dz u(z)e^{2i\lambda(y-z)}$. This expression holds true when $u \in L^1$, because the associated sequence converges absolutely. The function $\tilde{f}_1$ also admits a similar representation. Let us examine carefully the properties of $\tilde{f}$. If $y \mapsto |y|^n\lfloor u(y)\rfloor \in L^1$, then $\tilde{f}_1$ and $\tilde{f}_2$ are of class C^n over the real axis. If $u \in L^1$, then $\tilde{f}_1$ and $\tilde{f}_2$ can be analytically continued in the upper complex half-plane $\mathrm{Im}(\lambda) \geq 0$, where they have no singularity. Indeed, in view of the definition of M one can see that the exponential term has a norm equal to $e^{-2\mathrm{Im}\lambda(y-z)}$ (remember that we integrate over the domain $y - z > 0$), which decays faster than any polynomial term brought by the λ-derivatives.

Point Spectrum of the Operator $\mathbf{L}(u(t = t_0,.))$

From (19.24) we can define an analytic continuation of $a(\lambda)$ over the upper complex half-plane. A noticeable feature then appears. If λ_r is a zero of $a(\lambda)$, then f and g are linearly dependent, so there exists a coefficient ρ_r such that $g(x,\lambda_r) = \rho_r f(x,\lambda_r)$. The corresponding eigenfunction is bounded and decays exponentially as $x \to +\infty$ (because $|f| \sim e^{-\mathrm{Im}\lambda_r x}$) and as $x \to -\infty$ (because $|g| \sim e^{+\mathrm{Im}\lambda_r x}$). Thus λ_r is an element of the point spectrum of $\mathbf{L}(u)$. Moreover, we can compute from (19.20) and (19.24) the λ-derivative of a at $\lambda = \lambda_r$:

$$a'(\lambda_r) = -2i\rho_r \int_{-\infty}^{+\infty} dx\, f_1 f_2(x,\lambda_r). \tag{19.25}$$

It can then be proved that the set $(a(\lambda),b(\lambda),\lambda_r,\rho_r,a'(\lambda_r))$ characterizes the Jost functions f and g, as well as the solution u. The inverse transform is essentially based on the resolution of the linear integrodifferential Gelfand–Levitan–Marchenko equation, whose entries are defined by the set $(a,b,\lambda_r,\rho_r,a'(\lambda_r))$:

$$K_1(x,y) = \overline{\Phi}(x+y) - \int_x^\infty K_1(x,y'')\int_x^\infty \overline{\Phi}(y+y')\Phi(y'+y'')\,dy'\,dy'',$$

$$K_2(x,y) = -\int_x^\infty \overline{K_1}(x,y')\overline{\Phi}(y+y')\,dy', \tag{19.26}$$

where

$$\Phi(y) = -\sum_r \frac{i\rho_r}{a'(\lambda_r)}e^{i\lambda_r y} + \frac{1}{2\pi}\int_{-\infty}^{+\infty}\frac{b(\lambda)}{a(\lambda)}e^{i\lambda y}\,d\lambda.$$

We can get the eigenvector $f = (f_1,f_2)$ from the kernel $K = (K_1,K_2)$ solution of (19.26):

$$f(x, \lambda) = \mathbf{e_2} e^{i\lambda x} + \int_x^\infty K(x, y) e^{i\lambda y} \, dy \,. \tag{19.27}$$

We then obtain u by the formula $u(x) = -2i\overline{K_1}(x, x)$. The understanding of the inverse problem associated with the operator $\mathbf{L}(u)$ is not yet complete. In particular the precise characterization of the spectral data which lead to well-defined potentials u has not yet been completed. However, in the case that the initial condition u_0 is rapidly decaying, so that it satisfies $x \mapsto |x|^n |u_0|(x) \in L^1$ for any n, the inverse scattering can be rigorously obtained [2].

The great advantage of the method is that the evolution equations of the **scattering data** are uncoupled:

$$a(t, \lambda) = a(t_0, \lambda), \quad b(t, \lambda) = b(t_0, \lambda) e^{-4i\lambda^2(t-t_0)}, \quad \rho_r(t) = \rho_r(t_0) e^{-4i\lambda_r^2(t-t_0)} \,.$$

To sum up, the scattering transform involves the following operations:

$$
\begin{array}{ccc}
u(t_0, x) & \xrightarrow{\text{direct scatt.}} & (a, b, \lambda_r, \rho_r, a'(\lambda_r))\,(t_0) \\[2pt]
\text{NLS} \downarrow & & \downarrow \text{uncoupled evolution equations}\,. \\[2pt]
u(t, x) & \xleftarrow{\text{inverse scatt.}} & (a, b, \lambda_r, \rho_r, a'(\lambda_r))\,(t)
\end{array}
$$

What is striking is the remarkable analogy to Fourier analysis of the linear Schrödinger equation (see Section 19.1.1).

Conserved Quantities

There exists an infinite number of quantities that are preserved by the homogeneous nonlinear Schrödinger equation (19.19) [119]. They can be represented as functionals of the solution u or in terms of the scattering data. We present here only two of them, which are of physical interest.

- The mass of the wave $\mathcal{N} = \int |u|^2 dx$. With $n(\lambda) = -\pi^{-1} \ln |a(\lambda)|^2$, the mass is also given by

$$\mathcal{N} = \sum_r 2i(\overline{\lambda_r} - \lambda_r) + \int n(\lambda) \, d\lambda \,. \tag{19.28}$$

- The Hamiltonian $\mathcal{H} = \int |u_x|^2 - |u|^4 dx$, which can also be expressed as

$$\mathcal{H} = \sum_r \frac{8i}{3}(\overline{\lambda_r}^3 - \lambda_r^{\,3}) + 4 \int \lambda^2 n(\lambda) \, d\lambda \,. \tag{19.29}$$

Soliton

There exists a special solution of (19.19) with finite mass and Hamiltonian that is called a soliton:

$$u_0(t,x) = 2\nu_0 \frac{\exp i \left(2\mu_0(x - 4\mu_0 t) + 4(\nu_0^2 + \mu_0^2)t\right)}{\cosh\left(2\nu_0(x - 4\mu_0 t)\right)}. \tag{19.30}$$

This solution achieves a stable and perfect balance between dispersion and nonlinearity, so that its envelope (the modulus of the field) is traveling with constant velocity and shape. The mass and the velocity of the soliton are respectively $4\nu_0$ and $4\mu_0$. The width of the envelope of the soliton is conversely proportional to its mass. The soliton (19.30) is associated with the scattering data

$$a_0(\lambda) = \frac{\lambda - (\mu_0 + i\nu_0)}{\lambda - (\mu_0 - i\nu_0)}, \quad b_0(\lambda) = 0. \tag{19.31}$$

Here a_0 admits a unique zero in the upper complex half-plane denoted by $\lambda_0 = \mu_0 + i\nu_0$. The coefficient associated with the zero λ_0 is $\rho_0 = i \exp\left(-4i(\mu_0 + i\nu_0)^2 t\right)$. Figure 19.5 plots two different solitons at time $t = 0$. Both have the same mass, and consequently the same envelope, but they have different velocities. Note that in the case $\nu_0 \gg \mu_0$ (respectively $\nu_0 \ll \mu_0$), the soliton oscillates slowly (respectively quickly) within its envelope.

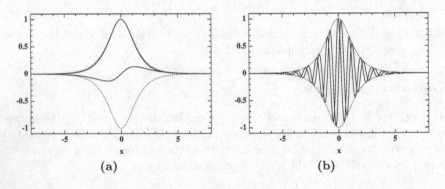

Fig. 19.5. Solitons at time $t = 0$. The dashed lines represent the envelopes (the moduli) of the solitons, while the solid lines represent the real and imaginary parts. Picture (a): $4\nu_0 = 2$, $4\mu_0 = 0.4$. Picture (b): $4\nu_0 = 2$, $4\mu_0 = 12.8$.

19.2.4 Soliton Propagation in Random Media

Competition Between Randomness and Nonlinearity

The main aim of the second part of this chapter is to discuss the stability of localization with respect to nonlinearities. More exactly, we want to know how the exponential decay of the transmission can be modified by a nonlinearity. The problem is much more difficult than for the linear case, and the methods used to study the linear case seem to fail completely. Some stability of localization has been conjectured. In particular, Fröhlich et al. have conjectured that solutions to the stationary nonlinear Schrödinger equation localize for sufficiently small initial data [70]. In the case of a general nonlinearity, results can be found in the literature for the stationary problem. It is very different from the linear case, since the transmission problem is no longer uniquely defined. Indeed, because of the nonlinearity, the transmitted intensity is not a linear function of the incident intensity. This phenomenon called bistability means that there may exist more than one output state for a given input state depending on hysteresis. The effect of randomness on bistability was addressed in [103]. The problem with fixed output was also considered in [47]. The authors show that for strong nonlinearity, the transmission coefficient cannot decay faster than a power law. These results show strong evidence that there exist delocalized transmission states. However, since only the time-harmonic problem has been addressed, not all of these states are physical [25], so that a complete study with a time-dependent model is required to understand this issue. This is the topic addressed in the second part of this chapter, where the propagation of a soliton through a slab of nonlinear and random medium is considered. Indeed, solitons can propagate without change of form or diminution of speed in homogeneous nonlinear dispersive media. Solitons are therefore candidates to test the stability of the exponential localization in nonlinear and random media. Physical [83], numerical [102], and experimental works [94] predict that for an NLS soliton propagating in a random medium, there exist two distinct regimes that depend on the soliton parameters. Furthermore, one of these regimes is expected to be very different from the localization regime in that the soliton retains its mass although it loses velocity. Using a perturbed version of the inverse scattering transform we can give a proof of this conjecture.

The Perturbed Nonlinear Schrödinger Equation

We consider a perturbed Schrödinger equation with a nonzero right-hand side:

$$iu_t + u_{xx} + 2|u|^2u = \varepsilon R(u)(t, x). \tag{19.32}$$

The small parameter $\varepsilon > 0$ characterizes the amplitude of the perturbation, whose spatial support is the interval $[0, L/\varepsilon^2]$. Outside this interval the perturbation is zero. The model of the perturbation is taken to be

$$R(u)(t, x) = V_1(x)u(t, x) + V_2(x)|u|^2u(t, x) + (V_3(x)u_x(t, x))_x, \tag{19.33}$$

for $x \in [0, L/\varepsilon^2]$. Here V_1, V_2, and V_3 are random, stationary, ergodic, zero-mean, and independent processes. We assume that a right-going soliton with parameters (ν_0, μ_0) is incoming from the homogeneous left half-space (Figure 19.6).

The Deterministic Asymptotic Regime

Let $L > 0$. We denote by Ω_L^ε the measurable set of realizations of the process $(V_j)_{j=1,2,3}$ such that the scattered wave consists of one soliton plus some

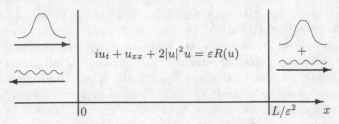

Fig. 19.6. Transmission of a soliton.

radiation. In terms of the spectral data it means that the Jost coefficient a admits a unique zero in the upper complex half-plane. We denote by ν^ε and μ^ε the rescaled processes defined on Ω_L^ε by $\nu^\varepsilon(x) = \nu(x/\varepsilon^2)$ and $\mu^\varepsilon(x) = \mu(x/\varepsilon^2)$ (i.e., the coefficients of the transmitted soliton in position x/ε^2), and on $\Omega_L^{\varepsilon\,c}$ by $\nu^\varepsilon(x) = 0$ and $\mu^\varepsilon(x) = 0$. We can now state our main convergence result (we then give a sketch of the proof).

Proposition 19.5.
1. $\lim\limits_{\varepsilon\to 0} \mathbb{P}\left(\Omega_L^\varepsilon\right) = 1.$
2. *The $\mathbb{R}^2$-valued process $(\nu^\varepsilon(x), \mu^\varepsilon(x))_{x\in[0,L]}$ converges in probability as a continuous process to the $\mathbb{R}^2$-valued deterministic function $(\nu_l(x), \mu_l(x))_{x\in[0,L]}$ that satisfies the system of ordinary differential equations*

$$\begin{cases} \dfrac{d\nu_l}{dx} = F(\nu_l, \mu_l), & \nu_l(0) = \nu_0\,, \\[2mm] \dfrac{d\mu_l}{dx} = G(\nu_l, \mu_l), & \mu_l(0) = \mu_0\,, \end{cases} \tag{19.34}$$

where the functions F and G are given by

$$F(\nu, \mu) = -\frac{1}{4\pi} \sum_{j=1}^{3} \int_{-\infty}^{\infty} |c_j|^2(\nu, \mu, \lambda) \gamma_j(2k(\nu, \mu, \lambda))\, d\lambda\,,$$

$$G(\nu, \mu) = -\frac{1}{8\pi} \sum_{j=1}^{3} \int_{-\infty}^{\infty} \left(\frac{\lambda^2}{\mu\nu} + \frac{\nu}{\mu} - \frac{\mu}{\nu}\right) |c_j|^2(\nu, \mu, \lambda) \gamma_j(2k(\nu, \mu, \lambda))\, d\lambda\,.$$

The coefficients γ_j and k are defined by:

$$\gamma_j(k) = \int_{-\infty}^{\infty} \mathbb{E}[V_j(0)V_j(x)] \cos(kx)\, dx\,, \qquad k(\nu, \mu, \lambda) = \frac{(\lambda - \mu)^2 + \nu^2}{2\mu}\,. \tag{19.35}$$

The functions c_j are written explicitly in [73, 1]. For consistency we write the full expression of c_1:

$$c_1(\nu, \mu, \lambda) = \frac{\pi}{2^4 \mu^3} \frac{(\lambda - \mu + i\nu)^2}{\cosh\left(\pi(\mu^2 - \nu^2 - \lambda^2)/(4\mu\nu)\right)}\,. \tag{19.36}$$

The first point means that the event "the scattered wave consists of one soliton plus some radiation" occurs with very high probability for small ε, while the second point gives the effective dynamics of the coefficients of the transmitted soliton, which turn out to be statistically stable quantities. This means that the shape of the transmitted soliton (and its velocity) are deterministic in our scaled regime. The situation is in some sense similar to the theory developed in Chapter 8 that describes the deformation of a linear pulse propagating in a random hyperbolic medium.

The analysis of the effective system (19.34) shows that there exist two main regimes up to transitory regimes.

1. If the mass of the incoming soliton is small enough, then the velocity of the soliton is almost constant, while its mass decreases to 0:
   - as $\exp(-L/L_{\text{loc}})$ (perturbation of the linear potential),
   - as $L^{-1/4}$ (perturbation of the nonlinear coefficient),
   - as $L^{-1/2}$, then as $\exp(-L/L'_{\text{loc}})$ (dispersive perturbation).
2. If the mass of the soliton is large enough, then it remains almost constant, while its velocity slowly decreases to 0. The decay rate depends on the tail of the spectrum of the perturbation, but we can state in great generality that it is at most logarithmic.

We discuss in more detail the case of a random potential in Section 19.2.5.

The Main Steps of the Proof

We now list the main steps of the proof [73, 1].

1. *A Priori Estimates*
   The following quantities (mass and Hamiltonian) are preserved by the perturbed Schrödinger equation (19.32):

$$\mathcal{N}_{\text{tot}} = \int |u|^2 \, dx, \qquad \mathcal{H}_{\text{tot}} = \int H_0(x) \, dx + \varepsilon \int H_1(x) \, dx, \qquad (19.37)$$

where $H_0(x) = |u_x|^2 - |u|^4$ and $H_1(x) = V_1(x)|u|^2 + \frac{1}{2}V_2(x)|u|^4 - V_3(x)|u_x|^2$. Assume that the V_j are bounded processes. Sobolev inequalities then prove that the H^1-norm, the L^4-norm, and the L^∞-norm of $u(t,.)$ are uniformly bounded with respect to $t \in \mathbb{R}$ and $\varepsilon \in (0,1)$. Furthermore, $\int H_1(x) \, dx$ can be bounded uniformly with respect to $t \in \mathbb{R}$ by a constant that depends only on $\mathcal{N}_{\text{tot}}$ and $\mathcal{H}_{\text{tot}}$.

2. *Stability of the Zero of the Jost Coefficient a*
   The zero of the Jost coefficient b corresponds to the soliton. This part strongly relies on the analytical properties of a in the upper complex half-plane. Basically, we apply Rouché's theorem so as to prove that the number of zeros is constant. This method is efficient to prove that the zero is preserved, but it does not give control on its precise location in the upper half-plane (remember that the real part of the zero is the parameter

μ and the imaginary part is ν). This step is not sufficient to compute the variations of the soliton parameters.

3. *Balance Between Radiation and Soliton*

The Jost coefficients a and b satisfy the coupled equations [93]

$$\frac{\partial a(\lambda, t)}{\partial t} = 0 + \varepsilon \left(a(\lambda, t)\check{\Gamma}(\lambda, t) + b(\lambda, t)\Gamma(\lambda, t) \right),$$

$$\frac{\partial b(\lambda, t)}{\partial t} = -4i\lambda^2 b(\lambda, t) - \varepsilon \left(a(\lambda, t)\overline{\Gamma}(\lambda, t) + b(\lambda, t)\check{\Gamma}(\lambda, t) \right),$$

where

$$\Gamma(\lambda, t) = -\int \left(R(u)f_2^2 + \overline{R(u)}f_1^2 \right) dx,$$

$$\check{\Gamma}(\lambda, t) = -\int \left(\overline{R(u)f_1}f_2 - R(u)f_1\overline{f_2} \right) dx.$$

From these equations we can estimate the amount of radiation emitted during some time interval in terms of mass and Hamiltonian thanks to (19.28–19.29). We are then able to deduce the evolution equations of the coefficients of the soliton by using the conservation of the mass and of the Hamiltonian, $\mathcal{N}_{\text{tot}}$ and $\mathcal{H}_{\text{tot}}$. Indeed, the total mass and Hamiltonian are conserved, but the discrete (soliton) and continuous (radiation) components vary along the propagation. The variations $\Delta(..)$ of these quantities are constrained by the relations

$$\Delta(\mathcal{N}_{\text{tot}}) = 0 = 4\Delta\nu + \int \Delta n(\lambda) \, d\lambda,$$

$$\Delta(\mathcal{H}_{\text{tot}}) = 0 = 16\Delta \left(\nu\mu^2 - \nu^3/3 \right) + 4 \int \lambda^2 \Delta n(\lambda) \, d\lambda + \varepsilon\Delta \left(\int_{\mathbb{R}} H_1(x) \, dx \right).$$

For times of order $O(\varepsilon^{-2})$, the radiated mass density $\Delta n(\lambda)$ is of order $O(1)$, while the last term in the expression of the total Hamiltonian is of order ε by the a priori estimates. Thus we can compute the long-time behavior of the coefficients of the soliton in the asymptotic framework $\varepsilon \rightarrow 0$. We can now in fact apply the diffusion approximation theorem presented in Chapter 6, and find that the coefficients of the soliton converge in probability to nonrandom functions that satisfy the system (19.34).

19.2.5 Reduction of Wave Localization by Nonlinearity

Localization Regime and Nonlinear Regime

In this subsection we describe in more detail the case of a random potential V_1 where $V_2 = V_3 = 0$ in (19.33). We study the asymptotic evolutions of the coefficients of the transmitted soliton as a function of the macroscopic length

L of the random slab, i.e., L/ε^2 in the microscopic scale. By Proposition 19.5 these evolutions are described by (19.34). We aim at exhibiting the relevant characteristics of this deterministic system of ordinary differential equations.

• In the regime $\nu_0 \ll \mu_0$, the system (19.34) can be simplified:

$$\begin{cases} \dfrac{d\nu_l}{dx} = -\dfrac{\gamma_1(4\mu_l)}{16}\dfrac{\nu_l}{\mu_l^2}, & \nu_l(0) = \nu_0\,, \\[2mm] \dfrac{d\mu_l}{dx} = -\dfrac{\gamma_1(4\mu_l)}{48}\dfrac{\nu_l^2}{\mu_l^3}, & \mu_l(0) = \mu_0\,. \end{cases} \tag{19.38}$$

This gives $\left(1 - 1/3\,(\nu_0/\mu_0)^2\right)^{1/2} \le \mu_l(x)/\mu_0 \le 1$, which shows that the velocity of the soliton is almost constant during the propagation, while the mass (equal to $4\nu_l$) decreases. The coefficient ν_l of the transmitted soliton decreases exponentially with the propagation distance:

$$\nu_l(x) \sim \nu_0 \exp\left(-\frac{x}{L_{\exp}}\right), \quad L_{\exp} = \frac{16\mu_0^2}{\gamma_1(4\mu_0)}. \tag{19.39}$$

Note that if the approximation $\nu \ll \mu$ holds true for the initial conditions, then it actually holds true during the whole propagation, since the velocity is almost constant while the mass decreases. The domain $\nu \ll \mu$ is therefore stable.

• In the regime $\mu_0 \ll \nu_0$, the system (19.34) can be simplified:

$$\begin{cases} \dfrac{d\nu_l}{dx} = -\dfrac{\pi\sqrt{2}\gamma_1(2\nu_l^2/(2\mu_l))}{2^8}\dfrac{\nu_l^{9/2}}{\mu_l^{11/2}}\exp\left(-\frac{\pi}{2}\frac{\nu_l}{\mu_l}\right), & \nu_l(0) = \nu_0\,, \\[2mm] \dfrac{d\mu_l}{dx} = -\dfrac{\pi\sqrt{2}\gamma_1(2\nu_l^2/(2\mu_l))}{2^9}\dfrac{\nu_l^{11/2}}{\mu_l^{13/2}}\exp\left(-\frac{\pi}{2}\frac{\nu_l}{\mu_l}\right), & \mu_l(0) = \mu_0\,. \end{cases} \tag{19.40}$$

It can be readily checked that $\left(1 - 2\,(\mu_0/\nu_0)^2\right)^{1/2} \le \nu_l(x)/\nu_0 \le 1$, which means that the mass of the soliton is almost constant during the propagation, while the velocity of the soliton decreases. The limit behavior for large x of the coefficient μ_l depends on the functions γ_1, more exactly on the high-frequency behaviors of the Fourier transforms of the autocorrelation functions of the process V_1. For instance, if $\mathbb{E}[V_1(0)V_1(x)] = \sigma^2\exp(-|x|/l_c)$, then

$$\lim_{x\to\infty} \mu_l(x) \times \ln(x) = \frac{\pi\nu_0}{2}\,,$$

which means that the velocity decreases as the logarithm of the length. This logarithmic rate actually represents the maximal decay of the velocity. Whatever the process V_1, the terms of the right-hand sides of (19.40) have at least an exponential decay of the type $\exp[-(\pi\nu)/(2\mu)]$, which implies $\liminf_{x\to\infty} \mu_l(x) \times \ln(x) \ge \pi\nu_0/2$. However, the decay rate may be much slower. As an example, if the autocorrelation functions are $\mathbb{E}[V_1(0)V_1(x)] =$

$\sigma^2 \exp(-x^2/l_c^2)$, then the velocity decreases as the square root of the logarithm of x:

$$\lim_{x \to \infty} \mu(x) \times \sqrt{\ln(x)} = \frac{\nu_0^2 l_c}{2}.$$

Note that the approximation $\mu \ll \nu$ actually holds true during the whole propagation, since the mass is almost constant while the velocity decreases, which shows that the domain $\mu \ll \nu$ is stable.

• The two previous stable regimes are actually the only ones that can be observed, up to transitory regimes. Indeed, for a fixed value of the initial velocity $4\mu_0$, there exists a threshold value $\mathcal{N}_c$ for the initial mass $4\nu_0$ above which we observe a slow decay of the velocity, while the mass goes to a constant value, and below which we observe an exponential decay of the mass, while the velocity goes to a constant value.

Asymptotics for a Small-Amplitude Soliton

It can be noted that in the limit case $\nu_0/\mu_0 \to 0$, the incoming soliton can be approximated by a linear pulse,

$$u_0(t, x) \sim \int_{-\infty}^{+\infty} \hat{f}(k) e^{ikx - ik^2 t} \, dk, \quad \text{with } \hat{f}(k) = \frac{1}{2} \cosh^{-1}\left(\frac{\pi}{4}\left(\frac{k - 2\mu_0}{\nu_0}\right)\right),$$

whose spectral content $\hat{f}$ is sharply peaked about the wave number $k_0 = 2\mu_0$. Furthermore, the spectrum of the emitted radiation is peaked around the wave number $-2\mu_0$ (there exists also a secondary peak about $+2\mu_0$, which is weaker).

These statements are in qualitative agreement with the linear approximation. The exponential decay length L_{exp} defined by (19.39) can be written in terms of the carrier wave number as $L_{\text{exp}} = 4k_0^2/\gamma_1(2k_0)$. It is equal to the localization length $L_{\text{loc}}(k_0)$ of a monochromatic wave with wave number k_0 scattered by a slab of linear random medium (see Proposition 19.1). However, the typical behavior of the transmitted energy (mass) for a linear pulse with carrier frequency k_0 is the exponential decay (19.15): $T(L) \sim \exp(-L/(4L_{\text{loc}}))$, whose decay rate is one fourth the decay rate $\exp(-L/L_{\text{loc}})$ of the soliton. This means that the transmitted wave consists of a soliton plus some radiation, and that for $L > L_{\text{loc}}$ most of the transmitted energy consists of radiation. This situation is in some sense close to the theory describing the stable propagation of the wave front in a random hyperbolic linear medium: as we have seen in Chapters 7 and 8, the energy of the wave front is decaying as $\exp(-L/L_{\text{loc}})$ (see (8.48), with $L_{\text{loc}} = 4\bar{c}^2/(\gamma\omega^2)$), while the total transmitted energy is decaying as $\exp(-L/(4L_{\text{loc}}))$ (see (7.77)).

Numerical Simulations

The results in the previous subsections hold true in the limit case $\varepsilon \to 0$, where the amplitudes of the perturbations go to zero and the length of the random slab goes to infinity. In this subsection we show that the asymptotic behaviors of the soliton can be observed in numerical simulations in the case that ε is small, more precisely, smaller than any other characteristic scale of the problem. We use a fourth-order split-step method to simulate the perturbed nonlinear Schrödinger equation (19.32). This numerical algorithm provides accurate and stable solutions [102]. For the sake of simplicity we consider only perturbations of the linear potential V_1 and take $V_2 = V_3 = 0$ in (19.33).

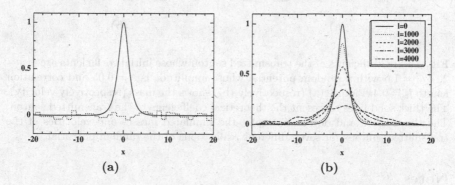

(a)   (b)

Fig. 19.7. Picture (a): Envelope of the initial soliton (solid line) with mass $4\nu_0 = 2$ and velocity $4\mu_0 = 1.6$. The dashed line plots a realization of the random potential εV_1 with $\varepsilon = 0.05$ and $l_c = 0.4$. Picture (b): Envelopes of the soliton when its center crosses different depth lines l for one of the realizations of the random potential. The coordinate x is normalized around the depth line l.

We assume in this subsection that the potential is constant over elementary intervals of length l_c and take independent random values over each interval that obey uniform distributions over $[-1, 1]$. In Figures 19.7–19.8, we present simulations in which the initial wave at time $t = 0$ is a soliton with mass $4\nu_0 = 2$ and velocity $4\mu_0 = 1.6$ centered at $x = 0$ (see Figure 19.7a). The simulated evolutions of the coefficients of the soliton are presented in Figure 19.8 for seven different realizations of the random potential with $\varepsilon = 0.05$ and $l_c = 0.4$. They are compared with the theoretical evolutions given by (19.34) in the scale x/ε^2. It thus appears that the numerical simulations are in very good agreement with the theoretical results. Figure 19.7b plots the envelopes of the solution at different depths corresponding to one of the simulations, which shows that the wave keeps the basic form of a soliton although it loses some mass. All these results confirm that the system (19.34) describes accurately the transmission of a soliton through a random slab in the regime of small perturbations and large slabs.

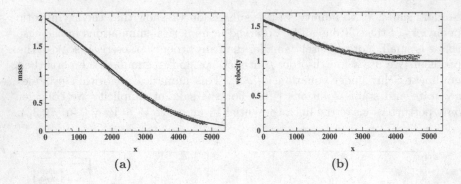

Fig. 19.8. Coefficients of the transmitted soliton whose initial coefficients are $4\nu_0 = 2$, $4\mu_0 = 1.6$ with a random potential whose amplitude is $\varepsilon = 0.05$ and correlation length $l_c = 0.4$. Picture (a) (respectively (b)) shows the mass (respectively velocity). The thick solid lines represent the theoretical coefficients of the transmitted soliton. The thin dashed and dotted lines plot the simulated masses and velocities of the transmitted solitons for seven different realizations of the random potential.

Notes

The results derived in the second part of this chapter and some extensions can be found in a series of papers [73, 1, 77]. Nonlinear and dispersive systems exhibit different behaviors, so it is worth considering other integrable systems. For instance, the Korteweg–de Vries equation, with a third-order dispersion, has been studied in [75], and the results turn out to be very different compared to the randomly perturbed NLS equation. Indeed, the scattering of the soliton generates not only continuous radiation during its motion, but also a soliton gas, that is, a collection of a very large number of solitons with very small masses, whose total mass is of order one. It would be interesting to get a classification of the integrable systems in terms of their respective behaviors with respect to random perturbations. Furthermore, the interaction of solitons in random media represents a great challenge for practical applications to communications or condensed matter. These issues are still the subject of intense research.

Propagation in Random Waveguides

In this chapter we consider wave propagation in an acoustic waveguide whose bulk modulus is a three-dimensional random function. This is the first time in this book that we consider such general random perturbations. However, using the propagating modes of the unperturbed waveguide we can reduce the three-dimensional wave propagation problem to the study of a system of ordinary differential equations with random coefficients. It is the Fourier transform of the mode amplitudes that satisfy these differential equations, to which we can apply the asymptotic theory of Chapter 6.

We deal with the usual scaled quantities, which are the propagation distance, the carrier wavelength, the bandwidth, and the standard deviation and the correlation lengths of the random perturbations in the different directions. We now encounter a new length scale: the width of the waveguide, which controls the number of propagating modes. We study the asymptotic behavior of the transmitted waves in the weakly heterogeneous regime, with a fixed number of propagating modes.

After deriving the coupled mode amplitude equations we consider the forward-scattering approximation in Section 20.2. In this approximation the coupling between forward- and backward-propagating modes is considered negligible compared to the coupling between only forward-propagating modes. The asymptotic theory of Chapter 6 can be applied in a straightforward way to the full forward and backward mode-coupling system. We focus our attention in this chapter on the forward-scattering approximation for two reasons. First, with the forward-scattering approximation we can calculate analytically many interesting quantities, such as the transmitted wave intensity and spatial focusing in time reversal. Without the forward-scattering approximation the reflection and transmission of energy can be analyzed with an infinite-dimensional system of transport equations. It is, however, much more complicated than the one in Chapter 9, which corresponds, essentially, to a single-mode waveguide. Second, from the asymptotic theory we see that the statistical coupling coefficient between a forward mode and a backward mode is proportional to the power spectral density of the random perturbations evaluated at the sum of their corresponding wave numbers. The statistical coupling coefficient between two forward modes is, however, proportional to the power spectral density of the random perturbations evaluated at the difference of the two wave numbers. This means that if the power spectral density has a cutoff wave number, or if it decays fast enough, which means that the random perturbations are not too rough, then the coupling between forward modes is dominant and the forward-scattering approximation is justified. We discuss further this important point in Section 20.2.6. The forward-scattering approximation is used widely in underwater acoustics, in fiber optics, and elsewhere. It is because of the use of the forward-scattering approximation that we restrict the analysis to the weakly heterogeneous regime introduced in Chapter 18.

In Section 20.4 we analyze the statistics of a transmitted broadband pulse. The main result is that the original pulse is decomposed into a sum of modal pulses that propagate with the different modal speeds and can be described by a front pulse stabilization theory, as in Chapter 8. The analysis is completed in Section 20.5 with a detailed study of time reversal for the transmitted field. The refocused field has a diffraction-limited transverse spatial profile that is independent of the size of the time-reversal mirror, provided that the propagation distance is large enough to ensure mode energy equipartition. The refocused field is, moreover, statistically stable. This is a consequence of the fact that in the broadband case the Fourier representation of the pulse is a superposition of many approximately uncorrelated components. This is the same phenomenon that we have seen in the one-dimensional case in Chapter 10.

The analysis of the pulse coda, which is the incoherent wave fluctuations in the broadband regime, as well as the study of pulse propagation in the narrowband regime, requires knowledge of the joint statistical distribution of the Fourier mode amplitudes at two nearby frequencies. In Section 20.6 we obtain the effective system of transport equations that describes the dynamics of the mode coupling in the narrowband case. We can then analyze the statistics, and in particular the autocorrelation function, of the incoherent wave fluctuations in Section 20.7.

Finally, in Sections 20.8–20.9 we study the transmission and time reversal of a narrowband pulse. An interesting result here is that statistical stability for the time-reversed refocused field can be achieved even in narrowband regimes when the refocused pulse is a superposition of a large number of modes and the time-reversal mirror is not too small. This statistical stability is different from that in the broadband regime, which is the result of decoherence in the frequency domain and there are no restrictions on the size of the time-reversal mirror.

20.1 Propagation in Homogeneous Waveguides

In this section we study wave propagation in an acoustic waveguide that supports a finite number of propagating modes. In an ideal waveguide the geometric structure and the medium parameters can have a general form in the transverse directions but must be homogeneous along the waveguide axis. There are two general types of ideal waveguides, those that surround a homogeneous region with a confining boundary and those in which the confinement is achieved with a transversely varying index of refraction. We will present the analysis of the effects of random perturbations on waveguides of the first type and we will illustrate specific results with a planar waveguide. The main difference in working with waveguides of the second type is that the transverse wave mode profiles depend on the frequency, but this does not affect the theory we present here.

20.1.1 Modeling of the Waveguide

As in most of this book, we consider linear acoustic waves propagating in three spatial dimensions modeled by the system of wave equations

$$\rho(\mathbf{r})\frac{\partial \mathbf{u}}{\partial t} + \nabla p = \mathbf{F},\tag{20.1}$$

$$\frac{1}{K(\mathbf{r})}\frac{\partial p}{\partial t} + \nabla \cdot \mathbf{u} = 0,\tag{20.2}$$

where p is the acoustic pressure, $\mathbf{u}$ is the acoustic velocity, ρ is the density of the medium, and K the bulk modulus. The source is modeled by the forcing term $\mathbf{F}(t, \mathbf{r})$.

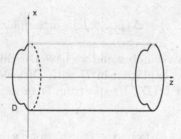

Fig. 20.1. Waveguide geometry.

We assume that the transverse profile of the waveguide is a simply connected region $\mathcal{D}$ in two dimensions. The direction of propagation along the waveguide axis is z, and the transverse coordinates are denoted by $\mathbf{x} \in \mathcal{D}$ (see Figure 20.1). In the interior of the waveguide the medium parameters are homogeneous,

$$\rho(\mathbf{r}) \equiv \bar{\rho}, \qquad K(\mathbf{r}) = \bar{K}, \qquad \text{for } \mathbf{x} \in \mathcal{D} \text{ and } z \in \mathbb{R}.$$

By differentiating with respect to time (20.2) and substituting (20.1) into it, we get the standard wave equation for the pressure field:

$$\Delta p - \frac{1}{\bar{c}^2} \frac{\partial^2 p}{\partial t^2} = \nabla \cdot \mathbf{F}, \qquad (20.3)$$

where $\Delta = \Delta_\perp + \frac{\partial^2}{\partial z^2}$ and $\Delta_\perp$ is the transverse Laplacian. The sound speed is $\bar{c} = \sqrt{K/\bar{\rho}}$.

We must now prescribe boundary conditions on the boundary $\partial\mathcal{D}$ of the domain $\mathcal{D}$. In underwater acoustics, or in seismic wave propagation, the density is much smaller outside the waveguide than inside. This means that we must use a pressure-release boundary condition since the pressure is very weak outside, and therefore, by continuity, the pressure is zero just inside the waveguide. Motivated by such examples, we will use Dirichlet boundary conditions

$$p(t, \mathbf{x}, z) = 0 \qquad \text{for } \mathbf{x} \in \partial\mathcal{D} \text{ and } z \in \mathbb{R}. \qquad (20.4)$$

We can also consider other types of boundary conditions if, for example, the boundary of the waveguide is a rigid wall, in which case the normal velocity vanishes. By (20.1) we obtain Neumann boundary conditions for the pressure.

20.1.2 The Propagating and Evanescent Modes

A waveguide mode is a monochromatic wave $p(t, \mathbf{x}, z) = \hat{p}(\omega, \mathbf{x}, z) e^{-i\omega t}$ with frequency ω, where $\hat{p}(\omega, \mathbf{x}, z)$ satisfies the time harmonic form of the wave equation (20.3) without a source term

$$\frac{\partial^2}{\partial z^2} \hat{p}(\omega, \mathbf{x}, z) + \Delta_\perp \hat{p}(\omega, \mathbf{x}, z) + k^2(\omega) \hat{p}(\omega, \mathbf{x}, z) = 0. \qquad (20.5)$$

Here $k = \omega/\bar{c}$ is the wave number and we have Dirichlet boundary conditions on $\partial\mathcal{D}$. The transverse Laplacian in $\mathcal{D}$ with Dirichlet boundary conditions on $\partial\mathcal{D}$ is self-adjoint in $L^2(\mathcal{D})$. Its spectrum is an infinite number of discrete eigenvalues

$$-\Delta_\perp \phi_j(\mathbf{x}) = \lambda_j \phi_j(\mathbf{x}), \; \mathbf{x} \in \mathcal{D}, \qquad \phi_j(\mathbf{x}) = 0, \; \mathbf{x} \in \partial\mathcal{D},$$

for $j = 1, 2, \ldots$. The eigenvalues are positive and nondecreasing, and we assume for simplicity that they are simple, so we have $0 < \lambda_1 < \lambda_2 < \cdots$. The eigenmodes are real and form an orthonormal set

$$\int_\mathcal{D} \phi_j(\mathbf{x}) \phi_l(\mathbf{x}) \, d\mathbf{x} = \delta_{jl}.$$

The modal wave numbers $\beta_j(\omega)$ are defined by

$$\beta_j(\omega) = \sqrt{\frac{\omega^2}{c^2} - \lambda_j}, \quad \lambda_j \le \frac{\omega^2}{c^2}, \tag{20.6}$$

and we denote by $N(\omega)$ the number of eigenmodes for which this inequality holds. The solutions

$$\hat{p}_j(\omega, \mathbf{x}, z) = \phi_j(\mathbf{x})e^{\pm i\beta_j(\omega)z}, \quad j = 1, \ldots, N(\omega),$$

of the wave equation (20.5) are the propagating waveguide modes. For $j > N(\omega)$ we define the modal wave numbers by

$$\beta_j(\omega) = \sqrt{\lambda_j - \frac{\omega^2}{c^2}}, \quad \lambda_j > \frac{\omega^2}{c^2}, \tag{20.7}$$

and the corresponding solutions

$$\hat{q}_j(\omega, \mathbf{x}, z) = \phi_j(\mathbf{x})e^{\pm \beta_j(\omega)z}, \quad j > N(\omega),$$

of the wave equation (20.5) are the evanescent waveguide modes.

The planar waveguide. This is the special case in which $\mathcal{D}$ is $(0, d) \times \mathbb{R}$ and we consider only solutions that depend on $x \in (0, d)$. In this case,

$$\lambda_j = \frac{\pi^2 j^2}{d^2}, \quad \phi_j(x) = \sqrt{\frac{2}{d}} \sin\left(\frac{\pi j x}{d}\right), \quad j \ge 1,$$

and the number of propagating modes is

$$N(\omega) = \left[\frac{\omega d}{\pi c}\right],$$

where $[x]$ is the integer part of x. The propagating modal wave numbers are now given by

$$\beta_j(\omega) = \sqrt{\frac{\omega^2}{c^2} - \frac{\pi^2 j^2}{d^2}}, \quad j = 1, 2, \ldots, N(\omega).$$

Variable index of refraction. When the wave mode structure is determined by a transversely varying index of refraction that is confining, then the eigenmodes are solutions of

$$\Delta_{\perp}\phi_j(\omega, \mathbf{x}) + k^2(\omega)n^2(\mathbf{x})\phi_j(\omega, \mathbf{x}) = \mu_j(\omega)\phi_j(\omega, \mathbf{x})$$

with or without boundary conditions, depending on the confining behavior of $n^2(\mathbf{x})$. The modal wave numbers are now given by $\beta_j(\omega) = \sqrt{\mu_j(\omega)}$ when $\mu_j \ge 0$ and by $\beta_j(\omega) = \sqrt{-\mu_j(\omega)}$ when $\mu_j < 0$. The main difference with the case above is that the mode profiles ϕ_j depend on the frequency ω.

20.1.3 Excitation Conditions for an Incoming Wave

For a right-propagating wave that is incoming from the left the pressure field at $z = 0$ can be written as a superposition of propagating modes

$$\hat{p}(\omega, \mathbf{x}, z = 0) = \sum_{j=1}^{N(\omega)} \frac{\hat{a}_j(\omega)}{\sqrt{\beta_j(\omega)}} \phi_j(\mathbf{x}),$$

where

$$\hat{a}_j(\omega) = \sqrt{\beta_j(\omega)} \int_{\mathcal{D}} \hat{p}(\omega, \mathbf{x}, z = 0) \phi_j(\mathbf{x}) \, d\mathbf{x}.$$

The factor $\sqrt{\beta_j}$ is introduced so that the differential equations derived in the next section will be symmetric. Since this is a right-propagating wave we know that the jth mode has modal wave number $+\beta_j$, so we can write the complete solution as

$$\hat{p}(\omega, \mathbf{x}, z) = \sum_{j=1}^{N(\omega)} \frac{\hat{a}_j(\omega)}{\sqrt{\beta_j(\omega)}} \phi_j(\mathbf{x}) e^{i\beta_j(\omega)z}.$$

20.1.4 Excitation Conditions for a Source

We consider a localized source in the plane $z = 0$ that emits a signal with orientation in the z-direction:

$$\mathbf{F}(t, \mathbf{x}, z) = f(t, \mathbf{x})\delta(z)\mathbf{e}_z.$$

Here $\mathbf{e}_z$ is the unit vector pointing in the z-direction. By (20.1), this source term implies that the pressure satisfies the following jump conditions across the plane $z = 0$,

$$\hat{p}(\omega, \mathbf{x}, z = 0^+) - \hat{p}(\omega, \mathbf{x}, z = 0^-) = \hat{f}(\omega, \mathbf{x}),$$

while (20.2) implies that there is no jump in the longitudinal velocity, so that the pressure field also satisfies

$$\frac{\partial \hat{p}}{\partial z}(\omega, \mathbf{x}, z = 0^+) - \frac{\partial \hat{p}}{\partial z}(\omega, \mathbf{x}, z = 0^-) = 0.$$

Here $\hat{f}$ is the Fourier transform of f with respect to time:

$$\hat{f}(\omega, \mathbf{x}) = \int f(t, \mathbf{x}) e^{i\omega t} \, dt, \qquad f(t, \mathbf{x}) = \frac{1}{2\pi} \int \hat{f}(\omega, \mathbf{x}) e^{-i\omega t} \, d\omega.$$

The pressure field can be written as a superposition of the complete set of modes

$$\hat{p}(\omega, \mathbf{x}, z) = \left[\sum_{j=1}^{N} \frac{\hat{a}_j(\omega)}{\sqrt{\beta_j(\omega)}} e^{i\beta_j z} \phi_j(\mathbf{x}) + \sum_{j=N+1}^{\infty} \frac{\hat{c}_j(\omega)}{\sqrt{\beta_j(\omega)}} e^{-\beta_j z} \phi_j(\mathbf{x}) \right] \mathbf{1}_{(0,\infty)}(z)$$

$$+ \left[\sum_{j=1}^{N} \frac{\hat{b}_j(\omega)}{\sqrt{\beta_j(\omega)}} e^{-i\beta_j z} \phi_j(\mathbf{x}) + \sum_{j=N+1}^{\infty} \frac{\hat{d}_j(\omega)}{\sqrt{\beta_j(\omega)}} e^{\beta_j z} \phi_j(\mathbf{x}) \right] \mathbf{1}_{(-\infty,0)}(z) \,,$$

where $\hat{a}_j$ is the amplitude of the jth right-going mode propagating in the right half-space $z > 0$, $\hat{b}_j$ is the amplitude of the jth left-going mode propagating in the left half-space $z < 0$, and $\hat{c}_j$ (respectively $\hat{d}_j$) is the amplitude of the j-th right-going (respectively left-going) evanescent mode. Substituting this expansion into the jump conditions, multiplying by $\phi_j(\mathbf{x})$, integrating with respect to $\mathbf{x}$ over $\mathcal{D}$, and using the orthogonality of the modes, we express the mode amplitudes in terms of the source:

$$\hat{a}_j(\omega) = -\hat{b}_j(\omega) = \frac{\sqrt{\beta_j(\omega)}}{2} \int_{\mathcal{D}} \hat{f}(\omega, \mathbf{x}) \phi_j(\mathbf{x}) \, d\mathbf{x} \,,$$

$$\hat{c}_j(\omega) = -\hat{d}_j(\omega) = -\frac{\sqrt{\beta_j(\omega)}}{2} \int_{\mathcal{D}} \hat{f}(\omega, \mathbf{x}) \phi_j(\mathbf{x}) \, d\mathbf{x} \,.$$

If we look at these waves in the region $z > 0$ far from the source, at a distance large compared to the wavelength, then the evanescent modes are negligible, and the waves have the form

$$\hat{p}(\omega, \mathbf{x}, z) = \sum_{j=1}^{N(\omega)} \frac{\hat{a}_j(\omega)}{\sqrt{\beta_j(\omega)}} \phi_j(\mathbf{x}) e^{i\beta_j(\omega)z} \,.$$

In the case of a point source $f(t, \mathbf{x}) = f(t)\delta(\mathbf{x} - \mathbf{x}_0)$, we simply have

$$\hat{a}_j(\omega) = \frac{1}{2} \sqrt{\beta_j(\omega)} \phi_j(\mathbf{x}_0) \hat{f}(\omega) \,.$$

20.2 Mode Coupling in Random Waveguides

A simple model for a randomly perturbed waveguide is one for which the density $\rho(\mathbf{r}) = \bar{\rho}$ is uniform, as in most of this book, and the reciprocal of the bulk modulus has the form

$$\frac{1}{K(\mathbf{r})} = \frac{1}{\bar{K}} (1 + \nu(\mathbf{r})) \,.$$

Here $\nu(\mathbf{r})$ is a three-dimensional stationary random field with mean zero, with a given covariance, and with some additional properties that are needed for the asymptotic analysis that are discussed later. We are not entering the theory of random fields but simply illustrate our model by the following example, which is a bounded random field:

$$\nu(\mathbf{r}) = \sum_j \nu_j \psi(\mathbf{r} - \mathbf{r}_j).$$

Here the $\{\nu_j\}$ are independent identically distributed bounded random variables with mean zero, and the points $\{\mathbf{r}_j\}$ are independent three-dimensional random variables that are uniformly distributed over the different cubes of a cubic tiling of $\mathbb{R}^3$. The function ψ is real-valued, smooth, bounded, and with compact support in $\mathbb{R}^3$.

Random perturbations of this kind are very different from the layered ones that we have considered up to now. The reason that we can use the asymptotic theory of Chapter 6 to analyze waves in randomly perturbed waveguides is that the mode amplitudes satisfy differential equations with z-dependent random coefficients that are projections of $\nu(\mathbf{x}, z)$ on the transverse eigenmodes $\{\phi_j(\mathbf{x})\}$.

We will use the weakly heterogeneous regime described in Chapter 18. We consider a randomly perturbed waveguide section occupying the region $z \in [0, L/\varepsilon^2]$, with two homogeneous waveguides occupying the two half spaces $z < 0$ and $z > L/\varepsilon^2$. The bulk modulus and the density have the form

$$\frac{1}{K(\mathbf{x}, z)} = \begin{cases} \frac{1}{K}(1 + \varepsilon\nu(\mathbf{x}, z)) & \text{for} \quad \mathbf{x} \in \mathcal{D}, \quad z \in [0, L/\varepsilon^2], \\ \frac{1}{K} & \text{for} \quad \mathbf{x} \in \mathcal{D}, \quad z \in (-\infty, 0) \cup (L/\varepsilon^2, \infty), \end{cases}$$
$$\rho(\mathbf{x}, z) = \bar{\rho} \quad \text{for} \quad \mathbf{x} \in \mathcal{D}, \quad z \in (-\infty, \infty).$$

The perturbed wave equation satisfied by the pressure field is

$$\Delta p - \frac{1 + \varepsilon\nu(\mathbf{x}, z)}{\bar{c}^2} \frac{\partial^2 p}{\partial t^2} = \nabla \cdot \mathbf{F}, \tag{20.8}$$

where the average sound speed is $\bar{c} = \sqrt{K/\bar{\rho}}$. The pressure field also satisfies the boundary conditions (20.4). We assume that a guided wave is incoming from the left through the perfect waveguide and (for $z < 0$) has in the Fourier domain the form

$$\hat{p}_{in}(\omega, \mathbf{x}, z) = \sum_{j=1}^{N(\omega)} \phi_j(\mathbf{x}) \hat{p}_j(\omega, z), \quad \hat{p}_j(\omega, z) = \frac{\hat{a}_{j,0}(\omega)}{\sqrt{\beta_j(\omega)}} e^{i\beta_j(\omega)z}, \tag{20.9}$$

where $\hat{a}_{j,0}(\omega)$ is the projection of the incident wave $\hat{p}_{in}(\omega, \mathbf{x}, z = 0)$ onto the jth mode:

$$\hat{a}_{j,0}(\omega) = \sqrt{\beta_j(\omega)} \int_{\mathcal{D}} \phi_j(\mathbf{x}) \hat{p}_{in}(\omega, \mathbf{x}, z = 0) \, d\mathbf{x}.$$

The weak fluctuations of the medium parameters induce a coupling between the propagating modes, as well as between propagating and evanescent modes, which build up and become of order one after a propagation distance of order ε^{-2}, as expected from the asymptotic theory of Chapter 6, which we will use below.

20.2.1 Coupled Amplitude Equations

We fix the frequency ω and expand the field $\hat{p}$ inside the randomly perturbed waveguide in terms of the transverse eigenmodes

$$\hat{p}(\mathbf{x}, z) = \sum_{j=1}^{N} \phi_j(\mathbf{x})\hat{p}_j(z) + \sum_{j=N+1}^{\infty} \phi_j(\mathbf{x})\hat{q}_j(z), \tag{20.10}$$

where $\hat{p}_j$ is the amplitude of the jth propagating mode and $\hat{q}_j$ is the amplitude of the jth evanescent mode. For $j \leq N$ we introduce the right-going and left-going mode amplitudes $\hat{a}_j$ and $\hat{b}_j$ defined by

$$\hat{p}_j(z) = \frac{1}{\sqrt{\beta_j}} \left(\hat{a}_j(z)e^{i\beta_j z} + \hat{b}_j(z)e^{-i\beta_j z} \right), \tag{20.11}$$

$$\frac{d\hat{p}_j(z)}{dz} = i\sqrt{\beta_j} \left(\hat{a}_j(z)e^{i\beta_j z} - \hat{b}_j(z)e^{-i\beta_j z} \right). \tag{20.12}$$

Then it follows that $\hat{a}_j$ and $\hat{b}_j$ have the form

$$\hat{a}_j(z) = \frac{i\beta_j \hat{p}_j + \frac{d\hat{p}_j}{dz}}{2i\sqrt{\beta_j}} e^{-i\beta_j z}, \qquad \hat{b}_j(z) = \frac{i\beta_j \hat{p}_j - \frac{d\hat{p}_j}{dz}}{2i\sqrt{\beta_j}} e^{i\beta_j z}.$$

The total field $\hat{p}$ satisfies the time-harmonic wave equation

$$\Delta\hat{p}(\omega, \mathbf{x}, z) + k^2(1 + \varepsilon\nu(\mathbf{x}, z))\hat{p}(\omega, \mathbf{x}, z) = 0. \tag{20.13}$$

Using (20.10) in this equation, multiplying it by $\phi_l(\mathbf{x})$, and integrating over $\mathbf{x} \in \mathcal{D}$, we deduce from the orthogonality of the eigenmodes $(\phi_j)_{j\geq 1}$ the following system of coupled differential equations for the mode amplitudes:

$$\frac{d\hat{a}_j}{dz} = \frac{i\varepsilon k^2}{2} \sum_{1\leq l\leq N} \frac{C_{jl}(z)}{\sqrt{\beta_j\beta_l}} \left(\hat{a}_l e^{i(\beta_l-\beta_j)z} + \hat{b}_l e^{-i(\beta_l+\beta_j)z} \right)$$

$$+ \frac{i\varepsilon k^2}{2\sqrt{\beta_j}} \sum_{l>N} C_{jl}(z)\hat{q}_l(z)e^{-i\beta_j z}, \qquad 1 \leq j \leq N, \tag{20.14}$$

$$\frac{d\hat{b}_j}{dz} = -\frac{i\varepsilon k^2}{2} \sum_{1\leq l\leq N} \frac{C_{jl}(z)}{\sqrt{\beta_j\beta_l}} \left(\hat{a}_l e^{i(\beta_l+\beta_j)z} + \hat{b}_l e^{i(\beta_j-\beta_l)z} \right)$$

$$- \frac{i\varepsilon k^2}{2\sqrt{\beta_j}} \sum_{l>N} C_{jl}(z)\hat{q}_l(z)e^{i\beta_j z}, \qquad 1 \leq j \leq N, \tag{20.15}$$

$$\frac{d^2\hat{q}_j}{dz^2} - \beta_j^2\hat{q}_j + \varepsilon g_j(z) = 0, \qquad j \geq N+1. \tag{20.16}$$

Here

$$C_{jl}(z) = \int_{\mathcal{D}} \phi_j(\mathbf{x}) \phi_l(\mathbf{x}) \nu(\mathbf{x}, z) \, d\mathbf{x}, \tag{20.17}$$

$$g_j(z) = k^2 \sum_{l>N} C_{jl}(z) \hat{q}_l + k^2 \sum_{1 \le l \le N} \frac{C_{jl}(z)}{\sqrt{\beta_l}} \left(\hat{a}_l e^{i\beta_l z} + \hat{b}_l e^{-i\beta_l z} \right). \tag{20.18}$$

Note that g_j is a linear function of $\hat{a}$, $\hat{b}$, and $\hat{q}$. The system (20.14–20.16) is complemented with the boundary conditions

$$\hat{a}_j(0) = \hat{a}_{j,0}, \qquad \hat{b}_j\left(\frac{L}{\varepsilon^2}\right) = 0, \tag{20.19}$$

for the propagating modes. The second condition indicates that no wave is incoming from the right. We also use the radiation conditions

$$\lim_{z \to \pm\infty} \hat{q}_j(z) = 0,$$

which imply that the evanescent modes are decaying. The solution of (20.16) that satisfies the radiation conditions is

$$\hat{q}_j(z) = \frac{\varepsilon}{2\beta_j} \int_{-\infty}^{\infty} g_j(z+s) e^{-\beta_j|s|} \, ds. \tag{20.20}$$

20.2.2 Conservation of Energy Flux

We saw in Chapter 4 that energy-flux conservation in the one-dimensional case implies a conservation relation for the right- and left-propagating wave amplitudes that has the form $|\hat{a}|^2 - |\hat{b}|^2 = $ constant. We now generalize this relation to waveguides.

Let $|\hat{a}|^2 = \sum_{j=1}^{N} |\hat{a}_j|^2$ and $|\hat{b}|^2 = \sum_{j=1}^{N} |\hat{b}_j|^2$. Using (20.14–20.15) we have

$$\frac{d}{dz}(|\hat{a}|^2 - |\hat{b}|^2) = -k^2 \varepsilon \, \mathrm{Im} \left[\sum_{j=1}^{N} \sum_{l>N} \frac{C_{jl}}{\sqrt{\beta_j}} (\overline{\hat{a}_j} e^{-i\beta_j z} + \overline{\hat{b}_j} e^{i\beta_j z}) \hat{q}_l \right].$$

By (20.18) the right-hand side can be rewritten as

$$\frac{d}{dz}(|\hat{a}|^2 - |\hat{b}|^2) = -\varepsilon \, \mathrm{Im} \left[\sum_{l>N} \left(\overline{\hat{g}_l} - k^2 \sum_{l'>N} C_{ll'} \overline{\hat{q}_{l'}} \right) \hat{q}_l \right] = -\varepsilon \, \mathrm{Im} \left[\sum_{l>N} \overline{\hat{g}_l} \hat{q}_l \right].$$

We use the integral representation (20.20) and we integrate over z to get

$$(|\hat{a}|^2 - |\hat{b}|^2)(z) - (|\hat{a}|^2 - |\hat{b}|^2)(0) = -\frac{\varepsilon^2}{2} \sum_{l>N} \frac{1}{\beta_l} \hat{G}_l(z), \tag{20.21}$$

where $\hat{G}_l(z)$ is defined by

$$\hat{G}_l(z) = \int_{-\infty}^{z} \int_{-\infty}^{\infty} \text{Im} \left[\overline{\hat{g}_l}(y) \hat{g}_l(y+s) \right] e^{-\beta_l |s|} \, ds \, dy \, .$$

The quantity $\hat{G}_l(z)$ can also be written as

$$2i\hat{G}_l(z) = \int_{-\infty}^{z} \int_{-\infty}^{\infty} \overline{\hat{g}_l}(y) \hat{g}_l(y+s) e^{-\beta_l |s|} \, ds \, dy$$

$$- \int_{-\infty}^{z} \int_{-\infty}^{\infty} \hat{g}_l(y) \overline{\hat{g}_l}(y+s) e^{-\beta_l |s|} \, ds \, dy \, .$$

We transform the second term as follows (by the simple changes of variables $y \mapsto y - s$ and then $s \mapsto -s$):

$$\int_{-\infty}^{z} \int_{-\infty}^{\infty} \hat{g}_l(y) \overline{\hat{g}_l}(y+s) e^{-\beta_l |s|} \, ds \, dy = \int_{-\infty}^{\infty} \int_{-\infty}^{z+s} \hat{g}_l(y-s) \overline{\hat{g}_l}(y) \, dy \, e^{-\beta_l |s|} \, ds$$

$$= \int_{-\infty}^{\infty} \int_{-\infty}^{z-s} \hat{g}_l(y+s) \overline{\hat{g}_l}(y) \, dy \, e^{-\beta_l |s|} \, ds \, ,$$

so that we get

$$\hat{G}_l(z) = \frac{1}{2} \int_{-\infty}^{\infty} \int_{z-s}^{z} \text{Im}[\overline{\hat{g}_l}(y) \hat{g}_l(y+s)] \, dy \, e^{-\beta_l |s|} \, ds \, .$$

The waveguide is homogeneous in the section $(L/\varepsilon^2, \infty)$, which implies that $\hat{g}_l(y) = 0$ if $y > L/\varepsilon^2$. In the previous double integral, we can observe that if $s < 0$, then $y > z$, while if $s > 0$, then $y + s > z$. As a consequence, if $z \geq L/\varepsilon^2$, then either y or $y + s$ is larger than L/ε^2, so that the product $\overline{\hat{g}_l}(y) \hat{g}_l(y+s)$ is always 0. This shows that $\hat{G}_l(z) = 0$ if $z \geq L/\varepsilon^2$. By (20.21) we thus obtain

$$|\hat{a}|^2 (L/\varepsilon^2) - |\hat{b}|^2 (L/\varepsilon^2) = |\hat{a}|^2 (0) - |\hat{b}|^2 (0) \, .$$

Using the boundary conditions (20.19) at 0 and L/ε^2, we get the conservation of energy flux

$$|\hat{a}|^2 (L/\varepsilon^2) + |\hat{b}|^2 (0) = |\hat{a}_0|^2 \, , \tag{20.22}$$

which is exact for any ε. The identity (20.22) expresses the conservation of the global flux. The local flux inside the inhomogeneous section of the waveguide is, however, not preserved except in the limit $\varepsilon \to 0$. The evanescent modes can store a small amount of energy during the mode-coupling process, which gives the $O(\varepsilon^2)$ correction term on the right side of (20.21). There cannot be any energy stored in the evanescent modes for $z \geq L/\varepsilon^2$ because it is not possible for them to give back this energy to the propagating modes in the homogeneous section $(L/\varepsilon^2, \infty)$. There is no coupling between modes in the homogeneous waveguide section.

20.2.3 Evanescent Modes in Terms of Propagating Modes

In this and the next subsection we show how to obtain asymptotically a closed
system of equations for the propagating-mode amplitudes while taking into
account coupling with the evanescent modes.

We first substitute (20.18) into (20.20) and rewrite it in the vector-matrix
form

$$(I_d - \varepsilon \Psi)\hat{q} = \varepsilon \tilde{q}.$$

Here the operator Ψ is defined by

$$(\Psi \hat{q})_j(z) = \frac{k^2}{2\beta_j} \sum_{l \geq N+1} \int_{-\infty}^{\infty} C_{jl}(z+s)\hat{q}_l(z+s)e^{-\beta_j|s|}\,ds, \quad j \geq N+1,$$

and the vector function $\tilde{q}$ is given by

$$\tilde{q}_j(z) = \frac{k^2}{2\beta_j} \sum_{1 \leq l \leq N} \int_{-\infty}^{\infty} \frac{C_{jl}(z+s)}{\sqrt{\beta_l}}$$

$$\times \left(\hat{a}_l(z+s)e^{i\beta_l(z+s)} + \hat{b}_l(z+s)e^{-i\beta_l(z+s)} \right) e^{-\beta_j|s|}\,ds, \quad j \geq N+1. \quad (20.23)$$

Introducing the norm $\|\hat{q}\| = \sum_{j \geq N+1} \beta_j^{-1} \sup_z |\hat{q}_j(z)|$, we have

$$\|\Psi \hat{q}\| \leq \sum_{j \geq N+1} \beta_j^{-1} \sup_z \frac{k^2}{2\beta_j} \sum_{l \geq N+1} \int_{-\infty}^{\infty} |C_{jl}(z+s)||\hat{q}_l(z+s)|e^{-\beta_l|s|}\,ds$$

$$\leq \frac{Kk^2}{2} \sum_{j \geq N+1} \beta_j^{-2} \sum_{l \geq N+1} \sup_z |\hat{q}_l(z)| \int_{-\infty}^{\infty} e^{-\beta_l|s|}\,ds$$

$$\leq KK'k^2 \sum_{l \geq N+1} \beta_l^{-1} \sup_z |\hat{q}_l(z)| = KK'k^2\|\hat{q}\|,$$

where $K = \sup_{j,l,z} |C_{jl}(z)| \leq \sup_{z,\mathbf{x}} |\nu(z,\mathbf{x})|$ and $K' = \sum_{j \geq N+1} \beta_j^{-2}$. Note
that K' is indeed finite for a broad class of waveguides. This is, for example, the
case in the planar waveguide because we have $\beta_j = \sqrt{\pi^2 j^2/d^2 - \omega^2/\bar{c}^2} \sim \pi j/d$
for $j \gg N$. Thus Ψ is a bounded operator, and so for ε small enough, $I_d - \varepsilon \Psi$
is invertible and can be approximated by $I_d + \varepsilon \Psi + O(\varepsilon^2)$. For $j \geq N+1$ we
can therefore write

$$\hat{q}_j(z) = \frac{\varepsilon k^2}{2\beta_j} \sum_{1 \leq l \leq N} \int_{-\infty}^{\infty} \frac{C_{jl}(z+s)}{\sqrt{\beta_l}}$$

$$\times \left(\hat{a}_l(z+s)e^{i\beta_l(z+s)} + \hat{b}_l(z+s)e^{-i\beta_l(z+s)} \right) e^{-\beta_j|s|}\,ds + O(\varepsilon^2).$$

Furthermore, over a distance of order 1, the variation of $\hat{a}$ and $\hat{b}$ is at most of
order ε, so we can substitute $\hat{a}_l(z)$ and $\hat{b}_l(z)$ for $\hat{a}_l(z+s)$ and $\hat{b}_l(z+s)$ up to
an error of order ε, and we obtain

$$\hat{q}_j(z) = \frac{\varepsilon k^2}{2\beta_j} \sum_{1 \le l \le N} \int_{-\infty}^{\infty} \frac{C_{jl}(z+s)}{\sqrt{\beta_l}}$$

$$\times \left(\hat{a}_l(z) e^{i\beta_l(s+z)} + \hat{b}_l(z) e^{-i\beta_l(s+z)} \right) e^{-\beta_j|s|} \, ds + O(\varepsilon^2) \,. \quad (20.24)$$

We have assumed in this analysis that the vector function $\tilde{q}$ defined by (20.23) is bounded in the norm $\| \cdot \|$ uniformly in ε. This will follow if the mode amplitudes $\hat{a}_j$ and $\hat{b}_j$ are bounded uniformly in ε. Using the methods of Chapter 6 it can be shown that such bounds hold in probability as $\varepsilon \to 0$.

20.2.4 Propagating-Mode-Amplitude Equations

Substituting (20.24) into (20.14–20.15), we have the following system of differential equations for the propagating-mode amplitudes

$$\frac{d\hat{a}}{dz} = \varepsilon[\mathbf{H}^{(aa)}(z)\hat{a} + \mathbf{H}^{(ab)}(z)\hat{b}] + \varepsilon^2[\mathbf{G}^{(aa)}(z)\hat{a} + \mathbf{G}^{(ab)}(z)\hat{b}] + O(\varepsilon^3) \,, \quad (20.25)$$

$$\frac{d\hat{b}}{dz} = \varepsilon[\mathbf{H}^{(ba)}(z)\hat{a} + \mathbf{H}^{(bb)}(z)\hat{b}] + \varepsilon^2[\mathbf{G}^{(ba)}(z)\hat{a} + \mathbf{G}^{(bb)}(z)\hat{b}] + O(\varepsilon^3) \,. \quad (20.26)$$

Here the matrices $\mathbf{G}^{(\cdot,\cdot)}(z)$ and $\mathbf{H}^{(\cdot,\cdot)}(z)$ are given by

$$H_{jl}^{(aa)}(z) = \frac{ik^2}{2} \frac{C_{jl}(z)}{\sqrt{\beta_j\beta_l}} e^{i(\beta_l - \beta_j)z} \,, \quad (20.27)$$

$$H_{jl}^{(ab)}(z) = \frac{ik^2}{2} \frac{C_{jl}(z)}{\sqrt{\beta_j\beta_l}} e^{-i(\beta_l + \beta_j)z} \,, \quad (20.28)$$

$$H_{jl}^{(ba)}(z) = -\frac{ik^2}{2} \frac{C_{jl}(z)}{\sqrt{\beta_j\beta_l}} e^{i(\beta_l + \beta_j)z} \,, \quad (20.29)$$

$$H_{jl}^{(bb)}(z) = -\frac{ik^2}{2} \frac{C_{jl}(z)}{\sqrt{\beta_j\beta_l}} e^{i(\beta_j - \beta_l)z} \,, \quad (20.30)$$

$$G_{jl}^{(aa)}(z) = \frac{ik^4}{4} \sum_{l'>N} \int_{-\infty}^{\infty} \frac{C_{jl'}(z)C_{ll'}(z+s)}{\sqrt{\beta_j\beta_{l'}^2\beta_l}} e^{i\beta_l(z+s) - i\beta_j z - \beta_{l'}|s|} \, ds \,, \quad (20.31)$$

$$G_{jl}^{(ab)}(z) = \frac{ik^4}{4} \sum_{l'>N} \int_{-\infty}^{\infty} \frac{C_{jl'}(z)C_{ll'}(z+s)}{\sqrt{\beta_j\beta_{l'}^2\beta_l}} e^{-i\beta_l(z+s) - i\beta_j z - \beta_{l'}|s|} \, ds \,, \quad (20.32)$$

$$G_{jl}^{(ba)}(z) = -\frac{ik^4}{4} \sum_{l'>N} \int_{-\infty}^{\infty} \frac{C_{jl'}(z)C_{ll'}(z+s)}{\sqrt{\beta_j\beta_{l'}^2\beta_l}} e^{i\beta_l(z+s) + i\beta_j z - \beta_{l'}|s|} \, ds \,, \quad (20.33)$$

$$G_{jl}^{(bb)}(z) = -\frac{ik^4}{4} \sum_{l'>N} \int_{-\infty}^{\infty} \frac{C_{jl'}(z)C_{ll'}(z+s)}{\sqrt{\beta_j\beta_{l'}^2\beta_l}} e^{-i\beta_l(z+s) + i\beta_j z - \beta_{l'}|s|} \, ds \,. \quad (20.34)$$

The matrices $\mathbf{G}^{(\cdot,\cdot)}$ represent the effective coupling of the propagating modes with the evanescent modes. We note that the rescaled processes $\hat{a}_j^\varepsilon$, $\hat{b}_j^\varepsilon$, $j = 1, \ldots, N$, given by

$$\hat{a}_j^\varepsilon(z) = \hat{a}_j\left(\frac{z}{\varepsilon^2}\right), \quad \hat{b}_j^\varepsilon(z) = \hat{b}_j\left(\frac{z}{\varepsilon^2}\right), \tag{20.35}$$

are solutions of

$$\frac{d\hat{a}^\varepsilon}{dz} = \left[\frac{1}{\varepsilon}\mathbf{H}^{(aa)}\left(\frac{z}{\varepsilon^2}\right) + \mathbf{G}^{(aa)}\left(\frac{z}{\varepsilon^2}\right)\right]\hat{a}^\varepsilon$$

$$+ \left[\frac{1}{\varepsilon}\mathbf{H}^{(ab)}\left(\frac{z}{\varepsilon^2}\right) + \mathbf{G}^{(ab)}\left(\frac{z}{\varepsilon^2}\right)\right]\hat{b}^\varepsilon, \tag{20.36}$$

$$\frac{d\hat{b}^\varepsilon}{dz} = \left[\frac{1}{\varepsilon}\mathbf{H}^{(ba)}\left(\frac{z}{\varepsilon^2}\right) + \mathbf{G}^{(ba)}\left(\frac{z}{\varepsilon^2}\right)\right]\hat{a}^\varepsilon$$

$$+ \left[\frac{1}{\varepsilon}\mathbf{H}^{(bb)}\left(\frac{z}{\varepsilon^2}\right) + \mathbf{G}^{(bb)}\left(\frac{z}{\varepsilon^2}\right)\right]\hat{b}^\varepsilon. \tag{20.37}$$

The propagating-mode amplitudes are completely determined by the system (20.36–20.37) as soon as we specify the two-point boundary conditions

$$\hat{a}_j^\varepsilon(0) = \hat{a}_{j,0}, \quad \hat{b}_j^\varepsilon(L) = 0, \tag{20.38}$$

which come from (20.19). This is the usual reflection-transmission setup that we follow in this book, where no wave enters the waveguide at $z = L$ (in the scaling of (20.36–20.37)) and the incident wave at $z = 0$ has the form (20.9).

The system (20.36–20.37) and the boundary conditions (20.38) determine the propagating-mode amplitudes. Even though this system involves only the propagating-mode amplitudes, it takes into account the effect of the evanescent modes as well. If one is interested in determining the evanescent modes, one should first integrate the system (20.36–20.37) and then substitute the solutions into the integral representations (20.24) of the evanescent modes.

20.2.5 Propagator Matrices

The two-point linear boundary value problem for (20.36–20.37) has the same structure as the one encountered in the one-dimensional case in Section 4.4.3. It can therefore be solved using propagator matrices, which are now $2N \times 2N$ random matrices. The system (20.36–20.37) can be put into full vector-matrix form

$$\frac{dX^\varepsilon}{dz} = \mathbf{H}^\varepsilon(z)X^\varepsilon,$$

where X^ε is the $2N$-vector obtained by concatenating the N-vectors $\hat{a}^\varepsilon$ and $\hat{b}^\varepsilon$,

$$X^\varepsilon(z) = \begin{bmatrix} \hat{a}^\varepsilon(z) \\ \hat{b}^\varepsilon(z) \end{bmatrix},$$

while $\mathbf{H}^\varepsilon$ is the $2N \times 2N$ matrix

$$\mathbf{H}^\varepsilon(z) = \frac{1}{\varepsilon}\begin{bmatrix} \mathbf{H}^{(aa)}\left(\frac{z}{\varepsilon^2}\right) & \mathbf{H}^{(ab)}\left(\frac{z}{\varepsilon^2}\right) \\ \mathbf{H}^{(ba)}\left(\frac{z}{\varepsilon^2}\right) & \mathbf{H}^{(bb)}\left(\frac{z}{\varepsilon^2}\right) \end{bmatrix} + \begin{bmatrix} \mathbf{G}^{(aa)}\left(\frac{z}{\varepsilon^2}\right) & \mathbf{G}^{(ab)}\left(\frac{z}{\varepsilon^2}\right) \\ \mathbf{G}^{(ba)}\left(\frac{z}{\varepsilon^2}\right) & \mathbf{G}^{(bb)}\left(\frac{z}{\varepsilon^2}\right) \end{bmatrix}.$$

The $2N \times 2N$ propagator matrices $\mathbf{P}^\varepsilon(z)$ are now solutions of the initial value problem

$$\frac{d\mathbf{P}^\varepsilon}{dz} = \mathbf{H}^\varepsilon(z)\mathbf{P}^\varepsilon ,$$

with the initial condition $\mathbf{P}^\varepsilon(z = 0) = \mathbf{I}$. The general solution of (20.36–20.37) satisfies, for any $0 \le z \le L$,

$$\begin{bmatrix} \hat{a}^\varepsilon(z) \\ \hat{b}^\varepsilon(z) \end{bmatrix} = \mathbf{P}^\varepsilon(z) \begin{bmatrix} \hat{a}^\varepsilon(0) \\ \hat{b}^\varepsilon(0) \end{bmatrix}, \qquad \mathbf{P}^\varepsilon(z) = \begin{bmatrix} \mathbf{P}^\varepsilon_{aa}(z) & \mathbf{P}^\varepsilon_{ab}(z) \\ \mathbf{P}^\varepsilon_{ba}(z) & \mathbf{P}^\varepsilon_{bb}(z) \end{bmatrix}.$$

When we specialize this relation to $z = L$ and use the boundary conditions (20.38), with $\hat{a}_0$ successively the unit vectors, we get

$$\begin{bmatrix} \hat{\mathbf{T}}^\varepsilon(L) \\ \mathbf{0} \end{bmatrix} = \mathbf{P}^\varepsilon(L) \begin{bmatrix} \mathbf{I} \\ \hat{\mathbf{R}}^\varepsilon(L) \end{bmatrix}. \tag{20.39}$$

Here we have defined the $N \times N$ random reflection and transmission matrices $\hat{\mathbf{R}}^\varepsilon(L)$ and $\hat{\mathbf{T}}^\varepsilon(L)$.

The matrices $\mathbf{H}^{(\cdot,\cdot)}$ and $\mathbf{G}^{(\cdot,\cdot)}$ satisfy the symmetry relation

$$\mathbf{H}^{(aa)}(z) = \overline{\mathbf{H}^{(bb)}(z)}, \qquad \mathbf{H}^{(ab)}(z) = \overline{\mathbf{H}^{(ba)}(z)},$$
$$\mathbf{G}^{(aa)}(z) = \overline{\mathbf{G}^{(bb)}(z)}, \qquad \mathbf{G}^{(ab)}(z) = \overline{\mathbf{G}^{(ba)}(z)},$$

so we can check that if $(\hat{a}^\varepsilon(z), \hat{b}^\varepsilon(z))$ is a solution, then $(\overline{\hat{b}^\varepsilon(z)}, \overline{\hat{a}^\varepsilon(z)})$ is also a solution. This imposes that the propagator have the form

$$\mathbf{P}^\varepsilon(z) = \begin{bmatrix} \mathbf{P}^\varepsilon_{aa}(z) & \mathbf{P}^\varepsilon_{ab}(z) \\ \overline{\mathbf{P}^\varepsilon_{ab}(z)} & \overline{\mathbf{P}^\varepsilon_{aa}(z)} \end{bmatrix}. \tag{20.40}$$

Note that the matrix $\mathbf{P}^\varepsilon_{aa}$ describes the coupling between different right-going modes, while $\mathbf{P}^\varepsilon_{ab}$ describes the coupling between right-going and left-going modes.

In addition to the general symmetry or reciprocity relation that gives the form (20.40) for the propagator matrices, we also have the global flux conservation relation (20.22). When this relation is used in conjunction with (20.39) we get the N-mode generalization of the reflection-transmission conservation relation

$$\hat{\mathbf{R}}^{\varepsilon\dagger}\hat{\mathbf{R}}^\varepsilon + \hat{\mathbf{T}}^{\varepsilon\dagger}\hat{\mathbf{T}}^\varepsilon = \mathbf{I}, \tag{20.41}$$

where the sign $\dagger$ stands for the conjugate transpose. This relation holds in general, with evanescent modes taken into consideration and for any ε. When the evanescent modes are not taken into consideration, so that the $\mathbf{G}$ terms in $\mathbf{H}^\varepsilon$ are identically zero, then there are additional symmetries satisfied by the propagator matrices $\mathbf{P}^\varepsilon$. We will not describe these symmetries here because we will use the forward-scattering approximation that is introduced in the next section.

Let us have a look at the set of random ordinary differential equations (20.36–20.37). This is a set of linear ordinary differential equations with boundary conditions and rapidly varying random coefficients. The situation is very similar to the one encountered in the previous chapters dealing with the one-dimensional case in the weakly heterogeneous regime:

- dimension of the system: Instead of having two modes (the right-going mode and the left-going mode in the one-dimensional case), we now have a set of $2N$ modes (the N right-going and the N left-going modes). This difference does not represent any theoretical difficulty, even though the computations are lengthy, but it will be almost impossible to extract information from the limit system comparable to what we can do in the $N = 1$ case.
- diffusive terms: The matrices $\mathbf{H}^{(\cdot,\cdot)}$ have exactly the same form as those encountered in the previous chapters, since they have the same symmetry properties and the same diffusive scaling.
- drift terms: A new feature in the waveguide case is the presence of the nonzero-mean terms $\mathbf{G}^{(\cdot,\cdot)}$. These terms can be dealt with according to the general theory of the diffusion approximation presented in Chapter 6. In the limit $\varepsilon \to 0$, they become equal to their average with respect to the distribution of ν and to the rapid phase (which we denote by $\langle \cdot \rangle$):

$$\left\langle \mathbb{E}[G_{jl}^{(aa)}(\cdot)] \right\rangle, \quad \left\langle \mathbb{E}[G_{jl}^{(ab)}(\cdot)] \right\rangle, \dots .$$

Considering the expression (20.31), because of the presence of the phases $e^{i(\beta_l - \beta_j)z}$ only the terms with $j = l$ are nonzero. As a result,

$$\left\langle \mathbb{E}[G_{jl}^{(aa)}(\cdot)] \right\rangle = i\kappa_j(\omega)\delta_{jl} ,$$

where

$$\kappa_j(\omega) = \sum_{l'>N(\omega)} \frac{\omega^4}{4\bar{c}^4 \beta_{l'}(\omega)\beta_j(\omega)}$$

$$\times \int_{-\infty}^{\infty} \mathbb{E}[C_{jl'}(0)C_{jl'}(s)] \cos(\beta_j(\omega)s)e^{-\beta_{l'}(\omega)|s|} \, ds . \quad (20.42)$$

Similarly,

$$\left\langle \mathbb{E}[G_{jl}^{(ab)}(\cdot)] \right\rangle = \left\langle \mathbb{E}[G_{jl}^{(ba)}(\cdot)] \right\rangle = 0 , \qquad \left\langle \mathbb{E}[G_{jl}^{(bb)}(\cdot)] \right\rangle = -i\kappa_j(\omega)\delta_{jl} .$$

The role of the evanescent modes in the limit $\varepsilon \to 0$ is completely characterized by the diagonal matrices $\langle \mathbb{E}[\mathbf{G}^{(aa)}] \rangle$ and $\langle \mathbb{E}[\mathbf{G}^{(bb)}] \rangle$. Note that the diagonal coefficients here are purely imaginary. This shows that the coupling with the evanescent modes does not remove energy from the propagating modes, nor does it affect the coupling between the propagating modes, but it introduces for each propagating mode a dispersive or frequency-dependent phase modulation.

20.2.6 The Forward-Scattering Approximation

The limit as $\varepsilon \to 0$ of $\mathbf{P}^{\varepsilon}(z)$ can be obtained and identified as a multidimensional diffusion process, meaning that the entries of the limit matrix satisfy a system of linear stochastic differential equations. The stochastic differential equations for the limit entries of $\mathbf{P}^{\varepsilon}_{ab}(z)$ are coupled to the limit entries of $\mathbf{P}^{\varepsilon}_{aa}(z)$ through the coefficients

$$\int_{-\infty}^{\infty} \mathbb{E}[C_{jl}(0)C_{jl}(z)] \cos\left((\beta_j(\omega) + \beta_l(\omega))z\right) dz \,, \quad j,l = 1, \dots, N(\omega)\,.$$

This is because the phase factors present in the matrices $\mathbf{H}^{(ab)}(z)$ and $\mathbf{G}^{(ab)}(z)$ are $\pm(\beta_j + \beta_l)z$. On the other hand, the stochastic differential equations for the limit entries of $\mathbf{P}^{\varepsilon}_{aa}(z)$ are coupled to each other through the coefficients

$$\int_{-\infty}^{\infty} \mathbb{E}[C_{jl}(0)C_{jl}(z)] \cos\left((\beta_j(\omega) - \beta_l(\omega))z\right) dz \,, \quad j,l = 1, \dots, N(\omega)\,.$$

This is because the phase factors present in the matrices $\mathbf{H}^{(aa)}(z)$ and $\mathbf{G}^{(aa)}(z)$ are $\pm(\beta_j - \beta_l)z$. If we assume that the power spectral density of the process ν (i.e. the Fourier transform of its z-autocorrelation function) possesses a cutoff frequency, then it is natural to consider the case in which

$$\int_{-\infty}^{\infty} \mathbb{E}[C_{jl}(0)C_{jl}(z)] \cos\left((\beta_j(\omega) + \beta_l(\omega))z\right) dz = 0\,, \quad j,l = 1, \dots, N(\omega)\,,$$

$$(20.43)$$

while (at least) some of the intracoupling coefficients (those with $|j - l| = 1$) are not zero. As a result of this assumption, the asymptotic coupling between $\mathbf{P}^{\varepsilon}_{aa}(z)$ and $\mathbf{P}^{\varepsilon}_{ab}(z)$ becomes zero. If we also take into account the initial condition $\mathbf{P}^{\varepsilon}_{ab}(z = 0) = \mathbf{0}$, then the limit of $\mathbf{P}^{\varepsilon}_{ab}(z)$ is $\mathbf{0}$.

In the forward-scattering approximation we neglect the left-going (backward) propagating modes. As we have just seen, it is valid in the limit $\varepsilon \to 0$ when the condition (20.43) holds. In this case we can consider the simplified coupled mode equation given by

$$\frac{d\hat{a}^{\varepsilon}}{dz} = \frac{1}{\varepsilon} \mathbf{H}^{(aa)}\left(\frac{z}{\varepsilon^2}\right) \hat{a}^{\varepsilon} + \mathbf{G}^{(aa)}\left(\frac{z}{\varepsilon^2}\right) \hat{a}^{\varepsilon}\,, \quad (20.44)$$

where $\mathbf{H}^{(aa)}$ and $\mathbf{G}^{(aa)}$ are the $N \times N$ complex matrices given by (20.27–20.31). The system (20.44) is supplemented with the initial condition $\hat{a}^{\varepsilon}_j(\omega, z = 0) = \hat{a}_{j,0}(\omega)$. In the forward-scattering approximation, conservation relation (20.22) becomes

$$\sum_{j=1}^{N} |\hat{a}^{\varepsilon}_j(L)|^2 = \sum_{j=1}^{N} |\hat{a}_{j,0}|^2\,. \quad (20.45)$$

As in the general case, we introduce the **transfer** or propagator matrix $\mathbf{T}^{\varepsilon}(\omega, z)$, which is the fundamental solution of (20.44). It is the $N(\omega) \times N(\omega)$ matrix solution of

$$\frac{d}{dz}\mathbf{T}^{\varepsilon}(\omega,z) = \left[\frac{1}{\varepsilon}\mathbf{H}^{(aa)}\left(\frac{z}{\varepsilon^2}\right) + \mathbf{G}^{(aa)}\left(\frac{z}{\varepsilon^2}\right)\right]\mathbf{T}^{\varepsilon}(\omega,z) \tag{20.46}$$

starting from $\mathbf{T}^{\varepsilon}(\omega,0) = \mathbf{I}$. The (j,l) entry of the transfer matrix is the transmission coefficient $T_{jl}^{\varepsilon}(\omega,L)$, i.e., the output amplitude of the mode j when the input wave is a pure l mode with amplitude one. The conservation relation (20.45) shows that $\mathbf{T}^{\varepsilon}(\omega,L)$ is a unitary matrix.

20.3 The Time-Harmonic Problem

In this section we consider the system of random differential equations (20.44) for a single frequency ω.

20.3.1 The Coupled Mode Diffusion Process

We will now apply the diffusion approximation results of Chapter 6 to the system (20.46). The limit distribution of $\hat{a}^{\varepsilon}$ as $\varepsilon \to 0$ is a diffusion on $\mathbb{C}^{N(\omega)}$. We will assume that the following nondegeneracy condition holds.

The longitudinal wave numbers β_j are distinct along with their sums and differences.

In this case the infinitesimal generator of the limit $\hat{a}$ has a simple form, provided we write it in terms of $\hat{a}$ and $\overline{\hat{a}}$, rather than in terms of the real and imaginary parts of $\hat{a}$. We thus get the following result.

Proposition 20.1. *The complex mode amplitudes* $(\hat{a}_j^{\varepsilon}(\omega,z))_{j=1,\ldots,N}$ *converge in distribution as* $\varepsilon \to 0$ *to a diffusion process* $(\hat{a}_j(\omega,z))_{j=1,\ldots,N}$ *whose infinitesimal generator is*

$$\mathcal{L} = \frac{1}{4}\sum_{j\neq l}\Gamma_{jl}^{(c)}(\omega)\left(A_{jl}\overline{A_{jl}} + \overline{A_{jl}}A_{jl}\right) + \frac{1}{2}\sum_{j,l}\Gamma_{jl}^{(1)}(\omega)A_{jj}\overline{A_{ll}}$$

$$+\frac{i}{4}\sum_{j\neq l}\Gamma_{jl}^{(s)}(\omega)(A_{ll} - A_{jj}) + i\sum_j \kappa_j(\omega)A_{jj}, \tag{20.47}$$

$$A_{jl} = \hat{a}_j\frac{\partial}{\partial\hat{a}_l} - \overline{\hat{a}_l}\frac{\partial}{\partial\overline{\hat{a}_j}} = -\overline{A_{lj}}. \tag{20.48}$$

Here we have defined the complex derivatives in the standard way: if $z = x+iy$, *then* $\partial_z = (1/2)(\partial_x - i\partial_y)$ *and* $\partial_{\bar{z}} = (1/2)(\partial_x + i\partial_y)$. *The coefficients* $\Gamma^{(c)}$, $\Gamma^{(s)}$, *and* $\Gamma^{(1)}$ *are given by*

$$\Gamma_{jl}^{(c)}(\omega) = \frac{\omega^4\gamma_{jl}^{(c)}(\omega)}{4\bar{c}^4\beta_j(\omega)\beta_l(\omega)} \quad \text{if } j \neq l, \tag{20.49}$$

$$\Gamma_{jj}^{(c)}(\omega) = -\sum_{n\neq j}\Gamma_{jn}^{(c)}(\omega), \tag{20.50}$$

$$\gamma_{jl}^{(c)}(\omega) = \int_{-\infty}^{\infty} \cos\left((\beta_j(\omega) - \beta_l(\omega))z\right) \mathbb{E}[C_{jl}(0)C_{jl}(z)]\, dz\,, \quad (20.51)$$

$$C_{jl}(z) = \int_{\mathcal{D}} \phi_j(\mathbf{x})\phi_l(\mathbf{x})\nu(\mathbf{x}, z)\, d\mathbf{x}\,, \quad (20.52)$$

$$\Gamma_{jl}^{(s)}(\omega) = \frac{\omega^4 \gamma_{jl}^{(s)}(\omega)}{4\bar{c}^4 \beta_j(\omega)\beta_l(\omega)} \ \ if \ j \neq l\,, \quad (20.53)$$

$$\Gamma_{jj}^{(s)}(\omega) = -\sum_{n \neq j} \Gamma_{jn}^{(s)}(\omega)\,, \quad (20.54)$$

$$\gamma_{jl}^{(s)}(\omega) = 2\int_{0}^{\infty} \sin\left((\beta_j(\omega) - \beta_l(\omega))z\right) \mathbb{E}[C_{jl}(0)C_{jl}(z)]\, dz\,, \quad (20.55)$$

$$\Gamma_{jl}^{(1)}(\omega) = \frac{\omega^4 \gamma_{jl}^{(1)}}{4\bar{c}^4 \beta_j(\omega)\beta_l(\omega)} \ \ for \ all \ j, l\,, \quad (20.56)$$

$$\gamma_{jl}^{(1)} = \int_{-\infty}^{\infty} \mathbb{E}[C_{jj}(0)C_{ll}(z)]\, dz\,. \quad (20.57)$$

The coefficients of the second derivatives of the generator $\mathcal{L}$ are homogeneous of degree two, while the coefficients of the first derivatives are homogeneous of degree one. As a consequence we can write closed differential equations for moments of any order. In the next section we shall introduce explicitly the equations for the first-, second-, and fourth-order moments. Before considering moments we will discuss some qualitative properties of the diffusion process $\hat{a}$.

The coefficients $\gamma_{jl}^{(c)}$, and thus $\Gamma_{jl}^{(c)}$, are proportional to the power spectral densities of the stationary process $C_{jl}(z)$ for $j \neq l$. They are therefore nonnegative. We shall assume that the off-diagonal entries of the matrix $\Gamma^{(c)}$ are positive.

The infinitesimal generator satisfies

$$\forall j, n, \quad A_{jn}\left(\sum_{l=1}^{N} |\hat{a}_l|^2\right) = 0 \implies \mathcal{L}\left(\sum_{l=1}^{N} |\hat{a}_l|^2\right) = 0\,.$$

This implies that the diffusion process is supported on a sphere of $\mathbb{C}^N$, whose radius R_0 is determined by the initial condition $R_0^2 = \sum_{l=1}^{N} |\hat{a}_{l,0}(\omega)|^2$. The operator $\mathcal{L}$ is not self-adjoint on the sphere, because of the term $\Gamma^{(s)}$ in (20.47). This means that the process is not reversible. However, the uniform measure on the sphere is invariant, and the generator is strongly elliptic. From the theory of irreducible Markov processes with compact state space, we know that the process is ergodic, which means in particular that for large z the limit process $\hat{a}(z)$ converges to the uniform distribution over the sphere of radius R_0. This fact can be used to compute the limit distribution of the mode powers $(|\hat{a}_j|^2)_{j=1,\ldots,N}$ for large z, which is the uniform distribution over $\mathcal{H}_N$,

$$\mathcal{H}_N = \left\{ (P_j)_{j=1,\ldots,N}, \; P_j \geq 0, \; \sum_{j=1}^N P_j = R_0^2 \right\}. \tag{20.58}$$

We carry out a more detailed analysis that is valid for any z in Section 20.3.3. We address now the computation of the first moment in the asymptotic limit $\varepsilon \to 0$.

20.3.2 Mean Mode Amplitudes

From Proposition 20.1 we get the following result.

Proposition 20.2. *The expected values of the mode amplitudes* $\mathbb{E}[\hat{a}_j^\varepsilon(\omega, z)]$ *converge as* $\varepsilon \to 0$ *to* $\mathbb{E}[\hat{a}_j(\omega, z)]$, *which is given by*

$$\mathbb{E}[\hat{a}_j(\omega, z)] = \exp\left(\frac{\Gamma_{jj}^{(c)}(\omega)z}{2} - \frac{\Gamma_{jj}^{(1)}(\omega)z}{2} + \frac{i\Gamma_{jj}^{(s)}(\omega)z}{2} + i\kappa_j(\omega)z \right) \hat{a}_{j0}(\omega).$$

The real part of the exponential factor is $[\Gamma_{jj}^{(c)}(\omega) - \Gamma_{jj}^{(1)}(\omega)]z/2$. The coefficient $\Gamma_{jj}^{(c)}(\omega)$ is negative. The coefficient $\Gamma_{jj}^{(1)}(\omega)$ is nonnegative because it is proportional to the power spectral density of C_{jj} at 0 frequency. As a result, the damping coefficient $[\Gamma_{jj}^{(c)}(\omega) - \Gamma_{jj}^{(1)}(\omega)]/2$ is negative, and therefore the mean mode amplitude decays exponentially with propagation distance. The exponential decay rate is given by

$$\sup_{j=1,\ldots,N} |\mathbb{E}[\hat{a}_j(\omega, z)]| \leq \sup_{j=1,\ldots,N} |\hat{a}_{j,0}| e^{-\lambda_1 z}$$

with

$$\lambda_1 = \frac{1}{2} \inf_{j=1,\ldots,N} \left\{ \Gamma_{jj}^{(1)} - \Gamma_{jj}^{(c)} \right\} = \inf_{j=1,\ldots,N} \left\{ \sum_{l=1}^N \frac{\omega^4 \gamma_{jl}^{(c)}(\omega)}{8\bar{c}^4 \beta_j(\omega)\beta_l(\omega)} \right\}.$$

This exponential decay rate implies that the transmitted field loses its coherence. The study of the incoherent field requires the analysis of the higher moments of the mode amplitude.

20.3.3 Coupled Power Equations

The generator of the limit process $\hat{a}$ possesses an important symmetry, which follows from noting that when applying the generator to a function of $(|\hat{a}_1|^2, \ldots, |\hat{a}_N|^2)$, we obtain another function of $(|\hat{a}_1|^2, \ldots, |\hat{a}_N|^2)$. This implies that the limit process $(|\hat{a}_1(z)|^2)_{j=1,\ldots,N}$ is itself a Markov process.

Proposition 20.3. *The mode powers* $(|\hat{a}_j^\varepsilon(\omega, z)|^2)_{j=1,\dots,N}$ *converge in distribution as* $\varepsilon \to$ *to the diffusion process* $(\check{P}_j(\omega, z))_{j=1,\dots,N}$ *whose infinitesimal generator is*

$$
\mathcal{L}_P = \sum_{j \neq l} \Gamma_{jl}^{(c)}(\omega) P_l \frac{\partial}{\partial P_j} \left[P_j \left(\frac{\partial}{\partial P_j} - \frac{\partial}{\partial P_l} \right) \right]
$$

$$
= \sum_{j \neq l} \Gamma_{jl}^{(c)}(\omega) \left[P_l P_j \left(\frac{\partial}{\partial P_j} - \frac{\partial}{\partial P_l} \right) \frac{\partial}{\partial P_j} + (P_l - P_j) \frac{\partial}{\partial P_j} \right] . \quad (20.59)
$$

As pointed out above, the diffusion process $(P_j(\omega, z))_{j=1,\dots,N}$ is supported in $\mathcal{H}_N$. As a first application of this result, we compute the mean mode powers:

$$
P_j^{(1)}(\omega, z) = \mathbb{E}[P_j(\omega, z)] = \lim_{\varepsilon \to 0} \mathbb{E}[|\hat{a}_j^\varepsilon(\omega, z)|^2] .
$$

Using the generator $\mathcal{L}_P$ we get the following proposition.

Proposition 20.4. *The mean mode powers* $\mathbb{E}[|\hat{a}_j^\varepsilon(\omega, z)|^2]$ *converge to* $P_j^{(1)}(\omega, z)$, *which is the solution of the linear system*

$$
\frac{\partial P_j^{(1)}}{\partial z} = \sum_{n \neq j} \Gamma_{jn}^{(c)}(\omega) \left(P_n^{(1)} - P_j^{(1)} \right) , \quad (20.60)
$$

starting from $P_j^{(1)}(\omega, z = 0) = |\hat{a}_{j,0}|^2$, $j = 1, \dots, N$.

The solution of this system can be written in terms of the exponential of the matrix $\Gamma^{(c)}$:

$$
P^{(1)}(\omega, z) = \exp\left(\Gamma^{(c)}(\omega) z \right) P^{(1)}(\omega, 0) .
$$

We note that the vector $P^{(1)}(\omega, z)$ has a probabilistic interpretation, which we consider in detail in Section 20.6.2. We give here some of its basic properties. First, the matrix $\Gamma^{(c)}$ is symmetric and real, its off-diagonal terms are positive, and its diagonal terms are negative. The sums over the rows and columns are all zero. As a consequence of the Perron–Frobenius theorem, $\Gamma^{(c)}$ has zero as a simple eigenvalue, and all other eigenvalues $\lambda_N \leq \cdots \leq \lambda_2 < 0$ are negative. The eigenvector associated with the zero eigenvalue is the uniform vector $(1, \dots, 1)^T$. This shows that

$$
\sup_{j=1,\dots,N} \left| P_j^{(1)}(\omega, z) - \frac{1}{N} R_0^2 \right| \leq C e^{-\lambda_2 z} ,
$$

where $R_0^2 = \sum_{j=1}^N |\hat{a}_{j,0}|^2$. In words, the mean mode powers converge exponentially fast to the uniform distribution, which means that we have asymptotic equipartition of mode energy. The Perron–Frobenius theorem also provides an expression for the rate of convergence

$$\lambda_2 = \inf \left\{ -B^T \Gamma^{(c)} B, \ B \in \mathbb{R}^N \text{ such that } \sum_{j=1}^{N} B_j = 0, \ \sum_{j=1}^{N} B_j^2 = 1 \right\}.$$

Using the relation $\Gamma_{jj}^{(c)} = -\sum_{l \neq j} \Gamma_{jl}^{(c)}$, we can get an estimate for λ_2 of the form

$$\lambda_2 \leq \inf_{j=1,\ldots,N} \left\{ -\Gamma_{jj}^{(c)} \right\} = \inf_{j=1,\ldots,N} \left\{ \sum_{l \neq j} \frac{\omega^4 \gamma_{jl}^{(c)}(\omega)}{4\bar{c}^4 \beta_j(\omega) \beta_l(\omega)} \right\}.$$

This estimate implies that $\lambda_2 \leq 2\lambda_1$, which means that the convergence rate to equilibrium of the average mode powers is slower than the convergence rate of the square mean amplitudes, although both are of the same order of magnitude. We shall denote by L_{equip} the inverse of λ_2, which is the equipartition distance for the mean mode powers.

20.3.4 Fluctuations Theory

Proposition 20.3 also allows us to study the fluctuations of the mode powers by looking at the fourth-order moments of the mode amplitudes:

$$P_{jl}^{(2)}(\omega, z) = \lim_{\varepsilon \to 0} \mathbb{E} \left[|\hat{a}_j^\varepsilon(\omega, z)|^2 |\hat{a}_l^\varepsilon(\omega, z)|^2 \right] = \mathbb{E}[P_j(\omega, z) P_l(\omega, z)].$$

Using the generator $\mathcal{L}_P$ we get a system of ordinary differential equations for limit fourth moments $(P_{jl}^{(2)})_{j,l=1,\ldots,N}$, which has the form

$$\frac{dP_{jj}^{(2)}}{dz} = \sum_{n \neq j} \Gamma_{jn}^{(c)} \left(4P_{jn}^{(2)} - 2P_{jj}^{(2)} \right),$$

$$\frac{dP_{jl}^{(2)}}{dz} = -2\Gamma_{jl}^{(c)} P_{jl}^{(2)} + \sum_n \Gamma_{ln}^{(c)} \left(P_{jn}^{(2)} - P_{jl}^{(2)} \right) + \sum_n \Gamma_{jn}^{(c)} \left(P_{ln}^{(2)} - P_{jl}^{(2)} \right),$$

for $j \neq l$. The initial conditions are $P_{jl}^{(2)}(z = 0) = |\hat{a}_{j,0}|^2 |\hat{a}_{l,0}|^2$. This is a system of linear ordinary differential equations with constant coefficients that can be solved by computing the exponent of the evolution matrix.

It is straightforward to check that the function $P_{jl}^{(2)} \equiv 1 + \delta_{jl}$ is a stationary solution of the fourth moment system. Using the positivity of $\Gamma_{jl}^{(c)}$, $j \neq l$, we conclude that this stationary solution is asymptotically stable, which means that the solution $P_{jl}^{(2)}(z)$ starting from $P_{jl}^{(2)}(z = 0) = |\hat{a}_{j,0}|^2 |\hat{a}_{l,0}|^2$ converges as $z \to \infty$ to

$$P_{jl}^{(2)}(z) \overset{z \to \infty}{\longrightarrow} \begin{cases} \dfrac{1}{N(N+1)} R_0^4 & \text{if } j \neq l, \\ \dfrac{2}{N(N+1)} R_0^4 & \text{if } j = l, \end{cases}$$

where $R_0^2 = \sum_{j=1}^{N} |\hat{a}_{j,0}|^2$. This implies that the normalized correlation

$$\text{Cor}(P_j, P_l)(z) := \frac{P_{jl}^{(2)}(z) - P_j^{(1)}(z)P_l^{(1)}(z)}{P_j^{(1)}(z)P_l^{(1)}(z)}$$

has the following asymptotic form:

$$\text{Cor}(P_j, P_l)(z) \xrightarrow{z \to \infty} \begin{cases} -\dfrac{1}{N+1} & \text{if } j \neq l, \\ \dfrac{N-1}{N+1} & \text{if } j = l. \end{cases}$$

We see from the $j \neq l$ result that if, in addition, the number of modes N becomes large, then the mode powers become uncorrelated.

The large-z behavior for $j = l$ shows that whatever the number of modes N, the mode powers P_j are not statistically stable quantities. The full distribution function of any mode power can be calculated in the asymptotic "z large" using the fact that the invariant measure of $(P_j)_{j=1,\dots,N}$ is the uniform measure over $\mathcal{H}_N$. Thus the probability distribution of P_1 is the marginal distribution obtained by integrating the normalized uniform distribution over $(P_j)_{j=2,\dots,N}$. By symmetry, all the P_j have the same asymptotic distribution, whose density is

$$\psi(p_1) = (N-1)R_0^2(R_0^2 - p_1)^{N-2}\mathbf{1}_{[0,R_0^2]}(p_1).$$

Therefore, in the limit $z \to \infty$ the moments of P_1 are given by

$$\mathbb{E}[P_1(z)^k] \xrightarrow{z \to \infty} \int p_1^k \psi(p_1)\, dp_1 = \frac{k!(N-1)!}{(N+k-1)!}R_0^{2k}.$$

When N is large, we have $k!(N-1)!/(N+k-1)! \sim k!/N^k$. This shows that for large N, the distribution of $P_1(z)$ is close to that of an exponential random variable with mean R_0^2/N.

20.4 Broadband Pulse Propagation in Waveguides

Bandwidth plays a basic role in the propagation of pulses in a waveguide because of dispersion. There are two types of dispersion. First, the modes travel with different group velocities $1/\beta_j'(\omega)$ for different modes $j = 1, \dots, N(\omega)$, which gives strong pulse spreading whatever the bandwidth. Second, each mode is dispersive because $\beta_j(\omega)$ is not linear in ω. This second type of dispersion depends strongly on the bandwidth.

20.4.1 Integral Representation of the Transmitted Field

We assume that a point source located inside the waveguide at $(z = 0, \mathbf{x} = \mathbf{x}_0)$ emits a pulse with carrier frequency ω_0 and bandwidth of order ε^q, $q > 0$,

$$\mathbf{F}^{\varepsilon}(t, \mathbf{x}, z) = f^{\varepsilon}(t)\delta(\mathbf{x} - \mathbf{x}_0)\delta(z)\mathbf{e}_z .$$

The pulse has the form

$$f^{\varepsilon}(t) = f(\varepsilon^q t)e^{i\omega_0 t} \tag{20.61}$$

in the time domain, and in the Fourier domain

$$\hat{f}^{\varepsilon}(\omega) = \frac{1}{\varepsilon^q}\hat{f}\left(\frac{\omega - \omega_0}{\varepsilon^q}\right) .$$

This point source generates evanescent modes, left-going propagating modes, which we do not need to consider, since they propagate in a homogeneous half-space, and right-going modes, which we analyze. As shown in Section 20.1.4, the interface conditions at $z = 0$, which are initial conditions in the forward-scattering approximation, have the form

$$\hat{a}_j^{\varepsilon}(\omega, 0) = \frac{1}{2}\sqrt{\beta_j(\omega)}\hat{f}^{\varepsilon}(\omega)\phi_j(\mathbf{x}_0) , \text{ for } j \leq N(\omega) .$$

The transmitted field observed at time t is therefore

$$p_{\mathrm{tr}}\left(t, \mathbf{x}, \frac{L}{\varepsilon^2}\right) = \frac{1}{4\pi}\int \sum_{j,l=1}^{N(\omega)} \frac{\sqrt{\beta_l(\omega)}}{\sqrt{\beta_j(\omega)}}\phi_j(\mathbf{x})\phi_l(\mathbf{x}_0)\hat{f}^{\varepsilon}(\omega)T_{jl}^{\varepsilon}(\omega)e^{i\beta_j(\omega)\frac{L}{\varepsilon^2} - i\omega t} \, d\omega .$$

$$\tag{20.62}$$

We assume in this section that $q = 1$. The analysis that follows can be carried out for any $0 < q < 2$. In this general case the pulse width is of order $\varepsilon^{-q} \ll \varepsilon^{-2}$, which means that it is much smaller than the propagation distance. As a result, the modes are separated in time by the modal dispersion during propagation. The restriction $q > 0$ allows us to simplify the expression for the transmitted field by freezing the number of modes to the one for the carrier frequency. The case $q = 2$, which is the narrowband case, is addressed in Section 20.8.

We observe the field in a time window of order $1/\varepsilon$, which is comparable to the pulse width, and centered at time t_0/ε^2, which is of order the travel time to go to distances of order $1/\varepsilon^2$:

$$p_{\mathrm{tr}}^{\varepsilon}(t_0, t, \mathbf{x}, L) := p_{\mathrm{tr}}\left(\frac{t_0}{\varepsilon^2} + \frac{t}{\varepsilon}, \mathbf{x}, \frac{L}{\varepsilon^2}\right) ,$$

$$p_{\mathrm{tr}}^{\varepsilon}(t_0, t, \mathbf{x}, L) = \frac{1}{4\pi\varepsilon}\int \sum_{j,l=1}^{N(\omega)} \frac{\sqrt{\beta_l(\omega)}}{\sqrt{\beta_j(\omega)}}\phi_j(\mathbf{x})\phi_l(\mathbf{x}_0)\hat{f}\left(\frac{\omega - \omega_0}{\varepsilon}\right)T_{jl}^{\varepsilon}(\omega)$$

$$\times e^{i\frac{\beta_j(\omega)L - \omega t_0}{\varepsilon^2}}e^{-i\frac{\omega t}{\varepsilon}} \, d\omega .$$

We change variables to $\omega = \omega_0 + \varepsilon h$ and we expand $\beta_j(\omega_0 + \varepsilon h)$ with respect to ε:

$$p_{\mathrm{tr}}^{\varepsilon}(t_0, t, \mathbf{x}, L) = \frac{1}{4\pi} \sum_{j,l=1}^{N(\omega_0)} \frac{\sqrt{\beta_l(\omega_0)}}{\sqrt{\beta_j(\omega_0)}} \phi_j(\mathbf{x})\phi_l(\mathbf{x}_0) e^{i\frac{\beta_j(\omega_0)L - \omega_0 t_0}{\varepsilon^2}} e^{-i\frac{\omega_0 t}{\varepsilon}}$$

$$\times \int \hat{f}(h) T_{jl}^{\varepsilon}(\omega_0 + \varepsilon h) e^{i\frac{[\beta_j'(\omega_0)L - t_0]h}{\varepsilon}} e^{i[\beta_j''(\omega_0)L\frac{h^2}{2} - ht]} \, dh \,. \quad (20.63)$$

Note that here we have also replaced $N(\omega_0 + \varepsilon h)$ by $N(\omega_0)$, which is a legitimate approximation for ε small.

20.4.2 Broadband Pulse Propagation in a Homogeneous Waveguide

We first consider the expression (20.63) in a homogeneous waveguide where $T_{jl}^{\varepsilon} = \delta_{jl}$. Because of the fast phase $\exp i[\beta_j'(\omega_0)L - t_0]h/\varepsilon$ in (20.63), the jth mode is different from zero only when the time window is centered at $t_0 = \bar{t}_j$,

$$\bar{t}_j := \beta_j'(\omega_0)L \,, \quad (20.64)$$

which is the travel time of the jth mode. This follows from the Riemann-Lebesgue lemma. Near this time, only one term survives in (20.63) and the transmitted field is given by

$$p_{\mathrm{tr}}^{\varepsilon}(\bar{t}_j, t, \mathbf{x}, L) = \frac{1}{4\pi} \phi_j(\mathbf{x})\phi_j(\mathbf{x}_0) e^{i\frac{\beta_j(\omega_0)L - \omega_0 t_0}{\varepsilon^2} - i\frac{\omega_0 t}{\varepsilon}} \int \hat{f}(h) e^{i[\beta_j''(\omega_0)L\frac{h^2}{2} - ht]} \, dh$$

$$= \frac{1}{2} \phi_j(\mathbf{x})\phi_j(\mathbf{x}_0) e^{i\frac{\beta_j(\omega_0)L - \omega_0 t_0}{\varepsilon^2} - i\frac{\omega_0 t}{\varepsilon}} K_{j,L} * f(t) \,, \quad (20.65)$$

where $K_{j,L}(t)$ is the function whose Fourier transform is

$$\hat{K}_{j,L}(\omega) = e^{i\beta_j''(\omega_0)L\frac{\omega^2}{2}} \,.$$

This means that for ε small we observe a train of transmitted pulses that are well separated from each other. Each pulse is that of a single mode, and it travels with the group velocity of this mode $1/\beta_j'(\omega_0)$. The support of the pulse is of order the original pulse width, that is to say, ε^{-1}. However, each pulse mode is also dispersed by the convolution kernel $K_{j,L}(t)$.

20.4.3 The Stable Wave Field in a Random Waveguide

We next consider the transmitted field (20.63) in a random waveguide. We fix again a time window whose width is comparable to that of the input pulse and centered at the travel time $t_0 = \bar{t}_j$ of the jth mode.

Proposition 20.5. *The transmitted field observed around time $t_0 = \bar{t}_j$ has the following form:*

$$p^{\varepsilon}(\bar{t}_j, t, \mathbf{x}, L) = \widetilde{p}_{\mathrm{tr},j}^{\varepsilon}(t, \mathbf{x}, L) e^{i\frac{\beta_j(\omega_0)L - \omega_0 t_0}{\varepsilon^2} - i\frac{\omega_0 t}{\varepsilon}} \,.$$

The field $\widetilde{p}_{\mathrm{tr},j}^{\varepsilon}(t,\mathbf{x},L)$ converges in distribution as $\varepsilon \to 0$ as a continuous function in $(t,\mathbf{x},L)$ to

$$\widetilde{p}_{\mathrm{tr},j}(t,\mathbf{x},L) = \frac{1}{2}\phi_j(\mathbf{x})\phi_j(\mathbf{x}_0)e^{iW^{(j)}(L)}K_{j,L} * f(t)\,, \qquad (20.66)$$

where $W^{(j)}(L)$ is a Brownian motion with variance

$$\mathbb{E}\left[\left(W^{(j)}(L)\right)^2\right] = \Gamma_{jj}^{(1)}(\omega_0)L\,,$$

and $K_{j,L}$ is a deterministic convolution kernel whose Fourier transform is

$$\hat{K}_{j,L}(\omega) = \exp\left(\frac{1}{2}\Gamma_{jj}^{(c)}(\omega_0)L + \frac{i}{2}\Gamma_{jj}^{(s)}(\omega_0)L + i\kappa_j(\omega_0)L + i\beta_j''(\omega_0)L\frac{\omega^2}{2}\right)\,.$$

The coefficients $\Gamma_{jj}^{(c)}$, $\Gamma_{jj}^{(s)}$, and $\Gamma_{jj}^{(1)}$ are defined in Proposition 20.1. In order to completely describe the joint distribution of the transmitted field for all the modes, we note that the random process $(W^{(j)}(L))_{j=1,\ldots,N}$ is an N-dimensional Brownian motion with covariance matrix

$$\mathbb{E}\left[W^{(j)}(L)W^{(l)}(L)\right] = \Gamma_{jl}^{(1)}(\omega_0)L\,.$$

This proposition looks somewhat similar to the pulse-stabilization theory presented in Chapter 8 in a one-dimensional random medium. We get stabilization of the pulse up to a random phase; that is, the pulse intensity observed near the travel time of each mode is deterministic. The random phase is characterized in terms of Brownian motions. However, the pulse intensities attenuate exponentially with propagation distance and spread dispersively through the kernel $K_{j,L}$. There is no diffusion for the deterministic pulse intensity as there is in the pulse-stabilization theory in the one-dimensional case. The proof of the proposition goes along the same lines as the one presented in detail in Chapter 8, so we do not give it here.

We note that we can identify the transmitted field as the solution of an effective stochastic equation. We write

$$p_{\mathrm{tr}}^{\varepsilon}\left(\bar{t}_j,t,\mathbf{x},L\right) \overset{\varepsilon\to 0}{\sim} \widetilde{p}_{\mathrm{tr},j}(t,\mathbf{x},z=L)e^{i\frac{\beta_j(\omega_0)L-\omega_0 t_0}{\varepsilon^2}-i\frac{\omega_0 t}{\varepsilon}}\,,$$

and note that the field components $\widetilde{p}_{\mathrm{tr},j}$ obey the stochastic system of Schrödinger equations

$$d\widetilde{p}_{\mathrm{tr},j} + \beta_j'(\omega_0)\frac{\partial\widetilde{p}_{\mathrm{tr},j}}{\partial t}dz = \left(\frac{\Gamma_{jj}^{(c)}(\omega_0)}{2} + i\frac{\Gamma_{jj}^{(s)}(\omega_0)}{2} + i\kappa_j(\omega_0)\right)\widetilde{p}_{\mathrm{tr},j}dz$$

$$+i\widetilde{p}_{\mathrm{tr},j}\circ dW^{(j)}(z) - \frac{i\beta_j''(\omega_0)}{2}\frac{\partial^2\widetilde{p}_{\mathrm{tr},j}}{\partial t^2}dz\,,$$

starting from $\widetilde{p}_{\mathrm{tr},j}(t, \mathbf{x}, z = 0) = \phi_j(\mathbf{x})\phi_j(\mathbf{x}_0)f(t)/2$. From this equation we get immediately the equation satisfied by the coherent field, or mean field

$$\mathbb{E}[p_{\mathrm{tr}}^{\varepsilon}(t, \mathbf{x}, L)] \overset{\varepsilon \to 0}{\sim} \sum_{j=1}^{N} p_{\mathrm{mf},j}(t, \mathbf{x}, z = L)e^{i\frac{\beta_j(\omega_0)L - \omega_0 t_0}{\varepsilon^2} - i\frac{\omega_0 t}{\varepsilon}}$$

with $p_{\mathrm{mf},j}(t, \mathbf{x}, z = L)$ the solution of the Schrödinger equation with damping

$$\frac{\partial p_{\mathrm{mf},j}}{\partial z} + \beta_j'(\omega_0)\frac{\partial p_{\mathrm{mf},j}}{\partial t} = \left(\frac{\Gamma_{jj}^{(c)}(\omega_0)}{2} - \frac{\Gamma_{jj}^{(1)}(\omega_0)}{2} + i\frac{\Gamma_{jj}^{(s)}(\omega_0)}{2} + i\kappa_j(\omega_0) \right) p_{\mathrm{mf},j}$$
$$- \frac{i\beta_j''(\omega_0)}{2}\frac{\partial^2 p_{\mathrm{mf},j}}{\partial t^2} .$$

The additional damping term $-\Gamma_{jj}^{(1)}(\omega_0)/2$ on the right-hand side is the Itô–Stratonovich correction that appears in the expectation of the previous stochastic Schrödinger equation. It corresponds to the averaging of the random phase given in terms of the Brownian motion $W^{(j)}$.

20.5 Time Reversal for a Broadband Pulse

20.5.1 Time Reversal in Waveguides

We will now consider time reversal in a waveguide in a setup similar to the one that we analyzed in Chapter 12. A point source located in the plane $z = 0$ at the lateral position $\mathbf{x}_0$ emits a pulse $f^{\varepsilon}(t)$ of the form (20.61). A time-reversal mirror is located in the plane $z = L/\varepsilon^2$ and occupies the subdomain $\mathcal{D}_M \subset \mathcal{D}$. The field is recorded for a time interval $[t_0/\varepsilon^2, t_1/\varepsilon^2]$ at the time-reversal mirror, and it is reemitted time-reversed into the waveguide toward the original source location at $z = 0$.

The transmitted wave observed at time t is

$$p_{\mathrm{tr}}\left(t, \mathbf{x}, \frac{L}{\varepsilon^2}\right) = \frac{1}{4\pi} \sum_{j,l=1}^{N} \frac{\sqrt{\beta_l}}{\sqrt{\beta_j}} \phi_j(\mathbf{x})\phi_l(\mathbf{x}_0) \int \hat{f}^{\varepsilon}(\omega)T_{jl}^{\varepsilon}(\omega)e^{i\beta_j(\omega)\frac{L}{\varepsilon^2} - i\omega t}\, d\omega ,$$

which is the same as (20.62). The time-reversal mirror records the field from time t_0/ε^2 up to time t_1/ε^2, and time-reverses it. The new source at the time-reversal mirror that will generate the back-propagating waves is

$$\mathbf{F}_{\mathrm{TR}}^{\varepsilon}(t, \mathbf{x}, z) = -f_{\mathrm{TR}}^{\varepsilon}(t, \mathbf{x})\delta\left(z - \frac{L}{\varepsilon^2}\right)\mathbf{e}_z ,$$

with

$$f_{\mathrm{TR}}^{\varepsilon}(t, \mathbf{x}) = p_{\mathrm{tr}}\left(\frac{t_1}{\varepsilon^2} - t, \mathbf{x}, \frac{L}{\varepsilon^2}\right)G_1(t_1 - \varepsilon^2 t)G_2(\mathbf{x}) ,$$

where G_1 is the time-window function, of the form $G_1(t) = \mathbf{1}_{[t_0,t_1]}(t)$, and G_2 is the spatial-window function $G_2(\mathbf{x}) = \mathbf{1}_{\mathcal{D}_M}(\mathbf{x})$. The source $\mathbf{F}^\varepsilon_{\text{TR}}$ points in the negative z-direction, toward the initial source location $z = 0$. The Fourier transform of the source intensity is

$$\hat{f}^\varepsilon_{\text{TR}}(\omega, \mathbf{x}) = \frac{1}{4\pi\varepsilon^2} \int \sum_{j,l=1}^{N} \frac{\sqrt{\beta_l}}{\sqrt{\beta_j}} G_2(\mathbf{x})\phi_j(\mathbf{x})\phi_l(\mathbf{x}_0)\overline{\hat{f}^\varepsilon(\omega')}\overline{T^\varepsilon_{jl}(\omega')}$$

$$\times \overline{\hat{G}_1}\left(\frac{\omega-\omega'}{\varepsilon^2}\right) e^{-i\beta_j(\omega')\frac{L}{\varepsilon^2}+i\omega\frac{t_1}{\varepsilon^2}}\, d\omega'\,.$$

This source generates both propagating and evanescent modes. The left-going propagating modes have amplitudes

$$\hat{b}_m(\omega) = \frac{\sqrt{\beta_m(\omega)}}{2} \int_{\mathcal{D}} \hat{f}^\varepsilon_{\text{TR}}(\omega, \mathbf{x})\phi_m(\mathbf{x})\, d\mathbf{x}\, e^{i\beta_m(\omega)\frac{L}{\varepsilon^2}}\,,$$

which when substituting the expression for the source intensity gives

$$\hat{b}_m(\omega) = \frac{1}{8\pi\varepsilon^2} \sum_{j,l=1}^{N} \frac{\sqrt{\beta_l\beta_m}}{\sqrt{\beta_j}} M_{mj}\phi_l(\mathbf{x}_0) \int \overline{\hat{f}^\varepsilon(\omega')}\overline{T^\varepsilon_{jl}(\omega')}$$

$$\times \overline{\hat{G}_1}\left(\frac{\omega-\omega'}{\varepsilon^2}\right) e^{i[\beta_m(\omega)-\beta_j(\omega')]\frac{L}{\varepsilon^2}+i\omega\frac{t_1}{\varepsilon^2}}\, d\omega'\,.$$

Here the coupling coefficients M_{jl} are given by

$$M_{jl} = \int_{\mathcal{D}} \phi_j(\mathbf{x})G_2(\mathbf{x})\phi_l(\mathbf{x})\, d\mathbf{x}\,. \tag{20.67}$$

We have explicit formulas for the coupling coefficients M_{jl} in two cases:

- If the mirror spans the complete cross section $\mathcal{D}$ of the waveguide, then we have $G_2(\mathbf{x}) = 1$ and $M_{jl} = 1$ if $j = l$ and 0 otherwise.
- If the mirror is pointlike at $\mathbf{x} = \mathbf{x}_1$, meaning $G_2(\mathbf{x}) = |\mathcal{D}|\delta(\mathbf{x}-\mathbf{x}_1)$, with the factor $|\mathcal{D}|$ added for dimensional consistency, then $M_{jl} = |\mathcal{D}|\phi_j(\mathbf{x}_1)\phi_l(\mathbf{x}_1)$.

The back-propagating wave is left-going, starting from L/ε^2, and it is given by

$$\hat{p}_{\text{inc(TR)}}\left(\omega, \mathbf{x}, \frac{L}{\varepsilon^2}\right) = \sum_{m=1}^{N} \frac{\hat{b}_m(\omega)}{\sqrt{\beta_m}}\phi_m(\mathbf{x})e^{-i\beta_m(\omega)\frac{L}{\varepsilon^2}}\,.$$

The refocused field at 0, in the Fourier domain, is given by

$$\hat{p}_{\text{ref(TR)}}(\omega, \mathbf{x}, 0) = \sum_{m,n=1}^{N} \frac{(\mathbf{T}^\varepsilon)^T_{n,m}(\omega)\hat{b}_m(\omega)}{\sqrt{\beta_n}}\phi_n(\mathbf{x})\,. \tag{20.68}$$

Here $(\mathbf{T}^\varepsilon)^T(\omega)$ is the transfer matrix for the left-going modes propagating from L/ε^2 to 0, and it is the transpose of $\mathbf{T}^\varepsilon(\omega)$. This follows from the unitarity of the transfer matrix $\mathbf{T}^\varepsilon(\omega)$. From (20.68) we obtain

$$\hat{p}_{\mathrm{ref(TR)}}(\omega, \mathbf{x}, 0) = \frac{1}{8\pi\varepsilon^2} \sum_{j,l,m,n=1}^{N} \frac{\sqrt{\beta_l \beta_m}}{\sqrt{\beta_j \beta_n}} M_{mj} \phi_n(\mathbf{x}) \phi_l(\mathbf{x}_0)$$

$$\times \int \overline{\hat{f}^\varepsilon(\omega')} \overline{T_{jl}^\varepsilon(\omega')} T_{mn}^\varepsilon(\omega) \overline{\hat{G}_1}\left(\frac{\omega - \omega'}{\varepsilon^2}\right) e^{i[\beta_m(\omega) - \beta_j(\omega')]\frac{L}{\varepsilon^2} + i\omega\frac{t_1}{\varepsilon^2}} d\omega'.$$

After the change of variable $\omega' = \omega - \varepsilon^2 h$, the refocused field observed at time t_{obs} reads

$$p_{\mathrm{ref(TR)}}(t_{\mathrm{obs}}, \mathbf{x}, 0) = \frac{1}{8\pi^2} \sum_{j,l,m,n=1}^{N} \frac{\sqrt{\beta_l \beta_m}}{\sqrt{\beta_j \beta_n}} M_{mj} \phi_n(\mathbf{x}) \phi_l(\mathbf{x}_0) \int \int \overline{\hat{f}^\varepsilon(\omega - \varepsilon^2 h)}$$

$$\times \overline{T_{jl}^\varepsilon(\omega - \varepsilon^2 h)} T_{mn}^\varepsilon(\omega) \overline{\hat{G}_1}(h) \, e^{i\beta_m(\omega)\frac{L}{\varepsilon^2} - i\beta_j(\omega - \varepsilon^2 h)\frac{L}{\varepsilon^2} + i\omega\frac{t_1}{\varepsilon^2} - i\omega t_{\mathrm{obs}}} \, dh \, d\omega.$$

In the forward-scattering approximation the power delay spread of transmitted signals is not very long because there is no backscattering to produce long codas. This is to be contrasted with what happens in one-dimensional random media as we saw in Chapter 12. Moreover, we concentrate our attention more on spatial effects in the chapter, so it is reasonable to assume that we record the field for all time at the time-reversal mirror. This means that we have $G_1 = 1$ and $\hat{G}_1(h) = 2\pi\delta(h)$. The expression for the refocused field now simplifies to

$$p_{\mathrm{ref(TR)}}(t_{\mathrm{obs}}, \mathbf{x}, 0) = \frac{1}{8\pi^2} \sum_{j,l,m,n=1}^{N} \frac{\sqrt{\beta_l \beta_m}}{\sqrt{\beta_j \beta_n}} M_{mj} \phi_n(\mathbf{x}) \phi_l(\mathbf{x}_0) \int \int \overline{\hat{f}^\varepsilon(\omega)}$$

$$\times \overline{T_{jl}^\varepsilon(\omega)} T_{mn}^\varepsilon(\omega) e^{i[\beta_m(\omega) - \beta_j(\omega)]\frac{L}{\varepsilon^2} + i\omega\frac{t_1}{\varepsilon^2} - i\omega t_{\mathrm{obs}}} \, d\omega. \qquad (20.69)$$

20.5.2 Integral Representation of the Broadband Refocused Field

In this section we consider a pulse of the form

$$f^\varepsilon(t) = f(\varepsilon t)e^{i\omega_0 t}, \qquad (20.70)$$

which is the same as (20.61) with $q = 1$, and in the Fourier domain

$$\hat{f}^\varepsilon(\omega) = \frac{1}{\varepsilon}\hat{f}\left(\frac{\omega - \omega_0}{\varepsilon}\right).$$

We observe the refocused field in a time window comparable to the width of the pulse and centered at a time $t_{\mathrm{obs}}/\varepsilon^2$, which is of the order of travel times into the waveguide

$$
p_{\mathrm{ref(TR)}}\left(\frac{t_{\mathrm{obs}}}{\varepsilon^2}+\frac{t}{\varepsilon},\mathbf{x},0\right)=\frac{1}{4\pi\varepsilon}\sum_{j,l,m,n=1}^{N}\frac{\sqrt{\beta_l\beta_m}}{\sqrt{\beta_j\beta_n}}M_{mj}\phi_n(\mathbf{x})\phi_l(\mathbf{x}_0)
$$

$$
\times\int\overline{\hat{f}\left(\frac{\omega-\omega_0}{\varepsilon}\right)\overline{T_{jl}^{\varepsilon}(\omega)}}T_{mn}^{\varepsilon}(\omega)e^{i[\beta_m(\omega)-\beta_j(\omega)]\frac{L}{\varepsilon^2}+i\omega\frac{t_1-t_{\mathrm{obs}}}{\varepsilon^2}-i\omega\frac{t}{\varepsilon}}\,d\omega\,.
$$

Here M_{mj} is defined by (20.67). We make the change of variable $\omega=\omega_0+\varepsilon h$ and expand $\beta_j(\omega_0+\varepsilon h)$ with respect to ε and get the following expression of the refocused field:

$$
p_{\mathrm{ref(TR)}}\left(\frac{t_{\mathrm{obs}}}{\varepsilon^2}+\frac{t}{\varepsilon},\mathbf{x},0\right)=\frac{1}{4\pi}\sum_{j,l,m,n=1}^{N}\frac{\sqrt{\beta_l\beta_m}}{\sqrt{\beta_j\beta_n}}M_{mj}\phi_n(\mathbf{x})\phi_l(\mathbf{x}_0)
$$

$$
\times e^{i[\beta_m-\beta_j](\omega_0)\frac{L}{\varepsilon^2}}e^{i\omega_0\frac{t_1-t_{\mathrm{obs}}}{\varepsilon^2}-i\frac{\omega_0 t}{\varepsilon}}
$$

$$
\times\int\overline{\hat{f}(h)\overline{T_{jl}^{\varepsilon}(\omega_0+\varepsilon h)}}T_{mn}^{\varepsilon}(\omega_0+\varepsilon h)e^{i\frac{[\beta_m'-\beta_j'](\omega_0)L+(t_1-t_{\mathrm{obs}})}{\varepsilon}h}e^{-iht}\,dh\,.\quad(20.71)
$$

20.5.3 Refocusing in a Homogeneous Waveguide

We first use the expression (20.71) for time reversal in a homogeneous wave-guide. In this case $T_{jl}^{\varepsilon}=\delta_{jl}$, and the refocused field is

$$
p_{\mathrm{ref(TR)}}\left(\frac{t_{\mathrm{obs}}}{\varepsilon^2}+\frac{t}{\varepsilon},\mathbf{x},0\right)=\frac{1}{2}\sum_{j,m=1}^{N}M_{mj}\phi_m(\mathbf{x})\phi_j(\mathbf{x}_0)e^{i[\beta_m-\beta_j](\omega_0)\frac{L}{\varepsilon^2}+i\omega_0\frac{t_1-t_{\mathrm{obs}}}{\varepsilon^2}}
$$

$$
\times e^{-i\frac{\omega_0 t}{\varepsilon}}f\left(\frac{[\beta_m'-\beta_j'](\omega_0)L+t_1-t_{\mathrm{obs}}}{\varepsilon}-t\right)\,.
$$

We observe a refocused pulse only for a discrete set of observation times parameterized by two integers $j,m\in\{1,\dots,N\}$

$$
t_{\mathrm{obs}}=\bar{t}_{jm}:=t_1+[\beta_m'-\beta_j'](\omega_0)L\,.
$$

These times are the differences of mode travel times from O to L and back. If we observe the refocused pulse at $t_{\mathrm{obs}}=\bar{t}_{jm}$ with $j\neq m$, then we see only one mode (the mode m) because the other contributions in $p_{\mathrm{ref(TR)}}$ are negligible as $\varepsilon\to 0$. Therefore the refocused field has the form

$$
p_{\mathrm{ref(TR)}}\left(\frac{\bar{t}_{jm}}{\varepsilon^2}+\frac{t}{\varepsilon},\mathbf{x},0\right)=\frac{1}{2}M_{mj}\phi_m(\mathbf{x})\phi_j(\mathbf{x}_0)
$$

$$
\times e^{i[\beta_m-\beta_j](\omega_0)\frac{L}{\varepsilon^2}-i\omega_0\frac{[\beta_m'-\beta_j'](\omega_0)L}{\varepsilon^2}-i\frac{\omega_0 t}{\varepsilon}}f(-t)\,.
$$

Note that the intensity of this field is independent of ε. If we observe the refocused field at $t_{\mathrm{obs}}=t_1$, which is the travel-time difference to the time-reversal mirror and back for a single mode, so that $m=j$ in $\bar{t}_{jm}$, we see a contribution from all modes

$$p_{\text{ref(TR)}}\left(\frac{t_1}{\varepsilon^2} + \frac{t}{\varepsilon}, \mathbf{x}, 0\right) = H_{\omega_0, \mathbf{x}_0}(\mathbf{x}) e^{-i\frac{\omega_0 t}{\varepsilon}} f(-t). \qquad (20.72)$$

Here the spatial profile $H_{\omega_0, \mathbf{x}_0}$ has the form

$$H_{\omega_0, \mathbf{x}_0}(\mathbf{x}) = \frac{1}{2} \sum_{j=1}^{N} M_{jj} \phi_j(\mathbf{x}) \phi_j(\mathbf{x}_0). \qquad (20.73)$$

In a homogeneous medium, spatial focusing of the time-reversed field at the source is determined by $H_{\omega_0, \mathbf{x}_0}(\mathbf{x})$. From (20.67) we see that spatial focusing depends on (a) the size of the time-reversal mirror, which is the support of G_2 in the matrix M, and (b) the number N of propagating modes. If the time-reversal mirror spans the width of the waveguide, then M_{jj} does not depend on j and $H_{\omega_0, \mathbf{x}_0}(\mathbf{x})$ becomes more focused for $\mathbf{x}$ near $\mathbf{x}_0$ as the number of propagating modes N increases. If the time-reversal mirror does not span the width of the waveguide, then the spatial focusing profile of the time-reversed field is more complicated. It is compared to focusing profiles in the random case in Figure 20.2 at the end of this chapter.

For a planar waveguide we have $\phi_j(x) = \sqrt{2/d}\sin(\pi j x/d)$ and $\beta_j = \sqrt{\omega^2/c^2 - \pi^2 j^2/d^2}$. For a time-reversal mirror that spans the width of the waveguide, $G_2(x) = \mathbf{1}_{[0,d]}(x)$, $M_{jj} = 1$ for all j. In the continuum limit $N \gg 1$ the spatial focusing profile at the source simplifies to

$$H_{\omega_0, x_0}(x) \xrightarrow{N \gg 1} \frac{1}{\lambda_0} \text{sinc}\left(\frac{2\pi(x - x_0)}{\lambda_0}\right),$$

where $\lambda_0 = 2\pi\bar{c}/\omega_0$ is the carrier wavelength and the sinc function is defined by $\text{sinc}(x) = \sin(x)/x$. In the limit $N \gg 1$ we interpret the sum in (20.73) as a Riemann sum and get the simplified expression above by computing the resulting integral. The limit sinc profile is the best transverse profile that we can obtain, because spatial focusing reaches the **diffraction limit** $\lambda_0/2$, which is the first zero of the sinc function. Note that the continuum limit $N \gg 1$ corresponds to $d \gg \lambda_0$.

20.5.4 Refocusing in a Random Waveguide

The time-reversed field at the plane of the source is given by (20.71). We will now use this expression when the transfer matrix $T_{jl}^{\varepsilon}(\omega)$ is the one for a random waveguide. We note first that the product of two elements of the matrix appears in (20.71), at the same frequency and with one of the two conjugated. We recall from (20.70) that in this broadband case the pulse width is of order $1/\varepsilon$ and the bandwidth is of order ε. We will see in the next section, in Proposition 20.7, that the decoherence frequency of $T_{jl}^{\varepsilon}(\omega)$ is of order ε^2. This means that in the integral over frequencies in (20.71) we are summing over a large number, of order $1/\varepsilon$, of approximately uncorrelated random variables. The expression for the time-reversed field in (20.71) is therefore self-averaging, by the law of large numbers. The refocused field does not differ much as $\varepsilon \to 0$ from its expected value, which is

$$\mathbb{E}\left[p_{\text{ref(TR)}}\left(\frac{t_{\text{obs}}}{\varepsilon^2} + \frac{t}{\varepsilon}, \mathbf{x}, 0\right)\right] = \frac{1}{4\pi} \sum_{j,l,m,n=1}^{N} \frac{\sqrt{\beta_l \beta_m}}{\sqrt{\beta_j \beta_n}} M_{mj} \phi_n(\mathbf{x}) \phi_l(\mathbf{x}_0)$$

$$\times e^{i[\beta_m - \beta_j](\omega_0)\frac{L}{\varepsilon^2}} e^{i\omega_0 \frac{t_1 - t_{\text{obs}}}{\varepsilon^2} - i\frac{\omega_0 t}{\varepsilon}}$$

$$\times \int \overline{\hat{f}(h)} \mathbb{E}[\overline{T_{jl}^{\varepsilon}(\omega_0 + \varepsilon h)} T_{mn}^{\varepsilon}(\omega_0 + \varepsilon h)] e^{i\frac{[\beta_m' - \beta_j'](\omega_0)L + (t_1 - t_{\text{obs}})}{\varepsilon}} h e^{-iht} \, dh \,. (20.74)$$

From this expression we see that we need to calculate the expected value $\mathbb{E}[\overline{T_{jl}^{\varepsilon}(\omega_0 + \varepsilon h)} T_{mn}^{\varepsilon}(\omega_0 + \varepsilon h)]$ as $\varepsilon \to 0$. The following propositions contain the information that we need.

Proposition 20.6. *The expectation of two transmission coefficients at the same frequency has a limit as $\varepsilon \to 0$, which is given by*

$$\mathbb{E}[T_{jj}^{\varepsilon}(\omega, L)\overline{T_{ll}^{\varepsilon}(\omega, L)}] \xrightarrow{\varepsilon \to 0} e^{Q_{jl}(\omega)L} \text{ if } j \neq l \,, \tag{20.75}$$

$$\mathbb{E}[T_{jl}^{\varepsilon}(\omega, L)\overline{T_{jl}^{\varepsilon}(\omega, L)}] \xrightarrow{\varepsilon \to 0} \mathcal{T}_j^{(l)}(\omega, L) \,, \tag{20.76}$$

$$\mathbb{E}[T_{jm}^{\varepsilon}(\omega, L)\overline{T_{ln}^{\varepsilon}(\omega, L)}] \xrightarrow{\varepsilon \to 0} 0 \text{ in the other cases} \,, \tag{20.77}$$

where $(\mathcal{T}_j^{(l)}(\omega, z))_{j=1,\dots,N(\omega)}$ is the solution of the system of linear equations

$$\frac{\partial \mathcal{T}_j^{(l)}}{\partial z} = \sum_{n \neq j} \Gamma_{jn}^{(c)}(\omega)\left(\mathcal{T}_n^{(l)} - \mathcal{T}_j^{(l)}\right) \,, \qquad \mathcal{T}_j^{(l)}(\omega, z = 0) = \delta_{jl} \,. \tag{20.78}$$

The coefficients $\Gamma_{jl}^{(c)}$ are given by (20.49–20.52). The damping factors Q_{jl} are

$$Q_{jl}(\omega) = \frac{\Gamma_{jj}^{(c)}(\omega) + \Gamma_{ll}^{(c)}(\omega)}{2} - \frac{\Gamma_{jj}^{(1)}(\omega) + \Gamma_{ll}^{(1)}(\omega) - 2\Gamma_{jl}^{(1)}(\omega)}{2}$$

$$+ i\frac{\Gamma_{jj}^{(s)}(\omega) - \Gamma_{ll}^{(s)}(\omega)}{2} + i[\kappa_j(\omega) - \kappa_l(\omega)] \,. \tag{20.79}$$

The real parts of the damping factors Q_{jl} are negative. This implies that the moments (20.75) decay exponentially with the length of the waveguide L.

From the analysis of Section 20.3.3, we know that

$$\sup_{j,l} \left| \mathcal{T}_j^{(l)}(\omega, z) - \frac{1}{N} \right| \leq Ce^{-z/L_{\text{equip}}} \,, \tag{20.80}$$

where L_{equip} is the equipartition distance for the mean mode powers introduced at the end of Section 20.3.3. Therefore, the expectation of the square moduli of the entries of the transfer matrix in (20.76) converge exponentially fast to the constant $1/N$.

By Proposition 20.6, the expectation of the product of transmission coefficients at the same frequency $\omega + \varepsilon h$ becomes independent of h as $\varepsilon \to 0$. Therefore the fast phase in h in (20.74) cannot be compensated for, unless it is zero. This shows that we can observe the refocused field only at time $t_{\text{obs}} = t_1$ or at times $t_{\text{obs}} = \bar{t}_{jm}$, $j \neq m$. Let us first consider the second case. For fixed $j \neq m$ we have

$$
\mathbb{E}\left[p_{\text{ref(TR)}} \left(\frac{\bar{t}_{jm}}{\varepsilon^2} + \frac{t}{\varepsilon}, \mathbf{x}, 0 \right) \right] = \frac{1}{4\pi} \sum_{l,n=1}^{N} \frac{\sqrt{\beta_l \beta_m}}{\sqrt{\beta_j \beta_n}} M_{mj} \phi_n(\mathbf{x}) \phi_l(\mathbf{x}_0)
$$
$$
\times e^{i[\beta_m - \beta_j](\omega_0) \frac{L}{\varepsilon^2}} e^{i\omega_0 \frac{t_1 - t_{\text{obs}}}{\varepsilon^2} - i \frac{\omega_0 t}{\varepsilon}} \int \hat{f}(h) \mathbb{E}[\overline{T_{jl}^{\varepsilon}(\omega_0 + \varepsilon h)} T_{mn}^{\varepsilon}(\omega_0 + \varepsilon h)] e^{-iht} \, dh \,.
$$

Proposition 20.6 shows that the limit of the expectation is nonzero only if $l = j$ and $n = m$, and then the limit becomes independent of h, so that

$$
\mathbb{E}\left[p_{\text{ref(TR)}} \left(\frac{\bar{t}_{jm}}{\varepsilon^2} + \frac{t}{\varepsilon}, \mathbf{x}, 0 \right) \right] = \frac{1}{2} M_{mj} \phi_m(\mathbf{x}) \phi_j(\mathbf{x}_0) e^{i[\beta_m - \beta_j](\omega_0) \frac{L}{\varepsilon^2}}
$$
$$
\times e^{i\omega_0 \frac{t_1 - t_{\text{obs}}}{\varepsilon^2} - i \frac{\omega_0 t}{\varepsilon}} e^{Q_{jm}(\omega_0) L} f(-t) \,.
$$

In this refocused field only one mode contributes, and its amplitude decays exponentially with propagation distance.

Let us next consider the refocused field at time $t_{\text{obs}} = t_1$. All modes contribute to it:

$$
\mathbb{E}\left[p_{\text{ref(TR)}} \left(\frac{t_1}{\varepsilon^2} + \frac{t}{\varepsilon}, \mathbf{x}, 0 \right) \right] = \frac{1}{4\pi} \sum_{j,l,n=1}^{N} \frac{\sqrt{\beta_l}}{\sqrt{\beta_n}} M_{jj} \phi_n(\mathbf{x}) \phi_l(\mathbf{x}_0) e^{-i \frac{\omega_0 t}{\varepsilon}} \int \hat{f}(h)
$$
$$
\times \mathbb{E}[\overline{T_{jl}^{\varepsilon}(\omega_0 + \varepsilon h)} T_{jn}^{\varepsilon}(\omega_0 + \varepsilon h)] e^{-iht} \, dh \,.
$$

By Proposition 20.6, only the terms with $l = n$ give nonzero values, and we therefore get the following expression for the mean refocused pulse:

$$
\mathbb{E}\left[p_{\text{ref(TR)}} \left(\frac{t_1}{\varepsilon^2} + \frac{t}{\varepsilon}, \mathbf{x}, 0 \right) \right] = \frac{1}{2} \sum_{j,l=1}^{N} M_{jj} T_j^{(l)}(\omega_0, L) \phi_l(\mathbf{x}) \phi_l(\mathbf{x}_0) e^{-i \frac{\omega_0 t}{\varepsilon}} f(-t) \,.
$$

$$(20.81)$$

This is the main result regarding time reversal in the broadband case. We will now consider it in some special cases in order to explain its implications for spatial focusing in time reversal.

For small propagation distances, $L \ll L_{\text{equip}}$, the transfer matrix is close to the identity matrix. We therefore recover the result (20.72) of the homogeneous case, where the distribution of the mirror-coupling coefficient matrix M, defined by (20.67), plays a central role.

For large propagation distances, $L \gg L_{\text{equip}}$, we use (20.80) and get

$$\mathbb{E}\left[p_{\text{ref(TR)}}\left(\frac{t_1}{\varepsilon^2} + \frac{t}{\varepsilon}, \mathbf{x}, 0\right)\right] \overset{L \gg L_{\text{equip}}}{\sim} \left(\frac{1}{N}\sum_{j=1}^{N} M_{jj}\right) H_{\omega_0, \mathbf{x}_0}(\mathbf{x}) e^{-i\frac{\omega_0 t}{\varepsilon}} f(-t),$$

(20.82)

where the spatial focusing profile $H_{\omega_0, \mathbf{x}_0}$ is given here by

$$H_{\omega_0, \mathbf{x}_0}(\mathbf{x}) = \frac{1}{2}\sum_{l=1}^{N(\omega_0)} \phi_l(\mathbf{x})\phi_l(\mathbf{x}_0).$$

(20.83)

In this case the distribution of the mirror-coupling coefficients M plays no role. Only the average of the diagonal entries M_{jj} appears as a multiplicative factor that does not affect the spatial focusing profile. This means that spatial focusing in time reversal in a random waveguide, in a regime of strong scattering, does not depend on the size of the time-reversal mirror.

In the case of a planar waveguide, we have $\phi_j(x) = \sqrt{2/d}\sin(\pi j x/d)$, $\beta_j = \sqrt{\omega^2/c^2 - \pi^2 j^2/d^2}$. In the continuum limit $N \gg 1$, we get the simplified focusing profile

$$H_{\omega_0}(x) \overset{N \gg 1}{\longrightarrow} \frac{1}{\lambda_0}\text{sinc}\left(\frac{2\pi(x - x_0)}{\lambda_0}\right),$$

which is the diffraction-limited focal spot in time reversal. The refocused pulse is therefore concentrated around the original source location x_0 with a resolution of half a wavelength.

As noted above, (20.81) is the main result of this section. It describes how in time-reversal focusing we go from (20.72) in a homogeneous medium to (20.82) for a random waveguide in the equipartition regime, $L \gg L_{\text{equip}}$.

To conclude this section we give some comments about the statistical stability of the refocused field. The self-averaging property can be justified with the same arguments that were used in the one-dimensional random media. As already noted, it comes from the broadband source pulse, meaning that the refocused field (20.71) is an integration over a frequency band that is much larger (of order ε) than the frequency correlation, or coherence, of the transfer matrix (of order ε^2). When we take the expectation of the square of (20.71) we get in the asymptotic limit $\varepsilon \to 0$ the square of the expectation, which means that the variance is going to zero and that the refocused field is self-averaging. The expression (20.81) is consequently valid not only in mean, but also in probability, that is, for any typical realization of the random medium.

20.6 Statistics of the Transmission Coefficients at Two Nearby Frequencies

20.6.1 Transport Equations for the Autocorrelation Function of the Transfer Matrix

Up to now we have been able to compute many quantities of interest, such as time-reversal focusing in the broadband case, with only single-frequency statistical properties of the transfer matrix, as $\varepsilon \to 0$. We did need the fact that frequency decoherence is of order ε^2 in order to have statistical stability, as in the previous section, but we did not need quantitative, two-frequency statistical information. However, in many physically interesting contexts, such as in calculating the mean transmitted intensity or the mean refocused field amplitude in narrowband cases, we do need this information. We now introduce a proposition that describes the two-frequency statistical properties that we will need in the applications in the remainder of this chapter.

Proposition 20.7. *The autocorrelation function of the transmission coefficients at two nearby frequencies admits a limit as $\varepsilon \to 0$:*

$$\mathbb{E}[T_{jj}^{\varepsilon}(\omega, L)\overline{T_{ll}^{\varepsilon}(\omega - \varepsilon^2 h, L)}] \xrightarrow{\varepsilon \to 0} e^{Q_{jl}(\omega)L} \text{ if } j \neq l, \tag{20.84}$$

$$\mathbb{E}[T_{jl}^{\varepsilon}(\omega, L)\overline{T_{jl}^{\varepsilon}(\omega - \varepsilon^2 h, L)}] \xrightarrow{\varepsilon \to 0} e^{-i\beta_j'(\omega)hL} \int \mathcal{W}_j^{(l)}(\omega, \tau, L)e^{ih\tau}\, d\tau, \tag{20.85}$$

$$\mathbb{E}[T_{jm}^{\varepsilon}(\omega, L)\overline{T_{ln}^{\varepsilon}(\omega - \varepsilon^2 h, L)}] \xrightarrow{\varepsilon \to 0} 0 \text{ in the other cases}, \tag{20.86}$$

where $(\mathcal{W}_j^{(l)}(\omega, \tau, z))_{j=1,\dots,N(\omega)}$ is the solution of the system of transport equations

$$\frac{\partial \mathcal{W}_j^{(l)}}{\partial z} + \dot{\beta}_j'(\omega)\frac{\partial \mathcal{W}_j^{(l)}}{\partial \tau} = \sum_{n \neq j} \Gamma_{jn}^{(c)}(\omega)\left(\mathcal{W}_n^{(l)} - \mathcal{W}_j^{(l)}\right), \quad z \geq 0, \tag{20.87}$$

$$\mathcal{W}_j^{(l)}(\omega, \tau, z = 0) = \delta(\tau)\delta_{jl}. \tag{20.88}$$

The coefficients $\Gamma_{jl}^{(c)}$ are given by (20.49–20.52). The damping factors Q_{jl} are

$$Q_{jl}(\omega) = \frac{\Gamma_{jj}^{(c)}(\omega) + \Gamma_{ll}^{(c)}(\omega)}{2} - \frac{\Gamma_{jj}^{(1)}(\omega) + \Gamma_{ll}^{(1)}(\omega) - 2\Gamma_{jl}^{(1)}(\omega)}{2}$$
$$+ i\frac{\Gamma_{jj}^{(s)}(\omega) - \Gamma_{ll}^{(s)}(\omega)}{2} + i[\kappa_j(\omega) - \kappa_l(\omega)]. \tag{20.89}$$

We note that the real parts of the damping factors Q_{jl} are negative and that the solutions of the transport equations are measures. If $j \neq l$, then $\mathcal{W}_j^{(l)}$ has a continuous density, but $\mathcal{W}_l^{(l)}$ has a Dirac mass at $\tau = \beta_l'(\omega)z$ with weight

$\exp(\Gamma_{ll}^{(c)}z)$, where $\Gamma_{ll}^{(c)}$ is given by (20.50), and a continuous density denoted by $\mathcal{W}_{l,c}^{(l)}(\omega, \tau, z)$:

$$\mathcal{W}_l^{(l)}(\omega, \tau, z)\, d\tau = e^{\Gamma_{ll}^{(c)}(\omega)z}\delta(\tau - \beta_l'(\omega)z)\, d\tau + \mathcal{W}_{l,c}^{(l)}(\omega, \tau, z)\, d\tau.$$

Note also that by integrating the system of transport equations with respect to τ, we recover the result of Proposition 20.6.

The system of transport equations describes the coupling between the N forward-going modes. It is in the form of a system of transport equations that we have encountered in previous chapters (Chapter 9), but it is has a very different interpretation. The indices $j = 1, \ldots, N$ play equivalent roles, for they represent the labels of the modes, while in the transport equations of Chapter 9 the indices stand for the order of the moments of the transmission or reflection coefficients.

The transport equations have quite the standard form. They describe the evolution of the coupled powers of the modes in space and time, with transport velocities equal to the group velocities of the modes $1/\beta_j'(\omega)$. Therefore, the transport equations (20.87) could have been written down as the natural space-time generalization of the coupled power equations (20.78). The mathematical content of Proposition 20.7 is that this simple and intuitive space-time extension of (20.78) does not connect accurately the quantities that satisfy the transport equation and the moments of the random transfer matrix. The two-frequency nature of the statistical quantities that satisfy the transport equations is clear in Proposition 20.7.

Proof. For fixed indices m and n we consider the product of two transfer matrices

$$U_{jl}^\varepsilon(\omega, h, z) = T_{jm}^\varepsilon(\omega, z)\overline{T_{ln}^\varepsilon(\omega - \varepsilon^2 h, z)},$$

and note that it is the solution of

$$\begin{aligned}
\frac{dU_{jl}^\varepsilon}{dz} =\ & \frac{ik^2}{2\varepsilon}\left(\frac{C_{jj}(\frac{z}{\varepsilon^2})}{\beta_j(\omega)} - \frac{C_{ll}(\frac{z}{\varepsilon^2})}{\beta_l(\omega - \varepsilon^2 h)}\right) U_{jl}^\varepsilon \\
& + \frac{ik^2}{2\varepsilon}\sum_{j_1 \neq j}\frac{C_{jj_1}(\frac{z}{\varepsilon^2})}{\sqrt{\beta_j\beta_{j_1}}(\omega)}e^{i(\beta_{j_1}-\beta_j)(\omega)\frac{z}{\varepsilon^2}}U_{j_1 l}^\varepsilon \\
& - \frac{ik^2}{2\varepsilon}\sum_{l_1 \neq l}\frac{C_{ll_1}(\frac{z}{\varepsilon^2})}{\sqrt{\beta_l\beta_{l_1}}(\omega - \varepsilon^2 h)}e^{i(\beta_l-\beta_{l_1})(\omega-\varepsilon^2 h)\frac{z}{\varepsilon^2}}U_{jl_1}^\varepsilon \\
& + \frac{ik^4}{4}\sum_{j_1}\sum_{l'>N}\int_{-\infty}^\infty \frac{C_{jl'}(\frac{z}{\varepsilon^2})C_{j_1 l'}(\frac{z}{\varepsilon^2}+s)}{\sqrt{\beta_j\beta_{l'}^2\beta_{j_1}}(\omega)}e^{i[\beta_{j_1}-\beta_j](\omega)\frac{z}{\varepsilon^2}} \\
& \qquad\qquad \times e^{i\beta_{j_1}(\omega)s - \beta_{l'}(\omega)|s|}\, ds\, U_{j_1 l}^\varepsilon \\
& - \frac{ik^4}{4}\sum_{l_1}\sum_{l'>N}\int_{-\infty}^\infty \frac{C_{ll'}(\frac{z}{\varepsilon^2})C_{l_1 l'}(\frac{z}{\varepsilon^2}+s)}{\sqrt{\beta_l\beta_{l'}^2\beta_{l_1}}(\omega - \varepsilon^2 h)}e^{i[\beta_l-\beta_{l_1}](\omega-\varepsilon^2 h)\frac{z}{\varepsilon^2}} \\
& \qquad\qquad \times e^{-i\beta_{l_1}(\omega-\varepsilon^2 h)s - \beta_{l'}(\omega-\varepsilon^2 h)|s|}\, ds\, U_{jl_1}^\varepsilon,
\end{aligned}$$

with the initial conditions $U_{jl}^{\varepsilon}(\omega, h, z = 0) = \delta_{mj}\delta_{nl}$. Expanding $\beta_l(\omega - \varepsilon^2 h)$ with respect to ε gives the simpler system

$$
\begin{aligned}
\frac{dU_{jl}^{\varepsilon}}{dz} = {}& \frac{ik^2}{2\varepsilon}\left(\frac{C_{jj}(\frac{z}{\varepsilon^2})}{\beta_j(\omega)} - \frac{C_{ll}(\frac{z}{\varepsilon^2})}{\beta_l(\omega)}\right)U_{jl}^{\varepsilon} \\
&+ \frac{ik^2}{2\varepsilon}\sum_{j_1 \neq j}\frac{C_{jj_1}(\frac{z}{\varepsilon^2})}{\sqrt{\beta_j\beta_{j_1}}(\omega)}e^{i(\beta_{j_1}-\beta_j)(\omega)\frac{z}{\varepsilon^2}}U_{j_1 l}^{\varepsilon} \\
&- \frac{ik^2}{2\varepsilon}\sum_{l_1 \neq l}\frac{C_{ll_1}(\frac{z}{\varepsilon^2})}{\sqrt{\beta_l\beta_{l_1}}(\omega)}e^{i(\beta_l-\beta_{l_1})(\omega)\frac{z}{\varepsilon^2}}e^{i(\beta_l'-\beta_{l_1}')(\omega)zh}U_{jl_1}^{\varepsilon} \\
&+ \frac{ik^4}{4}\sum_{j_1}\sum_{l'>N}\int_{-\infty}^{\infty}\frac{C_{jl'}(\frac{z}{\varepsilon^2})C_{j_1 l'}(\frac{z}{\varepsilon^2}+s)}{\sqrt{\beta_j\beta_{l'}^2\beta_{j_1}}}e^{i[\beta_{j_1}-\beta_j](\omega)\frac{z}{\varepsilon^2}} \\
&\qquad\qquad\qquad \times e^{i\beta_{j_1}(\omega)s-\beta_{l'}(\omega)|s|}\,ds\,U_{j_1 l}^{\varepsilon} \\
&- \frac{ik^4}{4}\sum_{l_1}\sum_{l'>N}\int_{-\infty}^{\infty}\frac{C_{ll'}(\frac{z}{\varepsilon^2})C_{l_1 l'}(\frac{z}{\varepsilon^2}+s)}{\sqrt{\beta_l\beta_{l'}^2\beta_{l_1}}}e^{i[\beta_l-\beta_{l_1}](\omega)\frac{z}{\varepsilon^2}} \\
&\qquad\qquad\qquad \times e^{-i\beta_{l_1}(\omega)s-\beta_{l'}(\omega)|s|}\,ds\,e^{i[\beta_{l_1}'-\beta_l'](\omega)hz}U_{jl_1}^{\varepsilon}.
\end{aligned}
$$

We can apply the limit theorems of Chapter 6 to this system of random differential equations or we can first introduce the Fourier transform

$$
V_{jl}^{\varepsilon}(\omega, \tau, z) = \frac{1}{2\pi}\int e^{-ih(\tau - \beta_l'(\omega)z)}U_{jl}^{\varepsilon}(\omega, h, z)\,dh,
$$

which is the solution of

$$
\begin{aligned}
\frac{\partial V_{jl}^{\varepsilon}}{\partial z} + \beta_l'(\omega)\frac{\partial V_{jl}^{\varepsilon}}{\partial \tau} = {}& \frac{ik^2}{2\varepsilon}\left(\frac{C_{jj}(\frac{z}{\varepsilon^2})}{\beta_j(\omega)} - \frac{C_{ll}(\frac{z}{\varepsilon^2})}{\beta_l(\omega)}\right)V_{jl}^{\varepsilon} \\
&+ \frac{ik^2}{2\varepsilon}\sum_{j_1 \neq j}\frac{C_{jj_1}(\frac{z}{\varepsilon^2})}{\sqrt{\beta_j\beta_{j_1}}(\omega)}e^{i(\beta_{j_1}-\beta_j)(\omega)\frac{z}{\varepsilon^2}}V_{j_1 l}^{\varepsilon} \\
&- \frac{ik^2}{2\varepsilon}\sum_{l_1 \neq l}\frac{C_{ll_1}(\frac{z}{\varepsilon^2})}{\sqrt{\beta_l\beta_{l_1}}(\omega)}e^{i(\beta_l-\beta_{l_1})(\omega)\frac{z}{\varepsilon^2}}V_{jl_1}^{\varepsilon} \\
&+ \frac{ik^4}{4}\sum_{j_1}\sum_{l'>N}\int_{-\infty}^{\infty}\frac{C_{jl'}(\frac{z}{\varepsilon^2})C_{j_1 l'}(\frac{z}{\varepsilon^2}+s)}{\sqrt{\beta_j\beta_{l'}^2\beta_{j_1}}}e^{i[\beta_{j_1}-\beta_j](\omega)\frac{z}{\varepsilon^2}} \\
&\qquad\qquad\qquad \times e^{i\beta_{j_1}(\omega)s-\beta_{l'}(\omega)|s|}\,ds\,V_{j_1 l}^{\varepsilon} \\
&- \frac{ik^4}{4}\sum_{l_1}\sum_{l'>N}\int_{-\infty}^{\infty}\frac{C_{ll'}(\frac{z}{\varepsilon^2})C_{l_1 l'}(\frac{z}{\varepsilon^2}+s)}{\sqrt{\beta_l\beta_{l'}^2\beta_{l_1}}}e^{i[\beta_l-\beta_{l_1}](\omega)\frac{z}{\varepsilon^2}} \\
&\qquad\qquad\qquad \times e^{-i\beta_{l_1}(\omega)s-\beta_{l'}(\omega)|s|}\,ds\,V_{jl_1}^{\varepsilon},
\end{aligned}
$$

with the initial conditions $V_{jl}^{\varepsilon}(\omega, h, z = 0) = \delta_{mj}\delta_{nl}\delta(\tau)$. We can now apply a variant of the diffusion approximation theorem of Chapter 6 and get the result stated in the proposition. The details of the proof follow closely that of Proposition 9.1. $\qquad\square$

20.6.2 Probabilistic Representation of the Transport Equations

The transport equations (20.87) have a probabilistic representation, which is similar to the one we introduced in Section 9.2.2. This probabilistic representation can be used for Monte Carlo simulations as well as for getting a diffusion approximation. It is primarily this diffusion approximation that we want to derive in this section. We will use it in the applications that follow in this chapter.

We introduce the jump Markov process $(J_z)_{z\geq 0}$ whose state space is $\{1, \ldots, N(\omega)\}$ and whose infinitesimal generator is

$$\mathcal{L}\phi(j) = \sum_{l \neq j} \Gamma_{jl}^{(c)}(\omega) \left(\phi(l) - \phi(j)\right) .$$

We also define the process $\mathcal{B}_z$ by

$$\mathcal{B}_z = \int_0^z \beta_{J_s} ds , \quad z \geq 0 ,$$

which is well defined because J_z is piecewise constant. As in Section 9.2.2, we get the probabilistic representation of the solutions of the system (20.78) and those of to the transport equations (20.87) in terms of the jump Markov process J_z:

$$T_j^{(n)}(\omega, L) = \mathbb{P}\left(J_L = j \mid J_0 = n\right) , \qquad (20.90)$$

$$\int_{\tau_0}^{\tau_1} \mathcal{W}_j^{(n)}(\omega, \tau, L) \, d\tau = \mathbb{P}\left(J_L = j , \ \mathcal{B}_L \in [\tau_0, \tau_1] \mid J_0 = n\right) . \quad (20.91)$$

The process J_z is an irreducible, reversible, and ergodic Markov process. Its distribution converges as $z \to \infty$ to the uniform distribution over $\{1, \ldots, N\}$. The convergence is exponential with a rate that is equal to the second eigenvalue of the matrix $\Gamma^{(c)} = (\Gamma_{jl}^{(c)})_{j,l=1,\ldots,N}$. The first eigenvalue of this matrix is zero, with eigenvector the uniform distribution over $\{1, \ldots, N\}$. The second eigenvalue can be written in the form $-1/L_{\mathrm{equip}}$, which defines the equipartition distance L_{equip}.

We next determine the asymptotic distribution of the process $\mathcal{B}_z$. From the ergodic theorem we have that with probability one,

$$\frac{\mathcal{B}_z}{z} \xrightarrow{z \to \infty} \widehat{\beta'(\omega)} ,$$

where

$$\widehat{\beta'(\omega)} = \frac{1}{N(\omega)} \sum_{j=1}^{N(\omega)} \beta_j'(\omega) .$$

We can interpret the z-large limit to mean that z is considerably larger than L_{equip}.

For a planar waveguide we have that $\beta_j = \sqrt{\omega^2/c^2 - \pi^2 j^2/d^2}$ and $N(\omega) = [(\omega d)/(\pi \bar{c})]$. In the continuum limit $N(\omega) \gg 1$ we obtain the expression

$$\widehat{\beta'(\omega)} = \frac{1}{\bar{c}} \int_0^1 \frac{1}{\sqrt{1-s^2}} \, ds = \frac{\pi}{2\bar{c}},$$

which is independent of ω. This ω-independence property is likely to hold for a broad class of waveguides.

By applying a central limit theorem for functionals of ergodic Markov processes, we find that in distribution,

$$\frac{B_z - \widehat{\beta'(\omega)}z}{\sqrt{z}} \xrightarrow{z\to\infty} \mathcal{N}(0, \sigma^2_{\beta'(\omega)}).$$

Here $\mathcal{N}(0, \sigma^2_{\beta'(\omega)})$ is a zero-mean Gaussian random variable with variance

$$\sigma^2_{\beta'(\omega)} = 2 \int_0^\infty \mathbb{E}_e \left[(\beta'_{J_0}(\omega) - \widehat{\beta'(\omega)})(\beta'_{J_s}(\omega) - \widehat{\beta'(\omega)}) \right] ds$$

$$= \frac{2}{N(\omega)} \sum_{j=1}^{N(\omega)} (\beta'_j(\omega) - \widehat{\beta'(\omega)}) \int_0^\infty \mathbb{E} \left[\beta'_{J_s}(\omega) - \widehat{\beta'(\omega)} \mid J_0 = j \right] ds$$

$$= -\frac{2}{N(\omega)} B^T (\Gamma^{(c)})^{-1} B,$$

where $B = (B_j)_{j=1,\dots,N(\omega)}$ with $B_j = \beta'_j(\omega) - \widehat{\beta'(\omega)}$ and $\mathbb{E}_e$ stands for expectation with respect to the stationary process J_z. We note here that $\Gamma^{(c)}$ is not an invertible matrix since it possesses zero as an eigenvalue. However, the equation $\Gamma^{(c)} \tilde{B} = B$ can be solved because B belongs to the orthogonal complement of the null space of $\Gamma^{(c)}$. The solution $\tilde{B}$ is uniquely defined up to a component that belongs to the null space of $\Gamma^{(c)}$. Therefore $B^T \tilde{B} = B^T (\Gamma^{(c)})^{-1} B$ is uniquely defined. Note also that the order of magnitude of $\sigma^2_{\beta'(\omega)}$ is

$$\sigma^2_{\beta'(\omega)} \approx 2 L_{\text{equip}} \times \frac{1}{N} \sum_{j=1}^N \left[\beta'_j(\omega) - \widehat{\beta'(\omega)} \right]^2.$$

This formula is only an order-of-magnitude estimate. The exact value of $\sigma^2_{\beta'(\omega)}$ is given above.

These limit theorems imply that, when $L \gg L_{\text{equip}}$, we have

$$T_j^{(n)}(\omega, L) \overset{L \gg L_{\text{equip}}}{\sim} \frac{1}{N(\omega)}, \tag{20.92}$$

$$\mathcal{W}_j^{(n)}(\omega, \tau, L) \overset{L \gg L_{\text{equip}}}{\sim} \frac{1}{N(\omega)} \frac{1}{\sqrt{2\pi \sigma^2_{\beta'(\omega)} L}} \exp\left(-\frac{(\tau - \widehat{\beta'(\omega)} L)^2}{2\sigma^2_{\beta'(\omega)} L} \right). \tag{20.93}$$

The asymptotic result (20.92) shows that $T_j^{(n)}(\tau)$ becomes independent of n, the initial mode index, and uniform over $j \in \{1, \ldots, N(\omega)\}$. This is the regime of energy equipartition among all propagating modes. The asymptotic result (20.93) is equivalent to the diffusion approximation for the system of transport equations (20.87). This system becomes asymptotically a system of uncoupled advection-diffusion equations that have the common form

$$\frac{\partial \mathcal{W}}{\partial z} + \overline{\beta'(\omega)} \frac{\partial \mathcal{W}}{\partial \tau} = \frac{1}{2} \sigma_{\beta'(\omega)}^2 \frac{\partial^2 \mathcal{W}}{\partial \tau^2} .$$

20.7 Incoherent Wave Fluctuations in the Broadband Case

In this section we complete the analysis carried out in Section 20.4, where we studied pulse propagation with a broadband source of the form (20.70). In Section 20.4.3 we showed that a coherent field, with a deterministic intensity, can be observed near the times $\bar{t}_j$ defined by (20.64). We now consider the statistics of the transmitted field at a time $t_0 \notin \{\bar{t}_j, j = 1, \ldots, N\}$.

We first consider the mean field

$$\mathbb{E}[p_{\mathrm{tr}}^\varepsilon(t_0, t, \mathbf{x}, L)] = \frac{1}{4\pi} \sum_{j,l=1}^N \frac{\sqrt{\beta_l}}{\sqrt{\beta_j}} \phi_j(\mathbf{x})\phi_l(\mathbf{x}_0) e^{i\frac{\beta_j(\omega_0)L - \omega_0 t_0}{\varepsilon^2} - \frac{\omega_0 t}{\varepsilon}}$$

$$\times \int \hat{f}(h) \mathbb{E}[T_{jl}^\varepsilon(\omega_0 + \varepsilon h)] e^{i\frac{[\beta_j'(\omega_0)L - t_0]h}{\varepsilon}} e^{i[\beta_j''(\omega_0)L\frac{h^2}{2} - ht]} \, dh .$$

By Proposition 20.2, the expected value of the transmission coefficient converges to a value independent of h. Therefore, the fast phase $e^{i\frac{[\beta_j'(\omega_0)L - t_0]h}{\varepsilon}}$ cannot be compensated with any other term of the integral, which is then proportional to $f(t + [t_0 - \bar{t}_j]/\varepsilon)$ and is asymptotically negligible. This means that the coherent field can be observed only near the times $\bar{t}_j$, $j = 1, \ldots, N$.

We compute next the mean transmitted intensity at a time $t_0 \notin \{\bar{t}_j, j = 1, \ldots, N\}$:

$$\mathbb{E}\left[|p_{\mathrm{tr}}^\varepsilon(t_0, t, \mathbf{x}, L)|^2\right] = \frac{1}{16\pi^2} \int \int \sum_{j,l,m,n=1}^N \frac{\sqrt{\beta_l \beta_n}}{\sqrt{\beta_j \beta_m}} \phi_j(\mathbf{x})\phi_l(\mathbf{x}_0)\phi_m(\mathbf{x})\phi_n(\mathbf{x}_0)$$

$$\times e^{i\frac{(\beta_j - \beta_m)(\omega_0)L}{\varepsilon^2}} \hat{f}(h)\overline{\hat{f}(h')} \mathbb{E}[T_{jl}^\varepsilon(\omega_0 + \varepsilon h)\overline{T_{mn}^\varepsilon(\omega_0 + \varepsilon h')}]$$

$$\times e^{i\frac{[\beta_j'(\omega_0)L - t_0](h - h')}{\varepsilon}} e^{i[\beta_j''(\omega_0)L\frac{h^2 - h'^2}{2} - (h - h')t]} \, dh \, dh' .$$

Because of the presence of the rapid phase $e^{i\frac{[\beta_j'(\omega_0)L - t_0](h - h')}{\varepsilon}}$, we make the change of variable $h' = h - \varepsilon\xi$:

$$\mathbb{E}\left[|p_{\text{tr}}^{\varepsilon}(t_0, t, \mathbf{x}, L)|^2\right] = \frac{\varepsilon}{16\pi^2} \int \int \sum_{j,l,m,n=1}^{N} \frac{\sqrt{\beta_l \beta_n}}{\sqrt{\beta_j \beta_m}} \phi_j(\mathbf{x}) \phi_l(\mathbf{x}_0) \phi_m(\mathbf{x}) \phi_n(\mathbf{x}_0)$$

$$\times \hat{f}(h) \overline{\hat{f}(h - \varepsilon\xi)} \mathbb{E}[T_{jl}^{\varepsilon}(\omega_0 + \varepsilon h) \overline{T_{mn}^{\varepsilon}(\omega_0 + \varepsilon h - \varepsilon^2 \xi)}]$$

$$\times e^{i\frac{(\beta_j - \beta_m)(\omega_0)L}{\varepsilon^2}} e^{i[\beta_j'(\omega_0)L - t_0]\xi} \, dh \, d\xi \,.$$

We can now apply Proposition 20.7 and obtain

$$\mathbb{E}\left[|p_{\text{tr}}^{\varepsilon}(t_0, t, \mathbf{x}, L)|^2\right] = \varepsilon I_1^{\varepsilon} + \varepsilon I_2^{\varepsilon} \,,$$

$$I_1^{\varepsilon} \overset{\varepsilon \to 0}{\sim} \frac{1}{16\pi^2} \int \int \sum_{j \neq m=1}^{N} \phi_j(\mathbf{x}) \phi_j(\mathbf{x}_0) \phi_m(\mathbf{x}) \phi_m(\mathbf{x}_0) e^{i\frac{(\beta_j - \beta_m)(\omega_0)L}{\varepsilon^2}}$$

$$\times |\hat{f}(h)|^2 e^{Q_{jm}(\omega_0)L} e^{i[\beta_j'(\omega_0)L - t_0]\xi} \, dh \, d\xi \,,$$

$$I_2^{\varepsilon} \overset{\varepsilon \to 0}{\sim} \frac{1}{16\pi^2} \int \int \sum_{j,l=1}^{N} \frac{\beta_l}{\beta_j} \phi_j^2(\mathbf{x}) \phi_l^2(\mathbf{x}_0)$$

$$\times |\hat{f}(h)|^2 \int \mathcal{W}_j^{(l)}(\omega_0, \tau, L) e^{i\xi\tau} e^{-it_0\xi} \, d\tau \, dh \, d\xi \,.$$

Carrying out the ξ integral in the limit expression for I_1^{ε} shows that it is concentrated at the times $t_0 = \bar{t}_j$, $j = 1, \ldots, N$. Since t_0 is different from this set of times, it follows that I_1^{ε} goes to zero as $\varepsilon \to 0$. The ξ integral can also be done in the limit expression for I_2^{ε}, and we have

$$\lim_{\varepsilon \to 0} I_2^{\varepsilon} = \frac{1}{8\pi} \int |\hat{f}(h)|^2 \, dh \times \sum_{j,l=1}^{N} \frac{\beta_l}{\beta_j} \phi_j^2(\mathbf{x}) \phi_l^2(\mathbf{x}_0) \int \mathcal{W}_j^{(l)}(\omega_0, \tau, L) \delta(\tau - t_0) \, d\tau \,.$$

The measure $\mathcal{W}_j^{(l)}$ contains a Dirac mass at $\tau = \bar{t}_j$ for $j = l$. Since t_0 is different from this set of times, only the absolutely continuous part of $\mathcal{W}_j^{(l)}$ contributes. The mean transmitted intensity is therefore locally stationary in time because it depends only on t_0 and not on t:

$$\lim_{\varepsilon \to 0} \frac{1}{\varepsilon} \mathbb{E}\left[|p_{\text{tr}}^{\varepsilon}(t_0, t, \mathbf{x}, L)|^2\right] = I(t_0, \mathbf{x}, L) \,, \tag{20.94}$$

where

$$I(t_0, \mathbf{x}, L) = \frac{F_0}{4} \sum_{j,l=1}^{N(\omega_0)} \frac{\beta_l(\omega_0)}{\beta_j(\omega_0)} \phi_j^2(\mathbf{x}) \phi_l^2(\mathbf{x}_0) \mathcal{W}_{j,c}^{(l)}(\omega_0, t_0, L) \,, \tag{20.95}$$

$$F_0 = \frac{1}{2\pi} \int |\hat{f}(\omega)|^2 \, d\omega \,.$$

In the same way we get the following asymptotic expression for the auto-correlation function of the transmitted field:

$$\lim_{\varepsilon \to 0} \frac{1}{\varepsilon} e^{i\frac{\omega_0(s-t)}{\varepsilon^2}} \mathbb{E}\left[p_{\mathrm{tr}}^\varepsilon(t_0, t, \mathbf{x}, L)\, p_{\mathrm{tr}}^\varepsilon(t_0, s, \mathbf{y}, L)\right] = c_{\omega_0, t_0}(\mathbf{x}, \mathbf{y}) F(t - s), \quad (20.96)$$

where

$$c_{t_0, \omega_0}(\mathbf{x}, \mathbf{y}) = \frac{1}{4} \sum_{j,l=1}^{N(\omega_0)} \frac{\beta_l(\omega_0)}{\beta_j(\omega_0)} \phi_j(\omega_0, \mathbf{x}) \phi_j(\mathbf{y}) \phi_l^2(\mathbf{x}_0) \mathcal{W}_{j,c}^{(l)}(\omega_0, t_0, L),$$

$$F(t) = \frac{1}{2\pi} \int |\hat{f}(\omega)|^2 e^{i\omega t}\, d\omega.$$

In the asymptotic equipartition regime $L \gg L_{\mathrm{equip}}$, the functions $\mathcal{W}_j^{(l)}$ become independent of j and l and are given by (20.93). The limit mean transmitted intensity becomes in this regime

$$I(t_0, \mathbf{x}, L) \overset{L \gg L_{\mathrm{equip}}}{\sim} F_0 H_{\omega_0, \mathbf{x}_0}(\mathbf{x}) K_{\omega_0}(t_0), \quad (20.97)$$

where $H_{\omega_0, \mathbf{x}_0}$ and K_{ω_0} are given by

$$H_{\omega_0, \mathbf{x}_0}(\mathbf{x}) = \frac{1}{4N(\omega_0)} \sum_{j=1}^{N(\omega_0)} \frac{\phi_j^2(\mathbf{x})}{\beta_j(\omega_0)} \times \sum_{l=1}^{N(\omega_0)} \phi_l^2(\mathbf{x}_0) \beta_l(\omega_0), \quad (20.98)$$

$$K_{\omega_0}(t) = \frac{1}{\sqrt{2\pi \sigma_{\beta'(\omega_0)}^2 L}} \exp\left(-\frac{(t - \widehat{\beta'(\omega_0)} L)^2}{2\sigma_{\beta'(\omega_0)}^2 L}\right). \quad (20.99)$$

The autocorrelation function becomes

$$c_{t_0, \omega_0}(\mathbf{x}, \mathbf{y}) \overset{L \gg L_{\mathrm{equip}}}{\sim} H_{\omega_0, \mathbf{x}_0}^{(2)}(\mathbf{x}, \mathbf{y}) K_{\omega_0}(t_0), \quad (20.100)$$

where $H_{\omega_0, \mathbf{x}_0}^{(2)}$ is given by

$$H_{\omega_0, \mathbf{x}_0}^{(2)}(\mathbf{x}, \mathbf{y}) = \frac{1}{4N(\omega_0)} \sum_{j=1}^{N(\omega_0)} \frac{\phi_j(\mathbf{x}) \phi_j(\mathbf{y})}{\beta_j(\omega_0)} \times \sum_{l=1}^{N(\omega_0)} \phi_l^2(\mathbf{x}_0) \beta_l(\omega_0). \quad (20.101)$$

These results show that in the equipartition regime the autocorrelation function has a universal form, in the sense that it depends only on the unperturbed waveguide, and not on the statistics of the random perturbations.

For the planar waveguide, and in the continuum limit $N(\omega_0) \gg 1$, we have

$$H_{\omega_0, x_0}(x) \overset{N(\omega_0) \gg 1}{\sim} \frac{\pi^2}{16\lambda_0 d}, \quad (20.102)$$

$$H_{\omega_0, x_0}^{(2)}(x, y) \overset{N(\omega_0) \gg 1}{\sim} \frac{\pi^2}{16\lambda_0 d} \times J_0\left(\frac{2\pi(x - y)}{\lambda_0}\right), \quad (20.103)$$

where J_0 is the zero-order Bessel function. We see from these results that the mean intensity becomes uniform across the waveguide cross-section, and that the spatial extent of the autocorrelation function is of the order of the wavelength. These are characteristic properties of the statistics of the speckle pattern of the transmitted field.

Summary. The transmitted field has a coherent part, which is a train of short pulses centered at the times $\bar{t}_j = \beta'_j(\omega_0)L$, $j = 1, \ldots, N$. The amplitudes of these pulses are of order one, their supports in time are of order ε, and they decay exponentially with propagation distance as described in Proposition 20.5.

The transmitted field has also an incoherent part, whose typical amplitude is of order $\sqrt{\varepsilon}$ and whose support in time is of order one. This field has zero mean and variance (20.94). It becomes dominant for long propagation distances, where its time profile becomes a Gaussian centered at $\widetilde{\beta'(\omega_0)}L$ and with a width increasing as $\sqrt{L}$.

20.8 Narrowband Pulse Propagation in Waveguides

We consider the same situation as in Section 20.4, with a source term of the form (20.61), but we assume here that $q = 2$, so that

$$f^\varepsilon(t) = f(\varepsilon^2 t)e^{i\omega_0 t} \tag{20.104}$$

in the time domain, and in the Fourier domain

$$\hat{f}^\varepsilon(\omega) = \frac{1}{\varepsilon^2}\hat{f}\left(\frac{\omega - \omega_0}{\varepsilon^2}\right).$$

A pulse width of order ε^{-2} is comparable to the travel time over the propagation distance. As a result, the modes overlap significantly during the propagation.

The transmitted field at time t/ε^2 has the form

$$p^\varepsilon_{\text{tr}}(t, \mathbf{x}, L) = p_{\text{tr}}\left(\frac{t}{\varepsilon^2}, \mathbf{x}, \frac{L}{\varepsilon^2}\right),$$

$$p^\varepsilon_{\text{tr}}(t, \mathbf{x}, L) = \frac{1}{4\pi\varepsilon^2}\int \sum_{j,l=1}^{N(\omega)} \frac{\sqrt{\beta_l(\omega)}}{\sqrt{\beta_j(\omega)}}\phi_j(\mathbf{x})\phi_l(\mathbf{x}_0)\hat{f}\left(\frac{\omega - \omega_0}{\varepsilon^2}\right)$$

$$\times T^\varepsilon_{jl}(\omega)e^{i\frac{\beta_j(\omega)L - \omega t}{\varepsilon^2}}\, d\omega.$$

We change variables $\omega = \omega_0 + \varepsilon^2 h$ and we expand $\beta_j(\omega_0 + \varepsilon^2 h)$ with respect to ε:

$$p^\varepsilon_{\text{tr}}(t, \mathbf{x}, L) = \frac{1}{4\pi}\int \sum_{j,l=1}^{N} \frac{\sqrt{\beta_l}}{\sqrt{\beta_j}}\phi_j(\mathbf{x})\phi_l(\mathbf{x}_0)\hat{f}(h)T^\varepsilon_{jl}(\omega_0 + \varepsilon^2 h)$$

$$\times e^{i\frac{\beta_j(\omega_0)L - \omega_0 t}{\varepsilon^2}}e^{i[\beta'_j(\omega_0)L - t]h}\, dh.$$

As before in this chapter, we do not show the dependence of N on ω_0 after we approximate $N(\omega_0 + \varepsilon^2 h)$ by $N(\omega_0)$.

20.8.1 Narrowband Pulse Propagation in a Homogeneous Waveguide

In a homogeneous waveguide we have that $T_{jl}^\varepsilon = \delta_{jl}$ and

$$
\begin{aligned}
p_{\mathrm{tr}}^\varepsilon(t, \mathbf{x}, L) &= \frac{1}{4\pi} \int \sum_{j=1}^{N} \phi_j(\mathbf{x})\phi_j(\mathbf{x}_0)\hat{f}(h) e^{i \frac{\beta_j(\omega_0)L - \omega_0 t}{\varepsilon^2}} e^{i[\beta_j'(\omega_0)L - t]h}\, dh \\
&= \frac{1}{2} \sum_{j=1}^{N} \phi_j(\mathbf{x})\phi_j(\mathbf{x}_0) e^{i \frac{\beta_j(\omega_0)L - \omega_0 t}{\varepsilon^2}} f\left(t - \beta_j'(\omega_0)L\right).
\end{aligned}
$$

The transmitted field is therefore a superposition of modes, each of which is centered at its travel time $\beta_j'(\omega)L$. The modal dispersion makes the overall spreading of the transmitted field linearly increasing with L.

20.8.2 The Mean Field in a Random Waveguide

The mean transmitted field is calculated using Proposition 20.2 with the special initial conditions $\hat{a}_j(0) = \delta_{jl}$. We express the result in the form of a new proposition.

Proposition 20.8. *The mean transmission coefficients* $\mathbb{E}[T_{jl}^\varepsilon(\omega, L)]$ *converge to zero as* $\varepsilon \to 0$ *if* $j \neq l$ *and to* $\bar{T}_j(\omega, L)$ *if* $j = l$, *where* $(\bar{T}_j(\omega, L))_{j=1,\ldots,N(\omega)}$ *is given by*

$$
\bar{T}_j(\omega, L) = \exp\left(\frac{\Gamma_{jj}^{(c)}(\omega)L}{2} - \frac{\Gamma_{jj}^{(1)}(\omega)L}{2} + \frac{i\Gamma_{jj}^{(s)}(\omega)L}{2} + i\kappa_j(\omega)L \right). \tag{20.105}
$$

In the asymptotic regime $\varepsilon \to 0$, the mean transmitted field is given by

$$
\begin{aligned}
\mathbb{E}[p_{\mathrm{tr}}^\varepsilon(t, \mathbf{x}, L)] &= \frac{1}{4\pi} \int \sum_{j=1}^{N} \phi_j(\mathbf{x})\phi_j(\mathbf{x}_0)\hat{f}(h)\bar{T}_j(\omega_0, L) e^{i \frac{\beta_j(\omega_0)L - \omega_0 t}{\varepsilon^2}} \\
&\qquad\qquad\qquad\qquad \times e^{i[\beta_j'(\omega_0)L - t]h}\, dh \\
&= \frac{1}{2} \sum_{j=1}^{N} \phi_j(\mathbf{x})\phi_j(\mathbf{x}_0)\bar{T}_j(\omega_0, L) e^{i \frac{\beta_j(\omega_0)L - \omega_0 t}{\varepsilon^2}} f\left(t - \beta_j'(\omega_0)L\right).
\end{aligned}
$$

The mean field is still a superposition of modes, but they are exponentially damped and vanish for L large, $L > L_{\mathrm{equip}}(\omega_0)$. Therefore, the mean field vanishes for large L. We now turn our attention to the mean intensity, which accounts for the conversion of the coherent field into incoherent wave fluctuations.

20.8.3 The Mean Intensity in a Random Waveguide

We express the transmitted intensity as the expectation of a double integral

$$
\mathbb{E}\left[|p_{\mathrm{tr}}^{\varepsilon}(t,\mathbf{x},L)|^2\right] = \frac{1}{16\pi^2}\sum_{j,l=1}^{N}\sum_{m,n=1}^{N}\frac{\sqrt{\beta_l\beta_n}}{\sqrt{\beta_j\beta_m}}\phi_j(\mathbf{x})\phi_l(\mathbf{x_0})\phi_m(\mathbf{x})\phi_n(\mathbf{x_0})
$$

$$
\times e^{i\frac{[\beta_j(\omega_0)-\beta_m(\omega_0)]L}{\varepsilon^2}}\int\int \hat{f}(h)\,\overline{\hat{f}(h')}\,\mathbb{E}[T_{jl}^{\varepsilon}(\omega_0+\varepsilon^2 h)\overline{T_{mn}^{\varepsilon}(\omega_0+\varepsilon^2 h')}]
$$

$$
\times e^{i[\beta_j'(\omega_0)L-t]h-[\beta_m'(\omega_0)L-t]h'}\,dh\,dh'.
$$

Using Proposition 20.7 we see that there are two contributions to this integral:

$$
\mathbb{E}\left[|p_{\mathrm{tr}}^{\varepsilon}(t,\mathbf{x},L)|^2\right] = I_1^{\varepsilon}(t,\mathbf{x},L) + I_2^{\varepsilon}(t,\mathbf{x},L), \qquad (20.106)
$$

where

$$
I_1^{\varepsilon}(t,\mathbf{x},L) \overset{\varepsilon\to 0}{\sim} \frac{1}{16\pi^2}\sum_{j\neq m=1}^{N}\phi_j(\mathbf{x})\phi_j(\mathbf{x_0})\phi_m(\mathbf{x})\phi_m(\mathbf{x_0})e^{i\frac{[\beta_j(\omega_0)-\beta_m(\omega_0)]L}{\varepsilon^2}}
$$

$$
\times\int\int \hat{f}(h)\,\overline{\hat{f}(h')}\,e^{Q_{jm}(\omega_0)L}e^{i[\beta_j'(\omega_0)L-t]h-i[\beta_m'(\omega_0)L-t]h'}\,dh\,dh',
$$

$$
I_2^{\varepsilon}(t,\mathbf{x},L) \overset{\varepsilon\to 0}{\sim} \frac{1}{16\pi^2}\sum_{j,l=1}^{N}\frac{\beta_l}{\beta_j}\phi_j^2(\mathbf{x})\phi_l^2(\mathbf{x_0})
$$

$$
\times\int\int \hat{f}(h)\,\overline{\hat{f}(h')}\int \mathcal{W}_j^{(l)}(\omega_0,\tau,L)e^{i(h-h')\tau}\,d\tau\,e^{i(h'-h)t}\,dh\,dh'.
$$

The limit of the first contribution is

$$
I_1^{\varepsilon}(t,\mathbf{x},L) \overset{\varepsilon\to 0}{\sim} \frac{1}{4}\sum_{j\neq m=1}^{N}\phi_j(\mathbf{x})\phi_j(\mathbf{x_0})\phi_m(\mathbf{x})\phi_m(\mathbf{x_0})e^{i\frac{[\beta_j(\omega_0)-\beta_m(\omega_0)]L}{\varepsilon^2}}
$$

$$
\times e^{Q_{jm}(\omega_0)L}f(t-\beta_j'(\omega_0)L)f(t-\beta_m'(\omega_0)L). \qquad (20.107)
$$

We see that it decays exponentially with the propagation distance because of the damping factors $\exp(Q_{jm}(\omega_0)L)$. We can therefore neglect this contribution for $L \gg L_{\mathrm{equip}}(\omega_0)$.

The limit of the second contribution is

$$
I_2^{\varepsilon}(t,\mathbf{x},L) \overset{\varepsilon\to 0}{\sim} \frac{1}{4}\sum_{j,l=1}^{N}\frac{\beta_l}{\beta_j}\phi_j^2(\mathbf{x})\phi_l^2(\mathbf{x_0})\int \mathcal{W}_j^{(l)}(\omega_0,\tau,L)f(t-\tau)^2\,d\tau.
$$

The measures $\mathcal{W}_j^{(l)}(\omega_0,\tau,L)$ admit densities for $j \neq l$, but the measure $\mathcal{W}_l^{(l)}(\omega_0,\tau,L)$ also possesses a Dirac mass with weight $\exp(\Gamma_{ll}^{(c)}L)$ at $\tau = \beta_j'(\omega)L$. As a result, we can write

$$\lim_{\varepsilon \to 0} I_2^\varepsilon(t, \mathbf{x}, L) = \frac{1}{4} \sum_{l=1}^{N} \phi_l^2(\mathbf{x}) \phi_l^2(\mathbf{x}_0) \exp(\Gamma_{ll}^{(c)}(\omega_0)L) f(t - \beta_l'(\omega_0)L)^2$$

$$+ \frac{1}{4} \sum_{j,l=1}^{N} \frac{\beta_l}{\beta_j} \phi_j^2(\mathbf{x}) \phi_l^2(\mathbf{x}_0) \int \mathcal{W}_{j,c}^{(l)}(\omega_0, \tau, L) f(t - \tau)^2 \, d\tau. \quad (20.108)$$

The first sum is exponentially decaying for large L because of the damping factor. The second term is the main contribution. In the asymptotic equipartition regime $L \gg L_{\text{equip}}(\omega_0)$ we use the diffusion approximation (20.93). We conclude that

$$\lim_{\varepsilon \to 0} \mathbb{E}\left[|p_{\text{tr}}^\varepsilon(t, \mathbf{x}, L)|^2 \right] \overset{L \gg L_{\text{equip}}}{\sim} H_{\omega_0, \mathbf{x}_0}(\mathbf{x}) \times [K_{\omega_0} * (f^2)](t), \quad (20.109)$$

where $H_{\omega_0, \mathbf{x}_0}$ and K_{ω_0} are given by (20.98–20.99).

Summary. The main results of this section are:

- The mean field decays exponentially with propagation distance.
- The mean transmitted intensity converges to a stationary transverse spatial profile $H_{\omega_0, \mathbf{x}_0}$.
- The mean transmitted intensity is concentrated around the time $\widehat{\beta'(\omega_0)}L$ with a spread that is of order $\sigma_{\beta'(\omega_0)}\sqrt{L} \sim \sqrt{L L_{\text{equip}}(\omega_0)}/\bar{c}$ for a pulse with carrier frequency ω_0. Note that $\sigma_{\beta'(\omega_0)}\sqrt{L} \ll L/\bar{c}$, which means that the time spread increases as $\sqrt{L}$ in a random waveguide, while it increases linearly in a homogeneous one. This is because the modes are strongly coupled together and propagate with the same "average" group velocity $1/\widehat{\beta'(\omega_0)}$ in the random waveguide. The "average" group velocity is actually the harmonic average of the group velocities of the modes $1/\beta_j'(\omega_0)$.

20.9 Time Reversal for a Narrowband Pulse

We consider the time-reversal setup of Section 20.5, but we now consider narrowband pulses of the form (20.104). We observe the refocused field at time $t_{\text{obs}}/\varepsilon^2$:

$$p_{\text{ref(TR)}}\left(\frac{t_{\text{obs}}}{\varepsilon^2}, \mathbf{x}, 0\right) = \frac{1}{4\pi\varepsilon^2} \sum_{j,l,m,n=1}^{N} \frac{\sqrt{\beta_l \beta_m}}{\sqrt{\beta_j \beta_n}} M_{mj} \phi_n(\mathbf{x}) \phi_l(\mathbf{x}_0) \int \overline{\hat{f}\left(\frac{\omega - \omega_0}{\varepsilon^2}\right)}$$

$$\times \overline{T_{jl}^\varepsilon(\omega)} T_{mn}^\varepsilon(\omega) e^{i[\beta_m(\omega) - \beta_j(\omega)]\frac{L}{\varepsilon^2} + i\omega \frac{t_1 - t_{\text{obs}}}{\varepsilon^2}} \, d\omega.$$

We make the change of variable $\omega = \omega_0 + \varepsilon^2 h$,

$$p_{\text{ref(TR)}}\left(\frac{t_{\text{obs}}}{\varepsilon^2}, \mathbf{x}, 0\right) = \frac{1}{4\pi} \sum_{j,l,m,n=1}^{N} \frac{\sqrt{\beta_l \beta_m}}{\sqrt{\beta_j \beta_n}} M_{mj} \phi_n(\mathbf{x}) \phi_l(\mathbf{x}_0) \int \overline{\hat{f}(h)}$$

$$\times \overline{T_{jl}^\varepsilon(\omega_0 + \varepsilon^2 h)} T_{mn}^\varepsilon(\omega_0 + \varepsilon^2 h) e^{i[\beta_m - \beta_j](\omega_0 + \varepsilon^2 h)\frac{L}{\varepsilon^2} + i\omega_0 \frac{t_1 - t_{\text{obs}}}{\varepsilon^2} + ih(t_1 - t_{\text{obs}})} \, dh,$$

and expand $\beta_j(\omega_0 + \varepsilon^2 h)$ with respect to ε,

$$p_{\text{ref(TR)}}\left(\frac{t_{\text{obs}}}{\varepsilon^2}, \mathbf{x}, 0\right) = \frac{1}{4\pi} \sum_{j,l,m,n=1}^{N} \frac{\sqrt{\beta_l \beta_m}}{\sqrt{\beta_j \beta_n}} M_{mj} \phi_n(\mathbf{x}) \phi_l(\mathbf{x}_0)$$

$$\times e^{i[\beta_m - \beta_j](\omega_0)\frac{L}{\varepsilon^2} + i\omega_0 \frac{t_1 - t_{\text{obs}}}{\varepsilon^2}}$$

$$\times \int \overline{\hat{f}(h)} \overline{T_{jl}^\varepsilon(\omega_0 + \varepsilon^2 h)} T_{mn}^\varepsilon(\omega_0 + \varepsilon^2 h) e^{i\{[\beta_m' - \beta_j'](\omega_0)L + (t_1 - t_{\text{obs}})\}h} \, dh \,.$$

20.9.1 Refocusing in a Homogeneous Waveguide

In the homogeneous case, $T_{jl}^\varepsilon = \delta_{jl}$ and the refocused field is

$$p_{\text{ref(TR)}}\left(\frac{t_{\text{obs}}}{\varepsilon^2}, \mathbf{x}, 0\right) = \frac{1}{2} e^{i\omega_0 \frac{t_1 - t_{\text{obs}}}{\varepsilon^2}} \sum_{j,m=1}^{N} e^{i[\beta_m - \beta_j](\omega_0)\frac{L}{\varepsilon^2}}$$

$$\times M_{mj} \phi_m(\mathbf{x}) \phi_j(\mathbf{x}_0) f\left([\beta_m' - \beta_j'](\omega_0)L + t_1 - t_{\text{obs}}\right) \,.$$

The refocused field is a weighted sum of modes. The weights depend on the size of the mirror through the coefficients M_{mj}.

20.9.2 The Mean Refocused Field in a Random Waveguide

The statistical stability that we have in the broadband case, discussed in Section 20.5.4, does not carry over to the narrowband case. This is because the decoherence frequency, which is of order ε^2, is comparable to the bandwidth, which is also of order ε^2 in the narrowband case. We compute first the mean refocused field and consider the statistical stability thereafter.

The mean refocused field is

$$\mathbb{E}\left[p_{\text{ref(TR)}}\left(\frac{t_{\text{obs}}}{\varepsilon^2}, \mathbf{x}, 0\right)\right] = \frac{1}{2} \sum_{j,l,m,n=1}^{N} \frac{\sqrt{\beta_l \beta_m}}{\sqrt{\beta_j \beta_n}} M_{mj} \phi_n(\mathbf{x}) \phi_l(\mathbf{x}_0).$$

$$\times e^{i[\beta_m - \beta_j](\omega_0)\frac{L}{\varepsilon^2} + i\omega_0 \frac{t_1 - t_{\text{obs}}}{\varepsilon^2}} \mathbb{E}[\overline{T_{jl}^\varepsilon(\omega_0)} T_{mn}^\varepsilon(\omega_0)] f\left([\beta_m' - \beta_j'](\omega_0)L + t_1 - t_{\text{obs}}\right) \,.$$

From Proposition 20.7 we have the limit values of the expectations of products of two transmission coefficients, so we can write

$$\mathbb{E}\left[p_{\text{ref(TR)}}\left(\frac{t_{\text{obs}}}{\varepsilon^2}, \mathbf{x}, 0\right)\right] = p_1^\varepsilon + p_2^\varepsilon \,,$$

$$p_1^\varepsilon \overset{\varepsilon \to 0}{\sim} \frac{1}{2} \sum_{j \neq m=1}^{N} M_{mj} \phi_m(\mathbf{x}) \phi_j(\mathbf{x}_0) e^{i[\beta_m - \beta_j](\omega_0)\frac{L}{\varepsilon^2} + i\omega_0 \frac{t_1 - t_{\text{obs}}}{\varepsilon^2}}$$

$$\times e^{Q_{jm}(\omega_0)L} f\left([\beta_m' - \beta_j'](\omega_0)L + t_1 - t_{\text{obs}}\right) \,,$$

$$p_2^\varepsilon \overset{\varepsilon \to 0}{\sim} \frac{1}{2} e^{i\omega_0 \frac{t_1 - t_{\text{obs}}}{\varepsilon^2}} f(t_1 - t_{\text{obs}}) \sum_{j,l=1}^{N} M_{jj} \phi_l(\mathbf{x}) \phi_l(\mathbf{x}_0) T_j^{(l)}(\omega_0, L) \,.$$

The term p_1^ε decays exponentially with propagation distance because of the damping factors coming from Q_{jm}. We can therefore neglect this term in the asymptotic equipartition regime. The term p_2^ε does contribute. It refocuses around the time $t_{\text{obs}} = t_1$ with the original pulse shape, time-reversed. The spatial focusing profile is a weighted sum of modes, with weights that depend on the size of the time-reversal mirror through the coefficients M_{jl} and on the mean square transmission coefficients $\mathcal{T}_j^{(l)}$.

In the asymptotic equipartition regime $L \gg L_{\text{equip}}$, the coefficients $\mathcal{T}_j^{(l)}(\omega_0, L)$ converge to $1/N$ for all j and l, which gives for the mean refocused field

$$\lim_{\varepsilon \to 0} \mathbb{E}\left[p_{\text{ref(TR)}}\left(\frac{t_{\text{obs}}}{\varepsilon^2}, \mathbf{x}, 0 \right) \right] \overset{L \gg L_{\text{equip}}}{\sim} e^{i\omega_0 \frac{t_1 - t_{\text{obs}}}{\varepsilon^2}} f\left(t_1 - t_{\text{obs}} \right)$$

$$\times \frac{1}{N(\omega_0)} \sum_{j=1}^{N(\omega_0)} M_{jj} \times \frac{1}{2} \sum_{l=1}^{N(\omega_0)} \phi_l(\mathbf{x})\phi_l(\mathbf{x}_0). \quad (20.110)$$

The spatial refocusing profile can then be computed explicitly because it does not depend on the mirror shape or size. In the case of a planar waveguide, we have $\phi_j(x) = \sqrt{2/d}\sin(\pi j x/d)$, $\beta_j = \sqrt{\omega^2/c^2 - \pi^2 j^2/d^2}$, and in the continuum limit $N \gg 1$ we have

$$\frac{1}{2} \sum_{l=1}^{N} \phi_l(x)\phi_l(x_0) \overset{N \gg 1}{\sim} \frac{1}{\lambda_0}\text{sinc}\left(2\pi \frac{x - x_0}{\lambda_0} \right).$$

The mean refocused pulse is therefore concentrated around the original source location x_0 with a resolution of half a wavelength, which is the diffraction limit.

20.9.3 Statistical Stability of the Refocused Field

As we noted already, we cannot claim that the refocused field is statistically stable by using the same argument as in the broadband case, in Section 20.5.4, because here we have a narrowband pulse. However, we can achieve statistical stability through the summation over the modes. We will show this result in the quasimonochromatic case in which the pulse envelope is $f(t) = 1$ and $\hat{f}(h) = 2\pi\delta(h)$. In this case, it is clear that statistical stability cannot arise from time averaging, and the refocused field is

$$p_{\text{ref(TR)}}\left(\frac{t_{\text{obs}}}{\varepsilon^2}, \mathbf{x}, 0 \right) = \frac{1}{2} \sum_{j,l,m,n=1}^{N} \frac{\sqrt{\beta_l \beta_m}}{\sqrt{\beta_j \beta_n}} M_{mj}\phi_n(\mathbf{x})\phi_l(\mathbf{x}_0)$$

$$\times e^{i[\beta_m - \beta_j](\omega_0)\frac{L}{\varepsilon^2} + i\omega_0 \frac{t_1 - t_{\text{obs}}}{\varepsilon^2}} \overline{T_{jl}^\varepsilon(\omega_0)} T_{mn}^\varepsilon(\omega_0).$$

From (20.110), the mean refocused pulse at $\mathbf{x} = \mathbf{x}_0$ is in the asymptotic equipartition regime

$$\lim_{\varepsilon \to 0} \mathbb{E} \left[p_{\text{ref(TR)}} \left(\frac{t_{\text{obs}}}{\varepsilon^2}, \mathbf{x}_0, 0 \right) \right] \overset{L \gg L_{\text{equip}}}{\sim} e^{i\omega_0 \frac{t_1 - t_{\text{obs}}}{\varepsilon^2}} \frac{1}{N} \sum_{j=1}^{N} M_{jj} \times \frac{1}{2} \sum_{l=1}^{N} \phi_l^2(\mathbf{x}_0).$$

(20.111)

We now compute the second moment of the refocused field observed at $\mathbf{x} = \mathbf{x}_0$:

$$\mathbb{E} \left[\left| p_{\text{ref(TR)}} \left(\frac{t_{\text{obs}}}{\varepsilon^2}, \mathbf{x}_0, 0 \right) \right|^2 \right] = \frac{1}{4} \sum_{j,m,j',m'=1}^{N} \frac{\sqrt{\beta_l \beta_m \beta_{l'} \beta_{m'}}}{\sqrt{\beta_j \beta_n \beta_{j'} \beta_{n'}}} M_{mj} M_{m'j'}$$

$$\times \phi_l(\mathbf{x}_0) \phi_n(\mathbf{x}_0) \phi_{l'}(\mathbf{x}_0) \phi_{n'}(\mathbf{x}_0) e^{i[\beta_m - \beta_j + \beta_{m'} - \beta_{j'}](\omega_0) \frac{L}{\varepsilon^2}} \mathbb{E}[\overline{T_{jl}^\varepsilon} T_{mn}^\varepsilon \overline{T_{j'l'}^\varepsilon} T_{m'n'}^\varepsilon].$$

Using the results of Section 20.3.4 regarding fourth-order moments of the transfer matrix, we have

$$\lim_{\varepsilon \to 0} \mathbb{E}[\overline{T_{jl}^\varepsilon} T_{mn}^\varepsilon \overline{T_{j'l'}^\varepsilon} T_{m'n'}^\varepsilon]$$

$$\overset{L \gg L_{\text{equip}}}{\sim} \begin{cases} \frac{2}{N(N+1)} & \text{if } (j,l) = (m,n) = (j',l') = (m',n'), \\ \frac{1}{N(N+1)} & \text{if } (j,l) = (m,n) \neq (j',l') = (m',n'), \\ \frac{1}{N(N+1)} & \text{if } (j,l) = (m',n') \neq (j',l') = (m,n), \\ 0 & \text{otherwise.} \end{cases}$$

Using these fourth-order moment results in the expression for the second moment of the refocused field, we see that in the limit $\varepsilon \to 0$ and in the asymptotic equipartition regime $L \gg L_{\text{equip}}$,

$$\lim_{\varepsilon \to 0} \mathbb{E} \left[\left| p_{\text{ref(TR)}} \left(\frac{t_{\text{obs}}}{\varepsilon^2}, \mathbf{x}_0, 0 \right) \right|^2 \right]$$

$$\overset{L \gg L_{\text{equip}}}{\sim} \frac{R_0^4}{4N(N+1)} \left[2 \sum_j M_{jj}^2 + \sum_{j \neq j'} M_{jj} M_{j'j'} + \sum_{j \neq j'} M_{jj'}^2 \right]$$

$$\overset{L \gg L_{\text{equip}}}{\sim} \frac{R_0^4}{4N(N+1)} \left[\left(\sum_j M_{jj} \right)^2 + \sum_{j,j'} M_{jj'}^2 \right],$$

where $R_0^2 = \sum_{l=1}^{N} \phi_l^2(\mathbf{x}_0)$. Let us introduce the relative standard deviation S of the refocused field amplitude

$$S^2 := \lim_{\varepsilon \to 0} \frac{\mathbb{E} \left[\left| p_{\text{ref(TR)}} \left(\frac{t_{\text{obs}}}{\varepsilon^2}, \mathbf{x}_0, 0 \right) \right|^2 \right] - \left| \mathbb{E} \left[p_{\text{ref(TR)}} \left(\frac{t_{\text{obs}}}{\varepsilon^2}, \mathbf{x}_0, 0 \right) \right] \right|^2}{\left| \mathbb{E} \left[p_{\text{ref(TR)}} \left(\frac{t_{\text{obs}}}{\varepsilon^2}, \mathbf{x}_0, 0 \right) \right] \right|^2}.$$

We have statistical stability when S is small. From (20.111) and the definition of R_0 we have

$$\lim_{\varepsilon \to 0} \left| \mathbb{E} \left[p_{\text{ref(TR)}} \left(\frac{t_{\text{obs}}}{\varepsilon^2}, \mathbf{x}_0, 0 \right) \right] \right| \overset{L \gg L_{\text{equip}}}{\sim} \frac{R_0^2}{2N} \sum_{j=1}^{N} M_{jj}.$$

We can therefore write S^2 in the asymptotic equipartition regime as

$$S^2 \overset{L \gg L_{\text{equip}}}{\sim} -\frac{1}{N+1} + \frac{N}{N+1}\frac{1}{Q_{\text{mirror}}}, \tag{20.112}$$

where the quality factor Q_{mirror} is defined by

$$Q_{\text{mirror}} = \frac{\sum_{j,l} M_{jj}M_{ll}}{\sum_{j,l} M_{jl}^2}.$$

This quality factor depends only on the time-reversal mirror. We will have statistical stability when the number of modes N is large and when the quality factor Q_{mirror} is large. We can consider two extreme cases:

- If the time-reversal mirror spans the waveguide cross-section, then $M_{jl} = \delta_{jl}$ and the quality factor is equal to N, which is optimal since the relative standard deviation is then zero for any N. This result is not surprising since the time-reversal mirror records the transmitted signal fully, in both time and space, which implies optimal refocusing.
- If the time-reversal mirror is pointlike at $\mathbf{x}_1$, then $M_{jl} = \phi_j(\mathbf{x}_1)\phi_l(\mathbf{x}_1)$ and the quality factor is 1, which is bad, because the relative standard deviation S is asymptotically equal to $\sqrt{N-1}/\sqrt{N+1}$. The fluctuations of the refocused field are therefore of the same order as the mean field, which means that there is no statistical stability.

In the next section, we address some particular cases where explicit calculations are possible.

20.9.4 Numerical Illustration of Spatial Focusing and Statistical Stability in Narrowband Time Reversal

We consider very narrowband quasimonochromatic pulses and compare the transverse profiles of the refocused fields for a homogeneous waveguide and for a random waveguide in the equipartition regime. We consider the planar waveguide where the modes are given by $\phi_j(x) = \sqrt{2/d}\sin(\pi j x/d)$ and the modal wave numbers by $\beta_j = \sqrt{\omega^2/c^2 - \pi^2 j^2/d^2}$. We also assume that $\lambda_0 \gg d$, so that the number of propagating modes is large.

Homogeneous Waveguide

In the homogeneous case of Section 20.9.1, the spatial profile of the refocused field is

$$|p_{\text{ref(TR)}}(x)| = \frac{1}{2}\left|\sum_{j,m=1}^{N} e^{i[\beta_m-\beta_j](\omega_0)\frac{L}{c^2}} M_{mj}\phi_m(x)\phi_j(x_0)\right|.$$

For a time-reversal mirror that spans the width of the waveguide this expression becomes, in the continuum limit $N \gg 1$, the sinc profile

$$|p_{\text{ref(TR)}}(x)| = \frac{1}{\lambda_0}\left|\text{sinc}\left(\frac{2\pi(x-x_0)}{\lambda_0}\right)\right|.$$

Let us consider a time-reversal mirror of size a located in $x \in [d/2-a/2, d/2+a/2]$: $G_2(x) = \mathbf{1}_{[d/2-a/2, d/2+a/2]}(x)$. We then have

$$M_{jl} = \frac{a}{d}\left[\cos\left(\frac{(j-l)\pi}{2}\right)\text{sinc}\left(\frac{(j-l)\pi a}{2d}\right)\right.$$
$$\left. - \cos\left(\frac{(j+l)\pi}{2}\right)\text{sinc}\left(\frac{(j+l)\pi a}{2d}\right)\right].$$

Using these formulas we plot in Figure 20.2 the spatial profile of the refocused field for different sizes a of the time-reversal mirror. The peak at the original source location is there in all cases, but for smaller time-reversal mirrors, large side lobes appear.

Random Waveguide

For a random waveguide in the equipartition regime $L \gg L_{\text{equip}}$, the mean spatial profile of the refocused field is, as in Section 20.9.2,

$$|\mathbb{E}[p_{\text{ref(TR)}}(x)]| = \frac{1}{N}\sum_{j=1}^{N}M_{jj} \times \frac{1}{2}\sum_{l=1}^{N}\phi_l(x_0)\phi_l(x).$$

In the continuum limit $N \gg 1$ we obtain the diffraction-limited sinc profile

$$|\mathbb{E}[p_{\text{ref(TR)}}(x)]| \overset{N \gg 1}{\approx} \frac{1}{N}\sum_{j=1}^{N}M_{jj} \times \frac{1}{\lambda_0}\text{sinc}\left(\frac{2\pi(x-x_0)}{\lambda_0}\right).$$

The mean spatial profile is, up to an amplitude factor, independent of the mirror size. However, the statistical statistical stability of the refocused field depends, in the narrowband case, on the size of the time-reversal mirror as shown in Figure 20.3, which is a plot of S in (20.112) for different sizes of the time-reversal mirror.

Notes

This chapter is devoted to the analysis of wave propagation in waveguides with random inhomogeneities in a regime in which there is significant mode coupling between a finite number of propagating modes. The statistical analysis of the coupled mode equations at a single frequency leads to the coupled power equations for the second moments of the mode amplitudes that were previously derived in ocean acoustics [105, 48] and used in fiber optics [121]. A recent treatment of propagation in random waveguides and mostly electromagnetic applications is [150]. The system of transport equations for the

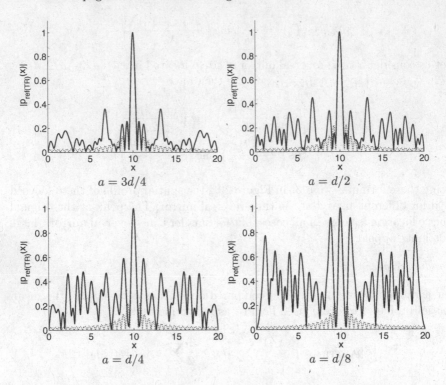

Fig. 20.2. Transverse profile of the refocused pulse in a homogeneous waveguide with diameter d. Here $d = 20$ and $\lambda_0 = 1$, so there are 40 modes. The original pulse location is $x_0 = d/2$. The dotted curve is the sinc profile, which is the focusing profile of a full-size time-reversal mirror. It is also the mean refocusing profile for a random waveguide, in the equipartition regime and for any size of time-reversal mirror. The solid curves are the refocusing profiles for time-reversal mirrors of different sizes a.

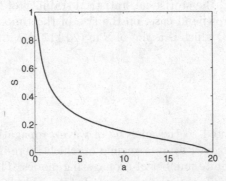

Fig. 20.3. The relative standard deviation S, from (20.112), of the refocused field as a function of the mirror size a. Here $d = 20$ and $\lambda_0 = 1$.

two-frequency autocorrelation function of the transfer matrix is derived from first principles here for the first time. The application of the transport equations to pulse propagation and time reversal is also new in this chapter.

A pulse propagating in a random waveguide acquires characteristic features that do not depend on the detailed statistics of the random perturbations. In the regime in which the transmitted field has lost memory of its original modal distribution because of strong mode coupling, the group velocity of the pulse becomes the harmonic average of the modal group velocities. In this regime, pulse spreading becomes proportional to the square root of the distance from the source, and not proportional to the distance, as it would be in a homogeneous waveguide. This effect has been known for more than thirty years [142], but we give here a complete and systematic presentation. The main mathematical tool for this is the two-frequency analysis of the mode transfer matrix.

Time reversal in waveguides is an important problem in many applications, in underwater acoustics and elsewhere, and several experiments have been carried out in media that can be modeled as randomly perturbed waveguides [113, 148]. Here we look carefully at the roles of the source bandwidth, the carrier frequency, which determines the number of propagating modes, and the size of the time-reversal mirror in the statistical stability of the refocused fields.

This chapter can be viewed as a transition from the main topic of this book, one-dimensional wave propagation or three-dimensional wave propagation in randomly layered media, to the general problem of wave propagation in three-dimensional random media. The general research area of waves in random media is huge, with dedicated journals and specialized literature. Indeed, in this chapter, we have considered waveguides with general three-dimensional random inhomogeneities, with a finite number of propagating modes. When the waveguide is very wide and the number of modes goes to infinity, then wave propagation in a random waveguide in the forward-scattering approximation tends to wave propagation in a general three-dimensional random medium in the paraxial regime [91, 16, 10]. Multimode propagation in random waveguides without the forward-scattering approximation [26] can be analyzed to some extent by the methods of this chapter. More generally, it can be analyzed in regimes other than the ones considered here, with applications in many areas of modern physics [12].

References

1. F. Kh. Abdullaev and J. Garnier, Solitons in media with random dispersive perturbations, Physica D **134** (1999), 303–315.
2. M. J. Ablowitz and H. Segur, *Solitons and the Inverse Scattering Transform*, SIAM, Philadelphia, 1981.
3. M. Abramowitz and I. Stegun, *Handbook of Mathematical Functions*, Dover Publications, New York, 1965.
4. D. G. Alfaro Vigo, J.-P. Fouque, J. Garnier, and A. Nachbin, Robustness of time reversal for waves in time-dependent random media, Stochastic Process. Appl. **111** (2004), 289–313.
5. P. W. Anderson, Absences of diffusion in certain random lattices, Phys. Rev. **109** (1958), 1492–1505.
6. L. Arnold, *Random Dynamical Systems*, Springer, Berlin, 2003.
7. L. Arnold, G. Papanicolaou, and V. Wihstutz, Asymptotic analysis of the Lyapounov exponent and rotation number of the random oscillator and applications, SIAM J. Appl. Math. **46** (1986), 427–450.
8. M. Asch, W. Kohler, G. Papanicolaou, M. Postel, and B. White, Frequency content of randomly scattered signals, SIAM Rev. **33** (1991), 519–625.
9. M. Asch, G. Papanicolaou, M. Postel, P. Sheng, and B. White, Frequency content of randomly scattered signals. Part I, Wave Motion **12** (1990), 429–450.
10. G. Bal, G. Papanicolaou, and L. Ryzhik, Self-averaging in time reversal for the parabolic wave equation, Stochastics and Dynamics **2** (2002), 507–531.
11. G. Bal and L. Ryzhik, Time reversal and refocusing in random media, SIAM J. Appl. Math **63** (2003), 1475–1498.
12. C. W. J. Beenakker, Random-matrix theory of quantum transport, Rev. Mod. Phys. **69** (1997), 731–808.
13. A. Bensoussan, J.-L. Lions, and G. Papanicolaou, *Asymptotic Analysis of Periodic Structures*, North Holland, Amsterdam, 1978.
14. G. Blankenship and G. Papanicolaou, Stability and control of systems with wide-band noise disturbances I, SIAM J. Appl. Math. **34** (1978), 437–476.
15. N. Bleisten and R. Handelsman, *Asymptotic Expansions of Integrals*, Dover, New York, 1986.
16. P. Blomgren, G. Papanicolaou, and H. Zhao, Super-resolution in time-reversal acoustics, J. Acoust. Soc. Am. **111** (2002), 230–248.

17. N. N. Bogoliubov and Y. A. Mitropolsky, *Asymptotic Methods in the Theory of Non-linear Oscillations*, Gordon and Breach, New York, 1961.
18. L. Borcea, G. Papanicolaou, and C. Tsogka, Coherent interferometry in finely layered random media, SIAM Multiscale Model. Simul. **5** (2006), 62–83.
19. L. Borcea, G. Papanicolaou, and C. Tsogka, Adaptive interferometric imaging in clutter and optimal illumination, Inverse Problems **22** (2006), 1405–1436.
20. M. Born and E. Wolf, *Principles of Optics*, Cambridge University Press, Cambridge, 1999.
21. R. Bouc and E. Pardoux, Asymptotic analysis of PDEs with wide-band noise disturbance expansion of the moments, Stochastic Anal. Appl. **2** (1984), 369–422.
22. J. Boussinesq, Théorie de l'intumescence liquide appelée onde solitaire ou de translation, se propageant dans un canal rectangulaire, C. R. Acad. Sci. Paris **72** (1871), 755–759; **73** (1871), 256–260; **73** (1871), 1210–1212.
23. L. Breiman, *Probability*, Addison Wesley, Reading, 1968.
24. L. M. Brekhovskikh, *Waves in Layered Media*, Academic Press, New York, 1980.
25. J. C. Bronski, D. W. McLaughlin, and M. J. Shelley, On the stability of time-harmonic localized states in a disordered nonlinear medium, J. Statist. Phys. **88** (1997), 1077–1115.
26. R. Burridge and G. Papanicolaou, The geometry of coupled mode propagation in one-dimensional random media, Comm. Pure Appl. Math. **25** (1972), 715–757.
27. R. Burridge, D. McLaughlin, and G. Papanicolaou, A stochastic Gaussian beam, J. Math. Phys., **14** (1973), 84–89.
28. R. Burridge and H. W. Chang, Multimode one-dimensional wave propagation in a highly discontinuous medium, Wave Motion **11** (1989), 231–249.
29. R. Burridge and K. Hsu, Effects of averaging and sampling on the statistics of reflection coefficients, Geophysics, **56** (1991), 50–58.
30. G. Papanicolaou, ed., *Wave Propagation in Complex Media*, The IMA Volumes in Mathematics and its Applications , Vol. 96, Springer, New York, 1998.
31. R. Burridge, G. Papanicolaou, P. Sheng, and B. White, Probing a random medium with a pulse, SIAM J. Appl. Math. **49** (1989), 582–607.
32. R. Burridge, G. Papanicolaou, and B. White, Statistics for pulse reflection from a randomly layered medium, SIAM J. Appl. Math. **47** (1987) 146–168.
33. R. Burridge, G. Papanicolaou, and B. White, One-dimensional wave propagation in a highly discontinuous medium, Wave Motion **10** (1988), 19–44.
34. R. E. Caflisch, M. J. Miksis, G. C. Papanicolaou, and L. Ting, Wave propagation in bubbly liquids at finite volume fraction, J. Fluid Mech. **160** (1985), 1–14.
35. M. Campillo and A. Paul, Long range correlations in the seismic coda, Science **29** (2003), 547–549.
36. R. Carmona, Random Schrödinger operators, in *École d'été de probabilités de Saint-Flour XIV*, Lecture Notes in Math., Vol. 1180, Springer, Berlin, 1986, pp. 1–124.
37. R. Carmona and J. Lacroix, *Spectral Theory of Random Schrödinger Operators*, Birkhäuser, Boston, 1990.
38. J. Chillan and J.-P. Fouque, Pressure fields generated by acoustical pulses propagating in randomly layered media, SIAM J. Appl. Math. **58** (1998) 1532–1546.
39. J.-F. Clouet and J.-P. Fouque, Spreading of a pulse traveling in random media, Ann. Appl. Probab. **4** (1994), 1083–1097.
40. J.-F. Clouet and J.-P. Fouque, A time-reversal method for an acoustical pulse propagating in randomly layered media, Wave Motion **25** (1997), 361–368.

41. H. Crauel, Lyapunov numbers of Markov solutions of linear stochastic systems, Stochastic **14** (1984), 11–28.

42. F. Dalfovo, S. Giorgini, L. P. Pitaevskii, and S. Stringary, Theory of Bose-Einstein condensation in trapped gases, Rev. Mod. Phys. **71** (1999), 463–512.

43. R. Dautray and J.-L. Lions, *Mathematical Analysis and Numerical Methods for Science and Technology*, Springer, Berlin, 2000 [*Analyse Mathématique et Calcul Numérique pour les Sciences et les Techniques*, Masson, Paris, 1987].

44. K. B. Davis, M. O. Mewes, M. R. Andrews, N. J. van Druten, D. S. Durfee, D. M. Kurn, and W. Ketterle, Bose-Einstein condensation in a gas of sodium atoms, Phys. Rev. Lett. **75** (1995), 3969–3971.

45. L. Debnath, *Nonlinear Partial Differential Equations for Scientists and Engineers*, Birkhäuser, Boston, 1997.

46. M. De Hoop, H.-W. Chang, and R. Burridge, The pseudo-primary field due to a point-source in a finely layered medium, Geophys. J. Int. **104** (1991), 489–506.

47. P. Desvillard and B. Souillard, Polynomially decaying transmission for the non-linear Schrödinger equation in a random medium, J. Statist. Phys. **43** (1986), 423–439.

48. L. B. Dozier and F. D. Tappert, Statistics of normal mode amplitudes in a random ocean, J. Acoust. Soc. Am. **63** (1978), 353–365; J. Acoust. Soc. Am. **63** (1978), 533–547.

49. V. A. Ditkin and A. P. Prudnikov, *Integral Transforms and Operational Calculus*, Pergamon Press, New York, 1965.

50. S. N. Ethier and T. G. Kurtz, *Markov Processes*, Wiley, New York, 1986.

51. L. C. Evans, *Partial Differential Equations*, American Mathematical Society, Providence, 2002.

52. A. C. Fannjiang, Self-averaging scaling limits for random parabolic waves, Arch. Rational Mech. Anal. **175** (2005), 343–387.

53. A. C. Fannjiang and K. Sølna, Superresolution and duality for time-reversal of waves in random media, Phys. Lett. A **352** (2005), 22–29.

54. W. Feller, *Introduction to Probability Theory and Its Applications*, Wiley, New York, 1971.

55. M. Fink, Time reversed acoustics, Physics Today **20** (1997), 34–40.

56. M. Fink, Time reversed acoustics, Scientific American **281**:5 (1999), 91–97.

57. M. Fink, D. Cassereau, A. Derode, C. Prada, P. Roux, M. Tanter, J.-L. Thomas, and F. Wu, Time-reversed acoustics, Reports on Progress in Physics **63** (2000), 1933–1995.

58. *Diffuse Waves in Complex Media (Les Houches, 1998)*, J.-P. Fouque, ed., NATO Sci. Ser. C Math. Phys. Sci., Vol. 531, Kluwer Acad. Publ., Dordrecht, 1999.

59. J.-P. Fouque, J. Garnier, J. C. Muñoz Grajales, and A. Nachbin, Time reversing solitary waves, Phys. Rev. Lett. **92** (2004), 094502.

60. J.-P. Fouque, J. Garnier, and A. Nachbin, Time reversal for dispersive waves in random media, SIAM J. Appl. Math. **64** (2004), 1810–1838.

61. J.-P. Fouque, J. Garnier, and A. Nachbin, Shock structure due to stochastic forcing and the time reversal of nonlinear waves, Physica D **195** (2004), 324–346.

62. J.-P. Fouque, J. Garnier, A. Nachbin, and K. Sølna, Time reversal refocusing for point source in randomly layered media, Wave Motion **42** (2005), 238–260.

63. J.-P. Fouque, J. Garnier, A. Nachbin, and K. Sølna, Imaging of a dissipative layer in a random medium using a time-reversal method, Proceedings of the

Conference MC2QMC 2004, H. Niederreiter and D. Talay, eds., Springer, Berlin, 2006, pp. 127–145.

64. J. P. Fouque and E. Merzbach, A limit theorem for linear boundary value problems in random media, Ann. Appl. Probab. **4** (1994), 549–569.

65. J.-P. Fouque and A. Nachbin, Time-reversed refocusing of surface water waves, SIAM Multiscale Model. Simul. **1** (2003), 609–629.

66. J.-P. Fouque and O. Poliannikov, Time reversal detection in one-dimensional random media, Inverse Problems **22** (2006), 903–922.

67. J.-P. Fouque and K. Sølna, Time-reversal aperture enhancement, SIAM Multiscale Model. Simul. **1** (2003), 239–259.

68. A. Friedman, *Partial Differential Equations of Parabolic Type*, Prentctice Hall, Englewood Cliffs, 1964.

69. H. Frisch and S. P. Lloyd, Electron levels in a one-dimensional random lattice, Phys. Rev. **120** (1960), 1175–1189.

70. J. Fröhlich, T. Spencer, and C. E. Wayne, Localization in disordered, nonlinear dynamical systems, J. Statist. Phys. **42** (1986), 247–274.

71. I. Gabitov and S. K. Turitsyn, Averaged pulse dynamics in a cascaded transmission system with passive dispersion compensation, Opt. Lett. **21** (1996), 327–329.

72. C. S. Gardner, J. M. Greene, M. D. Kruskal, and R. M. Miura, Method for solving the Korteweg-de Vries equation, Phys. Rev. Lett. **19** (1967), 1095–1097.

73. J. Garnier, Asymptotic transmission of solitons through random media, SIAM J. Appl. Math. **58** (1998), 1969–1995.

74. J. Garnier, Light propagation in square law media with random imperfections, Wave Motion **31** (2000), 1–19.

75. J. Garnier, Long-time dynamics of Korteweg-de Vries solitons driven by random perturbations, J. Statist. Phys. **105** (2001), 789–833.

76. J. Garnier, Imaging in randomly layered media by cross-correlating noisy signals, SIAM Multiscale Model. Simul. **4** (2005), 610–640.

77. J. Garnier, Solitons in random media with long-range correlation, Waves Random Media **11** (2001), 149–162.

78. P. Glotov, Time reversal for electromagnetic waves in randomly layered media, PhD dissertation, NC State University, 2006.

79. I. Goldsheid, S. Molchanov, and L. Pastur, A random homogeneous Schrödinger operator has a pure point spectrum, Functional Anal. Appl. **11** (1977), 1–10.

80. P. L. Goupillaud, An approach to inverse filtering of near-surface layer effects from seismic records, Geophysics, **26** (1961), 754–760.

81. S. A. Gredeskul and V. D. Freilikher, Waveguiding properties of randomly stratified media, Radiofiz. **31** (1988), 1210–1217 (861–867 in English).

82. S. A. Gredeskul and V. D. Freilikher, Localization and wave propagation in randomly layered media, Soviet Physics Usp. **33** (1990), 134–136.

83. S. A. Gredeskul and Yu. U. Kivshar, Propagation and scattering of nonlinear waves in disordered systems, Phys. Rep. **216** (1992), 1–61.

84. E. P. Gross, Structure of quantized vortex in boson systems, Nuovo Cimento **20**, 454–466 (1961); Hydrodynamics of a superfluid condensate, J. Math. Phys. **4**, 195–207 (1963); L. P. Pitaevskii, Vortex lines in an imperfect Bose gas, Sov. Phys. JETP **13**, 451–454 (1961).

85. M. A. Guzev and V. I. Klyatskin, Plane waves in a layered weakly dissipative randomly inhomogeneous medium, Waves Random Media **1** (1992), 7–19.

86. A. Hasegawa, *Optical Solitons in Fibers*, Springer, Berlin, 1989.

87. A. Hasegawa and F. Tappert, Transmission of stationary nonlinear optical pulses in dispersive dielectric fibers, I; Anomalous dispersion, Appl. Phys. Lett. **23** (1973), 142–144.

88. H. A. Haus and W. S. Wong, Solitons in optical communications, Rev. Mod. Phys. **68** (1996), 423–444.

89. M. H. Holmes *Introduction to Perturbation Methods*, Springer, New York, 1995.

90. A. Ishimaru, *Wave Propagation and Scattering in Random Media*, Academic Press, San Diego, 1978.

91. D. R. Jackson and D. R. Dowling, Phase-conjugation in underwater acoustics, J. Acoust. Soc. Am. **89** (1991), 171–181.

92. I. Karatzas and S. E. Shreve, *Brownian Motion and Stochastic Calculus, 2nd ed.*, Grad. Texts in Math. **113**, Springer-Verlag, New York, 1991.

93. V. I. Karpman, Soliton evolution in the presence of perturbations, Phys. Scr. **20** (1979), 462–478.

94. M. J. Kenna, R. L. Stanley, and J. D. Maynard, Effects of nonlinearity on Anderson localization, Phys. Rev. Lett. **69** (1992), 1807–1810; V. A. Hopkins, J. Keat, G. D. Meegan, T. Zhang, and J. D. Maynard, Observation of the predicted behavior of nonlinear pulse propagation in disordered media, Phys. Rev. Lett. **76** (1996), 1102–1104.

95. B. L. N. Kennett, *Seismic Wave Propagation in Stratified Media*, Cambridge University Press, Cambridge, 1983.

96. H. Kesten and G. Papanicolaou, A limit theorem for turbulent diffusion, Comm. Math. Phys. **65** (1979), 97–128.

97. R. Z. Khasminskii, On stochastic processes defined by differential equations with a small parameter, Theory Probab. Appl. **11** (1966), 211–228.

98. R. Z. Khasminskii, A limit theorem for solutions of differential equations with random right hand side, Theory Probab. Appl. **11** (1966), 390–406.

99. A. Kim, P. Kyritsi, P. Blomgren, and G. Papanicolaou, Spatial focusing and intersymbol interference in multiple-input single-output time-reversal communication systems, preprint 2006.

100. J.-H. Kim, Turning point problem in a one-dimensional refractive random multilayer, SIAM J. Appl. Math. **56** (1996), 1164–1180.

101. V. I. Klyatskin, *Stochastic Equations and Waves in Random Media*, Nauka, Moscow, 1980.

102. R. Knapp, Transmission of solitons through random media, Physica D **85** (1995), 496–508.

103. R. Knapp, G. Papanicolaou, and B. White, Transmission of waves by a nonlinear random medium, J. Statist. Phys. **63** (1991), 567–583.

104. W. Kohler and G. Papanicolaou, Power statistics for wave propagation in one dimension and comparison with transport theory, J. Math. Phys. **14** (1973), 1733–1745; **15** (1974), 2186–2197.

105. W. Kohler and G. Papanicolaou, Wave propagation in randomly inhomogeneous ocean, in Lecture Notes in Physics, Vol. 70, J. B. Keller and J. S. Papadakis, eds., *Wave Propagation and Underwater Acoustics*, Springer Verlag, Berlin, 1977.

106. W. Kohler, G. Papanicolaou, M. Postel, and B. White, Reflection of pulsed electromagnetic waves from a randomly stratified half-space, J. Opt. Soc. Am. A **8**, 1109–1125 (1991).

604   References

107. W. Kohler, G. Papanicolaou, and B. White, Reflection of waves generated by a point source over a randomly layered medium, Wave Motion **13** (1991), 53–87.
108. W. Kohler, G. Papanicolaou, and B. White, Localization and mode conversion for elastic waves in randomly layered media, I & II, Wave Motion **23**, 1–22 & 181–201 (1996).
109. W. Kohler, G. Papanicolaou, and B. White, Reflection and transmission of acoustic waves by a locally-layered slab, in: *Diffuse Waves in Complex Media (Les Houches, 1998)*, J.-P. Fouque, ed., NATO Sci. Ser. C Math. Phys. Sci., Vol. 531, Kluwer Acad. Publ., Dordrecht, 1999, pp. 347–382.
110. D. J. Korteweg and G. de Vries, On the change of form of long waves advancing in a rectangular canal, and on a new type of long stationary waves, Philos. Mag. Ser. **39** (1895), 422–443.
111. S. Kotani, Lyapunov indices determine absolutely continuous spectra of stationary random one-dimensional Schrödinger operators, in *Stochastic Analysis*, K. Ito ed., North Holland, Amsterdam, 1984, pp. 225–247.
112. H. Kunita, *Stochastic Flows and Stochastic Differential Equations*, Cambridge University Press, Studies in Advanced Mathematics **24**, Cambridge, 1990.
113. W. A. Kuperman, W. S. Hodgkiss, H. C. Song, T. Akal, C. Ferla, and D. R. Jackson, Phase conjugation in the ocean, experimental demonstration of an acoustic time-reversal mirror, J. Acoust. Soc. Am. **103** (1998), 25–40.
114. H. J. Kushner, *Approximation and Weak Convergence Methods for Random Processes*, MIT Press, Cambridge, 1984.
115. P. D. Lax, Integrals of nonlinear equations of evolution and solitary waves, Comm. Pure Appl. Math. **21** (1968), 467–490.
116. M. Lax and J. C. Phillips, One-dimensional impurity bands, Phys. Rev. **110** (1958), 41–49.
117. P. Lewicki, R. Burridge, and G. Papanicolaou, Pulse stabilization in a strongly heterogeneous medium, Wave Motion **20** (1994), 177–195.
118. P. Lewicki, R. Burridge, and M. De Hoop, Beyond effective medium theory: pulse stabilization for multimode wave propagation in high-contrast layered media, SIAM J. Appl. Math. **56** (1996), 256–276.
119. S. V. Manakov, S. Novikov, J. P. Pitaevskii, and V. E. Zakharov, *Theory of Solitons*, Consultants Bureau, New York, 1984.
120. L. Mandel and E. Wolf, *Optical Coherence and Quantum Optics*, Cambridge University Press, Cambridge, 1995.
121. D. Marcuse, *Theory of Dielectric Optical Waveguides*, Academic Press, New York, 1974.
122. G. Milton, *The Theory of Composites*, Cambridge University Press, Cambridge, 2001.
123. P. M. Morse and K. U. Ingard, *Theoretical Acoustics*, McGraw-Hill, New York, 1968.
124. A. Nachbin, A terrain-following Boussinesq system, SIAM J. Appl. Math. **63** (2003), 905–922.
125. A. C. Newell and J. V. Moloney, *Nonlinear Optics*, Addison-Wesley, Redwood City, 1992.
126. R. F. O'Doherty and N. A. Anstey, Reflections on amplitudes, Geophysical Prospecting **19** (1971), 430–458.
127. C. Oestges, A. D. Kim, G. Papanicolaou, and A. J. Paulraj, Characterization of space-time focusing in time reversed random fields, IEEE Trans. Antennas and Prop. **53** (2005), 283–293.

128. B. Øksendal, *Stochastic Differential Equations*, Springer, Berlin, 2000.

129. G. C. Papanicolaou, Wave propagation in a one-dimensional random medium, SIAM J. Appl. Math. **21** (1971), 13–18.

130. G. Papanicolaou, Diffusion in random media, in *Surveys in Applied Mathematics*, J. B. Keller, D. Mc Laughlin and G. Papanicolaou, eds., Plenum Press, New York, 1995, pp. 205–255.

131. G. Papanicolaou, Waves in one-dimensional random media, in *École d'été de Probabilités de Saint-Flour*, P. L. Hennequin, ed., Lecture Notes in Mathematics, Springer, 1988, pp. 205–275 .

132. G. Papanicolaou and W. Kohler, Asymptotic theory of mixing stochastic ordinary differential equations, Communications in Pure and Applied Mathematics, **27** (1974), 641–668.

133. G. Papanicolaou, M. Postel, P. Sheng, and B. White, Frequency content of randomly scattered signals. Part II: Inversion, Wave Motion, **12** (1990), 527–549.

134. G. Papanicolaou and K. Sølna, Ray theory for a locally layered random medium, Waves Random Media **10** (2000), 151–198.

135. G. Papanicolaou, D. W. Stroock, and S. R. S. Varadhan, Martingale approach to some limit theorems, in *Statistical Mechanics and Dynamical Systems*, D. Ruelle, ed., Duke Turbulence Conf., Duke Univ. Math. Series III, Part VI, 1976, pp. 1–120.

136. G. Papanicolaou, L. Ryzhik, and K. Sølna, Statistical stability in time reversal, SIAM J. Appl. Math. **64** (2004), 1133–1155.

137. G. Papanicolaou and S. Weinryb, A functional limit theorem for waves reflected by a random medium, Appl. Math. Optimiz. **30** (1991), 307–334.

138. E. Pardoux and A. Yu. Veretennikov, On the Poisson equation and diffusion approximation 1, Ann. Probab. **29** (2001), 1061–1085; ibidem 2, Ann. Probab. **31** (2003), 1166–1192; ibidem 3, Ann. Probab. **33** (2005), 1111–1133.

139. L. A. Pastur, Spectra of random self-adjoint operators, Russian Math. Surveys **28** (1973), 3–64.

140. L. Pastur and A. Figotin, *Spectra of Random and Almost Periodic Operators*, Springer Verlag, Heidelberg, 1992.

141. A. Paulraj, R. Nabar, and D. Gore, *Introduction to Space-Time Wireless Communications*, Cambridge University Press, Cambridge, 2003.

142. S. D. Personick, Time dispersion in dielectric waveguides, Bell Syst. Tech. J. **50** (1971), 843–859.

143. C. Pires and M. A. Miranda, Tsunami waveform inversion by adjoint methods, J. Geophys. Res. **106** (2001), 19733–19796.

144. C. Prada and M. Fink, Eigenmodes of the time reversal operator: A solution to selective focusing in multiple-target media, Wave Motion **20** (1994), 151–163.

145. C. Prada, S. Manneville, D. Spolianski, and M. Fink, Decomposition of the time reversal operator: Detection and selective focusing on two scatterers, J. Acoust. Soc. Am. **99** (1996), 2067–2076.

146. C. Prada, J.-L. Thomas, and M. Fink, The iterative time reversal mirror: Analysis of convergence, J. Acoust. Soc. Am. **97** (1995), 62–71.

147. C. Prada, F. Wu, and M. Fink, The iterative time reversal mirror: A solution to self-focusing in the pulse echo mode, J. Acoust. Soc. Am. **90** (1991), 1119–1129.

148. P. Roux and M. Fink, Time reversal in a waveguide: Study of the temporal and spatial focusing, J. Acoust. Soc. Am. **107** (2000), 2418–2429.

149. P. Roux and M. Fink, Greens function estimation using secondary sources in a shallow water environment, J. Acoust. Soc. Am. **133** (2003), 1406–1416.

150. H. E. Rowe, *Electromagnetic Propagation in Multi-mode Random Media*, Wiley, New York, 1999.

151. S. A. Shapiro, P. Hubral, and B. Ursin, Elastic wave reflectivity/transmissivity for randomly layered structures, Geophys. J. Int. **126** (1996), 184–196.

152. P. Sheng, Z.-Q. Zhang, B. White, and G. Papanicolaou, Multiple scattering noise in one dimension: universality through localization length scales, Phys. Rev. Lett. **57** (1986), 1000–1003.

153. P. Sheng, B. White, Z.-Q. Zhang, and G. Papanicolaou, Wave localization and multiple scattering in randomly-layered media, in *Scattering and Localization of Classical Waves in Random Media*, P. Sheng, ed., World Scientific, Singapore, 1990, pp. 563–619.

154. B. M. Shevtsov, Three-dimensional problem of inverse scattering in stratified randomly inhomogeneous media, Radiofiz. **25** (1982), 1032–1040 (742–749 in English).

155. B. M. Shevtsov, Statistical characteristics for wave packet scattering in a layered randomly inhomogeneous medium above a reflecting surface, Radiofiz. **30** (1987), 1007–1012 (750–754 in English).

156. B. M. Shevtsov, Backscattering of a wave in layered regularly and randomly inhomogeneous media, Radiofiz. **32** (1989), 1079–1083 (798–801 in English).

157. R. Snieder, Extracting the Greens function from the correlation of coda waves: A derivation based on stationary phase, Phys. Rev. E **69** (2004), 046610.

158. K. Sølna, Focusing of time-reversed reflections, Waves Random Media **12** (2002), 1–21.

159. K. Sølna and G. Milton, Can mixing materials make electromagnetic signals travel faster? SIAM J. Appl. Math. **62** (2002), 2064–2091.

160. K. Sølna and G. Papanicolaou, Ray theory for a locally layered medium, Waves Random Media **10** (2000), 155–202.

161. R. L. Stratonovich, *Topics in the Theory of Random Noise*, Gordon and Breach, New York, 1963.

162. T. Strohmer, M. Emami, J. Hansen, G. Papanicolaou, and A. Paulraj, Application of time-reversal with MMSE equalizer to UWB communications, in *Global Telecommunications Conference 2004*, IEEE, Vol. 5, Issue 29 Nov.-3 Dec. 2004, pp. 3123–3127.

163. D. W. Stroock and S. R. S. Varadhan, *Multidimensional Diffusion Processes*, Springer, Berlin, 1979.

164. M. Tanter, J. L. Thomas, F. Coulouvrat, and M. Fink, Breaking of time reversal invariance in nonlinear acoustics, Phys. Rev. E **64** (2001), 016602.

165. N. G. Van Kampen, *Stochastic Processes in Physics and Chemistry*, North Holland, Amsterdam, 1981.

166. R. Weaver and O. I. Lobkis, Ultrasonics without a source: Thermal fluctuation correlations at MHz frequencies, Phys. Rev. Lett. **84** (2001), 134301.

167. G. B. Whitham, *Linear and Nonlinear Waves*, Wiley, New York, 1974.

168. B. White, P. Sheng, and B. Nair, Localization and backscattering spectrum of seismic waves in stratified lithology, Geophysics **55** (1990), 1158–1165.

169. B. White, P. Sheng, M. Postel, and G. Papanicolaou, Probing through cloudiness: Theory of statistical inversion for multiply scattered data, Phys. Rev. Lett. **63** (1989), 2228–2231.

170. B. White, P. Sheng, Z.-Q. Zhang, and G. Papanicolaou, Wave localization characteristics in the time domain, Phys. Rev. Lett. **59** (1987), 1918–1921.
171. V. E. Zakharov and A. B. Shabat, Exact theory of two-dimensional self-focusing and one-dimensional self-modulation of waves in nonlinear media, Sov. Phys. JETP **34** (1972), 62–69.

Index

Stochastic Modelling and Applied Probability
formerly: Applications of Mathematics

(continued from page II)